T0350156

Advanced Information and Knowledge Processing

Series editors

Lakhmi C. Jain
Bournemouth University, Poole, UK, and
University of South Australia, Adelaide, Australia

Xindong Wu
University of Vermont

Information systems and intelligent knowledge processing are playing an increasing role in business, science and technology. Recently, advanced information systems have evolved to facilitate the co-evolution of human and information networks within communities. These advanced information systems use various paradigms including artificial intelligence, knowledge management, and neural science as well as conventional information processing paradigms. The aim of this series is to publish books on new designs and applications of advanced information and knowledge processing paradigms in areas including but not limited to aviation, business, security, education, engineering, health, management, and science. Books in the series should have a strong focus on information processing—preferably combined with, or extended by, new results from adjacent sciences. Proposals for research monographs, reference books, coherently integrated multi-author edited books, and handbooks will be considered for the series and each proposal will be reviewed by the Series Editors, with additional reviews from the editorial board and independent reviewers where appropriate. Titles published within the Advanced Information and Knowledge Processing series are included in Thomson Reuters' Book Citation Index.

More information about this series at http://www.springer.com/series/4738

Israël César Lerman

Foundations and Methods in Combinatorial and Statistical Data Analysis and Clustering

Springer

Israël César Lerman
Department of Data Knowledge
   and Management
University of Rennes 1, IRISA
Rennes, Ille-et-Vilaine
France

ISSN 1610-3947   ISSN 2197-8441   (electronic)
Advanced Information and Knowledge Processing
ISBN 978-1-4471-6791-4   ISBN 978-1-4471-6793-8   (eBook)
DOI 10.1007/978-1-4471-6793-8

Library of Congress Control Number: 2016931997

Printed on acid-free paper

This Springer imprint is published by SpringerNature
The registered company is Springer-Verlag London Ltd.

I dedicate this work to Rollande and our three daughters Sabine, Alix and Judith and their children

Preface

Relative to the basic notions of a descriptive attribute (variable) and an object described, there are two fundamental concepts in *Data Analysis*: association between attributes and similarity between objects. Given the description of objects by attributes the goal of data analysis methods is to propose a reduced representation of the data which preserves as accurately as possible the relationships between attributes and between objects. Mainly, there are two types of methods: *factorial* and *clustering*. *Factorial* methods are geometric. For these, the compression structure is obtained from a system of synthetic axes, called *factorial axes*. The most discriminant of them are retained in order to be substituted for the origin axes. Then, the set of data units (objects and also attributes) is represented by a cloud of points placed in the geometrical space, referring to the new system of axes. *Clustering* methods are combinatorial. The compression structure consists of an organized system of proximity clusters. In our terminology, an equivalent term for clustering is *classification*.

In our approach clustering (Classification) is considered as a central tool in data analysis. The extensive development of this principle has led to a very rich methodology. According to this standpoint the first facet of clustering concerns the organization of the attribute set. This enables us to discover the behavioural tendencies and subtendencies of the population studied from a sample of it, the latter defining the object set. The second facet concerns the proximity organization of the object set or a category set induced from it. Behaviour understanding is provided by the first facet and management control by the second facet. Geometrical factorial analysis is often considered as a special tool of data analysis for attribute set structuration. Clustering attributes is a non-classical subject in the literature on data analysis. Generally, the methods proposed for this problem consist of adapting methods created for clustering an object set. By distinguishing clearly the two dual problems: attribute clustering and object clustering, our approach is essentially different.

This book provides a large synthesis and systematic treatment in the area of clustering and combinatorial data analysis. A new vision of this very active field is given. The methodological principles are very new in the *data mining* field. All types of data structures are clearly represented and can be handled in a precise way: qualitative data of any sort, quantitative data and contingency data. The methods invented have been validated by many important and big applications. Their theoretical foundations are clearly and strongly established from three points of view: logical, combinatorial and statistical. In this way, the respective rationales of the distinct methods are clearly set up.

As expressed above, the special structure we are interested in for a reduced representation of the data is that obtained by clustering methods. A non-hierarchical clustering algorithm on a finite set E, endowed with a similarity index, produces a partition on E. Whereas a hierarchical clustering algorithm on E produces an ordered partition chain on E. This book is dominated by hierarchical clustering. However, methods of non-hierarchical clustering are also considered (see below).

In Chap. 1 we study some formal and combinatorial aspects of the sought mathematical structure: partition or ordered chain of partitions. More particularly, two sides are developed. The first is enumerative and consists of counting chains in the partition lattice or counting specific subsets in the partition set. In order to relate the partition type and the cardinality of the equivalence relation graph associated with it, we are led to address the set organized of an integer partition. The second important side concerns the mathematical representation of a partition and, more generally and importantly, an ordered chain of partitions on a finite set E. Thereby, the relationships between the latter structure and numerical (rep., ordinal) *ultrametric spaces* are established. In fact, all the algorithmic development of a given clustering method is dependent on the representation adopted. We end Chap. 1 by showing the transition between the formalization of symmetrical hierarchical clustering and that of directed hierarchical clustering, where junctions between clusters are directed according to a total (also said "linear") order on E.

Our method is focused on *ascendant agglomerative hierarchical clustering* (AAHC). However, non-hierarchical clustering plays an important role in the compression of data representation. This methodology addresses the problem of clustering an object set and not that of an attribute set. Its philosophy is different from that of hierarchical clustering. In these conditions, we describe in Chap. 2 two fundamental and essentially different methods of non-hierarchical clustering. These reflect two important families of no-hierarchical clustering algorithms. It is a matter of the *"central"* partitions of S. Régnier and that of "dynamic clustering" of E. Diday. The latter is derived from a generalization of the "allocating and centring" k-means algorithm, defined by D.J. Hall and G.H. Ball (see references of the chapter concerned). This method is discussed in this chapter. On the other hand, new theoretical and software developments are mentioned.

For the mathematical data representation the descriptive attributes are interpreted in terms of relations on the object set. Thereby, categorical attributes of any sort are represented faithfully. In these conditions, numerical attributes are defined as valued relations. Whereas classical approaches propose a converse reasoning by assigning, more or less arbitrarily, numerical values to categories.

In Chap. 3 we describe the set theoretic and relational representation of the data description. All types of data can be taken into account. Two description levels are considered: objects and categories. For each of the levels, object description and category description, two attribute types are considered depending on the arity of the representative relation on the object set, unary or binary. Notice that the arity of the representative relation associated with a given attribute can be greater than two. And this, is also considered in our development. Thus, in this framework, we define several structured attributes concerned by observation of real data.

The fundamental concept of resemblance between data units: attributes, objects or categories, is studied in Chaps. 4–7. It is based on a deep development of a similarity notion between combinatorial structures. Invariance properties of statistical nature are set up. These lead to a constructive and unified theory of the resemblance notion. Classical association coefficients such that the Goodman and Kruskal, Kendall and Yule coefficients are clearly stood in the framework of this theory. Two options are considered for normalization of the association coefficients between descriptive attributes: standard deviation and maximum. A probability scale, associated with the first normalization, is built in order to compare association coefficients between attributes or similarity indices between objects (resp., categories). This scale is obtained by associating independent random data with the observed one, the random model respecting the general characteristics of the data observed. This comparison technique is a part of the *likelihood linkage analysis* (LLA) clustering method where an observed value of a numerical similarity index is situated with respect to its unlikelihood bigness. Well-know non-parametric statistical theorems are needed for the application of this approach to the attribute comparison. New theorems are established. Based on the same principle an *index of implication* between Boolean attributes is set up. Also, we show how partial association coefficients between structured categorical attributes are built.

Comparing objects described is not equivalent to comparing descriptive attributes. We show in Chap. 7 how the LLA approach enables similarity indices between objects, described by heterogeneous attributes of different types, to be built. We also show how comparing categories is a specific task.

The fascinating concept of "natural" cluster of objects cannot be defined mathematically. Its realization in real cases is expected as a result derived from application of clustering algorithms. Such a cluster is interpreted intuitively. However, it is important to define it as accurately as possible. This definition is necessarily a statistical one. Nevertheless, statistical formalization of a "natural" cluster is very difficult. In Chap. 8 we address this concept. Statistical tools are established for understanding the meaning of such a cluster. For this purpose, initial description is examined for all types of data. Thus, the analysis of a "natural" cluster is essentially analytical. Another way consists of crossing with the target cluster associated with a "natural" cluster, known and discriminant clusters disjoint logically of it, but statistically linked. A "natural" cluster is a part of a "natural" clustering. Generally, this statistical structure sustains real data. However, it is important to test this hypothesis for the data treated. In these conditions, "classifiability" testing hypotheses are proposed and studied.

Whereas Chap. 8 is focused on the intrinsic analysis of clustering, Chap. 9 is devoted to comparing clusterings or clustering trees on the same finite set endowed with a similarity or dissimilarity index. In this chapter very powerful tools are established for this comparison. In this, the similarity data is either numerical or ordinal. A minute analysis of the comparison criteria for both types (numerical or ordinal) is provided. The criteria proposed have a combinatorial and non-parametric statistical nature and they are extremely general. They are established with respect to a probabilistic independence hypothesis between similarity and clustering structures. This enables us to establish significant and non-biased comparisons.

As mentioned above, AAHC is considered in this book as a main tool for *data analysis*. Starting with similarities or distances between data units (See Chaps. 4–7) we show in Chap. 10 how to build a classification tree on the data set corresponding to an agglomerative technique. Ordinal notion of pairwise similarities is treated first. Natural transition to a numerical version of this notion is shown. Defining a dissimilarity between disjoint subsets of the set to be clustered is a fundamental task in agglomerative hierarchical clustering. This dissimilarity is established from the pairwise dissimilarities of data units. Two families of dissimilarity indices are studied. The first is classical and employs distances and weightings. The second is defined from probabilistic indices obtained in the context of the LLA approach. The numerical dissimilarity indices between disjoint subsets of the data set enable comparisons between the clusters merged to be made. The algorithmic analysis of the clustering tree construction is a very important problem. Fundamental results for this problem are reported in this chapter. Thus, we describe some basic solutions provided for agglomerative hierarchical clustering of large data sets. Their computational complexities are expressed. We end this chapter by showing the transition between the usual symmetric hierarchical clustering and that directed where junctions between the branches of the hierarchical tree are compatible with a total order on the set clustered.

In Chap. 11 we begin by describing the *Classification Hiérarchique par Analyse de la Vraisemblance des Liens* (CHAVL) software. The address of a link is specified in the References section in order to access this software. The latter performs according to the LLA methodology, the AAHC of a descriptive attribute set or, dually, a described object (resp., category) set; and this, for a large family of data table structures. In this chapter the results obtained by the LLA method on many real cases are reported. These are provided from different areas: psychosociology, sociological surveys, biology, bioinformatics, image data processing, rural economy. The LLA hierarchical clustering method is applied in order to discover "natural" clusters and behavioural tendencies in the population observed. The cluster interpretation is based on the coefficients developed in Chaps. 4–8. In some of these cases, comparison of the LLA results with those of the Ward hierarchical clustering method, is expressed. In order to realize the different facets in applying the LLA method, some presentations of the processed real cases are detailed sufficiently.

The book ends with Chap. 12 devoted to a general conclusion in which several routes for future research works are outlined. Moreover, the contribution of the book to challenges and advances in cluster analysis is clearly specified. Further, in this chapter, the situation of the book content with respect to other books in the same field is described.

The starting point of the project of this book was a reviewed and completed English translation of the French book:

Classification et analyse ordinale des données

published—with the support of the CNRS—by Dunod (Paris) in 1981.

The progress of my research, the works I met and the considerable development of the field concerned have made that a single volume cannot suffice to cover the entire material expressed in the French book.

In the book we propose here, symmetrical synthetic structures for summarizing data are considered. For these structures—defined by partitions or partition chains— if x and y are two elements of the set E to be organized, the role of x with respect to y is identical to that of y with respect to x.

The different steps of the passage from the data table to the synthetic structure (partition or partition chain) on E are minutely studied. Recall that the set E to be clustered may be an attribute set or an object set (resp., a category set).

The book we propose is a *new* book. It corresponds with respect to the earlier French version, to a new writing, a new design and a much larger scope and potential. The intuitive introductions, the examples and the mathematical formalization and analysis of the subjects treated permit the reader to understand in depth the different approaches in data analysis and clustering. Special concern is devoted for expressing the relationships between these approaches. More precisely, the development provided in this book has the following general distinctive and related features:

1. Mathematical and statistical foundations of combinatorial data analysis and clustering;
2. Mathematical, formal conception and properties are set up in order to compare different approaches in the field concerned;
3. Definition of new methods, guided by a few fundamental principles taking into account the formal analysis;
4. Applying new methods to real data.

More specific distinctive features might be listed as follows:

- Formal descriptions and specific mathematical properties of the synthetic structures sought in clustering (partitions, partition chains (symmetrical and directed));
- Emphasizing data description by categorical attributes of different sorts (broad scope);
- Interpreting descriptive attributes in terms of relations on the object set described;
- Set theoretic representation of the relations defined by the descriptive attributes;
- Very clear typology of data description in the most general case;

- Development of a unified association coefficient notion (symmetrical and asymmetrical) between descriptive attributes of different sorts, including all types of categorical attributes;
- Development of a similarity notion between objects or categories for different types of description, including all types of categorical attributes;
- Probabilistic similarity measures between objects, object clusters, categories, category clusters, attributes, attribute clusters, ...;
- Clustering numerical or categorical descriptive attributes of different kinds;
- Clustering data units (objects or categories) described by a mixing of descriptive attribute types;
- Dual association between object clustering and attribute clustering;
- Seriation and clustering;
- Combinatorial and non-parametric statistical basis for the association coefficients, similarity indices and criteria in clustering;
- Algorithmic studies.

In the part of the French book not retaken here the synthetic structures summarizing the data are of *asymmetrical* nature. Ordinal considerations take part. The chapters concerned with the latter, which may constitute a second volume, are: 6–10. Let me give briefly the subject of each of them.

- Chapter 6: Principal component analysis and correspondence analysis;
- Chapter 7: Mathematical comparisons between *factorial* analysis and *classification* methods;
- Chapter 8: From combinatorial and statistical *seriation* methods to a family of cluster analysis methods;
- Chapter 9: Totally ordering the whole set of categories associated with a set of ordinal categorical attributes;
- Chapter 10: Assignation problems in *pattern recognition* between geometrical figures where the quality measure of the assignation has to be independent of specific geometrical transformations applied on the figures concerned.

As indicated in the title of the book, our work refers to *Combinatorial and Statistical Data Analysis*. The importance of this methodology has already been underlined in the well-known article "Combinatorial Data Analysis" by Phipps Arabie and Lawrence Hubert, published in 1992.

This book is not conceived *a priori* as a "text book". It is a result of my research led since 1966, with many collaborators (See below). Thus the main orientation is "research". However, the latter is placed in the framework of the entire domain concerned. Moreover, a very important part of this research is oriented towards the foundation and synthesis of different methods in combinatorial data analysis and clustering. Consequently, this book is a *reference book*. It will be very useful to master's and Ph.D. students. Wide parts of this book can be taught to students of computer science, statistics and mathematics. I did it.

Let me now cite, in alphabetic order, the names of different collaborators who have worked with me and participated in this research. Most often, but not always, they were around preparing theses and subsequent articles. I especially thank them. The theses defended at the University of Rennes 1 can be consulted at the link address: Sadoc.abes.fr/Recherche avancée.

Collaborators

Jérôme Azé, Helena Bacelar-Nicolaü, Kaddour Bachar, Jean-Louis Buard, Thierry Chantrel, Isaac Cohen-Hallaleh, Jean-Louis Cotrieux, François Daudé, Aziz Faraj, Jean-Paul Geffrault, Nadia Ghazzali, Régis Gras, Sylvie Guillaume, Ivan Kojadinovic, Pascale Kuntz, Jean-Yves Lafaye, Georges Lecalvé, Alain Léger, Henri Leredde, Jean Rémi Massé, Annie Moreau, Roger Ngouënet, Fernando Nicolaü Da Costa, Mohammed Ouali-Allah, Philippe Peter, Joaquim Pinto Da Costa, Annick Prod'Homme, Habibullah Rostam, Valérie Rouat, François Rouxel, Abdel Rahmane Sbii, Basavaneppa Tallur and Philippe Villoing.

Acknowledgements

Before proceeding, I express my gratitude to Dan A. Simovici, Professor at the University of Massachusetts Boston (Department of Computer Science) for having encouraged me to propose a recasted new English version of my book—mentioned above—"Classification et analyse ordinale des donées", published by Dunod (Paris) in 1981.

I especially thank Fionn Murtagh, Professor of Data Science at the University of Derby (Department of Computing and Mathematics), for having distributed the French book among classics in Clustering, in the framework of the *International Federation of Classification Societies* and the *Journal of Classification*. These books are now on the site:

http://www.brclasssoc.org.uk/books/index.html

I am grateful to the Institut de Recherche en Informatique et Systèmes Aléatoires (IRISA) institute where a great part of the research underlying the book, was carried out. Special thanks to the Directors Bruno Arnaldi and Jean-Marc Jezequel to have supported my project.

I also thank Gilles Lesventes, Director of the *ISTIC* UFR Informatique-Électronique to welcome me in his institute and to encourage me.

I owe a debt of gratitude to DUNOD publisher to have published the initial French book and to have retroceded me the copyright.

I also have immense gratitude to the editor of the series in which this book is published as well as to the editorial staff of Springer, particularly to Helen Desmond who has always been in my listening. I also thank James Robinson for taking care of the production process.

Validating new methods of data analysis on real cases is a fundamental task. In this regard, I especially thank researchers in computing science who have worked with me and built efficient and elegant softwares needed for applying the LLA methodology. Let me cite

Henri Leredde, Philippe Peter, Mohamed Ouali-Allah, Kaddour Bachar, Ivan Kojadinovic and Basavaneppa Tallur.

Philippe Louarn (INRIA-Rennes) has defined the general LATEX structure with respect to which I have composed this book. He helped me many times and his help was always valuable. I am very grateful to him.

I cannot conclude these acknowledgments without special thanks to my son-in-law Benjamin Enriquez (Professor of Mathematics at the University Louis Pasteur of Strasbourg). I regularly used to inform him about the progress of my writing. His encouragement and advice have always been very beneficial.

Contents

Chapter 1
On Some Facets of the Partition Set of a Finite Set

As indicated in the Preface, we shall start by describing mathematically the structure sought in *Clustering*. The latter is a partition or an ordered partition chain on a finite set E. A new structure has appeared these last years where each partition class is a subset of E, linearly ordered. Relative to the latter structure, an introduction will be given in Sect. 1.4.5.

The set E to be clustered may be a set of objects (resp., categories) or a set of descriptive attributes (see Chap. 3). According to the usage, without any generality restriction, we shall put E as a set $\mathcal{O}$ of objects. Nonetheless, the non-specified notation E will be maintained in Sect. 1.4.5.

1.1 Lattice of Partition Set of a Finite Set

1.1.1 Definition and General Properties

Let $\mathcal{O}$ be a finite set of n elements. A partition of $\mathcal{O}$ is a set of $\mathcal{O}$ subsets, mutually disjoint, whose union is the entire set $\mathcal{O}$. These subsets are called *classes* of the partition. Thereby, $\{o_a, o_b, o_c, o_d\}$, $\{o_e, o_f\}$ and $\{o_g\}$ are the classes of the partition $\{\{o_a, o_b, o_c, o_d\}, \{o_e, o_f\}, \{o_g\}\}$ of $\mathcal{O} = \{o_a, o_b, o_c, o_d, o_e, o_f, o_g\}$.

We shall use the notation $\mathcal{P}(\mathcal{O})$ to express the set of partitions of $\mathcal{O}$. To each element P of $\mathcal{P}(\mathcal{O})$ we will associate a binary relation on $\mathcal{O}$, which we can denote also, without ambiguity, by P. The latter is defined as follows:

$$(\forall (x, y) \in \mathcal{O} \times \mathcal{O}), x P y \Leftrightarrow x \text{ and } y \text{ are in the same class of the partition } P$$

© Springer-Verlag London 2016
I.C. Lerman, *Foundations and Methods in Combinatorial and Statistical Data Analysis and Clustering*, Advanced Information and Knowledge Processing, DOI 10.1007/978-1-4471-6793-8_1

P is an *equivalence relation*, that is to say,

- reflexive: $(\forall x \in \mathcal{O})$, xPx;
- symmetrical: $(\forall(x, y) \in \mathcal{O} \times \mathcal{O})$, $xPy \Leftrightarrow yPx$;
- and transitive: $(\forall(x, y, z) \in \mathcal{O} \times \mathcal{O} \times \mathcal{O})$, xPy and $yPz \Rightarrow xPz$.

Clearly, there is a bijective correspondence between $\mathcal{P}(\mathcal{O})$ and the set, designated by $\mathcal{E}_q(\mathcal{O})$ of equivalence relations on $\mathcal{O}$. To simplify notations, we will denote below by $\mathcal{P}$ the set introduced with the notation $\mathcal{P}(\mathcal{O})$.

The graph of a binary relation P on $\mathcal{O}$ is the subset of $\mathcal{O} \times \mathcal{O}$ defined by

$$Gr(P) = \{(x, y) | x \in \mathcal{O}, y \in \mathcal{O} \text{ and } xPy\} \tag{1.1.1}$$

When P is an equivalence relation associated with a partition $P = \{O_k | 1 \le k \le K\}$, $Gr(P)$ can be written as

$$Gr(P) = \bigcup_{1 \le k \le K} O_k \times O_k \tag{1.1.2}$$

Figure 1.1 shows $Gr(P)$ in the case of the example above.

Inclusion relation between subsets of the Cartesian product $\mathcal{O} \times \mathcal{O}$ provides an order relation $<$ on $\mathcal{P}$:

$$\big(\forall(P, P') \in \mathcal{P} \times \mathcal{P}\big), P < P' \Leftrightarrow Gr(P) \subset Gr(P') \tag{1.1.3}$$

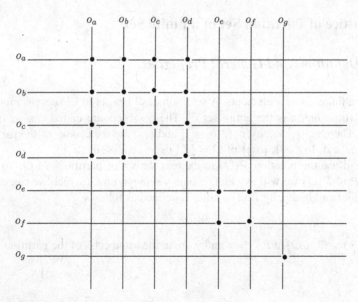

Fig. 1.1 Graph of the equivalence relation P

which means

$$\forall \big((x, y) \in \mathcal{O} \times \mathcal{O}\big), x P y \Rightarrow x P' y$$

In this case, P is said finer than P' or "P is a refinement of P'". Thus, relative to the set $\mathcal{O} = \{o_a, o_b, o_c, o_d, o_e, o_f, o_g\}$, the partition

$$P = \big\{\{o_a, o_b, o_c, o_d\}, \{o_e, o_f\}, \{o_g\}\big\}$$

is finer than

$$P' = \big\{\{o_a, o_b, o_c, o_d\}, \{o_e, o_f, o_g\}\big\}.$$

This order relation endows $\mathcal{P}$ with a *lattice* structure; that is to say, to every pair $\{P, P'\}$ of $\mathcal{P}$ elements corresponds in $\mathcal{P}$ a common *greatest lower bound* $P \wedge P'$ and a common *lowest upper bound* $P \vee P'$.

$P \wedge P'$ can be defined by the graph of the associated equivalence relation

$$Gr(P \wedge P') = Gr(P) \cap Gr(P')$$

where $Gr(P)$ (resp., $Gr(P')$) is the graph of the equivalence relation associated with P (resp., P').

$P \vee P'$ can also be defined from its graph. $Gr(P \vee P')$ is the graph of the transitive closure of the binary relation "P or P'". In more explicit words, for any $(x, y) \in \mathcal{O} \times \mathcal{O}$, $x P \vee P' y$, if and only if there exists a sequence $(z_0, z_1, \ldots, z_l)$, where $z_0 = x$, $z_l = y$ and such that $z_i P z_{i+1}$ or $z_i P' z_{i+1}$, for $i = 0, 1, \ldots, l - 1$.

Example Relative to above, consider $P = \big\{\{o_a, o_b, o_c, o_d\}, \{o_e, o_f\}, \{o_g\}\big\}$ and $P' = \big\{\{o_a, o_b\}, \{o_c, o_d\}, \{o_e, o_f, o_g\}\big\}$

$$P \wedge P' = \{\{o_a, o_b\}, \{o_c, o_d\}, \{o_e, o_f\}, \{o_g\}\}, P \vee P' = \{\{o_a, o_b, o_c, o_d\}, \{o_e, o_f, o_g\}\}$$

Clearly, the lattice $\mathcal{P}$ depends only on the cardinality n of $\mathcal{O}$. The smallest element of $\mathcal{P}$ is the finest partition, that for which each class is a "singleton" class, including exactly one element of $\mathcal{O}$. The biggest element of $\mathcal{P}$ is defined by the least fine partition of $\mathcal{O}$ comprising a single class which includes $\mathcal{O}$ in its totality. The finest and least fine partitions are considered as "trivial" partitions. They are called *partition of singletons* and *singleton partition*, and will be denoted below by P_s and P_t, respectively.

(a) A partition P' *covers* a partition P if and only if

1. $P < P'$;
2. $\{Q | Q \in \mathcal{P}, P < Q < P'\} =]P, P'[= \emptyset$.

where $]P, P'[$ designates the two sides open interval between P and P'. Item 2 means that there does not exist a partition Q which is strictly between P and P'. Obviously, P' covers P if and only if P' is obtained from P by merging two of its classes.

(b) The *atoms* of $\mathcal{P}$ are the elements of $\mathcal{P}$ which cover the partition of singletons P_s. Each of them identifies two elements of $\mathcal{O}$ in terms of the equivalence relation associated. Therefore, there are in all $\binom{n}{2}$ atoms. $\mathcal{P}$ is an *atomistic* lattice; that is to say, any element P of $\mathcal{P}$ can be written in the form

$$P = \bigvee_{1 \leq i \leq p} A_i \qquad (1.1.4)$$

where A_i, $1 \leq i \leq p$, are atoms of $\mathcal{P}$. There are several ways to carry out a decomposition such as (1.1.4). A systematic one consists of associating with each object pair joined by P, a single atom. More precisely and easier, consider a partition $P = \{O_1, O_2, \ldots, O_k, \ldots, O_K\}$ and denote by P_X the partition of $\mathcal{O}$ whose only class not reduced to a singleton class is X. We have

$$P = P_{O_1} \vee P_{O_2} \vee \cdots \vee P_{O_k} \vee \cdots P_{O_K} \qquad (1.1.5)$$

and this decomposition is unique. On the other hand, for any $\mathcal{O}$ subset X, with more than a single element ($X = \{x_1, x_2, \ldots, x_j, \ldots, x_m\}$ ($m \geq 2$)), we have

$$P_X = P_{\{x_1,x_2\}} \vee P_{\{x_1,x_3\}} \vee \cdots \vee P_{\{x_1,x_m\}}$$

or, also,

$$P_X = P_{\{x_1,x_2\}} \vee P_{\{x_2,x_3\}} \vee \cdots \vee P_{\{x_{(m-1)},x_m\}}$$

(c) A *chain* $C = (P_1, P_2, \ldots, P_i, \ldots, P_I)$ of the lattice $\mathcal{P}$ is a sequence totally (linearly) ordered of elements of $\mathcal{P}$:

$$P_1 < P_2 < \cdots < P_i < \cdots < P_I$$

The *length* $l(C)$ of a chain C is the number of its elements minus 1. With the notation above, $l(C) = I - 1$.

The *height* $h(P)$ of a partition P, belonging to $\mathcal{P}$, is defined by the *maximum* of the length chains connecting the partition P_s to P. It is given by $h(P) = n - c(P)$, where $c(P)$ is the class number of P.

The *height of the lattice* $\mathcal{P}$ $h(P)$ is the height of its greatest element, that is, the least fine partition P_t. Then, $h(P) = n - 1$.

(d) The height function h provides a gradation on $\mathcal{P}$, defined as follows:

1. $P' > P \Rightarrow h(P') > h(P)$;
2. If P' covers P, then $h(P') = h(P) + 1$.

In these conditions, the lattice $\mathcal{P}$ verifies the Jordan–Dedekind condition, that is, "For any P and P' in $\mathcal{P}$, such that $P < P'$, all of the maximal chains between P and P', have the same finite length". In fact, the length of any maximal chain between P and P' is $h(P') - h(P)$

Example

The diagram of the lattice $\mathcal{P}_4$ of the set $\mathcal{O} = \{o_a, o_b, o_c, o_d\}$ is depicted in Fig. 1.2. In this, a partition such that $\{\{o_a, o_b, o_d\}, \{o_c\}\}$ is written (abd, c). The straight line connecting (ab, c, d) to (abd, c) expresses that the former of both partitions is finer than the latter one.

(e) The lattice $\mathcal{P}$ graded by the height function h is *semimodular*, that is to say

$$\big(\forall (P, Q) \in \mathcal{P} \times \mathcal{P}\big), h(P) + h(Q) \geq h(P \vee Q) + h(P \wedge Q) \qquad (1.1.6)$$

The first step for proving this property consists of showing its equivalence with the following condition:

For any P, Q and R in $\mathcal{P}$, P and Q cover $R \Rightarrow P \vee Q$ covers P as well $P \vee Q$ covers Q.

(see [16] page 20 or [23]).

We can observe now that this condition is filled for $\mathcal{P}$. For this, to begin, denote the partition R as follows:

$$R = \{O_1, O_2, \ldots, O_k, \ldots, O_K\}$$

In these conditions and without loss of generality, the respective forms of P and Q may be either

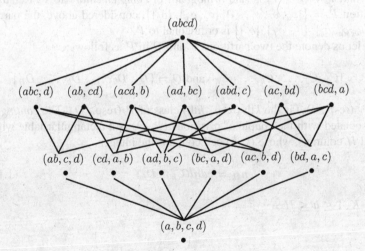

Fig. 1.2 Partition lattice on a set of four elements

$$P = \{O_1 \cup O_2, O_3, O_4, \ldots, O_k, \ldots, O_K\} \text{ and } Q = \{O_1, O_2 \cup O_3, O_4, \ldots, O_k, \ldots, O_K\}$$

$$\text{(1.1.7)}$$

or

$$P = \{O_1 \cup O_2, O_3, O_4, \ldots, O_k, \ldots, O_K\} \text{ and } Q = \{O_1, O_2, O_3 \cup O_4, \ldots, O_k, \ldots, O_K\}$$

$$\text{(1.1.8)}$$

In the first case (resp., second case) the class pair merged in P has (resp., has not) a common component with the class pair merged in Q.

In the first case (1.1.7)

$$P \vee Q = \{O_1 \cup O_2 \cup O_3, O_4, \ldots, O_k, \ldots, O_K\} \qquad \text{(1.1.9)}$$

And, in the second case (1.1.8)

$$P \vee Q = \{O_1 \cup O_2, O_3 \cup O_4, O_5, \ldots, O_k, \ldots, O_K\} \qquad \text{(1.1.10)}$$

In both cases, $P \vee Q$ covers P, as well as Q.

(f) A *geometric* lattice is defined as a *finite atomistic* (each element is a join of atoms) and *semimodular* lattice. In this definition, the atomistic property can be replaced by the *relative complementation* property [3, 23]. This property means that for any interval $\mathcal{I} = [P_1, P_2]$ in a geometric lattice $\mathcal{P}$ ($P_1, P_2 \in \mathcal{P}$ and $P_1 < P_2$) and for any P in $\mathcal{I}$, there exists at least one element Q in $\mathcal{I}$, such that $P \wedge Q = P_1$ and $P \vee Q = P_2$.

The geometric lattice $\mathcal{P}$ is geometric. P_s and P_t being the finest and coarsest partitions (see above), if P is a given partition in $\mathcal{P}$, a partition Q for which $P \wedge Q = P_s$ and $P \vee Q = P_t$ is said *orthogonal* or *complementary* to P. Relative to the partition $P = \big\{\{o_a, o_b, o_c, o_d\}, \{o_e, o_f\}, \{o_g\}\big\}$, considered above, the partition $Q = \big\{\{o_a, o_e, o_g\}, \{o_b, o_f\}, \{o_c\}\big\}$ is orthogonal to P.

Now, let us denote the two partitions P and Q in $\mathcal{P}$ as follows:

$$P = \{C_1, C_2, \ldots, C_k, \ldots, C_K\} \text{ and } Q = \{D_1, D_2, \ldots, D_h, \ldots, D_H\}$$

where C_k (resp., D_h) is the kth (resp., hth) class of P (resp., Q). The *contingency table* associated with the couple of partitions (P, Q) is a rectangular table with K rows and H columns, whose (k, h) cell contains the integer

$$n_{kh} = card(C_k \cap D_h) \qquad \text{(1.1.11)}$$

$1 \leq k \leq K, 1 \leq h \leq H$

This table gives the class cardinals of the partition

$$P \wedge Q = \{C_k \cap D_h | 1 \leq k \leq K, 1 \leq h \leq H\} \tag{1.1.12}$$

The *crossing grid* of (P, Q) is a rectangular table having the same respective sizes K and H as those of the associated contingency table. The content of the (k, h) cell of the new table is equal to 1 or 0, according to $n_{kh} \neq 0$ or $n_{kh} = 0$, respectively.

In these conditions, $P \wedge Q$ is the finest partition P_s if and only if the number of components equal to 1 in the kth row of the crossing grid table is exactly equal to $card(C_k)$, $1 \leq k \leq K$. In this situation we have necessarily $H \geq \max_{1 \leq k \leq K} card(C_k)$.

Now, designate by $C = \{c_1, c_2, \ldots, c_k, \ldots, c_K\}$ and $\mathcal{D} = \{d_1, d_2, \ldots, d_h, \ldots, d_H\}$ the row set and the column set of the crossing grid. Each column d_h can be interpreted as the indicator vector of a subset of C. The latter corresponds to the components of d_h equal to 1, $1 \leq h \leq H$. In the case where $P \wedge Q$ is the partition of singletons P_s, the number of components equal to 1 in the kth row is exactly equal to $n_{k.} = card(C_k)$, $1 \leq k \leq K$.

By considering the crossing grid representation, we can state that a necessary and sufficient condition that $P \vee Q$ is the singleton partition P_t is that for any column pair $\{d_h, d_{h'}\}$, there exists a chain of elements of $\mathcal{D}$ joining d_h to $d_{h'}$, such that the respective vectors of two consecutive elements have at least one common component equal to 1 (the intersection of the subsets indicated by two consecutive elements is non-empty).

In this case, we can say that the set of the grid column vectors is "weakly linked". Clearly, by interverting the respective roles of P and Q, this property is equivalent to that for which the grid rows are weakly linked. Therefore, we have not to distinguish between columns and rows for the property concerned and we can say globally, the grid is or is not weakly linked.

Given a partition P, building a partition Q orthogonal to P (see above for notations) assumes the following:

1. To fix a grid with K rows and H columns, such that $H \geq \max_{1 \leq k \leq K} n_{k.}$, where $n_{k.} = card(C_k)$, $1 \leq k \leq K$;
2. To fill the kth row with a Boolean vector having exactly $n_{k.}$ components equal to 1 $(1 \leq k \leq K)$ and such that the K vectors are weakly linked;
3. To give K injective mappings, where the kth one, which we denote by ϕ_k, maps C_k on the cells of the kth row including the value 1.

Example Let $P = \{\{o_a, o_b, o_c\}, \{o_d, o_e, o_f, o_g\}, \{o_h, o_i, o_j\}\}$ be a partition into three classes of $\mathcal{O} = \{o_a, o_b, o_c, o_d, o_e, o_f, o_g, o_h, o_i, o_j\}$ and let Table 1.1 be a cross grid sized 3×8, weakly linked.

On the other hand, consider the three following injective mappings ϕ_1, ϕ_2 and ϕ_3:

1. $\phi_1 : \{o_a, o_b, o_c\} \rightarrow \{d_2, d_5, d_6\}$;
2. $\phi_1 : \{o_d, o_e, o_f, o_g\} \rightarrow \{d_1, d_3, d_7, d_8\}$;
3. $\phi_1 : \{o_h, o_i, o_j\} \rightarrow \{d_2, d_4, d_7\}$.

Table 1.1 Crossing grid

	d_1	d_2	d_3	d_4	d_5	d_6	d_7	d_8
c_1	0	1	0	0	1	1	0	0
c_2	1	0	1	0	0	0	1	1
c_3	0	1	0	1	0	0	1	0

where

1. $\phi_1(o_a) = d_5$, $\phi_1(o_b) = d_2$, $\phi_1(o_c) = d_6$;
2. $\phi_2(o_d) = d_7$, $\phi_2(o_e) = d_3$, $\phi_2(o_f) = d_1$, $\phi_2(o_g) = d_8$;
3. $\phi_3(o_h) = d_4$, $\phi_3(o_i) = d_7$, $\phi_3(o_j) = d_2$.

Thereby, we get a partition $Q = \{\{o_f\}, \{o_b, o_j\}, \{o_e\}, \{o_h\}, \{o_a\}, \{o_c\}, \{o_d, o_i\}, \{o_g\}\}$, orthogonal to P.

Given a partition P, we may be asked how to enumerate or at least to generate the set of partitions Q orthogonal (complementary) to P. It is somewhat easy to count the partitions Q with H classes exactly, for which $P \wedge Q$ is the partition of singletons P_s (see Paragraph "Counting Partitions of n Objects: The Bell Numbers"). In the set of these partitions, those for which $P \vee Q$ is not reduced to a single class are characterized by the property that the crossing grid of (P, Q) can be brought down to the form of Fig. 1.3 where only the grey part can contain 1 values.

Let us express this latter property in other words. Consider the link binary relation between Boolean vector columns (resp., Boolean vector rows) of the grid, for which two vectors are related if and only if they have at least one component equal to 1 in common. Then, $P \vee Q$ can be reduced to a single class, if and only if the graph associated with the binary relation expressed is connected [1].

Relative to a partition P, it is not an easy task to count the number of partitions Q such that $P \vee Q$ is the trivial singleton partition P_t. However, we shall show that the proportion of partitions Q for which $P \vee Q$ is not the trivial partition P_t is negligible (see Paragraph "Enumerations Associated with Crossing Two Partitions").

Fig. 1.3 Not weakly linked crossing grid

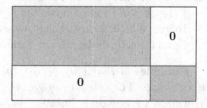

1.1.2 Countings

1.1.2.1 Counting Partition Chains in $\mathcal{P}$

Let $\mathcal{C}_0(\mathcal{O})$ be the chain set whose first element is the partition of singletons. C belongs to $\mathcal{C}_0(\mathcal{O})$, if and only if

$$C = (P_0, P_1, \ldots, P_k, \ldots, P_K) \qquad (1.1.13)$$

where $P_0 < P_1 < \cdots < P_k < \cdots < P_K$ and where P_0 is the partition of singletons, also denoted by P_s. Counting in $\mathcal{C}_0(\mathcal{O})$ does not restrict the generality with respect to counting partition chains in $\mathcal{P}$. In fact there is a clear bijective correspondence between $\mathcal{C}_0(\mathcal{O})$ and the set of partition chains, which we can denote by $\mathcal{C}_1(\mathcal{O})$, for which the first element is not P_0. This correspondence is defined by associating to each element C of $\mathcal{C}_0(\mathcal{O})$, the element C_1 of $\mathcal{C}_1(\mathcal{O})$, obtained from C by deleting its first component.

Recall that a chain of $\mathcal{C}_0(\mathcal{O})$ is *maximal* if and only if

1. $(\forall k, 1 \leq k \leq K - 1)$, P_{k+1} covers P_k;
2. P_K is the singleton partition P_t.

Such a chain will be called below *elementary*.

A *fineness* order relation can be defined as $\mathcal{C}_0(\mathcal{O})$. Given C and C' in $\mathcal{C}_0(\mathcal{O})$, C is finer than C', if and only if $C' \subset C$, or, more precisely, C' is a subsequence of C. $\mathcal{C}_0(\mathcal{O})$ endowed with this order is a semi-lattice for the intersection operation. Its smallest element is the sequence (P_0) reduced to a single partition, that is P_s, of singletons. In this semi-lattice the maximal elements are the elementary chains, whose common length being $n = card(\mathcal{O})$.

Clearly, the cardinal of $\mathcal{C}_0(\mathcal{O})$ depends uniquely on n. Let us denote it by $\psi(n)$. We have

Proposition 1

$$\psi(n) = \sum_{1 \leq k \leq n-1} S(n, k)\psi(k) \qquad (1.1.14)$$

where $S(n, k)$ designates the number of partitions of n objects into k non-empty classes. $S(n, k)$ that we will examine below is the Stirling number of the second kind.

For $k < n$, let $\mathcal{C}_{1k}(\mathcal{O})$ be the set $\{C | C \in \mathcal{C}_1(\mathcal{O}) \text{ and } c(P_1) = k\}$, where $c(P_1)$ is the number of classes of the partition P_1. Obviously, the $\mathcal{C}_{1k}(\mathcal{O})$, for $k = 1, 2, \ldots, n-1$, form a partition of the chain set $\mathcal{C}_1(\mathcal{O})$. Now, we shall calculate the cardinality of $\mathcal{C}_{1k}(\mathcal{O})$. For a given partition P_1 comprising k classes, there are $\psi(k)$ chains in $\mathcal{C}_{1k}(\mathcal{O})$ for which the first partition is P_1. There are $S(n, k)$ different partitions of n objects into k non-empty classes. Therefore, for $1 \leq k \leq n - 1$,

$$card\big(\mathcal{C}_{1k}(\mathcal{O})\big) = S(n, k)\psi(k)$$

Hence,

$$card\big(\mathcal{C}_0(\mathcal{O})\big) = card\big(\mathcal{C}_1(\mathcal{O})\big) = \sum_{1 \le k \le n-1} S(n, k)\psi(k)$$

$\square$

Example

For $n = 4$

$$\psi(4) = S(4, 1)\psi(1) + S(4, 2)\psi(2) = S(4, 3)\psi(3) = 1 \times 1 + 7 \times 1 + 6 \times 4 = 32$$

which can be directly verified on the lattice $\mathcal{P}_4$ shown in Fig. 1.2.

Proposition 2 *Let $\phi(k)$ designate the number of elementary chains of length n. We have*

$$\phi(n) = \frac{(n - 1)!n!}{2^{n-1}} \tag{1.1.15}$$

An elementary chain $C = (P_0, P_1, \ldots, P_k, \ldots, P_{n-1})$ of $\mathcal{C}_0(\mathcal{O})$ comprises exactly n components. P_0 (resp., P_{n-1}) is the partition of singletons (resp., the singleton partition). For a fixed k first terms of C, namely, $P_0, P_1, \ldots, P_{k-1}$, there are $\binom{n-k+1}{2}$ partitions P_k which cover P_{k-1}. Effectively, P_k is obtained from P_{k-1} by merging two classes of P_{k-1}. In these conditions, we have

$$\phi(n) = \binom{n}{2} \times \binom{n-1}{2} \times \cdots \times \binom{n-k+1}{2} \times \cdots \times \binom{2}{2}$$

Hence, the formula proposed (1.1.15). $\square$

Example

For $n = 4$, $\phi(4) = 3!4!/2^3 = 18$. This can be verified directly in Fig. 1.2.

Now, we shall be interested in enumerating the set of maximal partition chains joining two comparable vertices of the lattice $\mathcal{P}$. Suppose these are indicated as follows:

$$P = \{e_1, e_2, \ldots, e_j, \ldots, e_k\} \text{ and } P = \{E_1, E_2, \ldots, E_l, \ldots, E_m\} \tag{1.1.16}$$

where $P < Q$; that is to say and without restricting generality,

$$E_1 = \sum_{1 \le i \le h_1} e_i$$

$$E_2 = \sum_{h_1 < i \le h_2} e_i$$

$$\ldots$$

$$E_l = \sum_{h_{l-1} < i \le h_l} e_i$$

$$\cdots$$

$$E_m = \sum_{h_{m-1} < i \le h_m} e_i \qquad (1.1.17)$$

where the sum symbols indicate set sums, that is, unions of disjoint sets. By setting

$$k_1 = h_1, k_2 = h_2 - h_1, \ldots, k_l = h_l - h_{l-1}, \ldots, k_m = h_m - h_{m-1} \qquad (1.1.18)$$

E_1 appears as the union of the k_1 first classes of P, E_2 as the union of the next k_2 classes of P,..., E_m as the union of the k_m last classes of P.

Every maximal chain starting with P and ending with Q can be decomposed into m maximal chains. The lth one is on the subset set

$$P_l = \{e_{h_{l-1}+1}, \ldots, e_{h_l}\} \qquad (1.1.19)$$

$1 \le l \le m$, where h_0 is equal to 0. The lth chain starts with P_l and ends with the singleton partition $\{E_l\}$.

Clearly, the set of maximal partition chains starting with P_l and ending with $\{E_l\}$ is isomorphic to the set of maximal partition chains on a set of k_l elements. The cardinality of the latter is $\phi(k_l)$, where the function ϕ was specified in (1.1.15). To pass from P to Q there are in all

$$(k_1 - 1) + (k_2 - 1) + \cdots + \cdots + (k_l - 1) + \cdots + (k_m - 1) = k - m \qquad (1.1.20)$$

binary aggregations.

The binary aggregation-ordered sequence concerning P_l, $1 \le l \le m$ admits a representation by a classification tree (see Sect. 1.4.4.1), where each level of the latter is associated with one merging. In these conditions, associate one to one the label sequence

$$\left(\sum_{1 \le j \le l-1} k_j - (l - 2), \ldots, \sum_{1 \le j \le l} k_j - l \right) \qquad (1.1.21)$$

with the aggregation sequence on P_l, $1 \le l \le m$. Note that the number of elements of (1.1.21) is $k_l - 1$.

Now, to form a maximal chain going from P to Q, we have to build m maximal chains in the lattices $\mathcal{P}_{k_1}, \mathcal{P}_{k_2}, \ldots, \mathcal{P}_{k_l}, \ldots \mathcal{P}_{k_m}$ (where $\mathcal{P}_{k_l}$ concerns P_l (see (1.1.19))) and to form a permutation of $(1, 2, \ldots, k - m)$ such that its restrictions to each of the intervals (1.1.21) are identity. The purpose of this permutation consists of inserting from each other the respective levels of the different binary classification trees, established on the different sets P_l (see (1.1.19)), $1 \le l \le m$, respectively. There are in all

$$\frac{(k-m)!}{\Pi_{1\le l\le m}(k_l-1)!}$$

such permutations and then we have the following.

Proposition 3 *The number of maximal chains from P to Q (see (1.1.16)) is given by*

$$\frac{(k-m)!}{\Pi(k_l-1)!}\phi(k_1)\times\phi(k_2)\times\cdots\times\phi(k_m)=\frac{(k-m)!}{2^{k-m}}\Pi_{1\le l\le m}k_l! \qquad (1.1.22)$$

Example

Let P and Q be two partitions of the set $E=\{a,b,c,d,e,f,g,h,i,j\}$. $P=\{\{a\},\{b\},\{c\},\{d,e\},\{f\},\{g\},\{h,i,j\}\}$ and $Q=\{\{a,b,c,d,e\},\{f,g,h,i,j\}\}$. $Q=\{E_1,E_2\}$ includes two classes. The first one E_1 groups the classes $e_1=\{a\}$, $e_2=\{b\}$, $e_3=\{c\}$ and $e_4=\{d,e\}$. The second one groups the classes $e_5=\{f\}$, $e_6=\{g\}$ and $e_7=\{h,i,j\}$. In these conditions, consider an instance of two maximal chains on $\{e_1,e_2,e_3,e_4\}$ and $\{e_5,e_6,e_7\}$, respectively. These are shown in Fig. 1.4. The associated permutation with the label sequence of the first (resp., second) partition chain is $(1,2,3,4)$ (resp., $(5,6)$). The total permutation $(1,4,2,5,3)$ preserves each of the previous permutations. The associated labelled maximal chains inserting each from the other, the respective levels of both trees are figured in Fig. 1.5. In this particular example there are $5!4!3!/2^5$ maximal chains joining P to Q ($k=7$, $m=2$, $k_1=4$, $k_2=3$).

1.1.2.2   Counting Partitions

Counting Partitions of a Fixed Type

Let f be a surjective mapping of a set $\mathcal{O}$ of n objects onto the set $\{1,2,\ldots,k,\ldots,K\}$ of K integer labels (indices). f induces a partition of $\mathcal{O}$ into K non-empty classes. The kth class $\mathcal{O}_k$ is defined by

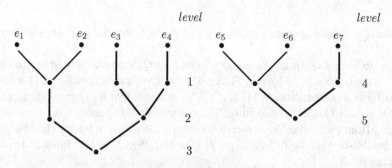

Fig. 1.4 Two maximal chains to be mutually inserted

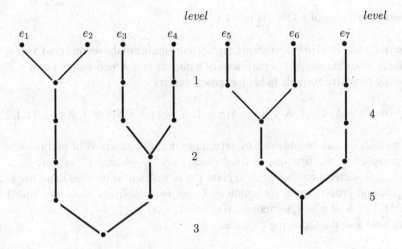

Fig. 1.5 A possible insertion of two maximal chains

$$O_k = f^{-1}(k) = \{x | x \in \mathcal{O} \text{ and } f(x) = k\},$$

$1 \le k \le K$. f determines a partition of $\mathcal{O}$ with *labelled classes*. If n_k is the cardinal of O_k, the *type* t of the partition is the ordered sequence $(n_1, n_2, \ldots, n_k, \ldots, n_K)$. $\mathcal{P}(n; t)$ will designate the set of partitions of $\mathcal{O}$ with labelled classes of type t. We have

$$card(\mathcal{P}(n; t)) = \frac{n!}{n_1! n_2! \ldots n_k! \ldots n_K!} \tag{1.1.23}$$

Now, divide the cardinal sequence t into subsequences where each of them is formed of integers, mutually equal. If the respective lengths of these are $k_1, k_2, \ldots, k_r$ $(k_1 + k_2 + \cdots + k_r = K)$, the number of partitions of $\mathcal{O}$, with non-labelled classes, whose cardinalities are $n_1, n_2, \ldots, n_k, \ldots, n_K$, is given by

$$card(\mathbb{P}(n; t)) = \frac{n!}{k_1! k_2! \ldots k_r! \times n_1! n_2! \ldots n_k! \ldots n_K!} \tag{1.1.24}$$

where $\mathbb{P}(n; t)$ designates the set of partitions concerned. In fact, a partition with non-labelled classes, belonging to $\mathbb{P}(n; t)$, gives rise to exactly $k_1! k_2! \ldots k_r!$ partitions with labelled classes in $\mathcal{P}(n; t)$. The latter are obtained by permuting the labels of the classes with the *same* cardinal.

The type of a partition with *non-labelled classes* is defined as the decreasing sequence of its class cardinals.

Counting Partitions of n Objects into k Classes

We have already used in Proposition 1 the Stirling number of the second kind $S(n, K)$. The latter gives the number of partitions of n objects into K non-empty classes. The following recursive formula holds for these numbers:

$$S(n, K) = S(n - 1, K - 1) = S(n - 1, K - 1) + K S(n - 1, K) \qquad (1.1.25)$$

This relation can be obtained by separating the set concerned of partitions into two groups. For the first one, a fixed object, say o_n—where $\mathcal{O} = \{o_1, o_2, \ldots, o_i, \ldots, o_n\}$—determines a singleton class (o_n is isolated in its class) and the second partition group is that for which o_n is not isolated in its class. We have for (1.1.25), $1 < K < n - 1$, moreover, $S(n, 1 = S(n, n) = 1)$.

We have also the following formulas:

$$l^n = \sum_{0 \le K \le l} \binom{l}{K} (K! S(n, K)) \qquad (1.1.26)$$

$$S(n, h) = \sum_{h-1 \le m \le n} \binom{n-1}{m} S(m, h - 1) \qquad (1.1.27)$$

The former of both previous equations is obtained by dividing the set of all mappings of a set of n objects into a set of l identified cells, $K! S(n, K)$ being the number of surjective mappings of a set of n objects into a set of K cells.

Equation (1.1.27) is obtained from dividing the partition set of n objects into h classes, according to the cardinal of the class to which a given object belongs. The mth term corresponds to the cardinal $(n - m)$.

Now, there remains to establish the explicit formula

$$K! S(n, K) = \sum_{0 \le j \le K} (-1)^j \binom{K}{K - j} (K - j)^n \qquad (1.1.28)$$

which can be derived from the inclusion and exclusion formula. This can be written as

$$\mu \left(\bigcup_{i \in \mathbb{K}} A_i \right) = \sum_{\mathbb{J} \subset \mathbb{K}} (-1)^{|\mathbb{J}|+1} \mu \left(\bigcap_{i \in \mathbb{J}} A_i \right) \qquad (1.1.29)$$

where $\mathbb{K} = \{1, 2, \ldots, i, \ldots, K\}$ and where $\{A_i | i \in \mathbb{K}\}$ is a set of K subsets of a set X endowed with a finite measure μ. For the formula to be established X is the mapping of $\mathcal{O}$ onto $\mathbb{K}$ and μ is the cardinal function. The previous sum on the right of (1.1.29) is extended over all the $\mathbb{K}$ subsets. $|\mathbb{J}|$ stands for $card(\mathbb{J})$.

Define A_i as the mapping set of $\mathcal{O}$ into $\mathbb{K} - \{i\}$. In this way, (1.1.29) gives the number of non-surjective mappings of $\mathcal{O}$ into $\mathbb{K}$. The right member can be written as

$$\mu\left(\bigcup_{i\in\mathbb{K}}\right) = \sum_{1\leq j\leq K} (-1)^{j+1}\binom{K}{h}(K-j)^n$$

The following relation

$$K!.S(n,K) = K^n - \mu\left(\bigcup_{i\in\mathbb{K}} A_i\right)$$

enables (1.1.28) to be get.

Counting Partitions of n Objects: The Bell Numbers

Let $B(n)$ designate the Bell number [2]. It is defined as the number of partitions of a set of n objects. We have the important recursive formula

$$B(n) = \sum_{0\leq m\leq n-1} \binom{n-1}{m} B(m) \tag{1.1.30}$$

where we set $B(0) = 1$. This equation is obtained by dividing the partition set according to the class cardinal to which a given object belongs. In (1.1.30), the class cardinal concerned is $n - m$. Notice the following relation:

$$B(n) = \sum_{1\leq k\leq n} S(n,k) \tag{1.1.31}$$

An additional formula to be mentioned, due to G. Dobinski (see in [2]), is

$$B(n) = \frac{1}{e} \sum_{k\geq 0} \frac{k^n}{k!} \tag{1.1.32}$$

where e is the Euler number defining the basis of natural logarithm.

Enumerations Associated with Crossing Two Partitions

Given a partition $P = \{C_1, C_2, \ldots, C_k, \ldots, C_K\}$ of an object set $\mathcal{O}$, we shall establish two countings. The first one is associated with the set composed of partitions Q such that the partition $P \wedge Q$ is reduced to the partition of singletons P_s. The second counting is associated with the partition set whose elements Q are such that $P \vee Q$ is the singleton partition P_t.

Q will be designated by $\{D_1, D_2, \ldots, D_{hj}, \ldots, D_H\}$. $t = (n_1, n_2, \ldots, n_k, \ldots, n_K)$ and $s = (m_1, m_2, \ldots, m_h, \ldots, m_H)$ denote the respective types of P and Q. In the second enumeration mentioned, we will start by the case where the type s is fixed. For the development below, keep in mind what we have expressed in **(f)** of Sect. 1.1.1.

To begin, let us consider the first counting. Clearly, if $H < \max_{1 \leq k \leq K} n_k$, there cannot exist a partition Q into H classes such that $P \wedge Q = P_s$. Consequently, we suppose $H \geq \max_{1 \leq k \leq K} n_k$. We have already observed (see Sect. 1.1.1) that the data of a partition Q into H classes *at most* for which $P \wedge Q = P_s$ can be given by a sequence of injective mappings $(\phi_1, \phi_2, \ldots, \phi_k, \ldots, \phi_K)$ of $C_1, C_2, \ldots, C_k, \ldots, C_K$, respectively, into the label set $\{1, 2, \ldots, h, \ldots, H\}$. The number of these sequences is given by

$$[n_1]_H \times [n_2]_H \times \cdots \times [n_k]_H \times \cdots \times [n_K]_H \qquad (1.1.33)$$

where

$$[n_k]_H = n_k \times (n_k - 1) \times \cdots \times (n_k - H + 1)$$

Now, we have to enumerate the set of sequences of injective mappings, as just considered, with respect to building a partition Q into H classes *exactly*, but such that

$$\bigcup_{1 \leq k \leq K} \phi(C_k) = \{1, 2, \ldots, h, \ldots, H\} \qquad (1.1.34)$$

For this enumeration purpose, begin by denoting F_j the set of sequences of injective mappings ϕ_k, for which

$$\bigcup_{1 \leq k \leq K} \phi(C_k) \subset \{1, 2, \ldots, h, \ldots, H\} - \{j\} \qquad (1.1.35)$$

Hence, the number desired is

$$[n_1]_H \times [n_2]_H \times \cdots \times [n_k]_H \times \cdots \times [n_K]_H - \mu \left(\bigcup_{1 \leq j \leq H} F_j \right) \qquad (1.1.36)$$

where μ is the cardinal function.

Inclusion and exclusion formula can be employed in order to evaluate the second term in (1.1.36). $\mathbb{H}$ denoting the set $\{1, 2, \ldots, h, \ldots, H\}$, we have

$$\mu \left(\bigcup_{j \in \mathbb{H}} F_j \right) = \sum_{G \subset \mathbb{H}} (-1)^{|G|+1} \mu \left(\bigcap_{j \in G} F_j \right) \qquad (1.1.37)$$

where $|\mathbb{G}|$ stands for $card(\mathbb{G})$. $\mu\left(\bigcap_{j\in\mathbb{G}} F_j\right)$ is the cardinal of injective mapping sequences which cannot reach $\mathbb{G}$. By setting $g = card(\mathbb{G})$ and $m = \max_{1\le k\le K} n_k$, we can observe that the latter cardinal is null if $m > H - g$. If not, this cardinal is equal to

$$[n_1]_{(H-g)} \times [n_2]_{(H-g)} \times \cdots \times [n_k]_{(H-g)} \times \cdots \times [n_K]_{(H-g)} \qquad (1.1.38)$$

and then, it depends only on $g = card(\mathbb{G})$.

In these conditions, the desired number (1.1.36) can be written as follows:

$$\sum_{1\le g\le H-m} (-1)^g \binom{H}{g} I(H-g) \qquad (1.1.39)$$

where $I(H - g)$ stands for (1.1.38).

The sequence of injective mappings $(\phi_1, \phi_2, \ldots, \phi_k, \ldots, \phi_K)$ of $C_1, C_2, \ldots,$ $C_k, \ldots$ and C_K, respectively, into the label set $\{1, 2, \ldots, h, \ldots, H\}$ determines a partition Q such that $P \wedge Q = P_s$. Every permutation σ of the sequence of the labels $(1, 2, \ldots, h, \ldots, H)$ gives the same partition Q from $(\psi_1 = \phi_1(\sigma), \psi_2 = \phi_2(\sigma), \ldots, \psi_k = \phi_k(\sigma), \ldots, \psi_K = \phi_K(\sigma))$. Reciprocally to a given Q, such that $P \wedge Q = P_s$, corresponds exactly $H!$ sequences of mappings $(\phi_1, \phi_2, \ldots, \phi_k, \ldots, \phi_K)$ which return Q. Hence, we have the following.

Proposition 4 *Let P be a partition of type $t = (n_1, n_2, \ldots, n_k, \ldots, n_K)$. The number of partitions Q into H non-empty classes such that $P \wedge Q = P_s$ is given by*

$$\sum_{1\le g\le H-m} \frac{(-1)^g}{g!(H-g)!} I(H-g) \qquad (1.1.40)$$

where $m = \max_{1\le k\le K} n_k$ and where $I(H - g)$ is given by (1.1.38).

Now, we shall enumerate the partition set composed of partitions Q such that $P \vee Q$ is the singleton partition P_t. Recall that $s = (m_1, m_2, \ldots, m_h, \ldots, m_H)$ indicates the type of a partition Q. First, we observe that in the set of partitions Q of type s, the proportion of those for which $P \vee Q$ is strictly finer than P_t is negligible.

Suppose, only to fix ideas the cardinals m_h, $1 \le h \le H$, are mutually distincts and consider the form to which the crossing grid of (P, Q) can be reduced—by permuting rows and columns—in the case where $P \vee Q < P_t$ (see Fig. 1.3). It follows that for a given partition P of type t, there corresponds at least one partition Q of type s, such that $P \vee Q < P_t$, if and only if there exists an ordered pair $(\mathbb{L}, \mathbb{M})$ of subsets of $\mathbb{H} = \{1, 2, \ldots, h, \ldots, H\}$ and $\mathbb{K} = \{1, 2, \ldots, k, \ldots, K\}$, respectively, satisfying the condition

$$\sum_{h\in\mathbb{L}} m_h = \sum_{k\in\mathbb{M}} n_k = l \qquad (1.1.41)$$

Relative to $\mathbb{M}$, consider the partition into two classes

$$P_{\mathbb{M}} = \left\{ \bigcup_{k \in \mathbb{M}} C_k, \bigcup_{k \notin \mathbb{M}} C_k \right\} \tag{1.1.42}$$

which is coarser than the partition P. To $\mathbb{M}$ we will associate the set of partitions Q for which $P \vee Q$ is at least as fine as the bi-partition (1.1.42). The proportion of such partitions is given by the formula

$$\frac{\frac{l!}{\Pi_{h \in \mathbb{L}} m_h!} \times \frac{(n-l)!}{\Pi_{h \notin \mathbb{L}} m_h!}}{\frac{n!}{\Pi_{1 \le h \le H}}} \tag{1.1.43}$$

In fact, a partition Q which takes part in the counting is equivalent to an ordered pair of partitions of respective types $(m_h | h \in \mathbb{L})$ and $(m_h | h \notin \mathbb{L})$ defined on $\bigcup_{k \in \mathbb{M}} C_k$ and $\bigcup_{k \notin \mathbb{M}} C_k$, respectively. Hence, (1.1.43) is deduced.

Notice that the proportion (1.1.43) can be brought down to the inverse of the binomial coefficient $\binom{n}{l}$. Hence the result.

Proposition 5 *Let P be a fixed partition of type $t = (n_1, n_2, \ldots, n_k, \ldots, n_K)$, the proportion of partitions Q into H non-empty classes such that $P \wedge Q = P_s$, of type $s = (m_1, m_2, \ldots, m_h, \ldots, m_H)$—where the m_h cardinals are mutually distinct—for which $P \vee Q < P_t$, is upper bounded by*

$$\sum_{(\mathbb{L}, \mathbb{M})} \frac{1}{\binom{n}{l}} \tag{1.1.44}$$

where the latter sum is extended over the ordered subset pairs $(\mathbb{L}, \mathbb{M})$, $\mathbb{L} \subset \mathbb{H}$ and $\mathbb{M} \subset \mathbb{K}$, such that the condition (1.1.41) is filled.

We can observe that (1.1.44) is negligible in the usual case where K is small with respect to n.

1.2 Partitions of an Integer

1.2.1 Generalities

The *type* of a partition P of an object set $\mathcal{O}$ of size n was defined above as the decreasing sequence of the class cardinals of P. By denoting this type

$$t = (n_1, n_2, \ldots, n_k, \ldots n_K) \tag{1.2.1}$$

we have

$$n_1 \geq n_2 \geq \cdots \geq n_k \geq \cdots \geq n_K \qquad (1.2.2)$$

The *order* of the partition is defined by the integer sequence $(r_1, r_2, \ldots, r_i, \ldots r_n)$ where r_i is the number of classes whose cardinality is i. We have

$$\sum_{1 \leq k \leq K} n_k = n \text{ and } \sum_{1 \leq i \leq n} i \times r_i = n \qquad (1.2.3)$$

A partition of n is defined as a decomposition of n into *parts*, such that their sum is equal to n. The order in which these parts are presented is irrelevant. Then, the partitions $(4, 2, 2, 1)$ and $(2, 1, 4, 2)$ of $n = 9$ are identical.

An integer partition of n can be brought down to be represented as in (1.2.1) and (1.2.2), by the decreasing sequence of its respective parts. In this way, the representation of a given integer partition is unique.

$\mathbb{P}(n)$ will designate the set of the partitions of an integer n. Each integer partition is represented by the decreasing sequence of its parts. Let $p = (n_1, n_2, \ldots, n_k, \ldots)$ and $q = (m_1, m_2, \ldots, m_k, \ldots)$ be two partitions of an integer n. The order relation $p \leq q$ we wish to introduce means that every part m_k of q is a sum of parts of p. Moreover, if $h \neq k$, the sum associated with the decomposition of m_h is disjoint from that associated with the decomposition of m_k. More formally, this can be written as follows:

$$(\forall h \neq k \geq 1), m_h = n_{i_1} + n_{i_2} + \cdots + n_{i_h}$$
$$m_k = n_{j_1} + n_{j_2} + \cdots + n_{j_k}$$
$$\text{and}$$
$$\{i_1, i_2, \cdots, i_h\} \cap \{j_1, j_2, \ldots, j_k\} = \emptyset \qquad (1.2.4)$$

$\mathbb{P}(n)$ can be expressed as the quotient set of the partition set $\mathcal{P}(\mathcal{O})$ of an object set of size n, by the equivalence relation corresponding to "have the same type". The mapping which associates to each partition in $\mathcal{P}(\mathcal{O})$, its type—defined as in (1.2.1) and (1.2.2)—is a surjective morphism between the ordered sets $\mathcal{P}(\mathcal{O})$ and $\mathbb{P}(n)$. For p in $\mathbb{P}(n)$ associated with P in $\mathcal{P}(\mathcal{O})$, n is partitioned into k parts by p, if P comprises k classes.

The ordered set $\mathbb{P}(n)$ has not a lattice structure. Two partitions of the integer n have not necessarily a single greatest lower bound. Therefore, the morphism mentioned just above between $\mathcal{P}(\mathcal{O})$ and $\mathbb{P}(n)$ is not a morphism between two lattice structures. In fact, the lowest upper bound operation $\vee$ of the $\mathcal{P}(\mathcal{O})$ lattice is not preserved in the correspondence concerned. Nonetheless, all maximal chains between two elements of $\mathbb{P}(n)$ have the same length.

1.2.2   Representations

A very known graphical representation of an integer partition is the Ferrer graph. Let $(n_1, n_2, \ldots, n_k, \ldots, n_K)$ be the type of the partition of an integer n:

$$n = n_1 + n_2 + \cdots + n_k + \cdots + n_K \text{ and } n_1 \geq n_2 \geq \cdots \geq n_k \geq \cdots \geq n_K \geq 1$$

The Ferrer graph is obtained by considering K equally spaced horizontal lines—justified left—and by drawing n_k equidistant points (or circles) on the kth line, from left to right (see Fig. 1.6).

Example

The integer partition $(4, 2, 2, 1)$ of $n = 9$ is represented by the Ferrer graph in Fig. 1.6. Notice that the two partitions $(5, 1)$ and $(4, 2)$ equally cover the two partitions $(4, 1, 1)$ and $(3, 2, 1)$. Hence, there is no *single* greatest lower bound.

The Ferrer graph representation visualizes an important property. If the Ferrer graph of a partition P is rotated of a right angle in the direction of clockwise, we obtain by symmetry with respect to a vertical line on the left of the diagram another partition of the integer concerned. The latter partition is called conjugate to P.

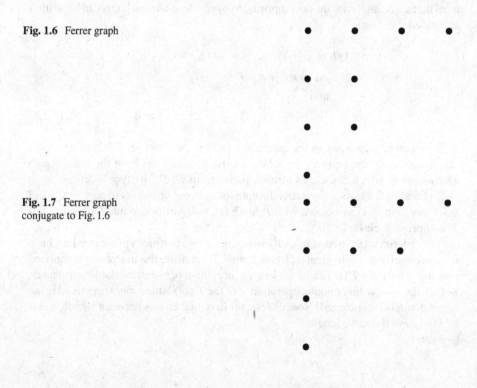

Fig. 1.6   Ferrer graph

Fig. 1.7   Ferrer graph
conjugate to Fig. 1.6

Example

Relative to the Ferrer graph depicted in Fig. 1.6, we get the graphical representation of Fig. 1.7 which corresponds to the partition $(4, 3, 1, 1)$ of $n = 9$.

1.3 Type of a Partition and Cardinality of the Associated Equivalence Binary Relation

Let F designate the set $P_2(\mathcal{O})$ of $\mathcal{O}$ subsets with two elements. F is also said the set of unordered pairs of $\mathcal{O}$:

$$F = \{\{x, y\} | x \in \mathcal{O}, y \in \mathcal{O}, x \neq y\} \tag{1.3.1}$$

A partition

$$P = \{O_1, O_2, \ldots, O_k, \ldots, O_K\} \tag{1.3.2}$$

into K non-empty classes induces a partition of F into two complementary classes $R(P)$ and $S(P)$:

$$R(P) = \{\{x, y\} \in F | \exists k, x \in O_k \text{ and } y \in O_k\}$$
$$S(P) = F - R(P) = \{\{x, y\} \in F | \exists j \neq k, x \in O_j \text{ and } y \in O_k\} \tag{1.3.3}$$

The graph data

$$Gr(P) = \{(x, y) \in \mathcal{O} \times \mathcal{O} | \exists k, x \in O_k \text{ and } y \in O_k\} \tag{1.3.4}$$

which is a subset of the Cartesian product $\mathcal{O} \times \mathcal{O}$—is equivalent to $R(P)$ and we have

$$card(Gr(P)) = 2card(R(P)) + n$$

where $n = card(\mathcal{O})$.

The purpose of this section is to study the relationship between the type of a partition P and the cardinal of $R(P)$ (see (1.2.1)).

Proposition 6

$$card(R(P)) = \sum_{1 \leq k \leq K} \frac{n_k(n_k - 1)}{2}$$

$$card(S(P)) = \sum_{1 \leq j < k \leq K} n_j n_k$$

$$card(R(P)) + card(S(P)) = card(F) = \frac{n(n-1)}{2} \tag{1.3.5}$$

where $(n_1, n_2, \ldots, n_k, \ldots, n_K)$ is the type of the partition P (see (1.2.1)).

Let R_k be the subset of F composed of object pairs joined in the kth class O_k of the partition P.

$$R_k = \{\{x, y\}|\{x, y\} \in F, x \in O_k \text{ and } y \in O_k\}$$

$1 \leq k \leq K$. The sets R_k constitute a partition of $R(P)$. The first of (1.3.5) is due to $card(R_k) = n_k(n_k - 1)/2$.

Let S_{jk} be the set of object pairs, such that for each pair, one component belongs to O_j and the other one to O_k:

$$S_{jk} = \{\{x, y\}|\{x, y\} \in F, x \in O_j \text{ and } y \in O_k\}$$

$1 \leq j < k \leq K$. The sets S_{jk} constitute a partition of $S(P)$. Due to the relation $card(S_{jk}) = n_j n_k$, the second of (1.3.5) is deduced.

The third of (1.3.5) is entailed by the fact that $R(P)$ and $S(P)$ form a partition of F.   □

Proposition 7 *Among all partitions of $\mathcal{O}$ into K classes, the type of those for which $card(R(P))$ is minimum is of the form*

$$(q + 1, q + 1, \ldots, q + 1, q, q, \ldots, q) \tag{1.3.6}$$

where q is the quotient of the division of n by K: $n = Kq + r$, where $r < K$.

In (1.3.6), K is the total number of components and r is the number of components equal to $q + 1$. The proof will be obtained recursively on n. Begin by noticing that

$$card(R(P)) = \frac{1}{2}\left(\sum_{1 \leq k \leq K} n_k^2 - n\right)$$

and then, minimizing $card(R(P))$ is equivalent to minimizing $\sum_{1 \leq k \leq K} n_k^2$.

For $n = K$, the property is obvious. Now, suppose the property true till a certain value of n strictly greater than K, that is,

$$r(q + 1)^2 = (K - r)q^2 \leq \sum_{1 \leq k \leq K} n_k^2 \tag{1.3.7}$$

for all types such that

$$\sum_{1 \leq k \leq K} n_k = n$$

In these conditions, we shall show that

$$(r + 1)(q + 1)^2 + \big(K - (r + 1)\big)q^2 \leq \sum_{1 \leq k \leq K} m_k^2 \tag{1.3.8}$$

for all types $(m_1, m_2, \ldots, m_k, \ldots, m_K)$ such that

$$\sum_{1 \leq k \leq K} m_k = n + 1$$

Notice that $r < k$ implies $r + 1 \leq K$. Equation (1.3.8) is equivalent to

$$r(q+1)^2 + (k-r)q^2 + 2q + 1 \leq (m_1 - 1)^2 + \sum_{2 \leq k \leq K} m_k^2 + 2m_1 - 1 \qquad (1.3.9)$$

m_1 is the greatest integer among the m_k, $1 \leq k \leq K$, then, necessarily, $m_1 \geq q + 1$, which is equivalent to

$$2q + 1 \leq 2m_1 - 1$$

Otherwise, $(m_1 - 1, m_2, \ldots, m_k, \ldots, m_K)$ defines a partition of the integer n into K parts. Then the recurrence hypothesis (1.3.7) applies and we have

$$r(q+1)^2 + (K-r)q^2 \leq (m_1 - 1)^2 + \sum_{2 \leq k \leq K} m_k^2$$

Adding member to member the two latter inequalities lead to the result desired.   □

Proposition 8 *Among all partitions of $\mathcal{O}$ into K classes, the type of those for which* $card\,(R(P))$ *is maximum is of the form*

$$(n - K + 1, 1, \ldots, 1, 1, 1, \ldots, 1) \qquad (1.3.10)$$

where the number of components equal to 1 is $K - 1$.

As for the preceding proposition the proof is based on a recursive reasoning with respect to n. It concerns maximization of $\sum_{1 \leq k \leq K} n_k^2$.

The property is trivial for $n = K$. Let us suppose that this property is true for a given n strictly greater than K, that is to say,

$$(n - K + 1)^2 + K - 1 \geq \sum_{1 \leq k \leq K} n_k^2 \qquad (1.3.11)$$

for any decomposition $n = \sum_{1 \leq k \leq K} n_k^2$ of n into K parts.

In these conditions, we have to prove

$$(n + 1 - K + 1)^2 + K - 1 \geq \sum_{1 \leq k \leq K} m_k^2 \qquad (1.3.12)$$

where $\sum_{1 \leq k \leq K} m_k = n + 1$.

Equation (1.3.12) is equivalent to

$$(n - K + 1)^2 + 2(n - K + 1) + 1 + K - 1 \geq (m_1 - 1)^2 + \sum_{2 \leq k \leq K} m_k^2 + 2(m_1 - 1) + 1$$

$$(1.3.13)$$

Now, $m_1 \leq (n + 1) - (K - 1)$, which can be written as follows:

$$2(n - K + 1) + 1 \geq 2(m_1 - 1) + 1$$

On the other hand, the recurrence hypothesis enables us to write

$$(n - K + 1)^2 + K - 1 \geq (m_1 - 1)^2 + \sum_{2 \leq k \leq K} m_k^2$$

Adding the associated members of the two previous inequalities ends the recursive proof.   □

Proposition 9 *In the partition set* $\mathcal{P}(\mathcal{O})$, *the partitions* P *which realize the maximum of* $card(R(P) \times S(P))$ *are those for which* $card(R(P)) = card(S(P))$.

By noticing the following relations

$$card(R(P) \times S(P)) = card(R(P)) \times card(S(P))$$
and
$$card(R(P)) + card(S(P)) = \frac{n(n - 1)}{2}$$

$$(1.3.14)$$

the result becomes obvious.   □'

Therefore, an upper bound of $card(R(P) \times S(P))$ is given by $(n(n - 1)/4)^2$. Moreover, the product expressed in (1.3.14) is all the more large so that the absolute value of the difference between $card(R(P))$ and $card(S(P))$ is small.

Illustration

We shall illustrate the previous proposition in the case of a partition with two classes. More precisely, the purpose consists of determining the type of such a partition, which maximizes $card(R(P) \times S(P))$. Designate by $(m, n - m)$ the type of this partition. We have $n - m \leq m \leq n - 1$.

The maximum value of $card(R(P) \times S(P))$ is get if there exists an integer m, $n - m \leq m \leq n - 1$, such that

$$m(n - m) = \frac{n(n - 1)}{4}$$

$$(1.3.15)$$

where the left member is nothing else than $card(S(P))$. However, such an integer m might not exist. In these conditions, we shall propose for m an integer value as near as possible a positive real value, solution of

$$\mu(n - \mu) = \frac{n(n-1)}{4} \tag{1.3.16}$$

where $n - \mu \le \mu \le n - 1$. Equation (1.3.16) leads to the quadratic equation

$$4\mu^2 - 4n\mu + n(n-1) = 0 \tag{1.3.17}$$

whose only admissible solution ($\mu \ge n - \mu$) is

$$\mu' = \frac{n + \sqrt{n}}{2} \tag{1.3.18}$$

In fact, the second solution of (1.3.17) is

$$\mu'' = \frac{n - \sqrt{n}}{2} = n - \mu' \tag{1.3.19}$$

(μ', μ'') corresponds to the type of a partition.

If n is a square of an integer, $\sqrt{n}$ is an integer. The type of the sought partition is then $((n + \sqrt{n})/2, (n - \sqrt{n})/2)$ (n and $\sqrt{n}$ have the same parity). If n is not an integer square, the partition type desired is

$$\left(\left[\frac{n + \sqrt{n}}{2} \right], \left[\frac{n - \sqrt{n}}{2} \right] \right) \tag{1.3.20}$$

where $[\mu]$ indicates the integer part of μ.

Proposition 10 *If the number of classes of the partition P is $K = 2$, the $card(R(P))$ data is equivalent to the partition type (n_1, n_2) data.*

As above, instead of working directly with $card(R(P))$, we can reason with respect the sum of squares $n_1^2 + n_2^2$. Consider $(x, n - x)$ and $(y, n - y)$ as two types of partitions with two classes ($x \ge n - x$ and $y \ge n - y$)

$$x^2 + (n - x)^2 = y^2(n - y)^2 \tag{1.3.21}$$

which is equivalent to the quadratic equation

$$x^2 - 2nx - y(y - 2n) = 0 \tag{1.3.22}$$

where x is expressed with respect to y. The two solutions of (1.3.22) are

$$x' = n - \sqrt{(n - y)^2} = y$$
$$\text{and}$$
$$x'' = n + \sqrt{(n - y)^2} = 2n - y, \tag{1.3.23}$$

respectively.

$x' = y$ gives the type $(y, n - y)$ which is identical to $(x, n - x)$ and $x'' = 2n - y$ *cannot* give a possible type since $n - (2n - y) = y - n < 0$.   □

The last result leads us to ask "In what extent the $card(R(P))$ data informs us on the type of the partition P?". More precisely, given a positive integer n, what are the partition types $(n_1, n_2, \ldots, n_k, \ldots, n_K)$ for which $\sum_{1 \leq k \leq K} n_k^2 = m$? Besides, what is the cardinality of all these partition types?

Proposition 11 *Let $\phi(n, m, K)$ designate the cardinal of the set of partition types $(n_1, n_2, \ldots, n_k, \ldots, n_K)$ into K non-empty classes for which $\sum_{1 \leq k \leq K} n_k^2 = m$. We have the recursive formula*

$$\phi(n, m, K) = \sum_{1 \leq l < \frac{n}{K}} \phi(n - lK - 1, m - 2ln + l^2 K - 1, K - 1) \qquad (1.3.24)$$

It is of importance to notice that the lth term of the latter sum corresponds to the set of the integer partitions of n for which the Kth part is $l + 1$. The other parts of a given element of this set are obtained by adding l to the corresponding parts of the associated partition of $n - lK - 1$. $\phi(n, m, K)$ is the cardinal of the set

$$\Phi(n, m, K) = \{(n_1, n_2, \ldots, n_k, \ldots, n_K) |$$

$$n_1 \geq n_2 \geq \ldots \geq n_k \geq \ldots \geq n_K, \sum_{1 \leq k \leq K} n_k = n \text{ and } \sum_{1 \leq k \leq K} n_k^2 = m\}$$

$$(1.3.25)$$

The formula (1.3.24) is obtained by partitioning $\Phi(n, m, K)$ according to the possible values of n_K, $1 \leq n_K \leq [n/K]$, where $[n/K]$ is the integer part of n/K. One possible element of $\Phi(n, m, K)$, for which $n_K = h$, satisfies

$$n_1 \geq n_2 \geq \cdots \geq n_k \geq \cdots \geq n_{K-1} \geq h,$$

$$h + \sum_{1 \leq i \leq K-1} n_i = n \text{ and } h^2 + \sum_{1 \leq i \leq K-1} n_i^2 = m \qquad (1.3.26)$$

Conditions (1.3.26) are equivalent to

$$n_1 - (h - 1) \geq n_2 - (h - 1) \geq \cdots \geq n_{K-1} - (h - 1) \geq 1$$

$$\sum_{1 \leq i \leq K-1} (n_i - (h - 1)) = n - (h - 1)K - 1$$

$$\sum_{1 \leq i \leq K-1} (n_i - (h - 1))^2 = m - 2(h - 1)n + K(h - 1)^2 - 1$$

$$(1.3.27)$$

Let $\nu_i = n_i - (h + 1)$, $1 \le i \le K - 1$. Equations (1.3.27) are reduced to

$$\nu_1 \ge \nu_2 \ge \cdots \ge \nu_{K-1} \ge 1$$

$$\sum_{1 \le i \le K-1} \nu_i = n - (h-1)K - 1$$

$$\sum_{1 \le i \le K-1} \nu_i = m - 2(h-1)n + K(h-1)^2 - 1 \qquad (1.3.28)$$

Therefore, the subset of $\Phi(n, m, K)$ (see (1.3.25)), for which $n_K = h$, can be associated bijectively with

$$\Phi\big(n - (h-1)K - 1, m - 2(h-1)n + K(h-1)^2 - 1, K - 1\big) \qquad (1.3.29)$$

By substituting l for $(h-1)$, we get the recurrence formula. $\qquad\square$

In order to determine the different elements of the set $\Phi(n, m, K)$, we proceed gradually. To begin, we define a partition of $\Phi(n, m, K)$ according to the value of n_K. The subset of $\Phi(n, m, K)$ for which $n_K = h$ is calculated from the set

$$\Phi\big(n - (h-1)K - 1, m - 2(h-1)n + (h-1)^2 K - 1, K - 1\big) \qquad (1.3.30)$$

as follows. To every element $\nu_1, \nu_2, \ldots, \nu_{K-1}$ of this set (see (1.3.28)) corresponds bijectively, an element $(n_1, n_2, \ldots, n_{K-1}, h)$ of $\Phi(n, m, K)$, where $n_k = \nu_k + h - 1$, $1 \le k \le K - 1$. Besides, to determine the set expressed in (1.3.30) we start by considering a partition of the latter, according to the value taken by ν_{K-1}, and so on. We shall make explicit this process on an example treated below. Clearly, the way by which the recursive formula (1.3.24) was established enables the set $\Phi(n, m, K)$ to be described.

The sets

$$\big\{\Phi(n, m, K) | 1 \le K \le n\big\}$$

allow computation of the set of all partitions of the integer n, for which $\sum_{1 \le k \le K} n_k^2$, to be performed. Designating by $\Psi(n, m)$ this set, we have

$$\Psi(n, m) = \sum_{1 \le K \le n} \Phi(n, m, K) \text{ (set sum)} \qquad (1.3.31)$$

Notice that $\phi(n, m, K)$ vanishes if n and m have not the same parity. In fact, $n + m$ is even since it is a sum of even integers, namely,

$$m + n = \sum_{1 \le k \le K} n_k^2 + \sum_{1 \le k \le K} n_k = \sum_{1 \le k \le K} n_k(n_k + 1) \qquad (1.3.32)$$

On the other hand, Propositions 7 and 8 entail

$$r(q+1)^2 + (K-r)q^2 \le m \le (n-K+1)^2 + K - 1 \qquad (1.3.33)$$

where q and r are the quotient and rest of the division of h by K.

Otherwise, we can notice that the problem developed above is a very specific version of the Waring problem [28]. In the framework of the latter, an asymptotic formula is established for the number of representations of a positive integer n as a sum of jth powers of K natural numbers:

$$n = n_1^j + n_2^j + \cdots + n_K^j$$

The case of interest considered above corresponds to $j = 2$.

Example

We shall now illustrate Eqs. (1.3.24) and (1.3.25) by determining the set $\Phi(20, 98, 5)$ (see (1.3.25)) whose cardinality is $\phi(20, 98, 5)$ (see (1.3.24)). This determination is based on recursive reductions associated with (1.3.24), followed by the use of Proposition 10. By applying (1.3.24), we have

$$\phi(20, 98, 5) = \phi(19, 97, 4) + \phi(14, 62, 4) + \phi(9, 37, 4) + \phi(4, 22, 4)$$

$$1 - \phi(19, 97, 4) = \phi(18, 96, 3) + \phi(14, 62, 3) + \phi(10, 36, 3) + \phi(6, 18, 3)$$

$$1.1 - \phi(18, 96, 3) = 0 \text{ because } 96 < 6^2 + 6^2 + 6^2 = 108 \text{ (see)} \quad (1.3.33)$$

$$1.2 - \phi(14, 62, 3) = 0 \text{ because } 62 < 5^2 + 5^2 + 4^2 = 66 \text{ (see)} \quad (1.3.33)$$

$$1.3 - \phi(10, 36, 3) = \phi(9, 35, 2) + \phi(6, 18, 2) + \phi(3, 7, 2)$$

$$1.3.1 - \phi(9, 35, 2) = 0 \text{ because } 35 < 5^2 + 4^2 = 41$$

$$1.3.2 - \phi(6, 18, 2) = 1 \text{ and the associated partition is } (\mathbf{3}, \mathbf{3})$$

$$1.3.3 - \phi(3, 7, 2) = 0 \text{ because } 7 > 2^2 + 1^2 = 5$$

$$1.4 - \phi(6, 18, 3) = \phi(5, 17, 2) + \phi(2, 8, 2)$$

$$1.4.1 - \phi(5, 17, 2) = 1 \text{ and the associated partition is } (\mathbf{4}, \mathbf{1})$$

$$1.4.2 - \phi(2, 8, 2) = 0$$

$$2 - \phi(14, 62, 4) = \phi(13, 61, 3) + \phi(9, 37, 3) + \phi(5, 21, 3) + \phi(1, 13, 3)$$

$$2.1 - \phi(13, 61, 3) = \phi(12, 60, 2) + \phi(9, 37, 2) + \phi(6, 20, 2) + \phi(3, 91, 2)$$

$$2.1.1 - \phi(12, 60, 2) = 0 \text{ because } 60 < 6^2 + 6^2 = 72$$

$$2.1.2 - \phi(9, 37, 2) = 0 \text{ because } 37 < 5^2 + 4^2 = 41$$

$$2.1.3 - \phi(6, 20, 2) = 1 \text{ and the associated partition is } (\mathbf{4, 2})$$

$$2.1.4 - \phi(3, 9, 2) = 0 \text{ because } 9 > 2^2 + 1^2 = 5$$

$$2.2 - \phi(9, 37, 3) = \phi(8, 36, 2) + \phi(5, 21, 2) + \phi(2, 12, 2)$$

Each term of the right member is null and then $\phi(9, 37, 3) = 0$

$$2.3 - \phi(5, 21, 3) = 0 \text{ because } 21 > 3^2 + 1^2 + 1^2 = 11$$

$$2.4 - \phi(1, 13, 3) = 0$$

Also, it is easy to verify that

$$3 - \phi(9, 37, 4) = 0$$

$$4 - \phi(4, 22, 4) = 0$$

As established above, in the case of partitions into two parts, the sum of squares $n_1^2 + n_2^2$ determines the integer partition type (n_1, n_2). Thereby, we have the following partition types:

$$\Phi(6, 18, 2) = \{(3, 3)\}, \ \Phi(5, 17, 2) = \{(4, 1)\} \text{ and } \Phi(6, 20, 2) = \{(4, 2)\}$$

associated, respectively, with the cases 1.3.2, 1.4.1 and 2.1.3.

Now, let us go back to the recursion. The sequences obtained from $(3, 3)$, $(4, 1)$ and $(4, 2)$ are

$$(3, 3), (4, 4, 2), (6, 6, 4, 3) \text{ and } (6, 6, 4, 3, 1),$$

$$(4, 1), (4, 1, 1), (7, 4, 4, 4) \text{ and } (7, 4, 4, 4, 1)$$

and

$$(4, 2), (6, 4, 3), (6, 4, 3, 1) \text{ and } (7, 5, 4, 2, 2)$$

Finally, we obtain

$$\Phi(20, 98, 5) = \{(6, 6, 4, 3, 1), (7, 4, 4, 4, 1), (7, 5, 4, 2, 2)\}$$

1.4   Ultrametric Spaces and Partition Chain Representation

1.4.1   Definition and Properties of Ultrametric Spaces

A *metric space* is defined by a pair $(\mathcal{O}, d)$ where $\mathcal{O}$ is a set and d a positive numerical function of $\mathcal{O} \times \mathcal{O}$ into the positive reals $\mathbb{R}_+$:

$$\mathcal{O} \times \mathcal{O} \longrightarrow \mathbb{R}_+,$$

satisfying the conditions

(M1) $\big(\forall (x, y) \in \mathcal{O} \times \mathcal{O}\big), d(x, y) = d(y, x)$ "symmetry"
(M2) $\big(\forall (x, y) \in \mathcal{O} \times \mathcal{O}\big), d(x, y) = 0 \Leftrightarrow$ "identification"
(M3) $\big(\forall (x, y, z) \in \mathcal{O} \times \mathcal{O} \times \mathcal{O}\big), d(x, z) \leq d(x, y) + d(y, z)$
"triangle inequality"   (1.4.1)

In a *ultrametric space* the condition (M3) is replaced by

(UM) $\big(\forall (x, y, z) \in \mathcal{O} \times \mathcal{O} \times \mathcal{O}\big), d(x, z) \leq \max\big(d(x, y), d(y, z)\big)$ (1.4.2)

Obviously, (UM) implies (M3) which becomes immaterial in the case of ultrametric space. Now, let us recall some definitions.

- A *circumference* of centre o and radius r in a metrical space $(\mathcal{O}, d)$ is the set

$$S(o; r) = \{x \,|\, x \in \mathcal{O} \text{ and } d(x, o) = r\} \qquad (1.4.3)$$

$S(o; r)$ is usually called *sphere*.
- A *closed ball* (resp., *open ball*) of centre o and radius r is the set

$$B(o; r) = \{x \,|\, x \in \mathcal{O} \text{ and } d(x, o) \leq r\}$$
$$(\text{resp., } B(o; r^-) = \{x \,|\, x \in \mathcal{O} \text{ and } d(x, o) < r\}) \qquad (1.4.4)$$

- A *divisor* of $(\mathcal{O}, d)$ is a binary equivalence relation on $\mathcal{O}$, satisfying

$$(\text{DIV}) \big(\forall (p, q, x, y) \in \mathcal{O}^4\big), pDq \text{ and } d(x, y) \leq d(p, q) \Rightarrow xDy \qquad (1.4.5)$$

- A *valuation* of a divisor D of the space $(\mathcal{O}, d)$ is the number

$$\nu(D) = sup\{d(x, y) \,|\, (x, y) \in \mathcal{O} \times \mathcal{O} \text{ and } xDy\} \qquad (1.4.6)$$

Theorem 1 *A metrical space* $(\mathcal{O}, d)$ *is ultrametric if and only if every triangle of this space is isosceles, the basis length being lower than the common length of the two other sides.*

Obviously, the latter geometrical property entails the condition (UM) (see (1.4.2)). Conversely, let (x, y, z) be a triangle of a ultrametric space $(\mathcal{O}, d)$ and assume (x, y) to be the lowest side. We have

$$d(x, z) \leq \max(d(x, y), d(y, z)) = d(y, z)$$

and

$$d(y, z) \leq \max(d(x, y), d(x, z)) = d(x, z)$$

Hence,

$$d(x, z) = d(y, z) \geq d(x, y)$$

$\square$

From now on, $(\mathcal{O}, d_u)$ will designate a ultrametric space.

Proposition 12 *Every point in a closed ball in a ultrametric space is a centre of this ball.*

Consider the notations in (1.4.4). Due to the ultrametric inequality, if $x \in B(o; r)$ and $p \in B(o; r)$, then $x \in B(p; r)$. Besides, if x is a point of the sphere $S(o; r)$ $(S(o; r) \subset B(o; r))$ and if p is a point of $B(o; r^-)$, we have $d(p, x) = d(o, x)$; that is to say, x is a point of the sphere $S(p; r)$.

Corollary 1 *Two non-disjoint balls* $B(o; r)$ *and* $B(p; s)$ *in a ultrametric space are concentric.*

In fact, every point of the intersection $B(o; r) \cap B(p; s)$ is a common centre of $B(o; r)$ and $B(p; s)$.

Corollary 2 *Two non-disjoint balls* $B(o; r)$ *and* $B(p; r)$ *of the same radius in a ultrametric space coincide.*

Corollary 3 *The balls of the same radius in a ultrametric space* $(\mathcal{O}, d_u)$ *constitute a partition of* $\mathcal{O}$. *The equivalence binary relation associated to the latter is a divisor of* $(\mathcal{O}, d_u)$.

Proposition 13 *Let* D *and* D' *be two divisors of a finite ultrametric space* $(\mathcal{O}, d_u)$. *If* D *is finer than* D' $(\forall (x, y) \in \mathcal{O} \times \mathcal{O}, x D y \Rightarrow x D' y)$, *then* $\nu(D) \leq \nu(D')$. *Reciprocally, given two divisors* D *and* D' *of* $(\mathcal{O}, d_u)$, *if* $\nu(D) \leq \nu(D')$, *then* D *is finer than* D'.

Proposition 14 *If $B(o; r)$ and $B(p; s)$ are two disjoint balls of $(\mathcal{O}, d_u)$, the distance between one element x of $B(o; r)$ and one element y of $B(p; s)$ depends only on the balls $B(o; r)$ and $B(p; s)$ and not on the specific points x and y.*

Let y' be an arbitrary element of $B(p; s)$, different from y. We have $d(y, y') < d(y, x)$. Theorem 1 entails

$$d(x, y) = d(x, y') > d(y, y') \qquad \square$$

Corollary 4 *The quotient $\mathcal{O}/D$ of a ultrametric space $(\mathcal{O}, d_u)$ by a divisor D is a ultrametric space. The distance between two elements (points) of the latter is strictly greater than $\nu(D)$ in the finite case.*

Proposition 15 *Every open ball $B(o; r^-)$ of radius r^- in $(\mathcal{O}, d_u)$, non-disjoint of a sphere $S(p; r)$ of radius r, is included in $S(p; r)$.*

Let x denote an element of $B(o; r^-)$. The set $S(p; r) \cap B(o; r^-)$ is non-empty. Therefore, there exists a point y which belongs to the latter set. By considering the triangle (x, p, y), we have $d(x, y) < r$. This is because y—which belongs to $B(o; r^-)$—is also the centre of $B(o; r^-)$. Consequently, by applying Theorem 1, we get $d(p, x) = d(p, y) = r$, and then, x belongs to $S(p; r)$. $\qquad \square$

Corollary 5 *A sphere of radius r in a ultrametric space is union of balls mutually disjoint of radius r^-.*

- A ultrametric proximity p_u is a positive numerical function (which might reach infinite value) defined by

$$p_u : \mathcal{O} \times \mathcal{O} \longrightarrow \mathbb{R}_+ \cup \{+\infty\}$$

verifying the following conditions:

(PUM1) $\big(\forall(x, y) \in \mathcal{O} \times \mathcal{O}\big)$, $p_u(x, y) = p_u(y, x)$ "symmetry"

(PUM2) $\big(\forall(x, y) \in \mathcal{O} \times \mathcal{O}\big)$, $p_u(x, y) = +\infty$ if and only if $x = y$ "identity"

(PUM3) $\big(\forall(x, y, z) \in \mathcal{O} \times \mathcal{O} \times \mathcal{O}\big)$, $p_u(x, z) \geq \min\big(p_u(x, y), p_u(y, z)\big)$ "ultrametric inequality"

$$(1.4.7)$$

Proposition 16 *If d_u is a ultrametric distance, then $-\log(d_u)$ is a ultrametric proximity. Conversely, if p_u is a ultrametric proximity, then $\exp(-p_u)$ is a ultrametric distance.*

- A ball of centre o and *smallness* s in the ultrametric space $(\mathcal{O}, p_u)$ is a set of the form

$$\{x \mid x \in \mathcal{O} \text{ and } p_u(o, x) \geq s \text{ (resp., } > s)\} \tag{1.4.8}$$

where s is a positive real value. The previous inequality is large (resp., strict) in the case of a closed (resp., open) ball.

- The order of a divisor D of $(\mathcal{O}, p_u)$ is defined as follows:

$$o(D) = inf\{p_u(x, y) \mid (x, y) \in \mathcal{O} \times \mathcal{O}\} \tag{1.4.9}$$

Clearly, $o(D) = -\log(\nu(D))$.

1.4.2 Partition Lattice Chains of a Finite Set and the Associated Ultrametric Spaces

We introduced in Sect. 1.1.1 the set $\mathcal{P}(\mathcal{O})$ of all partitions of $\mathcal{O}$, supposed here as a finite set. Relative to the lattice structure of $\mathcal{P}(\mathcal{O})$, described in Sect. 1.1.1, we shall consider totally ordered chains of partitions. A generic element of them can be written as follows:

$$C = (P_0, P_1, \ldots, P_i, \ldots, P_I) \tag{1.4.10}$$

where

- $\{0, 1, \ldots, i, \ldots, I\}$ is a begining segment of the natural integers $\mathbb{N}$;

- P_0 is the partition of singletons, for which each class includes exactly one element (see above);

- $(\forall i, 0 \leq i \leq I)$, $P_i < P_{i+1}$, which means P_i is finer than P_{i+1}.

Now, let us provide $\mathcal{O}$ with a distance function d associated with C (see (1.4.10)) as follows:

$$\big(\forall (x, y) \in \mathcal{O} \times \mathcal{O}\big),$$

$d(x, y)$ is the lowest integer i such that x and y are joined in a same class of P_i.

Proposition 17 *The distance defined above is ultrametric. Its divisors are the equivalence relations associated with the partitions P_i, $0 \leq i \leq I$.*

It is easy to verify this statement.

Proposition 18 *Conversely to Proposition 17, to every finite ultrametric space $(\mathcal{O}, d_u)$ corresponds in a unique way, a partition chain C in $\mathcal{P}(\mathcal{O})$ (see (1.4.10)) whose elements P_i, $0 \leq i \leq I$, are the divisors of $(\mathcal{O}, d_u)$.*

According to Corollary 3, consider the closed balls of radius r and let r scan increasingly $\mathbb{R}_+$ from 0. The finiteness of the set $\mathcal{O}$ makes that r will describe a finite set $r_0, r_1, \ldots, r_i, \ldots, r_I$ of r values:

$$r_0 < r_1 < \cdots < r_i < \cdots < r_I$$

where $I \le n-1$ and where the partition associated with r_i is finer than that associated with $r_{i+1}, 0 \le i \le I - 1$.

Propositions 17 and 18 correspond to the Johnson clustering scheme [11]. They have been established completely independently—with an ordinal version—in [14], Sect. IV and reported in [15].

Example

Let us consider an instance of a totally ordered chain of the partition lattice, defined on the set

$$\mathcal{O} = \{o_a, o_b, o_c, o_d, o_e, o_f, o_g, o_h, o_i, o_j\}$$

This chain which can be denoted by (P_0, P_1, P_2, P_3) is figured by the classification tree shown in Fig. 1.8. P_0 (resp., P_3) is the partition of singletons (resp., the singleton partition). P_1 and P_2 can be written as follows:

$$P_1 = \{\{o_a\}, \{o_b, o_c\}, \{o_d\}, \{o_e, o_f\}, \{o_g\}, \{o_h, o_i\}, \{o_j\}\}$$

and

$$P_2 = \{\{o_a, o_b, o_c\}, \{o_d, o_e, o_f, o_g\}, \{o_h, o_i, o_j\}\}$$

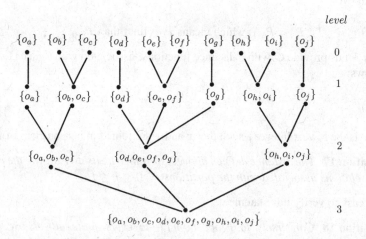

Fig. 1.8 Classification tree associated with a partition chain

Table 1.2 Ultrametric matrix associated with Fig. 1.8

	o_a	o_b	o_c	o_d	o_e	o_f	o_g	o_h	o_i	o_j
o_a	0	2	2	3	3	3	3	3	3	3
o_b	2	0	1	3	3	3	3	3	3	3
o_c	2	1	0	3	3	3	3	3	3	3
o_d	3	3	3	0	2	2	2	3	3	3
o_e	3	3	3	2	0	1	2	3	3	3
o_f	3	3	3	2	1	0	2	3	3	3
o_g	3	3	3	2	2	2	0	3	3	3
o_h	3	3	3	3	3	3	3	0	1	2
o_i	3	3	3	3	3	3	3	1	0	2
o_j	3	3	3	3	3	3	3	2	2	0

The level function defines a ultrametric distance. The matrix of the latter is given in Table 1.2. According to Proposition 17, the ultrametric distance associated with this partition chain is given by the square distance matrix of Table 1.2.

This matrix is symmetric with respect to its diagonal. In fact, we will consider below the upper part of this matrix, including its diagonal, namely,

$$\{d_u(x, y) \,|\, x \geq y\} \tag{1.4.11}$$

where x (resp., y) indicates the label column (resp., row) of the square table representing the ultrametric distance matrix.

Now, we shall give a characteristic property concerning a reduced form of a distance matrix, in order to be recognized as a *ultrametric* distance matrix. Initially, this property was proposed by M. Eytan (personal communication). However, the formulation we give makes clearer the algorithm which brought down a ultrametric matrix to the form mentioned. The latter will be specified by the hereafter Theorem 2

Lemma 1 *Let x_0 be an arbitrary point of a ultrametric space $(\mathcal{O}, d_u)$ where d_u takes its values in a beginning segment $\{1, 2, \ldots, i, \ldots, I\}$ of the natural integers $\mathbb{N}$. All of the points of the circular border $K(x_0; i) = B(x_0; i) - B(x_0; i - 1)$ are at equal distances of every point outside of $B(x_0; i)$.*

Notice that $K(x_0; i)$ is the sphere of centre x_0 and radius i. For x not belonging to $B(x_0; i)$, $d_u(x_0, x) > i$. Let y and z be two points of $K(x_0; i)$. We have

$$d_u(x_0, y) = d_u(x_0, z) = i$$

Therefore, by Theorem 1, $d_u(y, z) \leq i$. Since y belongs to $B(x_0; i)$, y is a centre of $B(x_0; i)$ (see Proposition 12). Then $d_u(y, z) = i$. And, for $x \notin B(x_0; i)$, $d_u(y, x) > i$. Finally, $d_u(x, y) = d_u(x, z) > d_u(y, z)$. $\qquad \square$

Algorithm

Let us label the elements of the object set $\mathcal{O}$. $\mathcal{O} = \{o_1, o_2, \ldots, o_i, \ldots, o_n\}$. To each ranking (permutation) $(o_{(1)}, o_{(2)}, \ldots, o_{(i)}, \ldots, o_{(n)})$ of the $\mathcal{O}$ elements, we will associate a square distance matrix where the ith column (resp., row) represents the $o_{(i)}$ object, $1 \leq i \leq n$. The distance value between $o_{(i)}$ and $o_{(j)}$ is set at the intersection of the ith row and the jth column, $1 \leq i, j \leq n$. We shall describe in the following an algorithm which determines a permutation of the sequence $(o_1, o_2, \ldots, o_i, \ldots, o_n)$ which gives a particular form, above mentioned, to the ultrametric distance matrix.

Let x_0 be an arbitrary element of a finite set $\mathcal{O}$, endowed with a ultrametric distance d_u. x_0 will be the first element of the permutation of $\mathcal{O}$ we have to build. Consider now the total preorder of $\mathcal{O}$, whose classes are the different circular borders $K(x_0, i) = B(x_0, i) - B(x_0, i - 1)$, for i describing a beginning section of natural integers, from $i = 1$ ($i \geq 1$). Building this preorder defines the first step of the algorithm

Let $K(x_0, i)$ and $K(x_0, j)$—with $i < j$—be the two concentric circular borders surrounding x_0. Every point x of $K(x_0, i)$ is at equal distances of all points of $K(x_0, j)$, because x is a centre of $B(x_0, j)$. Reciprocally, due to the previous lemma, any point of $K(x_0, j)$ is at equal distances of all points of $K(x_0, i)$. Consequently, a given class $K(x_0, i)$ of the total preorder defined cannot be refined by ranking its elements according to increasing or decreasing ultrametric proximities with respect to a point outside the circular border $K(x_0, i)$.

In these conditions, the next step consists of restricting to each of the classes $K(x_0, i)$, the application of the algorithm considered first at the level of the whole set, that is, ranking the elements of each $K(x_0, i)$ by decreasing ultrametric proximity with respect to one of the points of $K(x_0, i)$.

This refinement is taken up again, recursively, for each class defined by a circular border, until stability. This means that no new cutting of a class defined by a circular border is possible.

Theorem 2 *Let $D = \{d(i, j) | 1 \leq i, j \leq n\}$ be a distance matrix on a set $\mathcal{O} = \{o_i | 1 \leq i \leq n\}$ of n objects, where $d(i, j)$ stands for $d(o_i, o_j)$. The matrix D corresponds to a ultrametric distance if and only if the permutation obtained by the preceding algorithm gives to this matrix the following form:*

1. *Relative to the kth row, $1 \leq k \leq n$, from the diagonal element, which is equal to 0 ($d(k, k) = 0$), the successive elements from left to right go on increasing in a large sense;*
2. *For every index k, $1 \leq k \leq n$,*

$$if\, d(k, k + 1) = d(k, k + 2) = \cdots = d(k, k + l + 1)$$
$$then\, d(k + 1, j) \leq d(k, j)\, for\, k + 1 < j \leq k + l + 1$$
$$and\, d(k + 1, j) = d(k, j)\, for\, j > k + l + 1. \tag{1.4.12}$$

Condition 2 expresses that all triangles of the form $(k, k + 1, j)$, where $j > k + 1$, satisfy the conditions of Theorem 1. Therefore, if the distance d is ultrametric, conditions 1 and 2 above are satisfied.

Reciprocally, let (k, j, h) be a triangle where $k < j < h$. From condition 1 we have

$$d(k, j) \leq d(k, h)$$

On the other hand, condition 2 implies

$$d(j, h) \leq d(k, h)$$

If $d(k, j) < d(k, h)$, due to condition 2, we have

$$d(k + 1, j) \leq d(k, j) \text{ and } d(k + 1, h) = d(k, h)$$

Hence,

$$d(k + 1, j) < d(k + 1, h) = d(k, h)$$

because $d(k, j) < d(k, h) = d(k + 1, h)$.

By recurrence we obtain

$$d(j - 1, j) < d(j - 1, h) = d(k, h)$$

Consequently (condition 2)

$$d(j, h) = d(j - 1, h) = d(k, h)$$

Therefore, the triangle (k, j, h) is isosceles, the base (k, j) being the strictly lowest side.

Now, if $d(k, j) = d(k, h)$, the inequality $d(j, h) \leq d(k, h)$, obtained from condition 2, leads to the same result in terms of isosceles property.

1.4.3 Partition Lattice Chains and the Associated Ultrametric Preordonances

In the previous section we have seen that an ordered partition chain

$$\mathcal{C} = (P_0, P_1, \ldots, P_i, \ldots, P_I) \tag{1.4.13}$$

on a finite set $\mathcal{O}$ can be represented by a ultrametric space $(\mathcal{O}, d_\mathcal{C})$, where $d_\mathcal{C}$ is the ultrametric distance associated with $\mathcal{C}$. For (x, y) belonging to $\mathcal{O} \times \mathcal{O}$, $d_\mathcal{C}(x, y)$ is the lowest integer i, for which x and y are joined in a same class of P_i, $1 \leq i \leq I$.

We shall now give a representation of a partition chain where we emerge from the numerical distance concept. As in (1.3.1), let us consider the set F of unordered element pairs of the set $\mathcal{O}$:

$$F = \{\{x, y\}|x \in \mathcal{O}, y \in \mathcal{O}, x \neq y\}$$

F is assumed to be provided with a total preorder ϖ. A total preorder on F is called a *preordonance* on $\mathcal{O}$. The data of an ordered partition chain on $\mathcal{O}$ is equivalent to a total preorder on F, satisfying specific conditions which will be exposed hereafter. This preorder will be called "ultrametric preordonance" on $\mathcal{O}$.

A rank function ρ is naturally associated with a total preorder ϖ on F, as follows:

$$(\forall \{x, y\} \in F), \rho(\{x, y\}) = card\{\{x', y'\}|\{x', y'\} \in F \text{ and } \{x', y'\} \leq_\varpi \{x, y\}\} \tag{1.4.14}$$

Definition 1 A preordonance ϖ on $\mathcal{O}$ is a ultrametric one, if and only if the following condition is satisfied:

$$(\forall x, y, z \in \mathcal{O}), \rho(\{x, y\}) \leq r \text{ and } \rho(\{y, z\}) \leq r \Rightarrow \rho(\{x, z\}) \leq r \tag{1.4.15}$$

where r is a given integer and ρ is the rank function on F associated with the total preorder ϖ.

The terms of the latter definition are justified by the following.

Proposition 19 *A distance d, defined on $\mathcal{O}$, is ultrametric if and only if the associated preordonance ϖ_d is ultrametric.*

ϖ_d is defined as follows:

$$(\forall x, y, z \in \mathcal{O}), \{x, y\} \leq_{\varpi_d} \{z, t\} \Leftrightarrow d(x, y) \leq d(z, t) \tag{1.4.16}$$

If d is ultrametric

$$(\forall x, y, z \in \mathcal{O}), d(x, z) \leq \max(d(x, y), d(y, z)) \tag{1.4.17}$$

Such a condition implies for ϖ_d

$$(\forall x, y, z \in \mathcal{O}), \rho(x, z) \leq \max(\rho(x, y), \rho(y, z)) \tag{1.4.18}$$

which entails (1.4.15).

Conversely, if the preordonance associated with d is ultrametric, we have (1.4.18) which entails (1.4.17).

Representation

Strictly, beginning segments of an ultrametric preordonance ϖ_u are transitively closed. This means that if R is such a segment

$$(\forall x, y, z \in \mathcal{O}), \{x, y\} \in R \text{ and } \{y, z\} \in R \Rightarrow \{x, z\} \in R \qquad (1.4.19)$$

As a matter of fact a partition chain (see (1.4.13)) is equivalent to an increasing sequence (in the sense of set inclusion) of transitive closed subsets of F. If $(R_0, R_1, \ldots, R_i, \ldots, R_I)$ is such a sequence, we have

$$\emptyset = R_0 \subset R_1 \subset \cdots \subset R_i \subset \cdots \subset R_I = F \qquad (1.4.20)$$

where R_i is the set of element pairs joined by the partition $P_i, 0 \leq i \leq I$. $R_i \subset R_{i+1}$ results from P_i is finer than $P_{i+1}, 0 \leq i \leq I - 1$. R_0 and R_I correspond to the partition of singletons and to singleton partition, respectively.

In these conditions, the successive beginning sections of the ultrametric preordonance (which is a specific total preorder on F) associated with the partition chain $\mathcal{C}$ (see (1.4.13)) are

$$\emptyset = R_0, R_1, \ldots, R_i, \ldots, R_{I-1} \text{ and } R_I = F \qquad (1.4.21)$$

The ultrametric distance function $d_\mathcal{C}$ introduced above is in fact an ordinal function. It can be defined at the level of the sequence (1.4.21), where to each element p of F we will associate the lowest index i, for which $p \in R_i$. We have $I \leq n - 1$, because the maximum number of components of a partition chain is n.

We observe that the balls of the ultrametric space $(\mathcal{O}, d_\mathcal{C})$ are the sets $R_i, 0 \leq i \leq I$. On the other hand, the circular boundaries are $(R_i - R_{i-1}), 1 \leq i \leq I$.

The representation of a partition chain $\mathcal{C}$ (see (1.4.13)) by the total preorder

$$R_1 - R_0 < \cdots < R_i - R_{i-1} < \cdots < R_I - R_{I-1} \qquad (1.4.22)$$

is the one-to-one correspondence between the set of all partition chains on $\mathcal{O}$ and the set of all total ultrametric preordonances on $\mathcal{O}$. The represented and the representing sets have the same cardinality. This is an important feature of the representation we consider. A ultrametric preordonance is, in fact, the equivalence class in the set of ultrametric distances inducing the same partition chain.

1.4.4   Partition Hierarchies and Dendrograms

Fundamentally, two different representations of the structure desired in hierarchical clustering of a finite object set $\mathcal{O}$ are considered in the literature. The first one—examined above—is a classification tree associated with an ordered partition chain on $\mathcal{O}$. The structure underlying the second one is a hierarchy of parts (subsets) of $\mathcal{O}$. We reserve the designation of a dendrogram for the graphical representation of the latter structure. Both representations will be defined in this section and their respective relationships will be specified.

1.4.4.1 Partition Chains and Classification Trees

As just mentioned, the graphical representation of a classification tree associated with a partition chain was considered above (see Sect. 1.4.2 and Fig. 1.8). Each level of this graphical representation stands at one horizontal line. The level sequence is valuated by the increasing sequence of the first integer numbers. Each level i, $0 \le i \le I$, represents a partition P_i of the partition chain concerned. Each class of P_i is characterized by a point of the associated horizontal line. Let us designate by P_i and P_{i+1} two consecutive partitions, which we specify under the form

$$P_i = \{O_i^j | 1 \le j \le K_i\}$$
$$P_{i+1} = \{O_{i+1}^h | 1 \le h \le K_{i+1}\} \tag{1.4.23}$$

Let us now imagine that a given O_{i+1}^h results from merging a set of classes $\{O_i^j | j \in \mathbb{J}_h\}$, where $\mathbb{J}_h$ is a subset of $\{1, 2, \ldots, j, \ldots, K_i\}$. The points representing the latter classes form an interval of the horizontal level i. On the other hand $card(\mathbb{J}_h)$ branches link these points to the point of level $i + 1$, representing the class O_{i+1}^h. Thus, in the graphical representation of Fig. 1.8, the classes of level 1, $\{o_a\}$, $\{o_e, o_f\}$ and $\{o_g\}$ are connected with the class $\{o_a, o_e, o_f, o_g\}$ of level 2. It is of importance to notice that the classes of P_i which do not take part in any merging are transferred identically at the next level. Thus, in Fig. 1.8, the singleton classes $\{o_a\}$, $\{o_d\}$, $\{o_g\}$ and $\{o_j\}$ of level 0 are replicated at level 1.

Notice that in the drawing of the classification tree of Fig. 1.8, a top-down description of the different levels is carried out: starting with the 0 level (level of the leaves) and ending with the root level. Clearly, the reverse could have been done: the bottom (resp., upper) level becomes the leaves level (resp., the root level). Distinguishing these two drawing alternatives is immaterial.

An important property considered for a classification tree is its binariness. As usual, denote by $\mathcal{C}(\mathcal{O}) = (P_0, P_1, \ldots, P_i, P_{i+1}, \ldots, P_I)$ the ordered partition chain, represented by a classification tree $Tree(\mathcal{C}(\mathcal{O}))$.

Definition 2 *Tree($\mathcal{C}(\mathcal{O})$)* is *binary* if and only if each class fusion enabling the passage from a partition P_i to the next partition $P_{i+1}, 0 \le i \le I - 1$ includes exactly two arguments.

Usually in the literature, a single fusion of two classes is considered for the passage between two consecutive partitions of $\mathcal{C}(\mathcal{O})$. In the latter case, $I = n - 1$. However, several class pairs, without common component, might merge "at the same time". To be more explicit, if we consider that the partition P_i includes five classes,

$$P_i = \{O_i^1, O_i^2, O_i^3, O_i^4, O_i^5\}$$

a partition such as

$$P_{i+1} = \{O_i^1 \cup O_i^2, O_i^3, O_i^4 \cup O_i^5\}$$

results from binary fusions.

Proposition 2 gives the number of binary classification trees on a set of n objects, for which the two components of a single pair merge from one level to the next one. This result was established in [5, 15, 16, 26].

Proposition 1 gives the total number of classification trees on a set of n objects [12, 15, 16, 27].

In the development above, ordinal valuation is considered for the level sequence of a classification tree. However, positive numerical valuation can also be considered for the latter sequence. This is generally given by a positive numerical function over the partition sequence $\mathcal{C}(\mathcal{O}) = (P_0, P_1, \ldots, P_i, P_{i+1}, \ldots, P_I)$ (see for example Sect. 9.3.6 of Chap. 9). This form of valuation appears naturally in the ascendant agglomerative hierarchical construction of a binary classification tree (see Chap. 10). In this situation, the value of the associated numerical function for the level i is the distance or dissimilarity between the two offsprings of the node created at level i, $1 \leq i \leq n - 1$.

Consequently, the classification trees we consider are valuated by ordinal or numerical valuation. On the other hand, their leaves are labelled. In [24] one considers equivalence classes of classification trees in the case where no valuation or no leaves labelling hold. Such equivalence classes are called dendrograms. The case of no valuation but leaves labelling arises naturally in the definition of non-valuated hierarchy of set parts of a finite set $\mathcal{O}$.

1.4.4.2 Hierarchy of Parts of a Finite Set and Dendrograms

Definition 3 A hierarchy $\mathcal{H}(\mathcal{O})$ of parts of a set $\mathcal{O}$ is a set of $\mathcal{O}$ subsets satisfying the following conditions:

1. $\mathcal{O} \in \mathcal{H}(\mathcal{O})$
2. $(\forall o \in \mathcal{O}), \{o\} \in \mathcal{O}$
3. $(\forall X \in \mathcal{H}(\mathcal{O}), Y \in \mathcal{H}(\mathcal{O})), X \cap Y = X$ or $X \cap Y = Y$ or $X \cap Y = \emptyset$

$$(1.4.24)$$

In words, conditions 1, 2 and 3 can be expressed as follows:

1. The entire set $\mathcal{O}$ belongs to the hierarchy;
2. The singleton subsets belong to the hierarchy;

3. If X and Y are any two elements of the hierarchy, we have necessarily three alternatives: X is included in Y, Y is included in X or X and Y are disjoint.

Moreover, the hierarchy $\mathcal{H}(\mathcal{O})$ is *binary* if and only if every element of $\mathcal{H}(\mathcal{O})$, not reduced to a singleton, is the union of exactly two elements of $\mathcal{H}(\mathcal{O})$. Such a hierarchy will be denoted $\mathcal{H}_b(\mathcal{O})$. It comprises $2n - 1$ elements ($n = card(\mathcal{O})$).

Example

Let $\mathcal{O} = \{o_1, o_2, o_3, o_4, o_5, o_6, o_7, o_8, o_9, o_{10}\}$ be a set of ten objects, and an illustration of a binary hierarchy $\mathcal{H}_b(\mathcal{O})$ is given by the following:

$$\{\{o_1\}, \{o_2\}, \{o_3\}, \{o_4\}, \{o_5\}, \{o_6\}, \{o_7\}, \{o_8\}, \{o_9\}, \{o_{10}\},$$

$$\{o_2, o_3\}, \{o_5, o_6\}, \{o_8, o_9\},$$

$$\{o_1, o_2, o_3\}, \{o_4, o_5, o_6\}, \{o_8, o_9, o_{10}\},$$

$$\{o_4, o_5, o_6, o_7\}, \{o_1, o_2, o_3, o_4, o_5, o_6, o_7\},$$

$$\{o_1, o_2, o_3, o_4, o_5, o_6, o_7, o_8, o_9, o_{10}\}\}$$

Let $\mathbb{I}$ be a totally ordered set of values referring to an interval scale where the differences between pairs of values can be defined numerically. The more classical cases are those where $\mathbb{I}$ is included in positive reals $\mathbb{R}_+$ or in the natural integers $\mathbb{N}$. A mapping v of the $\mathcal{O}$ subset hierarchy $\mathcal{H}(\mathcal{O})$ onto $\mathbb{I}$, strictly increasing with respect to the inclusion relation between the elements of $\mathcal{H}(\mathcal{O})$, defines an *indexation* (a *valuation*) of $\mathcal{H}(\mathcal{O})$. The ordered pair $(\mathcal{H}(\mathcal{O}), v)$ is a hierarchy of $\mathcal{O}$ subsets endowed with a strictly increasing indexation. This indexation might be increasing in a large sense. In this case, the mapping v is increasing in a large sense (For $(X, Y) \in \mathcal{H}(\mathcal{O}) \times \mathcal{H}(\mathcal{O})$, $v(X) \leq v(Y) \Leftrightarrow X \subseteq Y$). Mostly, a numerical indexation is considered. For our concern, we focus on ordinal indexation taking values in the beginning segment of the integers (see [17]). In the latter case, the common value of the function v on each singleton subset is 0.

In these conditions, a directed and valuated graph, corresponding to the inclusion relation between the $\mathcal{O}$ subsets, elements of $\mathcal{H}(\mathcal{O})$, can be defined. Let us designate by $\mathcal{G} = (\mathcal{H}(\mathcal{O}), \Gamma, \nu)$ such a directed graph. Γ denotes the set of directed edges of $\mathcal{G}$ and ν, the valuation endowing Γ. If (X, Y) is an ordered pair of $\mathcal{H}(\mathcal{O})$, we have $\Gamma(X, Y)$ if and only if

1. $X \subset Y$ (strictly);
2. There does not exist an element Z of $\mathcal{H}(\mathcal{O})$ such that $X \subset Z \subset Y$ (strictly).

Otherwise, the valuation $\nu(X, Y)$ of the directed edge (X, Y) is defined by the difference $v(X) - v(Y)$, where v is the valuation function considered above on $\mathcal{H}(\mathcal{O})$.

Organizing the graph $\mathcal{G}$ level by level enables it to be graphically represented by a *tree diagram* called *dendrogram*. The algorithmic construction of the latter is carried out in a descendant way [19]. In the case where conditions 1 and 2 above are satisfied for an ordered pair (X, Y) of elements of $\mathcal{H}(\mathcal{O})$, Y and X play the respective roles of father and son nodes in the dendrogram. A directed edge connects Y to X. We may assume the valuation function ν established in such a way that all of the paths starting with the root—which represents the whole set $\mathcal{O}$—and ending with the leaves—which represent the singleton subsets—are of same length.

Example

The valuation function v is not necessarily injective on the set of internal nodes. In the illustration given (see Fig. 1.9) the value 4 is reached twice. The only mandatory requirement is that v increases strictly along every ascendant path from a leaf to the root.

The notion of a *binary* hierarchy of a finite set will be emphasized in the following Sect. 1.4.5. On the other hand, as mentioned above, ordinal valuation will be adopted for the binary hierarchy concerned.

The respective notions

- of a binary classification tree on an object set $\mathcal{O}$, associated with a totally ordered partition chain, where each partition covers the preceding one

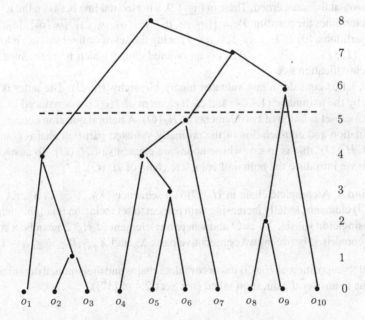

Fig. 1.9 Dendrogram associated with a valuated hierarchy

and that

- of a binary hierarchy of parts of $\mathcal{O}$, endowed with an ordinal valuation v, whose value set is $\{0, 1, \ldots, i, \ldots, n - 1\}$, where each value is reached,

are equivalent.

In other words, to a binary partition chain $C_b(\mathcal{O}) = (P_0, P_1, \ldots, P_i, P_{i+1}, \ldots, P_{n-1})$ on $\mathcal{O}$, where P_{i+1} covers P_i, $1 \leq i \leq n - 2$, there corresponds a unique binary hierarchy of parts $\mathcal{H}_b(\mathcal{O})$ of $\mathcal{O}$, endowed with an ordinal valuation, compatible with the inclusion relation between subsets, whose value set is $\{0, 1, \ldots, i, \ldots, n - 1\}$, where each value is reached.

Conversely, given a binary hierarchy $\mathcal{H}_b(\mathcal{O})$ of parts of a finite set $\mathcal{O}$, endowed with an ordinal valuation, compatible with the inclusion relation between subsets, whose value set is $\{0, 1, \ldots, i, \ldots, n - 1\}$, where each value is reached, there corresponds in a unique way an ordered partition chain $C_b(\mathcal{O})$, where each partition covers the preceding one.

Let us now indicate how to build $\mathcal{H}_b(\mathcal{O})$ from $C_b(\mathcal{O})$ ($C_b(\mathcal{O}) \to \mathcal{H}_b(\mathcal{O})$), and, reciprocally, $C_b(\mathcal{O})$ from $\mathcal{H}_b(\mathcal{O})$ ($\mathcal{H}_b(\mathcal{O}) \to C_b(\mathcal{O})$).

For the former correspondence ($C_b(\mathcal{O}) \to \mathcal{H}_b(\mathcal{O})$), every node (terminal, internal or root node) determines an element of $\mathcal{H}_b(\mathcal{O})$. A non-terminal node occurring at level i results from merging two classes of the partition P_{i-1}, $1 \leq i \leq n - 2$.

For the reverse correspondence ($C_b(\mathcal{O}) \to \mathcal{H}_b(\mathcal{O})$), where $\mathcal{H}_b(\mathcal{O})$ is ordinally valuated, consider the associated dendrogram as in Fig. 1.9. A cutting by a horizontal line situated between two consecutive nodes occurring at levels i and $i + 1$, $0 \leq i \leq n - 2$, determines a partition P_i. The latter is composed of the nodes appearing just beneath the horizontal line concerned. Thus, in Fig. 1.9, a horizontal line between the levels 5 and 6 determines the partition $P_5 = \{\{o_1, o_2, o_3\}, \{o_4, o_5, o_6, o_7\}, \{o_8, o_9\}, \{o_{10}\}\}$. In these conditions, $\{0, 1, 2, \ldots, i, \ldots, n - 1\}$ being the set of ordinal values endowing $\mathcal{H}_b(\mathcal{O})$, $(P_0, P_1, \ldots, P_i, \ldots, P_{n-1})$ is an ordered chain, which is represented by a binary classification tree.

Now, let us consider a non-valuated binary hierarchy $\mathcal{H}_b(\mathcal{O})$. The latter is then defined by the axiomatic (1.4.24) and each element of $\mathcal{H}_b(\mathcal{O})$ not reduced to a singleton $\mathcal{O}$ subset is union of two elements of $\mathcal{H}_b(\mathcal{O})$. A natural question concerns the determination and enumeration of the ordinally valuated partition chains compatible with $\mathcal{H}_b(\mathcal{O})$; that is to say, whose nodes are elements of $\mathcal{H}_b(\mathcal{O})$. To answer the question we introduce the notion of complete chain of $\mathcal{H}_b(\mathcal{O})$.

Definition 4 A complete chain in $\mathcal{H}_b(\mathcal{O})$ is a sequence $(X_0, X_1, \ldots, X_g, \ldots, X_h)$ of $\mathcal{H}_b(\mathcal{O})$ elements, strictly increasing with respect to set inclusion relation, such that X_0 is a singleton subset, $X_h = \mathcal{O}$ and where any element of $\mathcal{H}_b(\mathcal{O})$ can be a subset strictly comprised between two consecutive parts X_g and X_{g+1}, $0 \leq g \leq h - 1$

The decomposition of $\mathcal{H}_b(\mathcal{O})$ into n complete chains and their mutual organization enable us to answer the question asked (see Sect. 2.4 of [17]).

1.4.5 From a **Symmetrical** *Binary Hierarchy* to a **Directed** *Binary Hierarchy*

1.4.5.1 Introduction

As it will be seen in Chaps. 4–7, the condition of symmetry in the notion of resemblance is more often than not required. If x and y are the two arbitrary elements of the set E to be clustered and if S (resp., $\mathcal{D}$) is a numerical similarity (resp., dissimilarity) measure (see Chap. 4) on a set E of entities, mostly we assume that $S(x, y) = S(y, x)$ (resp., $\mathcal{D}(x, y) = \mathcal{D}(y, x)$). Comparison might concern objects, categories or descriptive attributes. In Chap. 3 different types of formal descriptions are given. Thus, the set E to be organized by clustering is generally a finite set of objects, categories or descriptive attributes. A classification tree on E adheres the respective resemblances between the elements of E if, as far as possible, the most similar entities are gathered in the classes merged at the first levels of the classification tree, whereas the most dissimilar ones are separated in different consistent classes appearing at high levels of this classification tree. In this section we consider the classification tree as binary and ordinally valuated (ranked). This case to which we refer to under certain conditions is mostly developed in the literature. As seen in Sect. 1.4.4.2, the structure concerned is equivalent to a binary hierarchy of parts of E, endowed with an ordinal valuation. In Chap. 10, methods are studied in order to build such classification trees.

Applications in different domains (e.g. social exchange, psychology, confusion matrices,…) require to define *asymmetrical* similarity measure on E. In Sect. 4.2.1.3 of Chap. 4 we will see how to derive in the case of Boolean data, an asymmetrical measure from a symmetrical one. On the other hand, we will see in Sect. 5.2.1.4 of Chap. 5 how the latter derivation is carried out for the probabilistic index of the *Likelihood of the Link*.

This measure applies very importantly in *Data Mining* field [7, 9, 18, 20, 21]. It allows *Interestingness Association rules* to be evaluated in a relative respect, in large data bases. A general form of an association rule can be written symbolically $a \rightarrow b$, where a and b are Boolean attributes. As in logical rules a and b play the respective roles of *premise* and *conclusion*. However, $a \rightarrow b$ expresses a *general tendency* and means that "when a is $TRUE$, then usually, b is also $TRUE$". Consequently, association rules tolerate statistically, a few number of counter examples, with respect to logical implications. Many measures have been proposed to quantify numerically the strength of the rule implicative tendency (see [10] for the state-of-the-art). Some of them are instances of those obtained by the general technique mentioned above and described in Sect. 4.2.1.3 of Chap. 4.

Clustering algorithms have been proposed in order to structure sets of rules provided from large data bases (e.g. [13]). Most of them build partitions. A binary partition chain on the rule set would have been a much richer structure. Even so, in the latter structure, the respective associations between clusters are *symmetrical*, whereas the rules to be represented are *asymmetrical* in nature.

In these conditions, the *symmetrical* structure of a binary classification tree makes it unsuitable for organizing directed links between Boolean attributes; or, more generally, elements of a set E endowed with an asymmetrical similarity measure. In [8] a structure of an oriented binary tree is proposed. The link between two classes merged is oriented. More precisely, if C and D denote two classes to be merged ($C \subset E$, $D \subset E$ and $C \cap D = \emptyset$), where C is at the left of D, the link is oriented from C to D and symbolized by a right arrow $\rightarrow: C \rightarrow D$. This structure was called "directed hierarchy" in [6]. Thus, the directed union between C and D can be written as $C \vec{\cup} D$.

Example

The set E to be organized is composed of four Boolean attributes: a, b, c and d. It is provided with an implication index Imp. Imp allows us to measure the strengths of the different rules, taken separately and also to compare two disjoint sets of rules organized each recursively as rules of rules. Let us assume that

$$Imp(a \rightarrow b) = 0.94, \, Imp(c \rightarrow d) = 0.81 \text{ and } Imp\big((a \rightarrow b) \rightarrow (c \rightarrow d)\big) = 0.70$$

Additionally, we suppose that the maximal value of Imp over other possible rules, namely,

$$b \rightarrow a, a \rightarrow c, c \rightarrow a, a \rightarrow d, d \rightarrow a, b \rightarrow c, c \rightarrow b, b \rightarrow d, d \rightarrow b, \text{ and } d \rightarrow c$$

is strictly lower than 0.81.

The graphical representation of the directed binary tree associated with the Imp function is given in Fig. 1.10. An equivalent version is given in Fig. 1.11, where the above right arrow indicates the common direction of the respective implications.

In the next section the formal expression of a directed binary hierarchy will be given. More particularly, we shall exhibit the needed changes for the *passage* from a symmetrical to a directed binary hierarchy. These aspects are substantially developed in [17, 22] and briefly retaken here.

Fig. 1.10 Directed binary classification tree

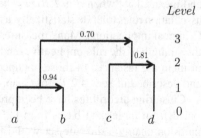

Fig. 1.11 Directed binary dendrogram

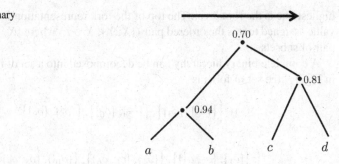

1.4.5.2 Binary Hierarchy: Set of *Forks*

Our objective is to carry out the transition from the formal expression of a *symmetric* binary hierarchy to that of a *directed* binary hierarchy. In other terms, we have to study how that becomes axiomatic and formal properties of symmetric binary hierarchy in the case of directed binary hierarchy.

The classical Definition 3 cannot be transposed in the case of *directed* binary hierarchy. A new definition is needed. The basic element considered for the latter is an unordered pair of subsets—called *fork*—instead of a subset. Precisely, this new version can be extended to the directed case by substituting *ordered* pairs of subsets (*directed* forks) for unordered ones.

Definition 5 A *fork* of E is an unordered pair $\{X, Y\}$ of non-empty disjoint subsets of E such that $X \neq E$, $Y \neq E$ and $X \cap Y = \emptyset$. X and Y are the two components of the fork $\{X, Y\}$. The *basis* of a fork $\{X, Y\}$ is the subset $X \cup Y$.

A fork $\{X, Y\}$ can be graphically represented by the simplest perfect binary tree with exactly three nodes (as in Fig. 1.12 which concerns a directed fork): two leaves associated with X and Y, respectively, and the root node associated with $X \cup Y$. Two levels are considered for this representation: the lowest one is the leaf level and the

Fig. 1.12 A directed fork

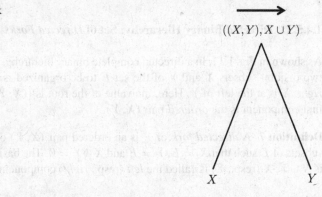

highest one is the root level. The top of the fork representation is the root node. The value assigned to it is the ordered pair $(\{X, Y\}, X \cup Y)$ where $\{X, Y\}$ is an *unordered* pair of subsets.

A complete binary hierarchy can be decomposed into a set of forks. For instance, in Fig. 1.9 the set of forks is

$$\Big\{ \{\{o_2\}, \{o_3\}\}, \{\{o_5\}, \{o_6\}\}, \{\{o_8\}, \{o_9\}\},$$

$$\{\{o_1\}, \{o_2, o_3\}\}, \{\{o_4\}, \{o_5, o_6\}\}, \{\{o_{10}\}, \{o_8, o_9\}\},$$

$$\{\{o_7\}, \{o_4, o_5, o_6\}\}, \{\{o_4, o_5, o_6, o_7\}, \{o_8, o_9, o_{10}\}\},$$

$$\{\{o_1, o_2, o_3\}, \{o_4, o_5, o_6, o_7, o_8, o_9, o_{10}\}\} \Big\}$$

Let us denote by $\mathcal{H}_f(E)$ a complete binary hierarchy described in terms of a set of forks.

Definition 6 A complete binary non-directed hierarchy $\mathcal{H}_f(E)$ is a set of forks which satisfies the following conditions:

1. $(\forall \ \{X, Y\} \in \mathcal{H}_f(E))$, if X (resp., Y) includes two elements at least, then X (resp., Y) is itself a basis of a fork;
2. $(\forall \ \{X, Y\}$ and $\{Z, T\} \in \mathcal{H}_f(E)$), such that $card(X \cup Y) \leq card(Z \cup T))$ we have one of the following properties:
   (i) $(X \cup Y) \cap (Z \cup T) = \emptyset$ or (ii) $X \cup Y \subset Z$ or (iii) $X \cup Y \subset T$;
3. $(\forall x \in E)$, there is a unique fork whose one component is equal to $\{x\}$;
4. There is a unique fork $\{X, Y\} \in \mathcal{H}_f(E)$ such that $X \cup Y = E$.

Clearly, a complete binary hierarchy on a set of cardinality n includes $n - 1$ forks. The equivalence between Definitions 3 and 6 is established in [17].

1.4.5.3 Directed Binary Hierarchy: Set of *Directed Forks*

As shown in Fig. 1.11, in a directed complete binary hierarchy, the junction between two disjoint subsets X and Y of the set E to be organized is directed from *left to right*. X is at the left of Y. Here, the value of the root is $((X, Y), X \cup Y)$ where the first component is the *ordered* pair (X, Y).

Definition 7 A *directed fork* of E is an ordered pair (X, Y) of non-empty disjoint subsets of E such that $X \neq E$, $Y \neq E$ and $X \cap Y = \emptyset$. The basis of the directed fork is $X \cup Y$. X (resp., Y) is called the *left* (resp., *right*) component of the directed fork.

For example in the two equivalent representations given in Figs. 1.10 and 1.11, we can distinguish the following directed forks:

$$(\{a\}, \{b\}), (\{c\}, \{d\}), (\{a, b\}\{c, d\})$$

Directed hierarchies might concern—with a specific interpretation—structuring a set of objects. However, mostly, they concern the structuration of a set of attributes. In these conditions, in order to get richer instances of directed forks, consider in the bottom of the diagram of Fig. 1.9 the replacement of the ordered sequence

$$(o_1, o_2, o_3, o_4, o_5, o_6, o_7, o_8, o_9, o_{10})$$

by

$$(a^2, a^7, a^1, a^6, a^4, a^5, a^9, a^8, a^{10}, a^3)$$

where $\mathcal{A} = \{a^j | 1 \leq j \leq 10\}$ is supposed to be a set of ten Boolean attributes. On the other hand, draw on the top of the diagram a right arrow, shown in Fig. 1.11, the common direction of the implications. In this way, each non-directed fork considered above corresponds to a directed fork on $\mathcal{A}$. Instances of these are $(\{a^8\}, \{a^{10}\})$, $(\{a^6\}, \{a^4, a^5\})$, $(\{a^4, a^5, a^6, a^9\}, \{a^3, a^8, a^{10}\})$,...

Definition 8 A *directed binary hierarchy* $\overrightarrow{\mathcal{H}}(E)$ is a set of directed forks of E which satisfies the following property: for each unordered pair $\{(X, Y), (Z, T)\}$ of distinct directed forks belonging to $\overrightarrow{\mathcal{H}}(E)$, where $card(X \cup Y) \leq card(Z \cup T)$, we have either $(X \cup Y) \cap (Z \cup T) = \emptyset$ or $X \cup Y \subseteq Z$ or $X \cup Y \subseteq T$. Moreover, a binary directed hierarchy is *complete* if the following conditions are satisfied:

1. $(\forall x \in E)$, there is a unique directed fork of $\overrightarrow{\mathcal{H}}(E)$ such that one component is $\{x\}$;
2. there is a directed fork in $\overrightarrow{\mathcal{H}}(E)$ such that $X \cup Y = E$;
3. for any directed fork (X, Y) in $\overrightarrow{\mathcal{H}}(E)$ such that $X \cup Y \neq E$, there exists a directed fork (X', Y') in $\overrightarrow{\mathcal{H}}(E)$ such that either $X' = X \cup Y$ or $Y' = X \cup Y$.

$(X \cup Y) \cap (Z \cup T) = \emptyset$ is called *exclusion* between the two directed forks (X, Y) and (Z, T). On the other hand, $X \cup Y \subseteq Z$ (resp. $X \cup Y \subseteq T$) is called a *left inclusion* (resp., a *right inclusion*) of the fork (X, Y) into the fork (Z, T).

The following properties are established in [22].

Proposition 20 *Let* $\overrightarrow{\mathcal{H}}(E)$ *be a directed binary hierarchy. For any directed fork* $(X, Y) \in \overrightarrow{\mathcal{H}}(E)$, *there is no other directed fork* $(Z, T) \in \overrightarrow{\mathcal{H}}(E)$ *such that* $X \cup Y = Z \cup T$. *In particular,* $(Y, X) \notin \overrightarrow{\mathcal{H}}(E)$.

Proposition 21 *Let* $\overrightarrow{\mathcal{H}}(E)$ *be a directed hierarchy. For any element x of E ($x \in E$), there is at most one directed fork with a component equal to $\{x\}$.*

Proposition 22 *Let $\overrightarrow{\mathcal{H}}(E)$ be a complete binary directed hierarchy. Given an unordered pair $\{x, y\}$ from E ($\{x, y\} \in P_2(E)$), there necessarily exists a directed fork in $\overrightarrow{\mathcal{H}}(E)$ such that one component contains x and the other one contains y. Moreover, this fork is unique.*

Theorem 3 *Let $\overrightarrow{\mathcal{H}}(E)$ be a complete binary directed hierarchy on E and R_H be the binary relation on E defined by*

$$\forall (x, y) \in E \times E, \, x \neq y, \, x R_H y \Leftrightarrow \exists! \, (X, Y) \in \overrightarrow{\mathcal{H}}(E), (x, y) \in X \times Y$$

R_H defines a strict total order on E.

Valued Directed Binary Hierarchies

The concept of a *valued* directed binary hierarchy is the same as in the classical case (see Definition 5). More explicitly, a numerical positive function ν defined on $\overrightarrow{\mathcal{H}}(E)$ is a *valuation* on $\overrightarrow{\mathcal{H}}(E)$ if and only if $(\forall ((X, Y), (Z, T)) \in \overrightarrow{\mathcal{H}}(E) \times \overrightarrow{\mathcal{H}}(E))$, such that $(X \cup Y) \cap (Z \cup T) \neq \emptyset$ and $card((X \cup Y)) < card((Z \cup T)) \Rightarrow \nu((X, Y)) < \nu((Z, T))$

1.4.5.4 Directed Ultrametrics

The classical notion of ultrametric distance d_u on a set $\mathcal{O}$ was defined in Sect. 1.4.1. It satisfies the so-called *ultrametric inequality* (see (1.4.2)). Remember the bijective correspondence established in Sect. 1.4.2 between a partition chain and a ultrametric distance on a finite set $\mathcal{O}$. We have considered more particularly the case where the ultrametric distance is defined ordinally from the level rank function of the classification tree associated with the partition chain. This correspondence applies in the case of binary trees associated with partition chains for which each partition covers the preceding one.

The transposition of the ultrametric inequality requires the integration of the total order R_H set up in Theorem 3. In these conditions, we are led to propose first a definition of a *directed dissimilarity* compatible with a total order relation R:

Definition 9 Let R be a total order on E. A directed dissimilarity d_R compatible with R is a mapping $E \times E \to \overline{\mathbb{R}}_+ = \mathbb{R}_+ \cup \{\infty\}$, where $\mathbb{R}_+$ is the set of positive numbers, of $E \times E$ on $\overline{\mathbb{R}}_+$ which satisfies the four following conditions:

1. $(\forall x \in E), d_R(x, x) = 0$;
2. $(\forall (x, y) \in E \times E), x \neq y, 0 < d_R(x, y) < +\infty$ if $x R y$;
3. $(\forall (x, y) \in E \times E), x \neq y, d_R(y, x) = \infty$ if $x R y$;
4. $(\forall$ triple $(x, y, z) \in E^3) \, x \neq y, x \neq z, y \neq z$, such that $x R y$ and $y R z$, then $d_R(x, z) \geq \max\{d_R(x, y), d_R(y, z)\}$.

Condition 3 restricts the directed dissimilarity d_R to the graph $\{(x, y)|x \in \mathcal{O}, y \in \mathcal{O}, xRy\}$ of the binary relation R, and Condition 4 is equivalent to the Robinson condition (see in [22]).

Therefore, in the associated geometrical space with d_R a new notion of directed ultrametricity will be introduced hereafter. It is characterized by the property that every *directed triangle* is isosceles with a basis length strictly lower than the common length of the two other sides.

Let us now illustrate the notion of directed isosceles triangle on the basis of the directed binary hierarchy, specified above, on the attribute set $\mathcal{A} = \{a^j | 1 \leq j \leq 10\}$. For this, consider the following linear order on the leaf set of the hierarchy concerned:

$$a^2 < a^7 < a^1 < a^6 < a^4 < a^5 < a^9 < a^8 < a^{10} < a^3.$$

Two instances of directed isosceles triangles will be examined: (a^6, a^4, a^5) and (a^4, a^5, a^9). Otherwise, we suppose the directed distance between two elements as given by the lowest level rank of the hierarchy at which the two elements are aggregated. We have

$$d_R(a^4, a^5) = 3 < d_R(a^6, a^4) = d_R(a^6, a^5) = 4$$

$$d_R(a^4, a^5) = 3 < d_R(a^4, a^9) = d_R(a^5, a^9) = 5$$

where d_R designates the directed ultrametric defined by the level rank function associated with the directed binary hierarchy.

More formally, we have the following.

Definition 10 Let E be a set (finite in our development), provided with a total order R and a directed dissimilarity d_R. A directed triangle of $\mathcal{O}$ is an ordered triplet (x, y, z) of elements of E, mutually distincts, such that xRy and yRz. The triangle (x, y, z) is isosceles if and only if

$$d_R(x, y) = d_R(x, z) \text{ or } d_R(x, z) = d_R(y, z).$$

In the former (resp., latter) case the basis of the triangle is (y, z) (resp., (x, y)).

Definition 11 Let us consider a total order R on E and a directed compatible dissimilarity d_R. d_R is called a *directed ultrametric* if, for any ordered triple $(x, y, z) \in E^3$ (xRy and yRz), the three following conditions are satisfied:

1. $d_R(x, y) \leq \text{Max}\{d_R(x, z), d_R(y, z)\}$;
2. $d_R(x, z) \leq \text{Max}\{d_R(x, y), d_R(y, z)\}$;
3. $d_R(y, z) \leq \text{Max}\{d_R(x, y), d_R(x, z)\}$.

The *directed ultrametric* is strict if 1 or, exclusively, 3 is strict. These inequalities cannot be strict simultaneously.

The three inequalities are required to define a directed ultrametric. For illustration, consider the above directed triangle (a^6, a^4, a^5). The three inequalities are

1. $d_R\left(a^6, a^4\right) \leq \text{Max}\left\{d_R\left(a^6, a^5\right), d_R\left(a^4, a^5\right)\right\}$;
2. $d_R\left(a^6, a^5\right) \leq \text{Max}\left\{d_R\left(a^6, a^4\right), d_R\left(a^4, a^5\right)\right\}$;
3. $d_R\left(a^4, a^5\right) \leq \text{Max}\left\{d_R\left(a^6, a^4\right), d_R\left(a^6, a^5\right)\right\}$.

and the third inequality is strict. Similar to the classical case, the directed triangles are isosceles in a directed ultrametric space. Moreover, for a *strict* ultrametric, the length of a directed triangle basis is strictly smaller than the length of its two equal sides. More formally, we have the following.

Theorem 4 *Let R be a total order on E and d_{uR} be a strict directed ultrametric compatible with R. For any directed triangle $(x, y, z) \in E^3$, $x \neq y, x \neq z, y \neq z$ we have either $d_{uR}(x, y) < d_{uR}(x, z) = d_{uR}(y, z)$ or $d_{uR}(y, z) < d_{uR}(x, y) = d_{uR}(x, z)$*

The reciprocal property is expressed by

Theorem 5 *If R is a total order on E and d_R is a directed dissimilarity such that each directed triangle (x, y, z) is isosceles such that the length of its basis, strictly smaller than the common length of the other sides, then d_R is strictly ultrametric.*

According to the so-called Johnson bijection theorem between a binary hierarchical classification tree and a ultrametric distance, we establish in [22] a one-to-one correspondence between a directed binary classification tree and a strict directed ultrametric. Moreover and importantly, an algorithm enabling the passage between a strict directed ultrametric and a directed binary classification tree is specified.

1.5   Polyhedral Representation of the Partition Set of a Finite Set

For[1] simplicity reasons the object set $\mathcal{O}$ is coded by the set $\mathbb{I} = \{1, 2, \ldots, i, \ldots, n\}$ of the first n integers. Thus, i indicates the ith object o_i, $1 \leq i \leq n$. A given element of the set $F = P_2(\mathcal{O})$ of distinct unordered object pairs of $\mathcal{O}$ (see (1.3.1)) is coded by a word ij, composed of two integers where the first one i is strictly lower than the second one j, $1 \leq i < j \leq n$. Ordering lexicographically the set of these words—that we designate also by F—gives

$$12, 13, \ldots, 1n, 23, 24, \ldots, 2n, \ldots, (n-1)n \tag{1.5.1}$$

A given partition P of $\mathcal{O}$ will be represented by the *indicator* function of the F subset $R(P)$, associated with unordered object pairs joined by the partition P

---

[1] This section could be examined after Sect. 2.2 of Chap. 2.

(see (1.3.3)). This indicator function that we designate also by P is a vertex of the cube $\Gamma = \{0, 1\}^m$ where $m = card(F) = n(n-1)/2$. The sequence of components $(\varpi_{ij} | 1 \leq i < j \leq n)$ of P, relative to F, ordered as indicated in (1.5.1), is such that $\varpi_{ij} = 1$ (resp., $\varpi_{ij} = 0$), if and only if the objects labelled i and j belong to a same class (resp., are separated into two different classes) of P.

Now, let us immerse the discrete cube $\Gamma = \{0, 1\}^m$ in the continuous one $[0, 1]^m$, where $[0, 1]$ is the closed beginning interval of length 1 of the positive reals $\mathbb{R}_+$. The set of the vertices associated with the set of partitions of $\mathcal{O}$ determines a polyhedral Π_n (Π_n is strictly included in Γ) such that all of its vertices belong to the sphere centred at the point whose all coordinates are equal to $1/2$ and of radius equal to $m/4$. This polyhedral is identical to its convex closure, because any point of $[0, 1]^m$ whose components are 0 or 1 can be expressed as a convex linear combination (linear combination with positive coefficients of sum unity) of points whose components are 0 or 1.

This geometrical representation was introduced by S. Régnier in order to analyse its "central partition method". The latter and the associated references are detailed in the next chapter. The problem addressed in this method consists of clustering a set of objects described by a set of nominal categorical attributes (see Sect. 3.3.1). Each of these attributes induces a partition of $\mathcal{O}$. Let $\{P^m | 1 \leq m \leq p\}$ designate the family of these partitions. A *central partition* P is defined as minimizing the average (mean) of the respective distances of P to the different partitions P^m, $1 \leq m \leq p$. The mean concerned might be weighed with positive coefficients of sum unity.

Precisely, the geometrical representation introduced enables the optimization criterion to be linearized. This criterion can be put as follows:

$$Arg\{\max_S (L(S))\}$$
$$\text{where}$$
$$L(S) = \sum_{ij \in F} t_{ij} s_{ij} \tag{1.5.2}$$

where $(s_{ij} | ij \in F)$ is the point of $[0, 1]^m$ which represents the target partition S and where t_{ij} is the value of a specific similarity function between the objects i and j, $1 \leq i < j \leq n$. These notations are the same as those adopted in Sect. 2.2.1.3 of Chap. 2 (see (2.2.24)). However, the reference space is here $F = P_2(\mathcal{O})$ and not the Cartesian product $\mathcal{O} \times \mathcal{O}$.

In order to resolve the optimization problem S. Régnier has proposed an heuristic called the "transfer algorithm" (see Sect. 2.2.2.1). In general, this algorithm does not lead to a global maximum, but to a local one. According to the principles of linear programming, a partition point which corresponds to a local maximum defines a global maximum if and only if the partition point concerned is better—in the sense of the criterion (1.5.2)—than all adjacent partition points in the polyhedra Π_n, expressed above. Therefore, recognizing adjacency in Π_n is a fundamental problem for the central partition method.

The development below is taken up again from the first chapter of [16]. It corresponds to a specific version that has been worked on, which refers to earlier works (1971–1975) of S. Régnier and W.F. De La Véga [4, 25].

Definition 12 Let X and Y be the two different vertices of the polyhedral Π_n. X and Y are adjacent if and only if the segment which connects them is an edge of Π_n. In this case we denote by XAY the adjacency binary relation between X and Y. The notation XDY expresses that X and Y are different and non-adjacent.

Two characteristic properties are given in the polyhedral theory relative to two different vertices X and Y of Π_n:

1. XAY if and only if $X \neq Y$ and there exists a linear form $\mathcal{L}$ on $\mathbb{R}^m$, such that $\mathcal{L}(X) = \mathcal{L}(Y) > \mathcal{L}(Z)$, for every vertex Z different from X and Y;
2. XDY if and only if there exists a set $\{Z_u | u \in U\}$ of vertices of Π_n and a system of real positive numbers $\{\lambda_u | u \in U\}$, such that

$$Y - X = \sum_{u \in U} \lambda_u (Z_u - X) \qquad (1.5.3)$$

where U indicates a subscript set.

XDY is the negation of XAY.

In [4], (1.5.3) is shown to be reduced to

$$\alpha X + \beta Y = \sum_{u \in U} \pi_u Z_u \qquad (1.5.4)$$

where

$$\alpha + \beta = \sum_{u \in U} \pi_u = 1$$

To establish (1.5.4), we begin by introducing $\Lambda = \sum_{u \in U} \lambda_u$. Using (1.5.3) and a given coordinate ij for which $|x_{ij} - y_{ij}| = 1$, Λ is proven to be strictly greater than 1. In these conditions, we put $\beta = 1/\Lambda$, $\alpha = 1 - \beta$ and $\pi_u = \lambda_u / \Lambda$, for all u in U.

Now, given a pair $\{X, Y\}$ of two distinct vertices of the polyhedra representing two distinct partitions, the matter consists of establishing a necessary and sufficient condition to have XAY. As above and in the following a partition and the vertex that represents it are identically denoted.

The characterization problem submitted can be reduced to the quotient set $\mathcal{O}' = \mathcal{O}/X \wedge Y$ whose elements are the classes of $X \wedge Y$. In fact,

$$XDY \Leftrightarrow X'DY' \qquad (1.5.5)$$

where X' and Y' are the partitions defined by X and Y on $\mathcal{O}'$. As a matter of fact, for each partition Z which takes part in (1.5.3), the components x_{ij}, y_{ij} and z_{ij} of X, Y

and Z are constant and equal to 1, for $\{i, j\}$ describing the element pairs of a same class of $X \wedge Y$. In these conditions (1.5.3) is equivalent to

$$Y' - X' = \sum_{u \in U} \lambda_u (Z'_u - X') \tag{1.5.6}$$

Thus, we characterize adjacency of a pair $\{X, Y\}$ of partitions for which $X \wedge Y$ is a partition of singletons. We denote the latter as $X \wedge Y = \mathbf{0}$.

An easy case is when X and Y are comparable. Suppose for example $X < Y$ (X is finer than Y). This implies $X = X \wedge Y = \mathbf{0}$. However, for $X = \mathbf{0}$, we can specify the set of all edges emerging from X. In the cube $\Gamma = [0, 1]^m$, the origin $\mathbf{0}$ is the extremity of m edges parallel to the different coordinate axes. All of them belong to the polyhedra Π_n. In fact, the other extremity of each of these edges is represented by a point of Γ such that exactly one coordinate is equal to 1, the others being equal to 0. Such a point corresponds to an *atomic* partition into $n - 1$ classes where only two objects are joined in one class; the other classes being singleton classes. There are in all m such partitions. On the other hand, the edges of Γ whose two extremities belong to Π_n are also edges of Π_n. Conversely, consider the convex cone generated by the m edges of Γ, all Γ is included in this cone. Since every extremal vertex of Π_n is also an extremal vertex of Γ, any other of the m atom partitions considered above can be adjacent to X. Therefore,

Proposition 23 *Relative to two comparable partitions X and Y, the necessary and sufficient condition that X and Y are adjacent is that one of both partitions covers the other one.*

Now, we shall consider the case where X and Y are non-comparable and satisfy the condition $X \wedge Y = \mathbf{0}$. This does not restrict the generality. In these conditions, notice that if a given $\mathcal{O}$ subset is both an X class and a Y class, it is also an $X \wedge Y$ class and then is reduced to a singleton class of the form $\{i\}$, where i codes an element of $\mathcal{O}$. For the latter we have

$$(\forall (g, h)1 \leq g < i < h \leq n), x_{gi} = y_{gi} = x_{ih} = y_{ih} = 0$$

where x_{gi} and x_{ih} (resp., y_{gi} and y_{ih}) are components of X (resp., Y). Hence,

$$(\forall (g, h)1 \leq g < i < h \leq n), z^u_{gi} = z^u_{ih} = 0$$

where z^u_{gi} and z^u_{ih} are the gi and ih components of Z_u in (1.5.3). Thus the contribution of each of the $n - 1$ unordered pairs $\{i, j\}$, where $j \in \mathbb{I}$, is null. It follows that $X D Y$ if and only if the respective restrictions of X and Y to $\mathcal{O} - \{o_i\}$ are non-adjacent.

Continuing step by step in an analogous way, we obtain

$$XDY \Leftrightarrow (X - G)D(Y - G)$$

where G is the part of $\mathcal{O}$ obtained as the union of all classes (necessarily singleton classes (see above)) which are both X and Y classes.

Now, consider (1.5.3) and a given pair ij ($ij \in F$). Denoting as above by z_{ij}^u the ijth component of Z_u, we have

$$x_{ij} = y_{ij} = 0 \Leftrightarrow z_{ij}^u = 0 \text{ for every } u \in U;$$
$$x_{ij} = 0 \text{ and } y_{ij} = 1 \Leftrightarrow z_{ij}^u = 0 \text{ or } 1, \text{ depending on } u \in U(x_{ij} \leq z_{ij}^u \leq y_{ij});$$
$$x_{ij} = 1 \text{ and } y_{ij} = 0 \Leftrightarrow z_{ij}^u = 0 \text{ or } 1, \text{ depending on } u \in U(y_{ij} \leq z_{ij}^u \leq x_{ij});$$
$$x_{ij} = y_{ij} = 1 \Leftrightarrow z_{ij}^u = 1 \text{ for every } u \in U. \tag{1.5.7}$$

Consequently, for a generic partition element Z intervening in the right member of (1.5.3), we have

$$X \wedge Y \leq Z \leq X \vee Y$$

A more precise relation is

$$(\forall i \in \mathbb{I}), Z(i) \subset X(i) \cup Y(i) \tag{1.5.8}$$

where $X(i)$, $Y(i)$ and $Z(i)$ are the classes of the respective partitions X, Y and Z to which o_i belongs.

Now, we shall establish that if none of both classes $X(i)$ and $Y(i)$ is included in the other one (i.e. if both of set differences $X(i) - Y(i)$ and $Y(i) - X(i)$ are non-empty), the class $Z(i)$ is exactly either $X(i)$ or $Y(i)$.

To begin, let us show that if $j \in X(i) - Y(i)$ and $k \in Y(i) - X(i)$, j and k cannot belong to the same Z class. In fact, according to (1.5.7)

$$z_{jk} = 1 \Leftrightarrow x_{jk} = 1 \text{ or } y_{jk} = 1$$

Suppose $x_{jk} = 1$, we have $x_{ij} = 1$ and then, $x_{ik} = 1$, which is contrary to $k \in Y(i) - X(i)$.

Similarly, if $y_{jk} = 1$, we have $y_{ik} = 1$ and then $y_{ij} = 1$, which is contrary to $j \in X(i) - Y(i)$.

Therefore, $Z(i)$ is necessarily included in either $X(i)$ or $Y(i)$. Let us examine more deeply the behaviour of (1.5.3).

For ij ($x_{ij} = 1$ and $y_{ij} = 1$), we have

$$-1 = \sum_{u \in U} \lambda_u (z_{ij}^u - 1)$$

For ik ($x_{ik} = 0$ and $y_{ik} = 1$), we have

$$1 = \sum_{u \in U} \lambda_u z_{ik}^u$$

Hence,

$$0 = \sum_{u \in U} \lambda_u (z_{ij}^u + z_{ik}^u - 1) \tag{1.5.9}$$

However, by transitivity we have

$$z_{ij}^u + z_{ik}^u - 1 \le z_{jk}^u = 0$$

Therefore, every term of the sum (1.5.8) cannot be positive. All of them are null. Thus, for Z participating to formula (1.5.3), we have

$$z_{ij} = 1 \text{ and } z_{ik} = 0 \text{ or (exclusively) } z_{ij} = 0 \text{ and } z_{ik} = 1$$

for all (j, k) chosen as indicated above.

Moreover, as we have $Z(i) \subset X(i)$ or (exclusively) $Z(i) \subset Y(i)$, necessarily, $Z(i) = X(i)$ or $Z(i) = Y(i)$.

From this result, we can see that if X and Y are two non-comparable partitions such that $X \vee Y$ is the singleton partition P_t with a single class, and every partition Z taking part in (1.5.3) brought down to X or to Y. This is provided from the property that we can associate a Y class to each X class, whose intersection and symmetrical difference with the X class concerned are non-empty. In this case necessarily X and Y are adjacent.

Reciprocally, in the case considered of a pair $\{X, Y\}$ of non-comparable partitions, where any X class is a Y class, the adjacency property between X and Y entails that $X \vee Y$ is the singleton partition. If not, $X \vee Y$ would be finer than a partition $\{O_1, O_2\}$ of $\mathcal{O}$ into two classes. In these conditions, let us consider the restrictions X_1 and Y_1 (resp., X_2 and Y_2) of the indicator functions X and Y to the pairs of F whose two components are in O_1 (resp., in O_2). Apart from the order of the F components, the associated vectors X and Y—representing the indicator functions X and Y—can be put as follows:

$$(X_1, X_2, \vec{0}) \text{ and } (Y_1, Y_2, \vec{0}) \tag{1.5.10}$$

where $\vec{0}$ represents a sequence of components that equal all to 0. Both vectors have the same size m.

Equation (1.5.10) is due to the property that any X_1 class (resp., Y_1 class) has a non-empty intersection with a X_2 class (resp., a Y_2 class). Consider now the vectors

$$V = (X_1, Y_2, \vec{0}) \text{ and } W = (Y_1, X_2, \vec{0}) \tag{1.5.11}$$

V and W represent two partitions, where V (resp., W) coincides with X on O_1 and with Y on O_2 (resp., coincides with Y on O_1 and with X on O_2). V and W represent intermediary partitions between X and Y; this because

$$X + Y = V + W \tag{1.5.12}$$

In accordance with (1.5.4), we have the following.

Theorem 6 *X and Y being non-comparable partitions, consider the restrictions $X|H$ and $Y|H$ of X and Y to the set H, which is the complement—with respect to O—of the union G of the O subsets which are equally an X class and a Y class. The necessary and sufficient condition to have XAY is that $(X|H) \vee (Y|H) = \{H\}$ which is the singleton partition of H.*

As we have seen in the last paragraph of Sect. 1.1.2.2 (see Proposition 5), the number of adjacent partitions to a given partition, as expressed in the latter theorem, is generally prohibitive.

References

1. Berge, C.: Principes de Combinatoire. Dunod, Paris (1970)
2. Berge, C.: Principles of Combinatorics. Academic Press, Cambridge (1971)
3. Birkhoff, N.: Lattice Theory. A.M.S., New York (1967)
4. De La Véga, W.F.: Caractérisation des sommets adjacents sur un plyèdre associé à l'ensemble des relations d'éqivalence sur un ensemble fini. Note Interne 47, Centre d'Analyse Documentaire pour l'Archéologie, C.N.R.S., Marseille (1971)
5. Frank, O., Svensson, K.: On probability distribution of single-linkage dendrograms. J. Statist. Comput. Simul. **12**, 121–131 (1981)
6. Gras, R., Kuntz, P.: Discovering r-rules with a directed hierarchy. Soft Comput. **10**, 453–460 (2005)
7. Gras, R., Kuntz, P.: An overview of the statistical implicative analysis development. In: Guillet, F., Gras, R., Suzuki, E., Spagnolo, F. (eds.) Statistical Implicative Analysis, pp. 11–40. Springer, New York (2008)
8. Gras, R., Larher, A.: L'implication statistique, une nouvelle méthode d'analyse des données. Mathématiques et Sciences Humaines, **120**, 5–31 (1993)
9. Guillaume, S., Lerman, I.C.: Analyse du comportement limite d'indices probabilistes pour une sélection discriminante. In: Khenchaf, A., Poncelet, P. (eds.) Revue de l'Information et des Nouvelles Technologies, RNTI E.20, EGC'2011, pp. 657–664. Hermann (2011)
10. Guillet, F., Hamilton, H.J. (eds.): Quality Measures in Data Mining, Studies in Computational Intelligence. Springer, New York (2007)
11. Johnson, S.C.: Hierarchical clustering schemes. Psychometrika **32**(3), 241–254 (1967)
12. Lengyel, T.: On the numbers of all agglomerative clustering hierarchies. In: COMPSTAT 1982, Part II, pp. 177–178. Physica-Verlag, Springer, Heidelberg (1982)
13. Lent, B., Swami, A.N., Widow, J.: Clustering association rules. In: Werner, B. (eds.) Proceedings of the 13th Conference on Data Engineering, pp. 220–231. IEEE Computer Society, Washington (1997)
14. Lerman, I.C.: Analyse du problème de la recherche d'une hiérarchie de classifications. Publication Interne 22, Maison des Sciences de l'Homme, Centre de Mathématiques Appliquées et de Calcul, February (1968)

15. Lerman, I.C.: Les bases de la classification automatique. Gauthier-Villars (1970)
16. Lerman, I.C.: Classification et analyse ordinale des données. Dunod and http://www.brclasssoc. org.uk/books/index.html (1981)
17. Lerman, I.C.: Analyse logique, combinatoire et statistique de la construction d'une hiérarchie binaire implicative; niveaux et noeuds significatifs. Mathématiques et Sciences humaines / Mathematics and Social Sciences, **184**, 47–103 (2008)
18. Lerman, I.C., Azé. A new measure of interestingness for association rules, based on the likelihood of the link. In: Guillet, F., Hamilton, H.J. (eds.) Quality Measures in Data Mining, Studies in Computational Intelligence, vol. 43, pp. 207–236. Springer, New York (2007)
19. Lerman, I.C., Ghazzali, N.: What do we retain from a classification tree? an experiment in image coding. In: Diday, E., Lecevallier, Y. (eds.) Symbolic-Numeric Data Analysis and Learning, pp. 27–42. Nova Science, New York (1991)
20. Lerman, I.C., Guillaume, S.: Analyse comparative d'indices discriminants fondés sur une échelle de probabilité. Research Report 7187, IRISA-INRIA, February (2010)
21. Lerman, I.C., Guillaume, S.: Comparing two discriminant probabilistic interestingness measures for association rules. In: Venturini, G., Guillet, F., Pinaud, B., Zighed, D.A. (eds.) Advances in Knowledge Discovery and Management, volume 471 of Studies in Computational Intelligence, pp. 59–83. Springer, New York (2013)
22. Lerman, I.C., Kuntz, P.: Directed binary hierarchies and directed ultrametrics. J. Classif. **28**, 272–296, October (2011)
23. Monjardet, B.: Note sur les treillis géométriques et l'axiomatique du treillis des partitions d'un ensemble fini. Mathématiques et Sciences Humaines, **22**, 23–26 (1968)
24. Murtagh, F.: Counting dendrograms: a survey. Discret. Appl. Math. **7**, 191–199 (1984)
25. Régnier, S.: Études sur le polyèdre des partitions. Mathématiques et Sciences Humaines, **82**, 85–111 (1983)
26. Saporta, G.: Théorie et Méthodes de la Statistique. Editions Technip (1978)
27. Volle, M.: Analyse des Données. Economica (1981)
28. Waring, E.: Meditationes Algebraicae, translation by Dennis Weeks of the 1782 edition. American Math. Soc., Providence (1991)

Chapter 2
Two Methods of Non-hierarchical Clustering

2.1 Preamble

As mentioned in the Preface, the development provided in this book is dominated by the potential of applying *ascendant agglomerative hierarchical clustering* to all types of data. Nonetheless, the specific methodology devoted to non-hierarchical clustering is also very important. In these conditions, we shall describe in this chapter two mutually very different methods of non-hierarchical clustering. The first one, called "Central Partition" method, is due to S. Régnier [35–37]. The second method called "méthode des Nuées Dynamiques" or "Dynamic cluster method" is due to E. Diday and collaborators [10, 11, 13, 16]. This approach corresponds to a vast generalization of the K-means method, initiated by Forgy [17] and Jancey [19].

A very important idea, common to both approaches, consists of summarizing a finite set E situated in a space endowed with a distance function, by a *centre*. For the central partition method, E is a set of partitions on a finite set of objects and for the dynamic cluster method, E is a cluster of elements of a finite set of objects, or, more generally, a finite set of structures on an object set $\mathcal{O}$. In the case presented here (see Sect. 2.3) the set $\mathcal{O}$ to be clustered is represented by points in the geometrical space $\mathbb{R}^p$, where $\mathbb{R}$ denotes the reals and p a given positive integer.

The presentation of the central partition method may partially require the formalism established in Chap. 3. Some recent theoretical and methodological developments of this method will be indicated, particularly, in Sect. 2.2.4. Otherwise, the K-means algorithm is certainly the basic technique which has received the greatest development in non-hierarchical clustering [3, 18]. We shall try to comment on some of their elements.

© Springer-Verlag London 2016
I.C. Lerman, *Foundations and Methods in Combinatorial and Statistical Data Analysis and Clustering*, Advanced Information and Knowledge Processing, DOI 10.1007/978-1-4471-6793-8_2

2.2 Central Partition Method

2.2.1 Data Structure and Clustering Criterion

2.2.1.1 The Data

A set $\mathcal{A} = \{a^m | 1 \leq m \leq p\}$ of nominal categorical attributes (see Sect. 3.3.1 of Chap. 3) is assumed established in order to describe a set $\mathcal{O}$ of objects. Let $\mathcal{C}_m$ be the value set of the attribute a^m, $1 \leq m \leq p$. Designating by a the vector attribute $(a^1, a^2, \ldots, a^m, \ldots, a^p)$, the description can be expressed in terms of the mapping

$$a \longrightarrow \mathcal{C} = \mathcal{C}_1 \times \mathcal{C}_2 \times \cdots \times \mathcal{C}_m \times \cdots \times \mathcal{C}_p$$
$$x \mapsto a(x) = \left(a^1(x), a^2(x), \ldots, a^m(x), \ldots, a^p(x)\right) \qquad (2.2.1)$$

where $x \in \mathcal{O}$ and where $a^m(x)$ is the category of a^m possessed by x, $1 \leq m \leq p$. The sets $\mathcal{C}_m$ are finite, without any specific structure. Thereby, each of the a^m attributes induces a partition P^m on the object set $\mathcal{O}$, $1 \leq m \leq p$. If K_m is the class number of P^m, the latter can be written as

$$P^m = \left\{O_1^m, O_2^m, \ldots, O_k^m, \ldots, O_{K_m}^m\right\} \qquad (2.2.2)$$

where O_k^m ($O_k^m \subset \mathcal{O}$) is composed of the set of objects having the kth value of the attribute a^m, $1 \leq k \leq K_m$, $1 \leq m \leq p$. Thus, the data are defined by the partition vector

$$(P^1, P^2, \ldots, P^m, \ldots, P^p) \qquad (2.2.3)$$

on $\mathcal{O}$.

As indicated above, this data structure will be expressed again in Sect. 3.3.1 of Chap. 3. Equally, it will intervene in Sect. 7.2.3 of Chap. 7 devoted to the construction of a similarity index between objects described by nominal categorical attributes. According to (2.2.3), a partition P that we have to compare with the family of partitions P^m, $1 \leq m \leq p$, will be represented as a partition vector having p components equal to the same partition P:

$$(P, P, \ldots, P, \ldots, P) \qquad (2.2.4)$$

2.2.1.2 The Notion of Central Partition

Let $\mathbb{P} = (P^m | 1 \leq m \leq p)$ be a sequence of partitions on a set $\mathcal{O}$ of objects. The notion of *central partition* of $\mathbb{P}$ assumes the definition of a distance δ on the set $\mathcal{P}$ of all partitions on $\mathcal{O}$. In these conditions, a central partition P of $\mathbb{P}$ is defined as

$$Arg \left\{ \min_{P \in \mathcal{P}} \left[\frac{1}{p} \sum_{1 \le m \le p} \delta(P^m, P) \right] \right\} \qquad (2.2.5)$$

Clearly, in the expression of this criterion, the multiplicative factor $\frac{1}{p}$ can be omitted. However, according to [35, 36], we will keep it. In fact, it refers to the *mean* notion.

If P and Q are two partitions of $\mathcal{O}$, S. Régnier considers for $\delta(P, Q)$ the cardinal of the symmetrical difference of the graphs in $\mathcal{O} \times \mathcal{O}$ of the equivalence relations associated with P and Q. Without risk of ambiguity, the latter relations can also be designated as P and Q.

Notice that the graph of the equivalence relation associated with P^m defined above (see (2.2.2)) is given as

$$gr(P^m) = \left\{ (x, y) | (x, y) \in \sum_{1 \le m \le K_m} O_k^m \times O_k^m \right\} \qquad (2.2.6)$$

Thereby, clearly, the representation of a partition of $\mathcal{O}$, by the graph of the associated equivalence relation, is situated in the Cartesian product $\mathcal{O} \times \mathcal{O}$.

In these conditions, the criterion to be minimized becomes

$$\delta[(P^1, P^2, \ldots, P^m, \ldots, P^p), (P, P, \ldots, P, \ldots, P)] = \frac{1}{p} \sum_{1 \le m \le p} card\big(gr(P^m) \Delta gr(P)\big)$$
$$(2.2.7)$$

where Δ designates the symmetrical difference.

Without loss of generality, we may consider the representation of the graph of an equivalence relation at the level of the set $F = P_2(\mathcal{O})$ of unordered distinct object pairs of $\mathcal{O}$, that is, by a subset of F (see Sect. 1.5 of Chap. 1 and Sect. 4.3.2 of Chap. 4). This representation at the F level is more reduced ($card(F) = n(n-1)/2$) than that in $\mathcal{O} \times \mathcal{O}$. The diagonal of the latter set, that is, the subset of pairs of the form (x, x) ($x \in \mathcal{O}$) does not take part in the new representation. Clearly, we can deduce one of both representations from the other.

Now, by considering the representation at the F level, we can observe that the distance index

$$\delta(P, Q) = card\big(gr(P) \Delta gr(Q)\big) \qquad (2.2.8)$$

where P and Q are two partitions of $\mathcal{O}$, can be reduced to the Rand similarity index expressed in Sect. 4.3.2 of Chap. 4. If $Rand(P, Q)$ designates this index, we can show easily that

$$\delta(P, Q) = n\big(1 - Rand(P, Q)\big) \qquad (2.2.9)$$

where $n = card(\mathcal{O})$.

2.2.1.3 Analysis of a Classification (Clustering) Criterion

As considered by S. Régnier [35, 36], let us return now to the representation in $\mathcal{O} \times \mathcal{O}$ of a partition P of an object set $\mathcal{O}$. To $gr(P)$ (where P denotes here the equivalence relation associated with a partition P) corresponds the indicator function ϖ defined as follows:

$$\left(\forall(x, y) \in \mathcal{O} \times \mathcal{O}\right) \varpi(x, y) = 1 \text{ if } x \text{ and } y \text{ are in a same class of } P$$
$$\left(\forall(x, y) \in \mathcal{O} \times \mathcal{O}\right) \varpi(x, y) = 0 \text{ if } x \text{ and } y \text{ are in different classes of } P$$
(2.2.10)

By coding the set $\mathcal{O}$ with the set $\mathbb{I} = \{1, 2, \ldots, i, \ldots, n\}$ of the first n integers, the partition P can be represented by the point S of the cube $\{0, 1\}^{n \times n}$, whose coordinate sequence is

$$\left(\varpi(x, y) | (x, y) \in \mathcal{O} \times \mathcal{O}\right)$$
(2.2.11)

We have

$$\left(\forall(x, y) \in \mathcal{O} \times \mathcal{O}\right), \ \varpi(x, y) = \varpi(y, x)$$
$$\left(\forall(x, y) \in \mathcal{O} \times \mathcal{O}\right), \ \varpi(x, y) = 1 \text{ and } \varpi(y, z) = 1 \Rightarrow \varpi(x, z) = 1 \quad (2.2.12)$$

Conversely, each point of the cube $\{0, 1\}^{n \times n}$, for which (2.2.11) and (2.2.12) are satisfied, is the representation of a partition of $\mathcal{O}$.

We will designate by ϖ^m and ϖ the indicator functions associated with the partitions P^m and P in (2.2.3) and (2.2.4), $1 \le m \le p$.

Notice that the logic cube $\{0, 1\}^{n \times n}$ represents all the binary relations on $\mathcal{O}$. By immersing it in the geometrical space $\mathbb{R}^{n \times n}$, endowed with the usual Euclidean metric, the square distance between two binary relations R and R', represented by their respective indicator functions ρ and ρ', is

$$d^2(\rho, \rho') = \sum_{(x,y) \in \mathcal{O} \times \mathcal{O}} [\rho(x, y) - \rho'(x, y)]^2$$
(2.2.13)

We have

$$card\left(gr(R) \triangle gr(R') = d^2(\rho, \rho')\right)$$
(2.2.14)

In the case where R and R' are partitions of $\mathcal{O}$, as in Eq. (2.2.7) for P^m and P, this index gives the number of ordered pairs (x, y) of $\mathcal{O} \times \mathcal{O}$ which are joined in the same class for one of both partitions and separated into two classes for the other.

In these conditions, the criterion defined in (2.2.7) becomes

$$\delta\big((P^1, P^2, \ldots, P^m, \ldots, P^p), (P, P, \ldots, P, \ldots, P)\big) = \frac{1}{p} \sum_{1 \leq m \leq p} d^2(S^m, S)$$

(2.2.15)

where S^m (resp., S) is the point of $\mathbb{R}^{n \times n}$ representing the equivalence relation P^m (resp., P), $1 \leq m \leq p$. The coordinates of S^m (resp., S) are given by the indicator function ϖ^m (resp., ϖ). More precisely, $\big(s_{ij}^m | (i, j) \in \mathbb{I} \times \mathbb{I}\big)$ (resp., $\big(s_{ij} | (i, j) \in \mathbb{I} \times \mathbb{I}\big)$) will represent the coordinate vector of S^m (resp., S). s_{ij}^m (resp., $s(i, j)$) is defined by $\varpi^m(x, y)$ (resp., $\varpi(x, y)$), where (x, y) is coded by $(i, j) \in \mathbb{I} \times \mathbb{I}$. Then, we will denote below indifferently, by s_{ij}^m or s_{xy}^m (resp., s_{ij} or s_{xy}), $1 \leq m \leq p$.

Equation (2.2.15) can be written as

$$\frac{1}{p} \sum_{1 \leq m \leq p} (s_{xy}^m - s_{xy})^2 = \frac{1}{p} \sum_{1 \leq m \leq p} \|S^m - S\|^2$$

(2.2.16)

where $\|\bullet\|$ designates the Euclidean norm in $\mathbb{R}^{n \times n}$.

The interest in immersing the cube of the binary relations on $\mathcal{O}$ in $\mathbb{R}^{n \times n}$ is due to the barycenter properties in the latter space. Let us designate by G the centre of gravity of the family of points $\{S^m | 1 \leq m \leq p\}$ of $\mathbb{R}^{n \times n}$, equally weighted as

$$G = \frac{1}{p} \sum_{1 \leq m \leq p} S^m$$

(2.2.17)

We have

$$\sum_{1 \leq m \leq p} \frac{1}{p} \|S^m - S\|^2 = \|S - G\|^2 + \sum_{1 \leq m \leq p} \frac{1}{p} \|S^m - G\|^2$$

(2.2.18)

This is a classical formula—easily demonstrable—concerning a cloud of points in an Euclidean space (see Sect. 10.3.3) where the moment of order 2 with respect to a given vertex S is decomposed relatively to the cloud inertia (moment of order 2 with respect to the gravity centre of the cloud).

In these conditions, a *central* partition can be defined as realizing the minimum of $\|S - G\|^2$, with respect to S. Considering the general property (2.2.10) in which s_{xy} can be substituted for ϖ_{xy}, the function $\|S - G\|^2$ of S can be simplified. In fact all of the points of the cube $\{0, 1\}^{n \times n}$ are at equal distance of the point H, whose coordinates are all equal to $\frac{1}{2}$. We have

$$\big(\forall (x, y) \text{ in } \mathcal{O} \times \mathcal{O}\big) \left(s_{xy} - \frac{1}{2}\right)^2 = \frac{1}{4}$$

and then,

$$\|S - H\|^2 = \sum_{(x,y) \in \mathcal{O} \times \mathcal{O}} \left(s_{xy} - \frac{1}{2}\right)^2 = \frac{n^2}{4}$$

(2.2.19)

We can say, in geometrical terms, that the cube of the binary relations on $\mathcal{O}$ is inscribed in the ball centred in H of radius $n^2/4$. It follows that

$$\|S - G\|^2 = \|(S - H) + (H - G)\|^2 = \|S - H\|^2 + \|H - G\|^2 + 2 \cdot \langle S - H, H - G \rangle \tag{2.2.20}$$

where $\langle \bullet, \bullet \rangle$ denotes the symmetrical bilinear form associated with the Euclidean norm. In these conditions, (2.2.20) can be written as

$$\|S - G\|^2 = \frac{n^2}{4} + \|G - H\|^2 - 2 \cdot \langle S - H, G - H \rangle \tag{2.2.21}$$

Therefore, minimizing $\|S - G\|^2$ is equivalent to maximizing the scalar product $\langle S - H, G - H \rangle$ which can be expanded as follows:

$$\sum_{(x,y) \in \mathcal{O} \times \mathcal{O}} \left(s_{xy} - \frac{1}{2} \right) \left(g_{xy} - \frac{1}{2} \right) \tag{2.2.22}$$

where the g_{xy}, $(x, y) \in \mathcal{O} \times \mathcal{O}$, are the coordinates of the point G, that is,

$$g_{xy} = \frac{1}{p} \sum_{1 \le m \le p} s_{xy}^m \tag{2.2.23}$$

By putting $t_{xy} = g_{xy} - \frac{1}{2}$, the central partition is defined as maximizing the linear form

$$L(S) = \sum_{(x,y) \in \mathcal{O} \times \mathcal{O}} t_{xy} . s_{xy} \tag{2.2.24}$$

For a partition $P = \{O_k | 1 \le k \le K\}$ of $\mathcal{O}$, whose classes are the O_k, $1 \le k \le K$, (2.2.24) can be written as

$$L(S) = \sum_{1 \le k \le K} \sum_{(x,y) \in O_k \times O_k} t_{xy} \tag{2.2.25}$$

where S is the point of $\{0, 1\}^{n \times n}$ representing the equivalence relation P, associated with the partition P. The kth term of the sum over $\{1, 2, \ldots, k, \ldots, K\}$ is a measure of cohesion or density of the class O_k. $L(S)$ is then the sum of densities of the different classes of P.

It is easy to see that $g_{xy} < 1/2$ for every (x, y) in $\mathcal{O} \times \mathcal{O}$, is a necessary and sufficient condition that the finest partition (each class is a singleton class) is optimal for $L(S)$. On the other hand, the coarsest partition into a single class can be optimal without having necessarily $g_{xy} > 1/2$ for all (x, y) in $\mathcal{O} \times \mathcal{O}$. Finally, we can notice that when g_{xy} is constant and equal to $1/2$, every partition of $\mathcal{O}$ is a central partition.

Example of Computing $L(S)$

Relative to the incidence data Table 4.1 given in Sect. 4.2.1.2 of Chap. 4, let us compare the partitions

$$P = \{C_1, C_2\} = \big\{\{o_a, o_d, o_f\}, \{o_b, o_c, o_e\}\big\}$$

and

$$Q = \{D_1, D_2\} = \big\{\{o_a, o_d, o_e, o_f\}, \{o_b, o_c\}\big\}$$

of the set $\mathcal{O} = \{o_a, o_b, o_c, o_d, o_e, o_f\}$, where we have $D_1 = C_1 + \{o_e\}$ and $D_2 = C_2 - \{o_e\}$.

This incidence data table illustrates the description of 6 objects by 18 Boolean attributes. In order to carry out the comparison mentioned, each Boolean attribute is interpreted here as a nominal categorical attribute with two values, 0 and 1. Table 2.1 gives the respective values of t_{xy} on $\mathcal{O} \times \mathcal{O}$. The rows of Table 2.1 are arranged in such a way that both partitions P and Q can clearly be distinguished.

For P, the criterion value is

$$\frac{41}{18} + \frac{37}{18} = \frac{13}{3} = \frac{39}{9}$$

and for Q,

$$\frac{46}{18} + \frac{24}{18} = \frac{35}{9}$$

Therefore, P is better than Q, according to the criterion concerned.

Admissible Metrics in $\mathbb{R}^{n \times n}$

The reason for which seeking for a central partition leads to maximizing a linear form (see (2.2.24)) is due to the fact that in $\mathbb{R}^{n \times n}$, all the points of the cube $\{0, 1\}^{n \times n}$

Table 2.1 Cohesion table

$\mathcal{O} \backslash \mathcal{O}$	o_a	o_d	o_f	o_e	o_b	o_c
o_a	$\frac{1}{2}$	$\frac{1}{6}$	$\frac{1}{9}$	$-\frac{1}{6}$	$-\frac{1}{9}$	$-\frac{1}{9}$
o_d	$\frac{1}{6}$	$\frac{1}{2}$	$\frac{1}{9}$	$\frac{1}{18}$	0	$-\frac{1}{9}$
o_f	$\frac{1}{9}$	$\frac{1}{9}$	$\frac{1}{2}$	0	$-\frac{1}{6}$	$-\frac{7}{18}$
o_e	$-\frac{1}{6}$	$\frac{1}{18}$	0	$\frac{1}{2}$	$\frac{1}{9}$	0
o_b	$-\frac{1}{9}$	0	$-\frac{1}{6}$	$\frac{1}{9}$	$\frac{1}{2}$	$\frac{1}{6}$
o_c	$-\frac{1}{9}$	$-\frac{1}{9}$	$-\frac{7}{18}$	0	$\frac{1}{6}$	$\frac{1}{2}$

are—for the ordinary Euclidean metric—equally distant from a specific point (called H above). We shall now characterize the symmetrical linear forms, giving metrics in $\mathbb{R}^{n \times n}$, which possess this property. In order to set forth the following proposition, let us designate by $\{E_I | 1 \leq I \leq n^2\}$ the canonical base of $\mathbb{R}^{n \times n}$. E_I is a vector with n^2 components for which the Ith component is equal to 1 and the others to 0. We have

Proposition 24 *Let $B(X, Y)$ $(X, Y \in \mathbb{R}^{n \times n})$ be a symmetrical bilinear form giving rise to a metric in $\mathbb{R}^{n \times n}$. A necessary and sufficient condition for which the metric associated with $B(X, Y)$ is such that there exists a point K $(K \in \mathbb{R}^{n \times n})$ equally distant from all points of the cube $\{0, 1\}^{n \times n}$, is that the canonical base $\{E_I | 1 \leq I \leq n^2\}$ is orthogonal.*

Clearly, the stated condition is sufficient. If the canonical base is orthogonal for $B(X, Y)$, we have

$$(\forall(I, J), 1 \leq I, J \leq n), B(E_I, E_J) = 1(\text{resp.}, 0) \text{ if } I = J(\text{resp.}, I \neq J)$$

If H is the point of $\mathbb{R}^{n \times n}$ whose coordinates are all equal to $\frac{1}{2}$, then, for any point X of $\{0, 1\}^{n \times n}$, we have

$$X - H = \sum_{1 \leq I \leq n^2} \frac{1}{2} E_I \qquad (2.2.26)$$

Therefore,

$$B(X - H, X - H) = \frac{1}{4} n^2 \qquad (2.2.27)$$

Conversely, to begin with, let us verify that if all points of the cube $\mathbb{C} = \{0, 1\}^{n \times n}$ are at equal distance of a given point H' of $\mathbb{R}^{n \times n}$, then, necessarily, $H' = H$. In fact, if X is an arbitrary point of the cube $\mathbb{C}$, associate the point X' with X, defined by $X' = U - X$, where U is the point of $\mathbb{C}$, whose coordinates are all equal to 1. Due to these assumptions, we have

$$Q(X - H) = Q(X' - H) \text{ and } Q(X - H') = Q(X' - H')$$

where $Q(Y) = B(Y, Y)$ is the square of the norm of Y, corresponding to B.

Now, let us expand $Q(X - X')$ of two different ways, by referring to H and K, respectively,

$$\begin{aligned} Q(X - X') &= B[(X - H) - (X' - H), (X - H) - (X' - H)] \\ &= 2B[(X - H), (X - H)] - 2B[(X - H), (X' - H)] \\ &= 2B[(X - H), (X - X')] \end{aligned} \qquad (2.2.28)$$

Similarly, we have

$$Q(X - X') = 2B[(X - H'), (X - X')] \tag{2.2.29}$$

From (2.2.28) and (2.2.29), we obtain

$$B(H' - H, X - X') = 0 \tag{2.2.30}$$

for all X and $X' = U - X$. Hence, $H' = H$.

H being the point of $\mathbb{R}^{n \times n}$ whose coordinates are all equal to $\frac{1}{2}$, if $Q(X - H)$ is a constant c^2, then, for every point Y whose coordinates are equal to 1 or -1, $Q(Y) = 4c^2$. In fact, there exists a single X of $\mathbb{C}$, for which $Y = 2(X - H)$. On the other hand, $Q(Y) = Q(-Y)$. It results that Q cannot include in its expression with respect to the canonical base, rectangle terms. If not, the value of $Q(Y)$ would change by substituting $Y' = -Y$ for Y. Therefore, $Q(X)$ is necessarily of the form

$$\sum_{(i,j) \in \mathbb{I} \times \mathbb{I}} q_{ij} x_{ij}^2$$

where the x_{ij}, $(i, j) \in \mathbb{I} \times \mathbb{I}$ are the coordinates of the vertex X in $\mathbb{C}$. $\qquad \square$

Following the end of this proof, notice that if X and X' are two points of the cube $\mathbb{C}$, representing two binary relations R and R' on $\mathcal{O}$, $Q(X - X')$ can be interpreted as a measure—defined by $\{q_{ij} | (i, j) \in \mathbb{I} \times \mathbb{I}\}$—of the symmetrical difference of the graphs of R and R' in $\mathcal{O} \times \mathcal{O}$.

2.2.2 Transfer Algorithm and Central Partition

2.2.2.1 Transfer Algorithm

The *transfer* algorithm is independent of the nature of the criterion to optimize. However, its expression is more consistent in the case of a *linear* criterion with respect to the set of object pairs. In these conditions, we pay attention to a criterion of the form (2.2.25). In the latter, $t_{xy} = g_{xy} - 0.5$ plays the role of a similarity index between x and y, where g_{xy} is a similarity index comprised between 0 and 1, $(x, y) \in \mathcal{O} \times \mathcal{O}$. In fact, up to a coherent scale transformation, the latter property can be filled by a very general family of similarity indices. Nevertheless, referring to the value 0.5 cannot be justified for most of the similarity indices on $\mathcal{O}$.

For a partition P on $\mathcal{O}$, let $c(P)$ be the value of a criterion c on P, which has to be maximized. From P, we can build a very near partition P' by moving one object of its class and putting it in another one. The latter might be a new class. To fix ideas, we consider the set $\mathcal{P}_K$ of partitions on $\mathcal{O}$ into K classes at most ($\mathcal{P}_n$ is the set of all partitions). By labelling the partition classes, a partition P can be represented by a mapping f which makes correspondence between $\mathcal{O}$ and the class labels:

$$f : \mathcal{O} \longrightarrow \{1, 2, \ldots, k, \ldots, K\} \tag{2.2.31}$$

The *transfer* of an object x from its class labelled l to the class labelled k, is defined by substituting for the mapping f, that f' of $\mathcal{O}$ into $\{1, 2, \ldots, k, \ldots, K\}$, defined as follows:

$$f'(x) = k \neq f(x) = l$$
$$f'(y) = f(y) \text{ if } y \neq x \tag{2.2.32}$$

This transfer is denoted in the following by $T_l^k(x)/f$.

This algorithm has to be initialized. A quick suggestion in the framework of the data structure presented in Sect. 2.2.1, consists of taking in (2.2.3) the partition P^m which maximizes $c(P)$.

By iterating the transformation (2.2.32) we can describe all the set $\mathcal{P}_K$ (see above). The algorithm consists of passing from one partition to the other (from P to P'), by transfer, choosing at each step the transfer $T_l^k(x)/f$ (see (2.2.32)), which maximizes the criterion c: $c(P') - c(P)$ maximum.

The development will be stopped when there does not remain any profitable transfer: $c(P') - c(P) < 0$ for all passages from P to P' (see (2.2.32)).

The algorithm definition above gives rise to a new distance between partitions. By denoting it τ, for a pair (P, Q) of partitions of $\mathcal{O}$, $\tau(P, Q)$ is the minimum number of transfers (2.2.32) needed to transform one of both partitions into the other. In [7, 8] some interesting formulas are given in certain conditions for the value of this distance.

2.2.2.2 Computing Central Partitions

As expressed above, the criterion to be maximized is linear with respect to the set $\mathcal{O} \times \mathcal{O}$ represented by the cube $\mathbb{C} = \{0, 1\}^{n \times n}$, see (2.2.25) which we again take up here:

$$L(S) = \sum_{1 \leq k \leq K} \left(\sum_{(x,y) \in O_k \times O_k} t_{xy} \right) \tag{2.2.33}$$

where S is the point of $\mathbb{C}$ representing the partition $P = \{O_k | 1 \leq k \leq K\}$. Besides, the notation P can replace S in $L(S)$. By considering the representation of the partition P by means of the mapping f (see (2.2.31)), $L(S)$ can be written as

$$L(S) = \sum_{1 \leq k \leq K} \left(\sum_{(x,y) \in \mathcal{O} \times \mathcal{O}, f(x)=f(y)=k} t_{xy} \right) \tag{2.2.34}$$

Let O_k and O_h be two classes of the partition P and let u be an element of O_k, the transfer of u from O_k to O_h modifies two elements of the first sum in (2.2.34), namely

$$\sum_{(x,y)\in O_k\times O_k} t_{xy} \text{ and } \sum_{(x,y)\in O_h\times O_h} t_{xy}$$

The sum of these two sums become

$$\sum_{(x,y)\in(O_k-\{u\})\times(O_k-\{u\})} t_{xy} + \sum_{(x,y)\in(O_h+\{u\})\times(O_h+\{u\})} t_{xy} \qquad (2.2.35)$$

The variation of $L(S)$ is

$$\Delta(L(S)) = -\sum_{y\in O_k} t_{uy} - \sum_{x\in O_k} t_{xu} + \sum_{x\in O_h} t_{xu} + \sum_{y\in O_h} t_{uy} \qquad (2.2.36)$$

By considering,

$$(\forall x, y, u \in \mathcal{O}), t_{ux} = t_{xu} \text{ and } t_{uy} = t_{yu}$$

and setting

$$A_u^l = \sum_{x|x\in O_l, x\neq u} t_{xu}: \text{attraction of the object } u \text{ by the class } l, 1 \leq l \leq K$$

we get

$$\Delta(L(S)) = 2(A_u^h - A_u^k) \qquad (2.2.37)$$

$\Delta(L(S))$ gives the gain in transferring the object u from the class O_k to the class O_h. As mentioned above, O_h might be a new class initially empty. In this case $A_u^h = t_{uu} = 0.5$.

Now, consider the table indexed by $\mathcal{O} \times \{1, 2, \ldots, l, \ldots, K\}$ with n rows and K columns

$$\mathbb{A} = \{A_u^l | u \in \mathcal{O}, 1 \leq l \leq K\} \qquad (2.2.38)$$

In the transfer $T_k^h(u)$, only the columns k and h change according to

$$A_u^l \longleftarrow A_u^k - t_{ux}$$
$$A_u^h \longleftarrow A_u^h + t_{ux} \qquad (2.2.39)$$

Example

Consider the partition $Q = \{\{o_a, o_d, o_e, o_f\}, \{o_b, o_c\}\}$ of the previous example. We can observe that the transfer of the object o_e from the cluster $D_1 = \{o_a, o_d, o_e, o_f\}$ to $D_2 = \{o_b, o_c\}$, is profitable.

2.2.3 Objects with the Same Representation

The transfer algorithm has been defined at the level of a set $\mathcal{O}$ of objects. A unit element corresponds to a single object of $\mathcal{O}$. However, there may exist objects of $\mathcal{O}$ having the same descriptive representation in

$$\mathcal{C} = \mathcal{C}_1 \times \mathcal{C}_2 \times \cdots \times \mathcal{C}_m \times \cdots \times \mathcal{C}_p \tag{2.2.40}$$

(See (2.2.1)). a designating the descriptive vector attribute $(a^1, a^2, \ldots, a^m, \ldots, a^p)$, two objects x and y have the same representation if and only if

$$(\forall m, 1 \leq m \leq p), a^m(x) = a^m(y) \tag{2.2.41}$$

If so, x and y are expected to be clustered together in the central partitions.

In fact, if we assume that x and y, for which $a(x) = a(y)$, are separated by a partition $P = \{O_1, O_2, \ldots, O_k, \ldots, O_K\}$, for example, $x \in O_j$ and $y \in O_h$, with $1 \leq j \neq h \leq K$, then transferring x in O_h or y in O_j increases strictly the linear form

$$L(P) = \sum_{1 \leq k \leq K} \left(\sum_{(u,v) \in O_k \times O_k} \right) t_{uv} \tag{2.2.42}$$

considered in (2.2.33).

Transferring x in O_h, entails the variation

$$\Delta_1(L(P)) = \frac{1}{2} + \sum_{v \in O_h} t_{xv} - \sum_{u \in O_j} t_{xu} \tag{2.2.43}$$

of $L(P)$. On the other hand, transferring y in O_j, gives the variation

$$\Delta_2(L(P)) = \frac{1}{2} + \sum_{u \in O_j} t_{yu} - \sum_{v \in O_h} t_{yv} \tag{2.2.44}$$

Now, the sum $\Delta_1(L(P)) + \Delta_2(L(P))$ is equal to 1 and then, necessarily $\Delta_1(L(P))$ or $\Delta_2(L(P))$ is strictly positive.

In these conditions, it comes to work at the level of the image $a(\mathcal{O})$ which defines a cloud of points in $\mathcal{C}$:

$$a(\mathcal{O}) = \{(\xi, n_\xi) | \xi \in \mathcal{C}\} \tag{2.2.45}$$

where n_ξ is the number of objects in $\mathcal{O}$ having the same representation ξ. Clearly, the cardinal of the set of ordered pairs

$$\{(\xi, n_\xi) | \xi \in \mathcal{C}, n_\xi \neq 0\} \tag{2.2.46}$$

is lower than $n = card(\mathcal{O})$, thereby we may denote by q ($q \leq n$), the cardinality of $a(\mathcal{O})$.

Therefore, it is advisable to work in the set $a(\mathcal{O})$. In these conditions, the criterion $L(P)$ can be written as

$$L(P) = \sum_{(\xi, \eta) \in a(\mathcal{O}) \times a(\mathcal{O})} n_\xi \cdot n_\eta (\varpi_{\xi\eta} - g_{\xi\eta})^2 \tag{2.2.47}$$

where $\varpi_{\xi\eta}$ is the common value of ϖ_{xy} for (x, y) in $a^{-1}(\xi) \times a^{-1}(\eta)$. Moreover,

$$g_{\xi\eta} = \frac{1}{p} \sum_{1 \leq m \leq p} s^m(x, y) \tag{2.2.48}$$

where (x, y) belongs to $a^{-1}(\xi) \times a^{-1}(\eta)$. Recall that $s^m(x, y)$ stands for $\varpi^m(x, y)$, where ϖ^m is the indicator function of the partition of $\mathcal{O}$ associated with the mth attribute a^m, $1 \leq m \leq p$.

To summarize, a partition of $\mathcal{O}$ is represented as a partition of $a(\mathcal{O})$ ($a(\mathcal{O}) \subseteq \mathcal{C}$). The relational representation of a partition of $\mathcal{O}$ is realized at the level of the cross product $a(\mathcal{O}) \times a(\mathcal{O})$:

$$(\forall (\xi, \eta) \in a(\mathcal{O}) \times a(\mathcal{O})), \varpi_{\xi\eta} = \varpi_{xy} \text{ for } (x, y) \in a^{-1}(\xi) \times a^{-1}(\eta) \tag{2.2.49}$$

The linear form of the criterion (see (2.2.24)) becomes

$$L(P) = \sum_{(\xi, \eta) \in a(\mathcal{O}) \times a(\mathcal{O})} t_{\xi\eta} s_{\xi\eta} \tag{2.2.50}$$

Due to the previous development and in particular to Eq. (2.2.47), we have

Proposition 25 *By adopting on $\mathbb{R}^{n \times n}$ the metric*

$$Q(R) = \sum_{(\xi, \eta) \in a(\mathcal{O}) \times a(\mathcal{O})} n_\xi \cdot n_\eta \rho_{\xi\eta}^2 \tag{2.2.51}$$

where $R = \{\rho_{\xi\eta} | (\xi, \eta) \in a(\mathcal{O}) \times a(\mathcal{O})\}$ is a valued binary relation on $a(\mathcal{O})$, central partitions defined at the level of $a(\mathcal{O})$ are identical to those defined at the level of $\mathcal{O}$.

This new metric has the same algebraic properties as those of the metric defined initially, which are expressed in Sect. 3.3.1; particularly, the barycenter property and that stated in Proposition 24. However, previously, the square distance between two relations X and Y on $\mathcal{O}$ was the cardinal of the symmetrical difference of their respective graphs in $\mathcal{O} \times \mathcal{O}$. Now, X and Y have to be defined on the representation set $\mathcal{C}$ (see (2.2.40)). Moreover, $Q(X - Y)$ represents the sum of the products $n_\xi.n_\eta$ over all the pairs (ξ, η) in $\mathcal{C} \times \mathcal{C}$ for which exactly one of both relations is satisfied. Consequently, it is a measure—with integer values—of the symmetrical difference of the graphs of X and Y in $\mathcal{C} \times \mathcal{C}$.

2.2.4 Statistical Asymptotic Analysis

2.2.4.1 Preamble

As it will be expressed in Sect. 3.1 of Chap. 3, the set $\mathcal{O}$ of objects is generally a sample of a *universe* (we can also say *population*) $\mathcal{U}$, much more vast than $\mathcal{O}$ and even infinite. Mostly, $\mathcal{U}$ is impossible to observe in its totality. And even if exhaustive observation is possible, it is of importance—particularly for computational complexity reasons—to know the validity of the inference obtained from a sample of $\mathcal{U}$.

On the other hand, dually, a complete description of the $\mathcal{O}$ elements is impossible. More often than not, the attributes retained for the description of $\mathcal{O}$, constitute a sample of a much larger set of attributes.

In these conditions, it is interesting to study some statistical asymptotic problems. We present two convergence statistical problems concerning the central partition method analysed by S. Régnier (1966), reported in [37] and worked on again in [23], Chap. 4, or [25], Chap. 1. In the case concerned, the descriptive attributes are nominal categorical (see Sect. 3.3.1 of Chap. 3). Extension of this analysis to other types of descriptions might be envisaged. The presentation of these asymptotic convergence problems require basic notions of probability theory, which can be found in [28, 33]. In the following development, only results obtained will be stated.

2.2.4.2 The First Convergence Problem

The object set $\mathcal{O}$ denoted here by $\{o_i | 1 \leq i \leq n\}$ is viewed as the realization of a random sample $\mathcal{O}^* = \{o_i^* | 1 \leq i \leq n\}$ of independent random elements provided from a universe $\mathcal{U}$ of objects. $\mathcal{U}$ might be infinite, nonetheless, generally, $\mathcal{U}$ is finite, its cardinality N $(N = \mathcal{U})$ being "very" large. $\mathcal{U}$ is supposed to be endowed with a probability measure that we designate by Φ. A realization $\mathcal{O}$ of $\mathcal{O}^*$ is obtained by a random drawing with replacement of a sequence of n objects. Due to the bigness of N with respect to n, the probability to get more than once the same object in $\mathcal{O}$, is extremely small.

The mapping $a = (a^1, a^2, \ldots, a^m, \ldots, a^p)$ of the universe $\mathcal{U}$ into the Cartesian product

$$\mathcal{C} = \mathcal{C}_1 \times \mathcal{C}_2 \times \cdots \times \mathcal{C}_m \cdots \times \mathcal{C}_p$$

defining the description of $\mathcal{U}$ (see (2.2.1)) is supposed fixed. $\mathcal{C}$ is finite and a is assumed measurable for Φ and for the uniform measure on $\mathcal{C}$. Thereby, every part of $\mathcal{C}$ is measurable and equally, the reverse image a^{-1} of the subset set of $\mathcal{C}$ into the subset set of $\mathcal{U}$ determines a set of $\mathcal{U}$ subsets, measurable for Φ (see [28, 33] for the definition of the measurability of a mapping).

A probability measure Ψ on $\mathcal{C}$ is induced from Φ as follows:

$$\Psi(\xi) = \Phi\big(a^1(\xi)\big) \tag{2.2.52}$$

The mapping a defines the representation of the random set $\mathcal{O}^\star$ of objects in $\mathcal{C}$. The sequence of images

$$\{\Omega_i = a(o_i^\star) | 1 \leq i \leq n\} \tag{2.2.53}$$

is a sequence of n independent random elements of $\mathcal{C}$, endowed with the probability measure Ψ

The realization of the sequence (2.2.53) in the form

$$\{\omega_i = a(o_i) | 1 \leq i \leq n\} \tag{2.2.54}$$

determines a discrete probability measure on $\mathcal{C}$, defined by

$$(\forall \omega \in \mathcal{C}),\ \Psi_n(\omega) = \frac{n_\omega}{n} \tag{2.2.55}$$

where n_ω is the number of objects of $\mathcal{O}$, whose representation is ω: $n_\omega = card$ $\big(a^{-1}(\omega) \cap \mathcal{O}\big)$.

According to the previous proposition, a central partition of $\mathcal{O}$ (resp., $\mathcal{O}^\star$) is a reciprocal image by a of a central partition of $\mathcal{C}$, when $\mathcal{C}$ is endowed with the probability measure Ψ_n (resp., Ψ). In these conditions, we consider the central partition search in $\mathcal{C}$.

Π designating the partition set of $\mathcal{C}$, let π be an arbitrary element of Π ($\pi \in \Pi$). According to (2.2.47), define

$$D_n(\pi) = \sum_{(\xi,\eta)in\mathcal{C}\times\mathcal{C}} (\varpi_{\xi\eta} - g_{\xi\eta})^2 \Psi_n(\xi) \cdot \Psi_n(\eta) \tag{2.2.56}$$

where $\varpi_{\xi\eta}$ is the indicator function of the equivalence relation on $\mathcal{C}$ associated with π ($\varpi_{\xi\eta} = 1$ if ξ and η are in the same class of π and 0 if not). On the other hand,

$$g_{\xi\eta} = \frac{1}{p} \sum_{1 \le m \le p} s_{\xi\eta}^m$$

where $s_{\xi\eta}^m$ is the value for (ξ, η) of the indicator function on C, associated with the image of the partition of $\mathcal{O}$, defined by the attribute a^m, $1 \le m \le p$.

The set of central partitions of C, provided with the probability measure Ψ_n, is defined as the following set C_n of partitions π where $D_n(\pi)$ (see Eq. (2.2.56)) reaches its lower bound:

$$C_n = \{\chi | D_n(\chi) \le D_n(\pi) \text{ for all } \pi \in \Pi\} \tag{2.2.57}$$

Similarly, putting

$$D(\pi) = \sum_{(\xi,\eta) in C \times C} (\varpi_{\xi\eta} - g_{\xi\eta})^2 \Psi(\xi).\Psi(\eta) \tag{2.2.58}$$

we have

$$C = \{\chi | D(\chi) \le D(\pi) \text{ for all } \pi \in \Pi\} \tag{2.2.59}$$

Relative to the latter equations, recall that $D_n(\pi)$ (resp., $D(\pi)$) is associated with $\mathcal{O} = \{o_i | 1 \le i \le n\}$ (resp., with $\mathcal{O}^* = \{o_i^* | 1 \le i \le n\}$). In the framework of C^n—realization of C on $\mathcal{O}$—endowed with its subset set and the probability measure induced by Ψ, the statistic $D_n(\pi)$, for a given π, is a realization of a random variable, also denoted—without ambiguity—by $D_n(\pi)$. To the random vector $(D_n(\pi) | \pi \in \Pi)$ corresponds the centre C_n (see (2.2.57)) which is a random subset of Π. The asymptotic behaviour of C_n ($n \to \infty$) is stated as follows:

Proposition 26 *Almost surely*

$$\bigcap_n \bigcup_{l \ge n} C_l \subset C \tag{2.2.60}$$

that is to say,

$$Prob\left\{\bigcap_n \bigcup_{l \ge n} C_l \subset C\right\} = 1$$

On the other hand,

$$lim_{n \to \infty} Prob\{C_n \subset C\} = 1 \tag{2.2.61}$$

The left member of (2.2.60) corresponds to $limsup_{n \to \infty} C_n$

2.2.4.3 The Second Convergence Problem

We assume here that the set $\mathcal{O}$ of objects is fixed. Each descriptive attribute a^m ($a^m : \mathcal{O} \to \mathcal{C}_m$) induces a partition P^m of $\mathcal{O}$, $1 \leq m \leq p$. The central partitions are those which minimize the function, already considered in (2.2.7):

$$\delta_p(P) = \frac{1}{p} \sum_{1 \leq m \leq p} \delta(P, P^m) \qquad (2.2.62)$$

where P is an element of the set $\mathcal{P}(\mathcal{O})$ of all partitions and where $\delta(P, P^m)$ designates the cardinal of the symmetrical difference between the graphs in $\mathcal{O} \times \mathcal{O}$ of the equivalence relations associated with P and P^m, respectively, $1 \leq m \leq p$.

In this problem of asymptotic behaviour, the sequence of attributes $(a^m | 1 \leq m \leq p)$ is regarded as a sample of a large set of attributes. More precisely, we suppose that the sequence of partitions $\{P^m | 1 \leq m \leq p\}$ is a sample of p independent random elements, provided from $\mathcal{P}(\mathcal{O})$, where the latter partition set is endowed with the algebra of the set of its subsets and with a probability measure Λ.

$\{P^m | 1 \leq m \leq p\}$ determines an empirical probability measure $\Lambda_p(X)$, for X belonging to $\mathcal{P}(\mathcal{O})$, which is the relative frequency (proportion) of partitions identical to X. Equation (2.2.62) becomes

$$\delta_p(P) = \sum_{X \in \mathcal{P}(\mathcal{O})} \delta(P, X) \Lambda_p(X) \qquad (2.2.63)$$

In this empirical case, the set denoted by C_p of central partitions of $\mathcal{O}$ is defined with respect to the family $\{P^m | 1 \leq m \leq p\}$ of $\mathcal{P}(\mathcal{O})$ elements, as follows:

$$C_p = \{P | P \in \mathcal{P}(\mathcal{O}) \text{ and } \delta_p(P) \leq \delta_p(Y) \text{ for all } Y \in \mathcal{P}(\mathcal{O})\} \qquad (2.2.64)$$

When p tends to infinity, the empirical distribution Λ_p tends to a distribution Λ. Putting

$$\delta(P) = \sum_{X \in \mathcal{P}(\mathcal{O})} \delta(P, X) \Lambda(X) \qquad (2.2.65)$$

the associated centre C, where $\delta(P)$ reaches its minimum, is written as

$$C = \{P | P \in \mathcal{P}(\mathcal{O}) \text{ and } \delta(P) \leq \delta(Y) \text{ for all } Y \in \mathcal{P}(\mathcal{O})\} \qquad (2.2.66)$$

For P fixed, by associating a family of random partitions

$$\{P^{1\star}, P^{2\star}, \ldots, P^{m\star}, \ldots, P^{p\star}\}$$

with the observed one, $\delta_p(P)$ becomes a random variable defined on $\mathcal{P}(\mathcal{O})^p$ endowed with the algebra of its subset set and the probability measure induced by Λ. In other words, the limit distribution

$$\{(X, \Lambda(X))|X \in \mathcal{P}(\mathcal{O})\}$$

is substituted for the empirical distribution

$$\{(X, \Lambda_p(X))|X \in \mathcal{P}(\mathcal{O})\}$$

In this situation

$$\left(\delta_p(P)|P \in \mathcal{P}(\mathcal{O})\right)$$

becomes a random vector and the associated centre C_p becomes a random subset of $\mathcal{P}(\mathcal{O})$.

Λ is assumed to be obtained as a limit of Λ_p, when p tends to infinity. C_p (resp., C) indicating the centre for Λ_p (resp., Λ), we have

Proposition 27 *Almost surely*

$$limsup_{p \to \infty} = \bigcap_p \bigcup_{l \geq p} C_l \subset C \tag{2.2.67}$$

that is to say,

$$Prob\left\{\bigcap_p \bigcup_{l \geq p} C_l \subset C\right\} = 1$$

On the other hand,

$$lim_{p \to \infty} Prob\{C_p \subset C\} = 1 \tag{2.2.68}$$

The proofs of the latter two propositions can be found in the references given in the introductory section (see Sect. 2.2.4.1).

2.2.5 Remarks on the Application of the Central Partition Method and Developments

In Sect. 2.2.1.3, relative to a pair of objects (x, y) in $\mathcal{O} \times \mathcal{O}$, we have seen introduced in a natural way the similarity index

$$g_{xy} = \frac{1}{p} \sum_{1 \leq m \leq p} s_{xy}^m$$

(See (2.2.23)) which represents the proportion of categorical attributes having the same value in both objects x and y. g_{xy} or more exactly $t_{xy} = g_{xy} - 0.5$ (see (2.2.24)) is the similarity index sustaining the *Central Partition Method*.

The weakest point of this technique is due to the fact that, in principle, all the descriptive categorical attributes are considered as equally discriminant. A consequence of this is the importance given to the reference value $\frac{1}{2}$ for g_{xy}. Thus, the value 0 of t_{xy} must correspond to the case for which x and y are moderately similar (middle resemblance).

Nonetheless, in the general case where the object set $\mathcal{O}$ is provided with a numerical similarity index $Sim(x, y)$, deduced from a description by attributes of any sort (see Chaps. 3 and 7), it is always possible to reduce the value scale of $Sim(x, y)$ to the interval $[0, 1]$, in such a way that the value 0.5 corresponds to a middle resemblance.

This reduction is in fact considered by S. Régnier in order to adapt the transfer algorithm to any similarity index Sim. The similarity function Sim is then represented by a point in the continuous cube $[0, 1]^{n \times n}$, included in $\mathbb{R}^{n \times n}$. On the other hand, the central partition P to be sought is represented in the latter cube by a point S whose coordinates ϖ_{xy} $((x, y) \in \mathcal{O} \times \mathcal{O})$ are 0 or 1 (0 (resp., 1) if x and y are separated (resp., joined) by P).

With this formalism, P is all the more preferred that the Euclidean distance $\|S - Sim\|$ is small ($\|\bullet\|$ is the Euclidean norm in $\mathbb{R}^{n \times n}$). Therefore, the criterion established in Sect. 2.2.1.3 applies. It becomes

$$L(S) = \sum_{(x,y) \in \mathcal{O} \times \mathcal{O}} \left(Sim(x, y) - \frac{1}{2} \right) \varpi_{xy} \qquad (2.2.69)$$

However, this criterion becomes independent of the central partition notion.

This linear form of an association coefficient between a similarity Sim and a partition P of $\mathcal{O}$ appears clearly in our statistical analysis of a criterion established to validate a clustering of $\mathcal{O}$. This criterion is written in the form

$$C(Sim, P) = \sum_{\{x,y\} \in F} \varpi_{xy}.c(x, y) \qquad (2.2.70)$$

(See Eq. 9.3.71 of Sect. 9.3.5.2 of Chap. 9) where $c(x, y)$ is a normalized version of $Sim(x, y)$. This normalization of statistical nature is carried out empirically with respect to the set F of unordered object pairs of $\mathcal{O}$. It plays a fundamental role in the *LLA* clustering approach. Therefore, $C(Sim, P)$ is not a linear function of the similarity table

$$\{Sim(x, y)|(x, y) \in \mathcal{O} \times \mathcal{O}\} \qquad (2.2.71)$$

In the method built by F. Marcotorchino and P. Michaud [31] the criterion to maximize takes the linear form

$$L(Sim, P) = \sum_{(x,y)\in\mathcal{O}\times\mathcal{O}} \varpi_{xy} \cdot Sim(x, y) = \sum_{1\leq k\leq K} \sum_{(i,j)\in\mathbb{I}_k\times\mathbb{I}_k} \varpi_{i,j} \cdot Sim_{ij} \quad (2.2.72)$$

where $\mathbb{I}_k$ is the index set coding the kth $\mathcal{O}_k$, $1 \leq k \leq K$.

The originality of the technique mentioned resides in using *linear programming under constraints* as follows:

Maximize $L(Sim, P)$ under the conditions

$$\varpi_{i,j} \in \{0, 1\}$$
$$\varpi_{i,i} = 1$$
$$\varpi_{i,j} = \varpi_{j,i}$$
$$(\forall(i, j, l) \in \mathbb{I} \times \mathbb{I} \times \mathbb{I}) \ \varpi_{i,j} + \varpi_{j,l} - \varpi_{i,l} \leq 1 \quad (2.2.73)$$

The solution provided corresponds to a *local* maximum.

In [30] an interesting and rich formalization is proposed for similarity indices on $\mathcal{O}$ in the case of a description by *Boolean* attributes. A large ensemble of similarity functions is mutually compared according to algebraic properties. We cannot here discuss this subject any further.

An important research direction indicated by A.K. Jain in an overview article [18] concerns the development of *ensemble clustering*. The aim consists of deriving a clustering on the set concerned $\mathcal{O}$ from a sequence of partitions of $\mathcal{O}$, obtained by applying sequentially a parametrized clustering algorithm on $\mathcal{O}$. In this process, the parametrization changes from one application to the next one. Iteration of the K-means algorithm is considered in [18]. In these conditions, each of the partitions obtained infers a nominal categorical attribute on $\mathcal{O}$. Consequently, the methods presented above apply and enable us to obtain a consensus partition of the different partitions obtained, in particular by the K-means algorithm. This algorithm is involved in the next section.

By an analogous way, we can also propose to summarize different partitions of $\mathcal{O}$, obtained from repeatable applications of a parametrized non-hierarchical clustering algorithm, by means of an *ascendant agglomerative hierarchical clustering*. For this purpose, the general method expressed in Sect. 7.2.1 of Chap. 7 can be followed. More particularly, we put for $S^j(x, y)$ of (7.2.1) the index corresponding to $S^j(o_i, o_{i'})$ of (7.2.15), $1 \leq j \leq p$, where p is the number of partitions obtained by repeating the non-hierarchical clustering algorithm (e.g. the K-means algorithm).

2.3 Dynamic and Adaptative Clustering Method

2.3.1 Data Structure and Clustering Criterion

As mentioned in the Preamble (see Sect. 2.1), we shall describe the initial version of the *dynamic cluster* method proposed by Diday [10, 11]. In this version the object set

$\mathcal{O}$ to be clustered is represented by a set of points in the geometrical space $\mathbb{R}^p$, where $\mathbb{R}$ is the reals and p, a positive integer, corresponding to the number of numerical attributes describing $\mathcal{O}$. To fix ideas, we may suppose $\mathbb{R}^p$ endowed with the ordinary Euclidean metric. On the other hand, if $P = \{O_1, O_2, \ldots, O_k, \ldots, O_K\}$ is a partition by proximity of $\mathcal{O}$, each of its classes is represented by a subset of $\mathcal{O}$, comprising a few elements, $1 \leq k \leq K$. This cluster representation, adequately determined, is expected to be more accurate than that defined by the centre of gravity (centroid) of the cluster concerned, as in the K-means algorithm (see Sect. 2.3.2). For simplicity, historical reasons and theoretical properties, the K-means version is mostly employed [18]. An interesting historical overview of this general approach is given in [3].

To begin let us specify our notations. $\mathcal{V} = \{v^j | 1 \leq j \leq p\}$ designates the set of numerical descriptive attributes and $\mathcal{O} = \{o_i | 1 \leq i \leq n\}$, the set of the objects described (see Sect. 3.2.2 of Chap. 3). The set $\mathbb{I} = \{1, 2, \ldots, i, \ldots, n\}$ codes $\mathcal{O}$ and can be decomposed, according to the partition P (see above), as $\{\mathbb{I}_k | 1 \leq k \leq K\}$, where $\mathbb{I}_k$ is the subset of $\mathbb{I}$, coding the class O_k, $1 \leq k \leq K$. n_k denoting the cardinality of O_k, or equivalently $\mathbb{I}_k$, we have

$$\sum_{1 \leq k \leq K} n_k = n = card(\mathcal{O}) \tag{2.3.1}$$

The point of $\mathbb{R}^p$, representing an object x of $\mathcal{O}$, can be written as

$$v(x) = \left(v^1(x), v^2(x), \ldots, v^j(x), \ldots, v^p(x)\right) \tag{2.3.2}$$

In the K-means algorithm, the definition of the criterion evaluating a partition P of the object set $\mathcal{O}$ depends *uniquely* on P. $P = \{O_1, O_2, \ldots, O_k, \ldots, O_K\}$, designating the partition of $\mathcal{O}$ into K non-empty classes, this criterion is a measure evaluating globally and additively the respective cohesions of the different classes O_k, $1 \leq k \leq K$. As mentioned above, each cluster O_k is represented by its centre of gravity, which we denote by g_k, $1 \leq k \leq K$. More precisely,

$$g_k = \frac{1}{n} \sum_{i \in \mathbb{I}_k} o_i \tag{2.3.3}$$

where—without ambiguity—o_i stands for the point of $\mathbb{R}^p$, representing the object o_i, $i \in \mathbb{I}$. Equation (2.3.3) can be written as

$$g_k - O = \frac{1}{n_k} \sum_{i \in \mathbb{I}_k} (o_i - O)$$

it reflects the vector writing

$$\vec{Og_k} = \frac{1}{n_k} \sum_{i \in \mathbb{I}_k} \vec{Oo_i}$$

where O designates here the origin point of $\mathbb{R}^p$.

The cohesion of a given cluster O_k is defined by the sum of squared distances of the different points of $\mathbb{R}^p$, representing O_k, to the gravity centre of these points, namely

$$W_k = \sum_{i \in \mathbb{I}_k} d^2(o_i, g_k) = \sum_{i \in \mathbb{I}_k} \|\overrightarrow{o_i g_k}\|^2 \tag{2.3.4}$$

Thereby, the cohesion of O_k is all the more strong as W_k is small. In these conditions, the global cohesion of the partition P is evaluated by the smallness of

$$W(P) = \sum_{1 \leq k \leq K} W_k \tag{2.3.5}$$

Consider now the cloud of points in $\mathbb{R}^p$ and designate it by $\mathcal{N}(\mathbb{I})$ (see Sect. 10.3.3 of Chap. 10). The inertia moment of $\mathcal{N}(\mathbb{I})$ can be written as

$$\mathcal{M}(\mathcal{N}(\mathbb{I})) = \sum_{i \in \mathbb{I}} d^2(o_i, G) \tag{2.3.6}$$

where G is the centre of gravity of $\mathcal{N}(\mathbb{I})$. We have

$$G = \frac{1}{n} \sum_{i \in \mathbb{I}} o_i = \frac{1}{n} \sum_{1 \leq k \leq K} n_k g_k \tag{2.3.7}$$

where n_k, specified above, is the cardinality of O_k, $1 \leq k \leq K$.

The general inertia decomposition formula with respect to the partition P is

$$\mathcal{M}(\mathcal{N}(\mathbb{I})) = \sum_{1 \leq k \leq K} (W_k + B_k) \tag{2.3.8}$$

where

$$B_k = n_k d^2(g_k, G) \tag{2.3.9}$$

In the previous formulas (2.3.3)–(2.3.9) all vertices in the cloud of $\mathbb{R}^p$, associated with $\mathcal{O}$, are equally weighted. It is easy to generalize the methods studied below, in the possible case of a representation by a cloud of points in $\mathbb{R}^p$, non-equally weighted. In the latter case, the inertia decomposition formula becomes (10.3.10) of Sect. 10.3.3.1 of Chap. 10.

The K-means algorithm described in the next section comprises successive steps of the same nature. In each of them, the criterion (2.3.5) is minimized at best. This criterion is interpreted as the "lost" (or "residual") inertia for the representation of $\mathcal{O}$ by the cloud

$$\mathcal{N}(\mathcal{G}) = \{(g_k, n_k) | 1 \leq k \leq K\} \tag{2.3.10}$$

Other criteria having the same general analytical form as (2.3.5) can be substituted for $W(P)$, that is,

$$\sum_{1 \le k \le K} \sum_{i \in \mathbb{I}_k} \delta(o_i, e_k) \tag{2.3.11}$$

where δ is a dissimilarity index on $\mathcal{O}$ and e_k, a prototype of the cluster O_k, $1 \le k \le K$.

Equation (2.3.11) can be denoted by $\Delta(L, P)$, where

$$L = (e_1, e_2, \ldots, e_k, \ldots, e_K) \tag{2.3.12}$$

is the prototype sequence, e_k, representing the cluster O_k, $1 \le k \le K$.

Equation (2.3.11) is the form of the criterion employed in dynamic cluster methods [3, 10, 11]. In order to make explicit this criterion, let us suppose the partition P with labelled classes. Hence, P is defined as the ordered sequence

$$P = (O_1, O_2, \ldots, O_k, \ldots, O_K) \tag{2.3.13}$$

$t(P) = (n_1, n_2, \ldots, n_k, \ldots, n_K)$, where $n_k = card(O_k)$, $1 \le k \le K$, is the partition *type* of P.

In this technique, the setting of the criterion depends on two arguments: the partition P (see (2.3.13)) and a sequence of integers

$$(m_1, m_2, \ldots, m_k, \ldots, m_K) \tag{2.3.14}$$

such that

$$(\forall k, 1 \le k \le K), m_k < n_k$$

where m_k specifies the cardinal of the subset e_k of the *entire* set $\mathcal{O}$, which represents the class O_k, $1 \le k \le K$. e_k will be called the *kernel* of O_k.

$\Delta(L, P)$ (see (2.3.11)) can be normalized in order to refer to a suitable scale taking into account the respective cardinalities of the e_k, $1 \le k \le K$. In practice, mostly, the different integers m_k are taken mutually equal in the form $m_k = [\alpha \times (n/K)]$, where α is a proportion chosen (e.g. $\alpha = 0.1$) and where $[\bullet]$ designates the integer part of $\bullet$. To fix ideas, without loss of generality, the latter option is considered in the following; that is, $m_k = m$ for all k, $1 \le k \le K$.

In these conditions, given P, the kernel e_k representing the kth class O_k, is defined as

$$Arg \left[\min_{e \in \mathcal{P}(m, \mathcal{O})} \left(\sum_{(x,y) \in O_k \times e} \delta(x, y) \right) \right] \tag{2.3.15}$$

where $\mathcal{P}(m, \mathcal{O})$ is the set of all subsets of cardinal m. There are in all $\binom{n}{m}$ such subsets.

In fact, e is obtained by carrying out a sort of the $m \mathcal{O}$ elements which are the nearest O_k cluster, according to the dissimilarity index

$$\delta(y, O_k) = \sum_{x \in O_k} \delta(x, y) \qquad (2.3.16)$$

Generally, only one solution exists for e, answering (2.3.15). If more than a single solution can be provided, the first encountered is selected.

The global fitting criterion between L (see (2.3.12)) and P (see (2.3.13)), corresponding to (2.3.11), can be written as

$$\Delta(L, P) = \sum_{1 \leq k \leq K} \sum_{(x,y) \in O_k \times e_k} \delta(x, y) \qquad (2.3.17)$$

Several types of δ dissimilarity indices can be considered [3]. The most classical one is associated with the Euclidean representation of $\mathcal{O}$ in $\mathbb{R}^p$. This index can be written as

$$d^2(x, y) = \sum_{1 \leq j \leq p} (\xi^j - \eta^j)^2 \qquad (2.3.18)$$

where $(\xi^j | 1 \leq j \leq p)$ and $(\eta^j | 1 \leq j \leq p)$ are the respective coordinates of the points denoted by x and y, representing the objects x and y in $\mathbb{R}^p$. Notice that (2.3.18) is directly derived from the distance index used in the K-means method.

Let us indicate here that the *LLA* approach enables us to get with a unique principle, dissimilarities called *informational* dissimilarities, for a very large scope of data structures. Different types of objects can be compared. The general form of the dissimilarity table concerned is given in (7.2.7) of Sect. 7.2.1 of Chap. 7.

2.3.2 The K-Means Algorithm

We consider the geometrical representation of the object set $\mathcal{O}$ by a cloud of points designated by $\mathcal{N}(\mathbb{I})$ ($\mathcal{N}(\mathbb{I}) = \{o_i | i \in \mathbb{I}\}$), equally weighted, in the space $\mathbb{R}^p$, endowed with the usual Euclidean metric (see (2.3.18)).

The initial state of the algorithm might be either a partition

$$P^0 = \{O_1^0, O_2^0, \ldots, O_k^0, \ldots, O_K^0\}$$

of $\mathcal{O}$, or a set

$$G^0 = \{g_1^0, g_2^0, \ldots, g_k^0, \ldots, g_K^0\}$$

of K distinct points. These define a system of K attraction centres. They are determined at random or from expert knowledge. Without loss of generality, we assume to start with

$$G^0 = \{g_1^0, g_2^0, \ldots, g_k^0, \ldots, g_K^0\} \tag{2.3.19}$$

In these conditions, the first step of the K-means algorithm consists of attaching each vertex of the cloud $\mathcal{N}(\mathbb{I})$ to the nearest attraction centre belonging to G^0. In the case where more than a single element satisfies the proximity requirement, the first encountered is selected. $\mathcal{O}$ is then divided into K classes. Let us denote by

$$P^0 = \{O_1^0, O_2^0, \ldots, O_k^0, \ldots, O_K^0\} \tag{2.3.20}$$

the partition obtained. O_k^0 is composed of the set of objects of $\mathcal{O}$ assigned to g_k^0, $1 \leq k \leq K$. We have

$$O_k^0 = \{x \mid x \in \mathcal{O}, d(x, g_k^0) \leq d(x, g_{k'}^0) \text{ for } k' \neq k, 1 \leq k, k' \leq K\} \tag{2.3.21}$$

As mentioned above, for a given x in $\mathcal{O}$, k is the lowest label for which the condition

$$d(x, g_k^0) \leq d(x, g_{k'}^0) \text{ for } k' \neq k$$

holds.

Notice that some of the classes of P^0 might be empty. In fact, some attraction centres in G^0—which do not correspond to elements of $\mathcal{N}(\mathbb{I})$—can remain isolated in the assignation process. In these conditions, let

$$P^1 = \{O_1^1, O_2^1, \ldots, O_k^1, \ldots, O_{K(1)}^1\} \tag{2.3.22}$$

be the partition of $\mathcal{O}$ into the non-empty classes of P^0, $K(1) \leq K$.

From P^1 a new system

$$G^1 = \{g_1^1, g_2^1, \ldots, g_k^1, \ldots, g_{K(1)}^1\} \tag{2.3.23}$$

of attraction centres is determined, where

$$g_k^1 = \frac{1}{n_k^1} \sum_{i \in \mathbb{I}_k^1} o_i \tag{2.3.24}$$

where $\mathbb{I}_k^1$ is the subset of $\mathbb{I}$ coding the class O_k^1, whose cardinality is denoted by n_k^1, $1 \leq k \leq K(1)$.

It is exceptional that two different gravity centres of G^1 coincide. However, mathematically, this case cannot be excluded. Hence, we may introduce

$$G^2 = \{g_1^2, g_2^2, \ldots, g_k^2, \ldots, g_{K(2)}^2\} \tag{2.3.25}$$

which is composed of mutually distinct elements of G^1, $K(2) \leq K(1)$.

It is easy to observe that each step of this algorithmic process decreases the criterion $W(P)$ (see (2.3.5)). Consequently, this process converges. In the case where this convergence is reached too slowly, the process is stopped when the relative decreasing of $W(P)$ becomes negligible.

The asymptotic convergence of this algorithm under different probabilistic or continuous models is studied in [2, 4, 26, 34]. This facet will be described more in Sect. 2.3.4.2.

To end this section let us recall the version of the K-means algorithm as it was named and defined by McQueen [29]. In this, the reactualization of the attraction centre (centre of gravity) of a given cluster is carried out as the class formation goes along. More precisely, suppose that at a given step, C is a cluster already constituted and denote by g the gravity centre of C. If at the next step an object x of $\mathcal{O}$ is assigned to C—because g is the x nearest gravity centre, among the different gravity centres of the clusters formed—then g is replaced by the centre of gravity of $C + \{x\}$. For more details we may refer to [1] pages 162–163. In the following, we will not be concerned in generalizing the McQueen version.

2.3.3 Dynamic Cluster Algorithm

Assume that K is the upper bound of the number of classes of the partition P to be sought. Moreover, suppose that each of the P classes is represented by a subset of the object set $\mathcal{O}$, whose cardinality is m. As expressed in the preceding section, m is generally a small number with respect to $n = card(\mathcal{O})$. As said, m might correspond to the integer part of $\alpha \times (n/K)$, where α is a small proportion, for example $\alpha = 0.1$.

Now, let $\mathcal{P}_K$ be the set of all labelled partitions of $\mathcal{O}$ into K non-empty classes at the most. We have

$$card(\mathcal{P}_K) = K^n \tag{2.3.26}$$

On the other hand let $\mathcal{L}_K$ designate the set of sequences of K kernels at the most, of equal size m. Thus, each kernel is defined by a subset of $\mathcal{O}$ whose cardinality is m. Therefore, there are $\binom{n}{m}$ choices for a given kernel e. The latter corresponds to a given non-empty component of a given L in $\mathcal{L}_K$. L will be denoted by $(e_1, e_2, \ldots, e_k, \ldots, e_K)$ and called a system of kernels.

2.3.3.1 Two Mappings ν and π: Allocating and Centring

ν makes correspondence from $\mathcal{P}_K$ to $\mathcal{L}_K$. To each element P of $\mathcal{L}_P$, ν will associate an element L of $\mathcal{L}_K$ which maximizes the fitting of L to P. More formally, ν is a mapping of $\mathcal{P}_K$ to $\mathcal{L}_K$:

$$\nu : \mathcal{P}_K \longrightarrow \mathcal{L}_K$$
$$P \mapsto \nu(P) = L \tag{2.3.27}$$

where L is a kernel system minimizing the criterion

$$\Delta(L, P) = \sum_{1 \leq k \leq K} \sum_{i \in \mathbb{I}_k} \delta(o_i, e_k) \tag{2.3.28}$$

considered in (2.3.11).

To fix ideas, assume in (2.3.27) that P is composed of K non-empty classes and follow the development of the previous section: Eqs. (2.3.11)–(2.3.17). The kernel e_k is obtained according to (2.3.15), as associated with the cluster O_k, $1 \leq k \leq K$. However, two distinct clusters might give rise to the same kernel. In these conditions, $\nu(P)$ is obtained by reducing the initial sequence $(e_1, e_2, \ldots, e_k, \ldots, e_K)$, preserving all mutually distinct kernels.

π is a mapping of $\mathcal{L}_K$ into $\mathcal{P}_K$:

$$\pi : \mathcal{L}_K \longrightarrow \mathcal{P}_K$$
$$L \mapsto \pi(L) = P \tag{2.3.29}$$

where the partition P is obtained by assigning each of the $\mathcal{O}$ objects to the nearest kernel of L, according to the criterion

$$\delta(e, x) = \sum_{y \in e} \delta(y, x) \tag{2.3.30}$$

In this way, each of the obtained P classes surrounds at the best one kernel of L. More precisely, the kth class of P, $1 \leq k \leq K$, is defined as

$$O_k = \{x | \delta(e_k, x) \leq \delta(e_{k'}, x) \text{ for } k' \neq k, 1 \leq k' \leq K\} \tag{2.3.31}$$

Now, if we consider the geometrical framework in which this algorithm lies, the dissimilarity $\delta(e, x)$ (see (2.3.30)) has to be defined by the sum of squared distances between x and the different elements of e, that is,

$$D(e, x) = \sum_{y \in e} d^2(y, x) \tag{2.3.32}$$

Fig. 2.1 A single kernel for
two clusters

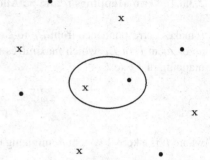

Fig. 2.2 A single cluster for
two kernels

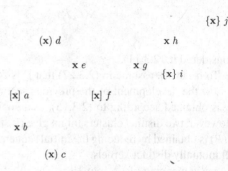

For this distance, as for possible others, a single P class might be obtained from two distinct kernels of L. In (2.3.32) we refer to (2.3.18) for $d^2(y, x)$.

A Geometrical Illustration

The distance between a set of points and a single point in a geometrical space is considered here defined as in (2.3.32). Two cases have to be illustrated.

1. A single kernel might correspond to two distinct classes;
2. Two distinct kernels might give rise to a single class.

For item 1, two clusters of points are considered in Fig. 2.1, where 10 points are involved. The points of one of both clusters are marked by "**x**" and those of the other one, by "•". Both clusters have the same kernel sized 2 which is surrounded.

For item 2, consider the set of 10 points in Fig. 2.2 and the kernel system $(\{a, f\}, \{c, d\}, \{i, j\})$. The elements of a given kernel are shown with brackets of the same type: (), [] or {}. Every point of the cluster $\{a, b, c, d, e, f\}$ is assigned to the kernel $\{a, f\}$; c and d are themselves assigned to $\{a, f\}$, according to the distance $D(e, x)$ (see (2.3.32))

2.3.3.2   Convergence Condition

Although the objective is to discover a "good" partition of $\mathcal{O}$, with clusters having strong cohesion, the dynamic cluster algorithm has to be viewed as searching a "good" kernel system. In fact, a single step of this algorithm consists of substituting for a given kernel an improved one. This is obtained from the partition associated from the last partition obtained by means of the criterion $\delta(e, x)$ (see (2.3.30) and (2.3.31)).

Let L^0 designate the initial kernel system. The latter might be chosen at random. We have for the following kernel systems

$$L^1 = \nu \circ \pi(L^0)$$
$$L^2 = \nu \circ \pi(L^1)$$
$$- - - - - - - -$$
$$L^i = \nu \circ \pi(L^{i-1})$$
$$- - - - - - - - \qquad (2.3.33)$$

where ν and π were defined in (2.3.27) and (2.3.29). L^i is the kernel system obtained after the ith step of the algorithm. The latter stops after the tth step if

$$\nu \circ \pi(L^t) = L^t \qquad (2.3.34)$$

L^t is a kernel system which cannot be enhanced. It defines a fixed point of the mapping $\nu \circ \pi$. We can write for the tth iteration

$$L^t = \nu \circ \pi(L^0) \qquad (2.3.35)$$

where t is the lowest integer for which

$$(\nu \circ \pi)^{t+1}(L^0) = (\nu \circ \pi)^t(L^0) \qquad (2.3.36)$$

Clearly, this algorithm is justified only if, for every i, the adequacy of L^i to the associated partition $\pi(L^i)$ is at least as good as that of L^{i-1} to $\pi(L^{i-1})$, that is, by referring to (2.3.28),

$$\Delta(L^i, \pi(L^i)) \leq \Delta(L^{i-1}, \pi(L^{i-1})) \qquad (2.3.37)$$

Proposition 28 *The sequence $\{\Delta(L^i, \pi(L^i)) | i \geq 0\}$ decreases if for any pair (L, M) of kernel systems, the following condition is satisfied:*

$$\Delta(L, \pi(M)) \leq \Delta(M, \pi(M)) \Rightarrow \Delta(L, \pi(L)) \leq \Delta(M, \pi(M)) \qquad (2.3.38)$$

In fact by construction (see (2.3.15)), we have

$$\Delta(L^i, \pi(L^{i-1})) \leq \Delta(L^{i-1}, \pi(L^{i-1})) \qquad (2.3.39)$$

The condition (2.3.38) implies

$$\Delta(L^i, \pi(L^i)) \leq \Delta(L^i, \pi(L^{i-1})) \qquad (2.3.40)$$

From (2.3.39) and (2.3.40) we obtain, by transitivity,

$$\Delta(L^i, \pi(L^i)) \leq \Delta(L^{i-1}, \pi(L^{i-1})) \qquad (2.3.41)$$

Intuitively, in geometrical context, condition (2.3.38) appears as very natural. It expresses that if the kernel system L does better than M, with respect to the partition $\pi(M)$, then L does better for $\pi(L)$, than M for $\pi(M)$.

Now, by considering the dissimilarity index $D(e, x)$ (see (2.3.32)), we shall show that the condition (2.3.38) is effectively satisfied for a set of objects $\mathcal{O}$, which is represented by a cloud of points in the geometrical space $\mathbb{R}^p$.

Let

$$\big(M, \pi(M)\big) = \big((c_1, c_2, \ldots, c_k, \ldots, c_K), (C_1, C_2, \ldots, C_k, \ldots, C_K)\big) \qquad (2.3.42)$$

be an ordered pair of a kernel system M and its associated partition $\pi(M)$. Recall that $card(c_k) = m$, for all k ($1 \leq k \leq K$) and suppose, to fix ideas, that all of the classes C_k are non-empty. In these conditions, consider a kernel system

$$(e_1, e_2, \ldots, e_k, \ldots, e_K)$$

and suppose satisfied the left side of the implication (2.3.38), that is,

$$\sum_{1 \leq k \leq K} \sum_{y \in C_k} D(y, e_k) \leq \sum_{1 \leq k \leq K} \sum_{y \in C_k} D(y, c_k) \qquad (2.3.43)$$

The partition

$$\pi(L) = (O_1, O_2, \ldots, O_k, \ldots, O_K)$$

where O_k is attracted by e_k, is necessary such that

$$\sum_{1 \leq k \leq K} \sum_{y \in O_k} D(y, e_k) \leq \sum_{1 \leq k \leq K} \sum_{y \in C_k} D(y, e_k) \qquad (2.3.44)$$

In fact, for a given k and $y \in C_k$, if $D(y, e_h) < D(y, e_k)$, for $k \neq h$, y will be assigned to O_h in $\pi(L)$ and the associated criterion diminution is $D(y, e_k) - D(y, e_h)$. Effectively, O_h is constituted of all elements y, such that $D(y, e_h)$ is minimum.

2.3.4 Following the Definition of the Algorithm

2.3.4.1 Clustering Structures Associated with the Repetition of the Algorithm Process

To fix ideas and without real loss of generality, we suppose in this section that the set of objects $\mathcal{O}$, as represented by a cloud of n points in the geometrical space $\mathbb{R}^p$, endowed with the ordinary Euclidean metric.

The most vulnerable point in the K-means or dynamic cluster algorithms consists of initial choice of attraction centres in the K-means, or kernels, in the dynamic cluster cases, respectively. Generally, these choices are carried out at random and then, repeating the algorithm process from different systems of attraction centres (for the K-means version) or kernels (for the dynamic cluster version), is needed. This repetition enables different algorithmic results, associated with different initializations, to be compared and synthesized. In the following, we shall focus on these aspects in the case of dynamic cluster algorithm.

Strong and Weak Patterns

Let

$$(L^1, L^2, \ldots, L^j, \ldots, L^J) \tag{2.3.45}$$

be a sequence of kernel systems. We may assume that the L^j, $1 \leq j \leq J$, are obtained randomly and independently. L^j is supposed to be written in the form

$$L^j = \left(e_1^j, e_2^j, \ldots, e_k^j, \ldots, e_K^j\right) \tag{2.3.46}$$

where $card(e_k^j) = m$, $1 \leq k \leq K$, $1 \leq j \leq J$.

Now, let

$$(P^1, P^2, \ldots, P^j, \ldots, P^J) \tag{2.3.47}$$

be the sequence of partitions obtained (after convergence) by the dynamic cluster algorithm from the kernel system (2.3.45). P^j is obtained from L^j, $1 \leq j \leq J$.

In relation to the partition sequence (2.3.47), Diday introduces [10], intuitively, the respective notions of "strong pattern" ("forme forte") and "weak pattern" ("forme faible"). In fact these notions can be formally expressed with respect to the lattice order endowing the partition set of $\mathcal{O}$ (see Sect. 1.1 of Chap. 1). For this, let us begin by defining the two following partitions:

$$P^\star = \bigwedge_{1 \leq j \leq J} P^j \tag{2.3.48}$$

$$Q^\star = \bigvee_{1 \leq j \leq J} P^j \tag{2.3.49}$$

where $\bigwedge$ and $\bigvee$ denote, respectively, the *greatest lower bound* (infimum) and the *least upper bound* (supremum) operations in the partition lattice.

We shall consider hereafter the partition $P^\star$. If we denote the partition P^j as

$$P^j = \left\{ O^j_k | 1 \leq k \leq K_j \right\} \tag{2.3.50}$$

where $K_j \leq K$, a given class of $P^\star$ can be written as follows:

$$\bigcap_{1 \leq j \leq J} O^j_{k_j} \tag{2.3.51}$$

where k_j is a subscript comprised between 1 and K_j, $1 \leq j \leq J$.

Thereby, we have

$$P^\star = \left\{ \bigcap_{1 \leq j \leq J} O^j_{k_j} | 1 \leq k_j \leq K_j \right\} \tag{2.3.52}$$

The theoretical number of classes of the partition $P^\star$ is

$$H = \prod_{1 \leq j \leq J} K_j \tag{2.3.53}$$

However, in practice, many of the $P^\star$ classes, as expressed in (2.3.52), are empty. This, all the more, as $\mathcal{O}$ is classifiable with respect to the distance, which is provided (see Sect. 8.7 of Chap. 8). The classes of $P^\star$ are called "strong patterns" ("formes fortes").

Note here that $Q^\star$ is the partition of $\mathcal{O}$ associated with the equivalence relation defined by the transitive closure of the binary relation on $\mathcal{O}$

$$Q = P^1 \vee P^2 \vee \cdots P^j \vee \cdots \vee P^J \tag{2.3.54}$$

where P^j indicates here the equivalence relation on $\mathcal{O}$, defined by the partition P^j of $\mathcal{O}$, $1 \leq j \leq J$

The classes of the partition Q are called "weak patterns" (formes faibles).

Facing to repetition of the dynamic cluster algorithm process, from a sequence of kernel systems (see (2.3.45)), strong and weak patterns enable the interpretation of the results (see (2.3.47)) to be ordered.

Statistical Significance of Strong Patterns

The strong patterns (formes fortes) are the more interesting to be handled and interpreted. There are two ways to observe their respective validities. The first one is static and can be expressed after J iterations of the algorithmic process (see (2.3.45)–(2.3.47)). The second one is recursive and is defined after each iteration of the algorithmic process.

For the first approach, it comes to measure the internal agreement of the different partitions of (2.3.47). The most natural is to refer mutual independence statistical hypothesis between the partitions observed at the end of the algorithmic process. In this context, the most classical measure is provided by the chi-square statistic [20] (see (6.2.43) of Sect. 6.2.3.5 of Chap. 6). In order to express this statistic here, let us introduce, relative to the partition P^j (see (2.3.50)), the parameters

$$(\forall k),\, 1 \le k \le K,\, 1 \le j \le J,\, n_k^j = card(O_k^j) \text{ and } p_k^j = \frac{n_k^j}{n} \qquad (2.3.55)$$

On the other hand, relative to the partition $P^\star$ (composed of strong patterns), introduce its class cardinals, that is,

$$n_{k_1 k_2 \dots k_j \dots k_J} = card \left(\bigcap_{1 \le j \le J} O_{k_j} \right) \qquad (2.3.56)$$

In these conditions, the chi-square statistic χ^2 can be written as follows:

$$\chi^2 = \sum_{(k_1, k_2, \dots, k_j, \dots, k_J)} \frac{\left(n_{k_1 k_2 \dots k_j \dots k_J} - np_{k_1} \times p_{k_2} \times \dots \times p_{k_j} \times \dots \times p_{k_J} \right)^2}{np_{k_1} \times p_{k_2} \times \dots \times p_{k_j} \times \dots \times p_{k_J}} \qquad (2.3.57)$$

where p_{k_j} is what we have denoted above by p_k^j (see (2.3.55)).

The independence hypothesis against to which the chi-square statistic (2.3.57) is established is far from the real nature of the problem in question. In fact, whatever is the relative arrangement of the respective points representing O in $\mathbb{R}^p$, the different partitions P^j, $1 \le j \le J$, are necessarily, near statistically, because they are obtained by the same algorithm, processed on the same set, endowed with a fixed distance.

In fact, the matter consists of studying the distribution of (2.3.57) in the case of a random class of n points in $\mathbb{R}^p$. The randomness might be defined by a multidimensional distribution (e.g. Gaussian distribution) adequately associated with the empirical distribution of the numerical attributes describing O. For simplicity reasons, the latter distribution can be defined as the product of p independent Gaussian probability laws, where the jth one is associated with the empirical distribution of the jth descriptive attribute, $1 \le j \le p$. In these conditions, the mean and variance of the jth probability law might be chosen as identical to those of the jth empirical distribution.

The analytical complexity for establishing the probability law of (2.3.57) under the probabilistic model we have just described leads us to propose to proceed by simulation. R independent random realizations of the cloud of n points associated with the observed one are carried out. For each realization, J independent repetitions of the dynamic cluster algorithm, from J kernel systems as (2.3.46) above, are performed and then, the chi-square statistic (2.3.57) is calculated. The sequence of the R values of χ^2, so obtained, enables the distribution of (2.3.57) to be observed under an adequate random model of no relation between the data units. Consequently, the observed value of (2.3.57) in a real case can be validated.

Notice that the coefficients described in Sect. 6.2.3.5 of Chap. 6 show that other indices than the χ^2 might be considered. For this, a certain research turn out to be needed.

As mentioned above, a second method, which permits the validity or prehension of the strong patterns to be established, has a sequential nature. It takes place after each iteration of the algorithm. More precisely, imagine the partition P^* (see (2.3.48) and (2.3.50)) get after J iterations and let

$$P = \{O_k | 1 \leq k \leq K\} \tag{2.3.58}$$

be the partition obtained after a new iteration (the $(J+1)$th one). If C is a class of P^*, the method evaluates how C is questioned by P.

For this evaluation—as suggested by Diday—the Shannon entropy can be used. It is written as follows:

$$\mathcal{H}(P/C) = -\sum_{1 \leq k \leq K} p(k/h)\log_K(p(k/h)) \tag{2.3.59}$$

where k and h stand for O_k and C and where

$$p(k/h) = \frac{card(O_k \cap C)}{card(C)} \tag{2.3.60}$$

$1 \leq k \leq K$.

A value 1 for one of the conditional probabilities (proportions) (2.3.60) means that the class C is included in one of the classes O_k. In this case, the class C is not questioned by the partition P and the value of $\mathcal{H}(P/C)$ is null. The opposite case is that for which C is equally distributed on the different classes O_k, $1 \leq k \leq K$. In this situation, each of the rations (2.3.60) is equal to $1/K$ and $\mathcal{H}(P/C) = 1$.

Consequently, the measure (2.3.59) permits us to evaluate the concentration of the class C with respect to the different classes of the partition P.

Now, to integrate the respective contributions of all the P^* classes, the following coefficient has to be adopted according to (2.3.59):

$$\mathcal{H}(P/P^*) = -\sum_{1 \leq h \leq H} \sum_{1 \leq k \leq K} p(k/h)\log_K(p(k/h)) \tag{2.3.61}$$

Other coefficients than (2.3.59) can be used for the same purpose. The most known is the Gini index. It can be written as follows:

$$\phi(P/C) = 1 - \sum_{1 \leq k \leq K} p(k/h)^2 \qquad (2.3.62)$$

Its value is comprised between 0 and $(1 - (1/K))$: 0 in the case where C is contained in one of the classes O_k and $(1 - (1/K))$ in the case where C is uniformly distributed among the different classes O_k, $1 \leq k \leq K$.

Consensus by a Partition or a Partition Chain

In terms of methodology, a few things have to be added to what was expressed at the end of Sect. 2.2.5. Nevertheless, the context here is very different. Let

$$(P^1, P^2, \ldots, P^j, \ldots, P^J) \qquad (2.3.63)$$

be the sequence of partitions obtained by the dynamic cluster algorithm from the kernel system (2.3.45). In (2.3.63), P^j is the partition obtained from L^j at the jth iteration of the algorithmic process, $1 \leq j \leq J$.

Every partition P^j can be assimilated to a partition induced by a nominal categorical attribute on $\mathcal{O}$, with K_j values, $1 \leq j \leq J$. The specificity of this interpretation is that these attributes are statistically very close. In fact, the different partitions P^j of $\mathcal{O}$, $1 \leq j \leq J$, are obtained by applying the *same algorithm*, but with *different initial states*. These partitions are all the more near that the set $\mathcal{O}$ endowed with its distance function is classifiable (see Sect. 8.7 of Chap. 8).

In these conditions, every clustering method of a set of objects described by nominal categorical attributes applies, in particular, the partition central method, which is conceived for this type of description (see Sect. 2.2). Also, we have point out at the end of Sect. 2.2.5 the usage of ascendant agglomerative hierarchical clustering, with the similarity index defined in Sect. 10.2.1 of Chap. 10.

In fact, following the definition of a similarity or dissimilarity index between objects described by categorical nominal attributes, every clustering algorithm which can handle the similarity or dissimilarity index concerned is able to built a consensus between the different partitions P^j, $1 \leq j \leq J$. Clearly, the consensus result is strongly dependent on the method used.

Notice that in our particular situation (see (2.3.63)), whatever the similarity (resp., dissimilarity) index chosen, the similarity (resp., dissimilarity) function is maximal (resp., minimal) for any pair of objects belonging to the same strong pattern. Thus, the strong patterns belonging to $P^\star$ (see (2.3.52)) will appear as subclasses of larger classes. It is interesting to observe how the strong patterns are organized for a given clustering consensus method.

2.3.4.2   Some Important Developments of the K-Means or Dynamic Cluster Methods

Due to the geometrical nature of the notions handled in these adaptive reallocating-recentring algorithms, the clearest expression of these concerns clustering a cloud of points in an Euclidean space. However, their flexibility, combined with the additive form of the adequation criterion (see (2.3.30)), make that they apply to a large extent of data structures and cluster representations [12]. In [14] each cluster is characterized by a factorial plan corresponding to a local component analysis [21, 25], Chap. 6.

In another statistical situation [15] each cluster is defined by a random model represented by a multidimensional distribution. The latter, denoted by F_k, is supposed characterized by a vector parameter θ_k, $1 \leq k \leq K$. The whole cloud in $\mathbb{R}^p$ associated with the representation of the set of objects $\mathcal{O}$ is seen as the realization of a probabilistic mixing of K distributions. It can be set in the form

$$F = \sum_{1 \leq k \leq K} p_k F_k \tag{2.3.64}$$

where p_k is the probability of the F_k occurrence, $1 \leq k \leq K$

$$\sum_{1 \leq k \leq K} p_k = 1$$

In this approach, the kernel system becomes

$$(\theta_1, \theta_2, \ldots, \theta_k, \ldots, \theta_K) \tag{2.3.65}$$

The criterion maximized is the likelihood logarithm of the observed sample $\{x_i | 1 \leq i \leq n\}$, where x_i represents the ith object in $\mathbb{R}^p$. This criterion comes down to the additive form (2.3.30) [16, 38].

This formalization of the clustering problem is the same as that of *Expectation-Maximisation (EM)* algorithm [5, 9, 32, 39]. Relative to the general principle, E and M, respectively, correspond to the allocation and centring steps of the general algorithm type considered in this section.

The development of the *EM* algorithm is more global than that resulting from the adaptation of dynamic cluster algorithm [16]. The same general form of the distributions F_k, $1 \leq k \leq K$, is assumed in the *EM* algorithm. Mostly, this form is Gaussian. In [6] *EM* algorithm is proved to be equivalent to the K-means algorithm—with the inertia criterion—under a spherical Gaussian mixture.

Let us return now to the initial representation of a class O_k of a partition $P = \{O_k | 1 \leq k \leq K\}$ of an object set $\mathcal{O}$ by a subset e of $\mathcal{O}$ (see Sect. 2.3.3.1). As already indicated, the most vulnerable feature of the dynamic cluster algorithm concerns the initial choice of the kernel system L^0 (see Sect. 2.3.3.2). For simplicity reasons and also because it is mostly practiced, we have supposed above a common size, denoted by m, for the different kernels. Generally, the order of m is a few units

($m = 1, 2$ or 3). If we accept the representation of each of the P classes by one of its elements, the "Attraction poles" method [22, 24, 27] and [25], Chap. 8, provides an objective statistical technique in order to determine a system of attraction poles, playing efficiently the role of an initial kernel system. This determination method is based on a variance analysis of the mutual distances between the elements of the object set $\mathcal{O}$. More precisely, suppose that we want a system of K attraction poles. If $\{e_1, e_2, \ldots, e_{k-1}\}$ designates the set of the $k - 1$ first poles extracted, $k \leq K$, the kth pole e_k is determined from a statistical coefficient combining two conditions:

1. The variance of the distances (or proximities) of e_k to the other elements of $\mathcal{O}$ is as large as possible;
2. e_k is as far as possible from $\{e_1, e_2, \ldots, e_{k-1}\}$.

The Pole attraction method was first conceived as a seriation method [24]. Afterwards, it was extended and gave rise to a large family of clustering methods [22, 27] and Chap. 8 of [25]. Let us precise briefly the general principle of these methods. An increasing sequence of the number K of classes is considered: $K = 2, 3, \ldots, l, \ldots, \mathbb{L}$. For a given value l of K, we determine a Pole attraction system $\{e_1, e_2, \ldots, e_l\}$, comprising l elements. Then, each of the objects of $\mathcal{O}$ is assigned to the (or one of the) nearest attraction pole among $e_1, e_2, \ldots, e_l$, creating thus a partition of $\mathcal{O}$ into exactly l clusters. In this way, e_i is naturally included in the ith cluster, $1 \leq i \leq l$.

Otherwise and importantly, this partition is evaluated according to the adequacy criterion studied in Sects. 9.3.5 and 9.3.6 of Chap. 9. Thereby, only some values of l are retained as corresponding to "natural" numbers of classes. Therefore, this method does not require to fix beforehand a number of classes. A top-down but *non-hierarchical* sequence of partitions is collected, each being accompanied by a high value of the adequacy criterion.

Determining an initial kernel system as the pole attraction system effectively contributes to any of the K-means or dynamic cluster algorithms. Experimentally, we observe a very quick convergence of these algorithms when the starting point is a system of attraction poles. On the other hand, theoretically, the attraction poles showed all interest in a continuous model of perfect classifiability [26].

This model is defined by a sequence of two consecutive intervals of an horizontal axis, I_1 and I_2, uniformly weighted with a density equal to 1, separated by an empty interval I. The respective lengths of I_1, I_2 and I are the parameters of the analysis of the K-means convergence algorithm studied in [26]. This analysis was discussed with great interest in [4].

The nature of the convergence problem considered in [34] is very different from that we have just mentioned. In [34] it is not a question of the convergence of the K-means algorithm towards an optimal solution. The matter is rather the convergence of an optimal solution of the K-means algorithm calculated for a random sample of infinite population towards the optimal solution of this algorithm, applied to the entire population. Therefore, the latter problem has a classical nature in terms of *Mathematical Statistics*. More precisely, consider the application of the K-means algorithm to the geometrical space $\mathbb{R}^p$, endowed with a probability measure P', such that the algorithm converges to a unique optimal solution defined by a system $L = (g_1, g_2, \ldots, g_k, \ldots, g_K)$ of centres of gravity. Now, consider a random sample $\mathcal{O}^{(n)}$ of size n provided from the probabilized space $\mathbb{R}^p$ and let us designate by $L^{(n)} = (g_1^{(n)}, g_2^{(n)}, \ldots, g_k^{(n)}, \ldots, g_K^{(n)})$ the unique optimal solution of the K-means algorithm applied on $\mathcal{O}^{(n)}$, the Pollard theorem specifies the conditions under which $L^{(n)}$ converges almost surely to L. This convergence is point by point, and it implies associated labellings between the elements of $L^{(n)}$ and those of L.

References

1. Anderberg, M.R.: Cluster Analysis for Applications. Academic Press, New York (1973)
2. Bock, H.-H.: On some significance tests in cluster analysis. J. Classif. **2**, 77–108 (1985)
3. Bock, H.-H.: Clustering methods: a history of k-means algorithms. In: Cucumel, G., Brito, P., Bertrand, P., de Carvalho, F. (eds.) Selected Contributions in Data Analysis and Classification, pp. 161–172. Springer, Berlin (2007)
4. Celeux, G.: Étude exhaustive de l'algorithme de réallocation-recentrage dans un cas simple. R.A.I.R.O. Recherche opérationnelle/Operations Research **20**(3), 229–243 (1986)
5. Celeux, G.: Reconnaissance de mélanges de densités de probabilité et applications à la validation des résultats en classification, thèse de doctorat ès sciences. Ph.D. thesis, Université de Paris IX Dauphine, September 1987
6. Celeux, G., Govaert, G.: A classification EM algorithm for clustering and two stochastic versions. Comput. Stat. Data Anal. **14**, 315–332 (1992)
7. Charon, I., Denoeud, L., Guénoche, A., Hudry, O.: Maximum transfer distance between partitions. J. Classif. **23**, 103–121 (2006)
8. Charon, I., Denoeud, L., Hudry, O.: Maximum de la distance de transfert à une partition donnée. Revue Mathématiques et Sciences **179**, 45–83 (2007)
9. Dempster, A.P., Laird, N.M., Rubin, D.B.: Maximum likelihood estimation from incomplete data via the EM algorithm (with discussion). JRSS **B. 39**, 1–38 (1977)
10. Diday, E.: Une nouvelle méthode de classification automatique et reconnaissance des formes: la méthode des nuées dynamiques. Revue de Statistique Appliquée **XIX**(2), 19–33 (1971)
11. Diday, E.: The dynamic clusters method in nonhierarchical clustering. J. Comput. Inf. Sci. **2**(1), 61–88 (1973)
12. Diday, E. and Collaborators: Optimisation en Classification Automatique, vol. I, II. Institut National de Recherche en Informatique et en Automatique (INRIA), Le Chesnay (1979)
13. Diday, E., Govaert, G.: Classification automatique avec distances adaptatives. R.A.I.R.O. Inf. Comput. Sci. **11**(4), 329–349 (1977)
14. Diday, E., Schroeder, A.: The dynamic clusters method in pattern recognition. In: Rosenfeld, J.L. (ed.) Information Processing 74, pp. 691–697. North Holland, Amsterdam (1974)
15. Diday, E., Schroeder, A.: A new approach in mixed distribution detection. Research Report 52, INRIA, January 1974
16. Diday, E., Schroeder, A.: A new approach in mixed distribution detection. R.A.I.R.O. Recherche Opérationnelle **10**(6), 75–1060 (1976)
17. Forgy, E.W.: Cluster analysis of multivariate data: efficiency versus interpretability of classifications. In: Biometrics, Biometric Society Meeting, vol. 21, p. 768 (1965)
18. Jain, A.K.: Data clustering: 50 years beyond k-means. Pattern Recognit. Lett. **31**, 651–666 (2010)
19. Jancey, R.C.: Multidimensional group analysis. Aust. J. Bot. **14**, 127–130 (1966)
20. Lancaster, H.O.: The Chi-Squared Distribution. Wiley, New York (1969)
21. Lebart, L., Fenelon, J.-P.: Statistique et Informatique appliquées. Dunod, Paris (1973)

22. Leredde. H.: La méthode des pôles d'attraction; la méthode des pôles d'aggrégation: deux nou-
    velles familles d'algorithmes en classification automatique et sériation. Ph.D. thesis, Université
    de Paris, 6 Oct 1979
23. Lerman, I.C.: Les bases de la classification automatique. Gauthier-Villars, Paris (1970)
24. Lerman, I.C.: Analyse du phénomène de la sériation. Revue Mathématiques et Sciences
    Humaines 38 (1972)
25. Lerman, I.C.: Classification et analyse ordinale des données. Dunod, Paris. http://www.
    brclasssoc.org.uk/books/index.html (1981)
26. Lerman, I.C.: Convergence optimale de l'algorithme de "réallocation-recentrage" dans le cas
    continu le plus simple. R.A.I.R.O. Recherche opérationnelle/Operations Research 20(1), 19–50
    (1986)
27. Lerman, I.C., Leredde, H.: La méthode des pôles d'attraction. In: IRIA (ed.): Analyse des
    Données et Informatique. IRIA (1977)
28. Loève, M.: Probability Theory. D. Van Nostrand Company, New Jersey (1963)
29. MacQueen, J.B.: Some methods for classification and analysis of multivariate observations.
    In: Proceedings of the 5th Berkeley Symposium on Mathematical Statistics and Probability,
    pp. 281–297. University of California Press, Berkeley (1967)
30. Marcotorchino, F.: Essai de Typologie Structurelle des Indices de Similarité Vectorielles par
    Unification Relationnelle. In: Benani, Y., Viennet, E. (eds.) RNTI A3, Revue des Nouvelles
    Technologies de l'Information, pp. 203–319. Cepaduès, Toulouse (2009)
31. Marcotorchino, F., Michaud, P.: Agrégation de similarités en classification automatique. Revue
    de Statistique Appliquée 2, 21–44 (1982)
32. McLachlan, G., Krishnan, T.: The EM Algorithm and Extensions. Wiley, New York (1997)
33. Neveu, J.: Bases Mathématiques du Calcul des Probabilités. Masson, Paris (1964)
34. Pollard, D.: Strong consistency of k-means algorithm. Ann. Stat. 9(1), 135–140 (1981)
35. Régnier, S.: Sur quelques aspects mathématiques des problèmes de la classification automa-
    tique. I.C.C. Bull. 4, 175–191 (1965)
36. Régnier, S.: Sur quelques aspects mathématiques des problèmes de la classification automa-
    tique. Revue Mathématiques et Sciences Humaines 22, 13–29 (1983)
37. Régnier, S.: Sur quelques aspects mathématiques des problèmes de la classification automa-
    tique. Revue Mathématiques et Sciences Humaines 22, 31–44 (1983)
38. Sclove, S.L.: Population mixture models and clustering algorithms. In: Commun. Stat. Theory
    Methods A6, 417–434 (1977)
39. Xu, R., Wunsh, D.: Survey of clustering algorithms. IEEE Trans. Neural Netw. 16(3), 645–678
    (2005)

Chapter 3
Structure and Mathematical Representation of Data

3.1 Objects, Categories and Attributes

In Chap. 2, two basic methods of non-hierarchical clustering were presented: the "transfer" method and the "dynamic adaptative method". Frequently, a clustering method is related to a given type of data. Thus, the first method is defined for a data description by nominal categorical attributes. The latter will be presented in Sect. 3.3.1. The second method is defined for a data description by numerical attributes. These will be presented in Sect. 3.2.2. The categorical attributes will play a very important part in our development. Consequently, we study in Sect. 3.2.3 different methods for transforming numerical attributes into categorical ones. A value of a categorical attribute is called "category". Conditions of exclusivity and exhaustivity are required for the different categories of a categorical attribute (see Sect. 3.3.1). A category is determined by the intuitive notion of "concept".

Formal definition, mathematically expressed, of a "concept" in real life is very difficult Sutcliffe (1992) [44]. It depends on a knowledge domain and on recognition techniques. For example, let us consider the cirrhosis concept defined in the hepatho-biliary pathology. For this pathology, we assume a universe $\mathcal{U}$ of liver ill persons. $\mathcal{U}$ is a real or hypothetical finite set and each of its elements defines an elementary and indivisible object interesting the domain studied. On a given element of $\mathcal{U}$, the concept may be *TRUE* or *FALSE*. A series of clinical tests are necessary in order to recognize if a given person is a cirrhosis ill. A concept is defined in "intension" (one can say in "comprehension") from a precise description using primitive concepts. It is detected by the domain expert for two reasons: first, its frequency is not negligible; and second, its consequences are globally comparable on all the objects where it is observed.

Let us denote by γ such a concept, for example the "liver cirrhosis". γ is defined with respect to the universe $\mathcal{U}$ of the ill liver persons. γ is represented by the subset of $\mathcal{U}$ that we denote by $\mathcal{U}(\gamma)$, whose disease is cirrhosis. Equivalently, a Boolean

© Springer-Verlag London 2016

I.C. Lerman, *Foundations and Methods in Combinatorial and Statistical Data Analysis and Clustering*, Advanced Information and Knowledge Processing, DOI 10.1007/978-1-4471-6793-8_3

attribute corresponds bijectively to the concept γ. This attribute, denoted by a_γ, is a mapping of $\mathcal{U}$ onto the set $\{0, 1\}$ comprising two codes 0 and 1:

$$a_\gamma : \mathcal{U} \to \{0, 1\} \tag{3.1.1}$$

For a given u in $\mathcal{U}$, $a_\gamma(u) = 1$ (resp., 0) if γ is *TRUE* (resp., *FALSE*) on u. In these conditions, the above subset $\mathcal{U}(\gamma)$ of $\mathcal{U}$ can be expressed as follows:

$$\mathcal{U}(\gamma) = a_\gamma^{-1}(1) \tag{3.1.2}$$

where $a_\gamma^{-1}(1)$ is the reciprocal image of 1.

$\mathcal{U}(\gamma)$ is called "extension" of the concept γ at the level of $\mathcal{U}$.

The introduced concept notion is Boolean. However, its definition may require non-Boolean attributes. Consider the following example taken in the framework of an epidemiological survey concerning a male adult population living in a given area and aged 18–60 years. The concept "smoking more than 20 cigarettes a day" is a Boolean concept. Nevertheless, its definition requires a non-Boolean attribute defined by "counting". The scale value of this attribute is the set of integer numbers.

The notion of a value scale associated with a descriptive attribute is preliminary to define descriptions in *Data Analysis* Suppes and Zinnes (1951) [43]. A given descriptive attribute a is defined at the level of a universe $\mathcal{U}$ of objects concerned by a set of concepts (see the above example where the concepts are those of the hepatho-biliary pathology). By denoting $\mathcal{E}$ the value scale of a, a is mathematically interpreted as a mapping of $\mathcal{U}$ in $\mathcal{E}$, associating a unique value in $\mathcal{E}$, denoted by $a(u)$, with each element u of $\mathcal{U}$.

$$a : \mathcal{U} \to \mathcal{E}$$
$$u \mapsto a(u) \tag{3.1.3}$$

Generally, in *data analysis* and *machine learning* the whole set $\mathcal{U}$ is not available. We only dispose a finite set $\mathcal{O}$ of objects representing $\mathcal{U}$. Mostly, $\mathcal{O}$ is a subset of $\mathcal{U}$, $\mathcal{O} \subset \mathcal{U}$. The observed results of a data analysis at the level of $\mathcal{O}$ are inferred at the level of $\mathcal{U}$. For this purpose, $\mathcal{O}$ has to be a statistical representative sample of $\mathcal{U}$. Usually, this condition is satisfied, and specially in *data mining* where the size of $\mathcal{O}$ is generally very large (many millions or even more) Therefore, for the following, our reference object set will be $\mathcal{O}$. Thus, the above attribute a will be regarded as a mapping of $\mathcal{O}$ into $\mathcal{E}$:

$$a : \mathcal{O} \to \mathcal{E}$$
$$u \mapsto a(u) \tag{3.1.4}$$

Now, let us define the scale $\mathcal{E} = \{e_1, e_2, \ldots, e_h, \ldots, e_k\}$ where e_h is one possible value of the attribute a, $1 \leq h \leq k$. The meaning of the field studied enables a structure on the value set $\mathcal{E}$ to be defined. This structure induces a mathematical relation on $\mathcal{O}$. In Sects. 3.2 and 3.3 we shall detail the different types of descriptive attributes and their formal representations at the level of the object set $\mathcal{O}$. The main type of a descriptive attribute is defined from the arity of the relation on $\mathcal{O}$, induced by the structure of the value set $\mathcal{E}$. We distinguish three main types I, II and III. For I the induced relation on $\mathcal{O}$ is unary. It comprises the Boolean attribute and the numerical one (Sects. 3.2.1 and 3.2.2). For II, the relation is binary, non-valuated or valuated. This type includes the nominal categorical attribute (Sect. 3.3.1), the ordinal categorical attribute (Sect. 3.3.2), the ranking attribute (Sect. 3.3.3) and the categorical attribute valuated by a numerical similarity (Sect. 3.3.4). These two types (I and II) cover a large range of formal descriptions in combinatorial data analysis (see Sects. 3.2, 3.3, and 3.5). Type III is defined when the attribute scale induces a binary relation—generally defined by a ranking (ordinal similarity) or a numerical valuation (numerical similarity)—on the set $P_2(\mathcal{O})$ (resp., $\mathcal{O} \times \mathcal{O}$) of unordered (resp., ordered) pairs of elements of $\mathcal{O}$. Then, the *preordonance categorical attribute*, the *taxonomic attribute* and the *taxonomic preordonance attribute* are presented in Sects. 3.4.1–3.4.3, respectively. In Sect. 3.4.4 preordonance representations of the different types of descriptive attributes are proposed. Section 3.4 is devoted for defining the representation of the different types of attributes when describing a set of categories instead of a set of objects.

Relational representation of descriptive attributes is clearly emphasized in our work. One of its specificities consists of highlighting the set theoretic representation sustaining the relational one Lerman (1970) [16], (1973) [17], (1981) [18], (1992) [20, 21], and (2009) [25]. In our work we have been very influenced by the M.G. Kendall work Kendall (1948) [12]. Several authors in *Combinatorial data analysis* interpret, implicitly or explicitly, the representation of a descriptive attribute of an object set $\mathcal{O}$ in terms of of a binary relation on $\mathcal{O}$. This is generally done for a given type of descriptive attribute [e.g. nominal (resp., ordinal) categorical attribute], in relation to a specific method to be developed Guénoche and Monjardet (1987) [10], Hubert (1987) [11], Marcotorchino and Michaud (1979) [35], Marcotorchino (2009) [34], Giakoumakis and Monjardet (1987) [8], Régnier (1965) [39]. In the following, we wish to present a general framework independent of a given methodology, in which the different types of descriptive attributes and their formal representations are expressed.

3.2 Representation of the Attributes of Type I

As mentioned above, an attribute of type I induces a unary relation on $\mathcal{O}$. Such an attribute can be called an "incidence" attribute. For this type, we distinguish exactly the "boolean" attribute and the "numerical" one.

3.2.1   The Boolean Attribute

This attribute is also called a "presence–absence" attribute. It has been already considered above in Sect. 3.1. As previously, let us denote it by a. Formally, a is a mapping of $\mathcal{O}$ onto the set $\{FALSE, TRUE\}$. Mostly, in *data analysis*, *FALSE* and *TRUE* are coded by the integer numbers 0 and 1, respectively. For the following mapping diagram, also considered above in a different context,

$$a : \mathcal{O} \to \{0, 1\}$$
$$x \mapsto a(x) \tag{3.2.1}$$

$a(x)$ denotes the value of a on x, $a(x) = 1$(resp., 0) if a is *TRUE* (resp., *FALSE*) for x.

a is represented by the subset $\mathcal{O}(a)$ of $\mathcal{O}$ constituted by those objects for which a is *TRUE*:

$$\mathcal{O}(a) = a^{-1}(1) \tag{3.2.2}$$

where a^{-1} denotes the reciprocal mapping of a.

Now, let us introduce the cardinalities of $\mathcal{O}$ and $\mathcal{O}(a)$ that we denote by n and $n(a)$: $n = card(\mathcal{O})$ and $n(a) = card(\mathcal{O}(a))$. Thus, the proportion or relative frequency $p(a)$ of objects for which a is *TRUE* is defined by $p(a) = n(a)/n$.

The negated Boolean attribute $\bar{a}$ is defined by

$$(\forall x \in \mathcal{O}), \bar{a}(x) = TRUE \text{ if and only if } a(x) = FALSE \tag{3.2.3}$$

Clearly, $\mathcal{O}(\bar{a})$ is associated with $\bar{a}$, the latter set is the complementary subset of $\mathcal{O}(a)$ in $\mathcal{O}$. We also can define $n(\bar{a}) = card(\mathcal{O}(\bar{a}))$ and $p(\bar{a}) = n(\bar{a})/n$. Obviously, we have $p(a) + p(\bar{a}) = 1$.

The couple $\{a, \bar{a}\}$ is the value set of a binary catégorical attribute. Let us denote this attribute by α. The empirical distribution of α is defined by the ordered pair $[p(a), p(\bar{a})]$.

Thus, binary categorical attributes can be associated with logically independent Boolean attributes.[1] Conversely, Boolean attributes can be associated with categorical binary attributes by retaining for every binary categorical attribute, one of its two possible values. Generally and for significant statistical reasons, the retained value is the least frequent among both values.

Boolean attributes occur very frequently in database descriptions. For the above example in Sect. 3.1 where $\mathcal{O}$ is defined by a sample of liver ill persons, the concept of liver cirrhosis specifies a Boolean attribute.

---

[1](I.e. no attribute can be derived from other).

3.2.2 The Numerical Attribute

The value scale of a numerical attribute is the set $\mathbb{R}$ of real numbers. Mostly, in *data analysis*, numerical descriptions are considered for measuring quantities (e.g. weight, size, ...). We shall assume a positive scale including 0 for the numerical attribute. In any case, this does not restrict the generality. In fact, by representing geometrically $\mathbb{R}$ with a horizontal axis directed from left to right, the scale origin can be moved to left in order to make positive the observed values of the numerical attribute on the object set $\mathcal{O}$. Let us denote by v the numerical attribute. v is a mapping of $\mathcal{O}$ onto the set that we denote by $\mathbb{R}_+$ of real positive numbers

$$v : \mathcal{O} \to \mathbb{R}_+ \qquad (3.2.4)$$

Associating a positive real number $v(o)$ with each object o of $\mathcal{O}$,

$$(\forall o \in \mathcal{O})\ o \mapsto v(o) \qquad (3.2.5)$$

v is interpreted as a valuated unary relation:

$$\{v(o)|o \in \mathcal{O}\} \qquad (3.2.6)$$

Thereby, Boolean and numerical attributes are considered at the same relational level. In fact, their extensions are represented in a set theoretic way at the level of the object set $\mathcal{O}$: a subset of $\mathcal{O}$ for the Boolean attribute, and a numerical valuation on $\mathcal{O}$ for the numerical attribute.

Taking into account the accuracy of measurement, the scale of decimal positive numbers is largely sufficient for defining a numerical attribute in *data analysis*. Therefore, by denoting $\mathbb{D}_+$ the latter scale, $\mathbb{D}_+$ can be substituted for $\mathbb{R}_+$ in the right member of (3.2.4).

As mentioned above, generally and more specially in *Data Mining* the size of the object set $\mathcal{O}$ is very large. Consequently, the size of the reached values by v on $\mathcal{O}$ is much smaller than the cardinality of $\mathcal{O}$. Let us designate by

$$[x_{(1)}, x_{(2)}, \ldots, x_{(l)}, \ldots, x_{(m)}] \qquad (3.2.7)$$

the increasing ordered sequence of reached values by v on $\mathcal{O}$ ($m < n$) and introduce the subset $\mathcal{O}_l = v^{-1}(x_{(l)})$ constituted by all objects whose v value is $x_{(l)}$, $1 \le l \le m$. Denote $n_l = card(\mathcal{O}_l)$, $1 \le l \le m$. The empirical distribution of v on $\mathcal{O}$ is defined by the sequence

$$\{(x_{(l)}, f_l)\} \qquad (3.2.8)$$

where $f_l = n_l/n$, f_l defines the relative frequency (proportion) of objects whose v value is x_l, $1 \le l \le m$.

As an example, let us consider the following increasing sequence of a numerical attribute v on a set of 10 objects

$$(1.5, 2.3, 3.4, 3.4, 3.4, 5.1, 7.2, 7.2, 8.5, 9.0)$$

According the accuracy measurement of v, only one decimal after the point is retained. The above distribution (3.2.8) becomes

$$[(1.5, 0.1), (2.3, 0.1), (3.4, 0.3), (5.1, 0.1), (7.2, 0.2), (8.5, 0.1), (9.0, 0.1)]$$
$$(3.2.9)$$

Thus, there are $m = 7$ distinct values.

In the geometrical methods of *data analysis*, a numerical attribute is represented by a *linear form* (see Chaps. 6 and 7 of [18]). This gives the projection measurement on a linear axis endowed with an origin and a unit vector. To fix idea, we assume a horizontal axis. Then, a given object o is represented by a point of the axis whose abscissa is $v(o)$. By considering the sequence (3.2.7) of the v values, n_l distinct objects are represented by the same point that we denote by M_l, whose abscissa is x_l, $1 \le l \le m$. A graphical representation, called "histogram" is obtained by drawing from each point M_l, $1 \le l \le m$, an ascendant vertical segment whose length is proportional to the relative frequency f_l, $1 \le l \le m$. Thus we obtain the distribution (3.2.9) (see Fig. 3.1).

In terms of component analysis, the sequence

$$\{(M_l, f_l)\} \tag{3.2.10}$$

defines one-dimensional cloud of points where the point M_l is endowed with the positive numerical value f_l, the latter being interpreted as a weight, $1 \le l \le m$.

This structure (see (3.2.10)) was exploited in image scalar quantization Ghazzali (1992) [7], Ghazzali et al. (1994) [6]. In this application, the numerical attribute v concerned is the luminance for which a scale of 256 grey levels, from 0 to 255, is established: 0 for the black and 255 for the blank. Here, the object set $\mathcal{O}$ is defined by the image pixels. Thus, for a squared image comprising 512 rows and 512 columns, the object set $\mathcal{O}$ includes $n = 512 \times 512 = 262144$ elements. The one-dimensional cloud (3.2.10) has at most 256 points. It can be put in the following form:

Fig. 3.1 Histogram

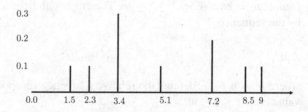

$$\{(l, f_l) | 0 \le l \le 255\} \tag{3.2.11}$$

where $f_l = n_l/n$ is the proportion of image pixels whose luminance is l, $0 \le l \le 255$.

3.2.3   Defining a Categorical Attribute from a Numerical One

As seen above, we distinguish in the framework of attributes of type I, two cases (see Sects. 3.2.1 and 3.2.2). The first one is defined by the Boolean attribute. The scale value of the latter is the poorest one. The second one is defined by the numerical attribute. Its scale value is the richest one.

There are several reasons for transforming a numerical attribute into a categorical one. One important reason may be to make homogeneous the description of the object set $\mathcal{O}$. Imagine that the vast majority of the description is provided by categorical attributes with a small number of categories attribute. In addition, suppose that a few descriptive attributes are numerical. In these conditions it might be appropriate to transform the numerical attributes into categorical ones (see the following Sects. 3.3.1 and 3.3.2). One more important reason related to the above first one, consists of retaining from the numerical value $v(o)$ of v on the object o, what can be significant in this value. In fact, knowing the exact value $v(o)$ might be less interesting than knowing that $v(o)$ is located inside a given interval. For this purpose, numerical attribute is categorized by dividing the interval variation of a numerical attribute into subintervals.

More precisely, regarding the sequence $[x_{(1)}, x_{(2)}, \ldots, x_{(l)}, \ldots, x_{(m)}]$ (see (3.2.7)), let us denote here $x_{(1)}$ by a, and $x_{(m)}^+$ by b, where $x_{(m)}^+$ is a value strictly greater than $x_{(m)}$ and as near as possible $x_{(m)}$. Therefore, the attribute v takes its values in the interval $[a, b[= \{x | x \in \mathbb{R}_+, a \le x < b\}$, where, as above, $\mathbb{R}_+$ denotes the real positive numbers. An increasing sequence, denoted by σ:

$$\sigma = (y_0 = a, y_1, \ldots, y_k, y_{k+1}, \ldots, y_l = b) \tag{3.2.12}$$

defines a subdivision of the interval $[a, b[$ into a sequence of l subintervals

$$\{[y_k, y_{k+1}[| 0 \le k \le l - 1\} \tag{3.2.13}$$

A category of a categorical attribute (see Sect. 3.3) is associated with each subinterval $[y_k, y_{k+1}[, 0 \le k \le l - 1$. By denoting c_k the category representing the interval $[y_k, y_{k+1}[, 0 \le k \le l - 1$, c_k is defined by the following Boolean attribute

$$(\forall o \in \mathcal{O}), c_k(o) = 1(\text{resp. } 0) \text{ iff } v(o) \in [y_k, y_{k+1}[(\text{resp. } v(o) \notin [y_k, y_{k+1}[)$$
$$\tag{3.2.14}$$

Now, the fundamental question is

How to define the subdivision σ?

There are a large variety of methods depending each on the objective considered. The simplest and the most direct one consists of defining σ by dividing the entire interval $[a, b[$ into subintervals with the same length. By denoting L the length of the interval $[a, b[$: $L = b - a$, we have

$$y_{k+1} = y_k + \frac{L}{l} \tag{3.2.15}$$

$0 \leq k \leq l-1$. Notice that for a given practical application, this technique requires an adequate value of l. On the other hand, for this technique, the statistical distribution of the attribute v on the object set $\mathcal{O}$ is not taken into account at all.

Another simple technique consists of interval division with as equal frequencies as possible. A given interval will be written $[x_i, x_{i'}[$ where $i' > i$ and where $n_i + \cdots + n_{i'}$ is adjusted to be as close as possible the ratio n/l. For this solution, in order to be consistent with the statistical distribution of the attribute v, we have also to determine in a non-arbitrary way an adequate value of l.

None of both previous techniques respect intimately the heterogeneity of the statistical distribution of v. Nevertheless, it might exist subintervals of $[a, b[$ such that two consecutive and near values of these, have comparable and high frequencies. Discovering such subintervals leads to the definition of "significant" categorical attribute (see (3,2.14)). Thus in Fig. 3.2 we recognize intuitively five subintervals.

These subintervals correspond to high concentration zones. Precisely, clustering methods enable to discover such subintervals. In these latter methods an objective criterion is optimized, locally or globally.

The interest of the well-known Fisher method Fisher(1958) [5] is to maximize *globally* the inter-cluster inertia in the case where the number of clusters is fixed. Each cluster is represented by an interval of the sought subdivision. The method is based on a dynamic programming algorithm.

Lafaye (1979) [14] analyzed and made effective a graphical method that we suggested the general idea. Let us consider the number of observations in a *small* interval window having a fixed length. This number defines a kind of "local density". By moving the window from left to right along the variation range of v, we can determine

Fig. 3.2 Histogram-2

the stable minima of this local density. This stability is obtained by varying in a suitable fashion the length of the window interval. The stable local minima define the cut points of the sought subdivision. For this method the number of the subdivision intervals is not fixed. It is a result of the algorithm employed. This method shown to be particularly effective for the treatment of small samples Kerjean (1978) [13]. In his paper [14] Lafaye gives a brief and interesting synthesis on the discretization of numerical attributes, with this respect, see also Rabaseda et al. (1995) [38].

In vector quantization (see the above example at the end of Sect. 3.2.2, considered in scalar quantization) Ghazzali (1992) [7], Ghazzali, Léger and Lerman (1994) [6] hierarchical ascendant (or agglomerative) clustering was applied. Two methods were processed: the *Likelihood Linkage Analysis* method Lerman (1993) [22] and the classical *Ward* method (1963) [48] (see Chap. 10 for a substantial development). In this case and for coding compression purpose the order of the number of clusters is given (for example $2^4 = 16$). Preliminary, it is established that each step of the aggregation algorithm joins two consecutive intervals of (3.2.7) obtained previously as clusters in the hierarchical process. A method of detecting the most consistent levels of a classification tree (see Sect. 9.3.6 of Chap. 9) enables us to obtain the subdivision wished.

A non-hierarchical clustering method as that presented in Sect. 2.2 of (Chap. 2 can also be applied in order to determine a subdivision σ. The method built by A. Schroeder (1976) [40] is based on a statistical criterion having a probabilistic nature: *the likelihood classifying criterion.*

For this criterion the observed cloud of points—one dimensional cloud in our case—is considered as provided from a mixing of probability laws, having distinct modes and distinct probability occurences, respectively, Symons (1981) [45]. Celeux and Govaert (1993) [3] employ a stochastic algorithm for separating the entire cloud into subclouds corresponding each to one of the probability laws. This separation enables an ordered sequence of intervals dominated each by one of the probability laws, to be detected. And then, a subdivision σ of the variation range of v can be deduced. This approach may lead to fine and sophisticated techniques. However, these assume probabilistic conditions on the shapes of the probability laws, difficult to validate in general. On the other hand, these methods may seem conceptually too heavy for the problem submitted of discretization of an observed numerical attribute.

3.3   Representation of the Attributes of Type II

As mentioned in Sect. 3.1, an attribute of type II induces a binary relation on $\mathcal{O}$. In this section we shall consider three sub-types of attributes. The nominal categorical attribute, the ordinal categorical attribute and the categorical attribute valuated by a numerical similarity.

3.3.1 The Nominal Categorical Attribute

Let $C = \{c_1, c_2, \ldots, c_h, \ldots, c_k\}$ be the value set of a categorical attribute c. No structure is assumed on C. The attribute c is defined by a mapping

$$c : \mathcal{O} \to \{c_1, c_2, \ldots, c_h, \ldots, c_k\} \qquad (3.3.1)$$

of $\mathcal{O}$ on C, such that for x in $\mathcal{O}$, $c(x) = c_h$ if and only if x possesses the value c_h, $1 \leq h \leq k$. $c_1, c_2, \ldots, c_h, \ldots$ and c_k are called the categories of the categorical attribute c. The value set C is assumed *exclusive* and *exhaustive*. That is to say: each object possesses necessarily and exactly one and only one categorical value. The attribute c induces a partition π on $\mathcal{O}$

$$\pi = \{\mathcal{O}_1, \mathcal{O}_2, \ldots, \mathcal{O}_h, \ldots, \mathcal{O}_k\} \qquad (3.3.2)$$

where $\mathcal{O}_h$ defined by $c^{-1}(c_h)$ is the reciprocal image of c_h, $1 \leq h \leq k$. Without loss of generality we assume this partition defined with labelled classes. Different equivalent expressions will be used for $\mathcal{O}_h$: $\mathcal{O}_h$ is the set of objects where c_h is *TRUE*, having the category c_h or belonging to the category c_h, ..., $1 \leq h \leq k$.

As just mentioned, in order to avoid ambiguity, we assume the classes of the partition π labelled. Then, to π corresponds the sequence $(n_1, n_2, \ldots, n_h, \ldots, n_k)$ of its cardinal classes. $n_h = card(\mathcal{O}_h)$, $1 \leq h \leq k$. This sequence defines the *type* of the partition π. We also associate the sequence of proportions or relative frequencies $(f_1, f_2, \ldots, f_h, \ldots, f_k)$ with π: $f_h = n_h/n$. This sequence defines the statistical distribution of c on $\mathcal{O}$.

The mapping (3.3.1) can be viewed as the representation of the nominal categorical attribute c at the level of the object set $\mathcal{O}$. It assigns to each elementary object o its value $c(o)$, the latter being a nominal code. It is important to realize that this representation does not allow us to compare directly two different nominal categorical attributes by comparing directly their respective values. In fact, given two nominal categorical attributes c and c' we cannot compare directly the statistical distributions of c and c' on $\mathcal{O}$. This comparison requires to cross the two respective partitions induced by c and c'.

There are different alternatives to carry out this comparison (see Chap. 5). That we adopt requires a higher representation level. By noticing that representing a categorical attribute c is equivalent to represent the associated partition π, we define the following binary relation P_π:

$$(\forall (x, y) \in \mathcal{O} \times \mathcal{O}), x P_\pi y \text{ iff } \exists h, 1 \leq h \leq k, \text{ s.t. } x \in \mathcal{O}_h, y \in \mathcal{O}_h \qquad (3.3.3)$$

This binary relation is an equivalence relation (reflexive, symmetrical and transitive relation). It can be represented at the level of the set

$$P = \{\{x, y\} | x \in \mathcal{O}, y \in \mathcal{O}, x \neq y\} \qquad (3.3.4)$$

of unordered object pairs. P is exactly the set $P_2(\mathcal{O})$ of all subsets with two elements of $\mathcal{O}$. Another notation we consider for $P_2(\mathcal{O})$ is $\mathcal{O}^{\{2\}}$. Two related representations denoted by $R(\pi)$ and $S(\pi)$ can be considered:

$$R(\pi) = \{\{x, y\} \in P_2(\mathcal{O}) | \exists h, 1 \leq h \leq k, x \in \mathcal{O}_h \text{ and } y \in \mathcal{O}_h\} \qquad (3.3.5)$$

and

$$S(\pi) = \{\{x, y\} \in P_2(\mathcal{O}) | \exists g \neq h, 1 \leq g, h \leq k, x \in \mathcal{O}_g \text{ and } y \in \mathcal{O}_h\} \qquad (3.3.6)$$

The indicator functions ρ_π and σ_π of $R(\pi)$ and $S(\pi)$, respectively, are defined as follows:

$$(\forall \{x, y\} \in P), \rho_\pi(\{x, y\}) = 1(resp., 0) \text{ iff } \{x, y\} \in R(\pi)(resp., \notin R(\pi)) \qquad (3.3.7)$$

and

$$(\forall \{x, y\} \in P), \sigma_\pi(\{x, y\}) = 1(resp., 0) \text{ iff } \{x, y\} \in S(\pi)(resp., \notin S(\pi)) \qquad (3.3.8)$$

The 1 and 0 values represent the logical values *TRUE* and *FALSE*, respectively. 1 and 0 can also be interpreted as specific integer values. In the latter case, we have

$$(\forall \{x, y\} \in P), \rho_\pi(\{x, y\}) + \sigma_\pi(\{x, y\}) = 1 \qquad (3.3.9)$$

In fact, $\{R(\pi), S(\pi)\}$ defines a bi-partition of P, that is to say, a partition of P into two classes. $R(\pi)$ [resp., $S(\pi)$] is the set of joined (resp. separated) distinct object pairs, by π.

The expressions of $R(\pi)$ and $S(\pi)$ with respect to the partition classes (see (3.3.2)), are respectively,

$$R(\pi) = \sum_{1 \leq h \leq k} P_2(\mathcal{O}_h)$$

and

$$S(\pi) = \sum_{1 \leq g < h \leq k} \mathcal{O}_g \star \mathcal{O}_h \qquad (3.3.10)$$

In these expressions, the sign Σ means a union of disjoint subsets. $P_2(\mathcal{O}_h)$ is the set of unordered pairs of elements of $\mathcal{O}_h$, or equivalently, the set of all 2-subsets of $\mathcal{O}_h, 1 \leq h \leq k$. $\mathcal{O}_g \star \mathcal{O}_h$ designates the set of all unordered pairs $\{x, y\}$ such that $x \in \mathcal{O}_g$ and $y \in \mathcal{O}_h, 1 \leq g \neq h \leq k$. We have

$$card[R(\pi)] = \sum_{1 \le h \le k} n_h \times (n_h - 1)/2$$

and

$$card[S(\pi)] = \sum_{1 \le g < h \le k} n_g \times n_h \qquad (3.3.11)$$

The following formula can be verified

$$card[R(\pi)] + card[S(\pi)] = n \times (n - 1)/2 \qquad (3.3.12)$$

where the right member is the cardinality of $P = P_2(\mathcal{O})$ (see (3.3.4)).

Now, let us consider an example given in the famous Goodman and Kruskal's paper (1954) [9], where the origin of the data is specified. The population is defined by white Protestant married couples living in Indianapolis, married in 1927, 1928 or 1929. Thus, each element u of our universe $\mathcal{U}$ is defined by such a married couple. The data sampling leads to an object set $\mathcal{O}$ comprising 1438 elements. One categorical attribute c defined is the "highest level of formal education of wife". Three categorical values are considered for this attribute. We take them as following

- c_1 = "less than three years high school";
- c_2 = "3 or 4 years high school";
- c_3 = "one year college or more".

The partition π (see (3.3.2)) includes here three classes $\mathcal{O}_1$, $\mathcal{O}_2$ and $\mathcal{O}_3$, where $\mathcal{O}_1$, $\mathcal{O}_2$ and $\mathcal{O}_3$ are the subsets of married couples whose formal education of wife are c_1, c_2 and c_3, respectively. We have according to the notations above, $n_1 = card[\mathcal{O}_1] = 591$, $n_2 = card[\mathcal{O}_2] = 608$ and $n_3 = card[\mathcal{O}_3] = 239$.
$P_2(\mathcal{O})$ is the set of unordered pairs of distinct married couples. $card[P_2(\mathcal{O})] = 1438 \times 1437/2 = 1033203$.

$P_2(\mathcal{O}_h)$ is the set of unordered pairs of distinct married couples whose formal education of wife is c_h, $1 \le h \le 3$.

- $card[P_2(\mathcal{O}_1)] = 591 \times 590/2 = 174345$;
- $card[P_2(\mathcal{O}_2)] = 608 \times 607/2 = 184528$;
- $card[P_2(\mathcal{O}_3)] = 239 \times 238/2 = 28441$.

Equations (3.3.10) and (3.3.11) give

$$card[R(\pi)] = 174345 + 184528 + 28441 = 387314$$

$\mathcal{O}_g \star \mathcal{O}_h$ is the set of unordered pairs of married couples whose formal education of wives are c_g and c_h, respectively, $1 \le g < h \le 3$.

- $card[\mathcal{O}_1 \star \mathcal{O}_2] = card(\mathcal{O}_1) \times card(\mathcal{O}_2) = 591 \times 608 = 359328$;
- $card[\mathcal{O}_1 \star \mathcal{O}_3] = card(\mathcal{O}_1) \times card(\mathcal{O}_3) = 591 \times 239 = 141249$;

- $card[\mathcal{O}_2 \star \mathcal{O}_3] = card(\mathcal{O}_2) \times card(\mathcal{O}_3) = 608 \times 239 = 145312$.

Equations (3.3.10) and (3.3.11) give

$$card[S(\pi)] = 359328 + 141249 + 145312 = 645889$$

We immediately verify the (3.3.12):

$$387314 + 645889 = 1033203$$

3.3.2 The Ordinal Categorical Attribute

As in the Goodman and Kruskal paper (1954) [9], we assume in the example above a total (i.e. linear) order on the category set $\{c_1, c_2, c_3\}$ for the categorical attribute c defined by the "Highest level of formal education of wife". We adopt the total order:

$$c_1 < c_2 < c_3$$

More generally, by considering the formalism introduced in Sect. 3.3.1 (see (3.3.1)), for an ordinal version of the categorical attribute c, the category set $\mathcal{C}$ is provided with a strict total order; that is, a ranking on $\mathcal{C}$. Assuming that the integer subscript h is the rank of the category c_h, $1 \leq h \leq k$, the latter total order is defined by

$$c_1 < c_2 < \cdots < c_h < \cdots < c_k \tag{3.3.13}$$

This structure induces a total preorder on the object set $\mathcal{O}$ that we denote by ω. By defining, as in Sect. 3.3.1, $\mathcal{O}_h = c^{-1}(c_h)$, $1 \leq h \leq k$, we have

$$\mathcal{O}_1 < \mathcal{O}_2 < \cdots < \mathcal{O}_h < \cdots < \mathcal{O}_k \tag{3.3.14}$$

and that means: for any (x, y) in the cartesian product $\mathcal{O}_g \times \mathcal{O}_h$, $c(x) < c(y)$ if and only if $g < h$, $1 \leq g < h \leq k$. The total order above on the set of classes $\mathcal{O}_1, \mathcal{O}_2, \ldots, \mathcal{O}_h, \ldots, \mathcal{O}_k$ is called the *quotient order* associated with ω. Given a total preorder on $\mathcal{O}$ consists of given a partition on $\mathcal{O}$ and, in addition, a total order on its classes. This mathematical expression formalizes the notion of a "ranking with ties" given in Kendall (1948) [12]. Now, the ordered sequence of the cardinal classes $(n_1, n_2, \ldots, n_h, \ldots, n_k)$, denoted as in Sect. 3.3.1, is called here *composition* of the total preorder ω. The statistical distribution of c on $\mathcal{O}$ has the same meaning as for the nominal case (Sect. 3.3.1). It is defined by $(f_1, f_2, \ldots, f_h, \ldots, f_k)$, where $f_h = n_h/n$, $1 \leq h \leq k$. Nevertheless, the labelling of the different classes satisfies here the linear order (3.3.14).

As for the nominal categorical attribute and in spite of the linear order defined on the category set $C = \{c_1, c_2, \ldots, c_h, \ldots, c_k\}$ the formal valuation on $\mathcal{O}$ defined by (3.3.1) does not allow ordinal categorical attributes to be compared (see Sects. 3.3.2, 3.3.4 and 3.4). As discussed in Sect. 3.3.3 a ranking attribute defining a linear order on the object set can be interpreted as a very specific case of an ordinal categorical attribute. However, the ranking function has a clear numerical interpretation and this allows us to obtain one version of comparison between two ranking attributes defining respectively, two specific numerical valuations on the object set $\mathcal{O}$.

In order to make possible the comparison between two ordinal categorical attributes in the most general case (see Chap. 6), let us define the following binary relation R_ω associated with the ordinal categorical attribute c:

$$(\forall((x, y) \in \mathcal{O} \times \mathcal{O})), \; xR_\omega y \text{ iff } c(x) < c(y) \tag{3.3.15}$$

Because of no symmetry and no reflexivity of R_ω, we consider the following representation set

$$C = \{(x, y) | x \in \mathcal{O}, y \in \mathcal{O}, x \neq y\} \tag{3.3.16}$$

In other words, C is the set of all ordered pairs of distinct objects from $\mathcal{O}$. Another notation we consider for C is $C^{[2]}$.

In C, R_ω is represented by the following subset:

$$R_\omega = \{(x, y) \in C | xR_\omega y\} \tag{3.3.17}$$

More explicitly R_ω can be put in the following form:

$$R(\omega) = \sum_{1 \leq g < h \leq k} \mathcal{O}_g \times \mathcal{O}_h \tag{3.3.18}$$

where Σ designates a union of disjoint subsets and where $\mathcal{O}_g \times \mathcal{O}_h$ is the cartesian product of $\mathcal{O}_g$ and $\mathcal{O}_h$.

C can be decomposed as follows:

$$C = R(\omega) + E(\omega) + S(\omega) \tag{3.3.19}$$

where

$$E(\omega) = \sum_{1 \leq h \leq k} \mathcal{O}_h^{[2]}$$

$$\text{and } S(\omega) = \sum_{1 \leq g < h \leq k} \mathcal{O}_h \times \mathcal{O}_g \tag{3.3.20}$$

The cardinalities of the three components of c (see (3.3.19)) are, respectively

$$card[R(\omega)] = \sum_{1 \leq g < h \leq k} n_g \times n_h$$

$$card(E(\omega)) = \sum_{1 \leq h \leq k} n_h \times (n_h - 1)$$

$$card(S(\omega)) = \sum_{1 \leq g < h \leq k} n_h \times n_g \qquad (3.3.21)$$

Clearly,

$$card[R(\omega)] = card(S(\omega)) \qquad (3.3.22)$$

on the other hand, we can verify the formula

$$card[R(\omega)] + card(E(\omega)) + card(S(\omega)) = n \times (n - 1) \qquad (3.3.23)$$

Let us take up again the above example in its ordinal version: $c_1 < c_2 < c_3$. We have

$$card[R(\omega)] = 591 \times 608 + 591 \times 239 + 608 \times 239$$
$$= 359328 + 141249 + 145312 = 645889$$
$$card(E(\omega)) = 591 \times 590 + 608 \times 607 + 239 \times 238$$
$$= 348690 + 369056 + 56882 = 774628$$

In these conditions, we verify the (3.3.23) where the right member is equal to $1438 \times 1437 = 2066406$.

By denoting ρ_ω the Boolean indicator function of $R(\omega)$, we have

$$(\forall (x, y) \in C), \ \rho_\omega(x, y) = 1 (resp., 0) \text{ iff } (x, y) \in R(\omega)(resp., \notin R(\omega)) \qquad (3.3.24)$$

As above (see what follows the Eqs. (3.3.7) and (3.3.8)), 1 and 0 are interpreted as the logical values *TRUE* and *FALSE*, respectively. However, the interpretation where 1 and 0 are numerical scorings holds also. Now, let $\mathcal{O}$ be an object set endowed with a ranking function R, inducing a linear order on $\mathcal{O}$. Consider a different linear order $R(\omega)$ on $\mathcal{O}$. In his work Kendall (1948) [12] uses a scoring numerical function with two values $+1$ and -1 in order to code $R(\omega)$ with respect to R. More clearly, for an R ordered pair (x, y) such that xRy, the scoring value is $+1$ (resp., -1) if $xR_\omega y$ (resp., $yR_\omega x$). The value $+1$ (resp., -1) indicates that $(x, y) \in R(\omega)$ (rep., $(y, x) \in R(\omega)$) (see (3.3.17)). Thus, our set theoretic representation is equivalent to the Kendall coding of a ranking. However, for our representation the initial ranking R is useless.

The set representation of the ordinal categorical attribute c by $R(\omega)$ (see (3.3.17)) does not take into account explicitly the set of ordered pairs (x, y) of distinct objects such that x and y belong to the same class $\mathcal{O}_h$ $c(x) = c(y) = h$, $1 \le h \le k$. This set is denoted by $E(\omega)$ in (3.3.19) and (3.3.20). However, due to (3.3.19) $E(\omega)$ is implicitly taken into account. In fact, $E(\omega)$ is deduced from $R(\omega)$. In order to weight the part of $E(\omega)$, the attribute c can be represented by the following numerical scoring function denoted $score_\omega$:

$$score_\omega(x, y) = \begin{cases} 1 & \text{if } (x, y) \in R(\omega) \\ 0.5 & \text{if } (x, y) \in E(\omega) \\ 0 & \text{if } (x, y) \in S(\omega) \end{cases} \qquad (3.3.25)$$

Let us now consider the more general case where the category set $\mathcal{C} = \{c_1, c_2, \ldots, c_h, \ldots, c_k\}$ is provided with a strict *partial* order. An instructive exercise we propose to the reader consists of taking up again the above set representation formalism for this general case. For our part, we reconsider the example above of the categorical attribute "Highest level of formal education of wife" where we only assume $c_1 < c_3$ and $c_2 < c_3$ for the categorical values (see above). Therefore, c_1 and c_2 are considered not comparable. This categorical scale induces a partial preorder on the object set $\mathcal{O}$, defined here by married couples. As above, denote ω this partial preorder. The expression of $R(\omega)$, $E(\omega)$ and $S(\omega)$ (see Eqs. 3.3.18–3.3.20) become

$$R(\omega) = \mathcal{O}_1 \times \mathcal{O}_3 + \mathcal{O}_2 \times \mathcal{O}_3$$
$$E(\omega) = \mathcal{O}_1^{[2]} + \mathcal{O}_2^{[2]} + \mathcal{O}_1^{[3]} + \mathcal{O}_1 \times \mathcal{O}_2 + \mathcal{O}_2 \times \mathcal{O}_1$$
$$S(\omega) = \mathcal{O}_3 \times \mathcal{O}_1 + \mathcal{O}_3 \times \mathcal{O}_2 \qquad (3.3.26)$$

where the sign $+$ indicates a set sum defined by a union of disjoint subsets.

The scoring function defined in (3.3.25) is appropriate to code the partial strict order ω.

3.3.3   The Ranking Attribute

In his book Kendall (1948) [12] uses the ability in a given subject taught (e.g. mathematics) to rank totally and strictly the described object set $\mathcal{O}$. Thus, a linear order that we denote by ω_l is induced on $\mathcal{O}$. For this, two conditions must be satisfied: (i) the cardinality of $\mathcal{O}$ is small enough; (ii) the scoring function is enough discriminating. The ranking attribute r is a bijective mapping of $\mathcal{O}$ onto the set of the first n integer numbers:

$$r : \mathcal{O} \to \{1, 2, \ldots, i, \ldots, n\} \qquad (3.3.27)$$

representing a given object x of $\mathcal{O}$, by its rank defined as follows:

$$(\forall x \in \mathcal{O}), x \mapsto r(x) = card\{y | y \in \mathcal{O} \text{ and } y \leq x \text{ for } \omega_l\} \quad (3.3.28)$$

Using the rank function r, the representation of the ranking attribute would have been considered in the framework of descriptive attributes of type I. In fact, r defines a numerical valuation on the set $\mathcal{O}$, given by the first n integer numbers. In fact, this coding of a linear order leads to the Spearman coefficient Spearman (1904) [41] (see Chap. 6). But, as expressed in the preceding Sect. 3.3.2, the originality of the Kendall representation is the level $\mathcal{O} \times \mathcal{O}$ of its definition. In fact, a ranking attribute can be seen as a very particular case of an ordinal categorical attribute. For this, each class of the induced total preorder on $\mathcal{O}$ becomes a singleton (i.e. comprises exactly one element) (see (3.3.14)). The representation set $R(\omega)$ (see (3.3.17)) becomes

$$R(\omega_l) = \{(x, y) | (x, y) \in \mathcal{O} \times \mathcal{O} \text{ and } r(x) < r(y)\} \quad (3.3.29)$$

where $r(x) < r(y)$ is equivalent to $x \leq y$ and not $y \leq x$ for ω_l.

We have

$$card[R(\omega_l)] = \frac{n \times (n - 1)}{2} \quad (3.3.30)$$

The "mean rank" function is an extension of the rank function (see (3.3.28)) used to code a total preorder (ranking with ties) on an object set $\mathcal{O}$. For the total preorder ω considered in (3.3.14), the mean rank function, denoted r_m, is defined as follows

$$(\forall h, 1 \leq h \leq k), (\forall x \in \mathcal{O}_h), r_m(x) = n_1 + n_2 + \cdots + n_{h-1} + \frac{1}{2} \times (n_h + 1) \quad (3.3.31)$$

where $n_g = card(\mathcal{O}_g)$, $1 \leq g \leq h$. The expression "mean rank" is explained by the fact that the mean rank of an arbitrary element x of $\mathcal{O}_h$, over all the linear orders compatible with the total preorder ω, is given by $r_m(x)$. The following formula is easy to verify:

$$\sum_{x \in \mathcal{O}} r_m(x) = \frac{n \times (n + 1)}{2} \quad (3.3.32)$$

In these conditions, a total preorder ω on $\mathcal{O}$ and then, an ordinal categorical attribute describing $\mathcal{O}$ can be represented as a particular numerical valuation on $\mathcal{O}$, given by the mean rank function r_m. However, the nature of this representation is very different from the relational one defined by $R(\omega)$ (see (3.3.17)). In fact, the latter representation is logical. It gives for a given ordered pair (x, y) in $\mathcal{O} \times \mathcal{O}$, a logical value.

For the above mentioned reasons, a ranking attribute inducing a linear (total and strict) order on $\mathcal{O}$ cannot occur in the case of description of large data sets. However, a realistic case where such a description occurs, concerns the problem of m rankings Kendall (1948) [12]. Let us imagine a few number of objects: $n = card(\mathcal{O})$ is relatively small. $\mathcal{O}$ is for example a set of manufactured products of a given type. We assume a set of m judges giving each his preferences by ranking (without ties) the n objects. Thus, each judge defines a ranking attribute on $\mathcal{O}$. In this type of data there is no restriction on how large is the number m of judges.

3.3.4 The Categorical Attribute Valuated by a Numerical Similarity

We consider here the case where the category set $\mathcal{C}$ (see (3.3.1)) is provided with a numerical similarity. The latter is supposed given *a priori*, for example, by an expert knowledge. Let us denote it by ξ. ξ is defined by a mapping of the cartesian product $\mathcal{C} \times \mathcal{C}$ onto the real numbers. Mostly, the set value of ξ is the *positive* reals $\mathbb{R}_+$:

$$\xi : \mathcal{C} \times \mathcal{C} \rightarrow \mathbb{R}_+$$
$$(c_g, c_h) \mapsto \xi(c_g, c_h) \qquad (3.3.33)$$

where $\xi(c_g, c_h)$ is the numerical similarity value between the categories c_g and c_h, $1 \leq g, h \leq k$.

Mostly, ξ is symmetrical:

$$(\forall(g, h), 1 \leq g, h \leq k), \xi(c_g, c_h) = \xi(c_h, c_g) \qquad (3.3.34)$$

However, real important cases of asymmetrical similarity may occur. This point will be mentioned below. On the other hand we assume

$$((\forall(g, h), 1 \leq g \neq h \leq k)), \min[\xi(c_g, c_g), \xi(c_h, c_h)] > \xi(c_g, c_h) \qquad (3.3.35)$$

Nevertheless, $\xi(c_h, c_h)$ is not necessarily invariant with respect to h, $1 \leq h \leq k$. ξ can be figured by the square matrix of Table 3.1.

Now, let us consider the example above of the categorical attribute "Highest level of formal education of wife", having the three categories denoted c_1, c_2 and c_3 (see Sect. 3.3.1). The matrix of ξ might be given by Table 3.2:

In fact, we have determined this matrix of numerical integer numbers from the following similarity ranking on $\mathcal{C} \times \mathcal{C}$

$$(c_1, c_3) < (c_1, c_2) < (c_2, c_3) < (c_2, c_2) < (c_1, c_1) < (c_3, c_3) \qquad (3.3.36)$$

Table 3.1 Matrix of ξ

$\mathcal{C}$	c_1	$\cdots$	c_h	$\cdots$	c_k
c_1					
$\vdots$					
c_g			$\xi(c_g, c_h)$		
$\vdots$					
c_k					

Table 3.2 Matrix of ξ for the example

$\mathcal{C}$	c_1	c_2	c_3
c_1	5	2	1
c_2	2	4	3
c_3	1	3	6

It may seem surprising that the similarity between a given category with itself is not the same whatever is this category. This can be justified by taking into account the specificity of the category concerned. In our example, we have considered that rarer is a category, more specific it is. Thereby, the category c_3 ("one year college or more") is more specific than c_2 ("3 or 4 years high school"). Intuitively, the resemblance between two different married couples whose common category is c_3, is stronger than that between two different married couples whose common category is c_2.

The valuation ξ (see (3.3.33)) induces a complete valuated symmetrical graph, without loops, on the object set $\mathcal{O}$. Explicitly

$$(\forall(x, y) \in \mathcal{O}^{[2]}), \xi(x, y) = \xi[c(x), c(y)] \qquad (3.3.37)$$

This graph is decomposed according to the partition π (see (3.3.2)) as follows:

$$(\forall(x, y) \in \mathcal{O}_g \times \mathcal{O}_h), \xi(x, y) = \xi(c_g, c_h)$$

and

$$(\forall(x, y) \in \mathcal{O}_h^{[2]}, \xi(x, y) = \xi(c_h, c_h) \qquad (3.3.38)$$

$1 \leq g \neq h \leq k$.

Now, let us illustrate this graph in the framework of the above example

$$(\forall(x, y) \in \mathcal{O}_1 \times \mathcal{O}_2 + \mathcal{O}_2 \times \mathcal{O}_1) \ \xi(x, y) = 2$$
$$(\forall(x, y) \in \mathcal{O}_1 \times \mathcal{O}_3 + \mathcal{O}_3 \times \mathcal{O}_1) \ \xi(x, y) = 1$$
$$(\forall(x, y) \in \mathcal{O}_2 \times \mathcal{O}_3 + \mathcal{O}_3 \times \mathcal{O}_2) \ \xi(x, y) = 3$$
$$(\forall(x, y) \in \mathcal{O}_1^{[2]}) \qquad\qquad \xi(x, y) = 5$$
$$(\forall(x, y) \in \mathcal{O}_2^{[2]}) \qquad\qquad \xi(x, y) = 4$$
$$(\forall(x, y) \in \mathcal{O}_3^{[2]}) \qquad\qquad \xi(x, y) = 6$$

Let us end this development by noticing that a nominal categorical attribute can be interpreted in terms of a very particular similarity categorical attribute as follows:

$$(\forall h, 1 \leq h \leq k), \xi(c_h, c_h) = 1$$
$$(\forall(g, h), 1 \leq g \neq h \leq k), \xi(c_g, c_h) = 0 \qquad (3.3.39)$$

Now, for an asymmetrical similarity providing the category set of a categorical attribute, (3.3.34) does not hold. On the other hand, $\xi(c_h, c_h)$, $1 \leq h \leq k$, may not be defined. Besides, Eqs. (3.3.37) and (3.3.38) are still valid. As an example, consider the traffic of cellular call phones between the towns of France. "Town of France" defines a categorical attribute. For a given ordered pair (A, B) of towns of France, let us consider the number, $\nu(A, B)$, of cellular call phones emitted from A to B during a given period. The ν function defines an asymmetrical similarity on the set of the towns of France (the category set in our example). In fact, generally $\nu(A, B) \neq \nu(B, A)$. The object set could be the set of antennae of cellular telephone located in the different towns.

3.3.5 The Valuated Binary Relation Attribute

The definition of this attribute is directly given at the level of the cartesian product $\mathcal{O} \times \mathcal{O}$ where $\mathcal{O}$ is the object set. Let us denote by B the binary relation defined by a given attribute. B is represented by the following subset of the cartesian product $\mathcal{O} \times \mathcal{O}$:

$$R(b) = \{(x, y)|(x, y) \in \mathcal{O} \times \mathcal{O} \text{ and } xBy\} \qquad (3.3.40)$$

Additionally, a valuation v on $R(b)$ is considered. v is a mapping of $R(b)$ on the value scale of v, denoted S in the following expression

$$v : R(b) \rightarrow S \qquad (3.3.41)$$

Generally S is the positive reals. The valuated binary relation can be represented by a valuated graph:

$$\{((x, y), v(x, y))|(x, y) \in R(b)\} \qquad (3.3.42)$$

The valuated binary relation attribute is a generalization of all the above attributes considered in this Sect. 3.3. However, it does not correspond to a categorical attribute. And, the specificity of the categorical structure is very important for comparing attributes or building similarity indices between objects or categories described (see Chaps. 4–7). The attribute type considered here occurs frequently in communication problems.

3.4 Representation of the Attributes of Type III

As mentioned in the introduction (Sect. 3.1) a categorical attribute is of type III if the similarity structure of its category set $C = \{c_1, c_2, \ldots, c_h, \ldots, c_k\}$ induces a binary relation on the set—denoted above by $\mathcal{O}^{\{2\}}$ (see (3.3.4))—of unordered pairs of distinct objects, or on the set denoted above by $\mathcal{O}^{[2]}$ (see (3.3.16)) of ordered pairs of distinct objects. Specific ordinal similarity structures on C occur importantly for describing data [28, 31, 36, 37]. Each of them defines a specific total preorder on $\mathcal{O}^{\{2\}}$ (ranking with ties). We shall distinguish below three versions of such a categorical attribute: "preordonance attribute", "taxonomic attribute" and "taxonomic preordonance attribute". A set theoretic representation at the level of the cartesian product $\mathcal{O}^{\{2\}} \times \mathcal{O}^{\{2\}}$ will be first given. Next, a representation by means of an adequate numerical valuation on $\mathcal{O}^{\{2\}}$ is proposed. The latter is easier and more efficient to handle for comparing the data described by such categorical attributes. As just mentioned, for the categorical attributes presented below, the induced total preorder is defined at the level of the set $\mathcal{O}^{\{2\}}$ of unordered object pairs. For this case, the ordinal similarity structure on C is symmetrical with respect to the categories to be compared. Nevertheless, extension can be envisaged for the case where the induced total preorder is defined at the level of $\mathcal{O}^{[2]}$ of ordered object pairs. For this case the ordinal similarity on C is asymmetrical.

3.4.1 The Preordonance Categorical Attribute

A "preordonance" categorical attribute is a categorical attribute whose category set C is provided with an ordinal similarity, called preordonance on C. Formally, as said above, a preordonance on C is a total preorder on a specific set of category pairs of C. For the categorical attributes to be introduced where the ordinal similarity is symmetrical,[2] the C category pairs to be considered is $\{(c_g, c_h) | 1 \leq g \leq h \leq K\}$. By denoting $\mathcal{K} = \{1, 2, \ldots, h, \ldots, K\}$ the category codes of the attribute concerned, the total preorder is defined on the following set

$$\mathcal{K}_2 = \{(g, h) | 1 \leq g \leq h \leq K\} \tag{3.4.1}$$

[19, 24, 28, 30, 36]. In fact, we have used already and implicitly such a categorical attribute in the example above of "Highest level of formal education of wife" (see (3.3.36)).

Let us now give an example concerning the problem of a database management of real estate advertisements [37]. One categorical attribute defined in this database is "subject of the transaction". Its categories are:

---

[2]The ordinal similarity between c_g and c_h is the same as that between c_h and c_g, $1 \leq g \leq h \leq K$.

$$c_1 = house, c_2 = villa, c_3 = apartment, c_4 = studio$$

$$apartment, c_5 = room, c_6 = garage \text{ and } c_7 = piece \text{ of } land.$$

By denoting gh the category pair (c_g, c_h), $1 \leq g \leq h \leq K$, the proposed preordonance is:

$$14 \sim 15 \sim 16 \sim 17 \sim 24 \sim 25 \sim 26 \sim 27 \sim$$
$$35 \sim 36 \sim 37 \sim 46 \sim 47 \sim 56 \sim 57 \sim 67 <$$
$$23 \sim 34 < 13 < 45 < 12 <$$
$$11 \sim 22 \sim 33 \sim 44 \sim 55 \sim 66 \sim 77 \qquad (3.4.2)$$

where the symbols $\sim$ and $<$ mean "equivalent" and "strictly lower than", respectively.

There are in all $K \times (K + 1)/2 = 7 \times (7 + 1)/2 = 28$ pairs. This total preorder comprises 6 classes. Their respective cardinalities (*composition* of the total preorder) are 16, 2, 1, 1, 1 and 7.

Generally, the expert establishes the total preorder on $\mathcal{K}_2$ (preordonance on $\mathcal{C}$), recursively, by sorting the pairs the most similar among pairs of $\mathcal{K}_2$ not yet sorted, at each step.

The preordonance categorical attribute has played an important part in our formalization work of data description (see the references indicated above). Let us mention that this type of attribute appeared also independently with a different expression and in a very different context in [4].

The preordonance categorical attribute concept does not require a notion of metrical difference between categories. Nevertheless, psychometric researchers consider a numerical scale measurement called "ordered metric scale" in which the differences between categories are defined and ordered Stevens (1951) [42].

Let us now define the set theoretic representation of a preordonance categorical attribute. Denote by $\omega(\mathcal{K}_2)$ the total preorder on $\mathcal{K}_2$ (see (3.4.1)) expressing this preordonance attribute.

$$(L_1, L_2, \ldots, L_q, \ldots, L_r) \qquad (3.4.3)$$

will designate the ordered sequence of the class preorder. L_1 (resp., L_r) comprises the most dissimilar (resp., similar) category pairs. To fix ideas and for consistent reasons, we can assume that the last classes are defined on the subset $\{(h, h) | 1 \leq h \leq K\}$. However, this point is out of our representation problem. The basic representation level is the cartesian product $\mathcal{K}_2 \times \mathcal{K}_2$. This representation has the same nature as that given in Sect. 3.3.2 for a total preorder on an object set (see Eq. (3.3.19)). Nevertheless, the context is different. In the latter, we define the following set sums of cartesian products

$$R(\omega(\mathcal{K}_2)) = \sum_{1 \leq p < q \leq r} L_p \times L_q \qquad (3.4.4)$$

$$E(\omega(\mathcal{K}_2)) = \sum_{1 \leq p \leq r} L_p^{[2]} \qquad (3.4.5)$$

$$S(\omega(\mathcal{K}_2)) = \sum_{1 \leq p < q \leq r} L_q \times L_p \qquad (3.4.6)$$

Now, let us denote by $\mathcal{I}_m(L_p \times L_q)'$ and $\mathcal{I}_m(L_p^{[2]})$ the respective representations of $L_p \times L_q$ and $L_p^{[2]}$ at the level of the components of the partition $\{\mathcal{O}_g | 1 \leq g \leq K\}$. We have

$$\mathcal{I}_m(L_p \times L_q) = \left(\sum_{(e,f) \in L_p} \mathcal{O}_e \star \mathcal{O}_f \right) \times \left(\sum_{(g,h) \in L_q} \mathcal{O}_g \star \mathcal{O}_h \right)$$

$$\mathcal{I}_m\left(L_p^{[2]}\right) = \left(\sum_{(g,h) \in L_p} \mathcal{O}_g \star \mathcal{O}_h \right)^2 \qquad (3.4.7)$$

$1 \leq p < q \leq r$, where as above the sums indicate union of disjoint subsets and where $\mathcal{O}_g \star \mathcal{O}_h$ is defined by the set of distinct unordered pairs whose components belong to $\mathcal{O}_g$ and $\mathcal{O}_h$, respectively, $1 \leq g \leq h \leq K$.

In order to avoid too big combinatorial complexity we code the total preorder $\omega(\mathcal{K}_2)$ with the "mean rank function" r_m (see (3.3.31) expressed in another context). Referring to (3.4.3), denote by l_q the cardinality of L_q, $1 \leq q \leq r$. Therefore, for a given pair (g, h) belonging to L_q we get

$$r_m(g, h) = \sum_{1 \leq p < q} l_p + \frac{l_q + 1}{2} \qquad (3.4.8)$$

Thus, we can verify that

$$\sum_{1 \leq q \leq r} l_q = \frac{K \times (K + 1)}{2}$$

For example, for the total preorder (3.4.2), the sequence of the values of r_m is

$$\frac{16 + 1}{2} = 8.5, 16 + \frac{2 + 1}{2} = 17.5, 19, 20, 21, 21 + \frac{7 + 1}{2} = 25$$

We verify that the sum of the r_m ranks is equal to $28 \times 29/2 = 406$.

Thus, a preordonance categorical attribute is coded as a specific categorical attribute valuated by a numerical similarity (see Sect. 3.3.4), given by the mean rank function.

Let us denote by R_q the right member of (3.4.8). R_q is the common value of the mean rank function r_m on the pairs (g, h) belonging to the qth preorder class L_q (see (3.4.3)). The empirical statistical distribution of R_q on $\mathcal{O} \star \mathcal{O}$ is given by

$$\{(R_q, \sum_{(g,h)\in L_q} (n_g \times n_h/n^{[2]}))|1 \leq q \leq r\} \tag{3.4.9}$$

where $n^{[2]} = n \times (n - 1)/2$.

We have considered above the case of symmetrical ordinal similarity for which the ordinal similarity for the ordered pair (c_g, c_h) is the same as that for (c_h, c_g), $1 \leq g \neq h \leq K$. As mentioned above, there might be data where the ordinal similarity is asymmetrical [26, 27]. In this case the total preorder comparing categories has to be established on

$$\mathcal{H}_2 = \{(g, h)|1 \leq g, h \leq K\} \tag{3.4.10}$$

instead of on $\mathcal{K}_2$ (see (3.4.1)).

In these conditions, an analogous development as above (see Eqs. (3.4.1) to (3.4.9)) has to be setup. In this, the non-ordered object pairs $\mathcal{O}^{[2]}$ has to be replaced by the ordered object pairs $\mathcal{O}^{[2]}$. We leave this development to be accomplished by the reader.

3.4.2 The Taxonomic Categorical Attribute

Let us start with an example taken from data we have processed Lerman and Peter (1988, 2007) [29], [31]. These data are provided from biological descriptions of phlebotomine sandfly species of French Guiana [15]. Descriptions are very complex. Relative to a descriptive categorical attribute and for subsets of category values, hierarchical logical dependencies associated with the *mother* → *daughter* relation, have to be taken into account. Consider the attributes 1, 18, 19 and 20 defined in this database [15] and retaken in [29]. We denote them by a^1, a^{21}, a^{31} and a^{32}, respectively. a^1 is the "Sex" attribute, a^{21} is defined by the "Number of style spines", a^{31} indicates the "Distribution of 4 style spines" and a^{32}, the "Distribution of 5 style spines". The code category sets of these attributes are: $\{1 : male\ 2 : female\}$, $\{1, 2, 3, 4, 5\}$, $\{1, 2, 3, 4, 5, 6\}$ and $\{1, 2, 3, 4, 5\}$, respectively. We obtain the following taxonomic structure organizing the different attributes according to the *mother* → *daughter* relation. Then, a daughter attribute might be associated with a categorical value of a given attribute. For example, the attribute a^{21} is associated with the value 1 of the attribute a^1 (see Fig. 3.3).

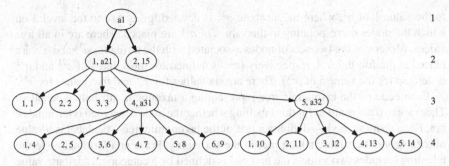

Fig. 3.3 Taxonomic attribute

Clearly, the attribute a^{21} is defined only when the a^1 value is 1. It is defined on the subset of objects whose a^1 value is 1 (male phlebotomine sandflies in our example). On the other hand, the attributes a^{31} and a^{32} are defined only when the values of a^{21} are 4 or 5. a^{31} (resp., a^{32}) is defined on the subset of objects where the value of a^1 is 1 and where the value of a^{21} is 4 (resp., 5). The common mother of the attributes a^{31} and a^{32} is a^{21}.

More generally, a taxonomic categorical attribute denoted τ, is defined by an organization of logically dependent attributes. It consists of a sequence of collections of categorical attributes of the following form:

$$\tau = (\{a^1\}, \{a^{21}, a^{22}, \ldots, a^{2k_2}\}, \ldots, \{a^{p1}, a^{p2}, \ldots, a^{pk_p}\},$$
$$\ldots, \{a^{q1}, a^{q2}, \ldots, a^{qk_q}\}) \qquad (3.4.11)$$

In the above example τ is instantiated as follows:

$$\tau = (\{a^1\}, \{a^{21}\}, \{a^{31}, a^{32}\})$$

Now, let us imagine two categorical attributes a^{41} and a^{42} defined respectively for $a^{31} = 4$ and for $a^{32} = 5$. By denoting A^{41} and A^{42} the respective category sets of a^{41} and a^{42}, A^{41} and A^{42} will be placed at the same 5th level of the above hierarchical structure (see Fig. 3.3). A^{41} and A^{42} will be instantiations of the above $\{a^{pi}, \ldots, a^{pi'}\}$ $(i' > i)$ subset of categories. Two categorical elements of A^{41} (resp., A^{42}) would have the same mother a^{41} (resp., a^{42}).

The construction of a taxonomic attribute must be done in a descendant manner, level by level. The first level 1 is assigned to the root of the taxonomy. Hence, in our example, the root is defined by the categorical attribute "Sex". Now, consider a given node ν defined at the lth level (l integer) and corresponding to a value of a preceding categorical attribute. The categorical attribute a^ν gives rise to a $(l+1)^{th}$ level, on which the values of a^ν are represented by different nodes. Directed descendant arrows join the node ν to the new nodes. Thus, in the example of Fig. 3.3, the node corresponding to the value 1 of a^1, where the attribute a^{21} is defined, gives rise to the level 3 on which the nodes corresponding to the values of a^{21} are placed. There are in all five values. Moreover, the two sets of nodes associated with the attributes a^{31} and a^{32} are placed at the fourth level, respectively (a^{31} is defined for the value 4 of a^{21} and a^{32} is defined for the value 5 of a^{21}). There are six values for a^{31} and five values for a^{32}.

Each node of the taxonomic tree representing a taxonomic attribute is labelled. There are two alternatives for this labelling whether the node concerned is terminal or not. For a terminal node—defining a leaf of the taxonomic tree—the label is a value of a categorical attribute taking part in the taxonomy. For a non-terminal node the labelling includes two values: the first one is defined by a categorical attribute value and the second one specifies the categorical attribute considered at this node. As an example, consider the level 3 of the taxonomic tree given in Fig. 3.3. From left to right, the first three nodes are terminal nodes and the two last nodes are non-terminal nodes.

A lateral ranking of leaves (terminal nodes) from left to right can be considered in the taxonomic structure. This ranking is given in the above pictured tree where there are in all 15 terminal nodes. Consider again the level 3 of the taxonomic tree. The extreme left node is a terminal node. It is labelled by the ordered pair $(1, 1)$. The first component is defined by the first value of the mother categorical attribute a^{21}. The second component is its lateral rank. The terminal node at the extreme left of level 4 is labelled $(1, 4)$. It corresponds to the value coded 1 of the categorical attribute a^{31} involved in its mother node. It is created directly after the terminal node $(3, 3)$ of level 3. Clearly, the lateral ranking depends on drawing options. At the 3 level, the terminal nodes are first placed. This could have been done for the level 2. The chosen alternative is due to legibility reason. In any case, a precise status could be defined for consistent and systematic technique of drawing the taxonomic tree associated with a taxonomic attribute. This is left to the reader as an exercise. In the following, each terminal node (leaf of the taxonomic tree) will be coded by its lateral rank. Notice that each terminal node can be identified with the associated complete chain of the taxonomic tree, starting with the root and ending with it.

Now, we shall associate bijectively a level labelled classification tree with the taxonomic structure defined by the taxonomic attribute (see (3.4.11) and the associated Fig. 3.3). This tree is established on the set $\mathcal{T}$ of the taxonomy terminal nodes. This is mathematically defined by a partition chain on $\mathcal{T}$ (see Chap. 1). Assume that k levels are comprised in the taxonomic structure ($k = 4$ for our example). The level 1 is the root level and the level k is determined by the terminal nodes corresponding to attributes introduced at the $(k-1)^{th}$ level. In our example these attributes are a^{31} and a^{32}. Denote by $(P_0, \ldots, P_{j-1}, P_j, \ldots, P_{k-1})$ the partition chain associated with the taxonomic structure. In order to specify this sequence of partitions, identify each non-terminal node with the subset of terminal nodes deriving from it. As an example, the node $(5, a^{32})$ is identified with the subset $\{10, 11, 12, 13, 14\}$. Besides, the node $(1, a^{21})$ is identified with the subset

$$\{1, 2, 3, 4, 5, 6, 7, 8, 9, 10, 11, 12, 13, 14\}.$$

P_0 is the finest partition. Each of its classes is a singleton including exactly one element. In our example P_0 comprises 15 classes. P_{k-1} is the least fine partition, it includes only one class grouping all elements. P_j is deduced from P_{j-1} by aggregating nodes (terminal or not) appearing at the the level $k - j + 1$ of the taxonomic structure and derived from a same categorical attribute attribute defined for a given node of the level $k - j$. In our example, the partition P_1 is obtained from the partition P_0 by aggregating $\{4\}, \{5\}, \{6\}, \{7\}, \{8\}, \{9\}$ on one side and by aggregating $\{10\}, \{11\}, \{12\}, \{13\}, \{14\}$ on the other side. Thus, we have

$$P_1 = \big\{\{1\}, \{2\}, \{3\}, \{4, 5, 6, 7, 8, 9\}, \{10, 11, 12, 13, 14\}, \{15\}\big\}$$

The partition P_2 into two classes is easily obtained

$$P_2 = \big\{\{1, 2, 3, 4, 5, 6, 7, 8, 9, 10, 11, 12, 13, 14\}, \{15\}\big\}$$

In Sect. 1.4.2 of Chap. 1, a bijective correspondence was established between a partition chain on a finite set E and a ultrametric distance on E given by a level notion of the classification tree associated with the partition chain. Here, equivalently, we shall define a bijective correspondence between a taxonomic structure (see (3.4.11)) and a ultrametric proximity on the set that denoted by T of its terminal nodes (see Fig. 3.3). Let us recall that this ultrametric proximity that we denote by p is characterized by the following property:

For any g, h and $i \in T$, $g \neq h$, $g \neq i$ and $h \neq i$
$$p(g, i) \geq \min[p(g, h), p(h, i)] \tag{3.4.12}$$

(see Chap. 1, Sect. 1.4.1).

Here, the node level is defined by 1 plus the number of branches joining the root of the taxonomic structure to the node concerned. As examples in Fig. 3.3, $level\big((4, a^{31})\big) = 3$ and $level\big((5, 8)\big) = 4$. For u and v belonging to T, $p(u, v)$ is defined by the highest node level of the taxonomic structure where u and v are aggregated. Thus, as an example relative to the figured taxonomic tree (see Fig. 3.3), $p(10, 14) = 3$ and $p(7, 13) = 2$; $p(10, 14) > p(7, 13)$.

The proof of the above condition (3.4.12) is analogous to that given for the ultrametric distance associated with a partition chain (see Chap. 1, Sect. 1.4.2). In fact, for a subset $\{x, y, z\}$ of three elements of T, suppose $p(g, h) > p(h, i)$: the highest aggregating level of g and h is strictly greater than that of h and i. Therefore, g and h belong to the same node—appearing at the level $p(g, h)$—excluding i. Otherwise $p(h, i) \leq p(g, h)$. Now, if $p(g, h) = p(h, i)$, the node highest level joining g and h is the same as that joining h and i. Hence and necessarily, the first node is identical to the second one, because both include h. Therefore, $p(g, h) = p(h, i) = p(g, i)$. This ends the proof of (3.4.12).

The proximity p function valuates a total preordonance on the set T of the terminal nodes; that is to say, a total preorder on the set

$$T_2 = \{(g, h) | g \in T, h \in T, 1 \leq g \leq h \leq |T|\} \qquad (3.4.13)$$

where $|T|$ is the cardinality of T ($|T| = 15$ in the above example).

This total preorder (ranking with ties) that we denote by $\omega(T_2)$ is established as follows

$$\text{(For any } (g, h) \text{ and } (k, l) \in T_2), (g, h) \leq (k, l) \text{ iff } p(g, h) \leq p(k, l) \qquad (3.4.14)$$

The total preordonance defined on T is ultrametric in the sense given in Sect. 1.4.3 of Chap. 1. More explicitly, if ρ is a ranking function on T_2 compatible with p (i.e. strictly increasing with respect to p), we have

$$\text{(For any } g, h, \text{ and } i \in T), \rho(g, h) \geq r \text{ and } \rho(h, i) \geq r \Rightarrow \rho(g, i) \geq r \qquad (3.4.15)$$

where r is an arbitrary positive integer.

We adopt for the ranking function ρ, the mean rank function r_m introduced above (see Eqs. (3.4.8) and (3.4.9)).

Finally, we represent a taxonomic attribute by a ultrametric preordonance on the set T of the terminal nodes of the associated taxonomic structure. This preordonance is numerically coded with the mean rank function defined on T_2 (see Eqs. (3.4.13) and (3.4.14)).

Now, let us illustrate how this representation is established in the framework of our example. Consider the diagram of Fig. 3.3 from down to top. The root node results from aggregating the two components of all ordered pairs of the form $(i, 15)$, $1 \leq i \leq 14$. Then, the first preorder class of $\omega(T_2)$ includes 14 pairs. Therefore, the common mean rank of each of them is $(14 + 1)/2 = 7.5$. The node $(1, a^{21})$ is obtained from three types of pair aggregations. These are

1. $\{(i, j) | 1 \leq i < j \leq 3\}$;
2. $\{(i, j) | 1 \leq i \leq 3, 10 \leq j \leq 14\}$;
3. $\{(i, j) | 1 \leq i \leq 9, 10 \leq j \leq 14\}$.

There are in all $3 + 3 \times 6 + 9 \times 5 = 66$ pairs. These constitute the second class of $\omega(T_2)$. In these conditions the mean rank of each of these pairs is $14 + (66 + 1)/2 = 47.5$.

The nodes $(4, a^{31})$ and $(5, a^{32})$ are constituted at level 3. The former aggregates the components of the set of pairs $\{(i, j) | 4 \leq i < j \leq 9\}$ and the latter those of $\{(i, j) | 10 \leq i < j \leq 14\}$. There are in all $((6 \times 5)/2) + ((5 \times 4)/2) = 25$ pairs. They constitute the third class preorder of $\omega(T_2)$. Consequently, the common mean rank of these pairs is $14 + 66 + (25 + 1)/2 = 93$. The fourth class preorder of $\omega(T_2)$ is defined by all ordered pairs of the form (i, i), $1 \leq i \leq 15$. The common mean rank assigned to each of these pairs is $14 + 66 + 25 + (15 + 1)/2 = 113$.

The expected value of the mean rank sum is easily obtained:

$$14 \times 7.5 + 66 \times 47.5 + 25 \times 93 + 15 \times 113 = 7260.$$

$\omega(\mathcal{T}_2)$ induces a total preorder on the set $O^{[2]}$ of unordered object pairs. In order to make explicit this total preorder (ranking with ties), denote $\mathcal{O}(i)$ the $\mathcal{O}$ subset defined by the value i of the taxonomic attribute, $1 \leq i \leq |\mathcal{T}|$. The set $\{\mathcal{O}(i) | 1 \leq i \leq |\mathcal{T}|\}$ defines a partition on $\mathcal{O}$. Now, for $i < j$, $1 \leq i < |\mathcal{T}|$, consider the unordered object pairs $\mathcal{O}(i) \star \mathcal{O}(j)$ (see what follows Eq. (3.3.10) for its definition) and substitute in $\omega(\mathcal{T}_2)$, $\mathcal{O}(i) \star \mathcal{O}(j)$ for (i, j). On the other hand, substitute in $\omega(\mathcal{T}_2)$, the unordered object pairs $P_2(\mathcal{O}(i))$ of $\mathcal{O}(i)$ for the ordered pair (i, i), $1 \leq i \leq |\mathcal{T}|$.

3.4.3 The Taxonomic Preordonance Attribute

Let us reconsider here the ordinal similarity structure provided by a taxonomic attribute τ organizing a set of logically dependent categorical attributes (see (3.4.11)). We further assume here that the category set denoted $\mathcal{C}(a^{pi})$ of a given categorical attribute a^{pi} is provided by a total preordonance (see Sect. 3.4.1), $1 \leq i \leq k_p$, $1 \leq p \leq q$. These preordonances are locally defined, attribute by attribute. They have to be integrated in the taxonomic structure. In these conditions, we have to build a total preordonance on the set of the taxonomy leaves, or, equivalently, on the set of the associated complete chains going from the root to the leaves. This preordonance must take into account both the preordonance defined by the taxonomic structure and those we have just mentioned.

Such preordonance is built step by step, decreasingly according to the taxonomic resemblance between terminal nodes (leaves of the taxonomy). The general principle consists of refining the ultrametric preordonance associated with the taxonomy (see Sect. 3.4.2) by means of the preordonances locally defined on the category sets of the different attributes.

In order to clarify this technique, let us begin by illustrating the refinement process in the framework of our example (see Fig. 3.3). By going from the deepest level (4 in our case) to the root 1, each refinement step concerns the terminal nodes aggregated at the first time at a given level. We begin by ordering the set

$$\Delta(\mathcal{T}) = \{(g, g) | g \in \mathcal{T}\} \tag{3.4.16}$$

according to the leaf depth in the taxonomy: in other words, the deeper the leaf the higher the ordinal similarity between the represented category and itself is. Thus, in the case of our example (see Fig. 3.3) we have

$$(4, 4) \sim (5, 5) \sim (6, 6) \sim (7, 7) \sim (8, 8) \sim (9, 9)$$
$$\sim (10, 10) \sim (11, 11) \sim (12, 12) \sim (13, 13) \sim (14, 14)$$
$$> (1, 1) \sim (2, 2) \sim (3, 3) > (15, 15)$$

By considering this ranking in a reverse manner, according to increasing ordinal similarity, the mean ranks assigned to the three preorder classes—that we can distinguish—are 106, 108 and 115, respectively. They replace the common rank 113 (see above). The sum rank concerned is preserved. In fact, $106 + 3 \times 108 + 11 \times 115 = 15 \times 113 = 1695$.

Now, let us consider the passage from level 4 to level 3. $A = \{4, 5, 6, 7, 8, 9\}$ and $B = \{10, 11, 12, 13, 14\}$ code the category sets of the attributes a^{31} and a^{32}, respectively. A and B are constituted by terminal nodes. According to the notations above (see Sects. 3.4.1 and 3.4.2), the set $P_2(A)$ (resp., $P_2(B)$) of unordered element pairs is defined by $\{(g, h) | 4 \leq g < h \leq 9\}$ (resp., $\{(g, h) | 10 \leq g < h \leq 14\}$). $P_2(A) \cup P_2(B)$ determines a unique class of the total preorder defined at the level 3 by the tree structure. This class comprises all the element pairs aggregated at the level 3 of the taxonomy. There are in all $card[P_2(A)] + card[P_2(B)] = 15 + 25$ pairs. Two preordonance structures on the category sets A and B of the attributes a^{31} and a^{32} provide total preorders on $P_2(A)$ and $P_2(B)$, respectively. These two total preorders must be employed consistently in order to define a unique total preorder on the entire set $P_2(A) \cup P_2(B)$. One option for this consists of requiring the expert knowledge for ranking the category pairs of A with respect to those of B. The option we adopt uses the mean rank functions, locally defined on $P_2(A)$ and $P_2(B)$ from the total preordonances on A and on B. These are interpreted as numerical similarity coefficients on A and on B, respectively. Denote r_A and r_B these mean rank functions. The value sum of r_A (rep., r_B) is $(15 \times 16)/2 = 120$ (rep., $(10 \times 11)/2 = 55$). In these conditions, we define a numerical function $r_{A \cup B}$ on $P_2(A) \cup P_2(B)$ directly deduced from r_A and r_B, as follows

$$r_{A \cup B} : P_2(A) \cup P_2(B) \rightarrow Val(r_A) \cup Val(r_B) \qquad (3.4.17)$$

where $Val(r_A)$ (resp., $Val(r_B)$) is the value set of r_A (resp., r_B) and where $r_{A \cup B}((g, h))$ is equal to $r_A((g, h))$ [resp., $r_B((g, h))$] if $\{g, h\} \in P_2(A)$ [resp., $P_2(B)$].

Therefore, according to the value scale of $r_{A \cup B}$, a total preorder on $P_2(A) \cup P_2(B)$ is established as follows

$$\forall \{g, h\} \text{ and } \{i, j\} \in P_2(A) \cup P_2(B),$$
$$\{g, h\} \leq \{i, j\} \text{ iff } r_{A \cup B}((g, h)) \leq r_{A \cup B}((i, j)) \qquad (3.4.18)$$

This total preorder is substituted for the unique class $P_2(A) \cup P_2(B)$. A new global mean ranking function refining the previous one is established on the set defined by this class. Recall that we had one common value 93 for all of its elements (see Sect. 3.4.2). The new assigned values belong necessarily to the interval $[14 + 66 + 1 + 81, 14 + 66 + 25]$.

The nature of the passage from level 3 to level 2 is somewhat different from the preceding one. All the nodes occupying the level 3 derive from only one mother node occupying the level 2, this being defined by the categorical attribute a^{21}. Recall that the unique preorder class of $\omega(\mathcal{T}_2)$ (see (3.4.13)) associated with the taxonomy is defined by the set of unordered leave pairs given in (3.4.16). For this, the mean rank of a given pair of such a preorder class is 47.5. The category set D of a^{21} is defined by $\{1, 2, 3, a^{31}, a^{32}\}$. A total preordonance on D enables us to determine a total preorder on the mentioned unique preorder class of $\omega(T_2)$. In order to build the latter, we have to do the following substitutions:

$$(\text{For any } g \in \{1, 2, 3\}), (g, 4) \leftarrow \{(g, h) \mid h \in A\}$$
$$(\text{For any } g \in \{1, 2, 3\}), (g, 5) \leftarrow \{(g, h) \mid h \in B\}$$
$$\text{for } \{4, 5\} \leftarrow \{(g, h) \mid (g, h) \in A \times B\} \tag{3.4.19}$$

where the different pairs included in a given class substitution are interpreted as equally similar, A and B have been defined above.

Notice that the new mean rank values in the class concerned of $\omega(T_2)$ are comprised between $14 + 1 = 15$ and $14 + 66 = 80$.

Now, let us give a general expression of the construction of a taxonomic preordonance attribute. For this purpose, we start with the Definition (3.4.11) of a taxonomic attribute τ. $\{a^{pi} \mid 1 \le i \le k_p\}$ is the set of the categories introduced at the pth level in a descendant way from the top to the bottom. Some of these categories define terminal nodes of the τ structure and some others define categorical attributes which are divided at the next $(p+1)^{th}$ level (see level $p = 3$ of the example). Denote $P_2[\mathcal{C}(a^{pi})]$ the set of unordered pairs of the category set $\mathcal{C}(a^{pi})$ of a^{pi}, $1 \le i \le k_p$. $\bigcup_i P_2[\mathcal{C}(a^{pi})]$ determines a new unique class—by going from down to top—of the taxonomic preordonance. This unique class is refined by means of the total preordonances defined on $\mathcal{C}(a^{pi})$, $1 \le i \le k_p$, respectively.

We begin by ordering the set $\Delta(T) = \{(g, g) \mid g \in T\}$ considered in (3.4.16), of category pairs of the form (g, g) where g is a terminal node (leaf) of the taxonomic structure (see above for the given example following (3.4.16)). Then and recursively, for $p = q$ to $p = 2$, the unique taxonomic class $\bigcup_i P_2[\mathcal{C}(a^{pi})]$ is refined. The following steps are needed for this refinement:

(i) Compute the local mean rank function r_m^i for the total preorder $P_2[\mathcal{C}(a^{pi})]$, defined from the preordonance attribute a^{pi};

(ii) Establish a global total preorder on $\bigcup_i P_2[\mathcal{C}(a^{pi})]$ compatible with the respective local mean ranking functions r_m^i, $1 \le i \le k_p$, that is to say

$$(\text{For any } (g, h) \in P_2[\mathcal{C}(a^{pi})], (i, j) \in P_2[\mathcal{C}(a^{pi'})]),$$
$$(g, h) \le (i, j) \text{ iff } r_m^i(g, h) \le r_m^{i'}(i, j) \tag{3.4.20}$$

(iii) Begin by associating the category set $C^t(a^{pi})$ corresponding to terminal nodes derived from a^{pi} with each category set $\mathcal{C}(a^{pi})$ and then, for any (g, h) belonging to $P_2[\mathcal{C}(a^{pi})]$ consider the subset $A(g)$ [resp., $A(h)$] of $C^t(a^{pi})$ defined by terminal nodes issued from g (resp., h). In these conditions, the set leaf pairs $A(g) \times A(h)$ is substituted for (g, h) (see (3.4.19)). All the pairs concerned are interpreted as equally similar and the mean rank function value $r_m^i(g, h)$ is applied to all of these pairs.

The set $\bigcup_{(g,h)} A(g) \times A(h)$ $((g, h) \in \bigcup_i P_2[\mathcal{C}(a^{pi})])$ of terminal node pairs constituted a unique class of the total preorder representing the taxonomic attribute. Now, this class is divided into subclasses depending on the respective preordonances defined on the category sets $\mathcal{C}(a^{pi})$ and the local mean rank functions r_m^i calculated for the respective total preorders on $P_2[\mathcal{C}(a^{pi})]$, $1 \leq i \leq k_p$.

When the process ends with $p = 2$, we obtain a total preorder on the entire set of terminal node pairs, comprising the pairs of $\Delta(\mathcal{T})$ (see (3.4.16)). This global preordonance is valuated by means of the mean rank function according to increasing ordinal similarity (see the example above).

3.4.4 Coding the Different Attributes in Terms of Preordonance or Similarity Categorical Attributes

We have just seen above (Sects. 3.4.2 and 3.4.3) how the taxonomic categorical attribute can be represented as a specific preordonance categorical attribute (see Sect. 3.4.1). By applying for each of them the mean rank function in order to code the associated total preorder on the category pairs (see (3.4.9)) we obtain categorical attributes whose value sets are evaluated by specific similarity measures (see Sect. 3.3.4), defined by the mean rank functions.

In fact, preordonance coding of the different attributes of types I or II can be considered. And this may contribute to enrich the scale value of the attribute concerned. Let us make clear this coding for the following cases: (i) *Boolean* (Sect. 3.2.1); (ii) *numerical* (Sect. 3.2.2); (iii) *nominal categorical* (Sect. 3.3.1); (iv) *ordinal categorical* (Sect. 3.3.2) and (v) *ranking* (Sect. 3.3.3).

3.4.4.1   Preordonance Coding of the Boolean Attribute

As expressed in Sect. 3.2.1, a binary categorical attribute is associated with a Boolean attribute a. The value set of the categorical attribute concerned is $\{a, \bar{a}\}$, where $\bar{a}$ is the negated Boolean attribute of a. Let us index by 1 (resp., 2) the category a (resp., $\bar{a}$) and consider the following set of ordered pairs

$$B_2 = \{(1, 2), (1, 1), (2, 2)\} \tag{3.4.21}$$

The preordonance is defined by a total preorder (ranking with ties) on B_2, and for this three alternatives can be envisaged:

1. $(1, 2) < (1, 1) \sim (2, 2)$;
2. $(1, 2) < (1, 1) < (2, 2)$;
3. $(1, 2) < (2, 2) < (1, 1)$.

There is no enrichment of the scale value in the first case. However, this representation is not equivalent to that given in Sect. 3.2.1.

Between 2 and 3 the chosen preordonance structure depends on the most significant category. If a (resp., $\bar{a}$) is the most important category according to expert knowledge, then the preordonance 3 (resp., 2) is adopted. For the example given in Sect. 3.1 the category $a = $ cirrhosis is more significant than $\bar{a} = $ non-cirrhosis. Therefore, the preordonance 3 is considered.

3.4.4.2 Preordonance Coding of the Numerical Attribute

Here we consider the presentation given in Sect. 3.2.2 from which the notations are retaken. In fact, the proposed representation is given by a categorical attribute valuated by a numerical similarity. In this case asymmetrical similarity is required. Denote by w the categorical attribute associated with the numerical attribute v (see Sect. 3.2.2). Consider the category set of w indexed by $\{1, 2, \ldots, i, \ldots, m\}$. Then, the ordered category pairs is represented by

$$C_2 = \{(i,j)|1 \leq i, j \leq m\} \tag{3.4.22}$$

it includes m^2 elements.

The valuation assigned to the ordered pair (i, j) is
$\xi(i,j) = x_{(j)} - x_{(i)}, 1 \leq i, j \leq m$ (see Sect. 3.2.2). It is asymmetrical and defined by the directed range—from left to right—of the interval $[x_{(i)}, x_{(j)}], 1 \leq i, j \leq m$. The empirical distribution of the valuation defined is

$$\{(\xi(i,j), f_i \times f_j)|1 \leq i, j \leq m\} \tag{3.4.23}$$

where f_i is defined in (3.2.8), $1 \leq i \leq m$.

As an example, consider the distribution given in (3.2.9). We have:

$$(\xi(2,5), f_2 \times f_5) = ((7.2 - 2.3), 0.1 \times 0.2) = (4.9, 0.02).$$

Notice that generally, a 0 value for $\xi(i, j)$ may occur for $i \neq j, 1 \leq i, j \leq m$. It occurs necessarily for $j = i, 1 \leq i \leq m$.

3.4.4.3 Preordonance Coding of the Nominal Categorical Attribute

The notations of Sect. 3.3.1 are taken up again here. $\{1, 2, \ldots, h, \ldots, k\}$ indexes the category set of a nominal categorical attribute c. We consider the ordered pairs of categories $\mathcal{K}_2 = \{(g, h)|1 \leq g \leq h \leq k\}$ (see (3.4.1)). If $g < h$, (g, h) designates the unordered pair of categories $\{g, h\}$ that we denote by gh (see (3.4.2)); if $g = h$, (h, h) denoted hh, is considered for the comparison between the category h with itself, $1 \leq g \leq h \leq k$.

In the established preordonance two distinct categories are considered as equally dissimilar. On the other hand, the similarity of a given category with itself is the same whatever is the category concerned. Consequently, the preordonance can be written as follows

$$12 \sim 13 \sim \cdots \sim 1k \sim 23 \sim 24 \sim \cdots \sim 2k$$
$$\sim \cdots \sim (k-1)k < 11 \sim 22 \sim \cdots \sim kk \qquad (3.4.24)$$

There are two classes for this total preorder on K_2. The first one comprises $k \times (k-1)/2$ elements and the second one, k elements. The corresponding mean ranks are $((k^2 - k + 2)/4)$ for one element of the first class and $(k^2 + 1)/2$, for one element of the second class. We verify easily that the mean rank sum is equal to $[(1/2) \times (k \times (k+1)/2)] \times [(k \times (k+1)/2) + 1]$.

Denote by $p = n \times (n-1)/2$ the cardinality of the set $P = \mathcal{O}^{\{2\}}$ of unordered object pairs of $\mathcal{O}$, the empirical statistical distribution of the categorical attribute whose value set is K_2 can be expressed as follows

$$\left\{ \left(gh, \frac{n_g \times n_h}{p} \right) | 1 \le g < h \le k \right\} \cup \left\{ hh, \frac{n_h \times (n_h - 1)}{2p} | 1 \le h \le k \right\} \quad (3.4.25)$$

(see (3.3.2) and following).

By denoting $r(\pi) = card[R(\pi)]$ and $s(\pi) = card[S(\pi)]$ (see (3.3.11)), the empirical statistical distribution of the mean rank function is given by

$$\left\{ \left(\frac{k^2 - k + 2}{4}, \frac{s(\pi)}{p} \right), \left(\frac{k^2 + 1}{2}, \frac{r(\pi)}{p} \right) \right\} \qquad (3.4.26)$$

Now, we shall illustrate the presentation above in the framework of the example given in Sect. 3.3.1, provided by Goodman and Kruskal (1954) [9]. Three categories c_1, c_2 and c_3 were defined in Sect. 3.3.1. The preordonance (3.4.24) becomes

$$12 \sim 13 \sim 23 < 11 \sim 22 \sim 33$$

The two mean rank values associated with the two preorder classes are 2 and 5. By recalling that $n_1 = 591$, $n_2 = 608$, $n_3 = 239$ and $p = 1033203$, the empirical statistical distribution (3.4.25) is instanciated as follows

$$\left\{ \left(12, \frac{591 \times 608}{1033203} = 0.3478 \right), \left(13, \frac{591 \times 239}{1033203} = 0.1367 \right), \right.$$
$$\left(23, \frac{608 \times 239}{1033203} = 0.1406 \right), \left(11, \frac{591 \times 590}{2066406} = 0.1687 \right),$$
$$\left. \left(22, \frac{608 \times 607}{2066406} = 0.1786 \right), \left(33, \frac{239 \times 238}{2066406} = 0.0275 \right) \right\}$$

We verify the value 1 for the sum of the relative frequencies. $r(\pi)$ and $s(\pi)$ are equal to 387314 and 645889, respectively (see the end of Sect. 3.3.1). Therefore, the empirical statistical distribution (3.4.26) of the mean rank function becomes

$$\{(2, 0.3749), (5, 0.6251)\}$$

3.4.4.4 Preordonance Coding of the Ordinal Categorical Attribute

Recall the notations of Sect. 3.3.2. The category set $\{c_h | 1 \leq h \leq k\}$ is indexed by $\{h | 1 \leq h \leq k\}$. The preordonance on the category set is defined by a total preorder on the set $H = \{(g, h) | 1 \leq g, h \leq k\}$ (see (3.4.10)) of all ordered category pairs. This will correspond to an asymmetrical ordinal similarity. More precisely, an ordered pair (g, h) is ranked according to the difference $h - g$ between the two integer codes h and g, $1 \leq g, h \leq k$. Therefore, a given class of the total preorder is defined by

$$D_e = \{(g, h) | 1 \leq g, h \leq k, h - g = e\} \tag{3.4.27}$$

where $1 - k \leq e \leq k - 1$. To each possible value of g corresponds a unique value of h. A value of h does exist if and only if

$$g \leq \min(k, k - e) \text{ and } g \geq \max(1, 1 - e)$$

In these conditions, the cardinality of D_e is

$$\min(k, k - e) - \max(1, 1 - e) + 1 \tag{3.4.28}$$

where the first (resp., the second) term correspond to the greatest (resp., the lowest) value of h. Therefore, the value of the mean rank function assigned to a current element of D_e is given by

$$\sum_{1-k \leq f \leq e-1} [\min(k, k - f) - \max(1, 1 - f) + 1]$$
$$+ \frac{1}{2}[\min(k, k - e) - \max(1, 1 - e) + 2] \tag{3.4.29}$$

Now, remind that $n \times (n - 1)$ is the cardinality of the set $\mathcal{O}^{[2]}$ of ordered distinct object pairs. For $e \neq 0$, the relative frequency of D_e is given by

$$\sum_{\{(g,h) \in D_e\}} \frac{n_g \times n_h}{n \times (n - 1)} \tag{3.4.30}$$

For $e = 0$, this relative frequency becomes

$$\sum_{1 \le g \le k} \frac{n_g \times (n_g - 1)}{n \times (n - 1)} \qquad (3.4.31)$$

(3.4.30) and (3.4.31) define the empirical statistical distribution of $h-g, 1 \le g, h \le k$.

In the previous treatment we have represented a given category c_h of an ordinal categorical attribute by the integer code $h, 1 \le h \le k$. And this is arbitrary. However, for comparing two ordered category pairs (c_g, c_h) and $(c_{g'}, c_{h'})$ of this type of attribute, it is intuitively acceptable to consider the ordinal dissimilarity between $(c_g$ and $c_h)$ strictly greater than that between $(c_{g'}$ and $c_{h'})$, if and only if $h - g > h' - g'$. Nevertheless, in the case where $h - g = h' - g' = e \ne 0$, assigning the same ordinal dissimilarity to (c_g, c_h) and to $(c_{g'}, c_{h'})$ may seem difficult to admit. One option might consist in refining the class D_e by associating the relative frequency $(n_g - n_h)/n = f_g - f_h$ with (g, h). Thus the ordinal dissimilarity on D_e becomes an increasing function of the latter relative frequency.

Now, let us illustrate this coding in our example considered in Sect. 3.3.2. The linear order defined for the three categories c_1, c_2 and c_3 is $c_1 < c_2 < c_3$. The total preorder on the set H of all ordered pairs $(g, h), 1 \le g, h \le 3$, is defined by

$$(3, 1) < (2, 1) \sim (3, 2) < (1, 1) \sim (2, 2) \sim (3, 3) <$$

$$(1, 2) \sim (2, 3) < (1, 3)$$

The associated mean rank values are 1, 4.5, 5, 7.5 and 9, respectively.

By recalling that $n_1 = 591, n_2 = 608$ and $n_3 = 239$, the second and the fourth classes can be refined as follows

$$(2, 1) < (3, 2) \text{ and } (2, 3) < (1, 2)$$

This, because, $n_1 - n_2 < n_2 - n_3$

Let us end this formal presentation by indicating that our focus in Chaps. 4 and 6 for comparing ordinal categorical attributes is the set theoretic representation given in Sect. 3.3.2.

3.4.4.5   Preordonance Coding of the Ranking Attribute

This coding can be interpreted as a particular case of the previous one. Each category is represented by only one object: $n_h = 1, 1 \le h \le k$. For this interpretation $k = n = card(\mathcal{O})$. In these conditions, the development above can be reconsidered. Two main approaches for building association coefficients between ranking attributes are derived from the two codings given in Sect. 3.3.3 and in this section, respectively (see Chap. 6).

3.5 Attribute Representations When Describing a Set $\mathcal{C}$ of Categories

3.5.1 Introduction

We have seen from the above sections that the basic data for describing an object set $\mathcal{O}$ by an attribute a is an ordered pair (a, o) where o is an element of $\mathcal{O}$. The value of a on o, denoted by $a(o)$ is unique. In this section the description concerns a set Γ categories. Let us designate by $\{C_1, C_2, \ldots, C_i, \ldots, C_I\}$ this set of categories:

$$\Gamma = \{C_1, C_2, \ldots, C_i, \ldots, C_I\} \tag{3.5.1}$$

Γ may be obtained from a nominal categorical attribute γ defined on a universe Ω of objects. We designate by Ω_i the subset of Ω constituted by the objects where C_i is *TRUE*: $\Omega_i = \gamma^{-1}(C_i)$, $1 \leq i \leq I$. For this description by an attribute a, the basic data is defined here by an ordered pair (a, C), where C belongs to Γ. Relative to a given ordered pair (a, C_i), $1 \leq i \leq I$, the description of C_i by a has necessarily a global and then, a statistical nature. To fix idea, but also because of effective reasons, a sample (set learning) $\mathcal{O}_i$, provided from Ω_i is substituted for C_i, $\mathcal{O}_i \subset \Omega_i$, $1 \leq i \leq I$. Nevertheless, cases may occur where the statistical distribution of a on C_i is directly estimated by expert knowledge, $1 \leq i \leq I$, Lebbe et al. (1987) [15].

For a given pair $(a, \mathcal{O})$ composed by a descriptive attribute a of a fixed type and by an object set $\mathcal{O}$ described by a we have previously (see the above sections) expressed the statistical distribution of a on $\mathcal{O}$. Here, instead of a single pair $(a, \mathcal{O})$, we have a sequence of such a pair, namely

$$\{(a, \mathcal{O}_i)|1 \leq i \leq I\} \tag{3.5.2}$$

where the different sets $\mathcal{O}_i$ are mutually disjoint.

Let us denote by D_a^i the statistical distribution of a on $\mathcal{O}_i$. On the other hand, define p_i as the relative frequency of the objects belonging to $\mathcal{O}_i$ with respect to all the objects belonging to the union of the $\mathcal{O}_i$. Explicitly,

$$p_i = \frac{card(\mathcal{O}_i)}{card(\bigcup_{1 \leq i' \leq I} \mathcal{O}_{i'})} \tag{3.5.3}$$

$1 \leq i \leq I$.

In these conditions, the data are defined here by the sequence

$$\{(D_a^i, p_i)|1 \leq i \leq I\} \tag{3.5.4}$$

Previously, we have observed that the statistical distribution of a descriptive attribute a on an object set $\mathcal{O}$, depends on the nature of the set theoretic representation of the attribute concerned. Consequently and in order to specify (3.5.4) for the different data description cases, we shall recall and highlight these statistical distributions. The attributes of type I (Boolean and numerical) will be considered in Sect. 3.5.2. The nominal and ordinal categorical attributes will be studied in Sect. 3.5.3 and finally, the preordonance attributes, in Sect. 3.5.4. In Sect. 3.5.5 the data table concept will be formalized by means of two relational systems: The Tarski system $\mathcal{T}$ and a statistical system that we designate by $\mathcal{S}$.

3.5.2   Attributes of Type I

Let us refer to Sect. 3.2. We take up again here the same notations. The attributes considered are the Boolean attributes (Sect. 3.2.1) and the numerical ones (Sect. 3.2.2). For the description of an object set $\mathcal{O}$ by a attribute, the relative frequency (proportion) $p(a) = n(a)/n$ has been defined and then, the sequence of distributions (3.5.4) becomes

$$\{(p_a^i, p_{i.})|1 \leq i \leq I)\} \tag{3.5.5}$$

where p_a^i is a proportion defined at the level of the object set $\mathcal{O}_{i.}$. More precisely, consider the subset $\mathcal{O}_{i.}(a)$ of $\mathcal{O}_{i.}$ where a is *TRUE* and denote by $n_{i.}$ and $n_{i.}(a)$ the cardinalities of $\mathcal{O}_{i.}$ and $\mathcal{O}_{i.}(a)$, respectively, we have:

$$p_a^i = n_{i.}(a)/n_{i.}, 1 \leq i \leq I.$$

Now, let us consider the case of a numerical attribute v whose distribution was expressed in the equation above (3.2.8). The distribution sequence can be written as

$$\{\{(x_{(l)}^i, f_{(l)}^i)|1 \leq i \leq I\}\} \tag{3.5.6}$$

where the distribution $(x_{(l)}^i, f_{(l)}^i)$ is defined at the level of the object set $\mathcal{O}_{i.}, 1 \leq i \leq I$.

3.5.3   Nominal or Ordinal Categorical Attributes

3.5.3.1   The Case of the Nominal Categorical Attribute

This case is particularly instructive. Different representation levels have been considered for this type of attribute (see Sect. 3.3.1). The most basic one is defined by (3.3.1). The latter gives a valuation at the level of the object set $\mathcal{O}$. This valuation assigns to each object o ($o \in \mathcal{O}$) a code representing the value $c(o)$ of the categorical attribute c. According to the notations of Sect. 3.3.1, we denote by $\{f_h^i|1 \leq h \leq k\}$ the statistical distribution of c on the set $\mathcal{O}_{i.}, 1 \leq i \leq I$. In other words, $f_h^i = n_{ih}/n_{i.}$ is the relative frequency (proportion) of objects from $\mathcal{O}_{i.}$ possessing the hth

category of the attribute c. As just defined above, $n_{i.} = card(\mathcal{O}_{i.})$. On the other hand, $n_{ih} = card(\mathcal{O}_{i.h})$ where $\mathcal{O}_{i.h} = c^{-1}(h) \cap \mathcal{O}_{i.}$, $1 \leq i \leq I$, $1 \leq h \leq k$. Thus, the distribution sequence (3.5.4) can be put in the following form

$$\{(\{f_h^i | 1 \leq h \leq k\}, p_{i.}) | 1 \leq i \leq I\} \tag{3.5.7}$$

Let us notice here that this data structure is exactly that addressed by *correspondence analysis and related methods* [1] (see Sect. 5.3.2.1 of Chap. 5 and its associated references). Using these techniques have meaning only when k and I are large enough. Now, let us consider the relational representation given by Eqs. (3.3.3)–(3.3.10). According to the notations of Sect. 3.4.4.3, $r(\pi)$ and $s(\pi)$ denote the cardinalities of the sets $R(\pi)$ and $S(\pi)$, respectively (see (3.3.11)). The statistical distribution of the indicator function ρ_π (see (3.3.7)) can be put in the following form

$$\left\{ \left(1, \frac{r(\pi)}{p}\right), \left(0, \frac{s(\pi)}{p}\right) \right\} \tag{3.5.8}$$

where $p = card(P) = n(n-1)/2$.

Let us denote by π^i the partition on $\mathcal{O}_{i.}$ induced by the categorical attribute c, namely

$$\pi^i = \{\mathcal{O}_{i.h} = c^{-1}(h) \cap \mathcal{O}_{i.} | 1 \leq h \leq k\},$$

$1 \leq i \leq I$.

$r(\pi^i)$ (resp., $s(\pi^i)$) designates the number of object pairs joined (resp., separated) by π^i, $1 \leq i \leq I$. Also, denote by $p^i = n_{i.} \times (n_{i.} - 1)/2$ the number of object pairs of $\mathcal{O}_{i.}$, $1 \leq i \leq I$. With these notations, the sequence of the statistical distributions (3.5.4) becomes

$$\left\{ \left(\left[\left(1, \frac{r(\pi^i)}{p^i}\right), \left(0, \frac{s(\pi^i)}{p^i}\right) \right], \frac{p^i}{sump} \right) | 1 \leq i \leq I \right\} \tag{3.5.9}$$

where $sump = \sum_{1 \leq i' \leq I} p^{i'}$.

In fact, the weighting $p^i/sump$ associated with the distribution D_c^i is obtained in a relative way from the cardinality of the set $\mathcal{O}_{i.}$ on which D_c^i is defined. However, this statistical distribution—defined by the previous equation—is too global in order to compare finely the different categories of Γ (see (3.5.1)), represented, respectively, by the sets $\mathcal{O}_{i.}$, $1 \leq i \leq I$. A discriminant representation has to take into account the decomposition of the sets $R(\pi)$ and $S(\pi)$ (see (3.3.10) and (3.3.11)). In these conditions, the sequence of the statistical distributions (3.5.4) becomes

$$\left\{ \left(\{f_{hh}^i | 1 \leq h \leq k\}, \{f_{gh}^i | 1 \leq g < h \leq k\}\}, \frac{p^i}{sump} \right) | 1 \leq i \leq I \right\} \tag{3.5.10}$$

where $f_{hh}^i = (n_{ih} \times (n_{ih} - 1))/(n \times (n - 1))$ and
$f_{gh}^i = (n_{ig} \times n_{ih})/(n \times (n - 1))$, $1 \le h \le k$, $1 \le g < h \le k$, $1 \le i \le I$. The relative
frequencies f_{hh}^i and f_{gh}^i are defined with respect to the set $\mathcal{O}_{i.}^{[2]}$ of unordered object
pairs.

In Sect. 3.4.4.3 a representation in terms of a preordonance attribute was proposed
for the nominal categorical attribute. This preordonance was coded by using the mean
rank function. The statistical distribution of this function (see (3.4.26)) refines that
(3.5.9). We obtain

$$\left\{ \left\{ \left(\frac{k^2 - k + 2}{4}, \frac{s(\pi^i)}{p^i} \right), \left(\frac{k^2 + 1}{2}, \frac{r(\pi^i)}{p^i} \right) \right\}, \frac{p^i}{sump} \, | 1 \le i \le I \right\} \qquad (3.5.11)$$

However, in spite of this refinement, comparing the categories C_i, $1 \le i \le I$ of
Γ, on the basis of such distributions, remains not discriminant enough. Finally, two
statistical relational representations have to be retained for this comparison (3.5.7)
and (3.5.10).

3.5.3.2   The Case of the Ordinal Categorical Attribute

The development of this case follows the same rationale as the preceding one. Assigning to each object of $\mathcal{O}$ a category (see (3.3.1)) provides a representation of the categorical attribute c at the level of $\mathcal{O}$. This representation is defined by a valuation
on $\mathcal{O}$. In these conditions, (3.5.7) can be reconsidered in its exact form. Therefore,
the linear order (3.3.13) on the category set $\mathcal{C} = \{c_1, c_2, \ldots, c_h, \ldots, c_k\}$ cannot be
taken into account.

Now, let us consider the relational representation of the categorical attribute c
(see Eqs. from (3.3.17) to (3.3.25)). Denote by $r(\omega^i) = card[R(\omega^i)]$, $e(\omega^i) =
card[E(\omega^i)]$ and $s(\omega^i) = card[S(\omega^i)]$ respectively, the cardinalities defined in
(3.3.21) and restricted to the set $\mathcal{O}_{i.}$. Then, denote $c_{i.}$ the cardinality of $\mathcal{O}_{i.}^{[2]}$ (set
of distinct object pairs from $\mathcal{O}_{i.}$), $c_{i.} = n_{i.} \times (n_{i.} - 1)$. By considering the $score_\omega$
function defined in (3.3.25), the sequence of the statistical distribution (3.5.4) can be
written

$$\left\{ \left(\left\{ \left(1, \frac{r(\omega^i)}{c_{i.}} \right), \left(0.5, \frac{e(\omega^i)}{c_{i.}} \right), \left(0, \frac{s(\omega^i)}{c_{i.}} \right) \right\}, \frac{c_{i.}}{sumc} \right) | 1 \le i \le I \right\} \qquad (3.5.12)$$

where $sumc = \sum_{1 \le i' \le I} c^{i'}$.

This sequence of distributions extends to the ordinal case that (3.5.9) considered
previously for the nominal case.

Now, let us consider the extension of the (3.5.10) expression in the ordinal case.
This expression is written here exactly in the same manner, namely

$$\left\{ \left(\{\{f^i_{hh}|1 \le h \le k\}, \{f^i_{gh}|1 \le g < h \le k\}\}, \frac{c^i}{sumc} \right) |1 \le i \le I \right\} \qquad (3.5.13)$$

where f^i_{hh} and f^i_{gh} have exactly the same meaning as for the nominal case; but here they derive from counting ordered pairs of distinct objects.

As an example and also as an exercise consider the table Table 3.3 below provided from Goodman and Kruskal (1954) [9]. These data have been already employed for illustration (see Sect. 3.3). Table 3.3 is a contingency table crossing two classifications. These are associated with two categorical attributes. In Sects. 3.3.1 and 3.3.2 we have considered the categorical attribute c: "Highest level of formal education of wife". Its values denoted c_1, c_2 and c_3 (see Sect. 3.3.1) index the columns of the contingency table (see Table 2.3). The rows of Table 3.3 are indexed by the values of the categorical attribute "Fertility planning status of couple". These values denoted D, C, B and A, are ranked from the lowest level D to the highest one A. Here, $I = 4$. On the other hand, $card(\mathcal{O}_{1.}) = 379$, $card(\mathcal{O}_{2.}) = 451$, $card(\mathcal{O}_{3.}) = 205$ and $card(\mathcal{O}_{4.}) = 403$.

Let us now illustrate the above distribution (3.5.12). We suppose as in Sect. 3.3.2, given the linear order $c_1 < c_2 < c_3$ for the values of the attribute c. For this illustration we shall give the contribution of $i = 2$ to the (3.5.12) expression. We have

$$r(\omega^2) = card[R(\omega^2)] = 168 \times 215 + 168 \times 68 + 215 \times 68 = 62164$$
$$s(\omega^2) = card[S(\omega^2)] = 215 \times 168 + 68 \times 168 + 68 \times 215 = 62164$$
$$e(\omega^2) = card[E(\omega^2)] = 168 \times 167 + 215 \times 214 + 68 \times 67 = 62164$$
$$c^2 = card(\mathcal{O}^{[2]}_{2.}) = 451 \times 450 = 202950$$
$$sumc = c^1 + c^2 + c^3 = 379 \times 378 + 451 \times 450 + 205 \times 204 = 388032$$
$$\frac{c^2}{sumc} = 0.5230$$
$$\frac{r(\omega^2)}{sumc} = 0.3063, \frac{s(\omega^2)}{sumc} = 0.3063 \text{ and } \frac{e(\omega^2)}{sumc} = 0.3874. \qquad (3.5.14)$$

Thus, the contribution of $i = 2$ to (3.5.12) is

$$(\{(1, 0.3036), (0.5, 0.3874), (0, 0.3063)\}, 0.5230)$$

The statistical distributions (3.5.9), (3.5.11) or (3.5.12) are too global for comparison purpose in the case of nominal or ordinal categorical attributes. The most accurate statistical distributions for comparing the categories of Γ ($\Gamma = \{A, B, C, D\}$) in the above example above), are (3.5.10) for the nominal case and (3.5.13) for the ordinal one. Illustrating these distributions in the framework of our example is left for the reader.

3.5.4 Ordinal (preordonance) or Numerical Similarity Categorical Attributes

In Sect. 3.4.1 the preordonance categorical attribute was defined. It was represented by using the mean rank function (see (3.4.8)) as a categorical attribute valuated by a specific numerical similarity. This type of attribute was presented in Sect. 3.3.4. We have Seen that the taxonomic categorical attribute (Sect. 3.4.3) is represented as a specific preordonance on the set of the leaves of the taxonomic tree. Consequently, it suffices, without any loss of generality, to give the form of the distribution sequence (3.5.4) in the case of a categorical attribute valuated by a numerical similarity. By reconsidering the notations of Sect. 3.3.4 with respect to the set of ordered distinct object pairs of $\mathcal{O}_{i.}$, $1 \leq i \leq I$, we obtain

$$\left\{ \left(\left\{ \{f_{hh}^i | 1 \leq h \leq k\}, \{f_{gh}^i | 1 \leq g \neq h \leq k\} \right\}, \frac{c^i}{sumc} \right) | 1 \leq i \leq I \right\} \qquad (3.5.15)$$

where $f_{hh}^i = (n_{ih} \times (n_{ih} - 1))/(n \times (n - 1))$ and
$f_{gh}^i = (n_{ig} \times n_{ih})/(n \times (n - 1))$, $1 \leq h \leq k$, $1 \leq g \neq h \leq k$, $1 \leq i \leq I$. c^i and $sumc$ have been defined above. The relative frequencies are computed for distinct object ordered pairs. More precisely, f_{hh}^i is the relative frequency in $\mathcal{O}_{i.}^{[2]}$ whose value is $\xi(c_h, c_h)$, $1 \leq h \leq k$, f_{gh}^i is the relative frequency in $\mathcal{O}_{i.}^{[2]}$ whose value is $\xi(c_g, c_h)$, $1 \leq g \neq h \leq k$.

Let us now illustrate the statistical distribution (3.5.14) in the case of our example (see Tables 3.2 and 3.3). Consider $i = 2$. We have $n_{i.} \times (n_{i.} - 1) = 451 \times 450 = 202950$.

$$f_{11}^2 = \frac{168 \times 167}{202950} = 0.1382 \ , f_{22}^2 = \frac{215 \times 214}{202950} = 0.2267$$

$$f_{33}^2 = \frac{68 \times 67}{202950} = 0.0224$$

$$f_{12}^2 = \frac{168 \times 215}{202950} = 0.1780 \ , f_{13}^2 = \frac{168 \times 68}{202950} = 0.0563$$

Table 3.3 Crossing between "Fertility planning" and "Highest level of formal education of wife"

$\mathcal{F} \backslash \mathcal{L}$	c_1	c_2	c_3	Row totals
D	223	122	34	379
C	168	215	68	451
B	90	80	35	205
A	110	191	102	403
Column totals	591	608	239	1438

$$f_{21}^2 = \frac{215 \times 168}{202950} = 0.1780 \, , f_{23}^2 = \frac{215 \times 68}{202950} = 0.0720$$

$$f_{31}^2 = \frac{68 \times 168}{202950} = 0.0563 \, , f_{32}^2 = \frac{68 \times 215}{202950} = 0.0720 \qquad (3.5.16)$$

3.5.5 The Data Table: A Tarski System $\mathcal{T}$ or a Statistical System $\mathcal{S}$

In Sects. 3.2–3.4, the description of a set $\mathcal{O}$ of objects by a descriptive attribute a has been defined. Relative to an ordered pair (a, o) constituted by a descriptive attribute a and by an object o, $a(o)$ designates the value of a on o. a is represented by a mapping of $\mathcal{O}$ onto a value scale $\mathcal{E}$. We assume that $\mathcal{E}$ is endowed with a relation r^a. This relation is unary in Sect. 3.2 (Boolean and numerical attributes), binary in Sect. 3.3 (nominal and ordinal categorical attributes, ranking attribute), binary on the set of object pairs in Sect. 3.4 (preordonance and taxonomic attributes). Thus, a very large range of attribute description in *combinatorial data analysis* and *machine learning* is covered. In addition, any arity of this relation can be handled [23] and [32]. r^a induces on $\mathcal{O}$ a relation that we denote by R^a. Thus, the description of $\mathcal{O}$ by a can be formalized by the ordered pair $(\mathcal{O}, R^a)$.

Generally, for the description of an object set $\mathcal{O}$ provided from a universe $\mathcal{U}$ of objects (see Sect. 3.1) the expert defines a vast set of descriptive attributes. Let us denote this set by $\mathcal{A} = \{a^j | 1 \leq j \leq p\}$ where p designates the number of attributes. Then, the very important notion of a data table T crossing an object set $\mathcal{O}$ with an attribute set $\mathcal{A}$ can be expressed (see Table 3.4). $\mathcal{O}$ that we denote by $\{o_i | 1 \leq i \leq n\}$ indexes the row set of T whereas, the attribute set $\mathcal{A}$ indexes the column set of T. A given cell is established at the intersection of the ith row and the jth column, it contains the value $a^j(o_i)$, $1 \leq i \leq n\}$, $1 \leq j \leq p$.

As said above, each descriptive attribute induces a relation on the object set $\mathcal{O}$. Let us designate by R^j the relation on $\mathcal{O}$ defined by the a^j attribute. Let us emphasize once again that this relation is induced from the relation which endows the value

Table 3.4 Data table T

$\mathcal{O} \backslash \mathcal{A}$	a^1	$\cdots$	a^j	$\cdots$	a^p
o_1	$a^1(o_1)$	$\cdots$	$a^j(o_1)$	$\cdots$	$a^p(o_1)$
$\vdots$	$\vdots$	$\vdots$	$\vdots$	$\vdots$	$\vdots$
o_i	$a^j(o_i)$	$\vdots$	$a^j(o_i)$	$\vdots$	$a^p(o_i)$
$\vdots$	$\vdots$	$\vdots$	$\vdots$	$\vdots$	$\vdots$
o_n	$a^1(o_n)$	$\cdots$	$a^j(o_n)$	$\cdots$	$a^p(o_n)$

scale of a^j, $1 \leq j \leq p$. Hence, we associate with the data table T the Tarski system Tarski (1954) [47].

$$T = (\mathcal{O}; R^1, R^2, \ldots, R^j, \ldots, R^p) \qquad (3.5.17)$$

where the different relations R^j do not necessarily have the same arity, $1 \leq j \leq p$.

In *clustering* there are two dual problems. The first one which is the most familiar and mostly the only considered consists of organizing by proximity into a classification structure the set $\mathcal{O}$ of objects. The proximity notion concerned is global with respect to the different relations R^j, $1 \leq j \leq p$, by integrating all of them. For a given relation R^j and two objects o_i and $o_{i'}$, the higher the proximity between o_i and $o_{i'}$, the more linked they are with respect to the relation R^j, $1 \leq j \leq p$, $1 \leq i < i' \leq n$ (see Chap. 7).

The second problem is dual of the first one. It consists of organizing by proximity into a classification scheme the set $\mathcal{A}$ of descriptive attributes; that is, and according to our formalism, the set $\{R^j | 1 \leq j \leq p\}$ of relations on $\mathcal{O}$, associated with the descriptive attributes. A fundamental task consists of building a proximity notion between two relations R^j and R^k on $\mathcal{O}$, $1 \leq j \leq p$. A numerical version of this gives rise to the notion of association coefficient between descriptive attributes (see Chaps. 4–6). A complementary and very instructive clustering analysis consists of organizing the whole set of the categories taking part in the different categorical attributes (Table 3.5).

For a classical data table indexed by the cartesian product $\mathcal{A} \times \mathcal{O}$, where $\mathcal{A}$ is the descriptive attribute set and $\mathcal{O}$ the object set described, a notion of "symbolic object" is introduced in [2]. Such a "symbolic object" is in fact defined by an area in the description space associated with $\mathcal{A}$. This zone, which is set *a priori* by the expert, can be expressed by the well-known notion of a Boolean attribute. The latter is present on a given object o ($o \in \mathcal{O}$), if and only if the description of o falls inside the zone concerned. Thereby, a set of "symbolic objects" can be liken to a set of Boolean attributes; each of the latter being sustained by a complex description. This complexity makes very difficult to recognize attribute clusters based on an association coefficient between attributes (see Sect. 4.2.1 of Chap. 4 and Sect. 5.2 of Chap. 5). On the other hand, cluster analysis of the object set $\mathcal{O}$, described by such boolean attributes (see Sect. 4.2.1 of Chap. 4 and Sect. 7.2.2 of Chap. 7) is very

Table 3.5 Data table S

$\Gamma \backslash \mathcal{A}$	a^1	$\cdots$	a^j	$\cdots$	a^p
C_1	$D^1(a^1)$	$\cdots$	$D^1(a^j)$	$\cdots$	$D^1(a^p)$
$\vdots$	$\vdots$	$\vdots$	$\vdots$	$\vdots$	$\vdots$
C_i	$D^i(a^j)$	$\vdots$	$D^i(a^j)$	$\vdots$	$D^i(a^p)$
$\vdots$	$\vdots$	$\vdots$	$\vdots$	$\vdots$	$\vdots$
C_I	$D^I(a^1)$	$\cdots$	$D^I(a^j)$	$\cdots$	$D^I(a^p)$

difficult to understand and to situate with respect to description by boolean attributes associated with "symbolic objects" (see Chap. 8).

The implicit principle of the *LLA* ascendant agglomerative hierarchical clustering method [22] (see Chaps. 5–7) is to cluster the set $\mathcal{A}$ of descriptive attributes before the set $\mathcal{O}$ of the described objects. An ultimate stage consists of interpreting by crossing techniques each clustering with respect to the other one (see Chap. 8). Condition of same arity for the different relations R^j, $1 \le j \le p$ is required in order to cluster them. In practice and mostly, this condition is brought back [20, 21, 36]. Moreover, statistical homogeneity in the description by the different attributes a^j, $1 \le j \le p$, is also an intuitive condition required. More explicitly and for example, for a description by nominal categorical attributes, the number of categories by attribute has to be of the same order.

In the *LLA* approach the two above requested conditions (same arity of the relations R^j ($1 \le j \le p$) and same statistical homogeneity of the different descriptive attributes) for clustering the set $\mathcal{A}$, are no more required for clustering the object set $\mathcal{O}$ (see Chap. 7).

We have presented above the description of a set $\Gamma = \{C_i | 1 \le i \le I\}$ of categories by an attribute interpreted as a relation on a set of objects. We have associated a learning set $\mathcal{O}_i$,—composed of objects belonging to the category C_i—with the category C_i, $1 \le i \le I$. If R denotes the relation endowing the value set of the attribute concerned, we have represented R by its statistical distribution on $\mathcal{O}_i$, $1 \le i \le I$. Thus Γ is represented by a sequence of statistical distributions whose general form is defined in (3.5.4) (see also (3.5.13)). Each distribution is computed for one category. It is weighted according to the relative frequency of the learning set that it represents.

Now, assume a set $\mathcal{A} = \{a^1, a^2, \ldots, a^j, \ldots, a^p, \}$ of p relational attributes. As expressed above, these define a sequence of p relations on each of the sets $\mathcal{O}_i$, $1 \le i \le I$. We denote by $\{R_i^1, R_i^2, \ldots, R_i^j, \ldots, R_i^p\}$ this sequence of relations on $\mathcal{O}_i$, $1 \le i \le I$. Notice that for category description, what is retained from R_i^j is its statistical distribution on $\mathcal{O}_i$, $1 \le j \le p$, $1 \le i \le I$ (see Sects. 3.5.1 to 3.5.4 when only one set is considered). Thus, we are led to define a system $\mathcal{S}$ as follows:

$$\mathcal{S} = \left(\Gamma; R^1, R^2, \ldots, R^j, \ldots, R^p\right) \tag{3.5.18}$$

where for each pair (C_i, R^j) $(C_i \in \Gamma)$, the statistical distribution of R^j on C_i is estimated on the basis of a learning set $\mathcal{O}_i$, $1 \le i \le I$, $1 \le j \le p$. In these conditions, the data table takes the form of Table 3.5.

As it was remarked above, the contingency table can be formalized by a system $\mathcal{S}_1 = (\Gamma, R^1)$ where only one relation is considered, the latter being defined by a nominal categorical attribute a^1. R^1 defines a partition on each of the sets $\mathcal{O}_i$, $1 \le i \le I$. Designate by $\mathcal{A}^1 = \{a_1^1, \ldots, a_g^1, \ldots, a_h^1, \ldots, a_k^1\}$ the category set of a^1. As said above, this data structure is the main matter of *correspondence analysis* [1] (see Sect. 5.3.2.1 of Chap. 5 and its associated references). In this approach, basically, the comparison between two categories γ_i and $\gamma_{i'}$ of Γ is evaluated from $\{|f_h^i - f_h^{i'}| | 1 \le h \le k\}$ where f_h^i (resp., $f_h^{i'}$) is the relative frequency in $\mathcal{O}_i$ (resp., $\mathcal{O}_{i'}$)

of objects having the category h, $1 \leq i < i' \leq I$. Let us suppose now a structure on the category set $\mathcal{C}^1$, given for example by a valuated binary relation (see Sect. 3.3.4). This structure cannot in anyway be taken into account in the mentioned approach. On the contrary, clustering approach as that given by the *LLA* method is able to integrate

$$\left\{ (val^1(g, h), f_g^i \times f_h^i) | 1 \leq g < h \leq k \right\}$$

and

$$\left\{ (val^1(g, h), f_g^{i'} \times f_h^{i'}) | 1 \leq g < h \leq k \right\} \tag{3.5.19}$$

As for the data defined by a Tarski system, it is matter for the data defined by a $\mathcal{S}$ system to organize by clustering the set Γ of categories and the set $\mathcal{A}$ of descriptive attributes. Clustering $\mathcal{A}$ consists of clustering the relation set $\{R^j | 1 \leq j \leq k\}$ on the basis of estimated statistical distributions on the different categories of Γ (see above). On the other hand and particularly for this structure, a fundamental problem for data analysis consists of clustering the whole set of the categories defined by the different attributes a^j, $1 \leq j \leq p$. Let us notice that we obtain the structure of a horizontal juxtaposition of contingency tables [33, 46], where the categorical attributes a^j, $1 \leq j \leq p$, are nominal.

To end, let us emphasize the importance of interpreting in a comparative manner a clustering of Γ, a clustering of $\mathcal{A}$ and also, a clustering of the set of categories taking part in the definition of $\mathcal{A}$. For this purpose specific tools are studied in Chaps. 4–7 and 9.

References

1. Benzécri, J.P.: L'analyse des données. tome II, Dunod (1973)
2. Billard, L., Diday, E.: Symbolic Data Analysis: Conceptual Statistics and Data Mining. Wiley, Hoboken (2006)
3. Celeux, G., Govaert, G.: Comparison of the mixture and the classification maximum likelihood in cluster analysis. J. Stat. Comput. Simul. **3–4**(47), 127–146 (1993)
4. Chah, S.: Critères de classification sur des données hétérogènes. Revue de Statistique Appliquée **33**(2), 19–36 (1985)
5. Fisher, W.D.: On grouping with maximum homogeneity. J. Am. Stat. Assoc. (53), 4–29 (1958)
6. Ghazzali, A., Léger, N., Lerman, I.C.: Rôle de la classification statistique dans la compression du signal image: panorama et une étude spécifique de cas. La Revue de Modulad. (14), 51–89 (1994)
7. Ghazzali, N.: Comparaison et réduction d' arbres de classification, en relation avec des problèmes de quantification en imagerie numérique. Ph.D. thesis, Université de Rennes 1, mai (1992)
8. Giakoumakis, V., Monjardet, B.: Coefficients d' accord entre deux préordres totaux. Mathématiques et Sciences Humaines. (98), 69–87 (1987)
9. Goodman, L.A., Kruskal, W.H.: Measures of association for cross classifications. J. Am. Stat. Assoc. (49), 732–764 (1954)
10. Guénoche, A., Monjardet, B.: Méthodes ordinales et combinatoires en analyse des données. Mathématiques et Sciences Humaines. (100), 5–47 (1987)

11. Hubert, L.J.: Assignment methods in combinatorial data analysis. Numerical Taxonomy. Marcel Dekker, New York (1987)
12. Kendall, M.G.: Rank correlation methods. Charles Griffin, 1st edn 1970 (1948)
13. Kerjean, A.M.: Tentative d'établissement de 100 typologies d'examens biologiques. Contribution à l'établissement du système A.D.M. Doctorat d'État. Ph.D. thesis, Université de Rennes 1 (1978)
14. Lafaye, J.Y.: Une méthode de discrétisation de variables continues. Revue de Statistique Appliquée, (27), 39–53 (1979)
15. Lebbe, J., Dedet, J.P., Vignes, R.: Identification assistée par ordinateur des phlébotomes de la Guyane Française. Publication Interne Version 1.02, Institut Pasteur de la Guyane Française, Juillet (1987)
16. Lerman, I.C.: Les bases de la classification automatique. Gauthier-Villars, Paris (1970)
17. Lerman, I.C.: Étude distributionnelle de statistiques de proximité entre structures finies de même type; application à la classification automatique. Cahiers du Bureau Universitaire de Recherche Opérationnelle. (19), 1–52 (1973)
18. Lerman, I.C.: Classification et analyse ordinale des données. Dunod. http://www.brclasssoc.org.uk/books/index.html (1981)
19. Lerman, I.C.: Construction d'un indice de similarité entre objets décrits par des variables d'un type quelconque. application au problème de consensus en classification. Revue de Statistique Appliquée **XXXV**(2), 39–60 (1987)
20. Lerman, I.C.: Conception et analyse de la forme limite d'une famille de coefficients statistiques d'association entre variables relationnelles, i. Revue Mathématique Informatique et Sciences Humaines (118), 35–52 (1992)
21. Lerman, I.C.: Conception et analyse de la forme limite d'une famille de coefficients statistiques d'association entre variables relationnelles, ii. Revue Mathématique Informatique et Sciences Humaines (119), 75–100 (1992)
22. Lerman, I.C.: Likelihood linkage analysis (lla) classification method (around an example treated by hand). Biochimie (75), 379–397 (1993)
23. Lerman, I.C.: Comparing classification tree structures: a special case of comparing q-ary relations. RAIRO-Ope. Res. (33), 339–365 (1999)
24. Lerman, I.C.: Comparing taxonomic data. Revue Mathématiques et Sciences Humaines(150), 37–51 (2000)
25. Lerman, I.C.: Analyse de la vraisemblance des liens relationnels une méthodologie d'analyse classificatoire des données. In Younès Benani and Emmanuel Viennet, editors, RNTI A3, Revue des Nouvelles Technologies de l'Information, pp. 93–126. Cèpaduès (2009)
26. Lerman, I.C., Guillaume, S.: Comparaison entre deux indices pour l'évaluation probabiliste discriminante des règles d'association. In Ali Khenchaf and Pascal Poncelet, editors, EGC'2011, RNTI E.20, pp. 647–656. Hermann (2011)
27. Lerman, I.C., Kuntz, P.: Directed binary hierarchies and directed ultrametrics. J. of Classif. (2011) (in press)
28. Lerman, I.C., Peter, Ph.: Organisation et consultation d'une banque de petites annonces à partir d'une méthode de classification hiérarchique en parallèle. In Data Analysis and Informatics IV, pp. 121–136. North Holland (1986)
29. Lerman, I.C., Peter, Ph.: Classification en présence de variables préordonnances taxonomiques à choix multiple. application à la structuration des phlébotomes de la Guyane Française. Publication Interne 426, IRISA-INRIA, Septembre (1988)
30. Lerman, I.C., Peter, Ph.: Indice probabiliste de vraisemblance du lien entre objets quelconques : analyse comparative entre deux approches. Revue de Statistique Appliquée, (LI(1)): 5–35 (2003)
31. Lerman, I.C., Peter, Ph: Representation of concept description by multivalued taxonomic preordonance variables. In: Cucumel, G., Brito, P., Bertrand, P., Carvalho, F. (eds.) Selected Contributions in Data Analysis and Classification, pp. 271–284. Springer, Berlin (2007)
32. Lerman, I.C., Rouxel, F.: Comparing classification tree structures: a special case of comparing q-ary relations ii. RAIRO-Oper. Res. (34), 251–281 (2000)

33. Lerman, I.C., Tallur, B.: Classification des éléments constitutifs d'une juxtaposition de tableaux de contingence. Revue de Statistique Appliquée (28), 5–28 (1980)
34. Marcotorchino, F.: Essai de typologie structurelle des indices de similarité vectoriels par unification relationnelle. In Younès Benani and Emmanuel Viennet, editors, RNTI A3, Revue des Nouvelles Technologies de l'Information, pp. 203–318. Cèpaduès (2009)
35. Marcotorchino, F., Michaud, P.: Optimisation en analyse ordinale des données. Masson (1979)
36. Ouali-Allah, M.: Analyse en préordonnance des données qualitatives. Application aux données numériques et symboliques. Ph.D. thesis, Université de Rennes 1, décembre (1991)
37. Peter, Ph.: Méthodes de classification hiérarchique et problèmes de structuration et de recherche d'informations assistée par ordinateur. Ph.D. thesis, Université de Rennes 1,mars (1987)
38. Rabaseda, S., Rakotomalala, R., Sebban, M.: Discretization of continuous attributes: a survey of methods. In Proceedings of the second Annual Joint Conference on Information Sciences, pp. 164–166 (1995)
39. Régnier, S.: Sur quelques aspects mathématiques des problèmes de la classification automatique. I.C.C. Bulletin (4), 175–191 (1965)
40. Schroeder, A.: Analyse d'un mélange de distributions de même type. Revue de Statistique Appliquée (24), 53–62 (1976)
41. Spearman, C.: The proof and measurement of association between two things. Am. J. Psychol. **15**(1), 72–101 (1904)
42. Stevens, S.S.: Mathematics, measurement and psychophysics. In Stevens, S.S. (ed.), Handbook of Experimental Psychology, pp. 1–49. New York, Wiley (1951)
43. Suppes, P., Zinnes, J.L.: Basic measurement theory. In: Bush, R.R., et al., Luce, R.D., Galanter, E.H. (eds.) Handbook of Mathematical Psychology, pp. 3–76. Wiley, New York (1951)
44. Sutcliffe, J.P.: Concept, class, and category in the tradition of aristotle. Categories and Concepts: Theoretical News and Inductive Data Analysis, pp. 35–65. Academic Press, London (1992)
45. Symons, M.J.: Clustering criteria and multivariate normal mixture. Biometrics (37), 35–43 (1981)
46. Tallur, B.: *Contribution à l'analyse exploratoire de tableaux de contingence par la classification, Doctorat dÉtat*. Ph.D. thesis, Université de Rennes 1, (1988)
47. Tarski, A.: Contribution to the theory of models. Indag. Math. (16), 572–588 (1954)
48. Ward, J.H.: Hierarchical grouping to optimize an objective function. J. Am. Stat. Assoc. (58), 236–244 (1963)

Chapter 4
Ordinal and Metrical Analysis
of the *Resemblance Notion*

4.1 Introduction

The concept of resemblance between data units (objects, categories or attributes) is the most important element in *data analysis* and *machine learning*. In Chap. 3, the description of a set of objects $\mathcal{O}$ (resp., categories $\mathcal{C}$) by a set of descriptive attributes $\mathcal{A}$ is formalized and a mathematical representation of this description is established. This will intimately influence the construction of the notion of resemblance, according to the elements to be compared (objects, attributes or categories). Let us denote by E the set of elements to be mutually compared. Mostly, the notion of resemblance in E is formalized by means of a numerical function of the Cartesian product $E \times E$ onto the reals, which we denote by $\mathcal{S}$. For any (x, y) belonging to $E \times E$, $\mathcal{S}(x, y)$ is a real number quantifying the extent to which x and y resemble one another. There are different synonymous names for such a function. Index (resp., measure or coefficient) of similarity (resp., of association) are expressions employed. In the development below, all of the six expressions "Similarity index", "Similarity measure", "Similarity coefficient", "Association index" "Association measure" and "Association coefficient" will be used. However and mainly, "Similarity index" or "Similarity measure" are more concerned for comparing objects or categories and "Association coefficient", for comparing descriptive attributes. The similarity function $\mathcal{S}(x, y)$ is mostly symmetrical. That is, for all (x, y) belonging to $E \times E$, $\mathcal{S}(x, y) = \mathcal{S}(y, x)$. However, there exist situations and problems requiring an asymmetrical association coefficient between descriptive attributes or an asymmetrical similarity index between objects (see Sect. 4.2.1.3). In this case, generally, $\mathcal{S}(x, y) \neq \mathcal{S}(y, x)$ for $(x, y) \in E \times E$.

The set E on which a similarity function has to be defined is generally represented by a subset (possibly weighted) of a geometrical or combinatorial space. Whatever the nature of this space, there is a large range of similarity indices (resp., association coefficients) proposed in the literature. In these conditions, a natural question concerns the stability of the results obtained by a given clustering technique when the similarity index (resp., association coefficient) involved is replaced by another

© Springer-Verlag London 2016
I.C. Lerman, *Foundations and Methods in Combinatorial and Statistical Data Analysis and Clustering*, Advanced Information and Knowledge Processing, DOI 10.1007/978-1-4471-6793-8_4

one. In practice, these results could change. Therefore and in order to obtain some invariance properties in relation to the choice of a numerical similarity S on E, an ordinal similarity could be substituted for it. The latter is defined by a total preorder (i.e. ranking with ties) on the set denoted by $F = P_2(E)$[1] of unordered element pairs of E, *compatible* with S. More precisely, there are two equivalent forms of this compatible total preorder denoted by ω and ϖ in the following:

$$\forall (p, q) \in F \times F, \, p \leq (\omega_S)q \Leftrightarrow S(p) \geq S(q)$$
$$\forall (p, q) \in F \times F, \, p \leq (\varpi_S)q \Leftrightarrow S(p) \leq S(q) \qquad (4.1.1)$$

ω_S (resp., ϖ_S) is called *preordonance* on E associated with the similarity S. This notion is the same as that expressed in Sect. 3.4.1 where it was defined by an ordinal similarity between the values of a categorical attribute. Here it is associated with a numerical similarity index on E. Depending on the context and for intuitive reasons, ω_S or ϖ_S will be used below. By substituting the associated preordonance for a numerical similarity, we may obtain a greater stability when a similarity index on E is replaced by another one. This problem is addressed in Sect. 4.2.2 in the context of Boolean data. More precisely, in this analysis the data structure is defined by a data table T (see Chap. 3, Sect. 3.5.4 and Table 3.4 where the attribute set $\mathcal{A}$ is constituted by Boolean attributes. Thus, the data table is indexed by the Cartesian product $\mathcal{O} \times \mathcal{A}$ where $\mathcal{O}$ denotes the described object set. In Chap. 3, $\mathcal{O}$ and $\mathcal{A}$ were designated as follows:

$$\mathcal{O} = \{o_1, o_2, \ldots, o_n\} \quad \text{and} \quad \mathcal{A} = \{a^1, a^2, \ldots, a^n\}$$

When the descriptive attributes are Boolean, the data table T has a perfect formal symmetry. Indeed, a given attribute a (resp., object o) can be faithfully represented by the subset $\mathcal{O}(a)$ of $\mathcal{O}$ (resp., $\mathcal{A}(o)$ of $\mathcal{A}$) where a is $TRUE$ (resp., which are $TRUE$ on o). We can also say $\mathcal{O}(a)$ is the $\mathcal{O}$ subset where a is present and $\mathcal{A}(o)$ is the $\mathcal{A}$ subset constituted by the attributes that are present on o. Therefore, the attribute set $\mathcal{A}$ is interpreted as a collection of $\mathcal{O}$ subsets. Notice that a given subset may be repeated in this collection. Similarly, the object set $\mathcal{O}$ is represented by a collection of $\mathcal{A}$ subsets. The underlying condition for this representation is that all of the attributes must be considered as of equal importance for the construction of a similarity index on $\mathcal{O}$. Many indices proposed admit this principle. However, semantically and then statistically, the meaning of a descriptive attribute is not the same as that of a described object. Indeed, all the objects generally have to be considered equally for the building of an association coefficient between descriptive attributes. On the contrary, it might be required not to attribute the same importance to the different descriptive attributes in order to establish a similarity index between objects. Indeed, a descriptive attribute a can be more or less statistically discriminant for estimating resemblance between objects. In this framework, the discriminant property of a Boolean attribute will depend on its relative frequency with respect to the observed object set $\mathcal{O}$ (see Sects. 3.2.1 and 7.2.2).

---

[1] $F = P_2(E) = \{\{x, y\} | x \in E, y \in E, x \neq y\}$.

In any case, in order to study the influence of a given similarity index on the associated preordonance in the case of Boolean data, we take for E (see above) the object set $\mathcal{O}$ and we adopt the principle of an equal importance of the different attributes of $\mathcal{A}$. The reason is first historical. Indeed was principally naturalists and life scientists who became interested in the classification of live animals and plants. Descriptive features were established for this purpose and equal weights were given to the different descriptive features. Different similarity indices were proposed. Some of them will be presented in Sect. 4.2.1.3. This view of the equal importance a priori of the different descriptive attributes is still held by many researchers in *numerical taxonomy* [33]. They claim that the relative importance of a given descriptive attribute has to be the a posteriori result of a data classification. In the case of Boolean data an axiomatic definition of a similarity index between objects is given in Sect. 4.2.1.2. This is based on the set theoretic representation of Boolean data, mentioned above. It covers all of the proposed indices. In addition, it enables us to study the influence of a given similarity index on its associated preordonance (see Sects. 4.2.1 and 4.2.2). A list of classic and representative indices is given in Sect. 4.2.1.3. Many other indices might be added to this list. However, this is not an objective in our development. The extension of the axiomatic definition to the numerical description of objects is given in a consistent way (Sect. 4.2.1.3). Conversely, fundamental indices established for comparing objects described by numerical attributes and representable in a geometrical space, are transposed for a description in a Boolean space (Sect. 4.2.1.3).

If the number of attributes present on a given object is constant in the object set $\mathcal{O}$, then all the similarity indices are equivalent with respect to the associated preordonance on $\mathcal{O}$ (i.e. determine the same compatible preordonance) (see Proposition 32). More generally, if this number of attributes present in a given object has a weak variance on $\mathcal{O}$, then the variation of the associated preordonance with a given similarity index is weak, when this index is replaced by another one. This variation is evaluated by a metric on the total preorders on the set $P = P_2(\mathcal{O})$ of unordered distinct object pairs of $\mathcal{O}$ (see Proposition 33). These results are important for data analysis of a population described by a questionnaire where a set of Boolean attributes is associated with each question, these attributes corresponding to the different categories of the categorical attribute defined by the question concerned.

As expressed above these results hold also in the case of association coefficients between descriptive Boolean attributes. However, in the case where a given Boolean attribute a ($a \in \mathcal{A}$) is represented by the subset of objects $\mathcal{O}(a)$ ($\mathcal{O}(a) \subset \mathcal{O}$) (see above), the variance of the number of objects where a is present ($\mathcal{O}(a)$) may be large. A theoretical calculation (see Sect. 4.2.2.2) and more importantly, a real example [20] show this property. In these conditions, the problem of building a relevant numerical association coefficient on a set $\mathcal{A}$ of Boolean attributes, remains. Nevertheless, the above stability study implicitly indicates that we should neutralize the effects of the cardinalities $n(a) = card[\mathcal{O}(a)]$ and $n(b) = card[\mathcal{O}(b)]$ in building an association coefficient between the two Boolean attributes a and b.

The association coefficients which have been considered for comparing Boolean attributes are directly derived from those proposed for comparing objects described by Boolean attributes (see Table 4.3 and Sect. 4.2.1.3). For a given pair (a, b) of Boolean attributes and referring to the basic index $s(a, b) = n(a \wedge b) = card[\mathcal{O}(a) \cap \mathcal{O}(b)]$, these different coefficients may be intuitively interpreted as deduced from different fashions of normalizing statistically this index. Indeed, $s(a, b)$ has a natural tendency to be "large" (resp., "small") if $n(a) = card(\mathcal{O}(a))$ and $n(b) = card(\mathcal{O}(b))$ are "large" (resp., "small") and this, independently of the relative positions of $\mathcal{O}(a)$ and $\mathcal{O}(b)$.

The association coefficients between Boolean attributes, obtained by transposition of similarity measures between objects described by Boolean attributes, are symmetrical. They are employed in the *data mining* field as *interestingness measures for association rules*. For a given ordered pair (a, b) of Boolean attributes, such a measure is asked to evaluate the implicative tendency $a \rightarrow b$, for which when a is *TRUE* on a given object x, then b is *generally*, but not absolutely, *TRUE* on x. For such a measure, an asymmetrical index seems to be more adequate. In Sect. 4.2.1.3, techniques for deriving asymmetrical measures from symmetrical ones, are proposed.

Taking into account the set theoretic representation of a categorical attribute of any type (see Chap. 3), the indices mentioned above (see Table 4.3 and Sect. 4.2.1.3) can be generalized for comparing categorical attributes of type II (see Chap. 3, Sect. 3.3). This generalization is carried out in Sect. 4.3. Thus, the famous Kendall τ_b index and the Goodman and Kruskal γ index, are clearly expressed and situated each with respect to the other in this new framework.

4.2   Formal Analysis in the Case of a Description of an Object Set $\mathcal{O}$ by Attributes of Type I; Extensions

4.2.1   Similarity Index in the Case of Boolean Data

4.2.1.1   Preliminaries

As defined in Sect. 3.2, type I attribute includes the Boolean and the numerical attributes. Let us remind that a Boolean attribute is represented by a subset of of the described object set $\mathcal{O}$ (see (3.2.2)). On the other hand, a numerical attributes is defined by a numerical valuation on $\mathcal{O}$ (see (3.2.6)). In this section we address the problem of defining a numerical similarity index on a set $\mathcal{O}$ of objects described by a set $\mathcal{A}$ of Boolean attributes. Classical and representative indices will be reported in Sect. 4.2.1.3. Their versions for comparing objects by numerical attributes will be defined in a consistent manner. On the other hand, extensions of these indices to the asymmetrical comparison between Boolean attributes will be proposed. Let us take again here the notations already introduced just above in Sect. 4.1:

$$\mathcal{O} = \{o_1, o_2, \ldots, o_n\} \quad \text{and} \quad \mathcal{A} = \{a^1, a^2, \ldots, a^n\}$$

According to the principle of a priori equal importance of the descriptive attributes, a given object o belonging to $\mathcal{O}$ will be represented by the subset $\mathcal{A}(o)$ constituted by the Boolean attributes which are $TRUE$ on o. As expressed in the General Introduction above, the formal symmetry of an incidence data table entails that the study holds for the dual case of comparing Boolean attributes.

Let us denote by x and y two given elements of the object set $\mathcal{O}$ and let $\mathcal{A}(x)$ and $\mathcal{A}(y)$ be the two subsets of the Boolean attribute set $\mathcal{A}$ representing x and y. As just defined $\mathcal{A}(x)$ (resp., $\mathcal{A}(y)$) is formed by all the Boolean attributes which are $TRUE$ on x (resp., y). Now, let us introduce the indicator functions ξ and η of the $\mathcal{A}$ subsets $\mathcal{A}(x)$ and $\mathcal{A}(y)$, respectively:

$$(\forall j, 1 \leq j \leq p)\xi(a^j) = 1(\text{resp.}, 0)$$
$$\text{if } a^j \text{ is } TRUE(\text{resp.}, FALSE) \text{ on } x$$
$$(\forall j, 1 \leq j \leq p)\eta(a^j) = 1(\text{resp.}, 0)$$
$$\text{if } a^j \text{ is } TRUE(\text{resp.}, FALSE) \text{ on } y \tag{4.2.1}$$

We consider a vectorial representation of ξ and η in the Cartesian space $\{0, 1\}^p$. More precisely, the Boolean vectors

$$(x_1, x_2, \ldots, x_j, \ldots, x_p) \quad \text{and} \quad (y_1, y_2, \ldots, y_j, \ldots, y_p)$$

denoted by $\alpha(x)$ and $\alpha(y)$ respectively, and where $x_j = \xi(a^j)$ and $y_j = \eta(a^j)$, $1 \leq j \leq p$, are associated with x and y. $\alpha(x)$ and $\alpha(y)$ are the representing points of x and y in the space $\{0, 1\}^p$. Let us indicate here that some of the below developments require a numerical interpretation of the values $x_j = \xi(a^j)$ and $y_j = \eta(a^j)$ (see Sect. 4.2.1.3).

Let us now designate by X and Y the subsets $\mathcal{A}(x)$ and $\mathcal{A}(y)$, respectively, and denote by X^c and Y^c the complementary subsets of X and Y in $\mathcal{A}$. In these conditions, introduce the following subsets

$$X \cap Y, \quad X \cap Y^c, \quad X^c \cap Y \quad \text{and} \quad X^c \cap Y^c \tag{4.2.2}$$

These define the relative cardinal positions of X with respect to Y, and vice versa.

4.2.1.2 Axiomatic Definition of a Symmetrical Similarity Index in the Case of Boolean Data

Let us take up again the vectors

$$\alpha(x) = (x_1, x_2, \ldots, x_j, \ldots, x_p) \quad \text{and} \quad (y_1, y_2, \ldots, y_j, \ldots, y_p)$$

representing the two objects to be compared. For a given attribute a^j, $1 \leq j \leq p$, x and y are said to have a *positive* (resp., *negative*) association if a^j is present (resp., absent) in both objects; that is equivalent to $x_j = 1$ and $y_j = 1$ (resp., $x_j = 0$ and $y_j = 0$), $1 \leq j \leq p$ (Fig. 4.1).

In this section the required conditions for a similarity index between two given objects are the following:

- The contribution of a positive association to the similarity measure is the *same* for each attribute of $\mathcal{A}$;
- The contribution of a negative association to the similarity measure is the *same* for each attribute of $\mathcal{A}$, in addition, it is *lower or equal* than the contribution of a positive association.

By considering the subsets defined in (4.2.2) let us introduce the following cardinalities:

$$s(x, y) = card(X \cap Y) = \sum_{1 \leq j \leq p} x_j \cdot y_j$$

$$t(x, y) = card(X^c \cap Y^c) = \sum_{1 \leq j \leq p} (1 - x_j) \cdot (1 - y_j)$$

$$u(x, y) = card(X \cap Y^c) = \sum_{1 \leq j \leq p} x_j \cdot (1 - y_j)$$

$$v(x, y) = card(X^c \cap Y) = \sum_{1 \leq j \leq p} (1 - x_j) \cdot y_j \qquad (4.2.3)$$

Fig. 4.1 Venn diagram

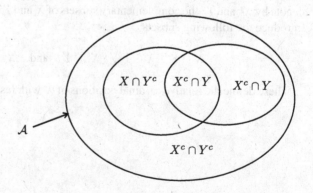

where, for any j, $1 \leq j \leq p$, the values 1 and 0 for x_j and y_j are interpreted numerically and where the symbol "$\cdot$" is defined by the numerical product.

s (resp., t) is the cardinal of the subset of attributes jointly present (resp., absent) in x and y.

u (resp., v) is the cardinal of the subset of attributes present in x (resp., y) and absent from y (resp., x).

We have:

$$s + u = card(X), s + v = card(Y), t + v = card(X^c), t + u = card(Y^c)$$

The symmetrical difference between X and Y is $(X \cap Y^c) + (X^c \cap Y)$. In this expression $+$ is a set sum defined by an union of disjoint subsets. The cardinality of the symmetrical difference between X and Y is $u + v$. On the other hand, we have: $s + u + v + t = card(A) = p$.

Definition 13 Let $I(x, y) = (s(x, y), u(x, y), v(x, y))$ be the triplet associated with a given object pair $\{x, y\}$ (see (4.2.3)). This triplet is called *Indicator* of $\{x, y\}$.

$I(x, y)$ is in our definition below the basis of building a similarity index (measure) between objects. Let us illustrate the values of this function in the following example.

The following data table describes an object set $\mathcal{O}$ including 6 elements $\mathcal{O} = \{o_a, o_b, o_c, o_d, o_e, o_f\}$ by an attribute set $\mathcal{A} = \{a^j | 1 \leq j \leq 18\}$ comprising 18 Boolean attributes. $\mathcal{O}$ (resp., $\mathcal{A}$) indexes the row set (resp., column set). A value 1 (resp., 0) in a given entry indicates the presence (resp., absence) of the column attribute in the row object concerned.

Now, let us consider the matrix of the indicator function $I(x, y)$, accompanied with the $u(x, y) + v(x, y)$ function on the attribute set corresponding to the incidence data in Table 4.1. We obtain (Table 4.2)

Definition 14 A similarity measure (or index) on $\mathcal{O}$ is a real positive function $\mathcal{S}$ on $\mathcal{O} \times \mathcal{O}$ which takes the following form:

$$\forall (x, y) \in \mathcal{O} \times \mathcal{O}, (x, y) \mapsto \mathcal{S}(x, y) = \mathcal{S}(I(x, y)) = \mathcal{S}(s, u, v)$$

where $\mathcal{S}(s, u, v)$ is a function defined on the subset $T = \{(s, u, v) | s + u + v \leq p\}$ of $\mathbb{N}^3$ ($\mathbb{N}$ is the integer set), such that the following properties are satisfied

1. For given u and v, $\mathcal{S}(s, u, v)$ increases with respect to s;
2. $\mathcal{S}(s, u, v)$ is symmetrical with respect to u and v:

$$(\forall (s, u, v) \in T), \mathcal{S}(s, u, v) = \mathcal{S}(s, v, u)$$

3. $\mathcal{S}(s, u, v)$ is increasing with respect to s and decreasing with respect to u. The increase with respect to s or (non-exclusively) the decrease with respect to u is necessarily strict.

Table 4.1 Incidence data table

$\mathcal{O}\backslash\mathcal{A}$	a^1	a^2	a^3	a^4	a^5	a^6	a^7	a^8	a^9
o_a	0	1	0	1	1	0	0	1	0
o_b	0	0	0	0	0	0	1	0	0
o_c	0	0	1	0	0	0	1	0	0
o_d	0	1	0	1	0	1	0	0	0
o_e	1	0	1	0	0	1	0	0	0
o_f	1	0	0	1	1	1	0	1	1
$\mathcal{O}\backslash\mathcal{A}$	a^{10}	a^{11}	a^{12}	a^{13}	a^{14}	a^{15}	a^{16}	a^{17}	a^{18}
o_a	1	1	0	1	0	0	1	1	0
o_b	1	0	0	0	1	1	0	1	1
o_c	1	1	1	1	0	1	0	0	1
o_d	0	1	0	1	1	0	0	1	0
o_e	1	0	0	0	1	0	0	0	0
o_f	0	1	0	0	1	0	1	1	0

Table 4.2 Values of $(I(x, y), u(x, y) + v(x, y))$ for Table 4.1

$\mathcal{O}\backslash\mathcal{O}$	o_a	o_b	o_c	o_d	o_e	o_f
o_a	(9, 0, 0), **0**	(2, 7, 4), **11**	(3, 6, 5), **11**	(5, 4, 2), **6**	(1, 8, 4), **12**	(6, 3, 4), **7**
o_b		(6, 0, 0), **0**	(4, 2, 4), **6**	(2, 4, 5), **9**	(2, 4, 3), **7**	(2, 4, 8), **12**
o_c			(8, 0, 0), **0**	(2, 6, 5), **11**	(2, 6, 3), **9**	(1, 7, 9), **16**
o_d				(7, 0, 0), **0**	(2, 5, 3), **8**	(5, 2, 5), **7**
o_e					(5, 0, 0), **0**	(3, 2, 7), **9**
o_f						(10, 0, 0), **0**

Let us compare this definition with a more analytical one proposed in [4].

Definition 15 A similarity measure (or index) on $\mathcal{O}$ is a real positive function $\mathcal{S}$ on $\mathcal{O} \times \mathcal{O}$, which satisfies the following conditions:

1. $(\forall (x, y) \in \mathcal{O} \times \mathcal{O})$, $\mathcal{S}(x, y) = \mathcal{S}(y, x)$ symmetry;
2. If x and y in $\mathcal{O}$ are identical except for one attribute a^k ($x_j = y_j$, for all j, $1 \leq j \leq p$, $j \neq k$ and $x_k \neq y_k$) and if z is such that $|z_k - x_k| < |z_k - y_k|$; then $\mathcal{S}(x, y) \geq \mathcal{S}(y, z)$

Note that in the above statement $|\bullet|$ designates the absolute value of $\bullet$.

Proposition 29 *If a similarity index (measure) $\mathcal{S}$ on $\mathcal{O}$ verifies the properties of Definition 14, then it verifies the properties of Definition 15. Reciprocally, a similarity index which can be put into the form $\mathcal{S}(s, u, v)$ and verifying the properties of Definition 15, is such that the function $\mathcal{S}$ decreases with respect to u and increases with respect to s, for $s + v$ constant.*

Proof Two cases have to be considered for x_k:

$$x_k = 1 \quad \text{and} \quad x_k = 0$$

For the first case, according to the conditions of Definition 14, $y_k = 0$ and $z_k = 1$. Consequently,

$$s(x, z) = s(y, z) + 1$$
$$u(x, z) = u(y, z)$$
$$v(x, z) = v(y, z) - 1 \tag{4.2.4}$$

Therefore

$$\mathcal{S}(s(x, z), u(x, z), v(x, z)) > \mathcal{S}(s(y, z), u(y, z), v(y, z))$$

Therefore, if a similarity measure satisfies the conditions of Definition 14, then it satisfies the conditions of Definition 15.

Reciprocally, suppose a similarity measure $\mathcal{S}(x, y)$ satisfying the conditions of Definition 15 and which can be put into the form $\mathcal{S}(s, u, v)$. The inequality $\mathcal{S}(x, z) \geq \mathcal{S}(y, z)$ gives

$$\mathcal{S}(s, u, v) \geq \mathcal{S}(s - \epsilon, u + 1 - \epsilon, v + \epsilon) \tag{4.2.5}$$

where (s, u, v) is defined for (x, z) and where $\epsilon = 1$ or 0.

$\epsilon = 1$ is concerned by the first case where $x_k = 1$, $y_k = 0$ and $z_k = 1$, and $\epsilon = 0$, by the second case where $x_k = 1$, $y_k = 1$ and $z_k = 0$. The $\epsilon = 0$ case entails the inequality

$$\mathcal{S}(s, u, v) \geq \mathcal{S}(s, u + 1, v)$$

which expresses the property of decreasing of the $\mathcal{S}$ function with respect to s for $s + v$ constant. □

Definition 14 is general enough in order to include all the numerical similarity indices proposed in the literature. These are mostly, to a 2 exponent rational fractions in s, u and v.

In Definition 14 the parameter t is interpreted as a function of the triple s, u and v. Indeed, $t = p - (s + u + v)$. Nevertheless, some of the above similarity measures are expressed without t or p (see (5), (6), (7), (8), (9) and (10) in Table 4.3), while others cannot be expressed without t or p (see (1), (2), (3), (4), (11), and (12) in Table 4.3). This formal distinction is highlighted in [23]. For the first type of similarity measure— expressed without t or p—the authors cited point out the following similarity measure

$$\mathcal{S}(x, y) = \frac{s}{s + \alpha \cdot u + \beta \cdot v}$$

Table 4.3 Classical similarity indices

Measure	Expression
(1) Russel and Rao (1940) [29]	$S(x, y) = \frac{s}{p}$
(2) Sokal and Michener (1958) [31]	$S(x, y) = 1 - \frac{u+v}{p} = \frac{s+t}{p}$
(3) Roger and Tanimoto (1960) [28]	$S(x, y) = \frac{p-(u+v)}{p+(u+v)}$
(4) Hamann (1961) [12]	$S(x, y) = 1 - \frac{2(u+v)}{p}$
(5) Jaccard (1908) [13]	$S(x, y) = \frac{s}{s+(u+v)}$
(6) Kulezynski(a) (1927) [15]	$S(x, y) = \frac{s}{u+v}$
(7) Dice (1945) [5]	$S(x, y) = \frac{2s}{2s+(u+v)}$
(8) Sokal and Sneath (1963) [32]	$S(x, y) = \frac{s}{s+2(u+v)}$
(9) Kulezynski(b) (1927) [15]	$S(x, y) = \frac{1}{2} \times \left(\frac{s}{s+u} + \frac{s}{s+v} \right)$
(10) Ochiai (1957) [25]	$S(x, y) = \frac{s}{\sqrt{(s+u)\times(s+v)}}$
(11) Yule (1912) [36]	$S(x, y) = \frac{s\cdot t - u\cdot v}{s\cdot t + u\cdot v}$
(12) Pearson (1926) [26]	$S(x, y) = \frac{s\cdot t - u\cdot v}{\sqrt{(s+u)\cdot(s+v)\cdot(t+u)\cdot(t+v)}}$

where α and β are interpreted for each of the similarity indices concerned.

4.2.1.3   Some Classical Indices (Measures), Their Versions
for Numerical Attributes; Extension
to the Asymmetrical Case

As mentioned above, the similarity indices proposed here are considered in the framework of comparing objects described by Boolean attributes. It is obvious (see General Introduction) that these indices can be defined in the context of comparing Boolean attributes describing a set of objects. This alternative will be considered in the third paragraph below.

Some Classical Similarity Measures for Boolean Data

Let us begin by some remarks on the respective forms and formal properties of these indices. The measure (1) is a function of the single parameter s. Each of the indices (2), (3) and (4) is a function of $d = (u + v)$. For these indices it is equivalent to say that each of them is a function of $a = (s + t)$, since $s + t = p - (u + v)$. The measures (5), (6), (7) and (8) are each a function of $\frac{s}{u+v}$. Finally, notice that the measures (9) an (10) are the algebraic mean and the geometrical mean of $\frac{s}{s+u}$ and $\frac{s}{s+v}$, respectively.

Now, consider the similarity measure $S(x, x)$ between a given object and itself. $S(x, x) = S(s, 0, 0)$ is the number of attributes possessed by x. This index can be put into the two equivalent forms: $card[\mathcal{A}(x)]$ and $\sum_{1 \leq j \leq p} x_j^2$ (see (4.2.3)). $S(s, 0, 0)$ is equal to 1 for all of the measures above, except for (1) and (6). For (1) $S(s, 0, 0)$ is the proportion of the attributes possessed by x and for (6), $S(s, 0, 0) = \infty$.

There are three different ranges for the values of a given similarity index of the above list (see Table 4.3): $[0, 1]$, $[-1, +1]$ and $[0, +\infty]$. The common range of the measures (1), (2), (3), (5), (7), (8), (9) and (10) is $[0, 1]$. $[-1, +1]$ is the common range of the indices (4), (11) and (12) and finally, resp., $[0, +\infty]$ is the range of the index (6).

Now, let us consider the following dissimilarity indices directly derived from (1), (2) and (5):

$$1 - \frac{s}{p}, \quad \frac{u+v}{p} \quad \text{and} \quad \frac{u+v}{s+u+v}$$

They define three distances on the set of all $\mathcal{A}$ subsets. That is the triangular inequality is satisfied for each of them. More precisely, let $\mathcal{D}$ be a dissimilarity index on the object set $\mathcal{O}$ which is represented by a sample in the set of all $\mathcal{A}$ subsets. $\mathcal{D}$ defines a distance on $\mathcal{O}$ if and only if

$$(\forall x, y \text{ and } z \text{ in } \mathcal{O}) \; \mathcal{D}(x, z) \leq \mathcal{D}(x, y) + \mathcal{D}(y, z)$$

In order to establish these results, the analytical expressions given in (4.2.3) have to be used. Let

$$(x_1, x_2, \ldots, x_j, \ldots, x_p), \quad (y_1, y_2, \ldots, y_j, \ldots, y_p) \quad \text{and} \quad (z_1, z_2, \ldots, z_j, \ldots, z_p)$$

be the three Boolean vectors associated with the objects x, y and z. For a given subscript j there are eight patterns of (x_j, y_j, z_j), namely: $(1, 1, 1)$, $(1, 1, 0)$, $(1, 0, 1)$, $(0, 1, 1)$, $(1, 0, 0)$, $(0, 1, 0)$, $(0, 0, 1)$ and $(0, 0, 0)$. The proofs of the triangular inequality for each of the three dissimilarity indices considered above can be obtained by distinguishing the respective contributions of each of these eight patterns. The detailed proof is left for the reader.

Extension of the Boolean Similarity Measures to the Case of Numerical Attributes

Now, let us propose an extension of the indices above (see Table 4.3) in the case of numerical data. For this purpose we take up again the representation of a given object x of $\mathcal{O}$ by means of a vector of Booleans denoted $\alpha(x) = (x_1, x_2, \ldots, x_j, \ldots, x_p)$ (see (4.2.1). The component x_j of this vector will be interpreted here as an intensity of presence of the attribute a^j on the object x, $1 \leq j \leq p$. It is equal to 0 or 1 if a^j is Boolean. In the case where a^j is a numerical attribute, we consider the distribution of the latter on the object set $\mathcal{O}$ and we determine a *minimal* value m_j and a *maximal* value M_j for this attribute, $1 \leq j \leq p$. In order to avoid outliers values such m_j

(resp., M_j) is not necessarily chosen as the lowest value (resp., the highest value) of a^j on $\mathcal{O}$. In these conditions we substitute the value of

$$a'^j(x) = \min \left\{ \max \left(\frac{a^j(x) - m_j}{M_j - m_j} \right), 1 \right\} \qquad (4.2.6)$$

for $a^j(x)$, and $a'^j(x)$ is included between 0 and 1. Notice that if m_j and M_j are, respectively, the minimal and maximal values of $a^j(x)$ on $\mathcal{O}$, the previous expression becomes

$$a'^j(x) = \frac{a^j(x) - m_j}{M_j - m_j} \qquad (4.2.7)$$

The reduction (4.2.6) is also proposed in [23]; but in a different background, namely, that of transforming a numerical similarity measure.

Now, let us consider the set $\{a'^j | 1 \leq j \leq p\}$ of the reduced attributes according to (4.2.6). Notice that there is no change for the values (0 and 1) of a Boolean attribute; except that these values are numerically interpreted. The extension of the different indices proposed (see Table 4.3) is based on the use of (4.2.3). These enable us to define the parameters

$$s(x, y) = \sum_{1 \leq j \leq p} a'^j(x) a'^j(y)$$

$$u(x, y) = \sum_{1 \leq j \leq p} a'^j(x)(1 - a'^j(y))$$

$$v(x, y) = \sum_{1 \leq j \leq p} (1 - a'^j(x)) a'^j(y)$$

$$t(x, y) = \sum_{1 \leq j \leq p} (1 - a'^j(x))(1 - a'^j(y)) \qquad (4.2.8)$$

Notice that the relation $s + u + v + t = p$ of the Boolean case holds. In these conditions, a similarity measure defined in the Boolean case and expressed by $\mathcal{S}(s, u, v)$ (see Definition 14) can be extended to the case where the descriptive attributes are constituted by a mixing of Boolean and numerical attributes. For this, we have to employ the previous expressions (see (4.2.8)).

Conversely, a similarity index defined for numerical description can be adapted to the case of Boolean data. Such a similarity index is generally derived from a distance function on which a decreasing function is applied. Let us recall three classical and well-known distance indices (Table 4.4).

In these notations

- x_j (resp., y_j) is the value of the jth numerical attribute a^j on the object x (resp., y), $1 \leq j \leq p$;
- $\alpha(x)$ (resp., $\alpha(y)$) denotes the vector representing x (resp., y) in the geometrical space $\mathbb{R}^p$;

Table 4.4 Classical distances measures for numerical data

Name	Expression
(1) Euclidean	$\sqrt{\sum_{1 \leq j \leq p}(x_j - y_j)^2}$
(2) Mahalanobis	$\sqrt{(\alpha(x) - \alpha(y))^t \Sigma^{-1}(\alpha(x) - \alpha(y))}$
(3) Minkowski	$(\sum_{1 \leq j \leq p}|x_j - y_j|^\gamma)^{\frac{1}{\gamma}}$

- $(\alpha(x) - \alpha(y))^t$ denotes the transpose of the vector $(\alpha(x) - \alpha(y))$ and Σ is the covariance matrix between attributes;
- γ is a real positive parameter defined by the user, the value 1 of this parameter gives the Manhattan distance.

Each of the above measures can be adapted for the case of Boolean description. For this, the value of a given Boolean attribute a on a given object x—which is 0 or 1—has to be interpreted numerically. In this interpretation, the assumed scale of a is the real interval $[0, 1]$. Therefore, $a(x)$ is viewed as the presence level of a on the object x. An interesting exercise consists of making explicit the adaptation of the Mahalanobis distance for the Boolean case. This is left for the reader.

Asymmetrical Versions of the Symmetrical Measures

Now, let us consider the dual case of comparing Boolean attributes. As expressed in the above introduction (see Sect. 4.1) and principally in Chap. 3, Sect. 3.2.1, the representation of the set of Boolean attributes is defined at the level of the set $\mathcal{P}(\mathcal{O})$ of subsets of the object set $\mathcal{O}$. As a consequence, for the comparison of two Boolean attributes a and b, the parameters s, u, v and t are defined as follows:

$$s(a, b) = card[\mathcal{O}(a) \cap \mathcal{O}(b)], \quad u(a, b) = card[\mathcal{O}(a) \cap \mathcal{O}(\bar{b})]$$
$$v(a, b) = card[\mathcal{O}(\bar{a}) \cap \mathcal{O}(b)] \quad \text{and} \quad t(a, b) = card[\mathcal{O}(\bar{a}) \cap \mathcal{O}(\bar{b})]$$

$$(4.2.9)$$

where $\mathcal{O}(a)$ (resp., $\mathcal{O}(b)$) is the $\mathcal{O}$ subset where the attribute a (resp., b) is present. $\bar{a}$ (resp., $\bar{b}$) is the a negated (resp., b negated) attribute. $\mathcal{O}(\bar{a})$ and $\mathcal{O}(\bar{b})$ are the complementary subsets in $\mathcal{O}$ of $\mathcal{O}(a)$ and $\mathcal{O}(b)$, respectively.

The parameters defined in (4.2.9) can be put in the following form:

$$s(a, b) = n(a \wedge b), \quad u(a, b) = n(a \wedge \bar{b})$$

$$v(a, b) = n(\bar{a} \wedge b) \quad \text{and} \quad t(a, b) = n(\bar{a} \wedge \bar{b})$$

where $\wedge$ designates the conjunction symbol.

In these conditions and clearly, a similarity measure $S(a, b)$ between two Boolean attributes a and b, satisfying the conditions of Definition 14 can be employed. More particularly, all the similarity measures of Table 4.3 can be considered for this comparison. These indices are symmetrical with respect to the two attribute components a and b. For most of them, generally and implicitly, they are built by starting with the "raw" index $s(a, b) = n(a \wedge b)$ and by normalizing it. This normalization is carried out in order to evaluate how large is $n(a \wedge b)$ with respect to the values of the other parameters (see (4.2.9)) and mainly, with respect to $n(a) = card(\mathcal{O}(a))$ and $n(b) = card(\mathcal{O}(b))$.

In the *data mining* field, an asymmetrical version of the similarity measure between a and b may be required [11]. In fact, an implicative tendency $a \rightarrow b$ has to be highlighted. It expresses that a $TRUE$ value of a observed on a given object o of $\mathcal{O}$ implies that *generally* but not *absolutely*, a $TRUE$ value of b on o is observed. The "raw" index becomes here $n(a \wedge \bar{b})$ and the evaluation problem becomes "how small is $n(a \wedge \bar{b})$ with respect to the values of the other parameters?". This form of evaluation asking appeared in [10], this in order to adapt for the implicative case the symmetrical probabilistic association coefficient of the *likelihood of the Link* between Boolean attributes established earlier [19, 20] (see Sect. 5.2.1.4 of Chap. 5).

Consequently, we propose to transpose the defined indices (see Table 4.3) using the basic substitution $(1 - y_j) \rightarrow y_j$ in (4.2.3). Thus, the parameters s, u, v and t become u, s, t and v, respectively. A given similarity measure $S(s, u, v)$ is transformed into $S(u, s, t)$. According to Definition 14, the obtained indices increase with respect to the parameter u. That is, are as more big as the implicative tendency is weak. For intuitive reasons it is wished to have the opposite behaviour where a value of a given measure is as more big as the implicative tendency is strong. Therefore, we will take an appropriate complement of the measure $S(u, s, t)$ depending on its analytical expression. More precisely and for the similarity measures derived from Table 4.3, we consider three cases:

1. The range of the similarity measure $S(u, s, t)$ is $[0, a]$, $(a = 1$ or $0.5)$;
2. The range of the similarity measure $S(u, s, t)$ is $[-1, +1]$;
3. The range of the similarity measure $S(u, s, t)$ is $[0, \infty]$.

Now, denote by $\mathcal{J}(s, u, v)$ the implicative similarity measure corresponding to $S(u, s, t)$. We will define

1. $\mathcal{J}(s, u, v) = a - S(u, s, t)$;
2. $\mathcal{J}(s, u, v) = -S(u, s, t)$;
3. $\mathcal{J}(s, u, v) = \frac{1}{S(u, s, t)}$.

Table 4.5 is associated with Table 4.1. It gives the implicative measures derived from those of the latter table (Table 4.1).

Notice that some of the similarity measures in Table 4.3 remain invariant when they are transformed into their asymmetrical versions. This occurs for the indices (2), (4), (11) and (12). Then for each of these indices, focusing on how large is $s(a, b)$ is equivalent to focusing on how small is $u(a, b)$. Let us now illustrate this

Table 4.5 Implicative similarity indices

Implicative measure	Expression
(1) Russel and Rao (1940) [29]	$\mathcal{J}(a \rightarrow b) = \frac{s+t+v}{n}$
(2) Sokal and Michener (1958) [31]	$\mathcal{J}(a \rightarrow b) = \frac{s+t}{n}$
(3) Roger and Tanimoto (1960) [28]	$\mathcal{J}(a \rightarrow b) = \frac{2(s+t)}{n+(s+t)}$
(4) Hamann (1961) [12]	$\mathcal{J}(a \rightarrow b) = 1 - \frac{2(u+v)}{n}$
(5) Jaccard (1908) [13]	$\mathcal{J}(a \rightarrow b) = \frac{s+t}{u+s+t}$
(6) Kulezynski(a) (1927) [15]	$\mathcal{J}(a \rightarrow b) = \frac{s+t}{u}$
(7) Dice (1945) [5]	$\mathcal{J}(a \rightarrow b)) = \frac{s+t}{2u+(s+t)}$
(8) Sokal and Sneath (1963) [32]	$\mathcal{J}(a \rightarrow b) = \frac{2(s+t)}{u+2(s+t)}$
(9) Kulezynski(b) (1927) [15]	$\mathcal{J}(a \rightarrow b) = \frac{1}{2} \times \left(\frac{s}{s+u} + \frac{t}{t+u} \right)$
(10) Ochiai (1957) [25]	$\mathcal{J}(a \rightarrow b) = 1 - \frac{u}{\sqrt{(s+u) \times (t+u)}}$
(11) Yule (1912) [36]	$\mathcal{J}(a \rightarrow b) = \frac{s \cdot t - u \cdot v}{s \cdot t + u \cdot v}$
(12) Pearson (1926) [26]	$\mathcal{J}(a \rightarrow b)) = \frac{s \cdot t - u \cdot v}{\sqrt{(s+u) \cdot (s+v) \cdot (t+u) \cdot (t+v)}}$

Table 4.6 Numerical illustration

Measure	Sym(a, b)	Imp(a → b)
(1) Russel and Rao	1/6	8/9
(2) Kendall, Sokal and Michener	1/2	1/2
(3) Roger and Tanimoto	1/2	2/3
(4) Hamann	0	0
(5) Jaccard	1/4	9/11
(6) Kulezynski(a)	1/6	3
(7) Dice	2/5	9/13
(8) Sokal and Sneath	1/7	9/10
(9) Kulezynski(b)	9/20	51/80
(10) Ochiai	$3 \times \sqrt{2}/10$	$1 - (\sqrt{10}/10)$
(11) Yule	1/8	1/8
(12) Pearson	$4 \times \sqrt{13}/260$	$4 \times \sqrt{13}/260$

transformation in the case of the following example: $n = 18, n(a) = 5, n(b) = 10$ and $s(a, b) = n(a \wedge b) = 3$. In these conditions we have: $u(a, b) = n(a \wedge \bar{b}) = 2$, $v(a, b) = n(\bar{a} \wedge b) = 7$ and $t(a, b) = n(\bar{a} \wedge \bar{b}) = 6$.

In Table 4.6 $Sym(a, b)$ and $Imp(a \rightarrow b)$ denote the symmetrical and the implicative asymmetrical forms, respectively.

In short, for comparing Boolean attributes, we have established in this section a technique for transposing a symmetrical similarity function (see Table 4.3) into an asymmetrical implicative function (see Table 4.5). Thus, we obtain new asymmetrical *interestingness measures for association rules*. Those considered in the *Data Mining* literature [7, 16, 34], were conceived directly without clear reference to the association coefficients (similarity measures) established between Boolean data in clustering and this, although many are symmetrical. It is interesting to situate them with respect to the above framework. Let us consider two of these measures which have an historical status: the Lœvinger index and the Confidence index of Agrawal et al. [1, 24].

For the Lœvinger index the conditional proportion $p(b|a) = n(a \wedge b)/n(a)$ is positioned with respect to the marginal proportion $p(b) = n(b)/n(a)$ by considering the difference $p(b|a) - p(b)$. This is normalized with respect to its maximal value $1 - p(b)$. Therefore, the resulting implicative index, that we denote by $Loe(a \rightarrow b)$, is expressed as follows:

$$Loe(a \rightarrow b) = \frac{[(p(a \wedge b)/p(a)) - p(b)]}{[1 - p(b)]}$$

$$= 1 - \frac{p(a \wedge \bar{b})}{p(a) \times p(\bar{b})} = 1 - \frac{p(\bar{b}|a)}{p(\bar{b})} = 1 - \frac{p(a|\bar{b})}{p(a)} \qquad (4.2.10)$$

The range of this index is $[-\infty, 1]$. Its value is 0 in the case of independence defined by $p(a \wedge b) = p(a) \times p(b)$ and 1 in the case of inclusion: $\mathcal{O}(a) \subset \mathcal{O}(b)$. We can easily realize that this index can reach a negative value. For this, let us imagine $p(a)$ and $p(\bar{b})$ tending to 0 in such a way that $p(a \wedge \bar{b})$ tends to a finite limit. For illustration, consider $n = 10^6$, $n(a) = 2000$, $n(b) = 8000$ and $n(a \wedge \bar{b}) = 1600$; then, the value calculated of the index concerned is -99.

By considering the above notations, $Loe(a \rightarrow b)$ can be put in the following form:

$$Loe(a \rightarrow b) = 1 - n \times \frac{u}{(s + u) \times (t + u)} \qquad (4.2.11)$$

In the list of implicative measures obtained by transposition of similarity measures (see Table 4.3), the analytically nearest (4.2.11) expression is (10) corresponding to the Ochiai index. Let us take up again its expression:

$$Ochiai(a \rightarrow b) = 1 - \frac{u}{\sqrt{(s + u) \times (t + u)}} \qquad (4.2.12)$$

This index varies between 0 and 1. However, it is not null in the case of independence between a and b. In this case, it is equal to $1 - \sqrt{p(a) \times p(\bar{b})}$.

As said above, the index $Loe(a \rightarrow b)$ is obtained by normalizing $p(b|a)$. Whereas, *Confidence* index $(Confidence(a \rightarrow b))$ proposed in 1993 [1] is symply defined by the conditional proportion $p(b|a)$. It varies between 0 and 1; 0, in the case of exclusion $(\mathcal{O}(a) \cap \mathcal{O}(b) = \emptyset)$, 1, in the case of inclusion $\mathcal{O}(a) \subset \mathcal{O}(b))$ and $p(b)$ in the case of independence. This index can be written as follows:

$$Confidence(a \rightarrow b) = \frac{s}{s+u} \qquad (4.2.13)$$

It corresponds to one of the two additive components of the Kulzynski(b) index.

4.2.2 Preordonance Associated with a Similarity Index in the Case of Boolean Data

4.2.2.1 Definition and General Properties

The object set is denoted here by $\mathcal{O} = \{o_i | 1 \leq i \leq n\}$ and $P = P_2(\mathcal{O}) = \{\{x, y\} | x \in \mathcal{O}, y \in \mathcal{O}, x \neq y\}$ designates the set of unordered object pairs (or 2subsets) of $\mathcal{O}$ (see (3.3.4)). Let S be a numerical similarity index on $\mathcal{O}$. As expressed in the general introduction (see Sect. 4.1) S induces a total preorder (i.e. ranking with ties) on P, called *preordonance* on $\mathcal{O}$. Two versions of this ranking can be considered depending on an increasing or a decreasing alternatives of the rank function associated with the total preorder (see Sect. 3.3.3 for the definition of the rank function in an other context). Let us take up again our notations of Sect. 4.1, ω_S and ϖ_S denote the decreasing and the increasing alternatives (see (4.1.1)), respectively. In this section the first alternative is considered, namely,

$$\forall (p, q) \in P \times P, \quad p \leq (\omega_S) q \Leftrightarrow S(p) \geq S(q)$$

Without risk of ambiguity, we will also denote by ω_S the graph of this total preorder in $P \times P$, as following:

$$gr(\omega_S) = \{(p, q) | (p, q) \in P \times P, p \leq q \text{ for } \omega_S\}$$

As expressed in the general introduction (see Sect. 4.1) and as observed in the similarity analysis of Boolean data, there is a large range of possible numerical similarity indices. In these conditions, a natural thinking consists of expecting more stable results with respect to the similarity index choice when we substitute the associated preordonance for the numerical similarity index, that is, when ω_S is substituted for S. And indeed, some combinatorial methods of data analysis are exclusively based on ordinal similarity.

Example 1 Consider the incidence data table and the indicator function matrix associated in Sect. 4.2.1.2. Let us express two preordonances associated with the similarity indices $S(s, u, v) = s$ and $S(s, u, v) = s + t$, respectively. A given object pair $\{o_g, o_h\}$ will be denoted by the two-letter word gh where g is lexicographically before h and where g and h are in $\{a, b, c, d, e, f\}$; thus, $\{o_b, o_f\}$ is denoted by bf. We have

$\omega_S : af < ad \sim df < bc < ac \sim ef < ab \sim bd \sim cd \sim be \sim ce \sim de \sim bf < cf \sim ae$

In this, $\sim$ denotes the equivalence symbol for the total preorder concerned. The respective values of s are: 6, 5, 4, 3, 2 and 1.

$\omega_{s+t} : ad \sim bc' < be \sim af \sim df < de < bd \sim ce \sim ef < ab \sim ac \sim cd < ae \sim bf < cf$

The respective values of $(s+t)$ are: 12, 11, 10, 9, 7, 6 and 2. Then, those of $(u+v)$ are 6, 7, 8, 9, 11, 12 and 16.

Let us now define the equivalence binary relation between similarity measures.

Definition 16 Two similarity measures S and S' defined on an object set $\mathcal{O}$ are equivalent on $\mathcal{O}$ if and only if the preordonances ω_S and $\omega_{S'}$ associated on $\mathcal{O}$, respectively, are identical.

To fix ideas suppose that S and S' satisfy the conditions of Definition 14 and then, $S(x, y)$ and $S'(x, y)$ can be written $S(s, u, v)$ and $S'(s, u, v)$ where (s, u, v) is the indicator of the object pair $\{x, y\}$. In these conditions, if there exists a strictly increasing numerical function f such that

$$S'(s, u, v) = f[S'(s, u, v)]$$

S and S' are equivalent on any object set $\mathcal{O}$ described by Boolean attributes.

Proposition 30 *If the number of attributes present on an object of $\mathcal{O}$ is constant in $\mathcal{O}$, then all the similarity measures are equivalent on $\mathcal{O}$.*

Proof Let l be the common number of attributes present in a object of $\mathcal{O}$. Consider for two given objects x and y their indicator function $I(x, y) = (s, u, v)$ and a similarity measure $S(x, y) = S(s, u, v)$. Because of the relations $u = l - s$ and $v = l - s$, $S(s, u, v)$ can be written $\mathcal{R}(s) = S(s, l - s, l - s)$ where $\mathcal{R}(s)$ is a positive numerical function defined on $\{0, 1, 2, \ldots, p\}$. Due to Definition 14, $\mathcal{R}(s)$ is strictly increasing with respect to s.   □

4.2.2.2   Disagreement Between Two Preordonances Associated with Two Similarity Measures

Let S and S' be two numerical similarity measures on the object set $\mathcal{O}$ and let ω_S and $\omega_{S'}$ be the preordonances associated with S and S', respectively. When these two indices are not equivalent, there is a matter to define the "Disagreement" between ω_S and $\omega_{S'}$. The following definitions enable us to obtain an objective criterion measuring this disagreement.

By denoting ω the graph in $P \times P$ of a given total preorder on P (see above the expression of $gr(\omega_S)$), let us introduce

$$\omega_1 = \{(p, q) | (p, q) \in \omega \text{ and not } ((q, p) \in \omega)\}$$
$$\omega_2 = \{(p, q) | (q, p) \in \omega_1\} \tag{4.2.14}$$

Notice that if ω is a strict total order on P, ω_1 coincides with ω and ω_2 corresponds to the complementary subset of ω_1 in $P \times P$.

Definition 17 Let ω and ω' be the graphs of two total preorders on P, the following subset of $P \times P$

$$\Delta(\omega, \omega') = \omega_1 \cap \omega_2' + \omega_2 \cap \omega_1'$$

is called "Disagreement" between ω and ω', where ω_1' and ω_2' are associated with ω' in the same way that ω_1 and ω_2 have been associated with ω. The cardinality of $\Delta(\omega, \omega')$, $card(\Delta(\omega, \omega'))$, defines the number of *inversions* between ω and ω'

Let us notice that

$$(p, q) \in \Delta(\omega, \omega') \Leftrightarrow (q, p) \in \Delta(\omega, \omega')$$

Therefore $card(\Delta(\omega, \omega'))$ is an even integer and then we define the "difference" between ω and ω' by the integer number: $\frac{1}{2} \times card(\Delta(\omega, \omega'))$. Notice that if ω and ω' are strict total orders on P, $\Delta(\omega, \omega')$ is the symmetrical difference set between ω and ω'.

When ω and ω' are associated with two similarity measures $\mathcal{S}$ and $\mathcal{S}'$ defined on the object set $\mathcal{O}$: $\omega = \omega(\mathcal{S})$ and $\omega' = \omega(\mathcal{S}')$, an ordered pair $(\{x, y\}, \{z, t\})$ of $P \times P$ gives rise to an inversion if and only if

$(\mathcal{S}(x, y) < \mathcal{S}(z, t)$ and $\mathcal{S}'(x, y) > \mathcal{S}'(z, t))$
or $(\mathcal{S}(x, y) > \mathcal{S}(z, t)$ and $\mathcal{S}'(x, y) < \mathcal{S}'(z, t))$

In the following we will denote by $\underline{\omega}$ and $\overline{\omega}$ the preordonances on $\mathcal{O}$ associated respectively with the similarity measures $\mathcal{S}(x, y) = s(x, y)$ and $\mathcal{S}'(x, y) = s(x, y) + t(x, y)$. On the other hand Δ will designate the disagreement set between $\underline{\omega}$ and $\overline{\omega}$, $\Delta = \Delta(\underline{\omega}, \overline{\omega})$. By taking up again the previous example we have

Example 2 $\Delta = \{(af, ad), (ad, af), (af, bc), (bc, af), (df, bc), (bc, df),$
$(ac, bd), (bd, ac), (ac, be), (be, ac), (ac, ce), (ce, ac),$
$(ac, de), (de, ac), (ef, be), (be, ef), (ef, de), (de, ef)\}$

In the following we will restrict our attention to the similarity measures which can be put into the general form $\mathcal{S}(s, u + v)$

Proposition 31 *Let $\Phi(s, u + v)$ and $\Psi(s, u + v)$ be two similarity measures on the object set $\mathcal{O}$ and let α and β denote the preordonances respectively associated. We have $\Delta(\alpha, \beta) \subseteq \Delta$.*

Proof Indeed, let $(\{x, y\}, \{z, t\})$ be a given element of $\Delta(\alpha, \beta)$ and let us denote by (s, u, v) and $(s + r, u + k, v + h)$ the indicators of the pairs $\{x, y\}$ and $\{z, t\}$, respectively. In these conditions, $\Phi(s + r, u + v + k + h) - \Phi(s, u + v)$ and $\Psi(s + r, u + v + k + h) - \Psi(s, u + v)$ are different from 0 and of opposite signs. Therefore, r and $k + h$ are necessarily different from 0 and of the same sign. Otherwise, inversion cannot occur for $(\{x, y\}, \{z, t\})$. Hence $(\{x, y\}, \{z, t\})$ belongs to Δ. $\square$

Corollary 6 *Among the similarity measures having the general form $\mathcal{S}(s, u+v)$, the two measures for which the preordonances respectively associated, have the largest "difference" are s and $s + t = p - (u + v)$.*

Proposition 32 *If the number of Boolean attributes present on a given object of $\mathcal{O}$ takes one of two consecutive values, l or $l + 1$, $0 \leq l \leq p - 1$, then the set Δ is empty: $\Delta = \emptyset$.*

Proof Indeed, if Δ is not empty, it would exist one element $(\{x, y\}, \{z, t\})$ in $P \times P$ such that:

$$\sum_{j \in J} x_j \cdot y_j < \sum_{j \in J} z_j \cdot t_j \text{ and } \sum_{j \in J} (x_j + y_j) > \sum_{j \in J} (z_j + t_j) + 2 \left(\sum_{j \in J} x_j \cdot y_j - \sum_{j \in J} z_j \cdot t_j \right)$$
$$(4.2.15)$$

In the above notations $J = \{1, 2, \ldots, j, \ldots, p\}$. These inequalities entail

$$\sum_{j \in J} (x_j + y_j) - \sum_{j \in J} (z_j + t_j) \geq 3 \qquad (4.2.16)$$

But we have

$$\sum_{j \in J} (x_j + y_j) \leq 2l + 2 \quad \text{and} \quad \sum_{j \in J} (z_j + t_j) \geq 2l$$

and these inequalities make impossible (4.2.16). $\qquad\square$

Obviously, under the conditions of the previous proposition, the inversion number between $\underline{\omega}$ and $\overline{\omega}$ is null.

Proposition 33 *Let V denote the variance in $\mathcal{O}$ of the number of attributes present on a given object, we have:*

$$\frac{card(\Delta)}{card(P \times P)} < \frac{4V}{9}$$

Proof Let us denote here by p_x the number of attributes possessed by x (or present on x). We have:

$$V = \frac{1}{n} \sum_{x \in \mathcal{O}} (p_x - \overline{p})^2$$

where $\overline{p} = \frac{1}{n} \sum_{x \in \mathcal{O}} p_x$.

The statement of the Proposition 33 expresses that the difference between $\underline{\omega}$ and $\overline{\omega}$ is strictly bounded by $card(P)^2 \times \frac{2V}{9}$.

Consider a partition of $P \times P$ into two classes denoted by H and G:

- H is the set of ordered pairs (p, q) of the form $(\{x, y\}, \{z, t\})$ where different letters designate different objects in $\mathcal{O}$

- G is the set of ordered pairs (p,q) of the form $(\{x, y\}, \{x, z\})$ where different letters designate different objects in $\mathcal{O}$

We have

$$card(H) = \frac{n \cdot (n-1) \cdot (n-2) \cdot (n-3)}{4}$$

$$card(G) = n \cdot (n-1) \cdot (n-2) \qquad (4.2.17)$$

Now, let us introduce the indicator function $\Phi(p,q)$ $(p,q) \in P \times P$ of Δ. We have

$$card(\Delta) = \sum_H \Phi(\{x, y\}, \{z, t\}) + \sum_G \Phi(\{x, y\}, \{x, z\}) \qquad (4.2.18)$$

By defining the proportions

$$\beta = \frac{card(H \cap \Delta)}{card(H)} \text{ and } \beta' = \frac{card(G \cap \Delta)}{card(G)}$$

the second member of (4.2.18) can be written as

$$card(\Delta) = \frac{n \cdot (n-1) \cdot (n-2) \cdot (n-3)}{4} \times \beta + n \cdot (n-1) \cdot (n-2) \times \beta' \quad (4.2.19)$$

In order to interpret probabilistically the parameters β and β', let us provide the object set $\mathcal{O}$ by an uniform probability measure (equal chance $1/n$ for each object) with mutual independence between different objects. For this model, the number p_x of attributes possessed by a given object x is a random integer variable whose mean and variance are equal to $\overline{p}$ and V respectively. In this framework let us consider a random element of H. Then β can be interpreted as the probability of the event: $(\{x, y\}, \{z, t\}) \in \Delta$. This event implies $\{|(p_x + p_y) - (p_z + p_t)| \geq 3\}$. Therefore,

$$\beta \leq Pr\{|(p_x + p_y) - (p_z + p_t)| \geq 3\}$$

Due to the Tchebycheff inequality, we have:

$$Pr\{|(p_x + p_y) - (p_z + p_t)| \geq 3\} \leq \frac{1}{3^2}\mathcal{E}(|(p_x + p_y) - (p_z + p_t)|^2) \approx \frac{4V}{9}$$

where $\mathcal{E}$ designates the mathematical expectation. More precisely, $variance((p_x + p_y) - (p_z + p_t)) = \frac{4n}{(n-1)} \times V$.

On the other hand, let us consider a random element in G. β' is the probability of the event $(\{x, y\}, \{x, z\}) \in \Delta$. This event implies $\{|p_y - p_z| \geq 3\}$. Therefore

$$\beta' \leq Pr\{|(p_y - p_z)| \geq 3\} \leq \frac{1}{3^2}\mathcal{E}((p_y) - p_z)^2) \approx \frac{2V}{9}$$

By considering account (4.2.19), $card(\Delta)$ is bounded by

$$\left[\frac{1}{9} \times n(n-1)(n-2)(n-3)+\frac{2}{9} \times n(n-1)(n-2)\right] V = \frac{1}{9} \times (n-2)(n-1)^2 nV)$$

Finally, by taking into account the cardinality $\frac{1}{4} \times (n-1)^2.n^2$ of $P \times P$, we have:

$$\frac{card(\Delta)}{card(P \times P)} \le \frac{4}{9} \times \frac{n-2}{n}V < \frac{4}{9}V \qquad (4.2.20)$$

$\square$

This result as that the used Tchebycheff one, has only a theoretical interest. It expresses that if the variance of the number of attributes possessed by a given object in $\mathcal{O}$ is small, then the inversion number between two preordonances associated respectively with two arbitrary similarity measures is also small. In practice, it happens frequently (questionnaire, descriptive object codes) that the attribute number possessed by a given object has a weak variance in the studied population. In this circumstance, the methods based on a preordonance associated with a similarity measure have an intrinsic property, that is, do not depend on a specific choice of a numerical similarity measure.

Probabilistic Model for Illustration

As seen above, the object set $\mathcal{O}$ is given as a sample in the set $\mathcal{P}(\mathcal{A})$ of all $\mathcal{A}$ subsets; or, equivalently, as a set of elements (points) in the Boolean vector space $\{0, 1\}^p$. Generally, $\mathcal{O}$ is obtained as a sample coming from a large universe Ω of objects and then, it can be viewed as a sample of n independent random elements in the set $\mathcal{P}(\mathcal{A})$ endowed with a probability measure. With this formalization, it is interesting to calculate the mathematical expectation of the inversion number between the preordonances $\underline{\omega}$ and $\overline{\omega}$ associated with the similarity measures s and $s + t$. We have just already seen that if the probability measure concerned is concentrated on one level or two consecutive levels of the simplex $\mathcal{P}(\mathcal{A})$, the inversion number between $\underline{\omega}$ and $\overline{\omega}$ is exactly null. Now, let us consider the random variable $card(X)$ associated with a random element X of the space probabilized $\mathcal{P}(\mathcal{A})$ ($X \subset \mathcal{A}$). Due to the previous proposition, a small value of the mathematical expectation mentioned, can be expected if the variance of $card(X)$ is small.

Let x, y, z and t be a sample of four independent objects taken in $\mathcal{P}(\mathcal{A})$ and let us denote by $\Phi(\{x, y\}, \{z, t\})$ the indicator function of the event

$$\sum_J (x_j - y_j)^2 > \sum_J (z_j - t_j)^2 \quad \text{and} \quad \sum_J x_j y_j > \sum_J z_j t_j \qquad (4.2.21)$$

which defines an inversion. In this equation $J = \{1, 2, \ldots, j, \ldots, p\}$.

By denoting ν the number of inversions, we have:

$$\nu = \sum_H \Phi(\{x, y\}, \{z, t\}) + \sum_G \Phi(\{x, y\}, \{x, z\}) \qquad (4.2.22)$$

where H and G have been defined above in the framework of the previous Proposition 33 (see (4.2.17)).

The mathematical expectation $\mathcal{E}(\nu)$ of ν can be written

$$\mathcal{E}(\nu) = \sum_H \mathcal{E}[\Phi(\{x, y\}, \{z, t\})] + \sum_G \mathcal{E}[\Phi(\{x, y\}, \{x, z\})] \qquad (4.2.23)$$

We have

$$\mathcal{E}[\Phi(\{x, y\}, \{z, t\})] = Prob\{\{\sum_J\}(x_j - y_j)^2 > \sum_J\}(z_j - t_j)^2$$

$$\text{and } \sum_J x_j y_j > \sum_J z_j t_j\}\} \qquad (4.2.24)$$

$$\mathcal{E}[\Phi(\{x, y\}, \{x, z\})] = Prob\{\{\sum_J\}(x_j - y_j)^2 > \sum_J\}(x_j - z_j)^2$$

$$\text{and } \sum_J x_j y_j > \sum_J x_j z_j\}\} \qquad (4.2.25)$$

By denoting π_1 the right member of (4.2.24) and by π_2, the right member of (4.2.25), we obtain

$$\mathcal{E}(\nu) = \frac{1}{4} \times n(n-1)(n-2)(n-3)\pi_1 + n(n-1)(n-2)\pi_2$$

$$= n(n-1)(n-2)(\frac{n-3}{4} \times \pi_1 + \pi_2) \qquad (4.2.26)$$

The total number of distinct ordered pairs is $w = \frac{1}{8} \times (n+1)n(n-1)(n-2)$ and then,

$$\frac{\mathcal{E}(\nu)}{w} = \frac{2(n-3)}{n+1} \times \pi_1 + \frac{8}{n+1} \times \pi_2 \qquad (4.2.27)$$

If n is large enough, $\frac{\mathcal{E}(\nu)}{w}$ is equal to $2\pi_1$, approximatively. In these conditions, only π_1 will interest us. With more synthetic notations we can write:

$$\pi_1 = Prob\{u + v > u' + v' \text{ and } s > s'\} \qquad (4.2.28)$$

where (s, u, v) and (s', u', v') are the indicators functions of (x, y) and (z, t), respectively, (see Definition 13).

Now, let us denote by B_k, B_h C_l and C_m the events $\{s = k\}$ $\{s' = h\}$ $\{u + v = l\}$ and $\{u' + v' = m\}$, respectively. We have

$$\pi_1 = \sum_{l>m} \sum_{k>h} Prob(B_k \wedge B_h \wedge C_l \wedge C_m) \qquad (4.2.29)$$

The events $B_k \wedge C_l$ and $B_h \wedge C_m$ are independent, because the first one is associated with the ordered pair (x, y) and the second one, with (z, t). Therefore

$$\pi_1 = \sum_{l>m} \sum_{k>h} Prob(B_k \wedge C_l) \times Prob(B_h \wedge C_m) \qquad (4.2.30)$$

Consequently, the probability π_1 can be determined if the model provides the probabilities $\{p_{kl} = Prob(B_k \wedge C_l) | 0 \leq k, l \leq p, k + l \leq p\}$

Example

Let us consider an illustrative example in which π_1 is calculated in the framework of a specific probabilistic model endowing the Cartesian space $\{0, 1\}^p$. For this, a given object x is defined as random element of $\mathcal{O}$ represented in $\{0, 1\}^p$ by a random Boolean vector $b(x) = (x_1, x_2, \ldots, x_j, \ldots, x_p)$, such that for any $j, 1 \leq j \leq p$

$$Prob\{x_j = 1\} = \alpha$$
$$Prob\{x_j = 0\} = 1 - \alpha \qquad (4.2.31)$$

where α is a real positive strictly number comprised between 0 and 1: $0 < \alpha < 1$. Moreover, we assume mutual independence between the different components of the random vector $b(x)$. Let us point out that these conditions cannot be satisfied by a realistic case.

Now, let us determine the probabilities $Prob(B_k \wedge C_l), 0 \leq k, l \leq p, k + l \leq p$. We have:

$$Prob(B_k \wedge C_l) = Prob(C_l | B_k) \times Prob(B_k)$$
$$Prob(B_k) = \binom{p}{k} \alpha^{2k} (1 - \alpha^2)^{p-k} \text{ and}$$
$$Prob(C_l | B_k) = \binom{p-k}{l} \left(\frac{2\alpha}{1+\alpha}\right)^l \left(\frac{1-\alpha}{1+\alpha}\right)^{p-k-l} \qquad (4.2.32)$$

Notice that the three largest values of each column occur for $0.2 \leq \alpha \leq 0.4$. On the other hand, these values are as more large as p is large (Table 4.7).

Table 4.7 π_1 values

$\alpha\backslash p$	$p = 10$	$p = 20$	$p = 30$	$p = 40$
$\alpha = 0.1$	0.029	0.058	0.081	0.099
$\alpha = 0.2$	0.079	0.120	0.140	0.151
$\alpha = 0.3$	0.099	0.128	0.141	0.150
$\alpha = 0.4$	0.091	0.114	0.125	0.132
$\alpha = 0.5$	0.072	0.092	0.102	0.108
$\alpha = 0.6$	0.049	0.066	0.075	0.081
$\alpha = 0.7$	0.026	0.046	0.048	0.054
$\alpha = 0.8$	0.009	0.017	0.023	0.028
$\alpha = 0.9$	0.001	0.002	0.004	0.006

Computing the Inversion Number

Concretely, by calculating the *distance* in terms of inversion number between $\underline{\omega}$ and $\overline{\omega}$ (see above), we can realize the extent to which the stability of a preordonance on $\mathcal{O}$, associated with a similarity measure on $\mathcal{O}$, is verified. For this purpose, let us associate with each pair $\{x, y\}$ of distinct objects of $\mathcal{O}$, the following point of $\mathbb{N}^2$

$$\left(\sum_J x_j y_j, \sum_J (x_j - y_j)^2\right) = (s, u + v)$$

where (s, u, v) is the indicator of $\{x, y\}$.

Now, let us introduce the set of object pairs

$$A_k^h = \{\{x, y\}|\{x, y\} \in P, s = k, u + v = h\} \tag{4.2.33}$$

an denote its cardinality by n_k^h ($h \le p - k$). Thus, by assigning the frequency n_k^h to the point (k, h), we define the empirical distribution of $(s, u + v)$ on the set P of unordered object pairs. Define

$$B_k = \{\{x, y\}|\{x, y\} \in P, s = k\} = \sum_{h=0}^{h=p-k} A_k^h$$

where $\sum$ designates a set theoretic sum (union of disjoint subsets).

In these conditions, the contribution of A_k^h to the inversion number—that we denote here by ν—is

$$n_k^h \times \sum_{l=0}^{l=k-1} \sum_{m=0}^{m=h-1} n_l^m \tag{4.2.34}$$

Thus, the latter sum is extended on the rectangle:

$$\{0 \leq l \leq (k-1), 0 \leq m \leq (h-1)\}$$

Therefore, the inversion number is given as

$$\nu = \sum_{k=0}^{k=p} \sum_{h=0}^{h=k-p} n_k^h \times \sum_{l=0}^{l=k-1} \sum_{m=0}^{m=h-1} n_l^m \qquad (4.2.35)$$

Let us take up again the above example (see Example 2). A graphic representation of the mentioned distribution above, is given as following:

The inversion number is then:

$$\nu = 2 + 4 + 1 + 2 = 9$$

From the above graphical representation the following statement can be derived: *A necessary and sufficient condition for no inversion ($\nu = 0$) is that there exists a non increasing function of $\mathbb{N}$ onto $\mathbb{N}$, whose graph contains all of the no-null frequency points of the triangle $\{(k, h)|k + h \leq p\}$ of $\mathbb{N} \times \mathbb{N}$.*

Let us now conclude by some bibliographic notes. The starting point of the work presented in Sects. 4.2.1 and 4.2.2, is the paper [17]. The latter was taken up again in [18, 21]. The development considered here is notably richer than the previous ones. Indeed, the cases of numerical attributes and asymmetrical association rules are studied. Two main and related objectives are considered in this treatment. The first one concerns a formal definition of the notion of a similarity measure and the second one consists of evaluating the extent to which two similarity measures are equivalent through their preordonances, respectively associated with the similarity measures. These two objectives underlie also many recent works [2, 3, 23, 30, 37]. Regarding the second objective, let us mention more particularly [37] where a notion of *topological graph* associated with a similarity measure, is proposed. This may extend in a some sense the notion of a preordonance associated with a similarity measure (Fig. 4.2).

4.2.2.3 On Two Particular Preordonances

Let us now consider two specific preordonances on $\mathcal{O}$ which have, in an algebraic sense, central properties with respect to similarity measures having the form $\mathcal{S}(s, u, v)$ and defined as *numerical* functions of the triplet (s, u, v) (see Definition 14). These preordonances are established directly from the indicator function $I(x, y) = (s, u, v)$ (see Definition 13). This function is defined on the set $\mathcal{O} \times \mathcal{O}$ of ordered object pairs. Adaptation of this function is here required on the set $P = P_2(\mathcal{O})$ of unordered object pairs. For this purpose, for any $\{x, y\} \in P$,

Fig. 4.2 Graphic representation of the distribution of n_k^h

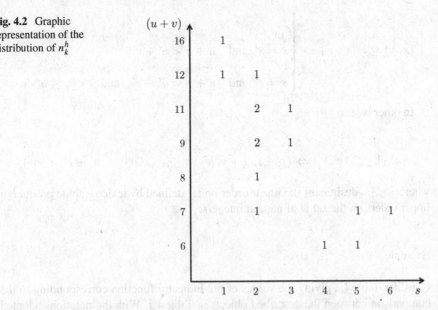

we set $I(\{x, y\}) = I(x, y)$ [resp., $= I(y, x)$] if and only if $u(x, y) \leq v(x, y)$ [resp;, $v(x, y) \leq u(x, y)$].

The principal interest of these two preordonances is formal. Thus, the high generality level of a preordonance as resemblance data for comparing objects is pointed out. For reasons which will appear below, the first of the two preordonances will be called *the most discriminant preordonance* and the second one, *the least discriminant preordonance*.

The Most Discriminant Preordonance

Let us work on again the general conditions for establishing a similarity index between two objects x and y described by Boolean attributes (see Sect. 4.2.1.2). We assume an additional condition for which *negative* associations can be neglected. This condition seems to be satisfied if the initial description is made with binary categorical attributes and if the set of the descriptive Boolean attributes $\mathcal{A}$ is built by retaining from each categorical attribute the categories (one or two) which are "significant" for the resemblance between objects, this significancy being evaluated by the expert knowledge.

The *most discriminant preordonance* that we denote by ω^1 is defined as follows.

Definition 18 Let $I(\{x, y\}) = (s, u, v)$ and $I(\{x', y'\}) = (s', u', v')$ be the indicators of two object pairs $\{x, y\}$ and $\{x', y'\}$ of P,

$$\{x, y\} \leq_{\omega^1} \{x', y'\} \Leftrightarrow \begin{cases} s > s' \\ \text{or} \\ s = s' \quad \text{and} \quad u + v < u' + v' \\ \text{or} \\ s = s' \quad \text{and} \quad u + v = u' + v' \quad \text{and} \quad u \cdot v \leq u' \cdot v' \end{cases}$$

In other words

$$\{x, y\} \leq_{\omega^1} \{x', y'\} \Leftrightarrow (p - s, u + v, u \cdot v) \leq_{lexico} (p - s', u' + v', u' \cdot v')$$

where $\leq_{lexico}$ designates the linear order on $\mathbb{N}^3$ defined by lexicographic product of linear orders on the set $\mathbb{N}$ of natural integers.

Example

Consider Table 4.2 giving the values of the Indicator function corresponding to the comparison between the described objects in Table 4.1. With the notations adopted above (see Example 2), we have

$$af < ad < df < bc < ef < ac < be < de < ce < bd$$

$$< ab < cd < bf < ae < cf$$

Now and as above, $\omega(S)$ denotes the total preorder on P associated with the similarity measure S, the latter being supposed satisfying the conditions of Definition 14. On the other hand, we denote by $\sim_{\omega(S)}$ (resp., $\sim_{\omega^1}$) the binary equivalence relation induced by the total preorder $\omega(S)$ (resp., ω^1) (let us recall that this equivalence relation is defined as follows: For any p and q in P,
$p \sim_{\omega(S)} q$ (resp., $p \sim_{\omega^1} q$) if and only if $p \leq q$ and $q \leq p$ for $\omega(S)$ (resp., for ω^1).

The most discriminant preordonance has a central property with respect to the similarity measures satisfying the conditions of Definition 14. This is expressed by the following proposition

Proposition 34 *The equivalence binary relation $\sim_{\omega^1}$ is finer than that $\sim_{\omega(S)}$ associated with an arbitrary similarity measure S which satisfies the conditions of Definition 2.*

Indeed, if $\{x, y\} \sim_{\omega^1} \{z, t\}$, we have $s = s', u + v = u' + v'$ and $u \cdot v = u' \cdot v'$. Therefore, $u = u'$ and $v = v'$. Consequently, $S(x, y) = S(z, t)$ and then $\{x, y\} \sim_{\omega(S)} \{z, t\}$.

The Least Discriminant Preordonance

A total preorder having a restrictive feature, on the set $P = P_2(\mathcal{O})$ of unordered object pairs of $\mathcal{O}$ will be proposed here. As for the previous one, its definition does not depend on a *numerical* function $\mathcal{S}(s, u, v)$. It is directly associated with the ordered pair of parameters $(s, u + v)$ or, equivalently, $(s, s + t)$. The definition of this total preorder is derived from the definition of a *strict* but *partial* order on P. Let us denote by ω^2 the latter.

Definition 19 Let p and q be two elements of P, with which are associated their two parameter vectors $(s, u + v)$ and $(s', u' + v')$, respectively.

$$p < q \quad \text{for } \omega^2 \Leftrightarrow s > s' \quad \text{and} \quad u + v < u' + v'$$

or equivalently

$$p < q \quad \text{for } \omega^2 \Leftrightarrow s > s' \quad \text{and} \quad s + t > s' + t'$$

ω^2 has a consensus property with respect to preordonances associated with similarity measures of the form $\mathcal{S}(s, u + v)$. Indeed, if $p < q$ for ω^2, then necessarily, by taking into account the conditions satisfied by $\mathcal{S}(s, u, v)$ (see Definition 14), $\mathcal{S}(s, u + v) > \mathcal{S}((s', u' + v'))$.

The total preorder proposed on P is, in a certain sense, compatible with ω^2. Let us denote it by ϖ^2. Its construction is obtained from the notion of an extremal elements in P or in a P subset, with respect to ω^2. Let Q be a subset of P. A given element p in Q is extremal in Q for the binary relation ω^2, if there does not exist p' in Q, such that $p' < p$ for ω^2. The construction of ϖ^2 is obtained step by step in a recursive process. The principle consists of evolving the set of object pairs by extracting at each step from the remaining P subset, its extremal elements for ω^2. More precisely, let us denote by $P^{(i)}$ the remaining object pairs before the ith stage. $P^{(1)} = P$. At the stage i, the set, denoted by $E^{(i)}$ of extremal elements of $P^{(i)}$ is determined. $E^{(i)}$ defines the ith class of the preorder ϖ^2 and then

$$P^{(i+1)} = P^{(i)} - E^{(i)}$$

The process ends for $P^{(imax+1)} = \emptyset$ and the last preorder class of ϖ^2 is $E^{(imax)}$.

Now, let us illustrate the result of the informal algorithm we have just outlined on the above example. We obtain

$$E^1 = \{ad, af, bc\}$$
$$E^2 = \{be, df\}$$
$$E^3 = \{ac, bd, ce, de, ef\}$$
$$E^4 = \{ab, bf, cd\}$$
$$E^5 = \{ae, cf\}$$

In these conditions the preordonance ϖ^2 is defined as follows:

$$ad \sim af \sim bc < be \sim df < ac \sim bd \sim ce \sim de \sim ef < ab \sim bf \sim cd <$$
$$ae \sim cf$$

In order to see this construction, it would be easier to visualize the respective positions of the different elements of P (there are in all 15) in a Cartesian diagram whose horizontal and vertical axis being sustained by the parameters s and $s + t$, respectively.

4.3 Extension of the Indices Defined in the Boolean Case to Attributes of Type II or III

4.3.1 Introduction

The principle of the extension of association coefficients between Boolean attributes (see Sect. 4.2.1.3) to the case of comparing attributes of types II or III (see Sects. 3.3 and 3.4 of Chap. 3), is based on the set theoretic representation of the relations on $\mathcal{O}$ induced by the descriptive attributes concerned. This representation was defined in Chap. 3 for different types of attributes describing $\mathcal{O}$. More precisely, we will consider the following types of attributes:

1. Nominal categorical attributes (Sect. 3.3.1, Chap. 3);
2. Ordinal categorical attributes (Sect. 3.3.2, Chap. 3);
3. Preordonance categorical attributes (Sect. 3.3.1, Chap. 3).

In each of the three previous cases, it comes to compare two categorial attributes. The difference between them is provided by the nature of the structure endowing the category set of each type of attribute (see Chap. 3). Let us denote by $\{c, d\}$ the categorical attribute pair of the same type to compare. $\{c_1, c_2, \ldots, c_h, \ldots, c_K\}$ (resp., $\{d_1, d_2, \ldots, d_l, \ldots, d_M\}$) denotes the value set of c (resp., d). $\pi = \{\mathcal{O}_1, \mathcal{O}_2, \ldots, \mathcal{O}_h, \ldots, \mathcal{O}_K\}$ (resp., $\chi = \{P_1, P_2, \ldots, P_l, \ldots, P_K\}$) denotes the partition on $\mathcal{O}$ induced by c (resp., d). The joint partition $\pi \wedge \chi$ is defined as follows:

$$\pi \wedge \chi = \{\mathcal{O}_{hl} = \mathcal{O}_h \cap P_l | 1 \leq h \leq K, 1 \leq l \leq M\} \qquad (4.3.1)$$

The contingency table associated with $\{c, d\}$ that we can designate by $Cont(c, d)$ is the table of the cardinal class of $\pi \wedge \chi$, namely,

$$Cont(c, d) = \{n_{hl} = card(\mathcal{O}_{hl}) = card(\mathcal{O}_h \cap P_l) | 1 \leq h \leq K, 1 \leq l \leq M\} \qquad (4.3.2)$$

Table 4.8 Contingency table

$c\backslash d$	d_1	…	d_l	…	d_M	
c_1	n_{11}	…	n_{1l}	…	n_{1M}	$n_{1.}$
⋮	⋮	⋮	⋮	⋮	⋮	⋮
c_h	n_{h1}	…	n_{hl}	…	n_{hM}	$n_{h.}$
⋮	⋮	⋮	⋮	⋮	⋮	⋮
c_K	n_{K1}	…	n_{Kl}	…	n_{KM}	$n_{K.}$
	$n_{.1}$	…	$n_{.l}$	…	$n_{.M}$	n

The general form of $Cont(c, d)$ is given by Table 4.8.

An example of such a contingency table is given in Table 3.3 of Chap. 3.

Clearly, the extension of the similarity coefficients given in Sect. 4.2.1.3 to the above three types of categorical attributes, is obtained when the parameters s, u, v and t defined for Boolean attributes are generalized. These generalizations will be given for each of the three types as specific functions of $Cont(c, d)$. For the first case of nominal categorical attributes the adapted expression of the Jaccard index [13] will be deduced. On the other hand, we will specify those coefficients of the Boolean case corresponding to the well known Rand (1971) and Fowlkes and Mallows (1983) coefficients [6, 27]. Moreover, a coefficient of the same type as that of Goodman and Kruskal (1954) [9] can be envisaged in this framework. We will illustrate on examples the calculation of some of these coefficients.

Nevertheless, not all of the association coefficients between nominal categorical attributes can be expressed as functions of the parameters s, u, v and t defined in this extended context. The structure of those coefficients—which cannot be expressed as a function of s, u, v and t—will be characterized. Mainly, they derive from the Chi square statistic associated with $Cont(c, d)$.

In Sect. 4.3.3 we will consider the case of ordinal categorical attributes. The generalized version of the parameters s, u, v and t will be specified for this case. The set theoretic representation defined in Sect. 3.3.2 of Chap. 3 will be employed. Thus, the famous coefficients γ of Goodman and Kruskal [9] and τ_b of Kendall [14] will be analyzed. We will see how they can be obtained from our formalization and we will specify the Boolean indices to which they correspond. Moreover, we will see how to transpose in the ordinal categorical case indices defined for Boolean attributes.

In Sect. 4.3.4, the transposition of association coefficients between Boolean attributes to those between preordonance categorical attributes, will be defined and specified (see Sect. 3.4.1 of Chap. 3). The coding in terms of the "mean rank" function of such categorical attributes leads us to consider the case of categorical attributes whose respective category sets are valuated by numerical similarity functions (see Sect. 3.3.4 of Chap. 3).

In each of the previous cases (Sects. 4.3.2 and 4.3.3), the meaning of an asymmetrical coefficient deduced from the transposition of an asymmetrical coefficient between Boolean attributes, will be clarified.

4.3.2 Comparing Nominal Categorical Attributes

As just expressed above taking into account the set theoretic representation of a nominal categorical attribute (see Sect. 3.3.1 of Chap. 3), the parameters s, u, v and t must be specified. As above, let us designate by $\{c, d\}$ the categorical attribute pair to be compared. π and χ are the partitions induced, respectively, by c and d on the object set $\mathcal{O}$. Let us recall that the set theoretic representation of a partition π of $\mathcal{O}$ is defined by a partition $\{R(\pi), S(\pi)\}$ of the set $P = P_2(\mathcal{O})$ of unordered object pairs of $\mathcal{O}$, into two complementary classes (see (3.3.4)–(3.3.6)). $\{R(\pi)$ (resp., $S(\pi))$ is the set of joined object pairs (resp., separated object pairs) by the partition π. The set theoretic representation of a given partition π may be $R(\pi)$ or $S(\pi)$. And in fact, some of the association coefficients between partitions are invariant whatever the representation chosen. In our treatment we emphasize the representation of a partition π by the subset $R(\pi)$ of P constituted by the object pairs joined by π.

To a pair $\{\pi, \chi\}$ of partitions corresponds a pair of bi-partitions (i.e. into two classes each), $\{\{R(\pi), S(\pi)\}, \{R(\chi), S(\chi)\}\}$, respectively. Let us consider the crossing between these two bi-partitions of P. Its classes are: $R(\pi) \cap R(\chi)$, $R(\pi) \cap S(\chi)$, $S(\pi) \cap R(\chi)$ and $S(\pi) \cap S(\chi)$. With the notations of Sect. 3.3.1 and those above (see (4.3.1)), we have:

$$R(\pi) \cap R(\chi) = \left(\sum_{h=1}^{h=K} P_2(O_h) \right) \cap \left(\sum_{l=1}^{l=M} P_2(P_l) \right)$$

$$= \sum_{h=1}^{h=K} \sum_{l=1}^{l=M} P_2(O_{hl})$$

$$R(\pi) \cap S(\chi) = \left(\sum_{h=1}^{h=K} P_2(O_h) \right) \cap \left(\sum_{1 \le e < l \le M} O_e \star O_l \right)$$

$$= \sum_{h=1}^{h=K} \sum_{1 \le e < l \le M} O_{he} \star O_{hl} \qquad (4.3.3)$$

By analogy, the reader will determine the set expressions of $S(\pi) \cap R(\chi)$ and $S(\pi) \cap S(\chi)$. The cardinalities of these different sets can be deduced immediately. Let us give one example, the rest is left for the reader.

$$card(R(\pi) \cap S(\chi)) = \sum_{h=1}^{h=K} \sum_{1 \le e < l \le M} n_{he} \times n_{hl}$$

In these conditions, the set expression of the parameters s, u, v and t (see (4.2.9)) becomes:

$$s(c, d) = card(R(\pi) \cap R(\chi)), u(c, d) = card(R(\pi) \cap S(\chi))$$
$$v(c, d) = card(S(\pi) \cap R(\chi)), t(c, d) = card(S(\pi) \cap S(\chi)) \qquad (4.3.4)$$

$R(\pi) \cap R(\chi)$ is the set of unordered object pairs joined by the joint partition $\pi \wedge \chi$:

$$R(\pi \wedge \chi) = R(\pi) \cap R(\chi)$$

Therefore, by referring to the parameter $s(c, d)$, we have

$$s(c, d) = \frac{1}{2} \sum_{h=1}^{h=K} \sum_{l=1}^{l=M} n_{hl} \cdot (n_{hl} - 1),$$
$$u(c, d) = card[R(\pi)] - s(c, d)$$
$$v(c, d) = card[R(\chi)] - s(c, d),$$
$$t(c, d) = card(P) - card[R(\pi)] - card[R(\chi)] + s(c, d). \qquad (4.3.5)$$

where $card[R(\pi)]$, $card[R(\chi)]$ and $card(P)$ were given in (3.3.11) and (3.3.12) of Chap. 3.

In these conditions and as announced above, every similarity measure between Boolean attributes satisfying the conditions of Definition 14 (see Sect. 4.2.1.2) can be transposed in the case of comparing two nominal categorial attributes inducing a couple of partitions on $\mathcal{O}$.

Some of the association coefficients between nominal categorical attributes were invented directly in the framework of comparing partitions. In fact as emphasized above and expressed in [22], they correspond to similarity indices established for comparing Boolean data. Thus, the Rand coefficient (1971) [27] is nothing else than the translation of the Sokal and Michener coefficient (1958) [31]. On the other hand, the Fowlkes and Mallows coefficient (1983) [6] can be defined as the transposition of the Ochiai index (1957) [25]. Moreover, one can establish that the Hamann coefficient (1961) [12] can take a form analogous to the Goodman and Kruskal coefficient (1954) [9]; that is, $[(s + t) - (u + v)]/(s + t + u + v)$.

Now, let us illustrate the computing of four coefficients, mutually different enough with respect to their analytical structures. For this we consider the contingency table Table 3.3 of Chap. 3. These coefficients are

$S_1(c, d) = \frac{s+t}{s+u+v+t}$ Sokal and Michener (1958) $\sim$ Rand (1971)

$S_2(c, d) = \frac{s}{s+u+v}$ Jaccard (1908)

$S_3(c, d) = \frac{s}{\sqrt{(s+u) \cdot (s+v)}}$ Ochiai (1957) $\sim$ Fowlkes and Mallows (1983)

$S_4(c, d) = \frac{st-uv}{st+uv}$ Yule (1911, 1912)

Clearly, both attributes c: "Fertility planning" and d: "Highest level of formal education of wife" are interpreted here as *nominal* categorial attributes. As above, π and χ designate the partitions induced by c and d, respectively. We have

$$card(P) = \frac{1}{2} \times 1438 \times 1437 = 1,033,203$$

$$card[R(\pi)] = \frac{1}{2} \times (379 \times 378 + 451 \times 450 + 205 \times 204 + 403 \times 402) = 275,019$$

$$card[R(\chi)] = \frac{1}{2} \times (591 \times 590 + 608 \times 607 + 239 \times 238) = 387,314$$

$$card[R(\pi \wedge \chi)] = \frac{1}{2} \times (223 \times 222 + 122 \times 121 + 34 \times 33$$
$$+168 \times 167 + 215 \times 214 + 68 \times 67$$
$$+90 \times 89 + 80 \times 79 + 35 \times 34$$
$$+110 \times 109 + 191 \times 190 + 102 \times 101) = 109,057$$

Thus,

$$s(c, d) = 109,057, u(c, d) = 165,962, v(c, d) = 278,257 \text{ and } t(c, d) = 479,927$$

In these conditions,

$$S_1(c, d) = \frac{109,057 + 479,927}{1,033,203} = 0.570$$

$$S_2(c, d) = \frac{109,057}{109,057 + 165,962 + 278,257} = 0.197$$

$$S_3(c, d) = \frac{109,057}{\sqrt{275,019 \times 387,314}} = 0.334$$

$$S_4(c, d) = \frac{109,057 \times 479,927 - 165,962 \times 278,257}{109,057 \times 479,927 + 165,962 \times 278,257} = 0.063$$

There is no reason to be surprised by the mutual differences between the values of these four indices. Indeed, their respective constructions follow very different intuitive logics. In any case, only comparing values of a same given index on different attribute pairs makes sense.

In Table 4.5 a transposition to asymmetrical implicative case of symmetrical association coefficients between Boolean attributes (see Table 4.3) has been proposed. By taking into account (4.3.5), the indices of Table 4.5 can be worked on again in the case of nominal categorical attributes. For a given pair (c, d) of such attributes and for a given implicative index $Imp(c \rightarrow d)$, a strong value of the latter index means, that for a given object pair $\{x, y\}$, generally but not absolutely, if x and y are joined by the partition π induced by c ($\{x, y\} \in R(\pi)$), then, x and y are joined by the partition χ induced by d ($\{x, y\} \in R(\chi)$). For illustration, we propose to the reader to calculate the values of the asymmetrical versions of the coefficients $S_2(c, d)$ and $S_3(c, d)$. We obtain 0.780 and 0.606, respectively. Notice that the asymmetrical versions of the indices $S_1(c, d)$ and $S_4(c, d)$ are identical to those symmetric ones.

A final point. For the nominal categorical attribute c (resp., d) let us introduce the indicator functions ρ_π and σ_π (resp., ρ_χ and σ_χ) of the subsets $R(\pi)$ and $S(\pi)$ [resp., $R(\chi)$ and $S(\chi)$] of $P = P_2(\mathcal{O})$ (see (3.3.7) and (3.3.8)). Using these Boolean functions, the parameters $s(c, d)$, $u(c, d)$, $v(c, d)$ and $t(c, d)$ (see 4.3.5) become

$$s(c, d) = \sum_{\{x, y\} \in P} \rho_\pi(\{x, y\}) \cdot \rho_\chi(\{x, y\})$$

$$u(c, d) = \sum_{\{x, y\} \in P} \rho_\pi(\{x, y\}) \cdot \sigma_\chi(\{x, y\})$$

$$v(c, d) = \sum_{\{x, y\} \in P} \sigma_\pi(\{x, y\}) \cdot \rho_\chi(\{x, y\})$$

$$t(c, d) = \sum_{\{x, y\} \in P} \sigma_\pi(\{x, y\}) \cdot \sigma_\chi(\{x, y\}) \tag{4.3.6}$$

where, obviously, we have $\rho_\pi(\{x, y\}) + \sigma_\pi(\{x, y\}) = 1$ (resp., $\rho_\chi(\{x, y\}) + \sigma_\chi(\{x, y\}) = 1$) for any $\{x, y\}$ in P.

4.3.3 Comparing Ordinal Categorical Attributes

Let ω and ϖ be two total preorders on the object set $\mathcal{O}$ induced respectively by two ordinal categorical attributes c and d. The contingency table crossing these two categorical attributes is represented exactly as for nominal case (see Table 4.8). The difference consists of supposing a linear (total) order on the category set of c (resp., d); from c_1 to c_K for c (resp., from d_1 to d_M for d). In Sect. 3.3.2 we have established the set theoretic representation of a total preorder ω induced by a categorical attribute c. We take again here this representation for each of both attributes c and d. The below notations for c are those of Sect. 3.3.2. For d, where we have

$$d_1 < d_2 < \cdots < d_l < \cdots < d_M \tag{4.3.7}$$

we obtain

$$P_1 < P_2 < \cdots < P_l < \cdots < P_M \tag{4.3.8}$$

where $P_l = d^{-1}(d_l)$ is the lth class of the total preorder ϖ.

According to (3.3.19) we have the following decomposition of the set C of ordered pairs of distinct objects from $\mathcal{O}$ (see (3.3.16))

$$C = R(\varpi) + E(\varpi) + S(\varpi) \tag{4.3.9}$$

Remind that in the case of comparing two *nominal* categorical attributes c and d (see the previous section) we were led to cross two bi-partitions (into *two* classes each) of the set $P = P_2(\mathcal{O})$, namely $\{R(\pi), S(\pi)\}$ and $\{R(\chi), S(\chi)\}$, where π and χ are the partitions of $\mathcal{O}$, induced by c and d, respectively. Here, where c and d are two *ordinal* categorical attributes, represented by two total preorders ω and ϖ (see Chap. 3, Sect. 3.3.2), we are led to cross two partitions of C (see (3.3.16)) into *three* classes each, namely $\{R(\omega), E(\omega), S(\omega)\}$ and $\{R(\varpi), E(\varpi), S(\varpi)\}$ (see (3.3.19)). The nine classes of the joint partition resulting of this crossing are

$$R(\omega) \cap R(\varpi), \ R(\omega) \cap \{E(\varpi), \ R(\omega) \cap S(\varpi),$$
$$E(\omega) \cap R(\varpi), \ E(\omega) \cap \{E(\varpi), \ E(\omega) \cap S(\varpi),$$
$$S(\omega) \cap R(\varpi), \ S(\omega) \cap \{E(\varpi), \ S(\omega) \cap S(\varpi). \tag{4.3.10}$$

Let us now give explicit expressions of the sets introduced in the previous equations using the classes of the total preorders ω and ϖ

$$R(\omega) \cap R(\varpi) = \left(\sum_{1 \le g < h \le K} \mathcal{O}_g \times \mathcal{O}_h \right) \cap \left(\sum_{1 \le e < l \le M} P_e \times P_l \right)$$
$$= \sum_{1 \le g < h \le K} \sum_{1 \le e < l \le M} \mathcal{O}_{ge} \times \mathcal{O}_{hl}$$

$$R(\omega) \cap E(\varpi) = \left(\sum_{1 \le g < h \le K} \mathcal{O}_g \times \mathcal{O}_h \right) \cap \left(\sum_{1 \le l \le M} P_l^{[2]} \right)$$
$$= \sum_{1 \le g < h \le K} \sum_{1 \le l \le M} \mathcal{O}_{gl} \times \mathcal{O}_{hl}$$

$$R(\omega) \cap S(\varpi) = \left(\sum_{1 \le g < h \le K} \mathcal{O}_g \times \mathcal{O}_h \right) \cap \left(\sum_{1 \le e < l \le M} P_l \times P_e \right)$$
$$= \sum_{1 \le g < h \le K} \sum_{1 \le e < l \le M} \mathcal{O}_{gl} \times \mathcal{O}_{he} \tag{4.3.11}$$

six other such sets have to be detailed. Let us give one more expression and leave the reader to specify the last five of them.

$$E(\omega) \cap E(\varpi) = \left(\sum_{1 \le h \le K} \mathcal{O}_h^{[2]} \right) \cap \left(\sum_{1 \le l \le M} P_l^{[2]} \right)$$
$$= \sum_{1 \le h \le K} \sum_{1 \le l \le M} (\mathcal{O}_h \cap P_l)^{[2]}$$
$$= \sum_{1 \le h \le K} \sum_{1 \le l \le M} \mathcal{O}_{hl}^{[2]} \tag{4.3.12}$$

In order to be guided in the construction of an association coefficient between ω and ϖ, let us follow the rationale of the elaboration of two famous coefficients: the τ_b of Kendall and the γ of Goodman and Kruskal [9, 14]. Our analysis will lead us to establish the *identity* of these two coefficients. For both coefficients we refer to the set of ordered object pairs which are not tied ranked for ω neither nor for ϖ. Denote by $C(\omega, \varpi)$ this set. It can be put into the following form:

$$C(\omega, \varpi) = C - (E(\omega) \cap E(\varpi))^c \qquad (4.3.13)$$

where the symbol '$-$' indicates the set difference and where $.^c$ indicates the complementary in C of $..$. In these conditions $\{R(\omega), S(\omega)\}$ (resp., $\{R(\varpi), S(\varpi)\}$) determines a bi-partition (i.e. into two classes) of $C - E(\omega)$ (resp., $C - E(\varpi)$). By considering the respective restrictions of the two bi-partitions to $C(\omega, \varpi)$, we define the following parameters:

$$s(c, d) = card[R(\omega) \cap R(\varpi)] = \sum_{1 \leq g < h \leq K} \sum_{1 \leq e < l \leq M} n_{ge} \times n_{hl}$$

$$u(c, d) = card[R(\omega) \cap S(\varpi)] = \sum_{1 \leq g < h \leq K} \sum_{1 \leq e < l \leq M} n_{gl} \times n_{he}$$

$$v(c, d) = card[R(\varpi) \cap S(\omega)] = \sum_{1 \leq g < h \leq K} \sum_{1 \leq e < l \leq M} n_{he} \times n_{gl}$$

$$t(c, d) = card[S(\omega) \cap S(\varpi)] = \sum_{1 \leq g < h \leq K} \sum_{1 \leq e < l \leq M} n_{hl} \times n_{ge} \qquad (4.3.14)$$

Clearly,

$$s(c, d) = t(c, d) \text{ and } u(c, d) = v(c, d) \qquad (4.3.15)$$

In these conditions, the coefficients τ_b of Kendall and γ of Goodman and Kruskal can be written as follows:

$$\tau_b = \frac{s - u}{\sqrt{(s + u) \cdot (s + v)}} = \frac{1}{2} \times \frac{(s + t) - (u + v)}{\sqrt{(s + u) \cdot (s + v)}} \qquad (4.3.16)$$

$$\gamma = \frac{(s + t) - (u + v)}{(s + t) + (u + v)} \qquad (4.3.17)$$

Thus, the numerators of τ_b and γ are identical. However the presentation of the denominator of τ_b is different from that of γ. Let us point out that the sets $E(\omega)$ and $E(\varpi)$ do not take part in the definition of the τ_b coefficient (see paragraph 3.16 of Chap. 3 in [14]).

It is interesting to notice that the γ index corresponds to the Hamann one [12] defined in the context of comparing Boolean data. On the other hand, the normalization of τ_b is structurally equivalent to that of the Ochiai index [25]. However, in the latter case, the numerator is defined by s instead of $s - u$ for the τ_b index.

Now, every index defined in the Boolean context and satisfying the conditions of Definition 14 (see Sect. 4.2.1.2) can be transposed in the case of ordinal categorical attributes, from the above definition of the parameters s, u, v and t (see (4.3.14)). Consider more particularly, the adaptation of the coefficients $S_1(c, d)$, $S_2(c, d)$, $S_3(c, d)$ and $S_4(c, d)$ mentioned in Sect. 4.2.2.2.

Let us consider again the denominators of the coefficients τ_b and γ. From the relations of (4.3.15) we have:

$$\sqrt{(s + u)(s + v)} = \sqrt{(s + u)(s + u)} = s + u \qquad (4.3.18)$$

$$(s + t) + (u + v) = 2(s + u) \qquad (4.3.19)$$

Therefore, τ_b and γ can be put in the common form as follows:

$$\frac{s - u}{s + u} = \frac{t - v}{t + v} \qquad (4.3.20)$$

Then we deduce

Theorem 7 *The coefficients τ_b and γ are identical.*

In order to illustrate the computing of the τ_b coefficient, take up again Table 3.3 of Chap. 3. We have

$$s(c, d) = 223 \times (215 + 68 + 80 + 35 + 191 + 102)$$
$$+ 122 \times (68 + 35 + 102)$$
$$+ 168 \times (80 + 35 + 191 + 102)$$
$$+ 215 \times (35 + 102)$$
$$+ 90 \times (191 + 102)$$
$$+ 80 \times 102 = 311{,}632$$

$$u(c, d) = 122 \times (168 + 90 + 110)$$
$$+ 34 \times (168 + 215 + 90 + 80 + 110 + 191)$$
$$+ 215 \times (90 + 110)$$
$$+ 68 \times (90 + 80 + 110 + 191)$$
$$+ 80 \times 110$$
$$+ 35 \times (110 + 191) = 168{,}295$$

With these values, we obtain the value of the τ_b index

$$\tau_b = \frac{311{,}632 - 168{,}295}{311{,}632 + 168{,}295} = 0.299$$

This value is also that of the index γ ; and indeed, it is given in [9].

For illustration, let us give here the values of the coefficients $S_1(c, d)$, $S_2(c, d)$, $S_3(c, d)$ and $S_4(c, d)$ mentioned in Sect. 4.2.2.2, in the context of comparing ordinal categorical attributes. We obtain:

$$S_1(c, d) = 0.649,\ S_2(c, d) = 0.481,\ S_3(c, d) = 0.649 \text{ and } S_4(c, d = 0.548$$

The set theoretic representation of an ordinal categorical attribute c might be considered at the level of the entire set C and not at that of $C(\omega, \varpi)$ (see (4.3.13)). However, the representation of c by $R(\omega)$ in C is asymmetrical. The structure of the complementary subset of $R(\omega)$ in C is different from that of $R(\omega)$. Indeed, $R(\omega)^c$ is defined by $S(\omega) + E(\omega)$ and then, it includes the subset $E(\omega)$ of object pairs tied ranked for ω.

For comparing two total preorders ω and ϖ on an object set $\mathcal{O}$, Giakoumakis and Monjardet [8] propose different generalizations of the coefficient τ established for comparing two total and strict orders (linear orders) on $\mathcal{O}$. Their analysis is based on the respective roles played by $E(\omega)$ and $E(\varpi)$. In order to take into account the latter sets constituted by tied ranked pairs for ω and for ϖ, respectively, the solution we propose is based on the coding of ω (resp., ϖ) by the function $score_\omega$ (resp., $score_\varpi$) (see (3.3.25)). Thus, a balanced representation is provided. With such a representation, the parameters $s(c, d)$, $u(c, d)$, $v(c, d)$ and $t(c, d)$ become:

$$s(c, d) = \sum_{(x,y)\in C} score_\omega(x, y) \cdot score_\varpi(x, y) = card[R(\omega) \cap R(\varpi)]$$

$$+\frac{1}{2}card[R(\omega) \cap E(\varpi)] + \frac{1}{2}card[E(\omega) \cap R(\varpi)] + \frac{1}{4}card[E(\omega) \cap E(\varpi)]$$

$$u(c, d) = \sum_{(x,y)\in C} score_\omega(x, y) \cdot [1 - score_\varpi(x, y)]$$

$$= card(R(\omega)) + \frac{1}{2}card(E(\omega)) - s(c, d)$$

$$v(c, d) = \sum_{(x,y)\in C} score_\varpi(x, y) \cdot [1 - score_\omega(x, y)]$$

$$= card(R(\varpi)) + \frac{1}{2}card(E(\varpi)) - s(c, d)$$

$$t(c, d) = \sum_{(x,y)\in C} [1 - score_\omega(x, y)] \cdot [1 - score_\varpi(x, y)]$$

$$= n(n-1) - \left[card(R(\omega)) + \frac{1}{2} card(E(\omega)) \right]$$
$$- \left[card(R(\varpi)) + \frac{1}{2} card(E(\varpi)) \right] + s(c,d)$$

$$(4.3.21)$$

These equations employ two decompositions of C according two partitions: $\{R(\omega), E(\omega), S(\omega)\}$ and $\{R(\varpi), E(\varpi), S(\varpi)\}$. For this representation we have neither the equality between $s(c,d)$ and $t(c,d)$ nor that between $u(c,d)$ and $v(c,d)$.

Obviously, every similarity measure defined in the Boolean context and satisfying the conditions of Definition 14 (see Sect. 3.2.1.2) can be transposed here with the expressions stated above of $s(c,d)$, $u(c,d)$, $v(c,d)$ and $t(c,d)$. The readers are invited to take up again the above similarity measures, $S_1(c,d)$, $S_2(c,d)$, $S_3(c,d)$ and $S_4(c,d)$ and to calculate this values in their new framework, for Table 3.3 of Chap. 3.

The previous codings obtained from the functions $score_\omega$ and $score_\varpi$ are binary. Let us end this section by considering a unary coding provided by the function "mean rank" applied on a total preorder on the object set $\mathcal{O}$, this being induced in our context, by an ordinal categorical attribute (see (3.3.31)). Denote by q_m and r_m the "mean rank" functions, coding the two total preorders on $\mathcal{O}$, induced by two ordinal categorical attributes c and d. We have:

$$(\forall h, 1 \leq h \leq K), \quad (\forall x \in O_h), \quad q_m(x) = N_{h-1.} + \frac{1}{2} \times (n_{h.} + 1) \qquad (4.3.22)$$

$$(\forall l, 1 \leq l \leq M), \quad (\forall x \in P_l), \quad r_m(x) = N_{.l-1} + \frac{1}{2} \times (n_{.l} + 1) \qquad (4.3.23)$$

where $N_{h-1.} = n_{1.} + n_{2.} + \cdots + n_{h-1.}$ (resp., $N_{.l-1} = n_{.1} + n_{.2} + \cdots + n_{.l-1}$). Thus, the maximal value of q_m (resp., r_m) is $N_{K-1.} + \frac{1}{2} \times (n_{K.} + 1)$ (resp., $N_{.M-1} + \frac{1}{2} \times (n_{.M} + 1)$). We denote by $Q_m(h)$ and $R_m(l)$ the right members of (4.3.22) and (4.3.23), respectively.

In order to adapt the formalism of the indices $s(c,d)$, $u(c,d)$, $v(c,d)$ and $t(c,d)$ (see above), let us introduce the following two functions, taking their values in the real interval [0, 1]:

$$(\forall x \in \mathcal{O}), \chi_m(x) = \frac{q_m(x)}{Q_m(K)} \qquad (4.3.24)$$

and

$$(\forall x \in \mathcal{O}), \rho_m(x) = \frac{r_m(x)}{R_m(M)} \qquad (4.3.25)$$

In these conditions, we can take up again the same form of the expressions considered in Sect. 4.2.1.3 as follows:

$$s(c,d) = \sum_{x \in \mathcal{O}} \chi_m(x) \cdot \rho_m(x)$$

$$u(c,d) = \sum_{x \in \mathcal{O}} \chi_m(x) \cdot [1 - \rho_m(x)]$$

$$v(c,d) = \sum_{x \in \mathcal{O}} [1 - \chi_m(x)] \cdot \rho_m(x)$$

$$t(c,d) = \sum_{x \in \mathcal{O}} [1 - \chi_m(x)] \cdot [1 - \rho_m(x)] \tag{4.3.26}$$

By considering the entries of the contingency table associated with the joint partition $\pi \wedge \chi$ (see (4.3.2)), we obtain:

$$s(c,d) = \sum_{1 \leq h \leq K} \sum_{1 \leq l \leq M} n_{hl} \chi_m(h) \cdot \rho_m(l)$$

$$u(c,d) = \sum_{1 \leq h \leq K} \sum_{1 \leq l \leq M} n_{hl} \chi_m(h) \cdot [1 - \rho_m(l)]$$

$$v(c,d) = \sum_{1 \leq h \leq K} \sum_{1 \leq l \leq M} n_{hl} [1 - \chi_m(h)] \cdot \rho_m(l)$$

$$t(c,d) = \sum_{1 \leq h \leq K} \sum_{1 \leq l \leq M} n_{hl} [1 - \chi_m(h)] \cdot [1 - \rho_m(l)] \tag{4.3.27}$$

where we have denoted by $\chi_m(h)$ (resp., $\rho_m(l)$) the common value of $\chi_m(x)$ (resp., $\rho_m(x)$) for x in $\mathcal{O}_h$, $1 \leq h \leq K$ (resp., for x in P_l, $1 \leq l \leq M$).

Therefore, every similarity coefficient defined in the Boolean context and satisfying the conditions of Definition 14, can be transposed in the ordinal categorical case by using the above expressions of (4.3.27). Let us see how these expressions are instantiated in the case of Table 3.3 of Chap. 3. The sequence of the values of $\chi_m(h)$, $1 \leq h \leq K$, becomes

$$\chi_m(1) = \frac{190}{1237} = 0.154, \quad \chi_m(2) = \frac{605}{1237} = 0.489$$

$$\chi_m(3) = \frac{933}{1237} = 0.754, \quad \chi_m(4) = 1$$

The sequence of the values of $\rho_m(l)$, $1 \leq l \leq M$, becomes

$$\rho_m(1) = \frac{295.5}{1318.5} = 0.224, \quad \rho_m(2) = \frac{895}{1318.5} = 0.679, \quad \rho_m(3) = 1$$

In these conditions, the value of $s(c, d)$ is obtained as follows:

$$s(c, d) = 223 \times 0.154 \times 0.224 + 122 \times 0.154 \times 0.679 + 34 \times 0.154 \times 1.$$
$$+168 \times 0.489 \times 0.224 + 215 \times 0.489 \times 0.679 + 68 \times 0.489 \times 1.$$
$$+90 \times 0.754 \times 0.224 + 80 \times 0.754 \times 0.679 + 35 \times 0.754 \times 1.$$
$$+110 \times 1. \times 0.224 + 191 \times 1. \times 0.679 + 102 \times 1. \times 1.$$
$$= 487.603$$

On the other hand, we have:

$$s(c, d) + u(c, d) = \sum_{1 \leq h \leq K} n_{h.} \chi_m(h)$$
$$= 379 \times 0.154 + 451 \times 0.489 \times +205 \times 0.754 + 403 \times 1.$$
$$= 836.475$$

$$s(c, d) + v(c, d) = \sum_{1 \leq l \leq M} n_{.l} \rho_m(l)$$
$$= 591 \times 0.224 + 608 \times 0.679 + 239 \times 1.$$
$$= 784.216$$

As a result we obtain

$$u(c, d) = 836.475 - 487.603 = 348.872$$
$$v(c, d) = 784.216 - 487.603 = 296.613$$
$$t(c, d) = 1438 - (487.603 + 348.872 + 296.613) = 304.912$$

Now and for illustration, we can calculate the values of the indices $S_1(c, d)$, $S_2(c, d)$, $S_3(c, d)$ and $S_4(c, d)$. Let us give the values of $S_1(c, d)$ and $S_2(c, d)$ and leave for the reader the calculation of $S_3(c, d)$ and $S_4(c, d)$.

$$S_1(c, d) = \frac{487.603 + 304.912}{1438} = 0.551$$
$$S_2(c, d) = \frac{487.603}{487.603 + 348.872 + 296.613} = 0.430$$

4.3.4   *Comparing Preordonance Categorical Attributes*

In Sect. 3.4.1 the mathematical representation of such categorical attributes was defined. Denote as above by $\{c, d\}$ the attribute pair to be compared. The following sets of category ordered pairs correspond to c and d, respectively.

$$C_2 = \{(c_g, c_h) | 1 \leq g \leq h \leq K\} \tag{4.3.28}$$

and

$$D_2 = \{(d_e, d_l) | 1 \leq e \leq l \leq M\} \tag{4.3.29}$$

Consider also the following sets of ordered category codes of c and d, namely,

$$K_2 = \{(g, h) | 1 \leq g \leq h \leq K\} \tag{4.3.30}$$

and

$$M_2 = \{(g, h) | 1 \leq e \leq l \leq M\} \tag{4.3.31}$$

The latter two sets are introduced in order to simplify our notations. K_2 and M_2 will represent the sets C_2 and D_2. $\omega(K_2)$ (resp., $\varpi(M_2)$) will designate a total preorder on K_2 (resp., on M_2) defining in an ordinal manner (by ranking) the similarities between the categories of the attribute c (resp., d). As in Sect. 3.4.1, denote by

$$(L_1, L_2, \ldots, L_p, \ldots, L_r) \tag{4.3.32}$$

and by

$$(M_1, M_2, \ldots, M_q, \ldots, M_s) \tag{4.3.33}$$

the ordered sequences of the $\omega(K_2)$ and $\varpi(M_2)$ preorder classes, respectively.

Two representations of a preordonance categorical attribute c were developed in Sect. 3.4.1 of Chap. 3. The first one is a set theoretic representation (see (3.4.4)–(3.4.7)) and the second one, is provided by means of a numerical valuation, defined from the notion of the "mean rank" function defined on the set of category pairs of c (see (3.4.8)). We have a clear analogy with the development in the previous section concerning the comparison of ordinal categorical attributes. However, the structure defined by an *ordinal* categorical attribute is a total preorder on the category set of this attribute, whereas, this total preorder is on the set of category *pairs* in the case of a *preordonance* categorical attribute.

4.3.4.1   Set Theoretic Representation

Let us work up again in this new framework the same type of decomposition into nine classes considered in (4.3.11). We use the same type of notations as in (3.4.4)–(3.4.7):

$$R[\omega(K_2)] \cap R[\varpi(M_2)] = \left(\sum_{1 \le p < p' \le r} L_p \times L_{p'} \right) \cap \left(\sum_{1 \le q < q' \le s} M_q \times M_{q'} \right)$$

$$R[\omega(K_2)] \cap E[\varpi(M_2)] = \left(\sum_{1 \le p < p' \le r} L_p \times L_{p'} \right) \cap \left(\sum_{1 \le q \le s} M_q^{[2]} \right)$$

$$R[\omega(K_2)] \cap S[\varpi(M_2)] = \left(\sum_{1 \le p < p' \le r} L_p \times L_{p'} \right) \cap \left(\sum_{1 \le q < q' \le s} M_{q'} \times M_q \right)$$

$$(4.3.34)$$

Here also, we leave for the reader to specify the detailed formulas for the six other classes of the decomposition concerned. These are

$$E[\omega(K_2)] \cap R[\varpi(M_2)], \quad E[\omega(K_2)] \cap E[\varpi(M_2)], \, E[\omega(K_2)] \cap S[\varpi(M_2)]$$
$$S[\omega(K_2)] \cap R[\varpi(M_2)], \quad S[\omega(K_2)] \cap E[\varpi(M_2)], \, S[\omega(K_2)] \cap S[\varpi(M_2)]$$

$$(4.3.35)$$

Building an association coefficient between the categorical attributes c and d, requires the calculation of the cardinalities of the sets we have just defined (see (4.3.34)–(4.3.35)). This must be done with respect to the class cardinals of the joint partition $\pi \wedge \chi$ (see (4.3.1) and (4.3.2)). Therefore, each of the above expressions must be detailed with respect to the classes $\mathcal{O}_{hl}$, $1 \le h \le K$, $1 \le l \le M$ of $\pi \wedge \chi$. For this purpose, we will employ the decompositions of the form (3.4.7). In order to be more explicit let us denote here by $\{\mathcal{O}_{h.}|1 \le h \le K\}$ and $\{\mathcal{O}_{.l}|1 \le l \le M\}$ the partitions π and χ, respectively. With these notations, one generic element of the sum (set sum) detailing $R[\omega(K_2)] \cap R[\varpi(M_2)]$ can be written as follows:

$$(\mathcal{O}_{e.} \star \mathcal{O}_{f.} \times \mathcal{O}_{g.} \star \mathcal{O}_{h.}) \cap (\mathcal{O}_{.i} \star \mathcal{O}_{.j} \times \mathcal{O}_{.k} \star \mathcal{O}_{.l}) = \mathcal{O}_{ei} \star \mathcal{O}_{fj} \times \mathcal{O}_{gk} \star \mathcal{O}_{hl} \quad (4.3.36)$$

where $ef \in L_p$, $gh \in L_{p'}$, $ij \in M_q$, $kl \in M_{q'}$. By referring to the contingency table $Cont(c, d)$ (4.3.2), the cardinality of the set (4.3.36) is given by

$$n_{ei} \times n_{fj} \times n_{gk} \times n_{hl} \quad (4.3.37)$$

The expression of one generic element of the sum defining $R[\omega(K_2)] \cap E[\varpi(M_2)]$ is analogous to the preceding generic element (4.3.36). But in this case $ef \in L_p$, $gh \in L_{p'}$, $(ij, kl) \in M_q^{[2]}$. The form of its cardinality is still given by (4.3.37).

Finally, the respective forms of the set theoretic and cardinal expressions of a generic element of $R[\omega(K_2)] \cap S[\varpi(M_2)]$ (see (4.3.36) and (4.3.37)) hold. Here, $ef \in L_p$, $gh \in L_q$, $ij \in M_{q'}$, $kl \in M_q$.

An exercise left for the reader consists of developing the 6 other expressions (4.3.35) and to write a program which enables us to compute the nine expressions (4.3.34) and (4.3.35). This program can be applied in the case of Table 3.3 of Chap. 3, where the two following preordonances may be considered:

$$(c_1, c_3) < (c_1, c_2) < (c_2, c_3) < (c_2, c_2) < (c_1, c_1) < (c_3, c_3) \cdot$$
$$(A, D) < (B, D) < (A, C) < (C, D) < (A, B) \sim (B, C)$$
$$< A, A) \sim (C, C) \sim (D, D) < (B, B) \quad (4.3.38)$$

Example

Let us indicate how to compute the value of $card(R[\omega(K_2)] \cap R[\varpi(M_2)])$ in the framework of Table 3.3 of Chap. 3.

- By taking $p = 3$ and $p' = 5$, $L_3 \times L_5$ has to take part.
- By taking $q = 4$ and $q' = 5$, $M_3 \times M_5$ has to take part.
- L_3 includes a single pair: (c_2, c_3) and then $ef = 23$
- L_5 includes a single pair: (c_1, c_1) and then $gh = 11$
- M_4 includes a single pair: (d_3, d_4) and then $ij = 34$
- M_5 includes a two pairs: (d_1, d_2) and (d_1, d_2),
  we take (d_1, d_2) and then, $kl = 12$

Therefore

$$n_{ei} \times n_{fj} \times n_{gk} \times n_{hl} = n_{23} \times n_{34} \times n_{11} \times n_{12} = 215 \times 34 \times 110 \times 90 = 72{,}369{,}000$$

The set theoretic representation given in (4.3.34)–(4.3.35) permits us to consider an analogous development to that given in Sect. 4.3.3, concerning the adaptation of every association coefficient between Boolean attributes in order to compare preordonance categorical attributes. For example, an extension of the Goodman and Kruskal coefficient can be expressed as follows:

$$\Gamma = \frac{R[\omega(K_2)] \cap R[\varpi(M_2)] - R[\omega(K_2)] \cap S[\varpi(M_2)]}{R[\omega(K_2)] \cap R[\varpi(M_2)] + R[\omega(K_2)] \cap S[\varpi(M_2)]}$$

4.3.4.2 Representation by the "Mean Rank" Function

Let us start with the example of both preordonances given above [see (4.3.38)]. The "mean rank" function was defined to code a categorical preordonance attribute in (3.4.8). As previously, c and d will designate two categorical preordonance attributes. q_m and r_m denote the "mean rank" functions defined on the set of category pairs of c and d, respectively. We have with understandable notations (see Table 4.9):

$$\chi_m(g, h) = \frac{q_m(g, h)}{\max\{q_m(g', h')|1 \le g' \le h' \le K\}} \quad (4.3.39)$$

Analogously (see Table 4.10),

Table 4.9 Values of the mean rank function for c

gh	11	12	13	22	23	33
$q_m(g,h)$	5	2	1	4	3	6
$\chi_m(g,h)$	5/6	1/3	1/6	2/3	1/2	1

Table 4.10 Values of the mean rank function for d

el	11	12	13	14	22	23	24	33	34	44
$r_m(e,l)$	8	5.5	3	1	10	5.5	2	8	4	8
$\rho_m(e,l)$	0.8	0.55	0.3	0.1	1	0.55	0.2	0.8	0.4	0.8

$$\rho_m(e,l) = \frac{r_m(e,l)}{\max\{r_m(e',l')|1 \leq e' \leq l' \leq M\}} \tag{4.3.40}$$

Now, let us determine how the expressions of $s(c,d)$, $u(c,d)$, $v(c,d)$ and $t(c,d)$ become for this coding. For simplicity, the representation set will be the set of ordered object pairs $\mathcal{O}^{[2]}$, instead of the set of unordered object pairs $\mathcal{O}^{\{2\}}$. The values obtained have to be divided by 2 in the case of referring to the latter set. More precisely, we have

$$s(c,d) = \sum_{(x,y)\in\mathcal{O}^{[2]}} \chi_m[c(x),c(y)] \cdot \rho_m[d(x),d(y)]$$

$$u(c,d) = \sum_{(x,y)\in\mathcal{O}^{[2]}} \chi_m[c(x),c(y)] \cdot (1 - \rho_m[d(x),d(y)])$$

$$v(c,d) = \sum_{(x,y)\in\mathcal{O}^{[2]}} (1 - \chi_m[c(x),c(y)]) \cdot \rho_m[d(x),d(y)]$$

$$t(c,d) = \sum_{(x,y)\in\mathcal{O}^{[2]}} (1 - \chi_m[c(x),c(y)]) \cdot (1 - \rho_m[d(x),d(y)]) \tag{4.3.41}$$

In order to compact the preceding expressions consider the partition of $\mathcal{O}^{[2]}$ according to the joint partition $\pi \wedge \chi$. We obtain

$$\mathcal{O}^{[2]} = \sum_{1\leq g\leq h\leq K, 1\leq e\leq l\leq M, (g,e)\neq(h,l)} \mathcal{O}_{ge} \times \mathcal{O}_{hl} + \sum_{1\leq g\leq K, 1\leq e\leq M} \mathcal{O}_{ge}^{[2]} \tag{4.3.42}$$

This partition comprises $(K \times M)^2$ classes. On each of its classes the argument of a given sum of (4.3.41) is invariant. Thus a class such that $\mathcal{O}_{ge} \times \mathcal{O}_{hl}$ for $(g,e) \neq (h,l)$, gives rise to the following contributions:

$$n_{ge} \times n_{hl} \chi_m(gh) \cdot \rho_m(el) \quad \text{for } s(c, d)$$
$$n_{ge} \times n_{hl} \chi_m(gh) \cdot [1 - \rho_m(el)] \quad \text{for } u(c, d)$$
$$n_{ge} \times n_{hl}[1 - \chi_m(gh)] \cdot \rho_m(el) \quad \text{for } v(c, d)$$
$$n_{ge} \times n_{hl}[1 - \chi_m(gh)] \cdot [1 - \rho_m(el)] \quad \text{for } t(c, d) \qquad (4.3.43)$$

On the other hand, a class such that $\mathcal{O}_{ge}^{[2]}$ gives rise to the following contributions:

$$n_{ge} \times (n_{ge} - 1)\chi_m(gg) \cdot \rho_m(ee) \quad \text{for } s(c, d)$$
$$n_{ge} \times (n_{ge} - 1)\chi_m(gg) \cdot [1 - \rho_m(ee)] \quad \text{for } u(c, d)$$
$$n_{ge} \times (n_{ge} - 1)[1 - \chi_m(gg)] \cdot \rho_m(ee) \quad \text{for } v(c, d)$$
$$n_{ge} \times (n_{ge} - 1)[1 - \chi_m(gg)] \cdot [1 - \rho_m(ee)] \quad \text{for } t(c, d) \qquad (4.3.44)$$

Now, let us illustrate these contributions in the case of Table 3.3 of Chap. 3. Take for example $g = 2, h = 3, e = 3$ and $l = 2$. The cardinal of the class $\mathcal{O}_{ge} \times \mathcal{O}_{hl} = \mathcal{O}_{23} \times \mathcal{O}_{32}$ is 215×35. On the other hand, $\chi_m(23) = 0.5$ and $\rho_m(3, 2) = \rho_m(2, 3) = 0.55$. Thus, the sequence of the values of the four contributions defined in (4.3.43) is

$215 \times 35 \times 0.55 = 2069.375$
$215 \times 35 \times 0.45 = 1693.125$
$215 \times 35 \times 0.55 = 2069.375$
$215 \times 35 \times 0.45 = 1693.125$

Let us continue our illustration by giving the respective contributions of $\mathcal{O}_{23}^{[2]}$ to the sequence defined in (4.3.44). The cardinal of $\mathcal{O}_{23}^{[2]}$ is 215×214. We obtain

$215 \times 214 \times \frac{2}{3} \times 0.8 = 24538.667$
$215 \times 214 \times \frac{2}{3} \times 0.2 = 6134.667$
$215 \times 214 \times \frac{1}{3} \times 0.8 = 12269.333$
$215 \times 214 \times \frac{1}{3} \times 0.2 = 3067.333$

Notice that we have worked here in the case where the notion of resemblance between two categories is symmetrical. The case where this notion is asymmetrical can be envisaged. In the latter case the preordonance associated with a given categorical attribute on the set $\mathcal{C}$ of its categories, is defined by a total preorder on the set $\mathcal{C} \times \mathcal{C}$ of ordered pairs of categories of $\mathcal{C}$.

A final point relative to the development above. The latter can be taken up again for comparing categorical attributes valuated by a numerical similarity notion (see Sect. 3.3.4). Indeed, given a categorical attribute, it suffices to substitute a numerical similarity function for the "mean rank" function, on the category set of the attribute concerned (see (3.3.33)).

Other very important facets of the resemblance notion will be addressed in the next two chapters. In Chaps. 5 and 6 a set theoretic, probabilistic and statistical methodology is carried out in order to build an association coefficient between descriptive attributes of any common type. This method which comprises the *LLA* approach, enables different famous coefficients established in the literature to be mutually compared.

The *LLA* approach is also concerned in Chap. 7. The objective consists in the latter of building a similarity index between objects or categories described by attributes of any types.

References

1. Agrawal, T., Imielinski, T., Swami, A.N.: Mining association rules between sets of items in large databases. In: Proceedings of the 1993 ACM SIGMOD International Conference on Management of Data, pp. 207–216 (1993)
2. Batagelj, V., Bren, M.: Comparing resemblance measures. J. Classif. **12**, 73–90 (1995)
3. Bouchon-Meunier, B., Rifqi, M., Bothorel, S.: Towards general measures of comparison of objects. Fuzzy Sets Syst. **2**, 143–153, 84 (1996)
4. de la Vega. W.F.: Techniques de classification automatique utilisant un indice de ressemblance. Revue Francaise de Sociologie (1967)
5. Dice, L.R.: Measures of the amount of ecologic association between species. Ecology **26**, 297–302 (1945)
6. Fowlkes, E.B., Mallows, C.L.: A method for comparing two hierarchical clusterings. J. Am. Stat. Assoc. J.A.S.A. **78**, 553–569 (1983)
7. Geng, L., Howard, J., Hamilton, J.: Choosing the right lens: Finding what is interesting in data mining. Studies in Computational Intelligence (SCI), vol. 43, pp. 3–24. Springer, New York (2007)
8. Giakoumakis, V., Monjardet, B.: Coefficients d ' accord entre deux préordres totaux. Mathématiques et Sciences Humaines **98**, 69–87 (1987)
9. Goodman, L.A., Kruskal, W.H.: Measures of association for cross classifications. J. Am. Stat. Assoc. **49**, 732–764 (1954)
10. Gras, R.: Contribution à l'étude expérimentale et à l'analyse de certaines acquisitions cognitives et de certains objectifs didactiques en mathématiques. Ph.D. thesis, Thèse de doctorat d'état, Université de Rennes 1 (1979)
11. Guillet, F., Hamilton, H.J. (eds.): Quality measures in data mining. Studies in Computational Intelligence, vol. 43. Springer, New York (2007)
12. Hamann, V.: Merkmalbestand und verwandtschaftsbeziehungen der farinosae. Beitragzum System der Monokotyledonen **2**, 639–768 (1961)
13. Jaccard, P.: Nouvelles recherches sur la distribution florale. Bulletin de la Société Vaudoise des Sciences Naturelles **44**, 223–270 (1908)
14. Kendall, M.G.: Rank Correlation Methods. Charles Griffin, New York (1970). (1st edn in 1948)
15. Kulczynski, S.: Die pflanzenassoziationen der pieninen [in polish, german summary]. Bull. Inter. Acad. Pol. Sci. Lett. Cl. Sci. Math. Nat (Sci. Nat) **2**, 57–203 (1927)
16. Lalich, S., Teytaud, O., Prudhomme, E.: Association rule interestingness: Measure and statistical validation. Studies in Computational Intelligence (SCI), vol. 43, pp. 251–275. Springer, New York (2007)
17. Lerman, I.C.: Indice de similarité et préordonnance associée. In: Barbut, M. (ed.) Ordres, Travaux du séminaire sur les ordres totaux finis. Aix-en-Provence, Mouton (1967)
18. Lerman, I.C.: Les Bases de la Classification Automatique. Gauthier-Villars, Paris (1970)
19. Lerman, I.C.: Sur l'analyse des données préalable à une classification automatique: proposition d'une nouvelle mesure de similarité. Mathématiques et Sciences Humaines **32**, 5–15 (1970)
20. Lerman, I.C.: Introduction à une méthode de classification automatique illustrée par la recherche d'une typolologie des personnages enfants à travers la littérature enfantine. Revue de Statistique Appliquée **XXI**(3), 23–49 (1973)

21. Lerman, I.C.: Classification et analyse ordinale des données (1981). Dunod and http://www.brclasssoc.org.uk/books/index.html
22. Lerman, I.C.: Comparing partitions (mathematical and statistical aspects). In: Bock, H.H. (ed.) Classification and Related Methods of Data Analysis, pp. 121–131. North-Holland, Amsterdam (1988)
23. Lesot, M.-J., Rifqi, M., Benhada, H.: Similarity measures for binary and numerical data. Int. J. Knowl. Eng. Soft Data Paradig. **1**, 63–84 (2009)
24. Loevinger, J.: A systematic approach to the construction and evaluation of tests of ability. Psychol. Monogr. **61**, 1–49 (1947)
25. Ochiai, A.: Zoogeographic studies on the soleoid fishes found in Japan and its neighbouring regions. Bull. Jpn. Soc. Sci. Fish. **22**, 526–530 (1957)
26. Pearson, K.: On the coefficient of racial likeness. Biometrika **18**, 105–117 (1926)
27. Rand, W.M.: Objective criteria for the evaluation of clustering methods. J. Am. Stat. Assoc. J.A.S.A. **66**, 846–850 (1971)
28. Rogers, D.J., Tanimoto, T.T.: A computer program for classifying plants. Science **132**, 1115–1118 (1960)
29. Russel, P.F., Rao, C.R.: On habitat and association of species of anopheline larvae in southeastern madras. J. Malar. Inst. India **T3**, 153–178 (1940)
30. Schneider, J., Borlund, P.: Matrix comparison, part 1: Motivation and important issues for measuring the resemblance between proximity measures or ordination results. J. Am. Soc. Inf. Sci. Technol. **58**(11), 1586–1595 (2007)
31. Sokal, R.R., Michener, C.: A statistical method for evaluating systematic relationships. Univ. Kans. Sci. Bull. **38**, 1409–1438 (1958)
32. Sokal, R.R., Sneath, P.H.A.: Principles of Numerical Taxonomy. W.H. Freeman, San Francisco (1963)
33. Sokal, R.R., Sneath, P.H.A.: Numerical Taxonomy. W.H. Freeman, San Francisco (1973)
34. Tan, P.-N., Kumar, V., Srivastava, J.: Selecting the right interestingness measure for association patterns. In: 8th ACM SIGKDD (ed.) Proceedings of the 8th ACM SIGKDD Conference on Knowledge Discovery and Data Mining (2002)
35. Yule, G.U.: An introduction of the theory of statistics. Charles Griffin, London (1911)
36. Yule, G.U.: On measuring association between attributes. J. Royal Statist. Soc. **75**, 579–642 (1912)
37. Zighed, D.A., Abdesselam, R., Bounekkar, A.: Equivalence topologique entre mesures de proximité. In: Khenchaf, A., Poncelet, P. (eds.) Revue de l'Information et des Nouvelles Technologies, RNTI 20, EGC'2011, pp. 53–64. Hermann (2011)

Chapter 5
Comparing Attributes by *Probabilistic and Statistical* Association I

5.1 Introduction

Data is defined by the observation of a set $\mathcal{A}$ of descriptive attributes on a set $\mathcal{O}$ of elementary objects or a set Γ of categories. In this and in the following chapters, we develop the construction of an association coefficient on $\mathcal{A}$. For this purpose, *Likelihood Linkage Analysis* approach is emphasized. It leads, in a unified process, to a very rich family of probabilistic association coefficients between descriptive attributes of any type. On the other hand, the principle of this method enables several association coefficients to be mutually compared.

This chapter is devoted to comparing descriptive attributes of type I, that is, Boolean and numerical attributes. Two cases will be distinguished. In the first one (see Sect. 5.2), description concerns a set $\mathcal{O}$ of elementary objects and in the second one (see Sect. 5.3), a set Γ of categories.

For comparing Boolean attributes observed on a set $\mathcal{O}$ of objects, the *LLA* approach was initiated in [23]. The starting point consists of defining a probabilistic coefficient for comparing two Boolean attributes [1, 23]. Denote as above a and b these two attributes observed on a set of objects $\mathcal{O}$. A "raw" coefficient between a and b is evaluated by means of a probability scale, with respect to a probabilistic independence hypothesis, taking into account the cardinalities $n(a)$ and $n(b)$. A very natural raw association coefficient is defined by $s(a, b) = card[\mathcal{O}(a) \cap \mathcal{O}(b)]$, also expressed in the preceding chapter, which represents the number of objects of $\mathcal{O}$ for which a and b are *TRUE*. Now, denote by $\mathcal{H}$ the independence hypothesis—also called *hypothesis of no relation* conditioned by the cardinalities $n(a)$ and $n(b)$. $\mathcal{H}$ enables us to establish a probabilistic reference scale for measuring resemblance between descriptive attributes. This construction leads to efficient normalization and the resulting probability scale has invariance properties. These are expressed in Sect. 5.2.1. Moreover, the probabilistic hypothesis of no relation $\mathcal{H}$ presents three distinct fundamental forms $\mathcal{H}_1$, $\mathcal{H}_2$ and $\mathcal{H}_3$ which will be defined in Sect. 5.2.1.2. Each of the three versions of the hypothesis of no relation associates to the observed attribute pair (a, b),

© Springer-Verlag London 2016

I.C. Lerman, *Foundations and Methods in Combinatorial and Statistical Data Analysis and Clustering*, Advanced Information and Knowledge Processing, DOI 10.1007/978-1-4471-6793-8_5

a random pair $(a^\star, b^\star)$, where $a^\star$ and $b^\star$ are independent in probability. In these conditions, $s(a, b)$ is located with respect to the distribution of the random association $s(a^\star, b^\star)$ by means of a probabilistic index. For this purpose, the normal distribution is used. In the case of attributes of type I this Gaussian reference is theoretically justified for each form ($\mathcal{H}_1$, $\mathcal{H}_2$ and $\mathcal{H}_3$) of the independence hypothesis [6, 26, 31]. Now, denote by $\mathcal{E}[s(a^\star, b^\star)]$ and $var(s[(\alpha^\star, \beta^\star)])$ the expectation and variance of the random raw coefficient. In the *LLA* approach, an important transformation in order to obtain normalized coefficients consists of centring and reducing $s(a, b)$ with respect to the mean and standard deviation of $s(a^\star, b^\star)$ [24, 27, 33, 34]. The obtained coefficient can be written as

$$Q(a, b) = \frac{\left(s(a, b) - \mathcal{E}[s(a^\star, b^\star)]\right)}{\sqrt{var(s(a^\star, b^\star))}}$$

At this level, indices derived from $Q(a, b)$ and neutralizing the effect of the object set size $n = card(\mathcal{O})$ can be proposed. If Φ denotes the cumulative normal distribution, a first version of a probabilistic coefficient is obtained by applying Φ to the normalized coefficient, namely $\Phi[Q(a, b)]$. The latter might not be discriminant enough. In these conditions, for mutual comparison a set $\mathcal{A}$ of descriptive attributes and a global normalization is in addition required, before applying Φ (see Sect. 5.2.1.5).

Asymmetrical probabilistic association coefficient between two Boolean attributes a and b is considered in Sect. 5.2.1.4. The latter defines an interestingness measure for the "association rule" $a \to b$ [36]. In this, the small size of $s(a, \bar{b}) = card[\mathcal{O}(a) \cap \mathcal{O}(\bar{b})]$—where $\bar{b}$ is the Boolean attribute opposite to b—is evaluated with respect to a probability scale built by the random model $\mathcal{H}_3$ defined in the symmetrical case, adopted in [11] after analysis led in [35].

Extension of the *LLA* approach for comparing numerical attributes observed on a set $\mathcal{O}$ of elementary objects is studied in Sect. 5.2.2.

The case of comparing descriptive attributes of type I, observed on a set Γ of categories (see Table 3.5 of Chap. 3), is studied in Sect. 5.3. To begin, we study the Boolean attribute case (see Sect. 5.3.2). Next, we consider the case where the attributes are numerical (see Sect. 5.3.3). The latter leads us to a direct extension for comparing ordinal and nominal categorical attributes distributed on Γ (see Sect. 5.3.3).

Generally, in the literature, the subject of clustering a set of objects is much more frequently considered than that of clustering a set of descriptive attributes. In this book, initially aggregating statistically near descriptive attributes is considered an important facet of a data clustering analysis. Consequently, a substantive part of the previous chapter, this chapter and the next (Chap. 6) are devoted to the problem of building an association coefficient between descriptive attributes. Establishing a similarity index between objects or categories is studied in the previous chapter and—according to the *LLA* method in Chap. 7. As already expressed and regardless of the nature of the data table, the problem consists of organizing both the row set and the column set by similarity according to a clustering scheme (hierarchical, for example). The interest in such a joint clustering will be discussed in Chap. 8. Cluster organization has to take into account in great detail the nature of the elements to be compared and their formal representation. The latter has been clearly explained in Chap. 3.

5.2 Comparing Attributes of Type I for an Object Set Description by the *Likelihood Linkage Analysis* Approach

5.2.1 The Boolean Case

5.2.1.1 Introduction

Let (a, b) be a couple of Boolean attributes represented by the couple $[\mathcal{O}(a), \mathcal{O}(b)]$ of subsets of the object set $\mathcal{O}$, where $\mathcal{O}(a)$ (resp., $\mathcal{O}(b)$) is the set of elements for which a (rep., b) is *TRUE* (see Sect. 3.2.1). Let us recall here the different cardinal parameters associated with $[\mathcal{O}(a), \mathcal{O}(b)]$, namely

$$n(a) = card(\mathcal{O}(a)), n(\bar{a}) = card(\mathcal{O}(\bar{a}))$$
$$n(b) = card(\mathcal{O}(b)), n(\bar{b}) = card(\mathcal{O}(\bar{b}))$$
$$s(a, b) = n(a \wedge b) = card(\mathcal{O}(a) \cap \mathcal{O}(b))$$
$$u(a, b) = n(a \wedge \bar{b}) = card(\mathcal{O}(a) \cap \mathcal{O}(\bar{b}))$$
$$v(a, b) = n(\bar{a} \wedge b) = card(\mathcal{O}(\bar{a}) \cap \mathcal{O}(b))$$
$$t(a, b) = n(\bar{a} \wedge \bar{b}) = card(\mathcal{O}(\bar{a}) \cap \mathcal{O}(\bar{b}))$$
$$p(a) = \frac{n(a)}{n}, p(\bar{a}) = \frac{n(\bar{a})}{n}, p(b) = \frac{n(b)}{n}, p(\bar{b}) = \frac{n(\bar{b})}{n}$$
$$p(a \wedge b) = \frac{n(a \wedge b)}{n}, p(a \wedge \bar{b}) = \frac{n(a \wedge \bar{b})}{n}$$
$$p(\bar{a} \wedge b) = \frac{n(\bar{a} \wedge b)}{n}, p(\bar{a} \wedge \bar{b}) = \frac{n(\bar{a} \wedge \bar{b})}{n} \tag{5.2.1}$$

where $n = card(\mathcal{O})$ and where $\bar{a}$ (resp., $\bar{b}$) is the Boolean attribute opposite to a (resp., b), for which we have: $\mathcal{O}(\bar{a}) = \mathcal{O} - \mathcal{O}(a)$ (resp., $\mathcal{O}(\bar{b}) = \mathcal{O} - \mathcal{O}(b)$) (*set difference*).

Clearly, whatever the association coefficient chosen between a and b on the basis of their observation on $\mathcal{O}$, $s(a, b) = card[\mathcal{O}(a) \cap \mathcal{O}(b)]$ has to play, explicitly or implicitly, an important role for measuring the resemblance between a and b. Indeed, the common presence of a and b on a given element of $\mathcal{O}$ has to contribute to the resemblance between a and b. $s(a, b)$ will play the role of a *raw* association coefficient. However, the only value of $s(a, b)$ provides a biased estimation of the resemblance between a and b. A *large* (resp., *small*) value of $n(a) = card(\mathcal{O}(a))$ and

$n(b) = card(\mathcal{O}(b))$ has a natural tendency to give rise to a *large* (resp., *small*) value of $s(a, b)$ and this, independently of the relative positions of $\mathcal{O}(a)$ and $\mathcal{O}(b)$. For comparing a and b, the basic idea of the *Likelihood Linkage Analysis* consists of locating $s(a, b)$ with respect to a probability scale provided by a random model associating with the triplet $[\mathcal{O}; \mathcal{O}(a), \mathcal{O}(b)]$ a random one $[\Omega; X, Y]$, where Ω is a random set of objects corresponding to $\mathcal{O}$ and where, for a given Ω, X and Y are two independent random subsets of Ω associated with $\mathcal{O}(a)$ and $\mathcal{O}(b)$, respectively. The model is defined such that the mathematical expectations $\mathcal{E}(card(\Omega))$, $\mathcal{E}(card(X))$ and $\mathcal{E}(card(Y))$ are equal to $n = card(\mathcal{O})$, $n(a) = card(\mathcal{O}(a))$ and $n(b) = card(\mathcal{O}(b))$, respectively. There are three fundamental random models denoted $\mathcal{H}_1$, $\mathcal{H}_2$ and $\mathcal{H}_3$, for the association

$$[\mathcal{O}; \mathcal{O}(a), \mathcal{O}(b)] \rightarrow [\Omega; X, Y]$$

these are defined from three random models of choosing a random set associated with an observed $\mathcal{O}$ subset. These, denoted $\mathcal{N}_1$, $\mathcal{N}_2$ and $\mathcal{N}_3$, respectively, will be expressed in Sect. 5.2.1.2. The expression of $\mathcal{N}_1$ is the easiest one. It is a pure combinatorial model. For this model, $\Omega = \mathcal{O}$. On the other hand, it consists of fixing $\mathcal{O}(a)$ (resp., $\mathcal{O}(b)$) and associating a random subset Y (resp., X) with $\mathcal{O}(b)$ (resp., $\mathcal{O}(a)$) in the set $\mathcal{O}$, endowed with a uniform probability measure of all subsets of $\mathcal{O}$ having the same cardinality $n(b)$ (resp., $n(a)$). Thus, two dual random variables (attributes) are defined as $S(a) = card(\mathcal{O}(a) \cap Y)$ and $S(b) = card(X \cap \mathcal{O}(b))$. Let us point out that these two random variables have exactly the same probability distribution, that of $S = card(X \cap Y)$, where X and Y are independent in probability. In these conditions, the probabilistic similarity coefficient we adopt between a and b is expressed as

$$P_1(a, b) = Prob^{\mathcal{H}_1}(S(a, b) \le s(a, b)) \tag{5.2.2}$$

where $Prob^{\mathcal{H}_1}$ is the probability defined under the random model $\mathcal{H}_1$. Thus, the similarity value between a and b is all the more large as the number of objects possessing a and b is unlikely to be big with respect to the no relation probabilistic hypothesis $\mathcal{H}_1$. We introduce here a notion of *likelihood* in the resemblance notion. By referring to the random variable $S(a)$ (resp., $S(b)$), the index $Prob^{\mathcal{H}_1}$, comprised between 0 and 1, is exactly defined by the proportion of $\mathcal{O}$ subsets of size $n(b)$ (resp., $n(a)$) whose cardinal of intersection with $\mathcal{O}(a)$ (resp., $\mathcal{O}(b)$) is lower or equal to $s(a, b)$. This probability index is exactly calculated by means of *hypergeometrical* distribution (see Sect. 5.2.1.2).

The idea of this association coefficient (similarity measure) is given in [23]. Its clear expression referring to the hypergeometric model is provided in [24]. It was worked on again in [27]. This similarity measure was retrieved in [51], where it was wrongly presented as a new one.

In most cases, the normal probability law gives a good approximation of the hypergeometrical distribution [9]. In these conditions, it is natural to consider the following coefficient obtained by centring and reducing $s(a, b)$ with respect $\mathcal{H}_1$, namely:

$$Q_1(a, b) = \frac{s(a, b) - \mathcal{E}(S(a, b))}{\sqrt{var(S(a, b))}} \qquad (5.2.3)$$

where $\mathcal{E}(S(a, b))$ and $var(S(a, b))$ are the mathematical expectation and variance of $S(a, b)$ under $\mathcal{H}_1$. Consequently, the following equality will be admitted:

$$P_1(a, b) = \Phi[Q_1(a, b)] \qquad (5.2.4)$$

where Φ is the cumulative normal distribution function. $P_1(a, b)$ defines—in our terminology—a "local" form of the probability index. In fact (see Sect. 5.2.1.2), its value tends quickly towards 0 or 1 when the number n of observations ($n = card(\mathcal{O})$) increases: 0, if $p(a \wedge b) < p(a) \times p(b)$ and 1, if $p(a \wedge b) > p(a) \times p(b)$, where $p(a \wedge b) = n(a \wedge b)/n, p(a) = n(a)/n$ and $p(b) = n(b)/n$. These latter Eqs. (5.2.2)–(5.2.4) will be worked on again in the next section.

Now let us consider the bilateral form of $\mathcal{H}_1$, that for which with the ordered pair $(\mathcal{O}(a), \mathcal{O}(b))$ of observed subsets is associated an ordered pair (X, Y) of independent random subsets. A different expression commonly used in mathematical statistics consists of saying: with the observed ordered pair (a, b) of Boolean attributes, is associated an ordered pair $(a^\star, b^\star)$ of independent random Boolean attributes. In these conditions, (X, Y) can be denoted $(\mathcal{O}(a^\star), \mathcal{O}(b^\star))$.

As mentioned above, there are two other forms of the hypothesis of no relation $\mathcal{H}_2$ and $\mathcal{H}_3$ associated with the way of specifying the random model. There are in all three random models denoted by $\mathcal{N}_1$, $\mathcal{N}_2$ and $\mathcal{N}_3$, sustaining, respectively, $\mathcal{H}_1$, $\mathcal{H}_2$ and $\mathcal{H}_3$. They will be developed in the next section (see Sect. 5.2.1.2). Let us give here an intuitive and easy expression of the *independence* random model $\mathcal{H}_2$. With the ordered pair of subsets $(\mathcal{O}(a), \mathcal{O}(b))$ is associated an ordered pair of random independent subsets (X, Y), from a sequence of n independent random objects taken in $\mathcal{O}$. In this sequence the probability for a given random object z to take part in the construction of X (resp., Y) is equal to $p(a) = n(a)/n$ (resp., $p(b) = n(b)/n$), the respective belonging to X and Y being independent. For the third form $\mathcal{H}_3$ of the hypothesis of no-relation, the random index $card(X \cap Y)$ follows a Poisson probability law whose parameter is $\mu = n(a) \times n(b)/n = n \times p(a) \times p(b)$. As it was for $\mathcal{N}_1$ above, $\mathcal{N}_2$ and $\mathcal{N}_3$ will be expressed in the next section at the level of a random choice of element in the subset set of a set.

5.2.1.2　The Three Versions of the Hypothesis of No-Relation

The Hypergeometric Model

As expressed above, for each of the three models of choosing a random element in the subset set of a set $\mathcal{N}_1$, $\mathcal{N}_2$ and $\mathcal{N}_3$, with the ordered pair of $\mathcal{O}$ subsets $(\mathcal{O}(a), \mathcal{O}(b))$, is associated an ordered pair of independent random subsets (X, Y) of a set Ω, corresponding to $\mathcal{O}$. The specification of Ω depends on the random model considered: $\mathcal{N}_1$, $\mathcal{N}_2$ or $\mathcal{N}_3$. X and Y are independent in probability for each of the three models. Consequently, for the model definition it suffices to clarify the association between an observed subset E of $\mathcal{O}$ and the random one, denoted by Z of Ω. Let us indicate below by e the cardinality of E.

The random model $\mathcal{N}_1$ has been introduced earlier (see Sect. 5.1). For this model $\Omega = \mathcal{O}$ and Z is a random element in the set—endowed with uniform probability measure—of all subsets whose cardinality is e. In other words, by considering the subset set $\mathcal{P}(\mathcal{O})$ of $\mathcal{O}$, represented by the simplex $2^{\mathcal{O}}$, this model concentrates all the probability on the level e of this simplex, $0 \le e \le n$. On the other hand, this probability is uniformly distributed on the different elements of this level. Under these conditions, if F is a given subset of $\mathcal{O}$

$$Prob\{Z = F \mid \mathcal{N}_1\} = \begin{cases} 0 & \text{if } card(F) \ne e \\ 1/\binom{n}{e} & \text{if } card(F) = e \end{cases} \tag{5.2.5}$$

where $\binom{n}{e}$ is the well-known binomial coefficient.

Now, let us consider again the ordered pair of subsets $[\mathcal{O}(a), \mathcal{O}(b)]$ associated with an ordered pair (a, b) of Boolean attributes. As mentioned in Sect. 5.1, the hypothesis of no relation $\mathcal{H}_1$ can be presented in unilateral form: for example by fixing $\mathcal{O}(a)$ [resp., $\mathcal{O}(b)$] and by associating with $\mathcal{O}(b)$ [resp., $\mathcal{O}(a)$] a random subset Y (resp., X) where Y (resp., X) follows the model defined in (5.2.5) with $e = n(b)$ (resp., $e = n(a)$). Thus, with the *raw* index $s(a, b) = n(a \wedge b)$ the random variables $S(a) = card(\mathcal{O}(a) \cap Y)$ and $S(b) = card(X \cap \mathcal{O}(b))$ are associated, respectively. Their probability laws can be directly calculated and obtained as

$$Prob^{\mathcal{H}_1}\{S(a) = k\} = \frac{\binom{n(a)}{k} \times \binom{n(\bar{a})}{n(b)-k}}{\binom{n}{n(b)}} \tag{5.2.6}$$

$$Prob^{\mathcal{H}_1}\{S(b) = k\} = \frac{\binom{n(b)}{k} \times \binom{n(\bar{b})}{n(a)-k}}{\binom{n}{n(a)}} \tag{5.2.7}$$

Proposition 35 *The probability distributions of $S(a)$ and $S(b)$ are identical.*

It is obvious to verify that the respective right members of (5.2.6) and (5.2.7) are identical.

Let us denote by $S(a, b)$ the common random variable $S(a)$ or $S(b)$. The *Likelihood of the Link* probabilistic measure

$$P_1(a, b) = Prob^{\mathcal{H}_1}(S(a, b) \leq s(a, b)) \tag{5.2.8}$$

which has been already introduced in (5.2.2) can be calculated for large enough values of n, $n(a)$ and $n(b)$. However, a more efficient and simple solution, especially for mutually comparing a set of several attribute pairs, is provided by the normal approximation of the hypergeometric probability law [9] (see (5.2.4), taken up again below).

The mathematical expectation $\mathcal{E}(S(a, b))$ (or *mean*) and the *variance* of $S(a, b)$, denoted μ_1 and σ_1^2, are equal to

$$\mu_1 = \frac{n(a) \cdot n(b)}{n} \text{ and } \sigma_1^2 = \frac{n(a) \cdot n(\bar{a}) \cdot n(b) \cdot n(\bar{b})}{n^2 \cdot (n-1)} \tag{5.2.9}$$

Statistical centring and standardizing the raw coefficient $s(a, b)$, obtained by the following formula:

$$Q_1(a, b) = \frac{s(a, b) - \mu_1}{\sigma_1} \tag{5.2.10}$$

leads to

$$Q_1(a, b) = \sqrt{n-1} \times \frac{[n(a \wedge b) \cdot n(\bar{a} \wedge \bar{b}) - n(a \wedge \bar{b}) \cdot n(\bar{a} \wedge b)]}{\sqrt{n(a) \cdot n(\bar{a}) \cdot n(b) \cdot n(\bar{b})}}$$

$$= \sqrt{n-1} \times \frac{[p(a \wedge b) \cdot p(\bar{a} \wedge \bar{b}) - p(a \wedge \bar{b}) \cdot p(\bar{a} \wedge b)]}{\sqrt{p(a) \cdot p(\bar{a}) \cdot p(b) \cdot p(\bar{b})}} \tag{5.2.11}$$

Therefore,

$$Q_1(a, b) = \sqrt{n-1} \times R_1(a, b) \tag{5.2.12}$$

where $R_1(a, b)$ is nothing else the Pearson coefficient (see (12) in Table 4.3 of Chap. 4). On the other hand, as expressed above $P_1(a, b)$ can be computed by means of the following equation:

$$P_1(a, b) = \Phi[Q_1(a, b)] \tag{5.2.13}$$

Table 5.1 2×2 contingency table

(a, b)	1	0
1	$n(a \wedge b))$	$n(a \wedge \bar{b})$
0	$n(\bar{a} \wedge b))$	$n(\bar{a} \wedge \bar{b})$

Under these conditions, for given $p(a \wedge b)$, $p(a)$ and $p(b)$ and for n tending to ∞, $P_1(a, b)$ tends towards 0 or 1, according to $p(a \wedge b) < p(a) \times p(b)$ or $p(a \wedge b) > p(a) \times p(b)$, respectively.

Now let us consider the well-known chi-square statistic [20] χ^2 associated with the following 2×2 contingency table.

The statistic can be written for this table, as follows:

$$\chi^2 = \sum_{(i,j)\in\{0,1\}^2} \left[\frac{n(i, j) - (n(i, .) \times n(., j)/n)}{\sqrt{(n(i, .) \times n(., j)/n)}} \right]^2 \qquad (5.2.14)$$

where $i = 1$ (resp., $i = 0$) codes a (resp., $\bar{a}$) and where $j = 1$ (resp., $j = 0$) codes b (resp., $\bar{b}$). On the other hand, $n(1, .) = n(a)$, $n(0, .) = n(\bar{a})$, $n(., 1) = n(b)$ and $n(., 0) = n(\bar{b})$.

For any $(i, j) \in \{0, 1\}^2$, $[n \cdot n(i, j) - n(i, .) \cdot n(., j)]^2$ is identical to

$$[n(a \wedge b) \cdot n(\bar{a} \wedge \bar{b}) - n(a \wedge \bar{b}) \cdot n(\bar{a} \wedge b)]^2$$

and then

$$\frac{\chi^2}{n} = R_1(a, b)^2$$

For n large enough, $n - 1$ and n may not to be distinguished. Consequently, the following equality holds (Table 5.1).

$$Q_1(a, b)^2 = \chi^2$$

Now, by considering the structure of the expression defining the chi-square statistic (see (5.2.14)) we are led to define the respective contributions of the different cells of the contingency table as follows:

Definition 20 The *oriented* contribution of the cell (i, j) to the χ^2 statistic is the coefficient

$$\chi(i, j) = \frac{n(i, j) - (n(i, .) \times n(., j)/n)}{\sqrt{(n(i, .) \times n(., j)/n)}}$$

$(i, j) \in \{0, 1\}^2$

Besides, denote by $\nu(i, j)$ the numerator of $\chi(i, j)$, $(i, j) \in \{0, 1\}^2$, and notice the following relations:

$$\nu(1, 1) = \nu(0, 0) = -\nu(1, 0) = -\nu(0, 1) \qquad (5.2.15)$$

The random model $\mathcal{H}_1$ will be called the *hypergeometric* model.

The Binomial Model

For this model denoted $\mathcal{N}_2$, with a given observed subset E of $\mathcal{O}$ is associated a random subset Z of $\mathcal{O}$. In the hypergeometric model, all the probability measure were concentrated on the single level defined by all the subsets of $\mathcal{O}$, having the same cardinality e $(e = card(E))$. Here, the probability measure is distributed on different levels of the simplex $2^{\mathcal{O}}$, in a more global fashion. Indeed, this model $(\mathcal{N}_2)$ includes two steps: the first step consists of randomly choosing a level of $2^{\mathcal{O}}$ and the second, in random choosing an element of this level.

• For the level random choice, consider a random integer variable K indicating a random level of $2^{\mathcal{O}}$, which is defined by all the $\mathcal{O}$ subsets having the same cardinality K. The probability law of K is defined as follows:

$$Prob\{K = k\} = \binom{n}{k} \eta^k \cdot (1 - \eta)^{n-k} \qquad (5.2.16)$$

$0 \leq k \leq n$, where η is the proportion $\frac{e}{n}$
• For the random choice of an element of a given level k, the probability (5.2.16) assigned to this level is uniformly distributed on the set of the $\binom{n}{k}$ elements of this level. Thus, the probability weight of each element of the level concerned is $\eta^k \cdot (1 - \eta)^{n-k}$. Hence, for a given subset F of $\mathcal{O}$, we have

$$Prob^{\mathcal{N}_2}\{Z = F \mid f = card(F)\} = \eta^f \cdot (1 - \eta)^{n-f} \qquad (5.2.17)$$

For an ordered subset pair $(\mathcal{O}(a), \mathcal{O}(b))$, representing an ordered attribute pair (a, b), corresponds—under the hypothesis of no relation—an ordered pair (X, Y) of independent random subsets, where X (resp., Y) is a random subset in the set $\mathcal{P}(\mathcal{O}) = 2^{\mathcal{O}}$, endowed with a probability measure defined by $\mathcal{N}_2$ and for which $\eta = n(a)/n$ (resp., $\eta = n(b)/n$) (see (5.2.16)). Therefore, in this case, the hypothesis of no relation has a bilateral form. The reference space of (X, Y) is $\mathcal{P}(\mathcal{O}) \times \mathcal{P}(\mathcal{O})$, where the probability measure endowing the first factor is different from that endowing the second factor.

Proposition 36 *The probability law of the random variable is binomial with the parameters n and $\pi = p(a) \times p(b)$.*

Proof According to the $\mathcal{H}_2$ model, we have

$$Prob\{card(X) = k, card(Y) = h\}$$

$$= \binom{n}{k}p(a)^k p(\bar{a})^{n-k} \times \binom{n}{h}p(b)^h \cdot p(\bar{b})^{n-h} \qquad (5.2.18)$$

On the other hand,

$$Prob\{card(X \cap Y) = s \mid card(X) = k, card(Y) = h\}$$

$$= \frac{\binom{k}{s}\binom{n-k}{h-s}}{\binom{n}{h}} = \frac{\binom{h}{s}\binom{n-h}{k-s}}{\binom{n}{k}} \qquad (5.2.19)$$

In fact, this comes from the hypergeometric probability defined in the framework of the $\mathcal{N}_1$ model. This probability holds if and only if $k \geq s, h \geq s$ and $h + k - s \leq n$. The sought probability can be put into the following form:

$$\sum_G \binom{k}{s}\binom{n-k}{h-s}\binom{n}{k} p(a)^k p(b)^h p(\bar{a})^{n-k} p(\bar{b})^{n-h} \qquad (5.2.20)$$

where $G = \{(k, h) | k \geq s, h \geq s, h + k - s \leq n\}$. By introducing new variables $u = k - s$ and $v = h - s$, G will be defined by $G = \{(u, v) | u \geq 0, v \geq 0$ and $u + v \leq n - s\}$. Consequently, the sought probability can be written as

$$\sum_{0 \leq u \leq n-s, \, 0 \leq v \leq n-s-u} \frac{n!}{s!u!v!(n-u-v-s)!} p(a)^{s+u} p(b)^{s+v} p(\bar{a})^{n-s-u} p(\bar{b})^{n-s-v} \qquad (5.2.21)$$

However, we have

$$\sum_{0 \leq v \leq n-s-u} \frac{(n-s-u)!}{v!(n-s-u-v)!} p(b)^v p(\bar{b})^{n-s-u-v} = 1 \qquad (5.2.22)$$

Therefore, the expression (5.2.21) can be simplified and written as

$$\sum_{0 \leq u \leq n-s} \frac{n!}{s!u!(n-s-u)!} p(a)^{s+u} p(b)^s p(\bar{a})^{n-s-u} p(\bar{b})^u \qquad (5.2.23)$$

This can be factorized as follows:

$$\Big(\sum_{0 \le u \le n-s} \frac{(n-s)!}{u!(n-s-u)!}[p(a)p(\bar{b})]^u p(\bar{a})^{n-s-u} \Big) \times \Big(\frac{n!}{(n-s)!s!}[p(a)p(b)]^s \Big)$$

The result expressed in the proposition is obtained by noticing that the first factor can be put in the following form:

$$[p(a)p(\bar{b}) + p(\bar{a})]^{n-s} = [1 - p(a)p(b)]^{n-s}$$

$\square$

The mean and the variance of the binomial random variable $card(X \cap Y)$ are, respectively,

$$\mu_2 = \frac{n(a)n(b)}{n} = np(a)p(b) \text{ and } \sigma_2^2 = \frac{n(a)n(b)}{n}[1 - p(a)p(b)] \qquad (5.2.24)$$

Under these conditions, the centred and standardized coefficient can be written as

$$Q_2(a, b) = \frac{n(a \wedge b) - (n(a)n(b)/n)}{\sqrt{(n(a)n(b)/n)[1 - p(a)p(b)]}} \qquad (5.2.25)$$

Since $\mu_1 = \mu_2$ its numerator is exactly the same as that of $Q_1(a, b)$ (see (5.2.9) and (5.2.10)). Otherwise,

$$\sigma_2^2 = \sigma_1^2 \times \frac{1 - p(a)p(b)}{p(\bar{a})p(\bar{b})} \qquad (5.2.26)$$

The Poisson Model

The hypothesis of no relation leading to this model has been denoted above by $\mathcal{N}_3$. As for the previous models $\mathcal{N}_1$ and $\mathcal{N}_2$, E will designate an observed $\mathcal{O}$ subset whose cardinality is e. In the preceding binomial model, the definition of the random subset Z associated with E required two steps. Here, three steps are needed for this definition. For the first step with the object set $\mathcal{O}$ is associated a random object set Ω from a real or a hypothetical universe from which $\mathcal{O}$ is provided. The only condition required for this step is that the cardinality of Ω, denoted by $\mathcal{M}$, follows a Poisson probability law, parametrized by $n = card(\mathcal{O})$. Namely

$$Prob\{\mathcal{M} = m\} = \frac{n^m}{m!} \cdot e^{-n} \qquad (5.2.27)$$

for any m in the integers $\mathbb{N}$.

Now the two following steps are analogous to those of the binomial model ($\mathcal{N}_2$). Therefore, for $\Omega = \Omega_0$, whose cardinality is m, $m \geq e$,

$$Prob^{\mathcal{N}_3}\{K = card(Z) = k \mid \Omega = \Omega_0\} = \binom{m}{k} \eta^k \cdot (1 - \eta)^{m-k} \qquad (5.2.28)$$

where $k \leq m$ and $\eta = \frac{e}{m}$. Moreover,

$$Prob\{Z = F \mid \Omega = \Omega_0, f = card(F)\} = \eta^f \cdot (1 - \eta)^{m-f} \qquad (5.2.29)$$

In this model as expressed above, with the observed triple $(\mathcal{O}, \mathcal{O}(a), \mathcal{O}(b))$ is associated a random triple (Ω, X, Y), where X and Y are random subsets of a random set Ω associated with $\mathcal{O}$. Consider $\Omega = \Omega_0$ with $card(\Omega_0) = m$ and $\pi = p(a) \times p(b)$ where $p(a) = n(a)/n$ and $p(b) = n(b/n)$. We have in the framework of the binomial model and for $s \leq m$

$$Prob\{card(X \cap Y) = s \mid \mathcal{M} = m\} = \binom{m}{s} \pi^s \cdot (1 - \pi)^{m-s} \qquad (5.2.30)$$

In these conditions, by taking into account (5.2.27), we obtain

$$Prob^{\mathcal{H}_3}\{card(X \cap Y) = s\} = \sum_{m \geq s} \binom{m}{s} \pi^s \cdot (1 - \pi)^{m-s} \cdot \frac{n^m}{m!} \cdot e^{-n}$$

$$= \frac{(n \cdot \pi)^s}{s!} \cdot e^{-n \cdot \pi} \left[\sum_{m \geq s} \frac{(n \cdot (1 - \pi))^{m-s}}{(m - s)!} \cdot e^{-n \cdot (1 - \pi)} \right] \qquad (5.2.31)$$

The last sum (between the right brackets) is equal to 1, because it represents the total probability sum of a Poisson law. Therefore,

Proposition 37 *Under the random model $\mathcal{H}_3$, $card(X \cap Y)$ follows a probability Poisson law parametrized by $n \cdot \pi$, where $\pi = p(a) \times p(b)$.*

In the framework of this random model, called the "Poisson" model, the $s(a, b)$ centered and standardized coefficient becomes

$$Q_3(a, b) = \frac{n(a \wedge b) - (n(a)n(b)/n)}{\sqrt{(n(a)n(b)/n)}} \qquad (5.2.32)$$

The numerator of the coefficient $Q_3(a, b)$ is the same as that of $Q_1(a, b)$ and $Q_2(a, b)$. On the other hand, for the variance σ_3^2, we have the following relations:

$$\sigma_3^2 = \sigma_1^2 \cdot p(\bar{a}) \cdot p(\bar{b}) = \sigma_2^2 \cdot [1 - p(a) \cdot p(b)] \qquad (5.2.33)$$

Let us notice that the $Q_3(a, b)$ coefficient is nothing else but the oriented contribution of the cell $(1, 1)$ to the chi-square statistic associated with the 2×2 contingency table, crossing the binary attributes a and b (see Definition 20).

Comparison of the Three Coefficients

Equations (5.2.26) and (5.2.27) give the relationships between the respective variances of the random index $s(a^\star, b^\star) = n(a^\star \wedge b^\star)$ for the three models $\mathcal{H}_1$, $\mathcal{H}_2$ and $\mathcal{H}_3$.

As for the probabilistic index $P_1(a, b)$, computing of the indices

$$P_2(a, b) = Prob^{\mathcal{H}_2}(S(a, b) \leq s(a, b)) \text{ and } P_3(a, b) = Prob^{\mathcal{H}_3}(S(a, b) \leq s(a, b)),$$

associated with the no relation models $\mathcal{H}_2$ and $\mathcal{H}_2$, can be carried out exactly. However, for the same reasons given above and concerning $P_1(a, b)$, the normal approximation will be used for large enough values of n, $n(a)$ and $n(b)$. More precisely, accurate approximations for the values of $P_2(a, b)$ and $P_3(a, b)$ are provided by means of the following formulas [9, 20]:

$$P_2(a, b) = \Phi[Q_1(a, b)] \tag{5.2.34}$$

$$P_3(a, b) = \Phi[Q_3(a, b)] \tag{5.2.35}$$

where Φ is the normal cumulative distribution function.

A correlative form, independent of n, associated with the hypothesis of no relation $\mathcal{H}_1$, is obtained, according to Eq. (5.2.12), by

$$R_1(a, b) = Q_1(a, b)/\sqrt{n - 1},$$

More precisely, we have

$$R_1(a, b) = \frac{p(a \wedge b) - p(a) \cdot p(b)}{\sqrt{p(a) \cdot p(\bar{a}) \cdot p(b) \cdot p(\bar{b})}} \tag{5.2.36}$$

Similarily, by considering for $\mathcal{H}_2$ and $\mathcal{H}_3$, the correlative forms $R_2(a, b) = Q_2(a, b)/\sqrt{n}$ and $R_3(a, b) = Q_3(a, b)/\sqrt{n}$, we obtain

$$R_2(a, b) = \frac{p(a \wedge b) - p(a) \cdot p(b)}{\sqrt{p(a) \cdot p(b) \cdot [1 - p(a) \cdot p(b)]}} \tag{5.2.37}$$

$$R_3(a, b) = \frac{p(a \wedge b) - p(a) \cdot p(b)}{\sqrt{p(a) \cdot p(b)}} \tag{5.2.38}$$

$R_1(a, b)$ has a correlation coefficient structure and then its value comprises between -1 and $+1$ (see Sect. 5.2.2). On the other hand, it is easy to show the inequalities

$$|R_3(a, b)| < |R_2(a, b)| < |R_1(a, b)|$$

where for a real x, $|x|$ indicates the absolute value of x.

Now, for mnemotechnic reasons, we may call Q_H, Q_B and Q_P the three coefficients previously denoted Q_1, Q_2 and Q_3 (H for "Hypergeometric", B for "Binomial" and "P" for Poisson). Obviously,

$$(\forall(a, b) \in \mathcal{A} \times \mathcal{A}), \, Q_P(a, b) < Q_B(a, b) < Q_H(a, b) \qquad (5.2.39)$$

This relation shows that Q_B is somewhat *intermediate* between $Q_P(a, b)$ and $Q_H(a, b)$. Consequently, Boolean data analysis has to be discussed with respect to $Q_P(a, b)$ and $Q_H(a, b)$. Indeed, these latter two coefficients are more differentiated than those $Q_B(a, b)$ and $Q_H(a, b)$ or those $Q_P(a, b)$ and $Q_B(a, b)$. The mathematician loving symmetry tends to prefer $Q_H(a, b)$ to $Q_P(a, b)$. In fact, this symmetry may induce a bias in the perception of similarity between Boolean attributes. Indeed, all things being equal, the association between rare Boolean attributes should be emphasized with respect to common attributes. This is precisely performed by the random model $\mathcal{H}_3$ which follows in its spirit the *Information theory* approach.

5.2.1.3 Invariance Properties of the *LLA* Probabilistic Coefficients

The different probabilistic indices $P_1(a, b)$, $P_2(a, b)$ and $P_3(a, b)$ have been obtained from the *raw* index $s(a, b) = n(a \wedge b)$, by changing the form of the hypothesis of no-relation: $\mathcal{H}_1$, $\mathcal{H}_2$ or $\mathcal{H}_3$. The latter respects strictly or in expectation to the cardinalities $n(a)$ and $n(b)$. Recall that many indices have been proposed in the literature and Table 4.3 of Chap. 4 gives a list of very representative ones of them. Consider the coefficients of this table in the framework of comparing Boolean attributes. A natural question arises: How are these indices transformed if a random model ($\mathcal{H}_1$, $\mathcal{H}_2$ or $\mathcal{H}_3$) is applied to them? In other words and to fix ideas, is the probabilistic index $P_i(a, b)$ ($i = 1, 2$ or 3) invariant by substituting for $s(a, b)$ another similarity measure? This issue was studied by M.H. Nicolaü at the level of the coefficients listed in Table 4.3 of Chap. 4 [38]. More generally the question is the invariance of the probabilistic index

$$Prob^{\mathcal{H}_i}\{\mathcal{S}^\star(s, u, v) \leq \mathcal{S}(s, u, v)\}$$

where $i = 1, 2$ or 3 and where $\mathcal{S}^\star(s, u, v)$ is the random similarity measure associated with an observed one, satisfying the conditions of Definition 14 of Chap. 4. The previous equation can be written as

Table 5.2 $[s, n(a), n(b)]$ Expression of classical binary association coefficients

Measure	Expression
(1) Russel and Rao (1940) [45]	$S(a, b) = \frac{s}{p}$
(2) Sokal and Michener (1958) [47]	$S(a, b) = \frac{[2s + (n - (n(a) + n(b)))]}{p}$
(3) Roger and Tanimoto (1960) [44]	$S(a, b) = \frac{2s + [n - (n(a) + n(b))]}{-2s + [n + (n(a) + n(b))]}$
(4) Hamann (1961) [16]	$S(a, b) = 1 - \frac{4s + [n - 2(n(a) + n(b))]}{p}$
(5) Jaccard (1908) [17]	$S(a, b) = \frac{s}{n(a) + n(b) - s}$
(6) Kulczynski(a) (1927) [18]	$S(a, b) = \frac{s}{n(a) + n(b) - 2s}$
(7) Dice (1945) [7]	$S(a, b) = \frac{2s}{n(a) + n(b)}$
(8) Sokal and Sneath (1963) [48]	$S(a, b) = \frac{s}{2[n(a) + n(b)] - s}$
(9) Kulczynski(b) (1927) [18]	$S(a, b) = \frac{1}{2} \times \left(\frac{s}{n(a)} + \frac{s}{n(b)} \right)$
(10) Ochiai (1957) [40]	$S(a, b) = \frac{s}{\sqrt{n(a) \times n(b)}}$
(11) Yule (1912) [52]	$S(a, b) = \frac{n.s - n(a).n(b)}{2s^2 + [n - 2(n(a) + n(b))].s + n(a).n(b)}$
(12) Pearson (1926) [41]	$S(a, b) = \frac{n.s - n(a).n(b)}{\sqrt{n(a).n(b).(n - n(a)).(n - n(b))}}$

$$Prob^{\mathcal{H}_i}\{\mathcal{S}^\star(s, n(a) - s, n(b) - s) \leq \mathcal{S}(s, n(a) - s, n(b) - s)\}$$

This form simplifies the answer to the invariance question since $n(a)$ and $n(b)$ are fixed for the random models $\mathcal{H}_1$ and $\mathcal{H}_2$. On the other hand, the mathematical expectations of $n(a^\star)$ and $n(b^\star)$ are equal to $n(a)$ and $n(b)$, respectively. Under these conditions, let us take up again Table 4.3 of Chap. 4 in the context of Boolean attribute comparison (see (4.2.30)) and express the similarity coefficients as functions of s, $n(a)$ and $n(b)$. For this purpose, take into account the following relations:

$$u = n(a) - s$$
$$v = n(b) - s$$
$$t = n - (u + v - s) = n - n(a) - n(b) + s$$
$$= s + [n - n(a) - n(b)] \qquad (5.2.40)$$

Clearly, for a given hypothesis of no relation $\mathcal{H}_i$ $(1 \leq i \leq 3)$, the centered and standardized coefficient

$$Q_i(a, b) = \frac{s(a, b) - \mathcal{E}(a^\star, b^\star)}{\sqrt{var(a^\star, b^\star)}}$$

Table 5.3 2 × 2 Cross classification between two binary attributes

$a \backslash b$	Not get disease	Got disease	
Vaccinated	42	9	51
Not vaccinated	28	17	45
	70	26	96

is invariant if we substitute for $s(a, b)$ a linear function of $s(a, b)$ depending on fixed values of n, $n(a)$ and $n(b)$. Therefore, the coefficients (1), (2), (4), (7), (9), (10) and (12) are equivalent with respect to the probability scale induced by $\mathcal{H}_i$ ($1 \le i \le 3$). Moreover, by using the δ method [5], M.H. Nicolaü showed in her Ph.D. [38] that all of the coefficients listed in Table 5.2 are—in a certain sense— asymptotically equivalent with respect to the probability scale induced and obtained by the normal approximation. Two random models were considered which correspond to $\mathcal{H}_1$ and $\mathcal{H}_2$ hypotheses of no-relation. $\mathcal{H}_3$ might also be considered. And, intuitively speaking, the asymptotic tendency towards the normal distribution of $Q_3(a^\star, b^\star)$ must be quicker than that of $Q_1(a^\star, b^\star)$.

Now, the general problem consists of determining the conditions on the analytical form of $\mathcal{S}(s(a, b), u(a, b), v(a, b))$ for which the exact or asymptotic invariance of the following probabilistic index holds.

$$Prob^{\mathcal{H}_i}\left(\mathcal{S}(s(a^\star, b^\star), u(a^\star, b^\star), v(a^\star, b^\star)) \le \mathcal{S}(s(a, b), u(a, b), v(a, b))\right)$$

Clearly, an exact invariance is obtained for a similarity coefficient satisfying the conditions of Definition 14 of Chap. 4, under the hypothesis of no relation $\mathcal{H}_1$, since, in this case, $\mathcal{S}(s, u, v)$ can be written as $\mathcal{S}(s, n(a) - s, n(b) - s)$ and then, becomes a strictly increasing function of s.

Illustration

Let us illustrate the calculation of the previous indices $Q_1(a, b)$, $Q_2(a, b)$ and $Q_3(a, b)$ (see (5.2.11), (5.2.25) and (5.2.32)) and those which are derived from them; namely $R_1(a, b)$, $R_2(a, b)$ and $R_3(a, b)$ (see (5.2.36–5.2.38)) and on the other hand, $P_1(a, b)$, $P_2(a, b)$ and $P_3(a, b)$ (see (5.2.13), (5.2.34) and (5.2.35)). For this purpose, we consider the 2 × 2 contingency table (Table 5.3) provided by the textbook "Theory and Problems of Statistics" by Murray R. Spiegel, McGray-Hill, 1961, page 214.

Now, let us define two Boolean attributes a and b as follows: $a = Vaccinated$ and $b = Not\ get\ disease$. In these conditions, we have:

$$n = 96, n(a) = 51, n(b) = 70 \text{ and } n(a \wedge b) = 42$$

It follows with the notations above (Table 5.3)

$$p(a \wedge b) = \frac{42}{96} = 0.438$$

$$p(a) = \frac{51}{96} = 0.531, \ p(b) = \frac{70}{96} = 0.729, \ p(\bar{a}) = \frac{45}{96} = 0.469, \ p(\bar{b}) = \frac{26}{96} = 0.271$$

We obtain

$$Q_1(a, b) = \sqrt{95} \times \frac{0.438 - 0.531 \times 0.729}{\sqrt{0.531 \times 0.729 \times 0.469 \times 0.271}} = 2.237$$

$$Q_2(a, b) = \sqrt{96} \times \frac{0.438 - 0.531 \times 0.729}{\sqrt{0.531 \times 0.729(1 - 0.531 \times 0.729)}} = 1.028$$

$$Q_3(a, b) = \sqrt{96} \times \frac{0.438 - 0.531 \times 0.729}{\sqrt{0.531 \times 0.729}} = 0.790$$

Therefore,

$$R_1(a, b) = 0.23 \qquad R_2(a, b) = 0.105 \qquad R_3(a, b) = 0.0806$$

Finally, using the normal approximation we obtain

$$P_1(a, b) = 0.9874 \qquad P_2(a, b) = 0.848 \qquad P_3(a, b) = 0.7852$$

5.2.1.4 Indices in Case of Asymmetrical Association

The construction of the probabilistic *LLA* asymmetrical association coefficient between Boolean attributes is analogous to the symmetrical one developed in Sects. 5.2.1.1 and 5.2.1.2. In Sect. 4.2.1.3 principles and techniques have been defined for the classical case in order to transform a symmetrical association coefficient of Table 4.3 of Chap. 4 onto an asymmetrical coefficient (see Table 4.5). The transposition of the *LLA* approach from the symmetrical version to the asymmetrical one is different in nature and a very specific one. The framework and the notations are those of Sect. 4.2.1.3. More particularly, take up again Eq. (4.2.9).

In the asymmetrical case, for a given ordered pair (a, b) of Boolean attributes, the smallness of $n(a \wedge \bar{b})$ has to be evaluated with respect to the cardinalities n, $n(a)$ and $n(b)$. In the *LLA* approach the question is to assess how unlikely is the smallness of $n(a \wedge \bar{b})$ by taking into account the parameters n, $n(a)$ and $n(b)$. The passage of the symmetrical version [23–25] to the asymmetrical one—in the context of the *LLA* approach—was first proposed in [11]. In this only the usual version of the Binomial model was taken into account. Our analysis [35] led to the more appropriate Poisson model. The latter was retained in [12]. For the moment let us not further specify the random model of no relation and denote it by $\mathcal{H}$ as above. Thus, the unlikelihood of the smallness of $n(a \wedge \bar{b})$ is measured by the probability

$$Prob^{\mathcal{H}}\{n(a^\star \wedge \bar{b}^\star) \leq n(a \wedge \bar{b})\} \tag{5.2.41}$$

Nevertheless, desiring an index which defines an increasing function of the strength of the implication $a \to b$, we adopt the complementary probability

$$\mathcal{I}^{\mathcal{H}}(a, b) = Prob^{\mathcal{H}}\{n(a^\star \wedge \bar{b}^\star) > n(a \wedge \bar{b})\} \tag{5.2.42}$$

Let us designate by $\mathcal{I}_1(a, b)$, $\mathcal{I}_2(a, b)$ and $\mathcal{I}_3(a, b)$ the three versions of $\mathcal{I}^{\mathcal{H}}(a, b)$ for $\mathcal{H} = \mathcal{H}_1$, $\mathcal{H}_2$ and $\mathcal{H}_3$, respectively. The exact values of the corresponding probabilities are calculated, respectively, by means of the hypergeometric, binomial and Poisson laws. The expressions needed can be derived from the development of Sect. 5.2.1.2.

Now, let us express the centered and standardized version of the coefficient $n(a \wedge \bar{b})$ according to each of the hypotheses of no relation $\mathcal{H}_1$, $\mathcal{H}_2$ and $\mathcal{H}_3$. Following (5.2.11), (5.2.25) and (5.2.32), we obtain

$$Q_1(a, \bar{b}) = \sqrt{n-1} \times \frac{[p(a \wedge \bar{b}) - p(a) \cdot p(\bar{b})]}{\sqrt{p(a) \cdot p(\bar{a}) \cdot p(b) \cdot p(\bar{b})}}$$

$$Q_2(a, \bar{b}) = \sqrt{n} \times \frac{p(a \wedge \bar{b}) - p(a) \cdot p(\bar{b}))}{\sqrt{(p(a) \cdot p(\bar{b})) \cdot [1 - p(a) \cdot p(\bar{b})]}}$$

$$Q_3(a, \bar{b}) = \sqrt{n} \times \frac{p(a \wedge \bar{b}) - p(a) \cdot p(\bar{b})}{\sqrt{p(a) \cdot p(\bar{b})}} \tag{5.2.43}$$

The correlative forms $R_1(a, \bar{b})$, $R_2(a, \bar{b})$ and $R_3(a, \bar{b})$ can be easily deduced as for (5.2.36–5.2.38), established for the symmetrical case. Relative to the hypothesis of no relation $\mathcal{H}_3$, a correlative implicative coefficient which increases with the strength of the association rule $a \to b$ is defined as

$$\mathcal{R}_3(a \to b) = -R_3(a, \bar{b}) = \frac{-p(a \wedge \bar{b}) + p(a) \cdot p(\bar{b})}{\sqrt{p(a) \cdot p(\bar{b})}} \tag{5.2.44}$$

Notice that we have the following symmetrical property of the Q_1 coefficient:

$$Q_1(a, \bar{b}) = Q_1(\bar{a}, b) = -Q_1(a, b) = -Q_1(\bar{a}, \bar{b}) \tag{5.2.45}$$

Therefore, for the random model $\mathcal{H}_1$, the implicative form of the coefficient is exactly equivalent to the symmetrical case. For $\mathcal{H}_3$, if $p(a) < p(b)$

$$|Q_3(a, \bar{b})| > |Q_3(b, \bar{a})| \tag{5.2.46}$$

A preliminary condition in order to consider the evaluation of the association rule $a \to b$ is a negative value of $[n(a \wedge \bar{b}) - (n(a).n(\bar{b})/n)]$. This expression represents the numerator of $Q_3(a, \bar{b})$. It is identical to the numerator of $Q_3(\bar{a}, b)$, which corresponds to the opposite implication: $b \rightarrow a$. However, since we have assumed the natural condition $n(b) > n(a)$, the latter implication is difficult to accept. The inequality (5.2.46) is a consistent statement since, for the probabilistic index associated with $\mathcal{H}_3$ (see (5.2.42)), we have

$$\mathcal{I}_3(a, b) > \mathcal{I}_3(b, a) \tag{5.2.47}$$

To be convinced by this property, consider the excellent normal approximation, for n, $n(a)$ and $n(b)$ enough large [9], of the probability Poisson laws of $n(a^\star \wedge \bar{b}^\star)$ and $n(b^\star \wedge \bar{a}^\star)$, under the $\mathcal{H}_3$ model:

$$\mathcal{I}_3(a, b) = 1 - \Phi(Q_3(a, \bar{b}))$$
$$\mathcal{I}_3(b, a) = 1 - \Phi(Q_3(b, \bar{a})) \tag{5.2.48}$$

where Φ—as above—is the cumulative distribution function of the standardized normal law.

Moreover, the necessary and sufficient condition to have (5.2.47) is $n(a) < n(b)$. For large databases, mostly we have $p(a) < p(b) < 0.5$. With this condition we obtain the following inequalities which include (5.2.47):

$$P_3(\bar{a}, \bar{b}) < \mathcal{I}_3(b, a) < \mathcal{I}_3(a, b) < P_3(a, b)$$

where P_3 is the likelihood of the link probabilistic index defined in (5.2.35).

However, if $0.5 < p(a) < p(b)$, we have

$$P_3(a, b) < \mathcal{I}_3(b, a) < \mathcal{I}_3(a, b) < P_3(\bar{a}, \bar{b})$$

Illustration

Here, the previous 2×2 contingency table is taken up again. Based on this data the computing of the indices $\mathcal{I}_1(a, b)$, $\mathcal{I}_2(a, b)$ and $\mathcal{I}_3(a, b)$ can be carried out. This is left for the reader who will have to employ (5.2.43). We are interested in the calculation of $\mathcal{I}_3(a, b)$, $\mathcal{I}_3(b, a)$, $P_3(a, b)$ and $P_3(\bar{a}, \bar{b})$. We obtain:

$$Q_3(a, \bar{b}) = -1.295, \ Q_3(\bar{a}, b) = -0.840, \ Q_3(a, b) = 0.784 \text{ and } Q_3(\bar{a}, \bar{b}) = 1.379$$

Therefore,

$$\mathcal{I}_3(a, b) = 0.9024, \ \mathcal{I}_3(b, a) = 0.7995, \ P_3(a, b) = 0.785 \text{ and } P_3(\bar{a}, \bar{b}) = 0.9160$$

In this case where $0.5 < p(a) = 51/96 < p(b) = 70/96$, the second version of the inequalities above, concerning these coefficients, holds.

5.2.1.5 Similarity Global Reduction for Mutual Comparison Between Attributes

Assume given in a contingency table the proportions (one says also *relative frequencies*) $p(a \wedge b), p(a), p(b)$ and to fix ideas consider the case of a symmetrical association between the Boolean attributes a and b. Clearly, if $p(a \wedge b) > p(a) \times p(b)$ (resp., $p(a \wedge b) < p(a) \times p(b)$) a probabilistic index such that $P_i(a, b)$ ($i = 1, 2$ or 3) is as large (resp., small) as $n = card(\mathcal{O})$ is large. More formally, assume a sequence $(\mathcal{O}_j | j \geq 1)$ of object sets—for example, provided from a universe $\mathcal{U}$ of objects—such that $n_j = card(\mathcal{O}_j)$ increases and tends to ∞. Let us denote by $p_j(a \wedge b), p_j(a)$ and $p_j(b)$ the above relative frequencies calculated at the level of $\mathcal{O}_j$. On the other hand, denote by $P_i^j(a, b)$ the probabilistic index $P_i(a, b)$ defined at the level of $\mathcal{O}_j$ ($i = 1, 2$ or 3). If we suppose the convergence of $(p_j(a \wedge b) - p_j(a) \times p_j(b))$ to a *positive* (resp., *negative*) finite limit, then $P_i^j(a, b)$ tends to 1 (resp., to 0). Under these conditions, the probability scale induced by $P_i(a, b)$ ($i = 1, 2$ or 3) becomes no discriminant. It only allows to distinguish between positive and negative associations.

In order to solve this *crucial* problem let us begin by recalling the classical approach of *Inferential Statistics*. In this, the object set $\mathcal{O}$ is regarded as the observation of a random sample $\mathcal{O}^*$, provided from a universe $\mathcal{U}$ of very large or infinity size. Denote by $\pi(a), \pi(b)$ and $\pi(a \wedge b)$ the presence of proportions of the Boolean attributes a, b and $a \wedge b$ at the level of $\mathcal{U}$. On the other hand, $F(a), F(b)$ and $F(a \wedge b)$ will designate the random relative frequencies of these Boolean attributes at the level of the random set $\mathcal{O}^*$. Thus, their observed values over $\mathcal{O}$ are $p(a), p(b)$ and $p(a \wedge b)$, respectively. In this random model with the correlative coefficient $R_i(a, b)$ for the symmetrical case (resp., $R_i(a, \bar{b})$) for the asymmetrical case], is associated a random coefficient $R_i^*(a, b)$ for the symmetrical case (resp., $R_i^*(a, \bar{b})$ for the asymmetrical case) defined at the level of $\mathcal{O}^*$, $i = 1, 2$ or 3, (see (5.2.36–5.2.38)). The coefficient corresponding to $R_i(a, b)$ (resp., $R_i(a, \bar{b})$ at the level of the universe $\mathcal{U}$ will be denoted as $\rho_i(a, b)$ (resp., $\rho_i(a, \bar{b})$). The latter coefficients are functions of $\pi(a), \pi(b)$ and $\pi(a \wedge b)$. Let us restrict our argument below to the symmetrical case, the asymmetrical one is analogous.

The strong law of large numbers [9] entails convergence in probability of $F(a), F(b)$ and $F(a \wedge b)$ to $\pi(a), \pi(b)$ and $\pi(a \wedge b)$. On the other hand, we show in [31] that, for n fixed, $R_1^*(a, b)$ (resp., $R_3^*(a, b)$) follows a normal distribution, of mathematical expectation $\rho_1(a, b)$ (resp., $\rho_3(a, b)$). The variance of this distribution can be put in the form C^2/n where the parameter C depends on the triplet $\pi(a), \pi(b), \pi(a \wedge b)$ and generally comprises between 0.75 and 1.25. Under these conditions, the symmetrical $1 - \alpha$ confidence interval for $\rho_i(a, b)$ can be written as

$$\left[r_i(a, b) - \sqrt{C^2/n} \times \Phi^{-1}\left(1 - \frac{\alpha}{2}\right), r_i(a, b) + \sqrt{C^2/n} \times \Phi^{-1}\left(1 - \frac{\alpha}{2}\right) \right]$$

For $n = 10^4$, $(1 - \alpha) = 0.99$ and $C^2 = 1$, we obtain $[r_1(a, b) - 0.025, r_1(a, b) + 0.025]$. Therefore, the estimation of $\rho_i(a, b)$ by $r_i(a, b)$ is accurate for large enough databases. However, the main paradigm of *Data Analysis* does not consist of *estimating* the values of association coefficients, but of *organizing* the elements of the data concerned according to their mutual relations. This can be performed much more accurately than estimating correlations. In order to illustrate this point, consider a couple of attribute pairs $(\{a, b\}, \{c, d\})$ where $\{a, b\}$ and $\{c, d\}$ are without a common component. Let us imagine that $\rho_1(a, b) = 0$ and $\rho_1(c, d) = 0.08$. These values reflect very weak dependencies between a and b on the one hand and between c and d on the other hand. Nevertheless, $\rho_1(a, b) < \rho_1(c, d)$. Under these conditions, $R_1^*(a, b) - R_1^*(c, d)$ asymptotically follows a normal distribution centred on $[\rho_1(a, b) - \rho_1(c, d)]$. As above for $n = 10^4$, we obtain

$$Prob\{R_1^*(a, b) < R_1^*(c, d)\} = \Phi(8/\sqrt{2}) \sim 1$$

Hence we observe very strong stability in the preservation of the inequality $\rho_1(a, b) < \rho_1(c, d)$.

Now, let us consider the main problem of mutual comparison of a set $\mathcal{A} = \{a^j | 1 \le j \le p\}$ of Boolean attributes observed on an object set $\mathcal{O}$. For this purpose, we consider the association coefficient matrix

$$\mathcal{Q} = \{Q(a^j, a^k) | 1 \le j < k \le p\} \tag{5.2.49}$$

where $Q(a, b)$ corresponds to one of the coefficients $Q_1(a, b)$, $Q_2(a, b)$ or $Q_3(a, b)$ (see (5.2.11), (5.2.25) and (5.2.32), respectively). With the above matrix is associated the following matrix of probabilistic association coefficients:

$$\mathcal{P} = \{P(a^j, a^k) | 1 \le j < k \le p\} \tag{5.2.50}$$

According to what was expressed at the beginning of this section, for a given attribute pair $\{a, b\}$ ($\{a, b\} \in P_2(\mathcal{A})$), it is difficult to distinguish $P(a, b)$ from 1 (resp., 0) in the case of a positive association (resp., in the case of a negative association) between a and b, for $n = card(\mathcal{O})$ large enough. But this should not surprise us. And indeed, if our universe is limited to a single attribute pair $\{a, b\}$, a binary answer, positive or negative, suffices for the nature of the association between a and b. Such an answer is precisely given by *Independence Statistical Tests*. In this approach an index such that $p = 1 - P(a, b)$ is called the $p - value$ of the test. Nevertheless, the probability scale defined by P remains discriminant for a moderate value of n, say of the order of several hundred (see the above illustration where $n = 96$). We call $P(a, b)$ a *local* probabilistic index, and (5.2.50) defines the matrix of the *local* probabilistic indices.

However large is n, the value of $Q(a, b)$ has to be evaluated in a relative way with respect to the empirical distribution defined in (5.2.49). The *Similarity Global Reduction* [31, 32] consists of substituting for (5.2.49), the matrix

$$\mathcal{Q}^g = \{Q^g(a^j, a^k) | 1 \le j < k \le p\} \tag{5.2.51}$$

where

$$Q^g(a^j, a^k) = \frac{Q(a^j, a^k) - mean_e(Q)}{\sqrt{var_e(Q)}} \qquad (5.2.52)$$

where $mean_e(Q)$ and $var_e(Q)$ are the empirical mean and variance of the distribution (5.2.49), namely

$$mean_e(Q) = \frac{2}{p \cdot (p-1)} \cdot \sum_{1 \leq j < k \leq p} Q(a^j, a^k)$$

$$var_e(Q) = \frac{2}{p \cdot (p-1)} \cdot \sum_{1 \leq j < k \leq p} [Q(a^j, a^k) - mean_e(Q)]^2$$

The matrix of *global* probabilistic indices is defined as follows:

$$\mathcal{P}^g = \{P^g(a^j, a^k) = \Phi(Q^g(a^j, a^k)) | 1 \leq j < k \leq p\} \qquad (5.2.53)$$

Other global reduction methods are envisaged in [31, 32]. The most elaborated of them uses the covariance matrix of a random Boolean vector (S, U, V, T) following a 4-nominal probability law of parameters n and $\pi(a \wedge b)$, $\pi(a \wedge \bar{b})$, $\pi(\bar{a} \wedge b)$, $\pi(\bar{a} \wedge \bar{b})$. In this book, we will adopt a global reduction as that defined above (see (5.2.52)). Nevertheless, this can be carried out with respect to a specific subset of $P_2(\mathcal{A})$.

In the case of asymmetrical associations between Boolean attributes (see Sect. 5.2.1.4), the simplest version of the empirical distribution with respect to which the normalization can be obtained, takes the following form:

$$\{Q_i(a, \bar{b}) | (a, b) \in \mathcal{A} \times \mathcal{A}, n(a) < n(b)\} \qquad (5.2.54)$$

This global reduction is studied in [21] where threshold conditions on *Support* and *Confidence* indices are taken into account. In this work the behaviour of the probabilistic indices (*local* and *global*) are investigated with respect to increasing models of size n.

As expressed above, a moderate size of the sample defined by the object set $\mathcal{O}$ yields a discriminant probability scale for the matrix $\mathcal{P}$ (see (5.2.50)) (see for example the illustration above where $n = 96$). The starting point of the approach proposed in [43], for a probabilistic measure of an association between two Boolean attributes is the notion of *Test Value*. In this, a specific normalization is considered in the case where n is large. It consists of reasoning as if the sample size is 100. The latter sample is hypothetical and assumed to summarize the initial observed sample. This approach is analysed in [22, 36] where different alternatives are discovered for such a reduction. On the other hand, this technique is compared theoretically and experimentally to the global reduction one presented above. The results obtained show the advantage and in particular, more stable behaviour of the "Similarity Global Reduction" method.

The latter will be adopted in our development for comparing attributes, objects or categories for attribute description of any type.

5.2.2 Comparing Numerical Attributes in the LLA approach

Boolean and numerical attributes define type I of the descriptive attributes (see Sect. 3.2). In the *LLA* approach three random models denoted $\mathcal{N}_1$, $\mathcal{N}_2$ and $\mathcal{N}_3$ were defined leading to three versions of the hypothesis of no relation between two Boolean attributes to be compared. $\mathcal{H}_1$, $\mathcal{H}_2$ and $\mathcal{H}_3$ were called *hypergeometric*, *binomial* and *Poisson* models, respectively (see Sect. 5.2.1.2). Our purpose here is to extend these three models in the numerical case for comparing two numerical attributes. Let us denote as v and w the numerical attributes to be compared. $R_1(a, b)$ of (5.2.12) or (5.2.36) defined under $\mathcal{H}_1$, becomes $R_1(v, w)$ which is the Bravais–Pearson correlation coefficient. New coefficients $R_2(v, w)$ and $R_3(v, w)$ corresponding to $R_2(a, b)$ (see (5.2.37)) and $R_3(a, b)$ (see (5.2.38)) will be obtained in the numerical case. Asymmetrical versions of these will also be obtained.

5.2.2.1 The Three Versions of the Hypothesis of No Relation Random Model

Let v be a numerical attribute defined on an object set $\mathcal{O}$ whose cardinality is n (see Sect. 3.2.2). Consider a set $I = \{1, 2, \ldots, i, \ldots, n\}$ of integer indices coding $\mathcal{O}$. i is an integer label coding a given element of $\mathcal{O}$, $1 \leq i \leq n$. v defines a valuation

$$x_I = \{x_i | i \in I\} \tag{5.2.55}$$

where x_i is the value of v on the object coded by i $1 \leq i \leq n$.

$\mathcal{N}_1$

For this model the basic random element is a permutation σ on I; that is, a self bijection on I. σ is an element of the set that we denote G_n, of all permutations on I, provided by a uniform probability measure. There are in all $n!$ permutations $[n! = card(G_n)]$ and each of them has an equal chance $(1/n!)$ to occur. The value of a given σ on I:

$$\sigma(I) = [\sigma(1), \sigma(2), \ldots, \sigma(i), \ldots, \sigma(n)] \tag{5.2.56}$$

is also called a *permutation* of I.

To the sequence $(x_1, x_2, \ldots, x_i, \ldots, x_n)$ of the values observed, corresponds the random sequence

$$x_{\sigma(I)} = [x_{\sigma(1)}, x_{\sigma(2)}, \ldots, x_{\sigma(i)}, \ldots, x_{\sigma(n)}] \tag{5.2.57}$$

$\mathcal{N}_2$

For this model, with the sequence of labels $(1, 2, \ldots, i, \ldots, n)$, a random sequence of n labels is associated. Denote the latter by

$$I^* = (i_1^*, i_2^*, \ldots, i_j^*, \ldots, i_n^*) \tag{5.2.58}$$

where the i_j^*, $1 \le j \le n$ are mutually independent and where for each j, $1 \le j \le n$, we have

$$\text{For any } i, 1 \le i \le n, Prob\{i_j^* = i\} = \frac{1}{n} \tag{5.2.59}$$

$\mathcal{N}_3$

The first step of this model consists of associating a random integer ν with the cardinal n. The probability law of ν is defined by the Poisson distribution of parameter n (see (5.2.27)).

For $\nu = m$, we consider the following random vector whose components are mutually independent

$$I_m^* = (i_1^*, i_2^*, \ldots, i_j^*, \ldots, i_m^*) \tag{5.2.60}$$

where for any j, $1 \le j \le m$, we have the (5.2.59), namely

$$\text{For any } i, 1 \le i \le n, Prob\{i_j^* = i\} = \frac{1}{n}$$

The third step of this model is analogous to $\mathcal{N}_2$, above. More precisely, for m fixed, with the sequence I_m^* of m independent random labels, we associate the sequence

$$x_{I_m^*} = (x_{i_1^*}, x_{i_2^*}, \ldots, x_{i_j^*}, \ldots, x_{i_m^*}) \tag{5.2.61}$$

which is twofold random: for ν and for I_m^*.

Now consider the representation of an $\mathcal{O}$ subset X as a valuation defined by means of a vector of zeros and ones, indicating X in $\mathcal{O}$ (see Sect. 4.2.1.1), the three random models defined in the Boolean case (see Sect. 5.2.1.2) can be retrieved from the above formulation. However, the Boolean case and the numerical one have to be clearly distinguished; otherwise, all the combinatorial phenomenons associated with the Boolean case will be hidden in the numerical formulation.

Notice that the random model $\mathcal{N}_2$ has the same nature as the "Bootstrap" resampling technique [8]. However, our way of deriving this model is completely independent. The point of view, the context and the objective of the approach concerned are completely different [33, 35].

5.2.2.2 Coefficients for Symmetrical or Asymmetrical Associations

As mentioned above, the two numerical attributes to associate are denoted v and w. According to the notations above, (x_I, y_I) designate the ordered pair of valuations on the object set $\mathcal{O}$ induced by v and w. More explicitly,

$$x_I = \{x_i | i \in I\} \text{ and } y_I = \{y_i | i \in I\} \tag{5.2.62}$$

where I codes $\mathcal{O}$.

Symmetrical Association

Let us denote by μ_v and σ_v^2 (resp., μ_w and σ_w^2) the mean and variance of the empirical distribution x_I (resp., y_I). Denote also as above by $\mathcal{H}_i$ the hypothesis of no relation associated with $\mathcal{N}_i$, $1 \leq i \leq 3$. The raw association coefficient can be written as

$$s(v, w) = <x_I, y_I> = \sum_{1 \leq i \leq n} x_i \cdot y_i \tag{5.2.63}$$

where $< x_I, y_I >$ denotes the inner product between the vectors x_I and y_I. This index is a natural generalization of the corresponding raw index defined in the Boolean case (see $s(a, b)$ in (5.2.1)). Indeed $s(a, b)$ and $< x_I, y_I >$ coincide if x_I (resp., y_I) is the indicator Boolean vector of the subset $\mathcal{O}(a)$ (resp., $\mathcal{O}$) (see (5.2.1)).

Under the hypothesis $\mathcal{H}_1$ of no relation, the random raw coefficient can be written as

$$s(\sigma(v), \tau(w)) = \sum_{1 \leq i \leq n} x_{\sigma(i)} \cdot y_{\tau(i)} \tag{5.2.64}$$

where σ and τ are two independent random permutations on I, according to the $\mathcal{N}_1$ model. In fact, the two dual random variables

$$s(v, \tau(w)) = \sum_{1 \leq i \leq n} x_i \cdot y_{\tau(i)} \text{ and } s(\sigma(v), w) = \sum_{1 \leq i \leq n} x_{\sigma(i)} \cdot y_i \tag{5.2.65}$$

have exactly the same distribution as that of $s(\sigma(v), \tau(w))$. Denote by $\mathcal{S}$ the common random variable. The mathematical expectation and variance of $\mathcal{S}$ are given as

$$\mathcal{E}(\mathcal{S}) = n\mu(v) \cdot \mu(w) \tag{5.2.66}$$

$$var_1(\mathcal{S}) = \frac{n^2}{n-1}\sigma^2(v) \cdot \sigma^2(w) \tag{5.2.67}$$

The centred and standardized raw coefficient $s(v, w)$ can be written as follows:

$$Q_1(v, w) = \frac{s - \mathcal{E}(\mathcal{S})}{\sqrt{var_1(\mathcal{S})}} = \sqrt{n-1} \cdot R_1(v, w) \tag{5.2.68}$$

where $R_1(v, w)$ is nothing else but the correlation coefficient of Bravais–Pearson, namely

$$R_1(v, w) = \frac{p(v, w) - \mu_v \cdot \mu_w}{\sigma_v \cdot \sigma_w} = \frac{\frac{1}{n} \cdot \sum_{1 \le i \le n}(x_i - \mu_v)(y_i - \mu_w)}{\sigma_v \cdot \sigma_w} \tag{5.2.69}$$

where $p(v, w) = \frac{s(v,w)}{n}$.

Now, let us take up again the Boolean case (see (5.2.1)). If we represent the $\mathcal{O}(a)$ (resp., $\mathcal{O}(b)$) subset by means of a valuation v (resp., w) defined by its indicator function, we retrieve $R_1(a, b)$ (see (5.2.12)) via $R_1(v, w)$. However, when σ (resp., τ) describes the set G_n of the $n!$ permutations on $I = \{1, 2, \ldots, i, \ldots, n\}$, a given subset X (rep., Y) corresponding to $\mathcal{O}(a)$ (resp., $\mathcal{O}(b)$) is repeated $(n(a)!) \times (n(\bar{a})!)$ (resp., $(n(b)!) \times (n(\bar{b})!)$) times.

Let us now consider the hypothesis $\mathcal{H}_2$ of no relation. $\nu = n$ and the random raw coefficient $\mathcal{S}$ can be written as

$$s(I^\star, I'^\star) = \; <x_{I^\star}, y_{I'^\star}> \; = \sum_{1 \le j \le n} x_{i_j^\star} \cdot y_{i_j'^\star} \tag{5.2.70}$$

where $I^\star = (i_1^\star, i_2^\star, \ldots, i_j^\star, \ldots, i_n^\star)$ and $I'^\star = (i_1'^\star, i_2'^\star, \ldots, i_j'^\star, \ldots, i_n'^\star)$ are two independent random sequences defined according to the $\mathcal{H}_2$ model. Mathematical calculation of the mean and variance of $\mathcal{S}$ gives

$$\mathcal{E}(\mathcal{S}) = n\mu_v \cdot \mu(w) \tag{5.2.71}$$

$$var_2(\mathcal{S}) = n(\sigma_v^2 \cdot \sigma_w^2 + \mu_v^2 \cdot \sigma_w^2 + \mu_w^2 \cdot \sigma_v^2) \tag{5.2.72}$$

We can notice that $\mathcal{E}(\mathcal{S})$ and $var_1(\mathcal{S})$ are reduced to

$$n \cdot p(a) \cdot p(b) \text{ and } n \cdot p(a) \cdot p(b) \cdot [1 - p(a) \cdot p(b)]$$

in the case where the valuations v and w define the indicator functions of $\mathcal{O}(a)$ and $\mathcal{O}(b)$, respectively, (see (5.2.43)). As for $\mathcal{N}_1$, we can compare here the respective cardinalities of the sample spaces in both Boolean and numerical cases. In the Boolean case there are 2^n possibilities for the choice of random subset X (resp., Y) associated with $\mathcal{O}(a)$ (resp., $\mathcal{O}(b)$), but there are n^n possible values for the choice of $x_I^\star$ (resp., $y_{I'}^\star$).

The centred and standardized version of the raw coefficient $s(v, w)$, with respect to the hypothesis of no relation $\mathcal{H}_2$, can be expressed as

$$Q_2(v, w) = \sqrt{n} \cdot R_2(v, w) \tag{5.2.73}$$

where

$$R_2(v, w) = \frac{p(v, w) - \mu_v \cdot \mu_w}{\sqrt{\sigma_v^2 \cdot \sigma_w^2 + \mu_v^2 \cdot \sigma_w^2 + \mu_w^2 \cdot \sigma_v^2}} \tag{5.2.74}$$

And in the case where v and w are Boolean valuations, $R_2(v, w)$ becomes $R_2(a, b)$ of (5.2.37).

It remains to be considered the case of the hypothesis of no relation $\mathcal{H}_3$, for which ν becomes a Poisson random variable parametrized by n (see (5.2.27)). Under these conditions, the raw random index S takes the following form:

$$s(I_\nu^\star, I_\nu'^\star) = \sum_{1 \le j \le \nu} x_{i^\star j} \cdot y_{i'^\star j} \tag{5.2.75}$$

Computing of the mathematical expectation and variance of S gives

$$\mathcal{E}(S) = n\mu_v.\mu_w \tag{5.2.76}$$

$$var_3(S) = n.(\sigma_v^2 + \mu_v^2).(\sigma_w^2 + \mu_w^2) \tag{5.2.77}$$

Then, the statistically normalized expression of the raw association coefficient $s(v, w)$ can be written as

$$Q_3(v, w) = \sqrt{n}.R_3(v, w) \tag{5.2.78}$$

where

$$R_3(v, w) = \frac{p(v, w) - \mu_v.\mu_w}{\sqrt{(\sigma_v^2 + \mu_v^2).(\sigma_w^2 + \mu_w^2)}} \tag{5.2.79}$$

We can verify that $R_3(v, w)$ becomes $R_3(a, b)$ of (5.2.38) in the case where v and w are Boolean valuations.

Asymmetrical Association

The retained hypothesis of no relation in the asymmetrical case for comparing Boolean attributes was $\mathcal{H}_3$, leading to the Poisson model (see Sect. 5.2.1.4) [12, 21, 35]. Therefore, we express the corresponding asymmetrical coefficient for comparing two numerical attributes v and w. This coefficient defines a measure of the implicative tendency $v \to w$. We derive it from that obtained in the symmetrical case [33] (see (5.2.78–5.2.79)) and by analogy with the Boolean case. As expressed in Sect. 3.2.2 we suppose the common value scale of v and w defined by the positive reals $\mathbb{R}_+$. Moreover, to be consistent with the Boolean case we substitute for v and w their reduced versions denoted by v' and w' and defined by (4.2.7) or (4.2.6). Thus, the common value scale of the new attributes is the interval $[0, 1]$. Under these conditions, the *raw* implicative index corresponding to $n(a \wedge \bar{b})$ becomes

$$s(v', \bar{w}') = \sum_{1 \le i \le n} x'_i \cdot [1 - y'_i] \qquad (5.2.80)$$

where x'_i (resp., y'_i) is the value of v' (rep., w') on the object coded by i, $1 \le i \le n$. The random raw index associated with $s(v', \bar{w}')$ can be written as follows:

$$s(v'^{\star}, \bar{w}'^{\star}) = \sum_{1 \le j \le \nu} x'_{i^{\star}j} \cdot (1 - y'_{i^{\star}j}) \qquad (5.2.81)$$

The local probabilistic index corresponding to (5.2.42) becomes

$$\mathcal{I}^{\mathcal{H}_3}(v, w) = Prob^{\mathcal{H}_3}\{s(v'^{\star}, \bar{w}'^{\star}) > s(v', \bar{w}')\} \qquad (5.2.82)$$

A mathematical calculation of the mean and variance of $s(v'^{\star}, \bar{w}'^{\star})$ yields

$$\mathcal{E}(s(v'^{\star}, \bar{w}'^{\star})) = n \cdot \mu_{v'} \cdot (1 - \mu_{w'}) \qquad (5.2.83)$$
$$var(s(v'^{\star}, \bar{w}'^{\star})) = n \cdot (o_{v'}^2 + \mu_{v'}^2) \cdot (o_{w'}^2 + (1 - \mu_{w'})^2) \qquad (5.2.84)$$

Therefore, the statistically normalized expression of the raw index can be written as

$$Q_3(v', \bar{w}') = \sqrt{n} \cdot R_3(v', \bar{w}') \qquad (5.2.85)$$

where

$$R_3(v', \bar{w}') = \frac{p(v', \bar{w}') - \mu_{v'} \cdot (1 - \mu_{w'})}{\sqrt{(o_{v'}^2 + \mu_{v'}^2) \cdot (o_{w'}^2 + (1 - \mu_{w'})^2)}} \qquad (5.2.86)$$

where $p(v', \bar{w}') = p(v', \bar{w}')/n$.

These results meet perfectly those obtained, independently in [19].

5.2.2.3 Asymptotic Distribution of the Random Raw Coefficient

Now we justify the normal asymptotic distribution of the random raw index $s(v^\star, w^\star)$ under each of the no relation (or independence) random models $\mathcal{H}_1$, $\mathcal{H}_2$ and $\mathcal{H}_3$.

$\mathcal{H}_1$ (The Wald and Wolfowitz Theorem)

Let us take up again the following unilateral version of the random raw coefficient expressed in the left part of (5.2.65)

$$s(v, \tau(w)) = \sum_{1 \leq i \leq n} x_i \cdot y_{\tau(i)} \tag{5.2.87}$$

Its centred and standardized version can be written as (see (5.2.66))

$$Q_1(v, w^\star) = \frac{s(v, \tau(w)) - n\mu_v \cdot \mu(w)}{\sqrt{\frac{n^2}{n-1}\sigma^2(v) \cdot \sigma^2(w)}} \tag{5.2.88}$$

Studying the limit distribution of $Q_1(v, w^\star)$ is the subject of a popular theorem of nonparametric statistics, the Wald and Wolfowitz theorem [15, 39, 50]. The normal tendency of this distribution is obtained under very general conditions. The version given below of this theorem is due to Noether.

Let $\{\zeta_n = (z_{1n}, z_{2n}, \ldots, z_{nn}) | n \text{ integer } \geq 1\}$ be a sequence of sequences of real numbers. Two conditions are considered for the $\{\zeta_n | n \text{ integer } \geq 1\}$ sequence, which we designate by $\mathbf{W}$ and $\mathbf{N}$. The former was set up by Wald and Wolfowitz and the latter by Noether.

Condition $\mathbf{W}$: For all $r = 3, 4, \ldots$

$$\frac{\frac{1}{n}\sum_{1 \leq i \leq n}(z_{in} - \bar{z}_n)^r}{\left[\frac{1}{n}\sum_{1 \leq i \leq n}(z_{in} - \bar{z}_n)^2\right]^{r/2}} = O(1)$$

Condition $\mathbf{N}$: For all $r = 3, 4, \ldots$

$$\frac{\sum_{1 \leq i \leq n}(z_{in} - \bar{z}_n)^r}{\left[\sum_{1 \leq i \leq n}(z_{in} - \bar{z}_n)^2\right]^{r/2}} = o(1)$$

where $\bar{z}_n = \frac{1}{n}\sum_{1 \leq i \leq n} z_{in}$ and where o and O are the Landau notations.

Condition $\mathbf{W}$ is more restrictive than $\mathbf{N}$ condition. By denoting W and N the coefficients concerned, these conditions (W for $\mathbf{W}$ and N for $\mathbf{N}$), we have $W = n^{(r-2)/2} \times N$.

Let $\xi_n = (x_{1n}, x_{2n}, \ldots, x_{nn})$ (resp., $\eta_n = (y_{1n}, y_{2n}, \ldots, y_{nn})$) be a sequence of sequences of real numbers and assume ξ_n (resp., η_n) defining the value sequence of a numerical attribute v_n (resp., w_n) on a sequence of n objects. Then, the Wald and Wolfowitz theorem can be expressed as follows:

Theorem 8 *If ξ_n satisfies the condition* **W** *and η_n satisfies condition* **N**, *then $Q_1(v_n, w_n^\star)$, where $Q_1(v, w^\star)$ is defined by (5.2.88) (using (5.2.87)), has a limiting normal distribution with mean 0 and variance 1.*

A detailed expression of the proof of this theorem is worked on again in Chap. 2 of [27]. It can be found in books of nonparametric statistics [10, 42].

Now let us consider the case where $\{\xi_n | n \text{ integer} \geq 1\}$ and $\{\eta_n | n \text{ integer} \geq 1\}$ are defined as sequences of Boolean sequences, and where a Boolean value is interpreted numerically. For such a sequence which we denote by $\{\alpha_n | n \text{ integer} \geq 1\}$ a mathematical computing shows that the coefficient W concerned by the **W** condition takes the following specific form:

$$\bar{a}_n \times \left(\frac{1 - \bar{a}_n}{\bar{a}_n} \right)^{\frac{r}{2}}$$

where $\bar{a}_n$ is nothing else than the proportion of 1 values in the Boolean vector α_n. Consider now two sequences of Boolean sequences $\{\alpha_n | n \text{ integer} \geq 1\}$ and $\{\beta_n | n \text{ integer} \geq 1\}$ and denote by $\overline{a_n}$ and $\overline{b_n}$ the respective proportions of 1 Boolean value in α_n and β_n. On the other hand, define two Boolean attributes a_n and b_n observed on an object set of size n and corresponding to α_n and β_n. Clearly, the random index $Q_1(v_n, w_n^\star)$ of the theorem above can be transposed in the case where it can be written $Q_1(a_n, b_n^\star)$. Therefore, we have the following corollary of the preceding theorem.

Corollary 7 *If both $\bar{a}_n$ and $\bar{b}_n$ tend towards finite limits, then the limiting distribution of $Q_1(a_n, b_n^\star)$ is normal with mean 0 and variance 1.*

This development meets the case of comparing independent random Boolean attributes under the hypergeometric model and justifies (5.2.1).

$\mathcal{H}_2$ Model

The random variables $\{x_{i_j^\star} . y_{i_j^\star} | 1 \leq j \leq n\}$ of which the sum define the random raw index $s(I^\star, I'^\star)$ (see (5.2.70)) are mutually independent. Therefore, the central limit theorem applies to $s(I^\star, I'^\star)$ and we have, for every fixed value q

$$Prob \left\{ \frac{s(I^\star, I'^\star) - n\mu_v \cdot \mu(w)}{\sqrt{n(\sigma_v^2 \cdot \sigma_w^2 + \mu_v^2 \cdot \sigma_w^2 + \mu_w^2 \cdot \sigma_v^2)}} \leq q \right\} \to \Phi(q)$$

where as usual, Φ designates the cumulative normal distribution.

$\mathcal{H}_3$ Model

The central limit theorem [9] applies also to the limit distribution of $s(I_\nu^\star, I_\nu'^\star) = \sum_{1 \le j \le \nu} x_{i^\star j} \cdot y_{i'^\star j}$ (see (5.2.75)). Indeed $\nu \to \infty$ with a probability tending to 1 (almost sure convergence in probability), and we have

$$Prob \left\{ \frac{s(I_\nu^\star, I_\nu'^\star) - n\mu_v \cdot \mu(w)}{\sqrt{n(\sigma_v^2 + \mu_v^2). + \mu_v^2 \cdot (\sigma_w^2 + \mu_w^2)}} \le q \right\} \to \Phi(q)$$

Obviously, these results extend to the case of asymmetrical association between numerical attributes. Otherwise, as mentioned above (see Sect. 5.2.1.5), "Similarity Global Reduction" is required for mutually comparing a set of numerical attributes, symmetrically or asymmetrically.

5.2.2.4 Extension to Partial Associations; the Boolean and Numerical Cases

The Boolean Case

An original method for building a partial association coefficient between descriptive attributes, observed on an object set $\mathcal{O}$, is developed in [28, 29]. Different types of attributes are treated: Boolean, numerical, ranking, categorical nominal and categorical ordinal types. This method involves two fundamental features of the *LLA* approach:

1. Set theoretic representation of descriptive attributes with respect to $\mathcal{O}$;
2. Centring and standardizing statistically an adequate raw index between combinatorial structures on $\mathcal{O}$ induced by the attributes to be compared.

As expressed above, this statistical normalization is carried out with respect to a hypothesis of no relation, respecting the cardinal characteristics of the induced structures. The hypothesis considered here is $\mathcal{H}_1$. The power of the method lies in how to deal with the attributes of type II (see Sect. 3.3, Chap. 3). To begin with, the cases of numerical and Boolean attributes will be considered here. The other cases will be presented in Sect. 6.2.6.

Let α, β and γ be three descriptive attributes of an object set $\mathcal{O}$. The general problem consists of measuring the association between β and γ, independently of the respective influences of α on β and γ. In other words, it is in question to neutralize the effect of α in the association measure between β and γ. Notice that β and γ are necessarily of the same type, but the type of α is not demanded to be the same as that of β and γ. For example, β and γ might be numerical attributes and β, a categorical one.

To start with, let us consider the same numerical type for the three attributes and denote as (u, v, w) a triplet of numerical attributes. The called "partial correlation

Fig. 5.1 Three Boolean
attributes: partial association

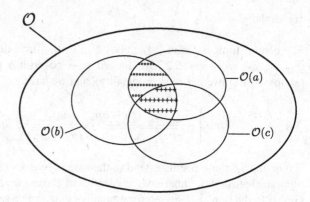

coefficient between v and w with u held constant", denoted by $\rho_{vw \cdot u}$ is provided
from a geometrical interpretation [2]. Let $\alpha_{11}u + \alpha_{12}$ be the least squares regression
line of v on u and $\alpha_{21}u + \alpha_{22}$ the least squares regression line of w on u. Let $v' =
v - (\alpha_{11}u + \alpha_{12})$ and $w' = w - (\alpha_{21}u + \alpha_{22})$. The correlation coefficient between
v' and w' is $\rho_{vw \cdot u}$.

$$\rho_{vw \cdot u} = \frac{\rho_{vw} - \rho_{uv} \cdot \rho_{uw}}{\sqrt{(1 - \rho_{uv}^2) \cdot (1 - \rho_{uw}^2)}} \qquad (5.2.89)$$

It is generally admitted to transpose such a coefficient in the Boolean case where
a Boolean vector of zeros and ones is interpreted numerically. We obtain

$$\rho_{bc \cdot a} = \frac{\rho_{bc} - \rho_{ab} \cdot \rho_{ac}}{\sqrt{(1 - \rho_{ab}^2) \cdot (1 - \rho_{ac}^2)}} \qquad (5.2.90)$$

where a, b and c are three Boolean attributes and where ρ_{bc}, ρ_{ab} and ρ_{ac} are the
Pearson correlation coefficients denoted above $R_1(b, c)$, $R_1(a, b)$ and $R_1(a, c)$ (see
(5.2.36)).

Let us now describe the combinatorial solution, corresponding to the *LLA*
approach, developed in [28] for establishing a partial association coefficient between
Boolean attributes b and c, neutralizing the influence of a. The situation can be
schematized by the diagram (see Fig. 5.1).

The raw index $s(b, c) = n(a \wedge b) = card(\mathcal{O}(b) \cap \mathcal{O}(c))$ is decomposed as follows:

$$s(b, c) = s((b, c)|a) + s((b, c)|\bar{a})$$
$$= card(\mathcal{O}(b) \cap \mathcal{O}(c) \cap \mathcal{O}(a)) + card(\mathcal{O}(b) \cap \mathcal{O}(c) \cap \mathcal{O}(\bar{a})) \qquad (5.2.91)$$

In order to build a partial association coefficient between b and c, neutralizing the
part of a, the hypothesis of no relation must preserve the cardinal relative positions
of $\mathcal{O}(b)$ and $\mathcal{O}(c)$ with respect to $\mathcal{O}(a)$). In this, it suffices to consider a unilateral

form of this hypothesis. Under these conditions, the random raw index is written as follows:

$$s^a(b, c^\star) = s((b, c^\star)|a) + s((b, c^\star)|\bar{a})$$
$$= card(\mathcal{O}(b) \cap \mathcal{O}(c^\star \wedge a)) + card(\mathcal{O}(b) \cap \mathcal{O}(c^\star \wedge \bar{a})) \qquad (5.2.92)$$

where $\mathcal{O}(c^\star \wedge a)$ (resp., $\mathcal{O}(c^\star \wedge \bar{a})$) is a random subset of $\mathcal{O}(a)$ (resp., $\mathcal{O}(\bar{a})$), whose cardinality is $n(a \wedge c)$ (resp., $n(\bar{a} \wedge c)$).

The random variables $s((b, c^\star)|a)$ and $s((b, c^\star)|\bar{a})$ are independent. Their distributions are hypergeometric parametrized by $(n(a), n(a \wedge b), n(a \wedge c))$ and $(n(\bar{a}), n(\bar{a} \wedge b), n(\bar{a} \wedge c))$, respectively. Identifying $n(a)$ with $n(a) - 1$ (resp., $n(\bar{a})$ with $n(\bar{a}) - 1$), the centred and standardized raw index can be written as $\sqrt{n} \cdot R_1^a(b, c)$, where

$$R_1^a(b, c) = \frac{p(b \wedge c) - \left(\frac{p(a \wedge b) \cdot p(a \wedge c)}{p(a)} + \frac{p(\bar{a} \wedge b) \cdot p(\bar{a} \wedge c)}{p(\bar{a})} \right)}{\sqrt{\frac{p(a \wedge b) \cdot p(a \wedge \bar{b}) \cdot p(a \wedge c) \cdot p(a \wedge \bar{c})}{p(a)^3} + \frac{p(\bar{a} \wedge b) \cdot p(\bar{a} \wedge \bar{b}) \cdot p(\bar{a} \wedge c) \cdot p(\bar{a} \wedge \bar{c})}{p(\bar{a})^3}}} \qquad (5.2.93)$$

Mathematical calculation enables us to establish that the numerator of $R_1^a(b, c)$ is identical to that of $\rho_{bc.a}$ (see (5.2.90)). However, the denominator of $R_1^a(b, c)$ is essentially different from that of $\rho_{bc.a}$. To verify this property consider the following example:

$$n(a \wedge b \wedge c) = 5, n(a \wedge b \wedge \bar{c}) = 15, n(a \wedge \bar{b} \wedge c) = 20, n(a \wedge b \wedge \bar{c}) = 60$$
$$n(\bar{a} \wedge b \wedge c) = 10, n(\bar{a} \wedge b \wedge \bar{c}) = 70, n(\bar{a} \wedge \bar{b} \wedge c) = 65, n(\bar{a} \wedge \bar{b} \wedge \bar{c}) = 55$$

We obtain 0.0475 and 0.0466 for the denominators of $R_1^a(b, c)$ and $\rho_{bc.a}$, respectively.

Substituting a nominal categorical attribute for the Boolean (or binary) attribute a leads to a natural generalization of the coefficient $R_1^a(b, c)$. The principle of this generalization will be given below in the case of comparing two numerical attributes.

Referring to the random model $\mathcal{H}_3$, a new version of the partial association coefficient, denoted $R_3^a(b, c)$ can be obtained. This is left for the reader.

Partial Association Between Two Numerical Attributes with Respect to a Nominal Categorical Attribute

As in Sect. 5.2.2.2, v and w will designate two numerical attributes. $I = \{1, 2, \ldots, i, \ldots, n\}$ denotes the indexation set of $\mathcal{O}$. $\{v(i)|1 \leq i \leq n\}$ (resp., $\{w(i)|1 \leq i \leq n\}$) is the sequence of the values of v (resp., w) on $\mathcal{O}$. Let c be a nominal categorical attribute defined on $\mathcal{O}$. As in Sect. 3.3.1, the category set is denoted here by $\mathcal{C} = \{c_1, c_2, \ldots, c_h, \ldots, c_k\}$. With each category c_h is associated a Boolean attribute a^h, defined as follows: For any x in $\mathcal{O}$, $a^h(x) = 1$ (resp., $a^h(x) = 0$) if and only if $c(x) = c_h$ (resp., $c(x) \neq c_h$), $1 \leq h \leq k$. Now let I_h be the object indices for which the value of c is c_h, $1 \leq h \leq k$.

As in the total case (see Eq. (5.2.63)) the raw association coefficient between v and w can be written as

$$s(v, w) = \sum_{i \in I} v(i) \cdot w(i) \tag{5.2.94}$$

Let us take up again here (5.2.65) to express the associated two dual random raw coefficients

$$s(v, \tau(w)) = \sum_{i \in I} x_i \cdot y_{\tau(i)} \text{ and } s(\sigma(v), w) = \sum_{i \in I} x_{\sigma(i)} \cdot y_i \tag{5.2.95}$$

where σ (resp., τ) is a random element of a probabilized space of permutations, chosen adequately. Here, for σ (resp., τ), the respective value sets of v and w on each I_h, $1 \leq h \leq k$, have to be held constant. Let us write $s(v, w)$ in the following form:

$$s(v, w) = \sum_{1 \leq h \leq k} \sum_{i \in I_h} v(i).w(i) \tag{5.2.96}$$

and consider one of the two dual random raw indices, for example, $s(v, \tau(w))$. Under these conditions, τ must be defined as a sequence $(\tau^1, \tau^2, \ldots, \tau^h, \ldots, \tau^k)$ of k independent random permutations on I_1, I_2, ..., I_h, ... and I_k, respectively. Therefore,

$$s(v, \tau(w)) = \sum_{1 \leq h \leq k} \sum_{i \in I_h} v(i) \cdot w(\tau^h(i)) \tag{5.2.97}$$

where the random permutation τ^h is equally distributed over all permutations on I_h, $1 \leq h \leq k$.

Clearly, the distribution of

$$s(\sigma(v), w) = \sum_{1 \leq h \leq k} \sum_{i \in I_h} v(\sigma^h(i)) \cdot w(i) \tag{5.2.98}$$

where the sequence $(\sigma^1, \sigma^2, \ldots, \sigma^h, \ldots, \sigma^k)$ has the same definition as that of $(\tau^1, \tau^2, \ldots, \tau^h, \ldots, \tau^k)$ —follows the same distribution as that of $s(v, \tau(w))$.

As in Sect. 3.3.1, n_h designates the cardinality of I_h, $1 \leq h \leq k$. According to (5.2.66) the mean and variance of $\sum_{i \in I_h} v(i) \cdot w(\tau^h(i))$ are

$$n_h \mu_h(v) \cdot \mu_h(w) \text{ and } \frac{n_h^2}{n_h - 1} \sigma_h^2(v) \cdot \sigma_h^2(w)$$

where $\mu_h(v)$ and $\sigma_h^2(v)$ (resp., $\mu_h(w)$ and $\sigma_h^2(w)$) are the mean and variance of v (resp., w) on I_h, $1 \leq h \leq k$.

By assuming, for any h $1 \leq h \leq k$, n_h large enough in order to identify n_h with $n_h - 1$, we obtain the following expression for the partial association coefficient between v and w, neutralizing the influence of the categorical attribute c:

$$R_1^c(v, w) = \frac{\sum_{1 \leq h \leq k} n_h \cdot \{\frac{1}{n_h} \cdot \sum_{i \in I_h} v(i)w(i) - \mu_h(v) \cdot \mu_h(w)\}}{\sqrt{\sum_{1 \leq h \leq k} n_h \sigma_h^2(v) \cdot \sigma_h^2(w)}} \qquad (5.2.99)$$

As for comparing Boolean attributes, a new version of the partial association coefficient, denoted $R_3^c(v, w)$ and referring to the random model $\mathcal{H}_3$, can be obtained. This is also left for the reader.

In Sect. 6.2.6 of Chap. 6, the generalization of partial association coefficient for comparing two nominal or ordinal categorical attributes is considered. In both cases, the attribute to be neutralized is necessarily of nominal categorical type. The case of partially comparing two ranking attributes is also expressed in this section.

5.3 Comparing Attributes for a Description of a Set of Categories

5.3.1 Introduction

So far, we have referred to a data table describing a set $\mathcal{O}$ of elementary objects (see Table 3.4, Sect. 3.5.4, Chap. 3). Now we pay attention to data table structure describing an exclusive and exhaustive set of categories, see Table 3.5, Sect. 3.5.4 of Chap. 3 and let us take up again the notations employed. $\Gamma = \{C_i | 1 \leq i \leq I\}$ denotes the set of categories concerned. The category C_i is represented by means of a learning set $\mathcal{O}_i$, $1 \leq i \leq I$. As above, the problem studied consists of building an association coefficient on the set $\mathcal{A} = \{a^j | 1 \leq j \leq p\}$ of descriptive attributes represented by the columns of Table 3.5. The dual problem of construction of a similarity index between elements of Γ will be studied in the following section (Sect. 7.2).

The general problem of comparing two given attributes a^j and a^k of $\mathcal{A}$, $1 \leq j < k \leq p$ returns to compare two sequences of distributions $(D^i(a^j) | 1 \leq i \leq I)$ and $(D^i(a^k) | 1 \leq i \leq I)$ where $D^i(a^j)$ and $D^i(a^k)$ are the respective statistical distributions of a^j and a^k on $\mathcal{O}_i$. Quantifying the latter similarity is a difficult problem. Although a^j and a^k are of the same type and have the same number of values (more than one single value), we cannot know how to match the respective values of a^j with those of a^k. Some possible solutions inspired from Mathematical Statistics contributions are proposed below.

The simplest case is where $\mathcal{A}$ is constituted by Boolean attributes. In this, two important subcases have to be distinguished. For the first one $\mathcal{A}$ is composed of the category set of a categorical attribute: each of the Boolean attributes is defined by one category. As expressed in Sect. 3.5.2, the table crossing Γ with $\mathcal{A}$, becomes a contigency table. In this alternative, the Boolean attributes a^j, $1 \leq j \leq p$, are logically dependent, they form an exclusive and exhaustive system: the different attributes are mutually exclusive and their union covers the whole set described. This case will be treated in Sect. 5.3.2. For the second subcase, the Boolean attributes are logically mutually independent. This case will also be addressed briefly in Sect. 5.3.2.

5.3.2 Case of a Description by Boolean Attributes

5.3.2.1 Case Where the Boolean Attributes Are Defined From Categories; the *Contingency Table*

Notations and Introduction

Let Γ and Λ denote the respective sets of categories both exclusive and exhaustive with respect to the description of a set of objects $\mathcal{O}$. The crossing of Γ and Λ produces a *contingency table* on $\mathcal{O}$. In the context of Sect. 3.5.4, this corresponds to a very particular case of the set $\mathcal{A}$ of descriptive attributes indexing the columns of Table 3.5. Let $\mathbb{I}$ and $\mathbb{J}$ be the respective sets of subscripts indexing the rows and columns of the contingency table:

$$\{n_{ij} | (i, j) \in \mathbb{I} \times \mathbb{J}\} \tag{5.3.1}$$

$I = card(\mathbb{I})$ and $J = card(\mathbb{J})$ will designate the number of rows and columns of this table. Writing $i \in \mathbb{I}$ (resp., $j \in \mathbb{J}$) is equivalent to write $1 \leq i \leq I$ (resp., $1 \leq j \leq J$). We put

$$n_{i.} = \sum_{j \in \mathbb{J}} n_{ij}, n_{.j} = \sum_{i \in \mathbb{I}} n_{ij}, 1 \leq i \leq I, 1 \leq j \leq J$$

$$n = n_{..} = \sum_{i \in \mathbb{I}} \sum_{j \in \mathbb{J}} n_{ij} \tag{5.3.2}$$

Now let us introduce the following proportions or relative frequencies:

$$f_{ij} = \frac{n_{ij}}{n}, p_{i.} = \frac{n_{i.}}{n} \text{ and } p_{.j} = \frac{n_{.j}}{n}, 1 \leq i \leq I, 1 \leq j \leq J.$$

The perfect symmetrical nature of a contingency table makes that the formalism needed to establish an association coefficient between the column categories can be transposed in order to establish an association coefficient between the row categories. Nevertheless, we are concerned here with the problem of establishing an association coefficient on the set Λ of column categories, interpreted as a set of descriptive attributes. For this interpretation, the set Γ of row categories is interpreted as a set of described objects. This point of view will be worked on again in the next section devoted to the construction of an *LLA* similarity index on an object set described by attributes of different types. In the formalism mentioned a specific geometrical representation of a contingency table will be needed.

Geometrical Representation of a Contingency Table

Correspondence Analysis method is based on this representation [4, 14]. With a given element i of $\mathbb{I}$ is associated its profile through $\mathbb{J}$, defined by the column vector

$$f_{\mathbb{J}}^i = (f_j^i | j \in \mathbb{J})^T$$

where $f_j^i = f_{ij}/p_{i.}$, $1 \leq i \leq I$ and where T indicates *transpose*.

$f_{\mathbb{J}}^i$ is represented in geometrical space $\mathbb{R}^J$ by a point whose coordinates are the f_j^i, $1 \leq j \leq J$. To this point is assigned the weight $p_{i.}$, $1 \leq i \leq I$. Thus, the whole representation is defined by the cloud

$$\mathcal{N}(\mathbb{I}) = \{(f_{\mathbb{J}}^i, p_{i.}) | i \in \mathbb{I}\} \tag{5.3.3}$$

Notice that this cloud is situated in the simplex defined by the convex hull of the canonical basis which we designate by $\{e_j | 1 \leq j \leq J\}$ of $\mathbb{R}^J$ (see Chap. 7). Furthermore, $\mathbb{R}^J$ is endowed with the following metric q:

$$q(e_j, e_k) = \frac{\delta_{jk}}{p_{.j}}$$

where δ_{jk} designates the δ of Kronecker ($\delta_{jk} = 0$ (resp. 1) if $j \neq k$ (resp., $j = k$)). This metric is called the Chi-square metric. The reason for this is that the square distance between two summits $f_{\mathbb{J}}^i$ and $f_{\mathbb{J}}^{i'}$ is the Chi-square distance between the distributions concerned with respect to the distribution $\{p_{.j} | 1 \leq j \leq J\}$, namely

$$d_{\chi^2}^2(f_{\mathbb{J}}^i, f_{\mathbb{J}}^{i'}) = \sum_{j \in \mathbb{J}} \frac{1}{p_{.j}} (f_j^i - f_j^{i'})^2 \tag{5.3.4}$$

Notice that the *center of gravity* (centroid) of $\mathcal{N}(\mathbb{I})$ can be written as

$$g_{\mathbb{I}} = \sum_{i \in \mathbb{I}} p_i f_{\mathbb{J}}^i = p_{.\mathbb{J}}$$

where the components of the point $p_{.\mathbb{J}}$ are the marginal proportions $p_{.j}$, $1 \le j \le J$.

Moreover, for this metric, the total inertia of $\mathcal{N}(\mathbb{I})$ is nothing else but the coefficient χ^2/n calculated for the whole contingency table crossing Γ with Λ:

$$\sum_{i \in \mathbb{I}} p_{i.} d_{\chi^2}^2 \left(f_{\mathbb{J}}^i, p_{.\mathbb{J}} \right) = \sum_{(i,j) \in \mathbb{I} \times \mathbb{J}} \frac{(f_{ij} - p_{i.} p_{.j})^2}{p_{i.} p_{.j}}$$

The introduction of the Chi-square metric is also justified by the so-called *principle of distributional equivalence* property. This means invariance of the mutual distances between elements of $\mathbb{I}$ (relative to the $\mathcal{N}(\mathbb{I})$ representation) and between elements of $\mathbb{J}$ (relative to the $\mathcal{N}(\mathbb{J})$ representation in its new space, see (7.2.35)) when two elements $i1$ and $i2$ of $\mathbb{I}$, having the same geometrical representation in $\mathbb{R}^J$ are replaced by one single new element $i0$, with the same representation as $i1$ and $i2$ and such that the weight of $i0$ is the sum of the weights of $i1$ and $i2$ ($p_{i0.} = p_{i1.} + p_{i2.}$).

The clustering problem of the rows or columns of a contingency table using an index as (5.3.4) and followed by a inertia criterion for comparing cluster formed on the set concerned ($\mathbb{I}$ or $\mathbb{J}$) is discussed in [13]. This approach will be detailed in Chap. 10.

We now illustrate the calculation of the cloud $\mathcal{N}(\mathbb{I})$ in the case of the *small* contingency table given by Table 3.3 of Chap. 3. As a matter of fact, analysis of the contingency table rows through its columns makes sense when the respective numbers of rows and columns are large enough. Let us specify that generally, in such analysis, the row number is greater than the column number. Now for Table 3.3, the lablels 1, 2, 3 and 4 of the set $\mathbb{I} = \{1, 2, 3, 4\}$ are for the rows indexed by D, C, B and A, respectively. On the other hand, the labels 1, 2, and 3 of the set $\mathbb{J} = \{1, 2, 3\}$ correspond to the columns c_1, c_2 and c_3, respectively. We have

1. $f_{\mathbb{J}}^1 = \left(\frac{223}{379}, \frac{122}{379}, \frac{34}{379} \right)^T$

2. $f_{\mathbb{J}}^2 = \left(\frac{168}{451}, \frac{215}{451}, \frac{68}{451} \right)^T$

3. $f_{\mathbb{J}}^3 = \left(\frac{90}{205}, \frac{80}{205}, \frac{35}{205} \right)^T$

4. $f_{\mathbb{J}}^4 = \left(\frac{110}{403}, \frac{191}{403}, \frac{102}{403} \right)^T$

where the T exponent indicates *transpose*. On the other hand, we have

- $p_{1.} = \frac{371}{1438}, p_{2.} = \frac{451}{1438}, p_{3.} = \frac{205}{1438}, p_{4.} = \frac{403}{1438}$

- $p_{.1} = \frac{591}{1438}, p_{.2} = \frac{608}{1438}, p_{.3} = \frac{231}{1438}$

In these conditions, according to the formula given in (5.3.4), we obtain

$$d^2_{\chi^2}(f^1_{\mathbb{J}}, f^2_{\mathbb{J}}) = \frac{1438}{591}\left(\frac{223}{379} - \frac{168}{451}\right)^2 + \frac{1438}{608}\left(\frac{122}{379} - \frac{215}{451}\right)^2 + \frac{1438}{239}\left(\frac{34}{379} - \frac{68}{451}\right)^2 = 0.1925$$

Correlation Coefficient Between Two Column Categories

The above geometrical representation makes that a given column Boolean attribute, indexed by j, associated with a category belonging to Γ, can be interpreted as a numerical attribute. On the other hand, for this representation, a given row Boolean attribute, indexed by i, is interpreted as an elementary object represented by the vector $f^i_{\mathbb{J}}$ (see above), the *value* of j on i, being f^i_j, $1 \le i \le I$, $1 \le j \le J$. Under these conditions the correlation coefficient between two columns Boolean attributes, indexed by j and k is

$$\rho(j, k) = \frac{\sum_{i \in \mathbb{I}} p_{i.}(f^i_j - p_{.j})(f^i_k - p_{.k})}{\sqrt{\left(\sum_{i \in \mathbb{I}} p_{i.}(f^i_j - p_{.j})^2\right)\left(\sum_{i \in \mathbb{I}} p_{i.}(f^i_k - p_{.k})^2\right)}} \tag{5.3.5}$$

Expressing this coefficient with respect to the following density table:

$$\left\{\frac{f_{ij}}{p_i.p_{.j}} | (i, j) \in \mathbb{I} \times \mathbb{J}\right\}$$

gives after a straightforward calculation

$$\rho(j, k) = \frac{\sum_{i \in \mathbb{I}} p_{i.} \frac{f_{ij}}{p_{i.}p_{.j}} \frac{f_{ik}}{p_{i.}p_{.k}} - 1}{\sqrt{\left(\sum_{i \in \mathbb{I}} p_{i.} \frac{f^2_{ij}}{p^2_i.p^2_{.j}} - 1\right)\left(\sum_{i \in \mathbb{I}} p_{i.} \frac{f^2_{ik}}{p^2_i.p^2_{.k}} - 1\right)}} \tag{5.3.6}$$

The design of this coefficient is the basis of [37, 49] works. Illustrating the computing of this coefficient by using the small example above, is left for the reader. An extension to multidimensional contingency table is provided in [49].

The geometrical representation adopted (see the preceding paragraph)—where a column category of a contingency table is assimilated to a numerical attribute—makes possible the extension of the partial association coefficient defined in the numerical case (see (5.2.89) in Sect. 5.2.2.4) for comparing column categories. More precisely, if h, j and k designate three column categories, the partial association coefficient $\rho_{jk.h}$ between j and k, neutralizing the influence of h, can be put as follows:

$$\rho_{jk.h} = \frac{\rho_{jk} - \rho_{hj} \cdot \rho_{hk}}{\sqrt{(1 - \rho_{hj}^2) \cdot (1 - \rho_{hk}^2)}} \tag{5.3.7}$$

where the ρ coefficient—when j and k are compared—is expressed in (5.3.5).

5.3.2.2   Extensions for Different Contingency Data Structures

Logically Independent Boolean Attributes

Let us consider now the case where none of the Boolean attributes of Λ is logically exclusive of another Λ attribute. That is, if a^j and a^k are two any distinct elements of Λ then, the sets of objects $\mathcal{O}(a^j)$ and $\mathcal{O}(a^k)$, where a^j and a^k are present, respectively, are not disjoint. We still denote by

$$(f_j^i | 1 \le i \le I)$$

the presence relative frequency vector of the Boolean attribute a^j on the sequence of disjoint subsets $\{O_{i.} | 1 \le i \le I\}$. In order to build an association coefficient on Λ, an option that comes naturally consists, as previously, of interpreting the elements of Λ as numerical attributes. In these conditions, f_j^i is the *value* of a^j on i, $1 \le i \le I$, $1 \le j \le J$. On the other hand, and as above, $p_{i.}$ is the weight assigned to i, $i \in \mathbb{I}$. The mean of a^j on $\mathbb{I}$, being $p_{.j}$, $1 \le j \le J$, the coefficient $\rho(j, k)$ can be taken up again, as it was in (5.3.5). New development for this comparison is proposed in the next paragraph.

Horizontal Juxtaposition of Contingency Tables

Let us consider again the data structure defined by the $\mathcal{S}$ system in Sect. 3.5.4 of Chap. 3 (see (3.5.18)). We assume $\mathcal{A} = \{a^1, a^2, \ldots, a^l, \ldots, a^p\}$ constituted by nominal categorical attributes. We denote by $\Lambda_l = \{a_l^1, a_l^2, \ldots, a_l^j, \ldots, a_l^{J_l}\}$ the category set of a^l. The latter will be represented by the set of subscripts $\mathbb{J} = \{1, 2, \ldots, j, \ldots, J_l\}$. Crossing Γ with Λ_l leads to a contingency table with I rows and J_l columns. Thus, we obtain a *horizontal juxtaposition of contingency tables* with I rows and $J = J_1 + J_2 + \cdots + J_l + \cdots + J_p$ columns. Under these conditions, consider the comparison between two column attributes a_l^j and $a_{l'}^k$. There are two cases: $l' = l$ or $l' \ne l$. In the former case, only one single contingency table has to be involved for evaluating the association between a_l^j and $a_{l'}^k$, that crossing Γ with Λ_l. In the latter case two contingency tables have to take part, giving rise to a horizontal juxtaposition of two contingency tables, indexed by $\Gamma \times \Lambda_l \cup \Lambda_{l'}$. If $l = l'$, we return to the case treated in the above section. Now, let us examine the case where $l' \ne l$ and consider the following two tables of relative frequencies:

$$\{f_{ij}(l)|(i, j) \in \mathbb{I} \times \mathbb{J}_l\} \text{ and } \{f_{ij}(l')|(i, j) \in \mathbb{I} \times \mathbb{J}_{l'}\} \qquad (5.3.8)$$

In [37] we propose to substitute for both tables, one single table

$$\{f'_{ij}|(i, j) \in \mathbb{I} \times \mathbb{J}_l \cup \mathbb{J}_{l'}\} \qquad (5.3.9)$$

where for any $(i, j) \in \mathbb{I} \times \mathbb{J}_l \cup \mathbb{J}_{l'}$

$$f'_{ij} = \frac{1}{2}f_{ij}(l) \text{ if } j \in \mathbb{J}_l (\text{resp. } \frac{1}{2}f_{ij}(l') \text{ if } j \in \mathbb{J}_{l'})$$

and to assimilate the latter table to a single contingency table. For this, due to $p_{i.}(l) = p_{i.}(l') = p_{i.} = card(\mathcal{O}_{i.}/card(\mathcal{O}))$, we have

$$p'_{i.} = \sum_{j \in \mathbb{J}_l \cup \mathbb{J}_{l'}} f'_{ij}$$

$$p'_{.j} = \frac{1}{2}p_{.j}$$

$$f'^i_j = \frac{f'_{ij}}{p_{i.}} = \frac{1}{2}f^i_j \text{ for } j \in \mathbb{J}_l \cup \mathbb{J}_{l'} \qquad (5.3.10)$$

and then

$$\text{For any } i \in \mathbb{I}, \sum_{j \in \mathbb{J}} f'^i_j = 1$$

Now let us return to the case discussed above where the Boolean attributes are logically mutually independent. Associate with each attribute a^j its opposite $\bar{a}^j$, $1 \leq j \leq p$. Crossing Γ with $\{a^j, \bar{a}^j\}$ gives a contingency table indexed by $\Gamma \times \{a^j, \bar{a}^j\}$ having I rows and 2 columns. Under these conditions, we can consider the following horizontal juxtaposition of contingency tables:

$$\Gamma \times \bigcup_{1 \leq j \leq p} \{a^j, \bar{a}^j\}$$

If the matter is to build an association coefficient between a^j and a^k $(k \neq j)$, then the juxtaposition of the contingency tables to consider is reduced to

$$\Gamma \times \{a^j, \bar{a}^j\} \cup \{a^k, \bar{a}^k\}$$

and the ith row of this data table is

$$(f_{ij}, p_{i.} - f_{ij}, f_{ik}, p_{i.} - f_{ik})$$

Therefore, the values of the f'_{ij} defined above become

$$\left(\frac{1}{2}f_{ij}, \frac{1}{2}(p_{i.} - f_{ij}), \frac{1}{2}f_{ik}, \frac{1}{2}(p_{i.} - f_{ik}) \right)$$

$1 \le i \le I, 1 \le j \ne k \le p$. In these conditions, it is easy to realize that the correlation coefficient calculated with the f'_{ij} values is exactly the same as that obtained above (see (5.3.5)).

Set Theoretic Approach for Establishing an Association Coefficient

The numerator of $\rho(j, k)$ in (5.3.5) can be put as

$$\sum_{1 \le i \le I} p_i f_j^i f_k^i - p_{.j} p_{.k}$$

where the second term is the mean of the first one under independence hypothesis. Consequently, in a consistent fashion with the construction of the coefficient $\rho(j, k)$, the *raw* association coefficient must be

$$s^c(a^j, a^k) = \sum_{1 \le i \le I} p_i f_j^i f_k^i$$

$$= \sum_{1 \le i \le I} \frac{n_{i.}}{n} \frac{n(i \wedge j)}{n_{i.}} \frac{n(i \wedge k)}{n_{i.}} = \frac{1}{n} \sum_{1 \le i \le I} \frac{n(i \wedge j)n(i \wedge k)}{n_{i.}} \qquad (5.3.11)$$

The first alternative of the hypothesis of no relation $\mathcal{H}_1$ consists of fixing the sets $P_{.j}$ and $P_{.k}$ and associating with the partition $\{O_{i.}|i \in \mathbb{I}\}$ a random partition with equal probabilities, in the set $\mathcal{P}(n; t)$ of all partitions of $\mathcal{O}$, with the same type $t = (n_{1.}, n_{2.}, \ldots, n_{i.}, \ldots, n_{I.})$. The associated random raw coefficient can be designated by (Fig. 5.2)

Fig. 5.2 Set association between categories

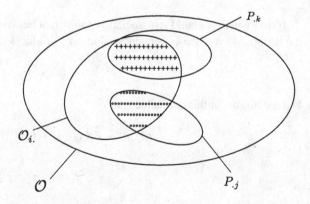

$$s^{c^*}(a^j, a^k)$$

The dual of this hypothesis of no relation consists of fixing $O_{i.}$ and associating with the partition $\{P_{.j} | j \in \mathbb{J}\}$ a random one in the set $\mathcal{P}(n; u)$, equally distributed, of all partitions of $\mathcal{O}$, with the same type $u = (n_{.1}, n_{.2}, \ldots, n_{.j}, \ldots, n_{.J})$.

$$s^c((a^*)^j, (a^*)^k)$$

It is shown in [30] that the distributions of the above two versions of the random raw coefficient are identical. Moreover, by denoting $S(c; (a^j, a^k))$ the common random variable associated with both random raw coefficients, we obtain the following results:

Theorem 9 *The mean and variance of $S(c; (a^j, a^k))$ are equal to*

$$\mathcal{E}(S(c; (a^j, a^k))) = \frac{(n-I)}{(n-1)} \frac{n(j)n(k)}{n}$$

$$var(S(c; (a^j, a^k))) = \sum_{1 \le i \le I} \frac{(n(i)-1)}{n(i)} \frac{n(j)n(h)}{n(n-1)}$$

$$\times \left(1 + \frac{(n(i)-2)(n(j)-1)}{n-2} + \frac{(n(i)-2)(n(k)-1)}{n-2}\right.$$

$$\left. + \frac{(n(i)-2)(n(i)-3)(n(j)-1)(n(k)-1)}{(n-2)(n-3)}\right)$$

$$+ 2 \sum_{1 \le i < i' \le I} \frac{(n(i)-1)(n(i')-1)n(j)(n(j)-1)n(k)(n(k)-1)}{n(n-1)(n-2)(n-3)}$$

$$- \left(\frac{(n-I)}{(n-1)}\right)^2 \left(\frac{n(j)n(k)}{n}\right)^2 \tag{5.3.12}$$

In the general case where I is small with respect to n the value of the ratio $(n-I)/n$ can be confused with unity. Thus, the centred raw coefficient can be written as

$$\sum_{1 \le i \le I} \frac{n(i \wedge j)n(i \wedge k)}{n(i)} - \frac{n(j)n(k)}{n}$$

and then, it is up to the multiplicative factor $1/n$, the numerator of the correlation coefficient $\rho(j, k)$ (see (5.3.5)). The denominator of $\rho(j, k)$ has a formal expression clearly different from the standard deviation of $S(c; (a^j, a^k))$.

This set theoretic approach can be extended for comparing column attributes of a horizontal juxtaposition of contingency tables [30].

5.3.3 Comparing Distributions of Numerical, Ordinal Categorical and Nominal Categorical Attributes

5.3.3.1 Ordinal Categorical Attributes

To begin with, we consider the case of category description by numerical attributes and this, for two related reasons. First, this case was worked in *Statistical Testing Hypotheses* theory. Next, the solution provided in the numerical case will guide us when the attributes to be compared are discrete, more precisely, in the cases of ordinal and nominal categorical attributes.

5.3.3.2 Comparing Distributions of Numerical Attributes

In the framework of the data structure S pictured in the data table Table 3.5 of Chap. 3, we consider two numerical attributes represented by two columns of Table 3.5. We designate these here by v and w. With the notations used in Table 3.5, v and w correspond to two given attributes a^j and a^k among those indexing the columns of Table 3.5, $1 \leq j \leq k \leq p$. We assume given, the distributions of v and w on the sequence $(O_{i.}|1 \leq i \leq I)$ of mutually disjoint sets. Recall that $O_{i.}$ is a training set (a sample) representing the category C_i which indexes the ith row of Table 3.5. Let us consider one generic element of this sequence which we denote by O.

The comparison of v and w on O is based on their cumulative empirical distribution functions, F and G, respectively. For a given real z, $F(z)$ (resp., $G(z)$) is the proportion (relative frequency) of objects in O for which the value of v (resp., w) is lower or equal to z.

$$F(z) = Prob\{X \leq z\}$$
$$G(z) = Prob\{Y \leq z\} \tag{5.3.13}$$

where X (resp., Y) is the random numerical attribute (variable) underlying F (resp., G). $Prob\{X \leq z\}$ (resp., $Prob\{Y \leq z\}$) corresponds to the probability of a random element o^* taken at random with equal chances in O to have a v value (resp., w value) lower or equal to z.

The statistic introduced by Smirnov in the context of Statistical Testing Hypotheses [46] is

$$Smir(F, G) = sup_{z \in \mathbb{R}}|F(z) - G(z)| \tag{5.3.14}$$

where $\mathbb{R}$ indicates the reals.

The probability law of this statistic has been established under the null hypothesis for which the empirical distributions F and G come from one theoretical cumulative distribution Ψ. In these conditions, a local version of an *LLA* probabilistic index can be provided as

$$Prob\{Smir(F^\star, G^\star) \geq Smir(F, G)\} \tag{5.3.15}$$

where F^* and G^* are two random empirical distributions coming from Ψ. However, the main problem here is not limited to the comparison of a single pair of distributions. It consists of mutually comparing a set of numerical attributes given by their distributions on a sequence of disjoint learning sets, denoted above $(O_{i.}|1 \leq i \leq I)$. For this, take up again Table 3.5 of Chap. 3, where the column attributes a^j, $1 \leq j \leq p$ are supposed numerical. Then, denote by F_i^j the empirical distribution of a^j on $O_{i.}$, $1 \leq i \leq I$. A distance coefficient between two column attributes a^j and a^k, $1 \leq j < k \leq p$, using the Smirnov Statistic, can be proposed as

$$Smir(a^j, a^k) = \sum_{1 \leq i \leq I} Smir(F_i^j, F_i^k) \qquad (5.3.16)$$

By denoting $mean_e(Smir)$ and $var_e(Smir)$ the empirical mean and variance of the following distance table:

$$\{Smir(a^j, a^k)|1 \leq j < k \leq p\}$$

the global reduction is applied (see (5.2.51)–(5.2.53) and (6.2.113)). From the following table:

$$\left\{ Smir^g(a^j, a^k) = \frac{Smir(a^j, a^k) - mean_e(Smir)}{\sqrt{var_e(Smir)}} \right\}$$

we get the probabilistic similarity table

$$\mathcal{P}_{Smir}^g = \{P_{Smir}^g(a^j, a^k) = \Phi(-Smir^g(a^j, a^k))|1 \leq j < k \leq p\} \qquad (5.3.17)$$

where, as usual, Φ designates the normal cumulative distribution. The adopted probabilistic reference does not assume that $Smir(F_i^{j^*}, F_i^{k^\star})$ follows a normal distribution under the null hypothesis for which $F_i^{j^*}$ and $F_i^{k^*}$ come from a single theoretical distribution. It corresponds rather to the third item expressed after (6.2.113). In this fashion, the *LLA* hierarchical agglomerative clustering can apply in order to organize by proximity the set

$$A = \{a^1, a^2, \ldots, a^k, \ldots, a^p\}$$

of numerical attributes according to their respective distributions on the sequence

$$(O_1, O_2, \ldots, O_{i.}, \ldots, O_{I.}).$$

The Cramér-Von Mises Two-Sample Test provides a different statistic than the Smirnov one. Its distribution is deeply studied in [3]. With the notations adopted above, this statistic can be expressed in the case where the empirical distributions of v and w are given on a same set O, by

$$Cram(F, G) = \sum_{o \in O} [F(v(o)) - G(v(o))]^2 + \sum_{o \in O} [F(w(o)) - G(w(o))]^2 \quad (5.3.18)$$

By substituting $Cram(F_i^j, F_i^k)$ for $Smir(F_i^j, F_i^k)$, a similar development as above can be led. Obviously, we have

$$Cram(a^j, a^k) = \sum_{1 \le i \le I} Cram(F_i^j, F_i^k) \quad (5.3.19)$$

The associated probabilistic similarity table on $\mathcal{A}$ that we consider can be written as

$$\mathcal{P}_{Cram}^g = \{P_{Cram}^g(a^j, a^k) = \Phi(-Cram^g(a^j, a^k)) | 1 \le j < k \le p\} \quad (5.3.20)$$

where, clearly, $Cram^g(a^j, a^k)$ is the centred and standardized coefficient $Cram(a^j, a^k)$ with respect to the distribution

$$\{Cram(a^j, a^k) | 1 \le j < k \le p\}.$$

5.3.3.3 Comparing Distributions of Ordinal Categorical and Nominal Categorical Attributes

Ordinal Categorical Attributes

Let c and d be two ordinal categorical attributes defined on a set O of objects. In our framework here O is a generic set corresponding to one of the sets $O_{i.}, 1 \le i \le I$ (see above). Let us recall our notations of Sects. 4.3.3 and 6.2.4. $\{c_1, c_2, \ldots, c_h, \ldots, c_K\}$ and $\{d_1, d_2, \ldots, d_l, \ldots, d_M\}$ designate the category sets of c and d, respectively. On the other hand, ω and ϖ indicate the respective total preorders on O induced by c and d. If $t = (n_1, n_2, \ldots, n_h, \ldots, n_K)$ and $u = (m_1, m_2, \ldots, m_l, \ldots, m_M)$ are the respective *compositions* of ω and ϖ, the empirical cumulative distributions of c and d are defined as follows:

$$F(c_h) = \frac{n_1 + n_2 + \cdots + n_h}{n}$$

$$G(d_l) = \frac{m_1 + m_2 + \cdots + m_l}{n} \quad (5.3.21)$$

where $n = card(O)$, $1 \le h \le K$, $1 \le l \le M$.

With a given h defined as a subscript of a c category, we associate $l(h)$ defined as one of the category subscripts of d such that

$$|F(c_h - G(d_{l(h)}))| \text{ minimum}$$

Similarly, with a given l defined as a subscript of a d category, we associate $h(l)$ defined as a category subscript of c, such that

$$|F(c_{h(l)} - G(d_l))| \text{ minimum}$$

In these conditions, a distance $\delta_{Smir}(F, G)$ inspired from the Smirnov statistic might be

$$\delta_{Smir}(F, G) = \max\{ \max_{1 \leq h \leq K} |F(c_h - G(d_{l(h)}))|,$$
$$\max_{1 \leq l \leq M} |F(c_{h(l)} - G(d_l))|\} \tag{5.3.22}$$

If we refer to the Cramér-von Mises statistic, the following distance could be proposed:

$$\delta_{Cram}(F, G) = \sum_{1 \leq h \leq K} |F(c_h - G(d_{l(h)}))|$$
$$+ \sum_{1 \leq l \leq M} |F(c_{h(l)} - G(d_l))|\} \tag{5.3.23}$$

The set $\mathcal{A} = \{a^1, a^2, \ldots, a^k, \ldots, a^p\}$ is now constituted by ordinal categorical attributes. The respective distributions of these are observed on each set $O_{i.}$ of the sequence $(O_{1.}, O_{2.}, \ldots, O_{i.}, \ldots, O_{I.})$, $O_{i.}$ being the sample representing the category C_i of Γ (see Sect. 3.5.4 of Chap. 3). For two given elements a^j and a^k of $\mathcal{A}$, $1 \leq j < k \leq p$, denote by F_j^i and F_k^i the distributions of a^j and a^k on $O_{i.}$, respectively, $1 \leq i \leq I$. In these conditions, the respective extensions of $\delta_{Smir}(F, G)$ and $\delta_{Cram}(F, G)$, for comparing the distributions of a^j and a^k on the overall Γ categories can be proposed as follows:

$$\delta_{Smir}(a^j, a^k) = \sum_{1 \leq i \leq I} \delta_{Smir}(F_j^i, F_k^i)$$
$$\delta_{Cram}(a^j, a^k) = \sum_{1 \leq i \leq I} \delta_{Cram}(F_j^i, F_k^i) \tag{5.3.24}$$

As for the previous numerical case (see (5.3.17) and (5.3.20)), global reduction of the distance tables

$$\{\delta_{Smir}(a^j, a^k)|1 \le j < k \le p\}$$
$$\{\delta_{Cram}(a^j, a^k)|1 \le j < k \le p\} \qquad (5.3.25)$$

has to be carried out before referring to a probability scale for probabilistic associations on $\mathcal{A}$.

Nominal Categorical Attributes

Now c and d are two nominal categorical attributes. The notations above are taken up again. Thus, $C = \{c_1, c_2, \ldots, c_h, \ldots, c_K\}$ and $D = \{d_1, d_2, \ldots, d_l, \ldots, d_M\}$ designate the respective category sets of c and d. Here, we do not suppose any structure on C and D and then, c and d induce two partitions $\pi = \{O_1, O_2, \ldots, O_h, \ldots, O_K\}$ and $\chi = \{P_1, P_2, \ldots, P_l, \ldots, P_M\}$ on the object set $\mathcal{O}$, respectively. For simplicity reasons, we assume these partitions with labelled classes. Their respective types are indicated by $t(\pi) = (n_1, n_2, \ldots, n_h, \ldots, n_K)$ and $t(\chi) = (m_1, m_2, \ldots, m_l, \ldots, m_M)$. Let us define, as in Sect. 3.3.1 of Chap. 3, the relative frequencies:

$$f_j = \frac{n_j}{n}, 1 \le j \le K \text{ and } g_l = \frac{m_l}{n}, 1 \le l \le M$$

For establishing indices such that Smirnov and Cramér-von Mises Statistics, it is essential to furnish total orders on C and D, respectively. Thus, the nominal categorical attributes c and d become ordinal categorical attributes. For clarity reasons, let us denote the latter by $\tilde{c}$ and $\tilde{d}$, respectively. In these conditions, $\tilde{c}$ and $\tilde{d}$ induce two total preorders on $\mathcal{O}$ that we denote by $\omega(\pi)$ and $\omega(\chi)$. $\omega(\pi)$ (resp., $\omega(\chi)$) have the same classes as π (resp, as χ). The general idea of both transformations π into $\omega(\pi)$ and χ into $\omega(\chi)$, is to establish the linear orders on C and D in such a way as to have the same basis for comparing the cumulative distributions of $\tilde{c}$ and $\tilde{d}$. We propose to build two category sequences $\tilde{C} = (c_{(1)}, c_{(2)}, \ldots, c_{(h)}, \ldots, c_{(K)})$ and $\tilde{D} = (d_{(1)}, d_{(2)}, \ldots, d_{(l)}, \ldots, d_{(M)})$, where the category sets C and $\tilde{C}$ (resp., D and $\tilde{D}$) are identical. Besides, $\big((1), (2), \ldots, (h), \ldots, (K)\big)$ (resp., $\big((1), (2), \ldots, (l), \ldots, (M)\big)$) is a permutation of the C subscript sequence $(1, 2, \ldots, h, \ldots, K)$ (resp., D subscript sequence $(1, 2, \ldots, l, \ldots, M)$) such that

$$f_{(1)} \le f_{(2)} \le \cdots \le f_{(h)} \le \cdots \le f_{(K)} \text{ and } g_{(1)} \le g_{(2)} \le \cdots \le g_{(l)} \le \cdots \le g_{(M)}$$

Clearly, we may denote by $\big(O_{(1)}, O_{(2)}, \ldots, O_{(h)}, \ldots, O_{(K)}\big)$ (resp., $\big(P_{(1)}, P_{(2)}, \ldots, P_{(l)}, \ldots, P_{(M)}\big)$) the ordered sequence of the $\omega(\pi)$ (resp., $\omega(\chi)$) classes. Thus,

$$(n_{(1)}, n_{(2)}, \ldots, n_{(h)}, \ldots, n_{(K)})$$

and

$$(m_{(1)}, m_{(2)}, \ldots, m_{(l)}, \ldots, m_{(M)})$$

are the *compositions* of $\omega(\pi)$ and $\omega(\chi)$, respectively. Obviously, the previous inequalities correspond to

$$n_{(1)} \leq n_{(2)} \leq \cdots \leq n_{(h)} \leq \cdots \leq n_{(K)} \text{ and } m_{(1)} \leq m_{(2)} \leq \cdots \leq m_{(l)} \leq \cdots \leq m_{(M)}$$

The transformations of the nominal categorical attributes c and d into the ordinal categorical attributes $\tilde{c}$ and $\tilde{d}$ enable the coefficients of the previous paragraph to be applied, namely (5.3.22) and (5.3.23). Clearly, the rest of the development above in the ordinal case holds here. More precisely, if $\mathcal{A} = \{a^1, a^2, \ldots, a^k, \ldots, a^p\}$ is constituted by nominal categorical attributes, with $\mathcal{A}$ is associated a set $\tilde{\mathcal{A}} = \{\tilde{a}^1, \tilde{a}^2, \ldots, \tilde{a}^k, \ldots, \tilde{a}^p\}$ where $\tilde{a}^k$ is associated with a^k in the same manner as $\tilde{c}$ or $\tilde{d}$ were associated with c and d, above, $1 \leq k \leq p$. Therefore, distance tables as (5.3.25) can be calculated followed by a global reduction in order to reach probabilistic association coefficients on $\tilde{\mathcal{A}}$. The latter will be interpreted as for comparing the elements of $\mathcal{A}$ according to their distributions on Γ.

References

1. Achard, P.: Choix d'un indice de similarité. Qual. Quant. **2**, 289–302 (1972)
2. Anderson, T.W.: An Introduction to Multivariate Statistical Analysis. Wiley, New York (1958)
3. Anderson, T.W.: On the distribution of the two-sample Cramér- Von mises criterion. Ann. Math. Stat. **3**, 1148–1159 (1962)
4. Benzécri, J.P.: L'analyse des données, tome II. Dunod, France (1973)
'5. Cramer, H.: Mathematical Methods of Statistics. Princeton University Press, USA (1946)
6. Daudé, F.: Analyse et justification de la notion de ressemblance entre variables qualitatives dans l'optique de la classification hiérarchique par *AVL*. Ph.D. thesis, Université de Rennes 1, June 1992
7. Dice, L.R.: Measures of the amount of ecologic association between species. Ecology **26**, 297–302 (1945)
8. Efron, B.: The Jacknife, the Boot-strap and other resampling plans. CBMS-NSF, Regional Conference series in Applied Mathematics (1994)
9. Feller, W.: An Introduction to Probability Theory and Its Applications. Wiley, New York (1968)
10. Fraser, D.A.S.: Nonparametric Methods in Statistics. Wiley, New York (1959)
11. Gras, R.: Contribution à l'étude expérimentale et à l'analyse de certaines acquisitions cognitives et de certains objectifs didactiques en mathématiques. Ph.D. thesis, Thèse de doctorat d'état, Université de Rennes 1 (1979)
12. Gras, R., Kuntz, P.: An overview of the statistical implicative analysis (SIA) development. In: Spagnolo, F., Suzuki, E., Gras, R., Guillet, F. (eds.) Statistical Implicative Analysis, Studies in Computational Intelligence, pp. 11–40. Springer, New York (2008)
13. Greenacre, M.: Clustering the rows and columns of a contingency table. J. Classif. **5**, 39–51 (1988)
14. Greenacre, M.: Correspondence Analysis in Practice. Chapman & Hall, London (2007)
15. Hajek, J.: Some extensions of the Wald-Wolfowitz-Noether theorem. Ann. Math. Stat. **32**, 506–523 (1961)

16. Hamann, V.: Merkmalbestand und verwandtschaftsbeziehungen der farinosae. Beitragzum System der Monokotyledonen **2**, 639–768 (1961)
17. Jaccard, P.: Nouvelles recherches sur la distribution florale. Bulletin de la Société Vaudoise des Sciences Naturelles **44**, 223–270 (1908)
18. Kulczynski, S.: Die pflanzenassoziationen der pieninen [in polish, german summary]. Bull. Inter. Acad. Pol. Sci. Lett. Cl. Sci. Math.Nat. (Sci. Nat) (suppl. 2), 57–203 (1927)
19. Lagrange, J.-B.: Analyse implicative d'un ensemble de variables numériques; application au traitement d'un questionnaire à réponses modales ordonnées. Revue de Statistique Appliquée **46**(1), 71–93 (1998)
20. Lancaster, H.O.: The Chi-squared Distribution. John Wiley, New Jersey (1969)
21. Lerman, I-C., Azé, J.: A new probabilistic measure of interestingness for association rules, based on the Likelihood of the Link. In: Guillet, F., Hamilton, H.J. (eds.) Quality Measures in Data Mining, Studies in Computational Intelligence, vol. 43, pp. 207–236. Springer, Heidelberg (2007)
22. Lerman, I-C., Guillaume, S.: Comparing two discriminant probabilistic interestingness measures for association rules. In: Pinaud, B. (ed.) Advances in Knowledge Discovery and Management 3, pp. 59–83. Springer, New York (2013)
23. Lerman, I.C.: Sur l'analyse des données préalable à une classification automatique; proposition d'une nouvelle mesure de similarité. Mathématiques et Sciences Humaines **32**, 5–15 (1970)
24. Lerman, I.C.: Étude distributionnelle de statistiques de proximité entre structures finies de même type; application à la classification automatique. Cahiers du Bureau Universitaire de Recherche Opérationnelle **19**, 1–52 (1973)
25. Lerman, I.C.: Introduction à une méthode de classification automatique illustrée par la recherche d'une typologie de personnages enfants à travers la littérature enfantine. Revue de Statistique Appliquée **XXI**(3), 23–49 (1973)
26. Lerman, I.C.: Formal analysis of a general notion of proximity between variables. In: Barra, J.R. et al., (ed.) Recent Developments in Statistics, pp. 787–795. North-Holland (1977)
27. Lerman, I.C.: Classification et analyse ordinale des données. Dunod, France (1981). http://www.brclasssoc.org.uk/books/index.html
28. Lerman, I.C.: Indices d'association partielle entre variables qualitatives nominales. RAIRO série verte **17**(3), 213–259 (1983)
29. Lerman, I.C.: Indices d'association partielle entre variables qualitatives ordinales. Publications Institut de Statistique des Universités de Paris, **XXVIII**(1, 2), 7–46 (1983)
30. Lerman, I.C.: Interprétation non-linéaire d'un coefficient d'association entre modalités d'une juxtaposition de tables de contingences. Mathématiques et Sciences Humaines **83**, 5–30 (1983)
31. Lerman, I.C.: Justification et validité d'une échelle [0, 1] de fréquence mathématique pour une structure de proximité sur un ensemble de variables observées. Publications de l'Institut de Statistique des Universités de Paris **29**, 27–57 (1984)
32. Lerman, I.C.: Rôle de l'inférence statistique dans une approche de l'analyse classificatoire des données. Journal de la Société de Statistique de Paris **127**, 233–238 (1986)
33. Lerman, I.C.: Conception et analyse de la forme limite d' une famille de coefficients statistiques d ' association entre variables relationnelles. i. Revue Mathématique Informatique et Sciences Humaines **118**, 35–52 (1992)
34. Lerman, I.C.: Conception et analyse de la forme limite d' une famille de coefficients statistiques d' association entre variables relationnelles. ii. Revue Mathématique Informatique et Sciences Humaines **119**, 75–100 (1992)
35. Lerman, I.C., Gras, R., Rostam, H.: Élaboration et évaluation d'un indice d'implication pour des données binaires i et ii. Revue Mathématique et Sciences Humaines, (**74** et **75**):5–35 et 5–47, 1981
36. Lerman, I.C., Guillaume, S.: Analyse comparative d'indices d'implication discriminants fondés sur une échelle de probabilité. Research Report 7187, IRISA-INRIA (2010)
37. Lerman, I.C., Tallur, B.: Classification des éléments constitutifs d' une juxtaposition de tableaux de contingence. Revue de Statistique Appliquée **28**, 5–28 (1980)

38. Nicolaü, M.H.: Contribuiçò es ao estudo dos coeficientes de comparação em análise classificatoria. Ph.D. thesis, University of Lisboa (1981)
39. Noether, G.: On a theorem by Wald and Wolfowitz. Ann. Math. Stat. **20**, 455–458 (1949)
40. Ochiai, A.: Zoogeographic studies on the soleoid fishes found in Japan and its neighbouring regions. Bull. Japan. Soc. Sci. Fish. **22**, 526–530 (1957)
41. Pearson, K.: On the coefficient of racial likeness. Biometrika **18**, 105–117 (1926)
42. Puri, M.L., Sen, P.K.: Nonparametric Methods in Multivariate Analysis. Wiley, New Jersey (1971)
43. Rakotomalala, R., Morineau, A.: The TVpercent principle for the counter examples statistic. In: Spagnolo, F., Suzuki, E., Gras, R., Guillet, F. (eds.) Statistical Implicative Analysis, Studies in Computational Intelligence, pp. 449–462. Springer, New York (2008)
44. Rogers, D.J., Tanimoto, T.T.: A computer program for classifying plants. Science **132**, 1115–1118 (1960)
45. Russel, P.F., Rao, C.R.: On habitat and association of species of anopheline larvae in southeastern madras. J. Malar. Inst. India **T3**, 153–178 (1940)
46. Smirnov, N.V.: Estimate of deviation between empirical distributrion functions in two independent samples (russian). Bulletin Moscow University, (**2:2**):3–16 (1939)
47. Sokal, R.R., Michener, C.: A statistical method for evaluating systematic relationships. Univ. Kans. Sci. Bull. **38**, 1409–1438 (1958)
48. Sokal, R.R., Sneath, P.H.A.: Principles of Numerical Taxonomy. Freeman (1963)
49. Tallur, B.: Contribution à l'analyse exploratoire de tableaux de contingence par la classification, Doctorat d' État. Ph.D. thesis, Université de Rennes 1 (1988)
50. Wald, A., Wolfowitz, J.: Statistical tests based on permutations of the observations. Ann. Math. Stat. **15**, 358–372 (1944)
51. Xiaobo, L., Dubes, R.C.: A probabilistic measure of similarity for binary data in pattern recognition. Pattern Recognit. **22**, 397–409 (1989)
52. Yule, G.U.: On measuring association between attributes. J. Royal Statist. Soc. **75**, 579–642 (1912)

Chapter 6
Comparing Attributes by a *Probabilistic and Statistical* Association II

6.1 Introduction

The data is defined by the observation of a set $\mathcal{A}$ of descriptive attributes on a set $\mathcal{O}$ of elementary objects. As indicated in the introduction of the preceding chapter (see Sect. 5.1 of Chap. 5) $\mathcal{A}$ is constituted of attributes of a same type belonging to the general type II (see Sect. 3.3 of Chap. 3). To fix ideas in this introduction, we may imagine $\mathcal{A}$ as composed of nominal categorical attributes. The different comparison cases are listed at the beginning of the following section (see Sect. 6.2). For this comparison, as expressed in the introductive Sect. 5.1 of Chap. 5, the *LLA* approach will be emphasized. It leads, in a unified process, to a very rich family of probabilistic association coefficients between descriptive attributes of any type. On the other hand, the principle of this method enables several association coefficients to be mutually compared.

The general principle is always the same. $\mathcal{A}$ is represented as a sample in the subset set of a set Ω adequately chosen. If α and β are the respective representative Ω subsets of two attributes a and b belonging to $\mathcal{A}$, a *raw* similarity index $s(\alpha, \beta)$ is defined. A random pair $(\alpha^\star, \beta^\star)$, composed of independent elements is associated with (α, β). The expectation $\mathcal{E}[s(\alpha^\star, \beta^\star)]$ and the variance $var(s[(\alpha^\star, \beta^\star)])$ of the random raw index $s(\alpha^\star, \beta^\star)$ are calculated mathematically (See Sects. 5.2 and 5.3 of Chap. 5 and Sect. 6.2 of this Chapter). These calculations [19, 21, 27, 28] reconsidered accurately here enable us to obtain standardized coefficients of the form

$$Q(\alpha, \beta) = \frac{(s(\alpha, \beta) - \mathcal{E}[s(\alpha^\star, \beta^\star)])}{\sqrt{var(s(\alpha^\star, \beta^\star))}}$$

If Φ denotes the cumulative normal distribution, a first version of a probabilistic coefficient is obtained by applying Φ to the normalized coefficient, namely, $\Phi[Q(\alpha, \beta)]$. The latter might be non-discriminant enough. In these conditions, for

© Springer-Verlag London 2016
I.C. Lerman, *Foundations and Methods in Combinatorial and Statistical Data Analysis and Clustering*, Advanced Information and Knowledge Processing, DOI 10.1007/978-1-4471-6793-8_6

mutual comparison a set $\mathcal{A}$ of descriptive attributes, as for the Boolean case, a global normalization is in addition required, before applying Φ (see Sect. 5.2.1.5 of Chap. 5).

The operation consisting of taking out from $s(\alpha, \beta)$ the expectation (mean) $\mathcal{E}[s(\alpha^\star, \beta^\star)]$ is often called "correction for chance" $s(\alpha, \beta)$ [1, 2, 14]. In order to obtain a coefficient comprised within the interval $[-1, +1]$ it is generally proposed to divide the centred coefficient $s(\alpha, \beta) - \mathcal{E}[s(\alpha^\star, \beta^\star)]$ by its maximum. This technique follows that applied by M.G. Kendall in 1948 (See [17]) (1970) for defining his famous τ coefficient for comparing two rankings without ties (i.e. total and strict orders) on an object set. But, determining an exact maximum for comparing two categorical attributes (nominal or ordinal) conditioned by fixed margins is a difficult problem. And generally, a too large bound for this maximum is proposed. Significant results for determining the exact maximum in the case of nominal or ordinal categorical attributes with fixed margins are provided in [25, 31, 44].

Let us once again point out that the normalization method required by the *LLA* approach is essentially different from those mentioned. The centred ("corrected for chance") coefficient $s(\alpha, \beta) - \mathcal{E}[s(\alpha^\star, \beta^\star)]$ is reduced by means of the standard deviation $\sqrt{var(s(\alpha^\star, \beta^\star))}$. The coefficient obtained lies in the interval $]-\infty, +\infty[$. By applying the normal cumulative distribution function to this coefficient or that deduced from an additional empirical global standardization (see Sect. 5.2.1.5), a new one, lying in the [0, 1] interval is obtained.

This general approach is studied in order to build *total* and *partial* association coefficients for different types of descriptive attributes. For type I (Boolean and numerical attributes) the partial case is considered in Sect. 5.2.2.4, and for type II (nominal categorical and ordinal categorical attributes) in Sect. 6.2.6.

In Sect. 6.2.6, the generalization of partial association coefficient for comparing two nominal or ordinal categorical attributes is considered. In both cases, the attribute to be neutralized has necessarily a nominal categorical type. The case of partially comparing two ranking attributes is also expressed in this section.

6.2 Comparing Attributes of Type II for an Object Set Description; the *LLA* Approach

6.2.1 Introduction; Alternatives in Normalizing Association Coefficients

So far, we have studied the comparison between descriptive attributes of type I, defined in Sect. 3.2. The latter comprises the Boolean and numerical cases (see Sects. 5.2.1 and 5.2.2 of Chap. 5). The subject of partial comparison between two Boolean or numerical attributes, neutralizing the role of a binary or nominal categorical attribute was addressed in Sect. 5.2.2.4. Now we shall generalize these comparisons for attributes of type II (see Sect. 3.3). More precisely, the following cases will be considered:

Fig. 6.1 Building an *LLA*
association coefficient

$$(a,b) \to (\alpha,\beta) \to ((R(\alpha), R(\beta))) \in \Omega_\alpha \times \Omega_\beta$$

$$s(a,b) = card(R(\alpha) \cap R(\beta)) \text{ (resp., } < R(\alpha), R(\beta) >)$$

$$(\alpha,\beta) \longrightarrow (\alpha^\star, \beta^\star) \in (\mathcal{A}, P_\mathcal{A}) \times (\mathcal{B}, P_\mathcal{B})$$

$$S = s(\alpha^\star, \beta^\star) = card(R(\alpha^\star) \cap R(\beta^\star)) \text{ (resp., } < R(\alpha^\star), R(\beta^\star) >)$$

$$Q(a,b) = \frac{s(a,b) - \mathcal{E}(S)}{\sqrt{var(S)}}$$

- Comparing two ranking attributes;
- Comparing two nominal categorical attributes;
- Comparing two ordinal categorical attributes;
- Comparing two valuated graphs attributes;
- Comparing two similarity categorical attributes;
- Comparing two taxonomic categorical attributes;
- Partial comparison between two categorical attributes (the nominal, ordinal and ranking cases).

In all cases the general scheme for establishing an association coefficient between two relational attributes, that we denote here by a and b, follows the diagram of Fig. 6.1.

a and b are two attributes of the same type. As above, to fix ideas, let us consider the example where a and b are nominal categorical attributes (see Sect. 3.3.1). α and β are the induced structures by a and b on the object set $\mathcal{O}$. For the illustrating example, α and β are two partitions on $\mathcal{O}$. $\mathcal{A}$ (resp., $\mathcal{B}$) is the set of all structures of the same type as α (resp., β). In the example considered, $\mathcal{A}$ (resp., $\mathcal{B}$) is the set of all partitions having the same type as α (resp., β). $R(\alpha)$ (resp., $R(\beta)$) defines the graph representation of α (resp., β) on $\mathcal{O}$ (see (3.3.5) for the illustrating example). Ω_α (resp., Ω_β) is the set of all graph representations associated bijectively with $\mathcal{A}$ (resp., $\mathcal{B}$). $s(a, b)$ defines the *raw* association coefficient. Its expression is given above for the example considered (see (4.3.4) and those above in Sect. 4.3.2). The raw index takes an inner product form $< R(\alpha), R(\beta) >$ in the case of comparing two valuated relations. The hypothesis of no relation that we denote by $\mathcal{H}$ is schematized in the rectangle. $\alpha^\star$ and $\beta^\star$ are the independent random structures associated with α and β, respectively. $\alpha^\star$ (resp., $\beta^\star$) is a random element of $\mathcal{A}$ (resp., $\mathcal{B}$) endowed with a probability measure $P_\mathcal{A}$ (resp., $P_\mathcal{B}$). $S = s(\alpha^\star, \beta^\star)$ is the random raw coefficient and $Q(a, b)$ is the centred and standardized coefficient. $s(a, b)$ and $Q(a, b)$ can also be denoted $s(\alpha, \beta)$ and $Q(\alpha, \beta)$, respectively.

As in (5.2.8), the *local* probabilistic coefficient can be written as

$$P(a, b) = Prob^{\mathcal{H}}\{s(\alpha^\star, \beta^\star) \leq s(\alpha, \beta)\} = Prob^{\mathcal{H}}\{Q(\alpha^\star, \beta)^\star \leq Q(\alpha, \beta)\}$$

$$(6.2.1)$$

As for comparing Boolean or numerical attributes there are three models of the hypothesis of no relation, $\mathcal{H}_1, \mathcal{H}_2$ and $\mathcal{H}_3$ corresponding to the random choice models $\mathcal{N}_1, \mathcal{N}_2$ and $\mathcal{N}_3$ (see Sects. 5.2.1.2 and 5.2.2.1). The two most different models are $\mathcal{H}_1$, and $\mathcal{H}_3$, corresponding to $\mathcal{N}_1$ and $\mathcal{N}_3$. $\mathcal{N}_1$ is the hypergeometric (or its gener- alization) model and $\mathcal{N}_3$ is the Poisson model. The respective means of $s(\alpha^\star, \beta^\star)$ under $\mathcal{H}_1$, and $\mathcal{H}_3$ are identical. But the respective variances are different. In most cases—but not always—the asymptotic normal tendency of $s(\alpha^\star, \beta^\star)$ is established [6, 11, 20, 36, 39, 47]. Normal asymptotic tendency is not verified for $\mathcal{H}_1$ model, in the case of comparing nominal categorical attributes having each a few number of categories and when the different categories of at least one of the two attributes to be compared, are more or less equally represented in the object set $\mathcal{O}$ [6, 36]. This circumstance occurs rarely. It is always possible to specify for these cases the limit distribution. The latter corresponds to a chi-square distribution. However, for sim- plicity reasons and a systematic treatment, data clustering requires a uniform process. In these conditions, the cumulative normal distribution will be used systematically in order to approximate the probability defined in (6.2.1). More precisely,

$$P(a, b) = Prob^{\mathcal{H}}\{Q(\alpha^\star, \beta)^\star \leq Q(\alpha, \beta)\} \approx \Phi(Q(\alpha, \beta))\qquad(6.2.2)$$

In the literature, mainly only the permutational model has been taken into account. The latter is clearly related to the $\mathcal{H}_1$ hypothesis of no relation. In the development below we focus on this model. However, important work was carried out [6] for comparing $\mathcal{H}_1, \mathcal{H}_2$ and $\mathcal{H}_3$ in the case of comparing nominal categorical attributes.

To derive a correlation coefficient from $Q(\alpha, \beta)$, let us refer to the case of compar- ing numerical attributes (see Sect. 5.2.2.2). For simplicity and a uniform treatment assume n large enough, in order to identify n and $n - 1$. There are three formal expres- sions which enable us to obtain the coefficient $R_1(v, w)$ from $Q_1(v, w)$. Regarding (5.2.68), the first one is defined by

$$R_1(v, w) = \frac{Q_1(v, w)}{\sqrt{n}} \qquad(6.2.3)$$

For the second expression, to begin, write the numerator of $Q_1(v, w)$ under the form $\{s(v, w) - \mathcal{E}(s(v^\star, w^\star))\}$, where $s(v^\star, w^\star)$ is the random raw coefficient asso- ciated with $s(v, w)$ (see (5.2.70)). Then the $R_1(v, w)$ coefficient can be obtained by the following formula

$$R_1(v, w) = \frac{s(v, w) - \mathcal{E}(s(v^\star, w^\star))}{\max\left(s(v^\star, w^\star) - \mathcal{E}(s(v^\star, w^\star))\right)} \qquad(6.2.4)$$

where the denominator is defined by the maximum of the random index associated with the numerator. This result is due to the Schwartz inequality for which the covariance between two numerical attributes is lower or equal than the product of the respective standard deviations of the two attributes.

The third formal expression leading to $R_1(v, w)$ from $Q_1(v, w)$ is

$$R_1(v, w) = \frac{Q_1(v, w)}{\sqrt{Q_1(v, v) \cdot Q_1(w, w)}} \qquad (6.2.5)$$

The formal expressions above (6.2.3)–(6.2.5) give exactly the same correlation coefficient in both cases: numerical and Boolean. However, different coefficients are obtained for attributes belonging to type II, specially, in the cases of nominal or ordinal categorical attributes. The case of comparing ranking attributes is specific. Versions corresponding to (6.2.4) and (6.2.5) give the same coefficient. Nevertheless, as we will see below, a version corresponding to (6.2.3) gives a different expression. Besides, it can be shown [28] that whatever is the common arity of the relations α and β to be compared, the coefficient $Q(\alpha, \beta)$ can be written as

$$Q(\alpha, \beta) = \sqrt{n} \cdot R(\alpha, \beta)$$

where $R(\alpha, \beta)$ is a *pure* coefficient not depending on n.

Let us express the normalization alternative proposed in (6.2.4) in the general context of the diagram above. The corresponding coefficient denoted by $\tau(\alpha, \beta)$ is

$$\tau(\alpha, \beta) = \frac{s(\alpha, \beta) - \mathcal{E}(s(\alpha^\star, \beta^\star))}{\max\left(s(\alpha^\star, \beta^\star)\right) - \mathcal{E}(s(\alpha^\star, \beta^\star))} \qquad (6.2.6)$$

The famous τ Kendall's coefficient [17], comparing two rankings on an object set, adopts this normalization principle. The latter inspired several researchers for comparing nominal categorical attributes [1, 12], inducing two partitions π and χ on the object set (see Sects. 3.3.1 and 4.3.2). In the *corrected for chance Rand coefficient* [12], the maximum proposed for $s(\pi^\star, \chi^\star)$ is $(card(R(\pi)) + card(R(\chi)))/2$. This bound is generally too large under the natural constraint of fixed types for both partitions, as it is required in the hypothesis of no relation. Determining the exact maximum of $s(\pi^\star, \chi^\star)$ is a very difficult problem for which an algorithmic solution using recursivity is provided in [31]. As illustration, consider a set of 70 elements, a couple of partitions π_0 and χ_0 whose respective types are $(24, 19, 18, 5, 4)$ and $(37, 12, 11, 7, 3)$. The bound calculated according to [12] gives 713.5. By substituting the geometrical mean $(\sqrt{card(R(\pi)) \cdot card(R(\chi))})$, a lower bound is necessarily obtained, its value for the illustration considered is 707.12. A more accurate but asymmetrical bound corresponds to $Min(card(R(\pi)), card(R(\chi)))$ and its value is 616. A maximal configuration of the contingency table concerned is computed by the algorithm defined in [31]. It gives a bound equal to 464. The algorithmic solution proposed has encouraged H. Messatfa [35] to work on again and to propose a treatment based on integer linear programming approach. The principle of the approach proposed in [44] is not new, it is very connected with the method proposed in [31], the later is not even mentioned.

In case of comparing two ordinal categorical attributes c and d, inducing two total preorders ω and ϖ on the object set $\mathcal{O}$, $card(R(\alpha) \cap R(\beta))$ of the diagram above becomes $card(R(\omega) \cap R(\varpi))$ (see (4.3.14)). The exact maximum of its random version $card(R(\omega^\star) \cap R(\varpi^\star))$, under the $\mathcal{H}_1$ model of no relation is given in [25], by means of a specific algorithm.

Let us now consider a set $\mathcal{A} = \{a^j | 1 \le j \le p\}$ of descriptive attributes having the same type, whatever is the latter. For mutually comparing the elements of $\mathcal{A}$, the *LLA* approach assumes the *Similarity Global Reduction* before obtaining the probability scale for evaluating in a relative fashion the associations concerned. Therefore, the Eqs. (5.2.51)–(5.2.53) can be taken up again in a very general framework. The justification of using cumulative normal distribution in (5.2.53) is established in [6, 24].

6.2.2   Comparing Two Ranking Attributes

Let ω_l and ϖ_l two linear orders on the object set $\mathcal{O}$ associated with two ranking attributes defined on $\mathcal{O}$ (see Sect. 3.3.3). In other words ω_l and ϖ_l are two total and strict orders associated with two ranking functions on $\mathcal{O}$. Applying the construction scheme of the diagram pictured in Fig. 6.1 starts with the set theoretic representation of $R(\omega_l)$ and $R(\varpi_l)$. These representations are defined at the level of the Cartesian product $\mathcal{O} \times \mathcal{O}$ (see (3.3.28) and (3.3.29)). The respective expressions of $R(\omega_l)$ and $R(\varpi_l)$ are the following:

$$R(\omega_l) = \{(x, y) | (x, y) \in \mathcal{O} \times \mathcal{O}, x <_{\omega_l} y\} \tag{6.2.7}$$

and

$$R(\varpi_l) = \{(x, y) | (x, y) \in \mathcal{O} \times \mathcal{O}, x <_{\varpi_l} y\} \tag{6.2.8}$$

where $<_{\omega_l}$ (resp., $<_{\varpi_l}$) is the order binary relation induced by ω_l (resp., ϖ_l).

Let us introduce the indicator functions ϕ and ψ of $R(\omega_l)$ and $R(\varpi_l)$, respectively: For any $(x, y) \in \mathcal{O} \times \mathcal{O}, x \ne y$,

$$\phi(x, y) = 1 \; (resp., 0) \text{ if and only if } x <_{\omega_l} y \; (resp., y <_{\omega_l} x)$$

and

$$\psi(x, y) = 1 \; (resp., 0) \text{ if and only if } x <_{\varpi_l} y \; (resp., y <_{\varpi_l} x)$$

In these conditions, the raw association coefficient

$$card(R(\omega_l) \cap R(\varpi_l)) = card\{(x, y) | (x, y) \in \mathcal{O} \times \mathcal{O}, x <_{\omega_l} y \text{ and } x <_{\varpi_l} y\}$$
$$\tag{6.2.9}$$

can be written as

$$card\big(R(\omega_l) \cap R(\varpi_l)\big) = \sum_{(x,y)\in\mathcal{O}\times\mathcal{O}} \phi(x,y) \cdot \psi(x,y) \qquad (6.2.10)$$

Now by considering an indexation of $\mathcal{O}$ such that each of its elements is coded by its rank for ω_l, the raw association coefficient can be put under the form

$$card\big(R(\omega_l) \cap R(\varpi_l)\big) = \sum_{1\le i<j\le n} \psi(i,j) \qquad (6.2.11)$$

Consequently, the associated random raw coefficient can be expressed as follows:

$$card\big(R(\omega_l) \cap R(\varpi_l^\star)\big) = \sum_{1\le i<j\le n} \psi(\sigma(i), \sigma(j)) \qquad (6.2.12)$$

where σ is a random permutation in the set of all permutations on $I = \{1, 2, \ldots, i,$ $\ldots, n\}$, endowed with a uniform probability measure (i.e. equal chance $1/n!$ for every permutation).

In order to follow the scheme of the diagram above, we have to calculate the mean and variance of the random raw coefficient. The mean can be written as

$$\sum_{I^{<2>}} Prob\{\psi(\sigma(i), \sigma(j)) = 1\} \qquad (6.2.13)$$

where $I^{<2>}$ denotes the set $\{(i,j)|1 \le i < j \le n\}$, whose cardinality being $n(n-1)/2$. The probability involved in the previous formula is equal to $1/2$, and then the mean is equal to $n(n-1)/4$.

A direct computing of the variance requires the development of the second absolute moment, namely

$$\mathcal{E}\left(\sum_{(i,j)\in I^{<2>}} \psi(\sigma(i), \sigma(j)) \right)^2$$

which is clearly equal to

$$\sum_{((i,j),(i',j'))\in I^{<2>}\times I^{<2>}} Prob\{\psi(\sigma(i), \sigma(j)) \cdot \psi(\sigma(i'), \sigma(j')) = 1\} \qquad (6.2.14)$$

The value of the probability involved in the previous equation depends on the structure of the ordered pair $((i,j), (i',j'))$ of elements of $I^{<2>} \times I^{<2>}$. Six configurations can be distinguished:

1. $\big((i,j), (i,j)\big)$
2. $\big((i,j), (i,h)\big)$
3. $\big((i,j), (h,i)\big)$
4. $\big((i,j), (h,j)\big)$
5. $\big((i,j), (j,h)\big)$
6. $\big((i,j), (h,k)\big)$

where different letters designate different indices. Let us denote by $\mathbb{I}, \mathbb{G}_1, \mathbb{G}'_1, \mathbb{G}_2, \mathbb{G}'_2$ and $\mathbb{H}$, the respective subsets of the Cartesian product $I^{<2>} \times I^{<2>}$, corresponding to the ordered sequence above of $\big((i,j), (i',j')\big)$ structure. The respective cardinalities of $\mathbb{I}, \mathbb{G}_1, \mathbb{G}'_1, \mathbb{G}_2, \mathbb{G}'_2$ and $\mathbb{H}$ are:

- $n(n-1)/2$;
- $n(n-1)(n-2)/3$;
- $n(n-1)(n-2)/6$;
- $n(n-1)(n-2)/3$;
- $n(n-1)(n-2)/6$;
- and $n(n-1)(n-2)(n-3)/4$.

We can verify that the sum of these is equal to $\big(n \cdot (n-1)/2\big)^2$. Now, the probability comprised in the sum of (6.2.14) is equal to $1/2, 1/3, 1/6, 1/3, 1/6$ and $1/4$ for $\big((i,j), (i',j')\big)$ having the structure 1, 2, 3, 4, 5 and 6 above, respectively. Therefore, the variance of the random raw coefficient (see (6.2.12)) can be written as

$$\frac{n(n-1)}{4} \cdot \left[1 + \frac{10}{9}(n-2) + \frac{1}{4}(n-2)(n-3) - \frac{n(n-1)}{4} \right]$$

and is equal to $(n-1)n(2n+5)/72$.

A linear expression of the random raw coefficient is considered in [10]. Then the use of the moment generating function and the Lindeberg–Feller convergence theorem enables the asymptotic normality of this random coefficient to be established. Finally, we have

Theorem 10 *The mean and variance of the random raw coefficient* $card\big(R(\omega_l) \cap R(\varpi_l^\star)\big)$ *are equal to* $n(n-1)/4$ *and* $(n-1)n(2n+5)/72$, *respectively. Moreover, the asymptotic distribution of the centred and standardized raw coefficient is the normal distribution.*

M.G. Kendall [17] defined his coefficient by

$$\tau(\omega_l, \varpi_l) = \frac{P - Q}{\frac{n(n-1)}{2}} \tag{6.2.15}$$

where with our formalism, $P = card\big(R(\omega_l) \cap R(\varpi_l)\big)$ and $Q = card\big(R(\omega_l) \cap S(\varpi_l)\big)$, where $S(\varpi_l)$ is the complementary subset of $R(\varpi_l)$ in $\mathcal{O}^{[2]}$. More explicitly, $S(\varpi_l) = \{(x, y)|(x, y) \in \mathcal{O}^{[2]}$ and $x >_{\varpi_l} y\}$. The denominator is the maximum of the numerator. It occurs when ω_l is identical to ϖ_l, that is, when $R(\omega_l)$ is identical to $R(\varpi_l)$. In this case, $R(\omega_l) \cap S(\varpi_l) = \emptyset$.

$$Q = \frac{n(n-1)}{2} - card\big(R(\omega_l) \cap R(\varpi_l)\big)$$

And then,

$$\frac{P - Q}{\frac{n(n-1)}{2}} = \frac{card\big(R(\omega_l) \cap R(\varpi_l)\big) - \frac{n(n-1)}{4}}{\frac{n(n-1)}{4}}$$

Consequently, we have the following:

Proposition 38 *The coefficient* $\tau(\omega_l, \varpi_l)$ *defined according to (6.2.6) is identical to the Kendall coefficient.*

Denoting as in the diagram by $Q(\omega_l, \varpi_l)$ the centred and standardized raw coefficient according to the preceding theorem, we have

Proposition 39 *The coefficient* $\dfrac{Q(\omega_l, \varpi_l)}{\sqrt{Q(\omega_l, \omega_l).Q(\varpi_l, \varpi_l)}}$ *is identical to the Kendall's coefficient.*

The third type of normalization that we can consider (see (6.2.3)) gives the coefficient

$$R(\omega_l, \varpi_l) = \sqrt{n} \cdot \frac{s(\omega_l, \varpi_l) - \frac{n(n-1)}{4}}{\sqrt{\frac{n^2(n-1)(2n+5)}{72}}}$$

whose maximum is equal to $3\sqrt{(n-1)/2(2n+5)}$. Therefore, it lies in the interval $[-1.5, 1.5]$. To obtain a coefficient comprised between -1 and $+1$, it suffices to multiply $R(\omega_l, \varpi_l)$ by the ratio $\dfrac{2}{3}$.

6.2.2.1 The Spearman Coefficient

Let us designate by q and r the two ranking functions associated with the linear orders ω_l and ϖ_l considered above. Definition of such ranking functions is provided in (3.3.28). $I = \{1, 2, \ldots, i, \ldots, n\}$ will denote an indexation set for the object set $\mathcal{O}$. We do not necessarily suppose here that i codes the object ranked at the ith position by ω_l, $1 \leq i \leq n$. q and r determine two numerical attributes, defining each, a bijective function of $\mathcal{O}$ on the set of the first n integer numbers. The commonly called ρ Spearman coefficient [43] is exactly the correlation coefficient $R_1(v, w)$ (see (5.2.69)), where q and r are substituted for v and w. Let us also denote the ρ Spearman coefficient by $R_S(q, r)$. We have

$$R_S(q, r) = \frac{\sum_i q(i)r(i) - \frac{1}{n} \left(\sum_i q(i) \right) \cdot \left(\sum_i r(i) \right)}{\sqrt{\left(\sum_i (q(i) - \bar{q})^2 \right) \cdot \left(\sum_i (r(i) - \bar{r})^2 \right)}} \qquad (6.2.16)$$

where the respective sums extend over the set I and where $\bar{q}$ (resp., $\bar{r}$) is the mean of q (resp., r).

However, there is a specificity of the Spearman coefficient with respect to a usual numerical correlation coefficient. This is related to the following equations satisfied by a ranking function p:

$$\sum_{1 \leq i \leq n} p(i) = \frac{n(n + 1)}{2}, \quad \sum_{1 \leq i \leq n} p(i)^2 = \frac{n(n + 1)(2n + 1)}{6}$$

Let us now introduce the attribute $d = r - q$, for which, $d(i) = q(i) - r(i)$, $1 \leq i \leq n$, the previous equations enable us to establish the well-known expression

$$R_S(q, r) = 1 - \frac{6 \sum_{1 \leq i \leq n} d(i)^2}{n^3 - n} \qquad (6.2.17)$$

Clearly, if ω_l and ϖ_l are identical, $q(i) = r(i)$, for all i, $1 \leq i \leq n$, and then $R_S(q, r) = 1$. Now, in the case where the linear order ϖ_l is inverse of ω_l, that is $r(i) = n - 2q(i) + 1$, $1 \leq i \leq n$, a calculation shows that $R_S(q, r) = -1 + (12/(n^2 - 1))$. Thus, the minimal value of the Spearman coefficient is slightly greater than -1.

In formula (6.2.17), a given absolute value of the difference $q(i) - r(i)$ is counted equally whatever the respective values of $q(i)$ and $r(i)$. In [41] it points out that there are applications where top ranks are much more important than lower ones. Weighting the distance between two associated ranks by means of a linear function is proposed. The new coefficient obtained has a Gaussian limit distribution.

Comparing the ρ Spearman coefficient and the τ Kendall coefficient is an intensive subject which has interested several researchers. Monjardet [37] provides a very interesting synthesis on the formal aspects of this comparison. In particular the Daniel's inequality

$$-1 \leq \frac{3(n + 2)}{n - 2} \tau - \frac{2(n + 1)}{n - 2} \rho \leq 1$$

is mentioned. Mainly, an expression for $\rho(\omega_l, \varpi_l) - \tau(\omega_l, \varpi_l)$ as a function of the parameters of the partial order defined by the intersection of ω_l and ϖ_l is given.

Besides, a product moment correlation coefficient between the Spearman's ρ and the Kendall's τ, under a permutational model is built in the pioneer paper of Daniels [5]. However, the proposed treatment is very specific. Generalization and less constrained expression will be given in Sect. 6.2.5.

Let us end this section with a claim of Kendall (see Chap. 1 of [17]): "from many practical and from most theoretical points of view τ is preferable to ρ". And in fact, with a set theoretic expression, the coefficients we will develop below, follow basic principles of the τ construction.

6.2.3 Comparing Two Nominal Categorical Attributes

6.2.3.1 Introduction

In Sect. 4.3.2 a natural way for extending association coefficients between Boolean attributes, to the case of comparing categorical nominal attributes was carried out. The notations of the latter section are taken up again here. c and d designate the two categorical attributes to be compared and π and χ are their corresponding partitions on the object set $\mathcal{O}$. The representation level of π and χ is the set P of unordered distinct object pairs (or 2-subsets) of $\mathcal{O}$. More precisely, as already expressed above, with a partition κ of $\mathcal{O}$ is associated a pair of P subsets $(R(\kappa), S(\kappa))$, where $R(\kappa)$ (resp., $S(\kappa)$) is the set of joined object pairs (resp., separated object pairs) by the partition κ. The parameters of the comparison between c and d, that is, between π and χ, are specified in (4.3.4). Here, in the framework of the *LLA* approach, the construction diagram of Fig. 6.1, will be followed. The general raw index $s(a, b) = card(R(\alpha) \cap R(\beta))$ becomes $s(a, b) = card(R(\pi) \cap R(\chi))$, its expression as a function of the contingency Table 4.8, is given in (4.3.5):

$$s(c, d) = \frac{1}{2} \sum_{h=1}^{h=K} \sum_{l=1}^{l=M} n_{hl} \cdot (n_{hl} - 1) \qquad (6.2.18)$$

To begin, let us specify what becomes the hypothesis of no relation $\mathcal{H}_1$. With the ordered pair (π, χ) is associated a random ordered pair $(\pi^\star, \chi\star)$, where $\pi^\star$ (resp., $\chi\star$) is a random partition taken according a multinomial model denoted by $\mathcal{N}_1$ and generalizing the hypergeometric model. It is very important for the following to specify this model.

The *type* of a partition was defined in Sect. 3.3.1. For a given partition, by assuming its respective classes labelled, this type is the sequence of its class cardinals. According to the contingency Table 4.8,

$$(n_{1.}, n_{2.}, \ldots, n_{h.}, \ldots, n_{K.})(\text{resp.}, (n_{.1}, n_{.2}, \ldots, n_{.l}, \ldots, n_{.M}))$$

will designate the type of the partition π (resp., χ). Let us introduce the set denoted by $\mathcal{P}(n; (n_{1.}, n_{2.}, \ldots, n_{h.}, \ldots, n_{K.}))$ (resp., $\mathcal{P}(n; (n_{.1}, n_{.2}, \ldots, n_{.l}, \ldots, n_{.M})))$ of all partitions having the type $(n_{1.}, n_{2.}, \ldots, n_{h.}, \ldots, n_{K.})$ (resp., $(n_{.1}, n_{.2}, \ldots, n_{.l}, \ldots, n_{.M}))$. For the random model $\mathcal{N}_1$, $\pi^\star$ (resp., $\chi\star$) is a random element in $\mathcal{P}(n; (n_{1.}, n_{2.}, \ldots, n_{h.}, \ldots, n_{K.}))$ (resp., $\mathcal{P}(n; (n_{.1}, n_{.2}, \ldots, n_{.l}, \ldots, n_{.M})))$, endowed with a uniform probability measure (equal chance for each of its elements). The respective types of π and χ will be denoted more briefly by $t(\pi)$ and $t(\chi)$.

With clear notations and in a generic way, we have

$$card(\mathcal{P}(n; (n_1, n_2, \ldots, n_j, \ldots, n_k))) = \frac{n!}{n_1! n_2! \ldots n_j! \ldots n_k!}$$

Notice that the centred and standardized coefficient obtained according to the construction diagram of Fig. 6.1 is invariant if we represent a partition by its separated object pairs. Thus, π and χ are represented by $S(\pi)$ and $S(\chi)$ (see Eq. (2.25)) and the raw coefficient can be written as $t(c, d) = card(S(\pi) \cap S(\chi))$. The last formula in (4.3.5) shows that $t(c, d) = s(c, d)$ up to $card(P) - card(R(\pi)) - card(R(\chi))$, which is constant under the independence random model $\mathcal{H}_1$.

In the following, the respective dual distributions of $card(R(\pi^\star) \cap R(\chi))$ and $card(R(\pi) \cap R(\chi^\star))$ will be examined. The respective moments of both distributions are shown to be equal, and then these distributions are identical. In a next step, the mean and variance of the common distribution are explicitly calculated. Therefore, the centred and standardized coefficient $Q(c, d)$ is obtained. It gives rise to association coefficients as those defined by (6.2.3) and (6.2.5).

As mentioned above (see what follows Fig. 6.1), in the case of comparing nominal categorical attributes, there are a few and rare situations for which the approximation given by the normal distribution as that given by (6.2.2) is not accurate enough. However, for mutually comparing a set of descriptive attributes of a given type, whatever the latter, the reference to the normal distribution is employed after the global reduction of the Q coefficients (see (5.2.51)–(5.2.53)). This is justified, theoretically and experimentally in most situations. In any case, this reference can set up a priori.

6.2.3.2 The Duality Properties

Let E be an arbitrary subset of $\mathcal{O}$ for which $card(E) = m$ $(m \le n)$. Let us introduce the set of all partitions we can obtain by associating with a m subset E of $\mathcal{O}$ the set of all of its partitions $(E_1, E_2, \ldots, E_e, \ldots, E_g)$ into classes labelled, the eth, having the cardinal m_e: $m_1 + m_2 + \cdots + m_e + \cdots + m_g = m$. We denote this set by $\mathcal{P}((n, m); (m_1, m_2, \ldots, m_e, \ldots, m_g))$. A given element of this set is defined by a m subset E of $\mathcal{O}$ and by a surjective mapping, that we denote by φ of E onto $\{1, 2, \ldots, e, \ldots, g\}$ for which $\varphi^{-1}(e) = E_e$ and such that $card(E_e) = m_e$, $1 \le e \le g$. The cardinality of $\mathcal{P}((n, m); (m_1, m_2, \ldots, m_e, \ldots, m_g))$ is

$$\binom{n}{m} \times \frac{m!}{m_1! m_2! \ldots m_e! \ldots m_g!} = \frac{n(n-1)(n-2) \ldots (n-m+1)}{m_1! m_2! \ldots m_e! \ldots m_g!}$$

Indeed, $\binom{n}{m}$ is the number of m-subsets of $\mathcal{O}$ and $m!/m_1! m_2! \ldots m_e! \ldots m_g!$ is the number of partitions of a given set with m elements, into classes labelled, having the cardinals $m_1, m_2, \ldots, m_e, \ldots, m_{g-1}$ and m_g, respectively.

Notice that

$$\mathcal{P}((n, n); (n_1, n_2, \ldots, n_j, \ldots, n_k))$$

is identical to

$$\mathcal{P}(n; (n_1, n_2, \ldots, n_j, \ldots, n_k))$$

defined in the previous section.

Let us now introduce the following theorem: Let E_0 be a fixed subset of $\mathcal{O}$ such that $m = card(E_0) \leq n = card(\mathcal{O})$. Consider the two integer sequences $u = (m_1, m_2, \ldots, m_e, \ldots, m_g)$ and $t = (n_1, n_2, \ldots, n_j, \ldots, n_k)$, for which, $g < k$ and $m_1 + m_2 + \cdots + m_e + \cdots + m_g = m$ (resp., $n_1 + n_2 + \cdots + n_j + \cdots + n_k = n$). Set $G = \{1, 2, \ldots, e, \ldots, g\}$ and $J = \{1, 2, \ldots, j, \ldots, k\}$ and assume an injective mapping τ of G onto J, such that, $m_e \leq n_{\tau(e)}$, for all e, $1 \leq e \leq g$. Finally, $\pi(u) = (E_1, E_2, \ldots, E_e, \ldots, E_g)$ and $\pi(t) = (O_1, O_2, \ldots, O_j, \ldots, O_k)$ (resp., $\pi_0(u) = (E_1^0, E_2^0, \ldots, E_e^0, \ldots, E_g^0)$ and $\pi_0(t) = (O_1^0, O_2^0, \ldots, O_j^0, \ldots, O_k^0)$) will designate two standard partitions (resp., two fixed partitions) belonging to $\mathcal{P}((n, m); u)$ and $\mathcal{P}(n; t)$, respectively. In these conditions we have the following:

Theorem 11 *The proportion of partitions $\pi(u)$ in $\mathcal{P}((n, m); u)$ for which $E_e \subset O_{\tau(e)}^0$, $1 \leq e \leq g$ is equal to the proportion of partitions $\pi(t)$ in $\mathcal{P}(n; t)$ for which $E_e^0 \subset O_{\tau(e)}$*

The value of the first proportion is

$$\frac{\Pi_{e \in G} \binom{n_{\tau(e)}}{m_e}}{\frac{n(n-1)\ldots(n-m+1)}{m_1! m_2! \ldots m_g!}} = \frac{\Pi_{e \in G} n_{\tau(e)}(n_{\tau(e)} - 1) \ldots (n_{\tau(e)} - m_e + 1)}{n(n-1) \ldots (n - m + 1)}$$

where Π denotes the product operation.

Indeed, $\binom{n_{\tau(e)}}{m_e}$ is the number of m_e subsets of $O_{\tau(e)}^0$. Hence $\Pi_{e \in G} \binom{n_{\tau(e)}}{m_e}$ is the number of g-uples $(E_1, E_2, \ldots, E_e, \ldots, E_g)$ of $\mathcal{O}$ subsets such that $card(E_e) = m_e$ and $E_e \subset O_{\tau(e)}^0$, for all $e \in G$. On the other hand, the number of elements of $\pi(t)$ in $\mathcal{P}(n; t)$ for which $E_e^0 \in O_{\tau(e)}$ for all $e \in G$ is equal to

$$\frac{(n - m)!}{\Pi_{e \in G}(n_{\tau(e)} - m_e)! \times \Pi_{j \in (\tau(G))^c} n_j!}$$

where $(\tau(G))^c$ is the complementary subset in J of the image of G by τ. Therefore, the proportion defined in the second part of the theorem is equal to

$$\frac{(n - m)!}{\Pi_{e \in G}(n_{\tau(e)} - m_e)!} \Big/ \frac{n!}{\Pi_{e \in G}(n_{\tau(e)})!}$$

which is identical to the proportion above, following the theorem statement.   □

Example 3 In order to illustrate the property above, let us consider an object set $\mathcal{O}$ of 10 objects and let $m = 5$, $u = (3, 2)$ and $t = (5, 3, 2)$. Besides, define an injective mapping τ of $G = \{1, 2\}$ onto $J = \{1, 2, 3\}$ defined by $\tau(1) = 1$ and $\tau(2) = 3$ and for which we have $m_e \leq n_{\tau(e)}$. The value of the proportion concerned is $1/504$

Let us introduce some basic notions required for the proof of the following theorem. Let $P^{[q]}$ the set of all q-uples of unordered object pairs distinct mutually. A standard element of $P^{[q]}$ can be written as $(p_1, p_2, \ldots, p_l, \ldots, p_q)$, where p_l is an unordered distinct object pair belonging to P, $1 \leq l \leq q$ (see (3.3.4)). Consider now a bijection of $\mathcal{O}$ into itself (an intra bijection) that we denote by σ. We extend σ

to $P^{[q]}$ by setting $\sigma(p_1, p_2, \ldots, p_l, \ldots, p_q) = (\sigma(p_1), \sigma(p_2), \ldots, \sigma(p_l), \ldots, \sigma(p_q))$, where $\sigma(\{x, y\}) = \{\sigma(x), \sigma(y)\}$. By this way, σ transforms $(p_1, p_2, \ldots, p_l, \ldots, p_q)$ into an element of $P^{[q]}$ having the same configuration denoted c, as that of $(p_1, p_2, \ldots, p_l, \ldots, p_q)$. That is, the object $\sigma(x)$ is substituted for x in all pairs, components of the q-uple where x was present. The set G_q^c of all pair q-uples having the same configuration c is the value set of $(\sigma(p_1), \sigma(p_2), \ldots, \sigma(p_l), \ldots, \sigma(p_q))$ when σ describes the $n!$ intra bijections of $\mathcal{O}$. Except for $q = 2$, we need not specify the different configurations c and the respective cardinalities of G_q^c. A specific case concerns the configuration for which any two pairs of a given q-uple have no common component.

Let $\pi_0(t)$ and $(p_1^0, p_2^0, \ldots, p_l^0, \ldots, p_q^0)$ be a fixed partition and a fixed q-uple in $\mathcal{P}(n; t)$ and G_q^c, respectively. Let us designate by ψ_0 (resp., ψ) the indicator function of the P subset $R(\pi_0(t))$ (resp., $R(\pi(t))$) corresponding to the equivalence relation associated with the partition $\pi_0(t)$ (resp., $\pi(t)$) (see (3.3.5)). $\pi(t)$ is a standard element of $\mathcal{P}(n; t)$.

Corollary 8 *The proportion of q-uples* $(p_1, p_2, \ldots, p_q)$ *in* G_q^c *for which* $\psi_0(p_1)\psi_0(p_2) \ldots \psi_0(p_q) = 1$ *is equal to the proportion of partitions of* $\mathcal{P}(n; t)$ *for which* $\psi(p_1^0)\psi_0(p_2^0) \ldots \psi(p_q^0) = 1$.

Let us associate with the set of pairs $\{p_1, p_2, \ldots, p_q\}$ involved in a given q-uple $(p_1, p_2, \ldots, p_q)$, its transitive closure. The latter determines a partition on a subset E of $\mathcal{O}$. We denote this partition by $\phi(p_1, p_2, \ldots, p_q) = (F_1, F_2, \ldots, F_f, \ldots, F_h)$. Getting $\psi_0(p_1)\psi_0(p_2) \ldots \psi_0(p_q) = 1$ is equivalent to have every class $F_f, 1 \leq f \leq h$, included in one class of $\pi_0(t)$ partition. Several F_f classes might be included in a unique class O_j^0. In these conditions, we have to consider all partitions $(E_1, E_2, \ldots, E_e, \ldots, E_g)$ on E resulting from aggregation of classes of $(F_1, F_2, \ldots, F_f, \ldots, F_h)$ for which there exists an injective mapping τ of $\{1, 2, \ldots, e, \ldots, g\}$ into $\{1, 2, \ldots, j, \ldots, k\}$ such that $m_e \leq n_{\tau(e)}$, where, as above, $m_e = card(E_e)$ and $n_j = card(O_j)$, $1 \leq e \leq g, 1 \leq j \leq k$.

Consider now the set of partitions $\mathcal{P}((n, m); u)$. Each of its elements can be obtained by the previous technique from elements of G_q^c and for symmetry reasons, the number of q-uples which determine a given partition of $\mathcal{P}((n, m); u)$ is the same. Consequently, for a given construction method, the first proportion stated in the corollary is equal to the first proportion stated in the previous theorem.

In analogous way, we consider the same construction from $(p_1^0, p_2^0, \ldots, p_q^0)$. It leads to a partition $(E_1^0, E_2^0, \ldots, E_e^0, \ldots, E_g^0)$ of the set E^0 of the objects appearing in the q-uple. For a given injective mapping τ, the second proportion stated in the corollary is equal to the second proportion stated in the theorem above.   □

Exercise 1 Give for $q = 3$ the different configurations c and for each of them calculate the proportion considered in the Corollary above.

6.2.3.3 Expression of the rth Moment Order of the Distribution of $s(\pi, \chi^\star) = card(R(\pi) \cap R(\chi^\star))$

As denoted above π and χ are two partitions of $\mathcal{O}$. In our context, π and χ are induced by two nominal categorical attributes. In order to simplify our notations with respect to those of Sect. 4.3.1, the respective types of π and χ are denoted here by $t = t(\pi) = (n_1, n_2, \ldots, n_i, \ldots, n_k)$ and $u = t(\chi) = (m_1, m_2, \ldots, m_i, \ldots, m_l)$. To fix ideas, we assume the classes of π (resp., χ) labelled. π (resp., χ) is an element of the set $\mathcal{P}(n; t)$ (resp., $\mathcal{P}(m; u)$) of partitions of $\mathcal{O}$, having the type t (resp., u). $\pi^\star$ (resp., $\chi^\star$) is a random element of $\mathcal{P}(n; t)$ (resp., $\mathcal{P}(m; u)$) endowed with a uniform probability measure.

Let us denote here by ϕ and ψ the indicator functions of $R(\pi)$ and $R(\chi)$ in the set of distinct unordered object pairs (or $\mathcal{O}$ 2-subsets) P. On the other hand, we denote by $\phi^\star$ and $\psi^\star$ the random indicator functions corresponding to the random subsets $R(\pi^\star)$ and $R(\chi^\star)$, respectively. We have

$$card(R(\pi) \cap R(\chi^\star)) = \sum_{p \in P} \phi(p)\psi^\star(p)$$

$$card(R(\pi^\star) \cap R(\chi)) = \sum_{p \in P} \phi^\star(p)\psi(p) \qquad (6.2.19)$$

The rth absolute moment of the first of the two previous random variables can be written as

$$\mathcal{E}\left(\sum coef(r; r_1, r_2, \ldots, r_q)(\phi(p_{i_1})^{r_1} \cdot \phi(p_{i_2})^{r_2} \ldots \phi(p_{i_q})^{r_q}\right)$$
$$\cdot (\psi^\star(p_{i_1})^{r_1} \cdot \psi^\star(p_{i_2})^{r_2} \ldots \psi^\star(p_{i_q})^{r_q})) \qquad (6.2.20)$$

where, as usual the $\mathcal{E}$ designates the expectation and where the sum is first extended over all partitions of r into q parts. Second, for a fixed partition $(r_1, r_2, \ldots, r_q)$, the sum is extended over all permutations on $(p_{i_1}, p_{i_2}, \ldots, p_{i_q})$ which can be obtained from a q-subset of P. Let us recall that

$$coef(r; r_1, r_2, \ldots, r_q) = \frac{r!}{l_1! l_2! \ldots l_h! \times r_1! r_2! \ldots r_q!}$$

where $r = r_1 + r_2 + \cdots + r_q$ and h the number of distinct r_j values, repeated l_1, $l_2, \ldots$ and l_h times, respectively.

Obviously we have

$$\phi(p_{i_1})^{r_1} \cdot \phi(p_{i_2})^{r_2} \ldots \phi(p_{i_q})^{r_q} = \phi(p_{i_1}) \cdot \phi(p_{i_2}) \ldots \phi(p_{i_q})$$

$$\psi^\star(p_{i_1})^{r_1} \cdot \psi^\star(p_{i_2})^{r_2} \ldots \psi^\star(p_{i_q})^{r_q} = \psi^\star(p_{i_1}) \cdot \psi^\star(p_{i_2}) \ldots \psi^\star(p_{i_q})$$

In these conditions, the expression concerned by (6.2.20) becomes

$$\sum coef(r; r_1, r_2, \ldots, r_q)(\phi(p_{i_1}) \cdot \phi(p_{i_2}) \ldots \phi(p_{i_q}))$$
$$\cdot \mathcal{E}(\psi^\star(p_{i_1}) \cdot \psi^\star(p_{i_2}) \ldots \psi^\star(p_{i_q})) \tag{6.2.21}$$

Similarly, the absolute rth moment of the random coefficient $card(R(\pi^\star) \cap R(\chi))$ can be reduced to

$$\sum coef(r; r_1, r_2, \ldots, r_q)\mathcal{E}(\phi^\star(p_{i_1}) \cdot \phi^\star(p_{i_2}) \ldots \phi^\star(p_{i_q}))$$
$$\cdot (\psi(p_{i_1}) \cdot \psi(p_{i_2}) \ldots \psi(p_{i_q})) \tag{6.2.22}$$

For a partition fixed $(r_1, r_2, \ldots, r_q)$ of r, consider the respective parts concerned in (6.2.21) and (6.2.22), namely

$$\sum (\phi(p_{i_1})\phi(p_{i_2}) \ldots \phi(p_{i_q})) \cdot (\psi^\star(p_{i_1}) \cdot \psi^\star(p_{i_2}) \ldots \psi^\star(p_{i_q})) \tag{6.2.23}$$

and

$$\sum (\phi^\star(p_{i_1})\phi^\star(p_{i_2}) \ldots \phi^\star(p_{i_q})) \cdot (\psi(p_{i_1}) \cdot \psi(p_{i_2}) \ldots \psi(p_{i_q})) \tag{6.2.24}$$

Now, decompose simultaneously the sums (6.2.23) and (6.2.24) according to the different sets G_q^c defined above. Due to the previous Corollary, we have

$$\mathcal{E}\left(\sum_{G_q^c}(\phi(p_{i_1}) \cdot \phi(p_{i_2}) \ldots \phi(p_{i_q})) \cdot (\psi^\star(p_{i_1}) \cdot \psi^\star(p_{i_2}) \ldots \psi^\star(p_{i_q}))\right)$$

$$= \mathcal{E}\left(\sum_{G_q^c}(\phi^\star(p_{i_1}) \cdot \phi^\star(p_{i_2}) \ldots \phi^\star(p_{i_q})) \cdot (\psi(p_{i_1}) \cdot \psi(p_{i_2}) \ldots \psi(p_{i_q}))\right)$$
$$\tag{6.2.25}$$

Consequently, the absolute rth moment of the distribution of $card(R(\pi) \cap R(\psi^\star))$ is equal to that of $card(R(\pi^\star) \cap R(\psi))$. Hence

Theorem 12 $\mathcal{P}(n; t)$ and $\mathcal{P}(n; u)$ being endowed with a uniform probability measure, the distributions of $s(\pi, \chi^\star) = card(R(\pi) \cap R(\psi^\star))$ and $s(\pi^\star, \psi) = card(R(\pi^\star) \cap R(\psi))$ are identical.

6.2.3.4   Mean and Variance of the Common Distribution of $s(\pi, \chi^\star)$ and $s(\pi^\star, \psi)$

To begin, we shall establish two lemmas which specify the previous Corollary for $q = 1$ and $q = 2$. In this, we will make explicit the different forms of G_q^c and we will calculate the common value of the proportion concerned by the statement of this

Corollary. Then the mean and variance of the distribution of $s(\pi^\star, \psi) = card(R(\pi^\star) \cap R(\psi))$ will be obtained.

Lemma 2 Let $p_0 = \{x_0, y_0\}$ a fixed pair in P and π_0 a fixed partition in $\mathcal{P}(n; t)$. The proportion of pairs in P whose two components are joined in a same class of π_0 is equal to the proportion of partitions in $\mathcal{P}(n; t)$ for which the two components of p_0 are joined in a same class.

The value of the first proportion indicated above, appears clearly. Indeed, the number of pairs joined by π_0 is $\sum_{1 \le i \le k} n_i(n_i - 1)/2$. Besides, $card(P) = n(n - 1)/2$. Now, the second proportion can be put in the form

$$\sum_{1 \le i \le k} \frac{\frac{(n-2)!}{n_1! \dots n_{i-1}!(n_i-2)!n_{i+1}! \dots n_k!}}{\frac{n!}{n_1!n_2! \dots n_k!}} = \sum_{1 \le i \le k} \frac{n_i(n_i - 1)}{n(n - 1)}$$

where a given term of the first sum represents the proportion of partitions in $\mathcal{P}(n; t)$ for which x_0 and y_0 are joined in the ith class of the partition.

Let us denote by G and H the set of ordered pairs of distinct object pairs (p, p') having the forms

$$(\{x, y\}, \{x, z\}) \text{ and } (\{w, x\}, \{y, z\})$$

where different letters indicate different $\mathcal{O}$ elements. For G (resp., H) the two pairs p and p' have one common component (resp., no common component). We have

$$card(G) = n(n - 1)(n - 2) \text{ and } card(H) = \frac{n(n - 1)(n - 2)(n - 3)}{4}$$

Lemma 3 Let π_0 be a fixed partition in $\mathcal{P}(n; t)$

1. If (p_0, p'_0) is an ordered pair of object pair fixed in G, the proportion of elements (p, p') in G such that both components of p and p' are joined by π_0 is equal to the proportion of partitions of $\mathcal{P}(n; t)$ for which the two components of $(p_0$ and $p'_0)$ are joined. The value of this proportion is

$$\sum_{1 \le i \le k} \frac{n_i(n_i - 1)(n_i - 2)}{n(n - 1)(n - 2)} \tag{6.2.26}$$

2. If (p_0, p'_0) is an ordered pair of object pair fixed in H, the proportion of elements (p, p') in H such that both components of p and p' are joined by π_0 is equal to the proportion of partitions of $\mathcal{P}(n; t)$ for which the two components of $(p_0$ and $p'_0)$ are joined. The value of this proportion is

$$\frac{\sum_{1 \le i \le k} n_i(n_i - 1)(n_i - 2)(n_i - 3) + \sum_{1 \le i \ne i' \le k} n_i(n_i - 1)n_{i'}(n_{i'} - 1)}{n(n - 1)(n - 2)(n - 3)} \tag{6.2.27}$$

Proof of the Part 1

The number of ordered pairs of the form $(\{x, y\}, \{x, z\})$ for which x, y and z are in the ith class denoted by $\mathcal{O}_i$, of π_0 is $n_i(n_i - 1)(n_i - 2)$ $(n_i = card(\mathcal{O}_i))$. Indeed, the number of 3-subsets of $\mathcal{O}_i$ is the binomial coefficient $\binom{n_i}{3} = n_i(n_i - 1)(n_i - 2)/6$ and a given 3-subset $\{a, b, c\}$ of a given set gives rise to six different ordered pairs of G, namely: $(\{a, b\}, \{a, c\})$, $(\{a, b\}, \{b, c\})$, $(\{b, c\}, \{a, b\})$, $(\{b, c\}, \{a, c\})$, $(\{a, c\}, \{a, b\})$ and $(\{a, c\}, \{b, c\})$. Then we get the proportion stated in the first part of 1.

Now, set $(p_0, p_0') = (\{x_0, y_0\}, \{x_0, z_0\})$ and assume $n_i \geq 3$. The proportion of partitions of $\mathcal{P}(n; t)$, for which x_0, y_0 and z_0 are joined in the ith class is given by

$$\frac{n_i(n_i - 1)(n_i - 2)}{n(n - 1)(n - 2)}$$

This proportion is obtained in an enumerative way.

Proof of the Part 2

The number of ordered pairs of the form $(\{w, x\}, \{y, z\})$ for which the 2-subset $\{w, x\}$ (resp., $\{y, z\}$) is included in the ith class (resp., in the i'th class) is equal to

$$\binom{n_i}{2} \times \binom{n_{i'}}{2} \text{ for } i \neq i' \text{ and } \binom{n_i}{2} \times \binom{n_i - 2}{2} \text{ for } i' = i$$

where we assume $n_i = card(\mathcal{O}_i) \geq 4$.

By considering the crossing of each 2-subset $\{w, x\}$ of a given class with the set of disjoint 2-subsets $\{y, z\}$, for which y and z are joined in a same class, we obtain the formula (6.2.27) for the first part of the statement 2.

Now, set $(p_0, p_0') = (\{w_0, x_0\}, \{y_0, z_0\})$. The proportion of partitions in $\mathcal{P}(n; t)$ for which the 2-subsets $\{w_0, x_0\}$ and $\{y_0, z_0\}$ are included in the ith class and the i'th class, respectively, is given by

$$\frac{n_i(n_i - 1)n_{i'}(n_{i'} - 1)}{n(n - 1)(n - 2)(n - 3)} \text{ if } i' \neq i \text{ and } \frac{n_i(n_i - 1)(n_i - 2)(n_i - 3)}{n(n - 1)(n - 2)(n - 3)} \text{ if } i' = i$$

Hence, we obtain the value (6.2.27) of the proportion defined in the second part of the statement 2, where we assume $n_i \geq 4$ for all i, $1 \leq i \leq k$.

Let us introduce for the statement of the next theorem the following parameters defined with respect to the type $t = (n_1, n_2, \ldots, n_i, \ldots, n_k)$

$$\lambda = \frac{\sum_{1 \leq i \leq k} n_i(n_i - 1)}{\sqrt{2n(n - 1)}}$$

$$\rho = \frac{\sum_{1 \leq i \leq k} n_i(n_i - 1)(n_i - 2)}{\sqrt{n(n - 1)(n - 2)}}$$

$$\theta = \frac{\left(\sum_{1 \leq i \leq k} n_i(n_i - 1)\right)^2 - 2\sum_{1 \leq i \leq k} n_i(n_i - 1)(2n_i - 3)}{2\sqrt{n(n - 1)(n - 2)(n - 3)}} \qquad (6.2.28)$$

Introduce also the parameters μ, σ and ζ which have the same forms as that of λ, ρ and θ, respectively, but with respect to the type $u = ((m_1, m_2, \ldots, m_j, \ldots, m_l))$

Theorem 13 *The mean and variance of the two identical dual distributions of $s(\pi, \chi^\star)$ and $s(\pi^\star, \chi)$, are $\lambda\mu$ and $\lambda\mu + \rho\sigma + \theta\zeta - \lambda^2\mu^2$, respectively.*

Let us take up again the first of both expressions (6.2.19). It would be equivalent for the following to consider the second expression. We have

$$\mathcal{E}\big(card(R(\pi) \cap R(\chi)^\star)\big) = \sum_{p \in P} \phi(p)\mathcal{E}\big(\psi^\star(p)\big) \qquad (6.2.29)$$

Lemma 1 enables the first stated result in the theorem above to be established. The development of the first term of the following expression of the variance

$$\mathcal{E}\left(\sum_{p \in P} \phi(p)\psi^\star(p)\right)^2 - \lambda^2\mu^2 \qquad (6.2.30)$$

$$\sum_{p \in P} \phi(p)\mathcal{E}\big(\psi^\star(p)\big) + \sum_{(p,p') \in P^{[2]}} \phi(p)\phi(p')\psi^\star(p)\psi^\star(p')$$

where $P^{[2]} = \{(p, p') | p \in P, p' \in P \text{ and } p \neq p'\}$. Equation (6.2.30) becomes

$$\lambda\mu - \lambda^2\mu^2 + \sum_{(p,p') \in P^{[2]}} \phi(p)\phi(p')\mathcal{E}\big(\psi^\star(p)\psi^\star(p')\big) \qquad (6.2.31)$$

By partitioning $P^{[2]}$ into the two disjoint subsets G and H (see Lemma 2) and by cutting the sum above into two parts, the first one over G and the second one over H, we get the stated result above from Lemma 2.

6.2.3.5 Proposition and Discussion Around Several Association Coefficients

Following above

$$Q_1(\pi, \chi) = \frac{s(\pi, \chi) - \lambda\mu}{\sqrt{\lambda\mu + \rho\sigma + \theta\zeta - \lambda^2\mu^2}} \qquad (6.2.32)$$

is the centred and standardized coefficient according to the diagram of Fig. 6.1.

Introduce the notations $r(\pi) = card(R(\pi))$, $r(\chi) = card(R(\chi))$ and notice that the previous numerator, defined by the centred raw coefficient, can be written as $s(\pi, \chi) - r(\pi)r(\chi)/card(P)$.

By analogy with the forms (6.2.3)–(6.2.5) of a pure association coefficient, not depending on the number of observations n, we may envisage the following coefficients:

$$R_1(\pi, \chi) = \frac{Q_1(\pi, \chi)}{\sqrt{n}} \tag{6.2.33}$$

$$R_1'(\pi, \chi) = \frac{s(\pi, \chi) - \mathcal{E}(s(\pi^\star, \chi^\star))}{max(s(\pi^\star, \chi^\star) - \mathcal{E}(s(\pi^\star, \chi^\star)))} \tag{6.2.34}$$

$$R_1''(\pi, \chi) = \frac{Q_1(\pi, \chi)}{\sqrt{Q_1(\pi, \pi) \cdot Q_1(\chi, \chi)}} \tag{6.2.35}$$

In the following theorem we give the limit form of $R_1(\pi, \chi)$ [26, 28] under natural asymptotic conditions. To begin set

$$f_i = \frac{n_i}{n}, g_j = \frac{m_j}{n} \text{ and } c_{ij} = \frac{n_{ij}}{n}$$

where n_{ij}—as in Table 4.8—is the cardinal of the intersection between the ith class of π and the jth class of χ, $1 \le i \le k$, $1 \le j \le l$.

Now, imagine, as in the strong law of large numbers, the convergence of the empirical frequencies f_i, g_j and c_{ij} to theoretical ones, denoted by π_i, χ_j and γ_{ij}, respectively, $1 \le i \le k$, $1 \le j \le l$. Then, define

$$p = \sum_{1 \le i \le k} \pi_i^2, t = \sum_{1 \le i \le k} \pi_i^3$$

$$q = \sum_{1 \le j \le l} \chi_j^2, t = \sum_{1 \le j \le l} \chi_j^3$$

$$\text{and } w = \sum_{1 \le i \le k, 1 \le j \le l} \gamma_{ij}^2 \tag{6.2.36}$$

In these conditions, we have the following theorem established in [28]

Theorem 14 *The limit form of* $R_1(\pi, \chi)$ *is*

$$\frac{\frac{1}{2}(w - pq)}{\sqrt{(t - p^2)(u - q^2) + \frac{1}{2n}((p - p^2) - 2(t - p^2))((q - q^2) - 2(u - q^2))}} \tag{6.2.37}$$

This result is obtained using a coding of π (resp., χ) with a $0 - 1$ valuation ϕ' (resp., ψ') deduced from the indicator function ϕ (resp., ψ) and defined at the level of the Cartesian product $\mathcal{O} \times \mathcal{O}$ (see Sect. 3.3.5). More precisely, for two distinct objects, x and y in $\mathcal{O}$, $\phi'(x, y) = 1$ (resp., $\phi'(x, y) = 0$) if and only if, x and y are joined by the partition π, Moreover, for any object x in $\mathcal{O}$, $\phi'(x, x) = 1$. Obviously, the definition of ψ' is analogous to that of ϕ', but with respect to the partition χ.

$(t - p^2)$ (resp., $(u - q^2)$) is in general strictly positive. Exceptionally it is null, the null value occurs when at least one of the two following conditions is satisfied:

- The π_i are equal for all i: $\pi_i = \frac{1}{k}$, $1 \le i \le k$.
- The χ_j are equal for all j: $\chi_j = \frac{1}{l}$, $1 \le j \le l$.

In these cases the magnitude order of the denominator of (6.2.37) changes. As a consequence, the Schwartz inequality type

$$Q_1(\pi, \chi) \le \sqrt{Q_1(\pi, \pi) \cdot Q_1(\chi, \chi)}$$

is not acquired for very specific cases when the entropy of either π or χ is too large. In any case, this has no effect for mutually comparing a set of nominal categorical attributes and besides, in real cases, each of the partitions compared has its cardinal classes mutually different enough.

At the end of the introductory Sect. 6.2.1, we have already presented and briefly discussed the coefficient $R_1'(\pi, \chi)$. Let us now consider the coefficient $R_1''(\pi, \chi)$ which is directly deduced from $Q_1(\pi, \chi)$. Assume, to fix idea, the most general case where the class cardinals of π (resp., χ) are not identical. Equation (6.2.37) shows that a new coefficient is obtained. For n large enough in order to render negligible the second term under the square root sign $\sqrt{\ }$ of (6.2.37), we get for $R_1''(\pi, \chi)$ a formal expression equivalent to the K. Pearson's coefficient between ϕ' and ψ', at the level of $\mathcal{O} \times \mathcal{O}$.

The coefficient we are going to define is closely related to the last mentioned one. Let us come back to the set theoretic representations of π and χ by the P-subsets $R(\pi)$ and $R(\chi)$ and consider the raw coefficient $s(\pi, \chi) = card(R(\pi) \cap R(\chi))$ (see Sects. 4.3.2 and 6.2.3.1). Now, and in order to define a less constrained hypothesis of no relation, let us "forget" the structural specificity of $R(\pi)$ and $S(\pi)$ as subsets of the set P of unordered distinct object pairs. A "free" version of the hypothesis of no relation consists of taken up again that defined in the Boolean case (see Sect. 5.2.1) and to substitute for $[\mathcal{O}, \mathcal{O}(a), \mathcal{O}(b)]$, $[P, R(\pi), R(\chi)]$. In these conditions and for the $\mathcal{H}_1$ model, with $R(\pi)$ (resp., $R(\chi)$) a random $r(\pi)$-subset of P, denoted by $R^\star(\pi)$, (resp., $r(\chi)$-subset of P, denoted by $R^\star(\chi)$,) is associated. For this model, the set of *all* $r(\pi)$-subsets (resp., $r(\chi)$-subsets) of P is endowed with a uniform probability measure. Notice that in the model presented in Sect. 6.2.3.1, only the $r(\pi)$-subsets (resp., $r(\chi)$-subsets) of P corresponding to a partition of type $t(\pi)$ (resp., $t(\chi)$) are reached. In these conditions, the coefficient obtained in transposing (5.2.10) is

$$\tilde{Q}_1(\pi, \chi) = \frac{card(R(\pi) \cap R(\chi)) - \frac{r(\pi)r(\chi)}{m}}{\sqrt{\frac{r(\pi)s(\pi)r(\chi)s(\chi)}{m^2(m-1)}}} \tag{6.2.38}$$

where $r(\pi) = card(R(\pi))$ and $s(\pi) = card(S(\pi))$ (resp., $r(\chi) = card(R(\chi))$ and $s(\chi) = card(S(\chi))$) and where $m = card(P) = r(\pi) + s(\pi)$ (see Sect. 4.3.2 for the notations).

By identifying $m - 1$ with m, the transposition of the K. Pearson coefficient gives

$$\widetilde{R}_1(\pi, \chi) = \frac{card(R(\pi) \cap R(\chi)) - \frac{r(\pi)r(\chi)}{m}}{\sqrt{\frac{r(\pi)s(\pi)r(\chi)s(\chi)}{m^2}}} \tag{6.2.39}$$

The numerator of this coefficient is exactly that of $Q_1(\pi, \chi)$ (see (6.2.32) and below). However and clearly, the denominator above is different. The variance of $card(R^\star(\pi) \cap R^\star(\chi))$ is much lower than that of $card(R(\pi^\star) \cap R(\chi^\star))$.

By referring to the Poisson model, the following coefficient can also be proposed

$$\widetilde{Q}_3(\pi, \chi) = \frac{card(R(\pi) \cap R(\chi)) - \frac{r(\pi)r(\chi)}{m}}{\sqrt{\frac{r(\pi)r(\chi)}{m}}} \tag{6.2.40}$$

The associated $\widetilde{R}_3(\pi, \chi)$ coefficient can be expressed by

$$\widetilde{R}_3(\pi, \chi) = \frac{card(R(\pi) \cap R(\chi)) - \frac{r(\pi)r(\chi)}{m}}{\sqrt{r(\pi)r(\chi)}} \tag{6.2.41}$$

According to the first of (5.2.11) and up to the multiplicative coefficient $1/m$, the common numerator of $\widetilde{Q}_1(\pi, \chi)$ and $\widetilde{Q}_3(\pi, \chi)$ can be put in the following form

$$card(R(\pi) \cap R(\chi)) \times card(S(\pi) \cap S(\chi)) - card(R(\pi) \cap S(\chi)) \times card(S(\pi) \cap R(\chi))$$

and this defines the numerator of a Goodman and Kruskal coefficient type. Indeed, the latter, designated generally by $\gamma(\pi, \chi)$ can be expressed as follows:

$$\frac{card(R(\pi) \cap R(\chi)) \cdot card(S(\pi) \cap S(\chi)) - card(R(\pi) \cap S(\chi)) \cdot card(S(\pi) \cap R(\chi))}{card(R(\pi) \cap R(\chi)) \cdot card(S(\pi) \cap S(\chi)) + card(R(\pi) \cap S(\chi)) \cdot card(S(\pi) \cap R(\chi))} \tag{6.2.42}$$

Let us now illustrate the calculation of $Q_1(\pi, \chi)$, $R_1(\pi, \chi)$, $R_1''(\pi, \chi)$, $\widetilde{Q}_1(\pi, \chi)$, $\widetilde{Q}_3(\pi, \chi)$ and $\gamma(\pi, \chi)$ on the basis of Table 3.3 of Chap. 3.

In the calculations below $\pi = \{\mathcal{O}_{1.}, \mathcal{O}_{2.}, \mathcal{O}_{3.}, \mathcal{O}_{4.}\}$ (resp., $\chi = \{\mathcal{O}_{.1}, \mathcal{O}_{.2}, \mathcal{O}_{.3}\}$) designates the partition expressed through the rows (resp., through the columns) of Table 3.3. We have

$card(R(\pi)) = \frac{1}{2}(379 \times 378 + 451 \times 450 + 205 \times 204 + 403 \times 402) = 275{,}019$

$card(S(\pi)) = 758{,}184$

$card(R(\chi)) = \frac{1}{2}(591 \times 590 + 608 \times 607 + 239 \times 238) = 387{,}314$

$card(S(\chi)) = 645{,}889$

Computing $Q_1(\pi, \chi)$

$$s(\pi, \chi) = card(R(\pi) \cap R(\chi)) = \frac{1}{2}((223 \times 222 + 122 \times 121 + 34 \times 33)$$
$$+ (168 \times 167 + 215 \times 214 + 68 \times 67) + (90 \times 89 + 80 \times 79 + 35 \times 34)$$
$$+ (110 \times 109 + 191 \times 190 + 102 \times 101)) = 109,057$$

The centred coefficient

$$s(\pi, \chi) - \frac{r(\pi) \times r(\chi)}{m} = 109,057 - \frac{275,019 \times 387,314}{1,033,203} = 5,961.375$$

The different parameters

$$\lambda = \frac{r(\pi)}{\sqrt{m}} = \frac{275,019}{\sqrt{1,033,203}} = 270.564$$

$$\mu = \frac{r(\chi)}{\sqrt{m}} = \frac{387,314}{\sqrt{1,033,203}} = 381.040$$

$$\rho = \frac{379 \times 378 \times 377 + 451 \times 450 \times 449 + 205 \times 204 \times 203 + 403 \times 402 \times 401}{\sqrt{1,438 \times 1,437 \times 1,436}} =$$
$$4,012.746$$

$$\sigma = \frac{591 \times 590 \times 589 + 608 \times 607 \times 606 + 239 \times 238 \times 237}{\sqrt{1,438 \times 1,437 \times 1,436}} = 8,123.356$$

By denoting $N^{[r]}$ the number $N(N - 1) \ldots (N - r + 1)$ where N is an integer number, we have

$$\theta = \frac{\left(\sum_i n_i^{[2]} - 1\right)^2 - 4 \sum_i n_i^{[3]}}{2\sqrt{n^{[4]}}}$$

We obtain $\sum_i n_i^{[2]} = 550,038$ and $\sum_i n_i^{[3]} = 218,588,190$ and then, $\theta = 73,094.698$

In an analogous way we obtain $\sum_j m_j^{[2]} = 774,628$ and $\sum_i m_j^{[3]} = 442,507,380$ and then, $\zeta = 144,964.374$

In these conditions, the value obtained for the variance is $119,318.21$, and consequently

$$\mathbf{Q}_1(\pi, \chi) = 17.258 \text{ and } \mathbf{R}_1(\pi, \chi) = 0.455$$

Computing $Q_1(\pi, \pi)$

$$s(\pi, \pi) = card(R(\pi) \cap R(\pi)) = card(R(\pi)) = 275,019$$

The centred coefficient

$$r(\pi)\left(1 - \frac{r(\pi)}{m}\right) = 275,019 \left(1 - \left(\frac{275,019}{1,033,203}\right)\right) = 201,814.170$$

The variance

$$\frac{r(\pi)^2}{m}\left(1 - \frac{r(\pi)^2}{m}\right) + \rho^2 + \theta^2$$

$$= \frac{275{,}019^2}{1{,}033{,}203}\left(1 - \frac{275{,}019^2}{1{,}033{,}203}\right) + 4{,}012.746^2 + 73{,}094.698^2 = 63{,}020.336$$

$$Q_1(\pi, \pi) = 803.917$$

Computing $Q_1(\chi, \chi)$
The centred coefficient

$$r(\chi)\left(1 - \frac{r(\chi)}{m}\right) = 387{,}314\left(1 - \left(\frac{387{,}314}{1{,}033{,}203}\right)\right) = 242{,}122.654$$

The variance

$$\frac{r(\chi)^2}{m}\left(1 - \frac{r(\chi)^2}{m}\right) + \sigma^2 + \zeta^2$$

$$= \frac{387{,}314^2}{1{,}033{,}203}\left(1 - \frac{387{,}314^2}{1{,}033{,}203}\right) + 8{,}123.356^2 + 144{,}964.374^2 = 276{,}785.89$$

$$Q_1(\chi, \chi) = 460.218$$

Therefore, we have

$$\mathbf{R}_1''(\pi, \chi) = \frac{Q_1(\pi, \chi)}{\sqrt{Q_1(\pi, \pi).Q_1(\chi, \chi)}} = \frac{17.258}{\sqrt{803.917 \times 460.218}} = 0.0284$$

Let us now calculate the K. Pearson coefficient type (see (6.2.39)). The numerator of this coefficient is the same as that of $Q_1(\pi, \chi)$ and has been calculated above. The denominator yields to

$$\frac{\sqrt{275{,}019 \times 387{,}314 \times 758{,}184 \times 645{,}889}}{1{,}033{,}203} = 221{,}051.538$$

And then,

$$\widetilde{R}_1(\pi, \chi) = \frac{5{,}961.375}{2{,}210{,}051.538} = 0.0270$$

the latter value is not very different from that of $\mathbf{R}_1''(\pi, \chi)$. Now, the value of a coefficient associated with $\widetilde{Q}_3(\pi, \chi)$ is equal to

$\widetilde{R}_3(\pi, \chi)$ expressed in (6.2.41) is equal to 0.0183.

To end, notice that $\gamma(\pi, \chi)$ is nothing else than the Yule coefficient (see also [9]). The latter was calculated in Sect. 4.3.2. Its value is 0.063.

Measures Based on Chi-Square Statistic

The well-known chi-square Statistic associated with the contingency Table 4.8 (see Sect. 4.3.1), crossing two categorical attributes c and d, is defined as

$$\chi^2(c, d) = \sum_{1 \leq h \leq K, 1 \leq l \leq M} \frac{(n_{hl} - \frac{n_{h.}n_{.l}}{n})^2}{\frac{n_{h.}n_{.l}}{n}} \tag{6.2.43}$$

It was conceived in order to test the hypothesis of no relation between c and d. Its form can be reduced to

$$\chi^2(c, d) = n \left(\sum_{1 \leq h \leq K, 1 \leq l \leq M} \frac{n_{hl}^2}{n_{h.}n_{.l}} - 1 \right)$$

Its distribution under the multinomial hypothesis of no relation $\mathcal{H}_1$ is—as so-called—*the chi-square distribution* with $(K - 1) \times (M - 1)$ degrees of freedom.

Various attempts were made to derive an association coefficient from it. Let us recall the most important of them

$$\phi^2(c, d) = \frac{\chi^2(c, d)}{n} \tag{6.2.44}$$

$$T(c, d) = \sqrt{\frac{\phi^2(c, d)}{\sqrt{(K - 1)(M - 1)}}} \tag{6.2.45}$$

$$C(c, d) = \sqrt{\frac{\phi^2(c, d)}{\min\{(K - 1), (M - 1)\}}} \tag{6.2.46}$$

All of them depend more or less on $(K - 1)$ and $(M - 1)$. $T(c, d)$ and $C(c, d)$ are the Tshuprow and the Cramér coefficients, respectively, [4, 45]. As emphasized by Goodman and Kruskal [9], it may be difficult to interpret directly the comparison between two different associations by means of these coefficients. In the *LLA* approach, locally applied, we refer to a probability scale by means of the coefficient

$$p_{\chi^2}(c, d) = Prob\{\chi_\nu^2 \leq \chi^2(c, d)\} \tag{6.2.47}$$

where $\nu = (K - 1) \times (M - 1)$ is the number of degrees of freedom. However, there are two important problems for applying this coefficient to mutually comparing a set of categorical attributes. First, each comparison between two categorical attributes requires to refer to a specific probability law. Second, for a number n of observations, large enough, the value of (6.2.47) is almost always very near of 1.

In order to refer to the same probability law, namely the Normal law, P. Villoing [46] employed the index $(\sqrt{2\chi_\nu^2} - \sqrt{2\nu - 1})$ whose probability law is pretty approximable by the Normal law when ν becomes larger than 10. He also employed the Statistic

$$W(\chi_\nu^2) = \left\{ \left(\frac{\chi_\nu^2}{\nu} \right)^{1/3} + \frac{2}{9\nu} - 1 \right\} / \sqrt{\frac{2}{9\nu}}$$

built by Wilson and Hilferty so that its probability law is the normal law [48]. For each of the both indices, the "Similarity Global Reduction" can be applied allowing discriminant probabilistic scale to be established for mutually comparing attributes.

Interesting comparative analysis between these coefficients and that derived from (6.2.32), in case of real data, was carried out in [46], through ascendant hierarchical clustering by the *LLA* method (see Chap. 10).

A final point. The formal conception of the chi-square Statistic is at the level of the object set $\mathcal{O}$ and not globally at the level of the Cartesian product $\mathcal{O} \times \mathcal{O}$. Notice in (6.2.43) that the contribution of the cell (h, l) is exactly the square of the coefficient (5.2.32) between the two Boolean attributes c_h and d_l. This coefficient that we denote here by $\chi(c_h, d_l)$ defines the *oriented* contribution of the cell (h, l) to the chi-square Statistic. This table will be generalized in Chap. 8. It will constitute a powerful tool for interpreting clustering results.

6.2.4 Comparing Two Ordinal Categorical Attributes

6.2.4.1 Introduction

This analysis follows the same logic as that concerning nominal categorical attributes (see Sect. 6.2.3)

Set theoretic representation of an ordinal categorical attribute was defined in Sect. 3.3.2. In Sect. 4.3.3, we showed how to transpose Boolean association coefficients in the case of comparing ordinal categorical attributes. Besides, the analysis of some famous coefficients as that the Goodman and Kruskal γ and the Kendall τ_b coefficients was carried out. Let us take up again the notations of Sect. 4.3.3. ω and ϖ designate the total preorders induced, respectively, by two ordinal categorical attributes c and d. Clearly, a total preorder of a set $\mathcal{O}$ is equivalent to a partition of $\mathcal{O}$ and a linear (total) and strict order on its classes. That is, an ordered sequence of disjoint and complementary subsets of $\mathcal{O}$. As in Sect. 4.3.3, we denote by $(O_1, O_2, \ldots, O_h, \ldots, O_K)$ and $(P_1, P_2, \ldots, P_l, \ldots, P_M)$ the ordered sequences associated with ω and ϖ, respectively. O_h (resp., P_l) is the hth (resp., lth) class of the total preorder ω (resp., ϖ), $1 \le h \le K$ (resp., $1 \le l \le M$). The *composition* of a total preorder is the ordered sequence of the respective cardinals of its classes. Thus, $t = (n_1, n_2, \ldots, n_h, \ldots, n_K)$ (resp., $u = (m_1, m_2, \ldots, m_l, \ldots, m_M)$) is the composition of ω (resp., ϖ). Let us designate by $\Omega(n; t)$ (resp., $\Omega(n; u)$) the set of all total preorders of $\mathcal{O}$, whose composition is t (resp., u). We have

$$card(\Omega(n; t)) = \frac{n!}{n_1! \, n_2! \ldots n_h! \ldots n_K!}$$

$$card(\Omega(n; u)) = \frac{n!}{m_1! \, m_2! \ldots m_l! \ldots m_M!} \tag{6.2.48}$$

In the *LLA* approach the raw association coefficient is $s(\omega, \varpi) = card(R(\omega) \cap R(\varpi))$ (see the first Equation of (4.3.13) and (4.3.14), respectively). Now, let us give the following form to the raw coefficient (see 4.8 for the notations):

$$s(\omega, \varpi) = card(R(\omega) \cap R(\varpi)) = \sum_{1 \leq h \leq K-1, 1 \leq l \leq M-1} n_{hl} \sum_{K \geq p > h, M \geq q > l} n_{pq}$$

$$(6.2.49)$$

This formula shows that the contribution of a given cell (h, l) is its content multiplied by the content the contingency table rectangle which is strictly to the right and below the cell concerned.

The hypothesis of no relation $\mathcal{H}_1$ associates with ω and ϖ an ordered pair (ω^*, ϖ^*) of independent random total preorders in $\Omega(n; t) \times \Omega(n; u)$, endowed with an uniform probability measure. The latter is a product of uniform probability measures on $\Omega(n; t)$ and $\Omega(n; u)$. In fact, we will establish in the following that the distributions of $card(R(\omega) \cap R(\varpi^*))$ and $card(R(\omega^*) \cap R(\varpi))$ are identical. The calculation of the mean and variance of this common distribution enables us to obtain a centred and standardized coefficient corresponding to the coefficient $Q(a, b)$ of the scheme expressed in Fig. 6.1.

6.2.4.2 The Duality Property

The framework of our development below is the set $(\mathcal{O}^{[2]})^{[q]}$ of q-uples of distinct ordered object pairs providing from $\mathcal{O}^{[2]}$, the latter being the set of ordered pairs of distinct objects. More precisely, a given element of $(\mathcal{O}^{[2]})^{[q]}$ can be written as $(p'_1, p'_2, \ldots, p'_l, \ldots, p'_q)$ where $p'_l = (x_l, y_l)$ is an ordered pair of distinct objects and where $p'_1, p'_2, \ldots, p'_l, \ldots, p'_q$ are mutually different. Now, let us consider a bijection σ of $\mathcal{O}$, σ can be represented by a permutation on $\mathcal{O}$. An extension of σ to $(\mathcal{O}^{[2]})^{[q]}$ is obtained by setting

$$\sigma(p'_1, p'_2, \ldots, p'_l, \ldots, p'_q) = (\sigma(p'_1), \sigma(p'_2), \ldots, \sigma(p'_l), \ldots, \sigma(p'_q)),$$

where $\sigma(x, y) = (\sigma(x), \sigma(y))$, for all (x, y) in $\mathcal{O}^{[2]}$.

$$(\sigma(p'_1), \sigma(p'_2), \ldots, \sigma(p'_l), \ldots, \sigma(p'_q))$$

has the same configuration as that $(p'_1, p'_2, \ldots, p'_l, \ldots, p'_q)$; that is, such that the object $\sigma(x)$ is substituted for x in all ordered pairs of $(p'_1, p'_2, \ldots, p'_l, \ldots, p'_q)$ where it appears. Denote by c this configuration and define the set G_q^c of all q-uples having the same configuration c. This set is the value set of $(\sigma(p'_1), \sigma(p'_2), \ldots, \sigma(p'_l), \ldots, \sigma(p'_q))$ when σ describes the $n!$ permutations $\mathcal{O}$.

Let ω_0 and $(p_1'^0, p_2'^0, \ldots, p_l'^0, \ldots, p_q'^0)$ be a fixed elements in $\Omega(n; t)$ and G_q^c, respectively. Besides, let us designate by ϕ_0 (resp., ϕ) the indicator function of the $\mathcal{O}^{[2]}$ subset $R(\omega_0)$ (resp., $R(\omega)$), where ω describes $\Omega(n; t)$. We have the following

Theorem 15 *The proportion of q-uples* $(p'_1, p'_2, \ldots, p'_l, \ldots, p'_q)$ *in* G^c_q *for which* $\phi_0(p'_1)\phi_0(p'_2)\ldots\phi_0(p'_l)\ldots\phi_0(p'_q) = 1$ *is equal to the proportion of total preorders in* $\Omega(n; t)$ *for which* $\phi(p'^0_1)\phi(p'^0_2)\ldots\phi(p'^0_l)\ldots\phi(p'^0_q) = 1$

Proof We have to establish

$$\mu\{\phi_0(p'_1)\phi_0(p'_2)\ldots\phi_0(p'_l)\ldots\phi_0(p'_q) = 1\} = \nu\{\phi(p'^0_1)\phi(p'^0_2)\ldots\phi(p'^0_l)\ldots\phi(p'^0_q) = 1\}$$

$$(6.2.50)$$

where μ and ν designate the first and second proportions stated in the theorem.

The proof is analogous to that of Corollary 3 above. To begin, notice the isomorphism between $\Omega(n; t)$ and the set of partitions with classes labelled $\mathcal{P}(n; t)$ where the class label is its rank for the linear order on the preorder classes (see (3.3.14) and (4.3.8)). As for the case of Corollary 3, we decompose both events under μ and ν in the previous equation, into respective and associated subevents. For a given q-uple $(p'_1, p'_2, \ldots, p'_l, \ldots, p'_q)$ of G^c_q, determine the subset $E = \{x_1, x_2, \ldots, x_i, \ldots, x_l\}$ of $\mathcal{O}$ elements which appear at least once in the mentioned q-uple. Then, consider an assignation function denoted here by γ of these different elements into the different classes of the total preorder ω_0 such that the event concerned by μ is satisfied. γ induces a partition of E whose classes are included in the ω_0 classes. γ can be represented by a mapping of the set of subscripts $\{1, 2, \ldots, i, \ldots, l\}$ in the class labels $\{1, 2, \ldots, h, \ldots, K\}$. Dually, let $E_0 = \{x^0_1, x^0_2, \ldots, x^0_i, \ldots, x^0_l\}$ be the subset whose elements appear at least once in $(p'^0_1, p'^0_2 \ldots, p'^0_l, \ldots, p'^0_q)$. We may assume that the E elements $x_1, x_2, \ldots, x_i, \ldots$ and x_l can be deduced from $x^0_1, x^0_2, \ldots, x^0_i, \ldots$ and x^0_l from a permutation σ on $\mathcal{O}$: $x_i = x^0_{\sigma(i)}, 1 \leq i \leq l$. In these conditions, we can consider a total preorder ω of $\Omega(n; t)$ such that the same mapping γ determines a partition of E_0 whose classes are included into the classes of ω in the same inclusion manner as previously. Therefore Theorem 4 completes the proof. □

The indicator functions ψ_0 and ψ of $R(\varpi_0)$ and $R(\varpi)$, respectively, can be introduced at the same time as those ϕ_0 and ϕ (see above). By a development analogous to that of Sect. 6.2.3.3, the following theorem can be established

Theorem 16 $\Omega(n; t)$ *and* $\Omega(n; u)$ *being endowed with a uniform probability measure, if* $\omega^\star$ *and* $\varpi^\star$ *are independent random elements in* $\Omega(n; t)$ *and* $\Omega(n; u)$, *respectively, the distributions of* $card(R(\omega_0) \cap R(\varpi^\star))$ *and* $card(R(\omega^\star) \cap R(\varpi_0))$ *are identical.*

6.2.4.3 Mean and Variance of $card(R(\omega_0) \cap R(\varpi^\star))$ and $card(R(\omega^\star) \cap R(\varpi_0))$ Distributions

To begin, we will establish two lemmas which specify the statement of Theorem 8 for $q = 1$ and $q = 2$. For these we will make explicit the different forms of G^c_q. Besides, the common value of the two proportions stated in this theorem, will be calculated. Hence, the mean and variance of $card(R(\omega_0) \cap R(\varpi^\star))$ (resp., $card(R(\omega^\star) \cap R(\varpi_0))$) will be deduced.

Lemma 4 *Let $p'_0 = (x_0, y_0)$ and ω_0 be a fixed ordered pair in $\mathcal{O}^{[2]}$ and ω_0 a fixed total preorder in $\Omega(n; t)$. The proportion of ordered pair (x, y) in $\mathcal{O}^{[2]}$ for which $x < y$ for ω_0 is equal to the proportion of total preorders in $\Omega(n; t)$ for which we have $x_0 < y_0$. The common value of this proportion is*

$$\frac{\sum_{1 \leq g < h \leq K} n_g n_h}{n(n-1)}$$

where $t = (n_1, n_2, \ldots, n_h, \ldots, n_K)$.

The value of the first of both proportions appears clearly. Indeed, we have

$$card(R(\omega_0)) = \sum_{1 \leq g < h \leq K} n_g n_h \text{ and } card(\mathcal{O}^{[2]}) = n(n-1)$$

The value of the second proportion can be expressed as follows:

$$\sum_{1 \leq g < h \leq K} \frac{\frac{(n-2)!}{n_1! n_2! \ldots (n_g-1)! \ldots (n_h-1)! \ldots n_K!}}{\frac{n!}{n_1! n_2! \ldots n_g! \ldots n_h! \ldots n_K!}} = \frac{\sum_{1 \leq g < h \leq K} n_g n_h}{n(n-1)}$$

where a generic term of the sum of the left member represents the number of total preorders whose composition is t, for which x_0 and y_0 belong to the gth and the hth classes, respectively, with $g < h$.

The next Lemma requires some definitions. We set

$$\mathcal{O}_4 = \mathcal{O}^{[2]} \times \mathcal{O}^{[2]} - \Delta$$

where $\Delta = \{(p', p') | p' \in \mathcal{O}^{[2]}\}$ is the diagonal of $\mathcal{O}^{[2]} \times \mathcal{O}^{[2]}$. $\mathcal{O}_4$ is the set of ordered pairs $((x, y), (x', y'))$ where (x, y) and (x', y') belong to $\mathcal{O}^{[2]}$ and where $(x, y) \neq (x', y')$. We have

$$card(\mathcal{O}_4) = n(n-1)[n(n-1) - 1] \tag{6.2.51}$$

A given element $((x, y), (x', y'))$ of $\mathcal{O}_4$ can take six forms denoted below by (0), (1), (1'), (2), (2') and (3). For (0), none of x and y is repeated in the second pair. For (3) both x and y are repeated. For (1) and (1') the first component x is repeated in its position or not, respectively. For (2) and (2'), the second component y is repeated in its position or not, respectively. More precisely, the six forms are listed as follows:

$$(0)\ ((x, y), (z, t))$$
$$(1)\ ((x, y), (x, t)),\ \text{·}(1')\ ((x, y), (z, x))$$
$$(2)\ ((x, y), (z, y)),\ (2')\ ((x, y), (y, t))$$
$$(3)\ ((x, y), (y, x)) \tag{6.2.52}$$

where different letters represent different objects.

Let $\{H, G_1, G_1', G_2, G_2', I\}$ be the partition of $\mathcal{O}_4$ according to the 6 previous forms. We have, respectively,

$$card(H) = n(n-1)(n-2)(n-3)$$
$$card(G_1) = card(G_1') = card(G_2) = card(G_2') = n(n-1)(n-2)$$
$$card(I) = n(n-1) \tag{6.2.53}$$

Lemma 5 *Let ω_0 and $\big((x_0, y_0), (x_0', y_0')\big)$ be two fixed elements in $\Omega(n; t)$ and $\mathcal{O}_4$, respectively. Whatever the form of $\big((x_0, y_0), (x_0', y_0')\big)$ (see (6.2.52)), the proportion of ordered pairs $\big((x, y), (x', y')\big)$ having the same form and for which $x < y$ and $x' < y'$, is equal to the proportion of total preorders in $\Omega(n; t)$ for which $x_0 < y_0$ and $x_0' < y_0'$.*

The common value of this proportion is

$$\frac{1}{n(n-1)(n-2)(n-3)} \sum_{1 \le g < h \le K} n_g n_h \sum_{1 \le g' < h' \le K} n_{g'}' n_{h'}'$$

if $\big((x_0, y_0), (x_0', y_0')\big) \in H$

where $n_{g'}' = n_{g'}$ (resp., $n_{g'}' = n_{g'} - 1$) if $g' \ne g$ and $g' \ne h$

(resp., $g' = g$ or $g' = h$)$n_{h'}' = n_{h'}$ (resp., $n_{h'}' = n_{h'} - 1$) if $h' \ne g$ and $h' \ne h$

(resp., $h' = g$ or $h' = h$)

$$\frac{1}{n(n-1)(n-2)} \sum_{1 \le g < h \le K} n_g n_h \sum_{K \ge j > g} n_j' = \frac{1}{n(n-1)(n-2)} \sum_{1 \le g \le K-1} n_g^f (n_g^f - 1)$$

if $\big((x_0, y_0), (x_0', y_0')\big) \in G_1$

where $n_j' = n_j$ (resp., $n_j' = n_j - 1$) if $j \ne h$ (resp., $j = h$) and where $n_g^f = \sum_{j > g} n_j$

$$\frac{1}{n(n-1)(n-2)} \sum_{1 \le g < h \le K} n_g n_h \sum_{j < h} n_j' = \frac{1}{n(n-1)(n-2)} \sum_{2 \le g \le K-1} n_g n_g^c n_g^f$$

if $\big((x_0, y_0), (x_0', y_0')\big) \in G_1'$

where $n_j' = n_j$ (resp., $n_j' = n_j - 1$) if $j \ne g$ (resp., $j = g$) and where $n_h^c = \sum_{j < h} n_j$

$$\frac{1}{n(n-1)(n-2)} \sum_{2 \le g \le K} n_g n_g^c (n_g^c - 1) \text{ if } \big((x_0, y_0), (x_0', y_0')\big) \in G_2$$

$$\frac{1}{n(n-1)(n-2)} \sum_{2 \le g \le K-1} n_g n_g^c n_g^f \text{ if } \big((x_0, y_0), (x_0', y_0')\big) \in G_2'$$

0 if $\big((x_0, y_0), (x_0', y_0')\big) \in I$ \hfill (6.2.54)

The number of ordered pairs of ordered $\mathcal{O}$ element pairs of the form (0), for which x, y, z and t belong to the gth, hth, g'th and h'th classes, respectively, is equal to $n_g n_h n_{g'}' n_{h'}'$, where $n_{g'}' = n_{g'}$ (resp., $(n_{g'} - 1)$) if $g' \ne g$ and $g' \ne h$ (resp., $g' = g$ or if

$g' = h$). In this, we assume $g \neq h$ and $g' \neq h'$. By taking into account the cardinality of H, we obtain the first proportion stated.

Dually, the number of total preorders of $\Omega(n; t)$ for which x_0, y_0 z_0 and t_0 occupy the gth, hth g'th and h'th classes, respectively, where $g \neq h$ and $g' \neq h'$ is equal to $(n - 4)! / \Pi_{1 \leq j \leq K} l_j!$, where

- $l_j = n_j$ if j is different from each of the subscripts g, h, g' and h'
- $l_g = n_g - 1$ (resp., $l_g = n_g - 2$) if $g' \neq g$ and $h' \neq g$ (resp., $g' = g$ or $h' = g$)
- $l_h = n_h - 1$ (resp., $l_h = n_h - 2$) if $g' \neq h$ and $h' \neq h$ (resp., $g' = h$ or $g' = g$).

Analogously,

- $l_{g'} = n_{g'} - 1$ (resp., $l_{g'} = n_{g'} - 2$) if $g \neq g'$ and $h \neq g'$ (resp., $g = g'$ or $h = g'$)
- $l_{h'} = n_{h'} - 1$ (resp., $l_{h'} = n_{h'} - 2$) if $g \neq h'$ and $h \neq h'$ (resp., $g = h'$ or $h = h'$).

Considering the cardinal of $\Omega(n; t)$ $(n! / \Pi_{1 \leq h \leq K} n_h!)$, we obtain the same proportion as above for the set of total preorders in $\Omega(n; t)$ for which $x_0 < y_0$ and $z_0 < t_0$.

Calculating the other proportions stated in the previous lemma is similar to the above computing. Nevertheless, let us establish it once again in the case where $((x_0, y_0), (x'_0, y'_0))$ has the form (1), that is $((x, y), (x, t))$. The number of ordered pairs of $\mathcal{O}^{[2]}$ having the latter form, for which x, y and t belong to the gth, hth and jth classes of the total preorder ω_0, is equal to $n_g n_h n'_j$, where $g \neq h$, $g \neq j$ and where $n'_j = n_j$ (resp., $n'_j = n_j - 1$) if $j \neq h$ (resp., $j = h$). By considering $card(G_1)$, we derive the proportion stated above, namely, that of the elements $((x, y), (x, t))$ for which $x < y$ and $x < t$ for ω_0.

Dually, the number of total preorders in $\Omega(n; t)$ for which x_0, y_0 and t_0 occupy the gth, hth and jth classes, respectively, where $g \neq h$ and $j > g$ is equal to

$$\frac{(n - 3)!}{\Pi_{1 \leq p \leq K} l_p!} \tag{6.2.55}$$

where

- $l_p = n_p$ if p is different from each of the subscripts g, h and j
- $l_h = (n_h - 1)$ (resp., $l_h = (n_h - 2)$) if $j \neq h$ (resp., $j = h$)
- $l_h = (n_j - 1)$ (resp., $l_h = (n_j - 2)$) if $h \neq j$ (resp., $h = j$).

Dividing (6.2.55) by $card(\Omega(n; t))$ gives the same proportion as that calculated previously, at the level of G_1 and concerning ω_0.

Theorem 17 *The common mean and variance of both dual identical distributions are*

$$\lambda\mu \text{ and } \lambda\mu + \rho_{ff}\sigma_{ff} + \rho_{cc}\sigma_{cc} + 2\rho_{cf}\sigma_{cf} + (\Lambda\Gamma - \lambda^2\mu^2), \tag{6.2.56}$$

respectively.

where

$$\lambda = \sum_{g<h} n_g n_h / \sqrt{n(n-1)}, \ \rho_{ff} = \sum_{g} n_g (n_g^f)^2 / \sqrt{n(n-1)(n-2)}$$

$$\rho_{cc} = \sum_{g} n_g (n_g^c)^2 / \sqrt{n(n-1)(n-2)}, \ \rho_{cf} = \sum_{g} n_g n_g^f n_g^c / \sqrt{n(n-1)(n-2)}$$

$$\Lambda = \sum_{g<h} n_g n_h \Big(\sum_{g'<h'} n_{g'} n_{h'} + n_g + n_h - 2n + 1 \Big) / \sqrt{n(n-1)(n-2)(n-3)}$$

$$(6.2.57)$$

The expressions of μ, σ_{ff}, σ_{cc}, σ_{cf} and Γ have the same forms as those of ρ, ρ_{ff}, ρ_{cf} and Λ, respectively, the n_g ($1 \le g \le K$) being replaced by the m_l ($1 \le l \le M$).

Lemma 3 above enables the result concerning the mean expression to be established. For the variance calculation, the same scheme as that considered for comparing a pair of partitions is followed. More precisely, we have to determine

$$\sum_{(q,q')\in\mathcal{O}_4} \phi_0(q)\phi_0(q')\mathcal{E}\big(\psi^\star(q)\psi^\star(q')\big) \qquad (6.2.58)$$

where ϕ_0 (resp., $\psi^\star$) is the indicator function of $R(\omega_0)$ (resp., $R(\varpi^\star)$) where ω_0 (resp., $\varpi^\star$) is a fixed element in $\Omega(n; t)$ (resp., a random element in $\Omega(n; u)$, endowed with a uniform probability measure).

By partitioning $\mathcal{O}_4$ into the classes H, G_1, G_1', G_2 G_2' and I (see above):

$$\mathcal{O}_4 = H + G_1 + G_1' + G_2 + G_2' + I \text{ (set sum)}$$

and by decomposing the sum (6.2.58) into 6 parts over H, G_1, G_1', G_2 G_2' and I, respectively, we deduce from Lemma 4 the result stated.

Discussion Around Several Association Coefficients Between Ordinal Categorical Attributes

As a consequence of the previous results, we obtain the centred and reduced association coefficient $Q_1(\omega, \varpi)$ corresponding to the general diagram of Fig. 6.1. It can be written as

$$Q_1(\omega, \varpi) = \frac{s(\omega, \varpi) - \lambda\mu}{\sqrt{\lambda\mu + \rho_{ff}\sigma_{ff} + \rho_{cc}\sigma_{cc} + 2\rho_{cf}\sigma_{cf} + (\Lambda\Gamma - \lambda^2\mu^2)}} \qquad (6.2.59)$$

Coefficients as those ((6.2.33)–(6.2.35)), previously defined for comparing partitions, can be proposed for comparing total preorders, namely

$$R_1(\omega, \varpi) = \frac{Q_1(\omega, \varpi)}{\sqrt{n}} \qquad (6.2.60)$$

$$R'_1(\omega, \varpi) = \frac{s(\omega, \varpi) - \mathcal{E}(s(\omega^\star, \varpi^\star))}{\max\left(s(\omega^\star, \varpi^\star) - \mathcal{E}(s(\omega^\star, \varpi^\star))\right)} \qquad (6.2.61)$$

$$R''_1(\omega, \varpi) = \frac{Q_1(\omega, \varpi)}{\sqrt{Q_1(\omega, \omega) \cdot Q_1(\varpi, \varpi)}} \qquad (6.2.62)$$

As mentioned at the end of the introductory Sect. 3.5.1, there is an optimal and very quick algorithm which provides the exact maximum concerned by the denominator of $R'_1(\omega, \varpi)$ [25].

Now, let us consider the centred raw coefficient $s(\omega, \varpi) - \lambda\mu$. It can be expressed as follows:

$$s - \frac{(s + u)(s + v)}{n^{[2]}} \qquad (6.2.63)$$

where as above $n^{[2]} = n(n - 1)$, where s, u and v were defined in (4.3.14).

Let us recall that in the case of comparing ranking attributes, the centred raw coefficient is identical to the numerator of the Kendall τ coefficient (see Proposition 38). This property does no more hold in the case of comparing total preorders. In fact, the τ_b coefficient can be written as (see (4.3.15) and (4.3.16)),

$$\tau_b(\omega, \varpi) = \frac{s - \frac{u+v}{2}}{\sqrt{(s + u)(s + v)}} \qquad (6.2.64)$$

And, $(u + v)/2$ is generally different from $(s + u)(s + v)/n^{[2]}$.

Now, let us introduce the parameters $r(\omega) = card(R(\omega)) = s + u$ and $r(\varpi) = card(R(\varpi)) = s + v$. As for comparing partitions (see the end of Sect. 6.2.3.5), let us consider a "free" hypothesis of no relation for which with $R(\omega)$ (resp., $R(\varpi)$) is associated a random $r(\omega)$ subset (resp., $r(\varpi)$) subset in the set of all $r(\omega)$ subsets (resp., $r(\varpi)$ subsets), endowed with a uniform probability measure. This model corresponds to the $\mathcal{H}_1$ hypergeometric model defined in the Boolean case (see Sect. 5.2.1.2). The corresponding coefficients to those defined by (6.2.38)–(6.2.41) can be easily formulated by analogy. We have,

$$\tilde{Q}_1(\omega, \varpi) = \frac{card(R(\omega) \cap R(\varpi)) - \frac{r(\omega)r(\varpi)}{m'}}{\sqrt{\frac{r(\omega)(m'-r(\omega))r(\varpi)(m'-r(\varpi))}{m'^2(m'-1)}}} \qquad (6.2.65)$$

where $r(\omega) = card(R(\omega))$ (resp., $r(\varpi) = card(R(\varpi))$) and where $m' = n^{[2]} = card(\mathcal{O}^{[2]})$.

By identifying $m' - 1$ with m', the transposition of the K. Pearson coefficient gives

$$\tilde{R}_1(\omega, \varpi) = \frac{card(R(\omega) \cap R(\varpi)) - \frac{r(\omega)r(\varpi)}{m'}}{\sqrt{\frac{r(\omega)(m'-r(\omega))r(\varpi)(m'-r(\varpi))}{m'^2}}} \qquad (6.2.66)$$

The numerator of this coefficient is exactly that of $Q_1(\omega, \varpi)$. However and clearly, the denominator above is different. The variance of $card(R^*(\omega) \cap R^*(\varpi))$ is much lower than that of $card(R(\omega^*) \cap R(\varpi^*))$.

By referring to the Poisson model the following coefficient can also be proposed

$$\tilde{Q}_3(\omega, \varpi) = \frac{card(R(\omega) \cap R(\varpi)) - \frac{r(\omega)r(\varpi)}{m'}}{\sqrt{\frac{r(\omega)r(\varpi)}{m'}}} \tag{6.2.67}$$

The associated $\tilde{R}_3(\omega, \varpi)$ coefficient can be expressed by

$$\tilde{R}_3(\omega, \varpi) = \frac{card(R(\omega) \cap R(\varpi)) - \frac{r(\omega)r(\varpi)}{m'}}{\sqrt{r(\omega)r(\varpi)}} \tag{6.2.68}$$

To continue in the vein of the example given in Sect. 6.2.3.4, for comparing two partitions associated, respectively, with two nominal categorical attributes, let us take up again Table 3.3 of Chap. 3 in order to illustrate the calculation of $Q_1(\omega, \varpi)$. Here, ω and ϖ denote, respectively, the total preorders induced by the categorical attributes "Fertility planning status of couple" and "Highest level of formal education of wife", interpreted as ordinal categorical attributes. We assume $D < C < B < A$ and $c_1 < c_2 < c_3$ for the categorical values of these attributes, respectively (see Table 3.3). By denoting as above $\{O_1, O_2, O_3, O_4\}$ (resp., $\{P_1, P_2, P_3\}$) the preorder classes of ω and ϖ respectively, we have:

$$O_1 < O_2 < O_3 < O_4 \text{ and } P_1 < P_2 < P_3$$

and

$r(\omega) = card(R(\omega)) = (379 \times (451 + 205 + 403) + 451 \times (205 + 403) + 205 \times 403) = 758,184$
$r(\varpi) = card(R(\varpi)) = (591 \times (608 + 239) + 608 \times 239) = 645,889$

Computing $Q_1(\omega, \varpi)$

$s(\omega, \varpi) = 223 \times (215 + 68 + 80 + 35 + 191 + 102) + 122 \times (68 + 35 + 102) + 168 \times (80 + 35 + 191 + 102) + 215 \times (35 + 102) + 90 \times (191 + 102) + 80 \times 102 = 311,632$

Computing $Q_1(\omega, \varpi)$
The parameters λ and μ

$\lambda = r(\omega)/\sqrt{1438 \times 1437} = 527.4324$
$\mu = r(\varpi)/\sqrt{1438 \times 1437} = 449.3141$

The centred coefficient

$s(\omega, \varpi) - \lambda\mu = 311,632 - 527.4324 \times 449.3141 = 74,649.1859$

Computing the variance; the different parameters

$\rho_{ff} = \big(379 \times (451 + 205 + 403)^2 + 451 \times (205 + 403)^2 451 \times (205 + 403)^2$
$+ 205 \times 403^2\big)/\sqrt{1438^{[3]}} = 11{,}474.4590$

where, according to above notations, $1438^{[3]} = 1438 \times 1437 \times 1436$

$\sigma_{ff} = \big(591 \times (608 + 239)^2 + 608 \times 239^2\big)/\sqrt{1438^{[3]}} = 8{,}420.9484$
$\rho_{cc} = \big(451 \times 379^2 + 205 \times (379 + 451)^2 + 403 \times (205 + 451 + 379)^2\big)/$
$\sqrt{1438^{[3]}} = 11{,}706.8044$
$\sigma_{cc} = \big(608 \times 591^2 + 239 \times (591 + 608)^2\big)/\sqrt{1438^{[3]}} = 10{,}205.876$
$\rho_{cf} = \cdot(451 \times 379 \times (205 + 403) + 205 \times (379 + 451) \times 403)/\sqrt{1438^{[3]}}$
$= 3{,}166.5925$
$\sigma_{cf} = 608 \times 591 \times 239/\sqrt{1438^{[3]}} = 1{,}576.5361$
$\Lambda = \Big(\big(379 \times (451 + 205 + 403) + 451 \times (205 + 403) + 205 \times 403\big)^2$
$+ (379 \times 451 + 379 + 451 - 2875) + (379 \times 205 + 379 + 205 - 2875)$
$+ (379 \times 403 + 379 + 403 - 2875) + (451 \times 205 + 451 + 205 - 2875)$
$+ (451 \times 403 + 451 + 403 - 2875) + (205 \times 403 + 205 + 403 - 2875)\Big)/$
$\sqrt{1438^{[4]}} = 278{,}572.8555$
$\Gamma = \Big(\big(591 \times (608 + 239) + 608 \times 239\big)^2 + (591{\times}608 + 591 + 608)$
$+ (591 \times 239 + 591 + 239 - 2875) + (608 \times 239 + 608 + 239 - 2875)\Big)/$
$\sqrt{1438^{[4]}} = 202{,}164.7557$
$\Lambda\Gamma - \lambda^2\mu^2 = 156{,}759{,}876.8$

We obtain for the value of the variance $275{,}616{,}911.2$ and therefore

$$Q_1(\omega, \varpi) = 4.4965 \approx 4.5$$

Some Comments

The value obtained for $Q_1(\omega, \varpi)$ is rather large. There are two reasons for this. On the one hand, there is a positive link between the categorical attributes "Fertility planning status of couple" and "Highest level of formal education of wife", ordinally interpreted. On the other hand, the sample size (1438) is large enough in order to clearly highlight the strength of the association. The probabilistic index $\Phi\big(Q_1(\omega, \varpi)\big)$ as expressed in the diagram of Fig. 6.1 is nearly equal 1. Therefore, according to *Statistical Independence Hypotheses*, the independence hypothesis between the two categorical attributes concerned is strongly rejected. However, as expressed previously, the fundamental problem in *Data Analysis* does not consist of pointing out one single link; but, of mutually comparing in a relative way many links. This can be carried out with a discriminant probabilistic index, after applying the *Global Reduction of the Similarities* (see Sect. 5.2.1.5 and Eqs. (5.2.51) and (5.2.52)). In the case here, we will have to mutually compare a set of ordinal categorical attributes.

The reader is asked to continue by illustrating the calculation of the coefficients (6.2.65)–(6.2.68) on the basis of the example above.

6.2.5   Comparing Two Valuated Binary Relation Attributes

6.2.5.1   Introduction

The attribute type concerned here was defined in Sect. 3.3.5. As in Sect. 5.2.2.1, let us code the object set $\mathcal{O}$ by means of the set $I = \{1, 2, \ldots, i, \ldots, n\}$ of integer subscripts. In this coding i indicates a specific element of $\mathcal{O}$, $1 \leq i \leq n$. Let us denote by $I^{[2]}$ the set of ordered pairs of distinct I elements:

$$I^{[2]} = \{(i,j)|1 \leq i \neq j \leq n\} \qquad (6.2.69)$$

A valuated binary relation on $\mathcal{O}$ is represented by a numerical valuation on $I^{[2]}$ or on the Cartesian product $I \times I$. Denoting by ξ and $\bar{\xi}$ these valuations we have

$$\xi = \{\xi_{ij}|(i,j) \in I^{[2]}\} \qquad (6.2.70)$$
$$\bar{\xi} = \{\xi_{ij}|(i,j) \in I \times I\} \qquad (6.2.71)$$

in the first or the second cases, respectively. In the second case, there are two specific structures which can be distinguished for ξ: symmetrical ($\xi_{ij} = \xi_{ji}$ for all $(i,j) \in I^{[2]}$) and antisymmetrical ($\xi_{ij} = -\xi_{ji}$ for all $(i,j) \in I^{[2]}$). For each of both structures, a common value has to be adopted in the diagonal $\Delta = \{(i,i)|i \in I\}$. This might be $\max\{\xi_{ij}|(i,j) \in I^{[2]}\}$ (resp., 0) for the symmetrical (resp., antisymmetrical) structure.

In Sect. 5.2.1.2 three versions denoted by $\mathcal{N}_1$, $\mathcal{N}_2$ and $\mathcal{N}_3$ of a random model associating with a given $\mathcal{O}$ subset, a random one, were specified. In Sect. 5.2.2.1 these models have been extended for associating with a $\mathcal{O}$ numerical valuation a random one. Principally, the random model $\mathcal{N}_1$ has been developed in the case where the given structure on $\mathcal{O}$ is a partition, a total preorder or a strict total order (ranking) on $\mathcal{O}$. For the case concerned here where the given structure is a valuated binary relation, the representation form (6.2.70) is the most adequate for extending the $\mathcal{N}_1$ model. Though, the expression of $\mathcal{N}_2$ and $\mathcal{N}_3$ models for this extension requires (6.2.71) representation [27]. $\mathcal{N}_1$ has a permutational nature. It associates with ξ the random valuation

$$\xi^\star = \{\xi_{\sigma(i)\sigma(j)}|(i,j) \in I^{[2]}\} \qquad (6.2.72)$$

where σ is a random permutation in the set, denoted by G_n, of all permutations of $I = \{1, 2, \ldots, i, \ldots, n\}$, provided by a uniform probability measure.

6.2.5.2   The Raw Coefficient and the Random Raw Coefficient
             Associated

Let

$$\xi = \{\xi_{ij}|(i,j) \in I^{[2]}\} \qquad (6.2.73)$$
$$\text{and } \eta = \{\eta_{ij}|(i,j) \in I \times I\} \qquad (6.2.74)$$

be two valuated binary relations of $\mathcal{O}$. The extension of a raw coefficient established in the case of comparing discrete combinatorial structures (e.g. partitions), to the case of comparing numerical valuations, has the same nature as that enabling two numerical attributes to be compared from the method of comparing two Boolean attributes (see Sect. 5.2.2.2). Hence, the raw coefficient between ξ and η is defined as an inner product. More precisely,

$$s(\xi, \eta) = <\xi, \eta> = \sum_{(i,j) \in I^{[2]}} \xi_{ij} \eta_{ij} \qquad (6.2.75)$$

The random raw coefficient defined under the hypothesis of no relation $\mathcal{H}_1$ can be written as

$$s(\xi^\star, \eta^\star) = \sum_{(i,j) \in I^{[2]}} \xi_{\sigma(i)\sigma(j)} \eta_{\tau(i)\tau(j)} \qquad (6.2.76)$$

where σ and τ are two independent random permutations taken according to the random model $\mathcal{N}_1$ in G_n.

By fixing ξ (resp., η) and by associating with η (resp., ξ) a permutational random valuation $\eta^\star$ (resp., $\xi^\star$) we obtain two dual random coefficients, defined as follows:

$$s(\xi, \eta^\star) = \sum_{(i,j) \in I^{[2]}} \xi_{ij} \eta_{\tau(i)\tau(j)} \text{ and } s(\xi^\star, \eta) = \sum_{(i,j) \in I^{[2]}} \xi_{\sigma(i)\sigma(j)} \eta_{ij} \qquad (6.2.77)$$

The duality property is expressed by

Proposition 40 *The distributions of $s(\xi, \eta^\star)$ and $s(\xi^\star, \eta)$ are identical.*

Proof The sum which defines $s(\xi, \eta^\star)$ (see the left side Equation in (6.2.77)) can also be written

$$\sum_{(\tau(i),\tau(j)) \in I^{[2]}} \xi_{ij} \eta_{\tau(i)\tau(j)}$$

Indeed, for a given permutation τ, $I^{[2]}$ is equivalently described by (i, j) or by $(\tau(i), \tau(j))$. By setting $i' = \tau(i)$ and $j' = \tau(j)$, the previous sum can be written as

$$\sum_{(i', j') \in I^{[2]}} \xi_{\tau^{-1}(i')\tau^{-1}(j')} \eta_{i'j'}$$

where τ^{-1} is the reciprocal permutation of τ and then τ^{-1} as τ describe uniformly G_n. Consequently, the latter random sum is identical to that of the right side Equation in (6.2.77), defining $s(\xi^\star, \eta)$.

Obviously, the common distribution expressed in the Proposition above is also that of $s(\xi^\star, \eta^\star)$ (see (6.2.76)).

The comparison of ordinal or categorical attributes (see Sects. 6.2.2, 6.2.3 and 6.2.4) may appear as specific cases of comparing binary valued attributes. In the case of comparing two totally ordinal attributes (ranking or ordinal categorical attributes (Sects. 6.2.2 and 6.2.4)), denoted here by ω and ϖ, by setting

$\xi_{ij} = 1$ (resp., $\xi_{ij} = 0$) if and only if $i < j$ (resp., $i \geq j$) for ω

$\eta_{ij} = 1$ (resp., $\eta_{ij} = 0$) if and only if $i < j$ (resp., $i \geq j$) for ϖ,

we obtain the raw association coefficients (6.2.10) or (6.2.49), depending on the strict or no strict ordinal natures of ω and ϖ, respectively.

Now, in the case of comparing two nominal categorical attributes inducing two partitions π and χ on the object set $\mathcal{O}$, we set

$\xi_{ij} = 1$ (resp., $\xi_{ij} = 0$) if and only if i and j are joined in a same π class (resp., are separated in different π classes)

$\eta_{ij} = 1$ (resp., $\eta_{ij} = 0$) if and only if i and j are joined in a same χ class (resp., are separated in different χ classes)

Thus and by taking into account our representation set ($I^{[2]}$), we obtain twice the raw coefficient (6.2.18). In this case of comparing partitions, the valuations ξ and η are *symmetrical*. It is not so in the case of comparing ordinal categorical attributes, inducing total preorders on $\mathcal{O}$ where ξ and η are neither symmetrical nor antisymmetrical.

The two dual raw random coefficients considered in (6.2.77) can be compared with those defined for comparing partitions (see Sect. 6.2.3) or total preorders (see Sect. 6.2.4). Let us take up again the set $\mathcal{P}(n; t)$ (resp., $\Omega(n; t)$) of all partitions of type (resp., of all total preorders of composition) $t = (n_1, n_2, \ldots, n_h, \ldots, n_K)$. To fix ideas, we begin with the partition case for which ξ and η have just been specified above. When σ describes the set G_n of $n!$ permutations on I, each of the partitions of $\mathcal{P}(n; t)$ is described exactly the same number of times, namely $n!/n_1!n_2!\ldots n_h!\ldots, n_K!$. Consequently, the distribution of $s(\xi^\star, \eta)$ (see the right side of (6.2.77)) is identical to that of $card(R(\pi^\star) \cap R(\chi))$ (see the second Equation of (6.2.19)). Similarly, the distribution of $s(\xi, \eta^\star)$ (see the left side of (6.2.77)) is identical to that of $card(R(\pi) \cap R(\chi^\star))$ (see the first Equation of (6.2.19)). Clearly, we obtain results of the same nature as the latter in the case of comparing total preorders.

The attribute structure considered here is very general. The different attribute types considered previously (see Sects. 6.2.2–6.2.4) may appear as particular cases of valuated binary relations. However, the specificity of the treatment devoted for comparing categorical or ordinal attributes is very important. It reveals all the combinatorial phenomenons underlying these comparisons. Besides, it enables us to visualize and analyze the construction of an association coefficient between two descriptive attributes of a given type. Consequently, the different coefficients proposed in the literature can be situated comparatively and new coefficients can be devised.

As expressed above, the coefficient $s(\xi^*, \eta^*)$ appeared in a very specific study of H.E. Daniels [5] concerning statistical association between the Spearman ρ and the Kendall τ coefficients. This work inspired G. Lecalvé [18] to extend the association coefficients between nominal or ordinal categorical attributes obtained in [19]. The product moment is used for the normalization process, whereas the standard deviation of the random raw coefficient is taken into account in our case. This extension was retaken an worked on again from a pure combinatorial approach [19, 21]. This leads to a clear formal expression of the moments of the random raw coefficient $s(\xi, \eta^*)$ (resp., $s(\xi^*, \eta)$). In [19] we were not aware of the Mantel contribution where an elegant expression of the variance of $s(\xi, \eta^*)$ (resp., $s(\xi^*, \eta)$) is given. However, in the Mantel development, considered with a view to regression approach, only symmetrical and antisymmetrical valuated relations are concerned. Our formal expression of this variance is essentially different. It is more structured and the most general one. The Mantel Statistic is emphasized in [13, 15] for testing the hypothesis of the conformity of proximity matrices.

6.2.5.3 The Centred and Standardized Coefficient $Q_1(\xi, \eta)$

Mean and Variance of $s(\xi, \eta^*)$

As mentioned above, the distributions of $s(\xi, \eta^*)$, $s(\xi^*, \eta)$ and $s(\xi^*, \eta^*)$ (see (6.2.76) and (6.2.77)) are identical. We will focus on the first of these random raw coefficients that we can designate by S_1, to carry out the calculations needed. Let us denote by μ_ξ and μ_η the means on $I^{[2]}$ of the valuations ξ and η, respectively, namely

$$\mu_\xi = \frac{1}{n^{[2]}} \sum_{(i,j) \in I^{[2]}} \xi_{ij}$$

$$\mu_\eta = \frac{1}{n^{[2]}} \sum_{(i,j) \in I^{[2]}} \eta_{ij} \tag{6.2.78}$$

It is straightforward to obtain

$$\mathcal{E}(s(\xi, \eta^*)) = n^{[2]} \mu_\xi \mu_\eta \tag{6.2.79}$$

For the variance calculation, as in (6.2.52), we have to decompose $(I^{[2]})^2$ into the following subsets, where different letters indicate different elements of I,

$$D = \{((i,j), (i,j))\}, E = \{((i,j), (j,i))\}$$
$$G_1 = \{((i,j), (i,k))\}, G_1' = \{((i,j), (h,i))\}$$
$$G_2 = \{((i,j), (h,j))\}, G_2' = \{((i,j), (j,k))\}$$
$$H = \{((i,j), (h,k))\} \tag{6.2.80}$$

$$var(S_1) = \frac{1}{n^{[2]}} \left(\sum_{I^{[2]}} \xi_{ij}^2 \right) \left(\sum_{I^{[2]}} \eta_{ij}^2 \right)$$

$$+ \frac{1}{n^{[2]}} \left(\sum_{I^{[2]}} \xi_{ij}\xi_{ji} \right) \left(\sum_{I^{[2]}} \eta_{ij}\eta_{ji} \right)$$

$$+ \frac{1}{n^{[3]}} \left(\sum_{G_1} \xi_{ij}\xi_{ik} \right) \left(\sum_{G_1} \eta_{ij}\eta_{ik} \right)$$

$$+ \frac{1}{n^{[3]}} \left(\sum_{G_1'} \xi_{ij}\xi_{hi} \right) \left(\sum_{G_1'} \eta_{ij}\eta_{hi} \right)$$

$$+ \frac{1}{n^{[3]}} \left(\sum_{G_2} \xi_{ij}\xi_{hj} \right) \left(\sum_{G_2} \eta_{ij}\eta_{hj} \right)$$

$$+ \frac{1}{n^{[3]}} \left(\sum_{G_2'} \xi_{ij}\xi_{jk} \right) \left(\sum_{G_2'} \eta_{ij}\eta_{jk} \right)$$

$$+ \frac{1}{n^{[4]}} \left(\sum_{H} \xi_{ij}\xi_{hk} \right) \left(\sum_{H} \eta_{ij}\eta_{hk} \right)$$

$$- \left(\frac{1}{n^{[2]}} \left(\sum_{I^{[2]}} \xi_{ij} \right) \left(\sum_{I^{[2]}} \eta_{ij} \right) \right)^2 \tag{6.2.81}$$

where for $r < n$, $n^{[r]} = n(n-1)(n-2)\ldots(n-r+1)$. This integer can be called the *r factorial* power of n.

Recall that we have

$$card(I) = n(n-1)$$
$$card(G_1) = card(G_1') = card(G_2) = card(G_2') = n(n-1)(n-2)$$
$$card(H) = n(n-1)(n-2)(n-3) \tag{6.2.82}$$

Now, in the case where ξ and η are both symmetrical or antisymmetrical the expression above of $var(S_1)$ becomes

$$var(S_1) = \frac{2}{n^{[2]}} \left(\sum_{I^{[2]}} \xi_{ij}^2 \right) \left(\sum_{I^{[2]}} \eta_{ij}^2 \right)$$

$$+ \frac{4}{n^{[3]}} \left(\sum_{G} \xi_{ij}\xi_{ik} \right) \left(\sum_{G} \eta_{ij}\eta_{ik} \right)$$

$$+ \frac{1}{n^{[4]}} \left(\sum_H \xi_{ij} \xi_{hk} \right) \left(\sum_H \eta_{ij} \eta_{hk} \right)$$

$$- \left(\frac{1}{n^{[2]}} \left(\sum_{I^{[2]}} \xi_{ij} \right) \left(\sum_{I^{[2]}} \eta_{ij} \right) \right)^2 \tag{6.2.83}$$

where G (resp., H) is the set of triples (i, j, k) (resp., quadruples (i, j, h, k)) with mutually distinct components.

The parameters introduced for $var(S_1)$ according to the Mantel expression are:

$$A_1 = \left(\sum_{I^{[2]}} \xi_{ij} \right)^2, A_2 = \sum_i \left(\sum_{j \in I - \{j\}} \xi_{ij} \right)^2, A_3 = \sum_{I^{[2]}} \xi_{ij}^2,$$

$$B_1 = \left(\sum_{I^{[2]}} \eta_{ij} \right)^2, B_2 = \sum_i \left(\sum_{j \in I - \{j\}} \eta_{ij} \right)^2, B_3 = \sum_{I^{[2]}} \eta_{ij}^2 \tag{6.2.84}$$

In the case of comparing two symmetrical (resp.antisymmetrical) valuated binary relations we have

$$var(S_1) = \frac{2}{n^{[2]}} A_3 B_3 + \frac{4}{n^{[3]}} (A_2 - A_3)(B_2 - B_3)$$

$$+ \frac{1}{n^{[4]}} (A_1 - 4A_2 + 2A_3)(B_1 - 4B_2 + 2B_3) - \frac{1}{(n^{[2]})^2} A_1 B_1 \tag{6.2.85}$$

The correspondence between formulas (6.2.83) and (6.2.85) is obtained by identifying in both expressions the multiplying factors of $\frac{2}{n^{[2]}}$, $\frac{4}{n^{[3]}}$, $\frac{1}{n^{[4]}}$ and $\frac{1}{(n^{[2]})^2}$. The expression (6.2.85) is simpler to compute than that (6.2.83). However, the latter—as said above—is formally clearer and this enables $q - ary$ $(q > 2)$ valuated relations on $\mathcal{O}$ to be compared [28].

In any case, the centred and standardized coefficient $Q_1(\xi, \eta)$ takes the following form

$$Q_1(\xi, \eta) = \frac{s(\xi, \eta) - n^{[2]} \mu_\xi \mu_\eta}{\sqrt{var(S_1)}} \tag{6.2.86}$$

Limit form of $Q_1(\xi, \eta)$

The limit form of $Q_1(\xi, \eta)$ has been studied in the symmetrical or exclusively antisymmetrical cases [28]. To fix ideas and without restricting generality, we may assume ξ and η taking their values in the real interval $[-1, +1]$. Then in the symmetrical (resp. antisymmetrical) cases we set, for any $i \in I$, $\xi_{ii} = \eta_{ii} = 1$ (resp., $\xi_{ii} = \eta_{ii} = 0$). To carry out the asymptotic analysis, it is easier to work at the level of the Cartesian product $I \times I$. In these conditions,

$$w(\xi, \eta) = \frac{1}{n^2} \sum_{(i,j) \in I \times I} \xi_{ij} \eta_{ij}$$

is substituted for $s(\xi, \eta)/n^{[2]}$. Now, by introducing the following parameters

$$p_1 = \frac{1}{n^2} \sum_{(i,j) \in I \times I} \xi_{ij}, r_1 = \frac{1}{n^2} \sum_{(i,j) \in I \times I} \eta_{ij}$$

$$p_2 = \frac{1}{n^2} \sum_{(i,j) \in I \times I} \xi_{ij}^2, r_2 = \frac{1}{n^2} \sum_{(i,j) \in I \times I} \eta_{ij}^2$$

$$q = \frac{1}{n^3} \sum_{(i,j,k) \in I \times I \times I} \xi_{ij} \xi_{ik}, s = \frac{1}{n^3} \sum_{(i,j,k) \in I \times I \times I} \eta_{ij} \eta_{ik} \qquad (6.2.87)$$

we get in [28] the following

Theorem 18 *For w as well as p_1, p_2 and q (resp., r_1, r_2 and s) tending almost surely to finite limits when n increases, the limit form of $Q_1(\xi, \eta)$ is*

$$\frac{\sqrt{n}}{2} \times \frac{w - p_1 r_1}{\sqrt{(q - p_1^2)(s - r_1^2) + \frac{1}{2n}\big((p_2 - p_1^2) - 2(q - p_1^2)\big)\big((r_2 - r_1^2) - 2(s - r_1^2)\big)}}$$
$$(6.2.88)$$

The instantiation of this result in the case where ξ and η code two partitions π and χ was given in (6.2.37). As expressed above, the coding of a total preorder ω on the object set $\mathcal{O}$ by the indicator function of its set theoretic representation $R(\omega)$ is neither symmetrical nor antisymmetrical (see Sect. 6.2.4). To exploit the previous general result in the case of comparing two total preorders ω and ϖ on the object set $\mathcal{O}$—induced by two ordinal categorical attributes—antisymmetrical codings for ω and ϖ have to be established. These can be done as follows:

For any $(i, j) \in I^{[2]}$, $\xi_{ij} = 1, 0$ or -1 (resp., $\eta_{ij} = 1, 0$ or -1) if and only if

- the class comprising i precedes strictly the class comprising j,
- the class comprising i is the same as the class comprising j,
- the class comprising i follows strictly the class comprising j,

for ω (resp., for ϖ).

Let us introduce the proportions $\omega_h = n_h/n$, $\varpi_l = m_l/n$ and $\nu_{hl} = n_{hl}/n$, $1 \le h \le K$, $1 \le l \le M$. In these conditions, we have

Proposition 41

$$p_1 = 0, \ p_2 = 2 \sum_{g<h} \omega_g \omega_h, \ q = \left(2 \sum_{f<g<h} \omega_f \omega_g \omega_h + \sum_{g<h} \omega_g \omega_h (\omega_g + \omega_h) \right)$$

$$r_1 = 0, \ r_2 = 2 \sum_{k<l} \varpi_k \varpi_l, \ s = \left(2 \sum_{j<k<l} \varpi_j \varpi_k \varpi_l + \sum_{k<l} \varpi_k \varpi_l (\varpi_k + \varpi_l) \right)$$

$$w = 2 \sum_{g<h} \sum_{k<l} (\nu_{gk} \nu_{hl} - \nu_{gl} \nu_{hk}) \tag{6.2.89}$$

The following result is got in [28]

Theorem 19 *If w and the vectors*

$$(\omega_1, \omega_2, \ldots, \omega_h, \ldots, \omega_K) \ and \ (\varpi_1, \varpi_2, \ldots, \varpi_l, \ldots, \varpi_M)$$

tend almost surely to finite limits when n increases, for the previous coding of ω and ϖ, the limit form of $Q_1(\omega, \varpi)$ is

$$\frac{\sqrt{n}}{2} \times \frac{w}{\sqrt{qs + \frac{1}{2n}(p_2 - 2q)(r_2 - 2s)}} \tag{6.2.90}$$

As for comparing partitions (see (6.2.33) and (6.2.35)) or total preorders (see (6.2.60) and (6.2.62)), the following coefficients, neutralizing the influence of the number n of observations can be proposed

$$R_1(\xi, \eta) = \frac{Q_1(\xi, \eta)}{\sqrt{n}} \tag{6.2.91}$$

$$R_1''(\xi, \eta) = \frac{Q_1(\xi, \eta)}{\sqrt{Q_1(\xi, \xi) Q_1(\eta, \eta)}} \tag{6.2.92}$$

The coefficients $\tilde{Q}_1(\pi, \chi)$ and $\tilde{Q}_1(\omega, \varpi)$ (see (6.2.38) and (6.2.65)) have been obtained with less restrictive hypothesis of no relation than that to obtain $Q_1(\pi, \chi)$ and $Q_1(\omega, \varpi)$ (see (6.2.32) and (6.2.59)). The K. Pearson coefficient built in the Boolean case was transposed in order to derive $\tilde{Q}_1(\pi, \chi)$ and $\tilde{Q}_1(\omega, \varpi)$. Let us now proceed with the same approach to establish an association coefficient, denoted by $\tilde{Q}_1(\xi, \eta)$, between two binary valuations ξ and η on the object set $\mathcal{O}$. In the latter case, the coefficient $Q_1(v, w)$ (see (5.2.68)) has to be transposed adequately. We may denote here by $\mathcal{H}'_1$ the hypothesis of no relation concerned. For this, the following random raw coefficient is associated with the raw coefficient $s(\xi, \eta)$ (see (6.2.75)).

$$s(\xi^\star, \eta^\star) = \sum_{ij \in I^{[2]}} \xi_{\sigma(ij)} \eta_{\tau(ij)} \tag{6.2.93}$$

where σ and τ are two independent random permutations on $I^{[2]}$. As in comparing numerical attributes (see (5.2.65)) the two dual random coefficients

$$s(\xi, \eta^{\star'}) = \sum_{ij \in I^{[2]}} \xi_{ij} \eta_{\tau(ij)} \text{ and } s(\xi^{\star'}, \eta) = \sum_{ij \in I^{[2]}} \xi_{\sigma(ij)} \eta_{ij} \qquad (6.2.94)$$

have exactly the same distribution as that of $s(\xi^{\star'}, \eta^{\star'})$. For the following and particularly to fix idea, we will focus on the random raw coefficient $s(\xi, \eta^{\star'})$. The *mean* and *variance* of the common distribution are

$$\mathcal{E}(s(\xi, \eta^{\star'})) = n^{[2]} \mu(\xi) \mu(\eta) \qquad (6.2.95)$$

$$var(s(\xi, \eta^{\star'})) = \frac{(n^{[2]})^2}{n^{[2]} - 1} var(\xi) var(\eta) \qquad (6.2.96)$$

where $\mu(\xi)$ and $var(\xi)$ (resp., $\mu(\eta)$ and $var(\eta)$) are the mean and variance of ξ (resp., η) over $I^{[2]}$.

Notice that $\mathcal{E}(s(\xi, \eta^{\star'})) = \mathcal{E}(s(\xi, \eta^{\star}))$ (see (6.2.79)). The centred and standardized coefficient can be written as

$$Q_1'(\xi, \eta) = \frac{s(\xi, \eta) - \mathcal{E}(s(\xi, \eta^{\star'}))}{\sqrt{var(s(\xi, \eta^{\star'}))}} = \sqrt{n^{[2]} - 1} R_1(\xi, \eta) \qquad (6.2.97)$$

where $R_1(\xi, \eta)$ is the correlation coefficient between ξ and η defined over $I^{[2]}$.

6.2.5.4 Comparing the Moments of the Distributions of $s(\xi, \eta^{\star})$ and $s(\xi, \eta^{\star'})$

To start with this comparison, notice that for fixed (i, j) in $I^{[2]}$, the distribution of $\eta_{\tau(i)\tau(j)}$ is the same as that of $\eta_{\sigma(ij)}$ where τ (resp., σ) is a random permutation in the set G_n (resp., G_m ($m = n^{[2]}$)) of all permutations on I (resp., $I^{[2]}$) provided by a uniform probability measure (i.e. all the permutations concerned have equal chances).

For the comparison we use

$$x_{ij} = \frac{\xi_{ij}}{\sqrt{var(\xi)}} \text{ and } y_{ij} = \frac{\eta_{ij}}{\sqrt{var(\eta)}} \qquad (6.2.98)$$

where $var(\xi)$ and $var(\eta)$ are the variances of ξ and η on $I^{[2]}$, namely, the variances of

$$\{\xi_{ij} | (i, j) \in I^{[2]}\} \text{ and } \{\eta_{ij} | (i, j) \in I^{[2]}\},$$

respectively, so that

$$\frac{1}{m} \sum_{(i,j) \in I^{[2]}} x_{ij}^2 = \frac{1}{m} \sum_{(i,j) \in I^{[2]}} y_{ij}^2 = 1 \qquad (6.2.99)$$

Comparing the respective behaviours of $s(\xi, \eta^\star)$ and $s(\xi, \eta^{\star'})$ is equivalent to compare the distributions of

$$S = \frac{1}{\sqrt{m}} \sum_{(i,j) \in I^{[2]}} x_{ij} y_{\tau(i)\tau(j)} \text{ and } U = \frac{1}{\sqrt{m}} \sum_{(i,j) \in I^{[2]}} x_{ij} y_{\sigma(ij)} \qquad (6.2.100)$$

The rth moment of U and S can be written, respectively, as follows:

$$m^{-r/2} \sum coef(r; r_1, r_2, \ldots, r_k) \sum_{(I^{[2]})^{[k]}} x_{q1}^{r_1} \ldots x_{qk}^{r_k} \times \frac{1}{m!} \sum_{\tau \in G_m} y_{\tau(q1)}^{r_1} \ldots x_{\tau(qk)}^{r_k}$$

$$(6.2.101)$$

and

$$m^{-r/2} \sum coef(r; r_1, r_2, \ldots, r_k) \sum_{(I^{[2]})^{[k]}} x_{q1}^{r_1} \ldots x_{qk}^{r_k}$$

$$\times \frac{1}{n!} \sum_{\sigma \in G_n} y_{\sigma(i_1), \sigma(j_1)}^{r_1} \ldots y_{\sigma(i_k), \sigma(j_k)}^{r_k} \qquad (6.2.102)$$

where $(q1, q2, \ldots, qk) = ((i_1, j_1), (i_2, j_2), \ldots, (i_k, j_k))$ is a k-uple of mutually distinct elements of $I^{[2]}$.

The nature of these expressions is similar to that (6.2.20). The first sum in both expressions extends over all partitions of r into k parts $r_1, r_2, \ldots$ and r_k: $r = r_1 + r_2 + \cdots + r_k$.

$$coef(r; r_1, r_2, \ldots, r_q) = \frac{r!}{l_1! l_2! \ldots l_h! \times r_1! r_2! \ldots r_q!}$$

was defined above (see (6.2.20) and below). Let us recall that h is the number of distinct r_j values, repeated $l_1, l_2, \ldots$ and l_h times, respectively.

The second summation extends over all the mutually distinct k-uples of elements of $I^{[2]}$. There are $m(m-1) \ldots (m-k+1)$ such k-uples in $(I^{[2]})^{[k]}$. This sum is split up according to a partition of $(I^{[2]})^{[k]}$ defined such that each of its classes is characterized by a configuration (c) which specifies for a given k-uple belonging to the class concerned the positions where an element of I is repeated. By denoting $G_k^{(c)}$ the class of the k-uples having the (c) configuration, if d is the number of distinct subscripts used in the configuration (c), we have

$$n(c) = card(G_k^{(c)}) = \binom{n}{d} \times d! = n(n-1) \ldots (n-d+1) \qquad (6.2.103)$$

When σ describes G_n, for $((i_1, j_1), (i_2, j_2), \ldots, (i_k, j_k))$ fixed

$$((\sigma(i_1), \sigma(j_1)), (\sigma(i_2), \sigma(j_2)), \ldots, (\sigma(i_k), \sigma(j_k)))$$

describes $(n - d)!$ times G_k^c. Hence, the expressions (6.2.101) and (6.2.102) can be put as

$$m^{-r/2} \sum coef(r; r_1, r_2, \ldots, r_k) \sum_{(c)} n(c) \bar{X}_c(r_1, r_2, \ldots, r_k) \bar{Y}'_k(r_1, r_2, \ldots, r_k)$$

$$(6.2.104)$$

and

$$m^{-r/2} \sum coef(r; r_1, r_2, \ldots, r_k) \sum_{(c)} n(c) \bar{X}_c(r_1, r_2, \ldots, r_k) \bar{Y}_c(r_1, r_2, \ldots, r_k)$$

$$(6.2.105)$$

where

$$\bar{X}_c(r_1, r_2, \ldots, r_k) = \frac{1}{card(G_k^{(c)})} \sum_{G_k^{(c)}} x_{q1}^{r_1} \ldots x_{qk}^{r_k},$$

$$\bar{Y}_c(r_1, r_2, \ldots, r_k) = \frac{1}{card(G_k^{(c)})} \sum_{G_k^{(c)}} y_{q1}^{r_1} \ldots y_{qk}^{r_k} \qquad (6.2.106)$$

and

$$\bar{Y}'_k(r_1, r_2, \ldots, r_k) = \frac{1}{card((I^{[2]})^{[k]})} \sum_{(I^{[2]})^{[k]}} y_{q1}^{r_1} \ldots y_{qk}^{r_k} \qquad (6.2.107)$$

where $card((I^{[2]})^{[k]}) = m(m - 1) \ldots (m - k + 1)$

Limit distribution of $s(\xi, \eta^\star)$

Let us consider a sequence $\{\mathcal{O}_n | n \geq 1\}$ of object sets increasing for the inclusion relation: $\mathcal{O}_n \subset \mathcal{O}_{n+1}$. Denote $n = card(\mathcal{O}_n)$ and $I_n = \{1, 2, \ldots, i, \ldots, n\}$ the integer set which codes $\mathcal{O}_n$. Now, suppose a sequence of pairs (x^n, y^n) of valuated binary relations where x^n and y^n are defined on $\mathcal{O}_n$. A more explicit expressions for these, are the following

$$x^n = \{x_{ij}^n | (i, j) \in I_n^2\} \text{ and } y^n = \{y_{ij}^n | (i, j) \in I_n^2\} \qquad (6.2.108)$$

Condition (6.2.99) is supposed satisfied for x^n and y^n, that is

$$\frac{1}{m} \sum_{(i,j) \in I^{[2]}} (x_{ij}^n)^2 = \frac{1}{m} \sum_{(i,j) \in I^{[2]}} (y_{ij}^n)^2 = 1 \qquad (6.2.109)$$

The definitions of the averages

$$\bar{X}_c^n(r_1, r_2, \ldots, r_k), \bar{Y}_c^n(r_1, r_2, \ldots, r_k) \text{ and } \bar{Y'}_k^n(r_1, r_2, \ldots, r_k)$$

are analogous to those

$$\bar{X}_c(r_1, r_2, \ldots, r_k), \bar{Y}_c(r_1, r_2, \ldots, r_k) \text{ and } \bar{Y'}_k(r_1, r_2, \ldots, r_k) \text{ (see (6.2.106) and}$$
(6.2.107)).

Consider, for a valued binary relation z^n on $\mathcal{O}_n$, the following limit condition when n increases and tends to infinity

$$\mathcal{E}\left(z_{\sigma(i_1)\sigma(j_1)}^n + \cdots + z_{\sigma(i_k)\sigma(j_k)}^n\right)^r = O(1) \text{ for all } r \geq 3 \text{ and } k \leq r \qquad (6.2.110)$$

where $((i_1, j_1), \ldots, (i_k, j_k))$ is an element in $(I^{[2]})^{[k]}$. O is the Landau notation ($O(1)$ indicates a finite limit) and $\mathcal{E}$ the permutational mean operator ($\frac{1}{n!} \sum_{\sigma \in G_n}$). The satisfaction of this condition by x^n and y^n ensures the convergence of $\bar{X}_c^n(r_1, r_2, \ldots, r_k)$, $\bar{Y}_c^n(r_1, r_2, \ldots, r_k)$ and $\bar{Y'}_k^n(r_1, r_2, \ldots, r_k)$ towards finite limits.

In the notations below, for simplicity reasons, the upper subscript n will be deleted. $n(c)$ is maximum for $d = 2k$. Denote by G_k^0 the set of k-uples of distinct elements of $I^{[2]}$ without common component, we have

$$card(G_k^0) = n(n-1) \ldots (n-2k+1) \qquad (6.2.111)$$

since

$$card((I^{[2]})^k) = m(m-1) \ldots (m-k+1) \qquad (6.2.112)$$

where $m = n(n-1)$, $card(G_k^0)/card((I^{[2]})^k)$ tends to unity when n tends to infinity.

Consequently, the magnitude order of the right sum of (6.2.104) is that of

$$n(0)\bar{X}_0(r_1, r_2, \ldots, r_k)\bar{Y'}_k(r_1, r_2, \ldots, r_k)$$

Similarly, the magnitude order of the right sum of (6.2.105) is

$$n(0)\bar{X}_0(r_1, r_2, \ldots, r_k)\bar{Y}_0(r_1, r_2, \ldots, r_k)$$

Due to the following equation

$$\bar{Y'}_k(r_1, r_2, \ldots, r_k) = \sum_c \frac{card(G_k^c)}{card((I^{[2]})^k)} \bar{Y}_c(r_1, r_2, \ldots, r_k)$$

the left member tends towards $\bar{Y}_0(r_1, r_2, \ldots, r_k)$.

Finally, by denoting μ_r and ν_r the rth absolute moments of S and U, respectively, we get the following:

Theorem 20 *Under the condition (6.2.110) the ratio μ_r/ν_r tends to 1 when n tends to infinity.*

As a result the distribution of S may be compared to that of U which is asymptotically Normal by virtue of the Wald and Wolfowitz theorem. This gives a general indication concerning the normal tendency of the random raw coefficient S, for most types of comparing attributes in real data with respect to the hypothesis of no relation $\mathcal{H}_1$. However, as mentioned above, there are particular structural situations where this tendency is not verified. One very known case concerns the comparison between partitions with a few number (e.g. 2 classes) of classes sized more or less equally [6, 36]. The theoretical and experimental analyses led in [6] shows that the Normal approximation is as more justified as

- the variability of the size class is greater;
- the number of classes is greater.

Besides, by considering the hypothesis of no relation $\mathcal{H}_3$, defined in [6, 28] and by substituting $\mathcal{H}_3$ for $\mathcal{H}_1$, the asymptotic normality of the distribution of the random raw coefficient between partitions is proven [6].

In the *LLA* hierarchical clustering of a set $\mathcal{A}$ of descriptive attributes of a given type, whatever the latter is (see Chap. 3, Sect. 5.2.53), a probabilistic association coefficient table analogous to (5.2.53) is established, namely,

$$\mathcal{P}^g = \{P^g(a^j, a^k) = \Phi(Q^g(a^j, a^k)) | 1 \leq j < k \leq p\} \qquad (6.2.113)$$

where the same cumulative normal distribution Φ is employed for all of the attribute pairs. The reasons are:

- For the vast majority of attribute pairs, Normal distribution limit holds;
- A systematic treatment is preferable to that for which specific probability laws are used locally—and in fact without real change—for a few attribute pairs;
- In any case, it is always possible to adopt a probabilistic association coefficients defined as in (6.2.113) and interpreted in terms of probabilities with respect to a no specified hypothesis of no relation.

Indeed, the alternative expressed by the third item is no more arbitrary than choosing a particular association coefficient.

In Chap. 3 Sect. 3.4.4, the representation of the different types of descriptive attributes: *Boolean, numerical, nominal categorical, ordinal categorical and ranking*, in terms of *preordonance* attributes, has been proposed. In the thesis work [40], this approach is developed in order to *LLA* cluster a set $\mathcal{A}$ of descriptive attributes. Important real applications were carried out in the work mentioned. In the case of preordonance representation of a categorical attribute, the total preorder on the set of category pairs is coded by the *mean rank function* (see (3.4.1) and (3.4.8)). This coding enables the preordonance representation of a categorical attribute to be interpreted as a particular case of a categorical attribute valuated by a numerical similarity (see Chap. 3, Sect. 3.3.4). In the next paragraph, we will give the calculations needed to obtain the centred an standardized association coefficient between two such attributes, according to the scheme of Fig. 6.1.

6.2.5.5   *LLA* Association Coefficient Between two Categorical Attributes Valued by Numerical Similarities

Let us denote by c and d two categorial attributes valued each by a numerical similarity. $C = \{c_1, c_2, \ldots, c_g, \ldots, c_K\}$ and $\mathcal{D} = \{d_1, d_2, \ldots, d_k, \ldots, d_M\}$ will designate the sets of categories of c and d, respectively.

$$\xi = \{\xi_{gh} | 1 \leq g \leq h \leq K\} \tag{6.2.114}$$

$$\text{and } \eta = \{\eta_{kl} | 1 \leq k \leq l \leq M\} \tag{6.2.115}$$

indicate the similarity valuations on C and $\mathcal{D}$, respectively.

As around (3.3.33) and (3.3.34) and Table 3.1 of Chap. 3, we assume, in particular to fix ideas, that ξ and η are two positive and symmetrical valuations.

Now and as above, denote by $\pi = \{O_g | 1 \leq g \leq K\}$ and $\chi = \{P_k | 1 \leq k \leq M\}$ the partitions of $\mathcal{O}$, induced by c and d, respectively. The cross partition is defined by

$$\pi \wedge \chi = \{O_g \cap P_k | 1 \leq g \leq K, 1 \leq k \leq M\} \tag{6.2.116}$$

and the associated contingency table by

$$Cont(\pi, \chi) = \{n_{gk} = card(O_g \cap P_k) | 1 \leq g \leq K, 1 \leq k \leq M\} \tag{6.2.117}$$

As above, we designate by $(n_1, n_2, \ldots, n_g, \ldots, n_K)$ (resp., $(m_1, m_2, \ldots, m_k, \ldots, m_M)$) the *type* of the partition π (resp., χ): $n_g = card(O_g, 1 \leq g \leq K$ (resp., $m_k = card(P_k, 1 \leq k \leq M)$.

$I = \{1, 2, \ldots, i, \ldots, n\}$ indexing the object set $\mathcal{O}$, let us designate by I_g (resp., J_k) the subset of I indexing O_g (resp., P_k), $1 \leq g \leq K$ (resp., $1 \leq k \leq M$). We can write

$$I = \sum_{1 \leq g \leq K, 1 \leq k \leq M} I_g \cap J_k \tag{6.2.118}$$

where the sum is a set theoretic one (i.e. union of mutually disjoint subsets). To simplify notations and without risk of ambiguity, we may write

$$I = \sum_{(g,k)} I_g \cap J_k \tag{6.2.119}$$

In these conditions, $I^{[2]}$ can be put under the following form

$$I^{[2]} = \left(\sum_{(g,k)} I_g \cap J_k \right) \times \left(\sum_{(g',k')} I_{g'} \cap J_{k'} \right) - \Delta \left(\sum_{(g,k)} I_g \cap J_k \times \sum_{(g',k')} I_{g'} \cap J_{k'} \right) \tag{6.2.120}$$

where Δ indicates the diagonal, that is the ordered pairs of I for which the two components are identical.

Let us extend now the definitions of ξ and η to $I \times I$ as follows:

$$\text{For any } (i, j) \in I \times I$$
$$\xi_{ij} = \xi_{(i,j)} = \xi_{(c(i),c(j))}$$
$$\eta_{ij} = \eta_{(i,j)} = \eta_{(d(i),d(j))} \tag{6.2.121}$$

where $c(i)$ (resp., $d(i)$) is the category of c (resp., d) possessed by i, $i \in I$.

In these conditions, the centred and standardized coefficient is obtained by applying in a specific way the coefficient $Q_1(\xi, \eta)$ (see (6.2.86)) built in the general case. The *raw* coefficient can be written as

$$s(c, d) = s(\xi, \eta) = \sum_{\big((g,l),(h,m)\big)} n_{gl} n_{hm} \xi_{gh} \eta_{lm} - \sum_{(g,l)} n_{gl} \xi_{gg} \eta_{ll} \tag{6.2.122}$$

Mean and Variance of the Random Raw Coefficient

$$\sum_{(i,j) \in I^{[2]}} \xi_{ij} = \sum_{(g,h)} n_g n_h \xi_{gh}$$

$$\sum_{(i,j) \in I^{[2]}} \eta_{ij} = \sum_{(l,m)} m_l m_m \eta_{lm} \tag{6.2.123}$$

(6.2.79) gives, under permutational model

$$\mathcal{E}[s(\xi, \eta^*)] = \frac{1}{n^{[2]}} \left(\sum_{(g,h)} n_g n_h \xi_{gh} - \sum_g n_g \xi_{gg} \right) \times \left(\sum_{(l,m)} m_l m_m \eta_{lm} - \sum_l m_l \eta_{ll} \right) \tag{6.2.124}$$

Taking into account the variance expression (6.2.85), we have to specify A_1, A_2, and A_3 on the one hand, B_1, B_2, and B_3 on the other hand. We get

$$A_1 = \left(\sum_{(g,h)} n_g n_h \xi_{gh} - \sum_g n_g \xi_{gg} \right)^2$$

$$A_2 = \sum_g \left(n_g (n_g - 1) \xi_{gg} + \sum_{h | h \neq g} n_g n_h \xi_{gh} \right)$$

$$A_3 = \sum_{(g,h)} n_g n_h \xi_{gh}^2 - \sum_g n_g \xi_{gg}^2 \tag{6.2.125}$$

let us specify how do we get the expression of A_2

$$A_2 = \sum_g \sum_{i \in I_g} \left(\sum_{j \in I - \{i\}} \xi_{ij} \right)^2 = \sum_g \sum_{i \in I_g} \left(\sum_{j \in I_g - \{i\}} \xi_{ij} + \sum_{\{h|h \neq g\}} \sum_{j \in I_h} \xi_{ij} \right)^2 \quad (6.2.126)$$

In these conditions we obtain

$$A_2 = \sum_g \sum_{i \in I_g} \left((n_g - 1)\xi_{gg} + \sum_{\{h|h \neq g\}} n_h \xi_{hh} \right)$$

$$= \sum_g \left(n_g(n_g - 1)\xi_{gg} + \sum_{\{h|h \neq g\}} n_g n_h \xi_{gh} \right) \quad (6.2.127)$$

The expressions of B_1, B_2 and B_3 are strictly analogous to those A_1, A_2 and A_3. They are obtained by substituting the η_{lm} for the ξ_{gh} ($1 \leq l, m \leq M$, $1 \leq g, h \leq K$) and $(m_1, m_2, \ldots, m_l, \ldots, m_M)$ for $(n_1, n_2, \ldots, n_g, \ldots, m_K)$.

The general centred and standardized coefficient $Q_1(\xi, \eta)$ (6.2.86) can be rewritten here by using the compact expressions above, we have just obtained.

6.2.5.6   The Z Score

An interesting and specific extension of the type of development considered above (see Sects. 5.2.2, 6.2.5 and 6.2.5.5) concerns association between two *aligned* proteic sequences of amino acids. Formally, one representation of such a proteic sequence is a word taking his letters in an alphabet $\mathbb{A}$ comprising 20 letters, representing the 20 amino acids. In fact, phylogeny biological considerations led to introduce in the alphabet an additional letter denoted by '-' which represents a *deletion*. The latter makes possible sequence alignment. Thus the new alphabet $\mathbb{A}'$ to which we refer comprises 21 letters. For the calculations below, we will designate it by

$$\mathcal{Z} = \{Z_i | 1 \leq i \leq 21\} \quad (6.2.128)$$

Given two proteic sequences s and s', as just mentioned, the introduction of the deletion, represented by the symbol '-' is due to biological phylogenetic justification for the alignment of s and s', position by position (site by site) in order to obtain the best matching (see Sect. 11.3.2 of Chap. 11). L will designate below the common length of the alignment.

To each site l of a given sequence s corresponds a valuated categorical attribute by a positive numerical similarity a^l (see Sect. 3.3.4 of Chap. 3). The value set of a^l is $\mathbb{A}'$, coded by $\mathcal{Z}$ (see (6.2.128)). The valuation is determined by a substitution matrix between amino acids (see Sect. 11.3.2 of Chap. 11). This positive numerical matrix is symmetrical with no constant value on its diagonal. We will designate it by

$$\Lambda = \{\lambda(i,j) = \lambda(Z_i, Z_j) | 1 \leq i \leq j \leq 21\} \quad (6.2.129)$$

The specification of λ for comparing amino acid with a deletion or two deletions between them is proposed in (11.4.104) and (11.4.105) of Sect. 11.4.6.3 of Chap. 11.

In Sect. 11.3 of Chap. 11, relative to a set S of aligned proteic sequences (multiple alignment) a *sequence* is represented by an object and a *site*, by a valuated categorical *attribute*. Here we consider the *dual* point of view. A *site* is represented by an *object* and a *sequence*, by a valuated categorical *attribute* whose value set is $\mathbb{A}'$, represented by $\mathcal{Z}$ (see (6.2.128)).

Let us designate by

$$\overline{L} = (1, 2, \ldots, l, \ldots, L) \tag{6.2.130}$$

the site sequence.

In the latter representation comparing two sequences s and s' comes down to compare two valuated categorical attributes whose respective sequences of values on the $\overline{L}$ sequence of sites are:

$$\text{For } s, (A_1, A_2, \ldots, A_l, \ldots, A_L)$$
$$\text{and for } s', (A'_1, A'_2, \ldots, A'_l, \ldots, A'_L) \tag{6.2.131}$$

where A_l (resp., A'_l) is the element of $\mathcal{Z}$ (see (6.2.128)) defining the lth letter of the word s (resp., s').

To carry out this comparison, we may adopt the formalization in Sect. 5 of [32], that is, to consider independently s and s' as two valuated categorical attributes to be associated statistically (see Equations (114) and (115) of the cited reference). In the latter conditions, we can use the coefficient developed above in Sect. 6.2.5.5.

However, here, s and s' are represented by two letter vectors in the same space $\mathbb{A}'^L$ (also designated for the calculations below by $\mathcal{Z}^L$) (see (6.2.131)) where the lth component A_l of s is comparable to the same lth component A'_l of s', this comparison being through the substitution matrix adopted. Thereby, the association value is

$$\lambda(A_l, A'_l) = \lambda(i, j) \text{ if } (A_l, A'_l) = (Z_i, Z_j) \tag{6.2.132}$$

$1 \leq l \leq L$, according to (6.2.129).

In these conditions, the *raw* association coefficient between s and s', denoted by $q(s, s')$ can be put in the following linear form

$$q(s, s') = \sum_{1 \leq l \leq L} \lambda(A_l, A'_l) \tag{6.2.133}$$

This linear form is that considered for comparing two numerical attributes (5.2.63) of Sect. 5.2.2 of Chap. 5; but, in the latter case, the association between the two values of the ith component x_i and y_i, has a multiplicative form $(x_i.y_i)$ with respect to the reals. In other words, $q(s, s')$ cannot be reduced to an inner product in the geometrical space $\mathbb{R}^L$.

According to previous cases, it suffices to consider a unilateral version of the permutational random model of the hypothesis of no relation for the definition of a random coefficient associated with $q(s, s')$. Therefore, the latter can be written as follows:

$$q(s, \chi(s)') = \sum_{1 \leq l \leq L} \lambda(A_l, A'_{\chi(l)}) \tag{6.2.134}$$

where χ is a random permutation in the set—provided with a uniform distribution— of the $L!$ permutations on the sequence $\overline{L}$ (see (6.2.130)).

This type of random variable was studied earlier, *completely independently*, in the context of asymptotic Mathematical Statistics by M. Motoo [38]. The form considered was

$$V = \sum_{1 \leq i \leq n} \gamma_{i\sigma(i)} \tag{6.2.135}$$

where σ is a random permutation, uniformly distributed, on $\mathbb{I} = \{1, 2, \ldots, i, \ldots, n\}$.

Now, consider a proteic sequence u denoted by

$$u = (B_1, B_2, \ldots, B_l, \ldots, B_L) \tag{6.2.136}$$

where $B_l \in \mathcal{Z}$, for all l, $1 \leq l \leq L$.

Let us introduce the "composition" of u. It is defined by

$$c(u) = (L_1(u), L_2(u), \ldots, L_i(u), \ldots, L_{21}(u)) \tag{6.2.137}$$

where $L_i(u)$ represents the number of letters B_l which represent the Z_i residue, $1 \leq i \leq 21$.

$c(u)$ represents the distribution in terms of absolute frequencies of $\mathbb{A}'$ on u.

Equally, we introduce the distribution of $\mathbb{A}'$ on u, in terms of relative frequencies:

$$f(u) = (p_1(u), p_2(u), \ldots, p_i(u), \ldots, p_{21}(u)) \tag{6.2.138}$$

where $p_i(u) = L_i(u)/L$ is defined by the proportion of times where Z_i occurs in u, $1 \leq i \leq 21$.

Now, let us designate by $\overline{L}_i(s)$ (resp., $\overline{L}_i(s')$) the $\overline{L}$ subsequence for which $A_l = Z_i$ (resp., $A'_l = Z_i$), $1 \leq i \leq 21$. We have $card(\overline{L}_i(s)) = L_i(s)$ (resp., $card(\overline{L}_i(s')) = L_i(s'))$, $1 \leq i \leq 21$.

In these conditions, the expression of $q(s, \chi(s'))$ (see (6.2.134)) can be written as follows:

$$q(s, \chi(s')) = \sum_{1 \leq i \leq 21} \sum_{l \in \overline{L}_i(s)} \lambda(Z_i, A'_{\chi(l)}) \tag{6.2.139}$$

According to the permutational random model, the probability for a given specific component Z_j of $\overline{L}_j(s')$ to face Z_i of $\overline{L}_i(s)$ through $A'_{\chi(l)}$ is

$$\frac{(L-1)!}{L!} = \frac{1}{L}$$

There are $L_j(s')$ such components Z_j, $1 \leq j \leq 21$. Consequently,

$$\mathcal{E}\left(\lambda(Z_i, A'_{\chi(l)})\right) = \frac{L_j(s')}{L}\lambda(i,j) = p_j(s') \qquad (6.2.140)$$

where, as usual, $\mathcal{E}$ denotes the mathematical expectation.

Therefore, the mean of $q(s, \chi(s'))$ can be written as follows:

$$M1 = \mathcal{E}\left(q(s, \chi(s'))\right) = L \sum_{1 \leq i \leq 21} \sum_{1 \leq j \leq 21} p_i(s) \cdot p_j(s')\lambda(i,j) \qquad (6.2.141)$$

Now, to get the variance of the random coefficient, we begin by calculating the absolute moment of order 2. We have:

$$M2 = \mathcal{E}\left((q(s, \chi(s')))^2\right) = \mathcal{E}\left[\left(\sum_{1 \leq l \leq L} \lambda(A_l, A'_{\chi(l)})\right)^2\right] \qquad (6.2.142)$$

The development of the right member gives

$$\mathcal{E}\left[\sum_{1 \leq l \leq L} \left(\lambda(A_l, A'_{\chi(l)})\right)^2\right]$$

$$+\mathcal{E}\left[\sum_{(l,l') \in \overline{L}^{[2]}} \lambda(A_l, A'_{\chi(l)}) \cdot \lambda(A_{l'}, A'_{\chi(l')})\right] \qquad (6.2.143)$$

where

$$\overline{L}^{[2]} = \{(l, l') | 1 \leq l \neq l' \leq L\}$$

Obviously, we have

$$card(\overline{L}^{[2]}) = L \times (L-1)$$

Taking into account the previous calculation, we have for the first term of (6.2.143)

$$\mathcal{E}\left[\sum_{1 \leq l \leq L} \left(\lambda(A_l, A'_{\chi(l)})\right)^2\right] = L \sum_{1 \leq i \leq 21} \sum_{1 \leq j \leq 21} p_i(s) \cdot p_j(s')(\lambda(i,j))^2 \qquad (6.2.144)$$

Now, it remains to evaluate the second term of (6.2.143). For this purpose, decompose $\overline{L}^{[2]}$ as follows:

$$\overline{L}^{[2]} = \sum_{1 \leq i \leq 21} (\overline{L}_i(s))^{[2]} + \sum_{(i,i') \in \underline{21}^{[2]}} \overline{L}_i(s) \times \overline{L}_{i'}(s) \qquad (6.2.145)$$

where

$$\underline{21}^{[2]} = \{(i, i') | 1 \leq i \neq i' \leq 21\}$$

The second term of (6.2.143) can be expressed as follows:

$$\mathcal{E}\left[\sum_{1 \leq i \leq 21} \sum_{(l,l') \in \overline{L}^{[2]}(s)} \alpha(l, l') + \sum_{(i,i') \in \underline{21}^{[2]}} \sum_{(l,l') \in \overline{L}_i(s) \times \overline{L}_{i'}(s)} \alpha(l, l') \right] \qquad (6.2.146)$$

where

$$\alpha(l, l') = \lambda(A_l, A'_{\chi(l)}) \cdot \lambda(A_{l'}, A'_{\chi(l')})$$

The linearity of the mathematical expectation makes that we have to specify $\mathcal{E}(\alpha(l, l'))$ for each of both sums under the right brackets of (6.2.146). For the first and second sums the respective forms of $\alpha(l, l')$ are

$$\lambda(Z_i, A'_{\chi(l)}) \cdot \lambda(Z_i, A'_{\chi(l')})$$
$$\lambda(Z_i, A'_{\chi(l)}) \cdot \lambda(Z_{i'}, A'_{\chi(l')}) \qquad (6.2.147)$$

where $i \neq i'$.

The probability for a fixed positions of $\chi(l)$ and $\chi(l')$ in $\overline{L}_j(s')^{[2]}$ or in the Cartesian product $\overline{L}_j(s') \times \overline{L}_{j'}(s')$, for $j \neq j'$ is

$$\frac{(L-2)!}{L!} = \frac{1}{L(L-1)}$$

Consequently, the mathematical expectation $\mathcal{E}(\alpha(l, l'))$ for the first form of $\alpha(l, l')$ given in (6.2.147) is

$$\frac{1}{L(L-1)} \left[\sum_{1 \leq j \leq 21} \sum_{(m,m') \in \overline{L}_j^{[2]}(s')} (\lambda(i,j))^2 \right.$$

$$\left. + \sum_{(j,j') \in \underline{21}^{[2]}} \sum_{(m,m') \in \overline{L}_j(s') \times \overline{L}_{j'}(s')} \lambda(i,j)\lambda(i,j') \right]$$

$$(6.2.148)$$

The structure of the calculation of for the second form of $\alpha(l, l')$ given in (6.2.147) is analogous to the preceding one. We have

$$\frac{1}{L(L-1)} \left(\sum_{1 \le j \le 21} \sum_{(m,m') \in \overline{L}_j^{[2]}(s')} \lambda(i,j) \cdot \lambda(i',j) \right.$$

$$\left. + \sum_{(j,j') \in \underline{21}^{[2]}} \sum_{(m,m') \in \overline{L}_j(s') \times \overline{L}_{j'}(s')} \lambda(i,j) \lambda(i',j') \right)$$

(6.2.149)

Finally, the absolute moment of order 2 can be written as

$$M2 = L \sum_{(i,j) \in \underline{21}^2} p_i(s) \cdot p_j(s')(\lambda(i,j))^2$$

$$+L^{[2]} \left[\left(\sum_{(i,j) \in \underline{21}^2} (p_i(s))^{[2]} \cdot (p_j(s'))^{[2]}(\lambda(i,j))^2 \right. \right.$$

$$+ \sum_{i \in \underline{21}} \sum_{(j,j') \in \underline{21}^{[2]}} (p_i(s))^{[2]} \cdot p_j(s') \cdot p_{j'}(s')\lambda(i,j) \cdot \lambda(i,j') \bigg)$$

$$+ \left(\sum_{(i,i') \in \underline{21}^{[2]}} \sum_{j \in \underline{21}} p_i(s) \cdot p_{i'}(s) \cdot (p_j(s))^{[2]}\lambda(i,j) \cdot \lambda(i',j) \right.$$

$$\left. \left. + \sum_{(i,i') \in \underline{21}^{[2]}} \sum_{(j,j') \in \underline{21}^{[2]}} p_i(s) \cdot p_{i'}(s) \cdot p_j(s') \cdot p_{j'}(s')\lambda(i,j) \cdot \lambda(i',j') \right) \right]$$

(6.2.150)

With these results, an exact mathematical formula is obtained for the well-known Z_{score} coefficient between two aligned proteic sequences s and s'. This coefficient is a statistical standardized version of the raw coefficient $q(s, s')$ (see (6.2.133)). It can be written as follows:

$$Z_{score}(s, s') = \frac{q(s, s') - M1}{\sqrt{M2 - M1^2}}$$

(6.2.151)

where, clearly, the denominator above is the standard deviation of $q(s, \chi(s'))$ (see (6.2.139)) and where $M1$ and $M2$ have been defined in (6.2.141) and (6.2.142), respectively.

In the bioinformatics literature (see for example [16]), the Statistics $M1$ and $M2$ are calculated approximatively, but with sufficient accuracy, by using Monte Carlo simulation procedure. For theoretical reasons and in practice, there is a clear advantage to substitute for simulations results the mathematical formulas proposed above. The calculations which enabled us to obtain these formulas were carried out several years ago. It is the first time where we publish them. They have to be compared analytically with those proposed in [3].

6.2.5.7   *LLA* Association Coefficient Between Two Taxonomic Attributes

In Sect. 3.4.1, the preordonance categorical attribute was defined. Coding this attribute in terms of a *mean rank* function on the category pairs $\mathcal{K}_2 = \{(g, h)|1 \leq g \leq h \leq K\}$, makes this attribute type as a specific one of a categorical attribute valuated by numerical similarity (see (3.4.1), (3.4.3) and (3.4.8)). Therefore, the development of the section above applies directly to this specific case. Nevertheless, this specificity has to be highlighted. For this purpose, some aspects of [30] are taken up again here. Recall that a taxonomic categorical attribute can be interpreted as a specific case of a preordonance categorical attribute (see (3.4.14) or (3.4.15)). In [29] we study the comparison between two taxonomic attributes with such a representation at the level of the set $P = P_2(\mathcal{O})$ of object pairs. In the work mentioned, these attributes are considered directly defined on the object set and not on categories predefined of the object set. A work to be done consists of specifying the expressions obtained in [29] when the taxonomic attributes are categorial ones, that is, are directly defined on an exhaustive set of categories. The nature of the passage to be built is the same as that considered for Sect. 6.2.5.5 from Sects. 6.2.5.2 and 6.2.5.3. Specifying this passage is left for the reader.

The categorical taxonomic attribute is a particular case of a categorical preordonance attribute. It defines a *ultrametric* preordonance (see (3.4.14) and (3.4.15)). A taxonomic attribute determines a *classification tree* on the object set $\mathcal{O}$. Mathematicaly, the defined structure is an ordered chain of partitions (see Chap. 1). In [29] we show how the ultrametric preordonance on $\mathcal{O}$ is built, step-by-step, from the sequence of partitions mentioned.

To be more explicit, let θ and ϑ be two taxonomic attributes defined on $\mathcal{O}$ and denote by λ_θ and λ_ϑ the two respective *mean rank* functions defined on P and coding the respective total preordonances associated with θ and ϑ. For P written as $P = \{(i, j)|1 \leq i < j \leq n\}$ and where $I = \{1, 2, \ldots, i, \ldots, n\}$ codes $\mathcal{O}$, the *raw* coefficient is

$$s(\theta, \vartheta) = \sum_P \lambda_\theta(i, j)\lambda_\vartheta(i, j) \tag{6.2.152}$$

The random *raw* coefficient can be expressed by

$$s(\theta, \vartheta^*) = \sum_P \lambda_\theta(i, j)\lambda_\vartheta(\tau(i), \tau(j)) \tag{6.2.153}$$

where τ is a random permutation in the set G_n of all permutations on I, endowed with a uniform probability measure. In these conditions and as usually, the centred and reduced coefficient is given by

$$Q_1(\theta, \vartheta) = \frac{s(\theta, \vartheta) - \mathcal{E}(s(\theta, \vartheta^*))}{\sqrt{var(s(\theta, \vartheta^*))}} \tag{6.2.154}$$

Several coefficients proposed in the literature—in particular that of Fowlks and Mallows (1983) [8]—are discussed in [29] and situated with respect to the *LLA* approach.

The representation of a preordonance categorical attribute at the level of the Cartesian product $P \times P$ is given in Sect. 3.4.1 (see (3.4.4) and (3.4.7)). This representation is specified and illustrated in [29] for a taxonomic attribute, directly defined on the object set $\mathcal{O}$. It is applied in [33] in the framework of the *LLA* approach. As previously, let us denote by θ a taxonomic attribute on $\mathcal{O}$. As just expressed θ defines a classification tree on $\mathcal{O}$. Let us denote by l_θ its level function: $l_\theta(i, j)$ is the lowest level where i and j are joined, for (i, j) belonging to P, written as $P = \{(i, j)|1 \le i < j \le n\}$ and where $I = \{1, 2, \ldots, i, \ldots, n\}$ codes $\mathcal{O}$. The set theoretic representation of θ at the level of $P \times P$ is

$$\mathcal{R}(\theta) = \{((i, j), (i', j'))|((i, j), (i', j')) \in P \times P \text{ and } l_\theta(i, j) < l_\theta(i', j')\} \tag{6.2.155}$$

Without ambiguity, we may also denote by θ the indicator function of $\mathcal{R}(\theta)$. Thus θ is represented as follows:

$$\theta((i, j), (i', j')) = 1 \text{ if } l_\theta(i, j) < l_\theta(i', j')$$
$$0 \text{ if not,} \tag{6.2.156}$$

for every $((i, j), (i', j')) \in P \times P$.

The *Raw* and the *Random Raw* Coefficient

Let θ and ϑ be two classification trees associated with two taxonomic attributes on $\mathcal{O}$. The *raw* association coefficient between θ and ϑ can be written as

$$\tilde{s}(\theta, \vartheta) = \sum_{P \times P} \theta((i, j), (i', j'))\vartheta((i, j), (i', j')) \tag{6.2.157}$$

Now, according to the general approach we consider two independent random trees θ^* and ϑ^* associated with θ and ϑ, respectively, under the permutational model. As previously and according to the duality property, $\tilde{s}(\theta, \vartheta^*)$, $\tilde{s}(\theta^*, \vartheta)$ and $\tilde{s}(\theta^*, \vartheta^*)$ are equivalent versions of the same random raw coefficient. $\tilde{s}(\theta, \vartheta^*)$ can be written as

$$\tilde{s}(\theta, \vartheta^*) = \sum_{\left((i,j),(i',j')\right) \in P \times P} \theta\left((i,j),(i',j')\right) \vartheta\left((\tau(i), \tau(j)), (\tau(i'), \tau(j'))\right) \quad (6.2.158)$$

where τ is a random permutation in the set G_n of all permutations on I, with equal chances.

To get the centred and standardized coefficient

$$\widetilde{Q}_1 = \frac{\tilde{s}(\theta, \vartheta) - \mathcal{E}(\tilde{s}(\theta, \vartheta^*))}{\sqrt{var(\tilde{s}(\theta, \vartheta^*))}} \quad (6.2.159)$$

the mathematical expectations $\mathcal{E}\left(\tilde{s}(\theta, \vartheta^*)\right)$ and $\mathcal{E}\left(\tilde{s}(\theta, \vartheta^*)\right)^2$ must be calculated exactly. The computing of the mathematical expectation requires the decomposition of $P \times P$ into classes characterized by the same configuration for each of their elements $\left((i,j),(i',j')\right)$, this configuration being specified by the components where subscripts are repeated. Similarly, the computing of the variance requires the same type of decomposition of $(P \times P) \times (P \times P)$. This calculation is carried out recursively [33] by using the tree *profiles* of θ and ϑ. More precisely, we introduce in [29] a notion of *indexed type of a classification tree*. This concept captures entirely the tree shape. It corresponds to the sequence of partitions types, associated in a specific way with the decreasing sequence of the level tree. The random trees θ^* and ϑ^* have the same indexed types as those of θ and ϑ, respectively.

Finally, let us indicate that simulations of the distributions of $Q_1(\theta, \vartheta^*)$ and $\widetilde{Q}_1(\theta, \vartheta^*)$ are provided in [33].

6.2.6 From the Total Association to the Partial One

6.2.6.1 Introduction

In Sect. 5.2.2.4, extension to partial association coefficients between descriptive attributes of type I (Boolean and numerical) was carried out. Now, we shall be interested in this extension for attributes of type II. More precisely, the categorical nominal, the categorical ordinal and the ranking attributes will be considered. The development of the following presentation is provided in a series of two articles [22, 23].

In his work Somers [42] studies mainly the ordinal categorical case. He situates the conception of a partial association coefficient in the framework of a general model based on the product moment. This approach leads—for total association—to the well-known and classical coefficients as those of Pearson, Spearman and Kendall. The normalization technique employed by Daniels [5] refers also to the Product–Moment model. In this approach, after representing descriptive attributes at the level of I or $I^{[2]}$, where $I = \{1, 2, \ldots, i, \ldots, n\}$ codes the object set $\mathcal{O}$ (see (6.2.69)), a geometrical representation is adopted. With respect to the latter, the inner product and the least squares method play an important role. The coefficients (5.2.89) and (5.2.90) may derive from this approach.

In [14] the permutational approach is emphasized using the tool provided by the Mantel Statistic [34] (see in Sect. 6.2.5 (6.2.84)–(6.2.86)). Let ζ, ξ and η be three binary valuated attributes on $\mathcal{O}$. Suppose we are interested in the role of ζ for the association between ξ and η. What is proposed in the latter work is to evaluate the extent to which a *large* (resp., *small*) value of ζ entails a *large* (resp., *small*) value of the product $\theta = \xi \times \eta$. Indeed, the *raw* association coefficient and the random associated one are

$$s(\zeta, \theta) = \sum_{(i,j) \in I^{[2]}} \zeta_{ij} \theta_{ij} \text{ and } s(\zeta, \theta^\star) = \sum_{(i,j) \in I^{[2]}} \zeta_{ij} \theta_{\tau(i)\tau(j)}$$

where for all $(i, j) \in I^{[2]}$, $\theta(ij) = \xi(ij) \times \eta(ij)$ and where as usually, τ is a random permutation in the set G_n of all permutations on I, uniformly distributed.

To begin, let us notice that a *large* (resp., *small*) value of θ_{ij} does not necessarily imply simultaneously a *large* (resp., *small*) values of both ξ_{ij} and η_{ij}. More importantly, the proposed notion of partial association does not correspond to that where the influence of the attribute ζ has to be neutralized in the relationship between ξ and η. In fact, as usually (see Eqs. (6.2.75)–(6.2.77)), the raw coefficient and its random version are

$$s(\xi, \eta) = \sum_{(i,j) \in I^{[2]}} \xi_{ij} \eta_{ij} \text{ and } s(\xi, \eta^\star) = \sum_{(i,j) \in I^{[2]}} \xi_{ij} \eta_{\tau(i)\tau(j)}$$

and $\eta^\star$ has *to be held constant* with respect to ζ. The *only* attribute type, defined at the level of the object set $\mathcal{O}$, which can be neutralized in the permutational approach is the nominal categorical attribute and this, without any valuation on its category set. Besides, an attribute defined by a subset of the distinct ordered object pairs $\mathcal{O}^{[2]}$ can also be neutralized in the association between ξ and η.

This consideration is clearly set up in the Davis paper [7] where a partial association coefficient for the Goodman and Kruskal γ is proposed. Nevertheless, if α designates the partition induced on the object set $\mathcal{O}$ by a nominal categorical attribute a to be neutralized for the association between two ordinal categorical attributes c and d, this neutralization is proposed in a global manner with respect to the set $R(\alpha)$ of unordered object pairs joined by α (see (3.3.5)). This technique corresponds to a second alternative in our approach. The first and more accurate one consists of neutralizing the respective influences of the different classes of α and this by preserving—for each of the partition classes of α—its relative positions with respect to c and d. To do so, we define in a specific fashion the hypothesis of no relation for comparing c and d. In the following, three cases will be considered:

1. c and d are nominal categorical attributes;
2. c and d are ordinal categorical attributes;
3. c and d are ranking attributes.

The latter case will appear as a very specific one.

6.2.6.2 The Nominal Categorical Case

Partial Association Coefficient According to the Chi-Square Statistic

To begin, let us consider a partial version of the most classical association coefficient between two nominal categorical attributes: the Chi-square Statistic (see (6.2.43)). c and d will designate two categorical attributes to be compared independently of the effect of a third attribute denoted by a. α, π and χ designate the partitions induced on $\mathcal{O}$ by a, c and d. More explicitly,

$$\alpha = \{A_1, A_2, \ldots, A_j, \ldots, A_J\}$$
$$\pi = \{O_1, O_2, \ldots, O_h, \ldots, O_K\}$$
$$\chi = \{P_1, P_2, \ldots, P_l, \ldots, P_M\} \tag{6.2.160}$$

The class cardinals and their respective frequencies will be denoted as follows:

$$n_\alpha(j) = card(A_j), \, n_\pi(h) = card(O_h) \text{ and } n_\chi(l) = card(P_l)$$
$$p_\alpha(j) = n_\alpha(j)/n, \, p_\pi(h) = n_\pi(h)/n \text{ and } p_\chi(l) = n_\chi(l)/n \tag{6.2.161}$$

where $1 \leq j \leq J$, $1 \leq h \leq K$ and $1 \leq l \leq M$.

The relative position of two partitions on an object set is determined by their contingency table (see Table 4.8). This is defined by the class cardinals of the corresponding joint partition. In these conditions, we must consider the following contingency tables associated with the joint partitions $\pi \wedge \chi$, $\alpha \wedge \pi$ and $\alpha \wedge \chi$, namely

$$n(\pi \wedge \chi) = \{n(h \wedge l) | 1 \leq h \leq K, 1 \leq l \leq M\}$$
$$n(\alpha \wedge \pi) = \{n(j \wedge h) | 1 \leq j \leq J, 1 \leq h \leq K\}$$
$$n(\alpha \wedge \chi) = \{n(j \wedge l) | 1 \leq j \leq J, 1 \leq l \leq M\} \tag{6.2.162}$$

where $n(h \wedge l) = card(O_h \cap P_l)$, $n(j \wedge h) = card(A_j \cap O_h)$ and $n(j \wedge l) = card(A_j \cap P_l)$, $1 \leq j \leq J$, $1 \leq h \leq K$ and $1 \leq l \leq M$.

We will also have to state with respect to the tables of proportions associated with the previous contingency tables (6.2.162), respectively:

$$p(\pi \wedge \chi) = \{p(h \wedge l)|1 \leq h \leq K, 1 \leq l \leq M\}$$
$$p(\alpha \wedge \pi) = \{p(j \wedge h)|1 \leq j \leq J, 1 \leq h \leq K\}$$
$$p(\alpha \wedge \chi) = \{p(j \wedge l)|1 \leq j \leq J, 1 \leq l \leq M\} \qquad (6.2.163)$$

where $p(h \wedge l) = n(h \wedge l)/n$, $p(j \wedge h) = n(j \wedge h)/n$ and $p(j \wedge l) = n(j \wedge l)/n$, $1 \leq j \leq J$, $1 \leq h \leq K$ and $1 \leq l \leq M$.

$\chi^2_{cd.a}$ will designate the partial association coefficient sought in accordance with the Chi-square statistic. This will be worked out by means of a consistent formal and statistical analogy with the case of total association. The main point concerns the manner by which the hypothesis of no relation is built. More precisely, the pair of random independent partitions $(\pi^\star, \chi^\star)$ associated with the pair of partitions observed (π, χ) must be such that the relative positions of $\pi^\star$ and $\chi^\star$ with respect to α are preserved and then, are the same as those of π and χ with respect to α. More explicitly, write

$$O_h = \sum_{1 \leq j \leq J} O_h \cap A_j \text{ and } P_l = \sum_{1 \leq j \leq J} P_l \cap A_j \qquad (6.2.164)$$

where $\sum$ is a set sum (i.e. union of disjoint sets). Denote by O_{hj} (resp., P_{lj}) the subset $O_h \cap A_j$ (resp., $P_l \cap A_j$), $1 \leq h \leq K$, $1 \leq l \leq M$ and $1 \leq j \leq J$.

Recall that in the case of total comparison between c and d, the hypothesis of no relation can be interpreted as associating independently with the content $n(h \wedge l)$ of every cell of the contingency table, a random one $n^\star(h \wedge l)$ ($1 \leq h \leq K$ and $1 \leq l \leq M$), according to the Poisson random model $\mathcal{H}_3$ (see remark following (5.2.33)). Here, with $(O_1, O_2, \ldots, O_h, \ldots, O_K)$ is associated a sequence $(O_1^\star, O_2^\star, \ldots, O_h^\star, \ldots, O_K^\star)$ of independent random subsets where $O_h^\star$ is defined from a sequence of random subsets $(O_{h1}^\star, O_{h2}^\star, \ldots, O_{hj}^\star, \ldots, O_{KJ}^\star)$ where $O_{hj}^\star$ is associated with O_{hj} in A_j, according to the Poisson random model $\mathcal{H}_3$, with the parameters $n(j)$ and $p(h \wedge j)/p(j)$, $1 \leq j \leq J$.

Independently, a sequence of random subsets $(P_1^\star, P_2^\star, \ldots, P_l^\star, \ldots, P_M^\star)$ is associated with $(P_1, P_2, \ldots, P_l, \ldots, P_M)$, in an analogous manner. Let us specify that the parameters concerning the random choice of $P_{lj}^\star$ are $n(j)$ and $p(l \wedge j)/p(j)$, $1 \leq l \leq M$ and $1 \leq j \leq J$. In these conditions, the random variable $card(O_h^\star \cap P_l^\star)$ can be put as

$$card(O_h^\star \cap P_l^\star) = card\left(\left(\sum_j O_{hj}^\star\right) \cap \left(\sum_j P_{lj}^\star\right)\right)$$
$$= \sum_{1 \leq j \leq J} card(O_{hj}^\star \cap P_{lj}^\star) \qquad (6.2.165)$$

since the random choice of $O_{hj}^\star$ (resp., $P_{lj}^\star$) is made with respect to A_j and the different A_j are disjoint mutually. Each of $card(O_{hj}^\star \cap P_{lj}^\star)$, $1 \leq j \leq J$, is a Poisson random variable, whose parameter being

$$\frac{n(j)p(h \wedge j)p(l \wedge j)}{p(j)^2} = \frac{np(h \wedge j)p(l \wedge j)}{p(j)} \qquad (6.2.166)$$

Hence, the expression (6.2.165) appears as a sum of J independent Poisson random variables. Then, it is a Poisson random variable whose parameter is the sum of the different parameters specified in (6.2.166). In these conditions, the centred and reduced coefficient associated with the content $n(h \wedge l)$ of the cell (h, l) of the contingency table crossing c and d, where the attribute a is neutralized, becomes

$$\frac{n(h \wedge l) - \sum_j \left(np(h \wedge j)p(l \wedge j)/p(j) \right)}{\sqrt{\sum_j \left(np(h \wedge j)p(l \wedge j)/p(j) \right)}} \qquad (6.2.167)$$

Consequently, the sum of squares of (6.2.167) for $1 \le h \le K, 1 \le l \le M$, can be considered, under the hypothesis of no relation expressed above, as an observed value of a Chi-square random variable with $(K - 1)(M - 1)$ degrees of freedom. Therefore, the Chi-square partial association coefficient between the categorical attributes c and d, neutralizing the effect of the categorical attribute a is

$$\chi^2_{cd.a} = \sum_{h,l} \frac{\left(n(h \wedge l) - \sum_j \left(np(h \wedge j)p(l \wedge j)/p(j) \right) \right)^2}{\sum_j \left(np(h \wedge j)p(l \wedge j)/p(j) \right)} \qquad (6.2.168)$$

The conditional independence between c and d with respect to a can be expressed by

$$\text{For any } j, h \text{ and } l, p(h \wedge l | j) = p(h | j)p(l | j) \qquad (6.2.169)$$

This equation can be written as

$$\text{For any } j, h \text{ and } l, p(j \wedge h \wedge l) - \frac{p(j \wedge h)p(j \wedge l)}{p(j)} = 0 \qquad (6.2.170)$$

Clearly, these relations imply the nullity of $\chi^2_{cd.a}$.

Reciprocally, the nullity of $\chi^2_{cd.a}$ implies that for every (h, l), the average of $\big(p(j \wedge h \wedge l) - (p(j \wedge h)p(j \wedge l)/p(j)) \big)$ weighted by the proportion $\{p(j) | 1 \le j \le J\}$ is null.

To end, notice that normalized coefficients can be derived from $\chi^2_{cd.a}$ in an analogous manner as that for total association (see Eqs. (6.2.44)–(6.2.46)).

Partial Association Coefficient According to the *LLA* Method

The Local Approach

The *raw* coefficient denoted by $s(c, d)$ is the same as that defined in the total case (see the first (4.3.3) and (6.2.18)). Let us take up again the (6.2.18)

$$s(c, d) = card\big(R(\pi) \cap R(\chi)\big) = \frac{1}{2} \sum_{1 \le h \le K} \sum_{1 \le l \le M} n(h \wedge l)\big(n(h \wedge l) - 1\big) \quad (6.2.171)$$

where, as above, $R(\pi)$ (resp., $R(\chi)$) is the set of unordered distinct object pairs joined by the partition π (resp., χ). $n(h \wedge l)$ has just been defined above (see (6.2.162)).

As in the total case, we fix one of both partitions π or χ and we associate with the not fixed partition a random one in the set of all partitions having its type and uniformly distributed. As in the total case, it is shown [22] that whatever is the fixed partition (π or χ), the distribution of the random raw coefficient, defined with respect to the partial version of the hypothesis of no relation, is the same. Partitions with labelled classes are considered for this study.

Fix π and associate with χ a random partition $\chi^* = \{P_1^*, P_2^*, \ldots, P_l^*, \ldots, P_M^*\}$, where P_l^* is the random subset associated with P_l under a hypothesis of no relation preserving its relative position with respect to the α partition classes (see (6.2.160)). Consequently, P_l^* will be defined as the following set sum

$$P_l^* = \sum_{1 \le j \le J} P_{lj}^* \quad (6.2.172)$$

$1 \le l \le M$, where $(P_{l1}^*, P_{l2}^*, \ldots, P_{lj}^*, \ldots, P_{lJ}^*)$ is a sequence of independent random subsets, where P_{lj}^* is taken at random and with equal chances in the $n(j \wedge l)$ subsets of A_j. In other words, P_{lj}^* is a random element in the $n(j \wedge l)$ subset set of A_j, endowed with a uniform probability measure, $1 \le j \le J$. This random model is a hypergeometric one.

Now, by denoting φ, ψ and ψ^* the indicator functions of $R(\pi)$, $R(\chi)$ and $R(\chi^*)$ in the set $\mathbb{P} = P_2(\mathcal{O})$ of unordered object pairs, the random raw association coefficient can be expressed as

$$s(\pi, \chi^*) = card\big(R(\pi) \cap R(\chi^*)\big) = \sum_{p \in \mathbb{P}} \varphi(p)\psi^*(p) \quad (6.2.173)$$

The calculation of the mean (expectation), variance or any other moment will be based on the following partition of P associated with the partition α. The latter can be put under the form

$$\{P_2(A_j) | 1 \le j \le J\} \cup \{A_i * A_j | 1 \le i < j \le J\} \quad (6.2.174)$$

where $A_i * A_j$ denotes the set of 2-subsets of $\mathcal{O}$ for which one element belongs to A_i and the other one to A_j, $1 \le i < j \le J$. Thus, the random coefficient $s(\pi, \chi^*)$ can be decomposed as follows:

$$s(\pi, \chi^*) = \sum_{1 \le j \le J} \sum_{p \in P_2(A_j)} \varphi(p)\psi^*(p) + \sum_{1 \le i < j \le J} \sum_{p \in A_i * A_j} \varphi(p)\psi^*(p) \quad (6.2.175)$$

Let us denote by $s(\pi, \chi^\star; \alpha)$ the previous random coefficient but where the hypothesis of no relation is defined as around (6.2.172). The calculation of the mean and variance of $s(\pi, \chi^\star; \alpha)$ is detailed in [22], and then the centred and standardized partial association coefficient is obtained. We leave the reader interested to discover this development by himself.

The principle of the construction of this coefficient underlies the following notion of conditional independence

- For any j and (h, l), $1 \leq j \leq J$, $1 \leq h \leq K$ and $1 \leq l \leq M$, the proportion of object pairs joined in the class $A_j \cap O_h \cap P_l$ of the joint partition $\alpha \wedge \pi \wedge \chi$ is equal to the product of the proportion of object pairs joined in the class $A_j \cap O_h$ of the partition $\alpha \wedge \pi$ by the proportion of object pairs joined in the class $A_j \cap P_l$ of the partition $\alpha \wedge \chi$, these proportions being conditional and relative to the set $P_2(A_j)$:

For any j, $1 \leq j \leq J$, and (h, l), $1 \leq h \leq K$ and $1 \leq l \leq M$,

$$\frac{n(j \wedge h \wedge l)}{n(j)(n(j) - 1)} = \frac{n(j \wedge h)(n(j \wedge h) - 1)}{n(j)(n(j) - 1)} \times \frac{n(j \wedge l)(n(j \wedge l) - 1)}{n(j)(n(j) - 1)};$$

- For any (i, j) and (h, l), $1 \leq i < j \leq J$, $1 \leq h \leq K$ and $1 \leq l \leq M$, the proportion of object pairs of the class $O_h \cap P_l$ of $\pi \wedge \chi$ whose two components belong to the classes $A_i \cap O_h \cap P_l$ and $A_j \cap O_h \cap P_l$, respectively, of $\alpha \wedge \pi \wedge \chi$ is equal to the product of the proportion of the object pairs of O_h whose two components belong to the classes $A_i \cap O_h$ and $A_j \cap O_h$, respectively, of $\alpha \wedge \pi$ by the proportion of object pairs of P_l whose two components belong to the classes $A_i \cap P_l$ and $A_j \cap P_l$, respectively, of $\alpha \wedge \chi$, these proportions being conditional and relative to the set $A_i \star A_j$:

For any (i, j), $1 \leq i < j \leq J$, (h, l), $1 \leq h \leq K$ and $1 \leq l \leq M$,

$$\frac{n(i \wedge h \wedge l)n(j \wedge h \wedge l)}{n(i)n(j)} = \frac{n(i \wedge h)n(j \wedge h)}{n(i)n(j)} \times \frac{n(i \wedge l)n(j \wedge l)}{n(i)n(j)}.$$

The Global Approach

The construction of a partial association coefficient between c and d neutralizing the influence of a can be worked out in the framework of a more global hypothesis of no relation (see Sect. 4.3 of [22]). The calculations of the conditional mathematical expectations remain at the level of the object set $\mathcal{O}$. However, the only information retained for a given object pair $\{x, y\}$ is the set $R(\alpha)$ or $S(\alpha)$ to which it belongs (recall that $R(\alpha)$ (resp., $S(\alpha)$) is the set of joined pairs (resp., separated pairs) by α). This manner ignores how x and y are situated with respect to the different classes A_j, $1 \leq j \leq J$. Similarly, for a given element $(\{x, y\}, \{x, z\})$ of the set G of ordered pairs with one common component (resp., $(\{w, x\}, \{y, z\})$ of the set H of ordered pairs without common component) the only information required is its belonging to the Cartesian product $R(\alpha) \times R(\alpha)$, $R(\alpha) \times S(\alpha)$, $S(\alpha) \times R(\alpha)$ or $S(\alpha) \times S(\alpha)$. By this manner, the respective positions of the objects included in the ordered pairs of unordered object pairs are ignored. In fact, the general principle of this method is implicitly used in the the Kendall and Davis works [7, 17]. We will go over again this point below. In this case, we obtain different expressions of the moments of the random raw coefficient. The calculation of the mean leads to the following notion of conditional independence:

- The proportion of object pairs of the class $O_h \cap P_l$ of $\pi \wedge \chi$, $1 \leq l \leq M$, whose two components are joined in a same but not specified class of the partition α, is equal to the product of the proportion of object pairs of the class O_h whose components are joined in a same but not specified class of the partition α, by the proportion of object pairs of the class P_l whose two components are joined in a same but not specified class of the partition α, these proportions being conditional and relative to the set $R(\alpha)$ of object pairs joined by α:

For any (h, l), $1 \leq h \leq K$, $1 \leq l \leq M$,

$$\frac{\sum_{1 \leq j \leq J} n(j \wedge h \wedge l)\big(n(j \wedge h \wedge l) - 1\big)}{\sum_{1 \leq j \leq J} n(j)\big(n(j) - 1\big)}$$

$$= \frac{\sum_{1 \leq j \leq J} n(j \wedge h)\big(n(j \wedge h) - 1\big)}{\sum_{1 \leq j \leq J} n(j)\big(n(j) - 1\big)} \times \frac{\sum_{1 \leq j \leq J} n(j \wedge l)\big(n(j \wedge l) - 1\big)}{\sum_{1 \leq j \leq J} n(j)\big(n(j) - 1\big)};$$

- The proportion of object pairs of the class $O_h \cap P_l$ of $\pi \wedge \chi$, $1 \leq l \leq M$, whose two components are separated into two distinct but not specified classes of the partition α is equal to the product of object pairs of the class O_h whose two components are separated into two but not specified classes of the partition α, by the proportion of object pairs of P_l whose two components are separated into two distinct but not specified classes of the partition α ; all these proportions being conditional and relative to the set $S(\alpha)$ of object pairs separated by α :

For any (h, l), $1 \leq h \leq K$, $1 \leq l \leq M$, $\dfrac{\sum_{1 \leq i \leq j \leq J} n(i \wedge h \wedge l)n(j \wedge h \wedge l)}{\sum_{1 \leq i \leq j \leq J} n(i)n(j)}$

$$= \frac{\sum_{1 \leq i \leq j \leq J} n(i \wedge h)n(j \wedge h)}{\sum_{1 \leq i \leq j \leq J} n(i)n(j)} \times \frac{\sum_{1 \leq i \leq j \leq J} n(i \wedge l)n(j \wedge l)}{\sum_{1 \leq i \leq j \leq J} n(i)n(j)}$$

6.2.6.3 The Ordinal Categorical Case

This study [23] is parallel to the preceding one. Here, a, c and d are ordinal categorical attributes (see Sect. 3.3.2) and we are interested in neutralizing the effect of a in the association between c and d. Three total preorders on $\mathcal{O}$, denoted by α, ω and ϖ, are induced by a, c and d. As for total association (see (6.2.49)), the *raw* association coefficient and the random associated one can be written as, respectively,

$$s(\omega, \varpi) = card(R(\omega) \cap R(\varpi)) \text{ and } s(\omega, \varpi^\star) = card(R(\omega) \cap R(\varpi^\star)) \quad (6.2.176)$$

The Local Approach

What changes is the hypothesis of no relation. The latter becomes more restrictive in order to preserve the relative position of $\varpi^\star$ with respect to α. In fact, what is preserved in the calculations is the contingency table crossing α and ϖ. Finally and as expressed in the previous Introduction (Sect. 6.2.6.1), the retained structure for α which can be neutralized in this *local* approach of the association between ω and ϖ, is a partition. The hypothesis of no relation concerned is analogous to that considered for partial comparison between partitions. By denoting $(P_1^\star, P_2^\star, \ldots, P_l^\star, \ldots, P_M^\star)$ the ordered sequence of the $\varpi^\star$ total preorder classes, (5.2.68) can be taken up again. Nevertheless, the nature of the calculations changes. These will be based on the following decomposition of the set $\mathcal{O}^{[2]}$ of distinct object ordered pairs, into disjoint subsets

$$\mathcal{O}^{[2]} = \sum_{1 \leq j \leq J} A_j^{[2]} + \sum_{1 \leq i \neq j \leq J} A_i \times A_j$$

In these conditions, the random raw coefficient $s(\omega, \varpi^\star; \alpha)$ related to (5.2.68) can be put as

$$s(\omega, \varpi^\star; \alpha) = \sum_{1 \leq j \leq J} \sum_{q \in A_j^{[2]}} \varphi(q)\psi^\star(q) + \sum_{1 \leq i \neq j \leq J} \sum_{q \in A_i \times A_j} \varphi(q)\psi^\star(q) \quad (6.2.177)$$

The calculation of the mean and variance of $s(\omega, \varpi^\star; \alpha)$ is detailed in [23], and therefore the centred and standardized partial association coefficient between two ordinal categorical attributes c and d, neutralizing the influence of a categorical attribute a is obtained. In the reference mentioned, the reader interested will discover this development.

The principle of the hypothesis of no relation for this construction of partial association underlies the following notion of conditional independence

- For any j, $1 \leq j \leq J$, and a set of four distinct subscripts h, k, l and m, such that $1 \leq h < k \leq K$ and $1 \leq l < m \leq M$, the conditional proportions of ordered object pairs (x, y) belonging to $(A_j \cap O_h \cap P_l) \times (A_j \cap O_k \cap P_m)$ is equal to the conditional proportion of ordered pairs (x, y) belonging to $(A_j \cap O_h) \times (A_j \cap O_k)$ multiplied by the conditional proportion of ordered pairs (x, y) belonging to $(A_j \cap P_l) \times (A_j \cap P_m)$, these conditional proportions being relative to $A_j^{[2]}$:

  For any j, (h, k), (l, m), $1 \leq j \leq J$, $1 \leq h < k \leq K$ and $1 \leq l < m \leq M$,

$$\frac{n(j \wedge h \wedge l)n(j \wedge k \wedge m)}{n(j)\big(n(j)-1\big)} = \frac{n(j \wedge h)n(j \wedge k)}{n(j)\big(n(j)-1\big)} \times \frac{n(j \wedge l)n(j \wedge m)}{n(j)\big(n(j)-1\big)}.$$

- For any (i,j), $1 \le i \ne j \le J$ and a set of four distinct subscripts h, k, l and m, such that $1 \le h < k \le K$ and $1 \le l < m \le M$, the conditional proportions of ordered object pairs (x, y) belonging to $(A_i \cap O_h \cap P_l) \times (A_j \cap O_k \cap P_m)$ is equal to the conditional proportion of ordered pairs (x, y) belonging to $(A_i \cap O_h) \times (A_j \cap O_k)$ multiplied by the conditional proportion of ordered pairs (x, y) belonging to $(A_i \cap P_l) \times (A_j \cap P_m)$, these conditional proportions being relative to $A_i \times A_j$:

For any $(i,j), (h,k), (l,m), 1 \le i \ne j \le J, 1 \le h < k \le K$ and $1 \le l < m \le M$,

$$\frac{n(i \wedge h \wedge l)n(j \wedge k \wedge m)}{n(i)n(j)} = \frac{n(i \wedge h)n(j \wedge k)}{n(i)n(j)} \times \frac{n(i \wedge l)n(j \wedge m)}{n(i)n(j)}.$$

The Global Approach

This development follows in its spirit that called identically and established in the nominal categorical case (see the previous section). For this global version of the hypothesis of no relation between two *ordinal* categorical attributes c and d, neutralizing the influence of an ordinal categorical attribute a, the ordinal nature of a is taken into account. Therefore, the induced structure α by a on the object set $\mathcal{O}$ is interpreted faithfully as a total preorder on $\mathcal{O}$, the sequence of its classes being $(A_1, A_2, \ldots, A_j, \ldots, A_J)$. However, the representation which can be retained here is restricted to be directly the subset $R(\alpha)$ of $\mathcal{O}^{[2]}$:

$$R(\alpha) = \sum_{1 \le i < j \le J} A_i \times A_j$$

(see (3.3.18)).

Consequently, for a given ordered pair (x, y), the only information taken into account is its belonging to $R(\alpha)$ or to the complementary of $R(\alpha)$ in $\mathcal{O}^{[2]}$, denoted here by $T(\alpha)$. $T(\alpha) = E(\alpha) + S(\alpha)$, according to the notations of (3.3.19). Besides, concerning an ordered pair of ordered object pairs $\big((x, y), (x', y')\big)$ in $\mathcal{O}^{[2]} \times \mathcal{O}^{[2]}$, two items must be considered

- the structure of $\big((x, y), (x', y')\big)$ according to repetition of its different components;
- the belonging of $\big((x, y), (x', y')\big)$ to $R(\alpha) \times R(\alpha)$, $R(\alpha) \times T(\alpha)$, $T(\alpha) \times R(\alpha)$ or $T(\alpha) \times T(\alpha)$.

The calculations of the mean and variance of the random raw coefficient under these conditions are carried out in average, they are minutely reported in [23]. The calculation of the mean is related to the following notion of conditional independence where we denote by O_{hl} the (h, l) class of the intersection between the total preorders ω and ϖ, $1 \le h \le K$, $1 \le l \le M$.

- The conditional proportion of ordered pairs belonging to $(O_{hl} \times O_{km}) \cap R(\alpha)$ is equal to the product of the conditional proportion of ordered pairs belonging to $(O_h \times O_k) \cap R(\alpha)$ by the conditional proportion of ordered pairs belonging to $(P_l \times P_m) \cap R(\alpha)$, these conditional proportions being relative to $R(\alpha)$, namely

$$\text{For any } (h, k, l, m), 1 \le h < k \le K, 1 \le l < m \le M,$$

$$\frac{\sum_{1 \le h < k \le K} n(i \wedge h \wedge l) n(j \wedge k \wedge m)}{\sum_{1 \le i < j \le J} n(i) n(j)} =$$

$$\frac{\sum_{1 \le i < j \le J} n(i \wedge h) n(j \wedge k)}{\sum_{1 \le i < j \le J} n(i) n(j)} \times \frac{\sum_{1 \le i < j \le J} n(i) n(j) n(i \wedge l) n(j \wedge m)}{\sum_{1 \le i < j \le J} n(i) n(j)}.$$

- The conditional proportion of ordered pairs belonging to $(O_{hl} \times O_{km}) \cap T(\alpha)$, $1 \le h < k \le K$, $1 \le l < m \le M$ is equal to the product of the conditional proportion of ordered pairs belonging to $(O_h \times O_k) \cap T(\alpha)$ by the conditional proportion of ordered pairs belonging to $(P_l \times P_m) \cap T(\alpha)$, namely

$$\text{For any } (h, k, l, m), 1 \le h < k \le K, 1 \le l < m \le M,$$

$$\frac{\sum_{1 \le j \le i \le J} n(i \wedge h \wedge l) n(j \wedge k \wedge m)}{card(T(\alpha))} =$$

$$\frac{\sum_{1 \le j \le i \le J} n(i \wedge h) n(j \wedge k)}{card(T(\alpha))} \times \frac{\sum_{1 \le j \le i \le J} n(i \wedge l) n(j \wedge m)}{card(T(\alpha))}.$$

where (see (3.3.21))

$$card(T(\alpha)) = \sum_{1 \le j < i \le J} n(i) n(j) + \sum_{1 \le j \le J} n(i) (n(i) - 1)$$

6.2.6.4 The Ranking Case

Let α_l, ω_l and ϖ_l be three linear orders on $\mathcal{O}$ induced by three ranking attributes a, b and c, where a has to be neutralized in the association between b and c. In this case, the set $T(\alpha)$ is reduced to

$$S(\alpha) = \{(x, y) | (x, y) \in \mathcal{O}^{[2]} \text{ and } y < x \text{ for } \alpha_l\}$$

The calculations carried out in the ordinal categorical case can be retaken here leading to a new partial association coefficient [23]. As usually, the random raw association coefficient $s(\omega_l, \varpi_l^*; \alpha_l)$ is centred and reduced with its mean and standard deviation, respectively. The hypothesis of no relation considered here is the global one (see the preceding section). The mean of $s(\omega_l, \varpi_l^*; \alpha_l)$ is

$$\mathcal{E}\big(s(\omega_l, \varpi_l^*; \alpha_l)\big) = \frac{card\big(R(\omega) \cap R(\alpha)\big)card\big(R(\varpi) \cap R(\alpha)\big)}{card\big(R(\alpha)\big)} +$$
$$\frac{card\big(R(\omega) \cap S(\alpha)\big)card\big(R(\varpi) \cap S(\alpha)\big)}{card\big(S(\alpha)\big)} \tag{6.2.178}$$

where $\mathcal{E}$ designates as usually the mathematical expectation. We refer to [23] for the calculation of the variance of $s(\omega_l, \varpi_l; \alpha_l)$. The conditional independence notion—for which the mean of the random raw coefficient is null—becomes

$$\frac{card\big(R(\omega_l) \cap R(\varpi_l) \cap R(\alpha_l)\big)}{card\big(R(\alpha_l)\big)} = \frac{card\big(R(\omega_l) \cap R(\alpha_l)\big)}{card\big(R(\alpha_l)\big)} \times \frac{card\big((R(\varpi_l) \cap R(\alpha_l)\big)}{card\big(R(\alpha_l)\big)}$$

and

$$\frac{card\big(R(\omega_l) \cap R(\varpi_l) \cap S(\alpha_l)\big)}{card\big(S(\alpha_l)\big)} = \frac{card\big(R(\omega_l) \cap S(\alpha_l)\big)}{card\big(S(\alpha_l)\big)} \times \frac{card\big(R(\varpi_l) \cap S(\alpha_l)\big)}{card\big(S(\alpha_l)\big)}$$

Let us now examine the technique proposed by M.G. Kendall (1942) (see in [17]). This starts by representing $R(\omega_l)$ and $R(\varpi_l)$ at the level of $R(\alpha_l)$, as if the total space is defined by $R(\alpha_l)$. In these conditions, the 2×2 contingency table for comparing $R(\omega_l)$ and $R(\varpi_l)$ in $R(\alpha_l)$ becomes

$((b, c)/a)$	$R(\varpi_l) \cap R(\alpha_l)$	$S(\varpi_l) \cap R(\alpha_l)$
$R(\omega_l) \cap R(\alpha_l)$	$s = R(\omega_l) \cap R(\varpi_l) \cap R(\alpha_l)$	$u = R(\omega_l) \cap S(\varpi_l) \cap R(\alpha_l)$
$S(\omega_l) \cap R(\alpha_l)$	$v = S(\omega_l) \cap R(\varpi_l) \cap R(\alpha_l)$	$t = S(\omega_l) \cap S(\varpi_l) \cap R(\alpha_l)$

Then the partial association coefficient proposed is the K. Pearson coefficient between $R(\omega_l) \cap R(\alpha_l)$ and $R(\varpi_l) \cap R(\alpha_l)$ in $R(\alpha_l)$, namely, by using the notations in the contingency table

$$\tau(\omega_l, \varpi_l; \alpha_l) = \frac{st - uv}{\sqrt{(s+u)(s+v)(t+u)(t+v)}} \tag{6.2.179}$$

M.G. Kendall notes with surprise by assigning the result to a coincidence that the coefficient can be put under the form

$$\tau(\omega_l, \varpi_l; \alpha_l) = \frac{\tau(\omega_l, \varpi_l) - \tau(\omega_l, \alpha_l)\tau(\varpi_l, \alpha_l)}{\sqrt{1 - \tau(\omega_l, \alpha_l)^2}\sqrt{1 - \tau(\varpi_l, \alpha_l)^2}} \tag{6.2.180}$$

where τ designates the rank correlation coefficient he has established. We show in [23] that this result is due to the fact that the contribution of the centred coefficient in $R(\alpha_l)$ is identical to that in $S(\alpha_l)$. Besides, we may notice the similarity of the form of this coefficient with that (5.2.90) considered in the Boolean case.

On the other hand, the calculations required to obtain the coefficient (6.2.180) assume a hypothesis of no relation even more global than that mentioned above. These calculations do not distinguish the different structures of an ordered pair of ordered distinct object pairs $\big((x, y), (x', y')\big)$. The denominator of the coefficient obtained in the more restrictive global option is essentially different from that (6.2.180) and this is very important for the discriminant property of the measure concerned.

References

1. Albatineh, A.N., Niewiadomska-Bugaj, M.: Correcting jaccard and other similarity indices for chance agreement in cluster analysis. Adv. Data Anal. Class. **5**, 179–200 (2011)
2. Albatineh, A.N., Niewiadomska-Bugaj, M., Mihalko, D.: On similarity indices and correction for chance agreement. J. Class. **23**, 301–313 (2006)
3. Booth, H.S., Maindonald, J.H., Wilson, S.R., Gready, J.E.: An efficient z-score algorithm for assessing sequence alignments. J. Comput. Biol. **11**(4), 616–625 (2004)
4. Cramer, H.: The Elements of Probability Theory and Some of Its Applications. Wiley, New York (1946)
5. Daniels, H.E.: The relation between measures of correlation in the universe of sample permutations. Biometrika **33**, 129–135 (1944)
6. Daudé, F.: Analyse et justification de la notion de ressemblance entre variables qualitatives dans l'optique de la classification hiérarchique par *AVL*. Ph.D. thesis, Université de Rennes 1, June 1992
7. Davis, J.A.: A partial coefficient for goodman and kruskal's gamma. J. Am. Stat. Assoc. **62**(317), 189–193 (1967)
8. Fowlkes, E.B., Mallows, C.L.: A method for comparing two hierarchical clusterings. J. Am. Stat. Assoc. **78**, 553–569 (1983)
9. Goodman, L.A., Kruskal, W.H.: Measures of association for cross classifications. J. Am. Stat. Assoc. **49**, 732–764 (1954)
10. Haigh, J.: A neat way to prove asymptotic normality. Biometrika **3**, 677–678 (1971)
11. Hajek, J.: Some extensions of the Wald-Wolfowitz-Noether theorem. Ann. Math. Stat. **32**, 506–523 (1961)
12. Hubert, L., Arabie, P.: Comparing partitions. J. Classif. **2**, 193–218 (1985)
13. Hubert, L.J.: Inference procedures for the evaluation and comparison of proximity matrices. In: Felsenstein, J. (ed.) Numerical Taxonomy. Springer, Berlin (1983)
14. Hubert, L.J.: Combinatorial data analysis: association and partial association. Psychometrika **50**(4), 449–467 (1985)
15. Hubert, L.J.: Assignment methods in combinatorial data analysis. Numerical Taxonomy. Marcel Dekker, New York (1987)
16. Hulsen, T., de Vlieg, J., Leunissen, J., Groenen, P.: Testing statistical significance with structure similarity. BMC Bioinf. **7**(444), 1 (2006). Online
17. Kendall, M.G.: Rank correlation methods. Charles Griffin, London (1970). First edition in 1948
18. Lecalvé, G.: Un indice de similarité pour des variables de types quelconques. Statistique et Analyse des Données **01–02**, 39–47 (1976)
19. Lerman, I.C.: Étude distributionnelle de statistiques de proximité entre structures finies de même type; application à la classification automatique. Cahiers du Bureau Universitaire de Recherche Opérationnelle **19** 1–52 (1973)

20. Lerman, I.C.: Formal analysis of a general notion of proximity between variables. In: Barra, J.R., et al. (eds.) Recent Developments in Statistics, pp. 787–795. North-Holland, New York (1977)
21. Lerman, I.C.: Classification et analyse ordinale des données. Dunod and http://www.brclasssoc.org.uk/books/index.html (1981)
22. Lerman, I.C.: Indices d'association partielle entre variables qualitatives nominales. RAIRO série verte 17(3), 213–259 (1983)
23. Lerman, I.C.: Indices d'association partielle entre variables qualitatives ordinales. Publications Institut de Statistique des Universités de Paris, (XXVIII, 1,2), 7–46 (1983)
24. Lerman, I.C.: Justification et validité d'une échelle [0, 1] de fréquence mathématique pour une structure de proximité sur un ensemble de variables observées. Publications de l'Institut de Statistique des Universités de Paris 29, 27–57 (1984)
25. Lerman, I.C.: Maximisation de l'association entre deux variables qualitatives ordinales. Mathématiques et Sciences Humaines 100, 49–56 (1987)
26. Lerman, I.C.: Comparing partitions (mathematical and statistical aspects). In: Bock, H.H. (ed.) Classification and Related Methods of Data Analysis, pp. 121–131. North-Holland, Amsterdam (1988)
27. Lerman, I.C.: Conception et analyse de la forme limite d'une famille de coefficients statistiques d'association entre variables relationnelles, i. Revue Mathématique Informatique et Sciences Humaines 118, 35–52 (1992)
28. Lerman, I.C.: Conception et analyse de la forme limite d'une famille de coefficients statistiques d'association entre variables relationnelles, ii. Revue Mathématique Informatique et Sciences Humaines 119, 75–100 (1992)
29. Lerman, I.C.: Comparing classification tree structures: a special case of comparing q-ary relations. RAIRO-Oper. Res. 33, 339–365 (1999)
30. Lerman, I.C.: Comparing taxonomic data. Revue Mathématiques et Sciences Humaines 150, 37–51 (2000)
31. Lerman, I.C., Peter, P.: Structure maximale pour la somme des carrés d'une contingence aux marges fixées; une solution algorithmique programmée. Revue française d'automatique, d'informatique et de recherche opérationnelle 22(2), 83–136 (1988)
32. Lerman, I.C., Peter, P., Risler, J.L.: Matrices AVL pour la classification et l'alignement de séquences protéiques. Research Report 2466, IRISA-INRIA, September 1994
33. Lerman, I.C., Rouxel, F.: Comparing classification tree structures: a special case of comparing q-ary relations ii. RAIRO-Oper. Res. 34, 251–281 (2000)
34. Mantel, N.: Detection of disease clustering and a generalized approach. Cancer Res. 27(2), 209–220 (1967)
35. Messatfa, H.: An algorithm to maximize the agreement between partitions. J. Classif. 9(1), 5–15 (1992)
36. Mielke, P.W.: On asymptotic non-normality of null distributions of MRPP statistics. In: Communications in Statistics, Theory and Methods, pp. A8:1541–1550 (1979)
37. Monjardet, B.: Concordance between two linear orders: The Spearman and Kendall coefficients revisited. J. Classif. 14, 269–295 (1997)
38. Motoo, M.: On the Hoeffding's combinatorial central limit theorem. Ann. Inst. Stat. Math. 8, 145–154 (1957)
39. Noether, G.: On a theorem by Wald and Wolfowitz. Ann. Math. Stat. 20, 455–458 (1949)
40. Ouali-Allah, M.: Analyse en préordonnance des données qualitatives. Application aux données numériques et symboliques. Ph.D. thesis, Université de Rennes 1, Decembre 1991
41. Pinto Da Costa, J.F., Roque, L.A.C.: Limit distribution for the weighted rank correlation coefficient, r_W. REVSTAT - Stat. J. 3, 189–200 (2006)
42. Somers, R.H.: Analysis of partial rank correlation measures based on the product-moment model: Part one. Social Forces 53(2), 229–246 (1974)
43. Spearman, C.: The proof and measurement of association between two things. Am. J. Psychol. 15(1), 72–101 (1904)

44. Steinley, D., Hendrickson, G., Brusco, M.J.: A note on maximizing the agreement between partitions: a stepwise optimal algorithm and some properties. J. Classif. **32**, 114–126 (2015)
45. Tshuprow, A.A.: Principles of the Mathematical Theory of Correlation (trans: Kantorowitsch, M). W. Hodge and Co, London (1939)
46. Villoing, P.: Classification ascendante hiérarchique et indices de similarité sur données qualitatives nominales selon l'algorithme de la vraisemblance de la vraisemblance du lien. Ph.D. thesis, Université de Rennes 1, December 1980
47. Wald, A., Wolfowitz, J.: Statistical tests based on permutations of the observations. Ann. Math. Stat. **15**, 358–372 (1944)
48. Wilson, E.B., Hilferty, MM: The distribution of chi-square. In: Proceedings of the National Academy of Sciences of the United States of America, vol. 17, pp. 684–688 (1931)

Chapter 7
Comparing Objects or Categories Described by Attributes

7.1 Preamble

Let us recapitulate briefly the development of the previous chapters. As emphasized above (see Sect. 3.5.4 of Chap. 3), given a set $\mathcal{A} = \{a^1, a^2, \ldots, a^j, \ldots, a^p\}$ of descriptive attributes, description may concern a set $\mathcal{O} = \{o_1, o_2, \ldots, o_i, \ldots, o_n\}$ of elementary objects or a set $\Gamma = \{C_1, C_2, \ldots, C_i, \ldots, C_I\}$ of categories or classes. In Sect. 4.2 of Chap. 4 we studied the problem of the definition of a similarity index between objects or attributes when $\mathcal{A}$ is constituted by Boolean attributes. In this analysis several facets are considered: *combinatorial, metrical, ordinal and statistical*. We saw in Sect. 4.2.1.3, how to transpose a similarity index between objects defined in the framework of Boolean description, to the case of description by numerical attributes. Conversely, we have transposed a similarity or distance function defined on $\mathcal{O}$ in the case of numerical description to that where $\mathcal{A}$ is a set of Boolean attributes.

Up to now, our development has been dominated by the problem of constructing an association coefficient between descriptive attributes. The symmetrical and asymmetrical cases have been dealt with. The asymmetrical version is especially important in the case where $\mathcal{A}$ is composed of Boolean attributes.

In Sect. 4.3 of Chap. 4, extension of classical association coefficients between Boolean attributes to attributes of types II and III, was carried out.

In Sect. 5.2 of Chaps. 5 and 6 we have developed the *LLA* approach for building a probabilistic association coefficient on the set $\mathcal{A}$ of descriptive attributes. The approach concerned is combinatorial and probabilistic. The case of the attributes of type I is treated in Sect. 5.2, and that of the attributes of types II and III, in Sect. 6.2. Partial association coefficients are defined and extensively studied for all types of attributes (see Sects. 5.2.2.2 and 6.2.6). The new coefficients obtained are comparatively situated in relation to classically proposed coefficients.

In Sect. 5.3 of Chap. 5 we assume that set $\mathcal{A}$ of the descriptive attributes is observed on a set Γ of categories. In this situation, the distribution of every element of $\mathcal{A}$ is

© Springer-Verlag London 2016
I.C. Lerman, *Foundations and Methods in Combinatorial and Statistical Data Analysis and Clustering*, Advanced Information and Knowledge Processing,
DOI 10.1007/978-1-4471-6793-8_7

available on each of the categories of Γ. An important case is that where $\mathcal{A}$ is defined by an exclusive and exhaustive set of Boolean attributes associated with the category set of a nominal categorial attribute. This case gives rise to a contingency table, crossing $\mathcal{A}$ with Γ.

Concerning the description of a set $\mathcal{O}$ of objects by a set $\mathcal{A}$ of descriptive attributes, represented by a data table such as Table 3.4 of Chap. 3, we shall now consider the construction of a similarity index (one can also say a proximity index) on $\mathcal{O}$. The nature of building such a similarity index is conceptually different from the construction of an association coefficient on $\mathcal{A}$. In fact, a data table such as Table 3.4 is essentially dissymmetrical. There is as much difference between an element indexing a column and an element indexing a row as between a measurement scale and the object which is measured. To visualize this point, we may imagine the case where $\mathcal{A}$ is composed by ordinal categorical attributes.

The construction of a similarity index between objects may appear as more problematic than that of building an association coefficient between attributes. Indeed, to compare two descriptive attributes a^j and a^k, $1 \le j < k \le p$, the different objects have—a priori—to participate equally, with the same importance. Whereas, if the matter is to compare two objects o_i and $o_{i'}$, $1 \le i < i' \le n$, we cannot know if a priori the same importance (the same weight) has to be assigned to the different descriptive attributes. The respective scales and discriminant abilities of the different attributes may not be mutually comparable.

Now, let us refer to the most classical case where the attributes are numerical. In this, we put the data table in the following form

$$T_{num} = \{x_i^j = a^j(o_i) | 1 \le i \le n, 1 \le j \le p\} \tag{7.1.1}$$

In the geometrical representation of this data table, a given attribute a^j is assimilated as a *coordinate linear form* of the space $(\mathbb{R}^p)^\star$ (see Sect. 3.2.2 of Chap. 3). Thus, the jth coordinate linear form e_j^* is associated with a^j, $1 \le j \le p$. On the other hand, the vector of $\mathbb{R}^p$ whose components are x_i^j is associated with the object o_i, $1 \le i \le n$, $1 \le j \le p$. More precisely, it is the vector

$$\overrightarrow{OO_i} = \sum_{1 \le j \le p} x_i^j \vec{e}_j$$

where $\{\vec{e}_j | 1 \le j \le p\}$ is the canonical base of the geometrical space $\mathbb{R}^p$. In fact, the affine space is considered and the representation of the object o_i is assimilated to the point O_i given by the previous equation. Thus the vector $\overrightarrow{OO_i}$ is expressed as the difference $O_i - O$ between two points: O_i and O, $1 \le i \le n$. In general, we may suppose the object set $\mathcal{O}$ endowed with a system $M = \{\mu_i | 1 \le i \le n\}$ of weights, where μ_i is a positive real defining the weight of o_i. $\mathbb{I}$ designating the index set $\{i | 1 \le i \le n\}$, we associate the following cloud of points with the data table T_{num} (see (7.1.1)) and the weight system M,

$$\mathcal{N}(\mathbb{I}) = \{(O_i, \mu_i) | i \in \mathbb{I}\} \tag{7.1.2}$$

in the geometrical space $\mathbb{R}^p$ endowed with a positive definite metric, adequately chosen in order to evaluate the mutual distances between the points O_i, these, representing the respective objects o_i, $1 \le i \le n$. The centroid (centre of gravity) G of $\mathcal{N}(\mathbb{I})$ is defined by

$$G = \frac{1}{\mu} \sum_{i \in \mathbb{I}} \mu_i O_i$$

where $\mu = \sum_{i \in \mathbb{I}} \mu_i$.

Generally, we may assume—if not, we have to return to the fact—that the different numerical scales of the different attributes are of the same order of magnitude: it then makes sense to compare the values of two different attributes on the same object. In these conditions, there are two basic concepts in measuring the resemblance between two given objects o_i and $o_{i'}$. The first one, corresponding to a dissimilarity, is the Euclidean distance—according to a metric defined on $\mathbb{R}^p$—between the points O_i and $O_{i'}$ (which represent the objects o_i and $o_{i'}$) and the second, corresponding to a similarity, is the cosine of the angle $\widehat{O_i O O_{i'}}$, $1 \le i < i' \le n$. In the first notion the respective sizes of both vectors $(O_i - O)$ and $(O_{i'} - O)$ play an important role. Whereas, it is the similarity of the respective directions of $(O_i - O)$ and $(O_{i'} - O)$ which plays the more important role in the second notion. Methods based on distance and weight concepts employ the first notion of dissemblance. The *LLA* approach is focused on the notion of correlative similarity. Therefore, this method is more concerned by the second notion of similarity. In the similarity indices proposed below, statistical reduction is applied, attribute by attribute. Thus, all of the attributes become equally discriminant for measuring resemblance between objects. Other transformations can be considered [14, 15].

Clearly, in *Data Analysis*, it is always possible to derive a similarity index on a set E endowed with a distance index, by applying to the latter a decreasing function of the positive reals $\mathbb{R}_+$ onto itself. For a given table of distance indices

$$\mathcal{D} = \{D(x, y) | \{x, y\} \in P_2(E)\}$$

where $P_2(E)$ is the set of distinct object pairs of E, the most usual transformations are $D_{\max} - D(x, y)$ and $D_{\max}/D(x, y)$, where $D_{\max}$ is the maximal value of $\mathcal{D}$.

More sophisticated transformations of a distance index into a similarity index can be considered. For this, we may use as a point of departure [14] where the converse transformation of a similarity onto a distance is proposed.

The geometrical representation of a numerical data table T_{num} (see (5.3.25)) by means of a cloud of points in an euclidean space (see (7.1.1)) enables the distance or similarity notion to be visualized. As said above, this assumes—after possible normalization of the numerical scales of the different descriptive attributes—that there makes sense in comparing values of different attributes on a same object. This sense is lost for categorical attributes. Indeed, if c_l^j and c_m^k are two categorical values of two different categorical attributes a^j and a^k of the attribute set $\mathcal{A}$, observed on the same object o of $\mathcal{O}$, we cannot compare, without any reference, c_l^j and c_m^k. Thus, a Cosine index type, such as that considered above in the numerical case, cannot be envisaged in the categorical case.

Precisely, the *LLA* approach enables us to address the latter case for all types of categorical attributes.

7.2   Comparing Objects or Categories by the *LLA* Method

7.2.1   The Outline of the LLA *Method for Comparing Objects or Categories*

As in Sect. 4.2 of Chap. 4, x, y, z, ... will designate generic elements of

$$\mathcal{O} = \{o_1, o_2, \ldots, o_i, \ldots, o_I\}.$$

On the other hand, suppose all of the attributes of $\mathcal{A}$, having the same type. The basic argument to define a similarity measure on the object set $\mathcal{O}$ is the ordered pair

$$(\{x, y\}, a^j)$$

where $\{x, y\}$ is a unordered object pair of $\mathcal{O}$ and a^j an element of $\mathcal{A}$. In the categorical case, the comparison between the values $a^j(x)$ and $a^j(y)$ enables a similarity index between x and y to be produced. We will designate it by $s^j(x, y)$. This defines the *raw* contribution of the attribute a^j to the resemblance between x and y. In the numerical and Boolean cases (see Sect. 7.2.2) the definition of $s^j(x, y)$ may be relative to the whole attribute set and then, may require global measures of $\mathcal{A}$ on x and y, respectively. In fact, in our approach below, $s^j(x, y)$ is taken as an adequate contribution of a^j to a Cosine similarity index between the vectors X and Y which represent the objects x and y in the geometrical space.

Whatever the type of the data table, $s^j(x, y)$ is centred and standardized on the set $\mathcal{O} \times \mathcal{O}$ of ordered object pairs. More precisely, we define the *normalized* contribution of a^j to the similarity between x and y, as

$$S^j(x, y) = \frac{s^j(x, y) - mean_e(s^j)}{\sqrt{var_e(s^j)}} \tag{7.2.1}$$

where $mean_e(s^j)$ and $var_e(s^j)$ are the mean and variance of the empirical distribution of $s^j(x', y')$ on $\mathcal{O} \times \mathcal{O}$.

The sum of the normalized contributions for the different attributes is then

$$Q(x, y) = \sum_{1 \le j \le p} S^j(x, y) \tag{7.2.2}$$

For any j, $1 \leq j \leq p$, the mean and variance of $S^j(x, y)$ are equal to 0 and 1, respectively. By referring to the independence model, independent random attributes are associated with the observed ones. Then, the following reduced index might be proposed:

$$S(x, y) = \frac{1}{\sqrt{p}} Q(x, y)$$

However, this reduction has no effect on the following.

The distribution

$$Q = \{Q(x, y) | \{x, y\} \in P_2(\mathcal{O})\} \qquad (7.2.3)$$

of the Q index on the set $P_2(\mathcal{O})$ of distinct object pairs, is considered. The determination of its empirical mean and variance, $mean_e(Q)$ and $var_e(Q)$ respectively, leads to the following table of globally normalized indices

$$Q^g = \{Q^g(x, y) | \{x, y\} \in P_2(\mathcal{O})\} \qquad (7.2.4)$$

where

$$Q^g(x, y) = \frac{Q(x, y) - mean_e(Q)}{\sqrt{var_e(Q)}} \qquad (7.2.5)$$

Explicitly

$$mean_e(Q) = \frac{2}{n(n-1)} \sum_{1 \leq i < i' \leq n} Q(o_i, o_{i'})$$

and

$$var_e(Q) = \frac{2}{n(n-1)} \sum_{1 \leq i < i' \leq n} \left(Q(o_i, o_{i'}) - mean_e(Q) \right)^2$$

where I denotes here the cardinality of $\mathcal{O}$.

Consequently, we propose a table of probabilistic indices as follows

$$\mathcal{P}^g = \{P^g(x, y) = \Phi(Q^g(x, y)) | \{x, y\} \in P_2(\mathcal{O})\} \qquad (7.2.6)$$

where as usual, Φ denotes the cumulative distribution function of the normal probability law. This scheme is analogous to that followed through (5.2.51)–(5.2.53).

Referring to the normal distribution can be justified in the framework of no relation hypothesis where a set $\{a^{j^*} | 1 \leq j \leq p\}$ of independent random attributes is associated with the set $\{a^j | 1 \leq j \leq p\}$ of observed attributes, the random attributes having the same types and the same marginal distributions as those observed. This justification refers to the *central limit theorem*. For this, we may assume the number of attributes p to be large enough.

One final point. We propose to associate the following table of dissimilarities with the table $\mathcal{P}^g$ of probabilistic indices.

$$\mathcal{D} = \{D_{inf}(x, y) = -\text{Log}_2(P^g(x, y)) | \{x, y\} \in P_2(\mathcal{O})\} \qquad (7.2.7)$$

We call $\mathcal{D}$, table of *informational dissimilarities*. In fact, $-\text{Log}_2(P^g(x, y))$ is the entropy of the event whose probability is $P^g(x, y)$.

Our coefficients are oriented towards the *LLA* methodology. However, their extreme generality makes it possible to adapt and to apply them in the framework of any clustering method. One issue could be to employ the previous table in clustering methods which requires dissimilarity on the set to be analyzed.

Let us summarize. In relation to a data table crossing a set $\mathcal{A}$ of descriptive attributes with a set $\mathcal{O}$ of objects (see Table 3.4 of Chap. 3), for each attribute a^j of $\mathcal{A}$ and an object pair $\{x, y\}$ of $\mathcal{O}$, we have to calculate the *raw* contribution $s^j(x, y)$ of a^j to the similarity between x and y. In the case where $\mathcal{A}$ is composed of attributes of type I (numerical or Boolean), $s^j(x, y)$ of a^j depends on $a^j(x)$, $a^j(y)$ and possibly, a global measure associated with the values of the different attributes of $\mathcal{A}$ on x and y. In the case where $\mathcal{A}$ is composed by attributes of type II or III, $s^j(x, y)$ associated with a^j depends only on the values $a^j(x)$ and $a^j(y)$. The case of description by attributes of type I will be examined in Sect. 7.2.2. In Sect. 7.2.3 we will consider two subcases of description by categorical attributes: nominal and ordinal. In Sect. 7.2.4 we consider the case where descriptive attributes are categorical preordonance or categorical such that whose category set is valuated by a numerical similarity. The specific case where $\mathcal{A}$ includes taxonomic attributes, will provide—via the *LLA* clustering method— an elegant, simple and efficient solution to the well known consensus problem in classification [4, 5, 18]. This will be seen in Sect. 7.2.5. The flexibility of the method employed enables the problem of building a similarity index between objects, in the case of heterogenous description, to be treated (see Sect. 7.2.6). Thus, whatever the description nature of the object set by descriptive attributes, we end in defining a probabilistic similarity index of the *Likelihood of the Link* on $\mathcal{O}$. This index has proved its relevance through cluster analysis of many and difficult real data.

In the Goodall work as well and independently, a probabilistic similarity index between objects is proposed [6, 7]. Following our comparative analysis [12], some comments on the latter index will be given in Sect. 7.2.7. This index proved to be very complex to calculate and seems to be problematic to handle with real data of large size.

To end, we will consider in Sect. 7.2.8, the problem of construction of a similarity index between rows of an horizontal juxtaposition of contingency tables. For this purpose, we will start with one single table. Defining similarity index on the row set is interpreted here as comparing objects numerically described.

Recall that the dual point of view where comparing columns of a contingency table is assimilated to comparing numerical attributes, was developed in Sect. 5.3.2.1 (5.3.5). The association coefficient then proposed derives from the correlation coefficient between numerical attributes. And, taking into account the perfect symmetry structure of a contingency table, we have expressed that this association coefficient applies for comparing the rows of the contingency table.

Here and once again, the point of view which prevails consists of comparing rows of a contingency table as objects represented by a set $\mathbb{I}$ of points in an affine Eucledian space. This was defined by (7.1.1) given in Sect. 5.3.2.1. Besides, in this section the chi-square distance on $\mathbb{I}$ has been specified. Another interesting index in this context is the Matusita coefficient [1, 3, 16]. We will mention it in Sect. 7.2.8.

Now, let us come back to the general case of describing an object set $\mathcal{O}$ by an attribute set $\mathcal{A}$. The general process adopted for establishing probabilistic similarities on $\mathcal{O}$ (see (7.2.1)–(7.2.7)) shows that for each attribute a^j of $\mathcal{A}$ and an object pair $\{x, y\}$ of $P_2(\mathcal{O})$, it suffices to define the *raw* similarity coefficient $s^j(x, y)$ between x and y with respect to a^j. This definition will be discussed below for a large range of attribute types.

7.2.2 Similarity Index Between Objects Described by Numerical or Boolean Attributes

To begin, we consider the case of description by numerical attributes. As specified above, the respective scales of the different attributes are assumed to be—possibly after an adequate reduction—comparable. The notations are those considered above (see (7.1.1) and (7.1.2)). In these conditions, let us consider the raw similarity index $s^j(o_i, o_{i'}) = (x_i^j - \bar{x}^j)(x_{i'}^j - \bar{x}^j)$ which compares o_i and $o_{i'}$ with respect to the attribute a^j, $\bar{x}^j$ being the mean over $\mathcal{O}$ of x_i^j, $1 \le i < i' \le n$, $1 \le j \le p$. The geometrical and statistical nature of the index $Q(o_i, o_{i'})$ (see (7.2.2)), resulting from the sum of the normalized contributions of the different attributes of $\mathcal{A}$ is given by the following property

Proposition 42 *The similarity index $Q(o_i, o_{i'})$ is the inner product, according to the normalized principal components metric, between the vectors $(O_i - G)$ and $(O_{i'} - G)$, where G is the centroid of the cloud of points representing $\mathcal{O}$.*

Proof The proof is straightforward. The empirical mean and variance of $s^j(o_i, o_{i'})$ over $\mathcal{O} \times \mathcal{O}$ are 0 and $(var_e(a^j))^2$, respectively, $var_e(a^j)$ being the empirical variance of a^j over $\mathcal{O}$. On the other hand (see Chap. 6 of [10] and Sect. 10.3.3 of Chap. 10), the q metric of the *normalized principal component analysis* is diagonal, with $q(\vec{e}_j, \vec{e}_j) = 1/var_e(a^j)$.

LLA approach is focused on the detection of the similarity of forms and then, our reference—for comparing the objects o_i and $o_{i'}$—will be the cosine of the angle $\widehat{O_i O O_{i'}}$. This index is invariant if we substitute for the vector $(O_i - O)$ (resp., $(O_{i'} - O)$) a homothetic one. It can be written as follows:

$$Cos(\widehat{O_iOO_{i'}}) = \frac{\sum_j x_i^j x_{i'}^j}{\sqrt{\sum_j \left(x_i^j\right)^2 \sum_j \left(x_{i'}^j\right)^2}}$$

$$= \sum_{1 \le j \le p} \xi_i^j \xi_{i'}^j \qquad (7.2.8)$$

where $\xi_i^j = x_i^j / \sqrt{\sum_j \left(x_i^j\right)^2}$ $\left(\text{resp., } \xi_{i'}^j = x_{i'}^j / \sqrt{\sum_j \left(x_i^j\right)^2}\right), 1 \le j \le p.$

The raw contribution of the component j to the similarity cosine index between o_i and $o_{i'}$ could be

$$\xi_i^j \cdot \xi_{i'}^j \qquad (7.2.9)$$

However, if $O_i = O_{i'}$, the value of this contribution is dependent on $\xi_i^j = \xi_{i'}^j$. Preferably, the raw similarity contribution of j is chosen as

$$s^j(o_i, o_{i'}) = \frac{1}{p} - \frac{1}{2}\left(\xi_i^j - \xi_{i'}^j\right)^2 \qquad (7.2.10)$$

With this definition we have

- $\sum_{1 \le j \le p} s^j(o_i, o_{i'}) = Cos(\widehat{O_iOO_{i'}})$

- $s^j(o_i, o_{i'})$ is maximal and equal to $1/p$ if the vectors $(O_i - O)$ and $(O_{i'} - O)$ are homothetic.

Consequently, by applying the sequence of equations from (7.2.1) to (7.2.6), the probability scale is reached through the last Eq. (7.2.6).

Clearly, other options might be considered for the raw similarity contribution of the jth component. For example

$$s^j(o_i, o_{i'}) = d_j^2 - \left(x_i^j - x_{i'}^j\right)^2$$

where

$$d_j^2 = \max\left\{\left(x_i^j - x_{i'}^j\right)^2 | (i, i') \in \mathbb{I} \times \mathbb{I}\right\}$$

$1 \le j \le p.$ In this, the contribution of the attribute a^j is disconnected from all of other attributes. In fact, there is a large variety of dissimilarity or distance indices between objects described by numerical attributes. Important works were carried out in order to analyze the *metrical* or *Euclidean* natures of these indices and to try to study the impact of these properties on an adequate choice of a dissimilarity index facing data of a given type [8]. Whatever the index chosen and for a given object pair

$\{o_i, o_{i'}\}$, it is always possible to determine a suitable raw similarity contribution of the jth attribute a^j to this index, $1 \leq i < i' \leq n, 1 \leq j \leq p$. In these conditions, *likelihood of the link* probabilistic index can be derived by following the process defined by the sequence of equations from (7.2.1) to (7.2.6). As in our development, the Boolean description is related to the numerical one in [8]. In the case of description by Boolean attributes we obtain for ξ_i^j (see (7.2.8))

$$\xi_i^j = \frac{\epsilon_i^j}{\sqrt{n(o_i)}}$$

where $n(o_i)$ denotes the number of attributes possessed by the object o_i and where $\epsilon_i^j = 1$ (resp., $\epsilon_i^j = 0$) if the jth attribute a^j is *TRUE* (resp., *FALSE*) on o_i, $1 \leq i \leq n$, $1 \leq j \leq p$. As previously discussed, the above development from (7.2.1) to (7.2.6) is applied to

$$s^j(o_i, o_{i'}) = \frac{1}{p} - \left(\xi_i^j - \xi_{i'}^j\right)^2$$

with the latter expression of ξ_i^j, in order to get the probabilistic similarity index between o_i and $o_{i'}$, $1 \leq i < i' \leq n$.

7.2.2.1   Valuation on the Attribute Set

We have already expressed in the introduction of this chapter (see Sect. 4.1) that methods for introducing an a priori numerical valuation (one can also say "weighting") on the attribute set $\mathcal{A}$ are more or less difficult to justify. This valuation is supposed to give the respective importances with which the different attributes have to take part in the evaluation of similarity between objects. Nevertheless, analyzing the variation results with respect to such a valuation might be instructive. The latter can be defined by a sequence $\alpha = (\alpha_1, \alpha_2, \dots, \alpha_j, \dots, \alpha_p)$ of positive reals, whose sum is unity. In our similarity index (see (7.2.1) and (7.2.2)), as for the metric of *normalized component analysis*, all of the descriptive attributes are reduced in order to have—in a certain sense—the same variability. The form of the Q coefficient (see (7.2.2)) enables the valuation to be integrated with the following formula

$$Q(\alpha; x, y) = \sum_{1 \leq j \leq p} \alpha_j S^j(x, y) \tag{7.2.11}$$

Now, let us give some objective strategies to determine an valuation. These are clearer to define in the case of a description by numerical attributes.

For clustering a set $\mathcal{O}$ of objects, some experts in Component Factorial Analysis (*CFA*) (see Chap. 6 of [10]) advice to adopt for these a description by the first and most discriminating factors of a *CFA* on the cloud of points (see (7.1.2)) associated with the initial description of $\mathcal{O}$. In fact, this technique employs an implicit weighting of the initial attributes according to the factorial model. This valuation could be highlighted more explicitly if in (7.2.11) α_j is related to the sum of correlation squares between a^j and the factors retained for a reduced description.

Other objective valuations on the attribute set $\mathcal{A}$ can be envisaged. In Chap. 8, a notion of *projective importance* of a given attribute a^j is defined. This corresponds to the variance of the association coefficient between a^j and the set $\{a^{j'} | 1 \leq j' \neq j \leq p\}$ of the other attributes, $1 \leq j \leq p$.

This weighting problem does not depend on the attribute types of $\mathcal{A}$. However, notice that in the case where description is made by categorical attributes, *CFA* is difficult to define. Thus, we will no more mention this problem in the following.

7.2.3 Similarity Index Between Objects Described by Nominal or Ordinal Categorical Attributes

7.2.3.1 The Nominal Case

As in Sect. 3.5.4 of Chap. 3, the attribute set is denoted by

$$\mathcal{A} = \{a^1, a^2, \ldots, a^j, \ldots, a^p\}$$

where for every j, a^j is a nominal categorical attribute.

$$C_j = \left\{ d^j_1, d^j_2, \ldots, d^j_h, \ldots, d^j_{K_j} \right\}$$

will designate the category set of a^j, $1 \leq j \leq p$. The partition of $\mathcal{O}$ induced by a^j is indicated as

$$\pi(a^j) = \left\{ O^j_1, O^j_2, \ldots, O^j_h, \ldots, O^j_{K_j} \right\}$$

where $O^j_h = a^{j-1} \left(d^j_h \right)$ (see (3.3.2) of Chap. 3). $\mathbb{I} = \{1, 2, \ldots, i, \ldots, I\}$ is the subscript set indexing $\mathcal{O}$ and $\mathbb{I}^j_h$ is the $\mathbb{I}$ subset indexing O^j_h, $1 \leq h \leq K_j$.

For a given object pair $\{o_i, o_{i'}\}$ of $P_2(\mathcal{O})$, the raw contribution of a^j to the similarity between o_i and $o_{i'}$ is set as follows

$$s^j(o_i, o_{i'}) = \begin{cases} 1 & \text{if } a^j(o_i) = a^j(o_{i'}) \\ 0 & \text{if } a^j(o_i) \neq a^j(o_{i'}) \end{cases} \tag{7.2.12}$$

For the empirical mean of s^j over $\mathbb{I} \times \mathbb{I}$ we have

$$mean(s^j) = \frac{1}{I^2} \sum_{(i,i') \in \mathbb{I} \times \mathbb{I}} s^j(o_i, o_{i'})$$

$$= \frac{1}{I^2} \sum_{1 \le h \le K_j} \sum_{(i,i') \in \mathbb{I}_h^j \times \mathbb{I}_h^j} 1$$

$$= \sum_{1 \le h \le K_j} \left(f_h^j \right)^2 \tag{7.2.13}$$

where f_h^j is the relative frequency $card(\mathbb{I}_h^j)/card(\mathbb{I})$, $1 \le h \le K_j$.

The calculation of the empirical variance of s^j over $\mathbb{I} \times \mathbb{I}$ follows the same principle and we get

$$var(s^j) = \sum_{1 \le h \le K_j} \left(f_h^j \right)^2 \left[1 - \sum_{1 \le h \le K_j} \left(f_h^j \right)^2 \right] \tag{7.2.14}$$

In these conditions, the normalized contribution of a^j to the similarity between o_i and $o_{i'}$ can be written

$$S^j(o_i, o_{i'}) = \frac{s^j(o_i, o_{i'}) - \sum_{1 \le h \le K_j} \left(f_h^j \right)^2}{\sqrt{\sum_{1 \le h \le K_j} \left(f_h^j \right)^2 \left[1 - \sum_{1 \le h \le K_j} \left(f_h^j \right)^2 \right]}} \tag{7.2.15}$$

And then, the rest of the process defined by (7.2.2)–(7.2.6), is to be applied in order to reach the probabilistic similarity indices given in (7.2.6).

The called *complete disjunctive coding* of a data table concerned by nominal categorical attributes, is obtained by substituting for each of the nominal categorical attribute a^j, its category set $C_j = \{a_1^j, a_2^j, \ldots, a_h^j, \ldots, a_{K_j}^j\}$, where a_h^j is interpreted as a Boolean attribute, $1 \le j \le p$. In these conditions, the data table includes $K_1 + K_2 = \cdots + K_j + \cdots + K_p$ columns. It comprises only 0 and 1 values, where the absence (resp., the presence) of a given category on a given object is coded by 0 (resp., by 1). Hence, the representation of a given object is a row Boolean vector containing exactly p components equal to 1. The jth one is situated between the positions $K_1 + K_2 = \cdots + K_{j-1}$ and $K_1 + K_2 = \cdots + K_j$.

Clearly, the index $s^j(o_i, o_{i'})$ can be conceived in the framework of *complete disjunctive coding*. Denoting by $a_h^j(o_i)$ the Boolean value (0 or 1) of a_h^j on o_i, $1 \le h \le K_j$, $1 \le i \le n$, we have

$$s^j(o_i, o_{i'}) = \sum_{1 \le h \le K_j} a_h^j(o_i) a_h^j(o_{i'})$$

on the other hand, the fact that

$$\sum_{1 \leq j \leq p} \sum_{1 \leq h \leq K_j} d_h^j(o_i) = p$$

for any i, makes that there is no sense to reduce the value of $a^j(o_i)$ with respect to

$$\{a^{j'}(o_i) | 1 \leq j' \leq p, 1 \leq h' \leq K_{j'}\}$$

7.2.3.2 The Ordinal Case

The notations are exactly the same as those considered above for the nominal case. But here, the value set $C_j = \{a_1^j, a_2^j, \ldots, a_h^j, \ldots, a_{K_j}^j\}$ of a given attribute a^j is totally ordered (see Sect. 3.3.2 of Chap. 3). Consequently, the raw contribution $s^j(o_i, o_{i'})$ to the similarity index between o_i and $o_{i'}$ has to take into account this ordinal structure. Assume this linear order defined as follows

$$a_1^j < a_2^j < \cdots < a_h^j < \cdots < a_{K_j}^j \tag{7.2.16}$$

A possible and natural definition of $s^j(o_i, o_{i'})$ derives from assimilating the respective values of a^j to their ranks according the linear order (7.2.16). More precisely, $s^j(o_i, o_{i'})$ is set as

$$s^j(i, i') = K_j - |a^j(i) - a^j(i')| \tag{7.2.17}$$

where i stands for o_i, $i \in \mathbb{I}$. The value of this index lies in the interval $[1, K_j]$. The 1 value is obtained when $(a^j(i) = K_j$ and $a^j(i') = 1)$ or $(a^j(i) = 1$ and $a^j(i') = K_j)$. The K_j value is obtained when $a^j(i) = a^j(i')$.

$s^j(i, i')$ can be retrieved with the first of (4.2.3) from a specific coding of the rank values $1, 2, \ldots, h, \ldots, K_j$ by means of Boolean vectors, including each, $2K_j$ components. K_j components are equal to 1 and K_j components to 0. The components equal to 1 are situated at the extremities of the vector concerned. The number of the components equal to 1 which are situated a the right extremity represents the rank value coded. As an example, let us consider $K_j = 5$, the sequence of integer codes $(1, 2, 3, 4, 5)$ is represented by the sequence of vectors

1. $(1, 1, 1, 1, 0, 0, 0, 0, 0, 1)$
2. $(1, 1, 1, 0, 0, 0, 0, 0, 1, 1)$
3. $(1, 1, 0, 0, 0, 0, 0, 1, 1, 1)$
4. $(1, 0, 0, , 0, 0, 1, 1, 1, 1)$
5. $(0, 0, 0, 0, 0, 1, 1, 1, 1, 1)$

The components b_h, $1 \leq h \leq 2K_j$ of a given the Boolean vector considered above can be expressed as follows:

- For $h \leq K_j$, $b_h(i) = 1 \leq a^j(i) \leq (K_j - h)$
- For $K_j < h \leq 2K_j$, $b_h(i) = (2K_j - h) < a^j(i) \leq K_j$

The sum of $|a^j(i) - a^j(i')|$ which appears in (7.2.17) is equal to

$$\sum_{1 \leq h < k \leq K_j} \sum_{(i,i') \in \mathbb{I}_h \times \mathbb{I}_k} |a^j(i) - a^j(i')| = \sum_{1 \leq h < k \leq K_j} n_h n_k (k - h)$$

Therefore, by denoting $(i^\star, i'^\star)$ a random element in $\mathbb{I} \times \mathbb{I}$, endowed with equal probabilities, the mean and variance of $s^j(i^\star, i'^\star)$ are equal to

$$mean(s^j(i^\star, i'^\star)) = K_j - \sum_{1 \leq h < k \leq K_j} \frac{n_h n_k}{n^2} (k - h)$$

$$var(s^j(i^\star, i'^\star)) = \sum_{1 \leq h < k \leq K_j} \frac{n_h n_k}{n^2} (k - h)^2 -$$

$$\left(\sum_{1 \leq h < k \leq K_j} \frac{n_h n_k}{n^2} (k - h) \right)^2 \qquad (7.2.18)$$

In these conditions, the normalized contribution of a^j to the similarity between o_i and $o_{i'}$ is given by

$$S^j(o_i, o_{i'}) = \frac{s^j(i, i') - mean(s^j(i^\star, i'^\star))}{\sqrt{var(s^j(i^\star, i'^\star))}} \qquad (7.2.19)$$

An then, the rest of the process defined by (7.2.2)–(7.2.6), is to be applied in order to reach the probabilistic similarity indices given in (7.2.6).

The Case of *Ranking* Attributes

As expressed in Sect. 3.3.3 of Chap. 3, this case can be interpreted as a very specific case of the preceding one where the descriptive attributes are ordinal categorical. In practice, this case arises when the size of the object set is moderate. A clear and known example, given in Sect. 3.3.3, is when the object set $\mathcal{O}$ is defined by a set of products and when, the attribute set $\mathcal{A}$ corresponds to a set of judges, giving each— according to its preference—a linear order on $\mathcal{O}$. The raw similarity contribution of the jth attribute a^j to the similarity between two objects o_i and $o_{i'}$ becomes

$$s^j(i, i') = n - |r^j(i) - r^j(i')| \qquad (7.2.20)$$

where $r^j(i)$ (resp., $r^j(i')$) corresponds to the rank of $a^j(o_i)$ (resp., $a^j(o_{i'})$). $1 \leq j \leq p$, $1 \leq i < i' \leq n$. For simplicity notations, we will omit in the following the upper subscript j. The value of $s(i,i')$ is comprised between 1 and n: 1, if $|r(i) - r(i')| = (n-1)$ and n if $r(i) = r(i')$. As previously, let us consider the random element (i^*, i'^*) equally distributed in $\mathbb{I} \times \mathbb{I}$. The probability distribution of $d(i^*, i'^*) = |r(i^*) - r(i'^*)|$ is given by

$$Prob\{d(i^*, i'^*) = m\} =$$
$$\frac{2(n - m)}{n^2} \text{ for } 1 \le m \le (n - 1) =$$
$$\frac{1}{n} \text{ for } m = 0 \tag{7.2.21}$$

Consequently, the *mean* M_1 and *absolute moment 2* M_2 of $d(i^*, i'^*)$ are given by

$$M_1 = \sum_{1 \le m \le (n-1)} \frac{2(n - m)}{n^2} \times m = \frac{n^2 - 1}{3n}$$
$$M_2 = \sum_{1 \le m \le (n-1)} \frac{2(n - m)}{n^2} \times m^2 = \frac{1}{6}(n^2 - 1) \tag{7.2.22}$$

Establishing these equations require the following formulas:

$$\sum_{1 \le m \le (n-1)} m2 = \frac{(n - 1)n(2n - 1)}{6}, \quad \sum_{1 \le m \le (n-1)} m^3 = \frac{(n - 1)^2 n^2}{4}$$

In these conditions, we have for the variance of $d(i^*, i'^*)$

$$var\big(d(i^*, i'^*)\big) = \frac{(n^2 - 1)(n^2 + 2)}{18n^2}$$

Finally, the normalized raw contribution of a^j to the similarity between o_i and $o_{i'}$ can be written as

$$S^j(i, i') = \frac{(n^2 - 1)/3n - |a^j(i) - a^j(i')|}{\sqrt{(n^2 - 1)(n^2 + 2)/18n^2}} \tag{7.2.23}$$

7.2.4 Similarity Index Between Objects Described by Preordonance or Valuated Categorical Attributes

The notion of a *preordonance categorical attribute* was defined in Sect. 3.4.1 of Chap. 3. We suppose here that $\mathcal{A} = \{a^1, a^2, \dots a^j, \dots, a^p\}$ is constituted by such attributes. As above, denote by $C_j = \{d_1^j, d_2^j, \dots, d_h^j, \dots, d_{K_j}^j\}$ the category set of a^j which is supposed to be coded with the integer subscript set $\mathcal{K}^j =$

$\{1, 2, \ldots, h, \ldots, K_j\}$. The preordonance, associated with a^j, is defined by a total preorder $\omega(\mathcal{K}^j)$ on the following set which represents the ordered category pairs of a^j

$$\mathcal{K}_2^j = \{(g, h) | 1 \leq g \leq h \leq K_j\} \tag{7.2.24}$$

Two mathematical representations for this type of attribute were proposed (see Sect. 3.4.1 of Chap. 3). The first one is at the level of the cartesian product $\mathcal{K}_2^j \times \mathcal{K}_2^j$ (see (3.4.4)–(3.4.6)). The second one is at the level of $\mathcal{K}_2^j$, using the *mean rank function* (see (3.4.8)). The latter will be employed in the development below.

Given two objects o_i and $o_{i'}$ of $\mathcal{O}$, the raw contribution of a^j to the similarity between o_i and $o_{i'}$ is defined by

$$s^j(o_i, o_{i'}) = r^j\left(a^j(o_i), a^j(o_{i'})\right) \tag{7.2.25}$$

where r^j is the mean rank function associated with $\omega(\mathcal{K}^j)$. For notation simplicity, $s^j(i, i')$ will be substituted for $s^j(o_i, o_{i'})$, $1 \leq i, i' \leq n$.

As for the nominal case above, denote by $\{\mathbb{I}_h^j | 1 \leq h \leq \mathcal{K}^j\}$ the partition of the subscript set $\mathbb{I}$, according to the partition of $\mathcal{O}$, induced by the categorical attribute a^j. Now and as above, define $(i^\star, i'^\star)$ as a random ordered pair in $\mathbb{I} \times \mathbb{I}$ endowed with a uniform probability measure. The mean and variance of $s^j(i^\star, i'^\star)$ have to be calculated in order to determine the normalized contribution of a^j to the similarity between o_i and $o_{i'}$. The first and second absolute moments M_1 and M_2 are given by

$$M_1 = \mathcal{E}\left(s^j(i^\star, i'^\star)\right) = \sum_{1 \leq g, h \leq K_j} f_g f_h r^j(g, h)$$

$$M_2 = \mathcal{E}\left(s^j(i^\star, i'^\star)\right) = \sum_{1 \leq g, h \leq K_j} f_g f_h (r^j(g, h))^2$$

where $\mathcal{E}$ designates the mathematical expectation and where f_h is the proportion n_h/n, $1 \leq h \leq K_j$. Therefore, the variance is

$$var\left(s^j(i^\star, i'^\star)\right) = M_2 - M_1^2 = \sum_{(g,h)} f_g f_h (r^j(g, h))^2 - \left(\sum_{(g,h)} f_g f_h r^j(g, h)\right)^2$$

In these conditions, the normalized contribution of the attribute a^j to the similarity between o_i and $o_{i'}$ can be written

$$S^j(o_i, o_{i'}) = \frac{r^j(a^j(o_i), a^j(o_{i'})) - M_1}{\sqrt{M_2 - M_1^2}} \tag{7.2.26}$$

And then, as previously, the rest of the process defined by the sequence (7.2.2)–(7.2.6) has to be applied in order to obtain the table of probabilistic similarity indices given in (7.2.6).

Representing the preordonance attribute a^j as a total preorder $\omega(\mathcal{K}^j)$ on $\mathcal{K}_2^j$ (see (7.2.24)) assumes implicitly the symmetrical nature of the ordinal similarity considered on the category set C_j. That is, if a_g^j and a_h^j are two distinct categories, the ordinal similarity of the ordered pair (a_g^j, a_h^j) is the same as that of (a_h^j, a_g^j). However, the asymmetrical case where the latter condition is not satisfied can occur in real data. For this, the total preorder has to be defined on

$$\mathcal{K}_{(2)}^j = \{(g, h) | 1 \leq g, h \leq K_j\} \tag{7.2.27}$$

Moreover, a categorical attribute valuated by a numerical similarity is generally defined by a valuation on the set $\mathcal{K}_{(2)}^j$ of all ordered category pairs. In Sect. 3.4.4 the coding of different attribute types in terms of preordonances or valuated similarity categorical attributes have been proposed. Coding a categorical attribute in terms of a preordonance on its category set enables us to add more structural information from the expert perception of the similarity between the categories and then, to enrich the attribute scale. When this preordonance coding does not introduce any new scale structural information, it can be shown that there is a large stability between the initial coding and the new one for the definition of the normalized similarity contribution of a given attribute in comparing two objects. More precise study of this point is left for the reader. Let us initialize this analysis by considering the Boolean attribute case. In this, the set above (7.2.24) becomes

$$\mathcal{K}_2 = \{(1, 1), (1, 2), (2, 2)\}$$

where the attribute value 1 (resp., 2) indicates the *absence* (resp., *presence*) of the binary attribute, that we denote by a. Let us set out for the raw similarity contribution between these values

$$s(1, 2) = p, s(1, 1) = s(2, 2) = q$$

where p and q are two positive real numbers such that $p < q$. The partition of $\mathbb{I}$ induced by a includes two classes $\mathbb{I}_1$ and $\mathbb{I}_2$, whose cardinalities are denoted by n_1 and n_2, $n_1 + n_2 = n = card(\mathbb{I})$. The absolute moments M_1 and M_2 of the random raw similarity index are obtained by the following formulas

$$M_1 = 2f_1 f_2 p + \left(f_1^2 + f_2^2\right) q$$
$$M_2 = 2f_1 f_2 p^2 + \left(f_1^2 + f_2^2\right) q^2 \tag{7.2.28}$$

where f_1 and f_2 stand for the relative frequencies n_1/n and n_2/n, respectively.

Introducing $\phi = 2f_1f_2$, we have

$$M_1 = \phi p + (1 - \phi)q$$
$$M_2 = \phi p^2 + (1 - \phi)q^2 \tag{7.2.29}$$

Therefore, the variance is equal to

$$\phi(1 - \phi)(q - p)^2$$

And then, the normalized contributions for a p and q values are

- $-\sqrt{\frac{(1-\phi)}{\phi}}$
- $\sqrt{\frac{\phi}{(1-\phi)}}$

respectively. Notice that they do not depend on the p and q values. This property may be surprising. In fact, it is a natural and suitable property. The resemblance between objects described by a binary attribute has not to depend on a numerical valuation assigned to both values of this dichotomy.

Notice that the range variation of ϕ is the interval $[0, 0.5]$, the value 0.5 being reached for $f_1 = f_2 = 0.5$. The function of ϕ in the first (resp., second) item above, increases (resp., decreases) when ϕ varies from 0 to 0.5. The maximal opposition is obtained for $\phi = 0.5$ where the respective values are equal to -1 and $+1$.

A final point. In order to illustrate the possible enrichment of the descriptive scale when preordonance coding is adopted, let us take as an example the case of a binary attribute. Now, suppose that the expert considers the resemblance between one given category and itself not identical for each of both categories. If he perceives the similarity between the category 2 and itself stronger than the similarity between the category 1 and itself. Then, the preordonance is established as: $(1, 2) < (1, 1) < (2, 2)$.

7.2.5 Similarity Index Between Objects Described by Taxonomic Attributes. A Solution for the Classification Consensus Problem

In Sect. 3.4.2 the notion of categorical taxonomic attribute was presented. This type of attribute was interpreted and represented in a specific manner as a preordonance categorical attribute. The value set of this is defined by the leaf set T of the taxonomy. The manner by which this preordonance is established on T will be once more expressed here with some more details. This preordonance is ultrametric in a sense already given (see (3.4.15)). Each internal node ν of the taxonomy corresponds to a cluster of leaves. If ν_1 and ν_2 are two nodes such that one descendant branch relates ν_1 to ν_2 ($\nu_1 \rightarrow \nu_2$), then ν_1 is a direct parent of ν_2. Recall that an integer level index is assigned to every node of the taxonomy, including the root and the leaf nodes. The root index is 1. The index of a given node is 1 plus the number of descendant branches from the root to the node concerned (see Fig. 3.3 of Chap. 3). The depth of the taxonomy is the maximum number of descendant branches relating the root to a leaf. If we designate by $k - 1$ this value, k will be the maximal index of a leaf in this taxonomy. Associated with the ultrametric preordonance a different formal representation of the taxonomic tree was given in Sect. 3.4.2. It is expressed in terms of an ordered sequence of partitions $\mathcal{P} = (P_0, P_1, \ldots, P_{h-1}, P_h, \ldots, P_{k-1})$ on the category set. P_0 is the finest partition, that, whose classes are *singletons* including each, exactly one single element of the category set and P_{k-1} is the least fine partition, comprising exactly one single class grouping the whole category set. The new classes of the partition P_h with respect to those of P_{h-1}, correspond to the nodes of the taxonomy, indexed by $(k - h)$ and obtained by merging nodes or leaves indexed by $(k - h + 1)$ (see Fig. 3.3).

Now, let us specify how is built the class sequence $\mathcal{L} = (L_1, L_2, \ldots, L_q, \ldots, L_r)$ of the total preorder on the following set of ordered category pairs

$$\mathcal{K}_2 = \{(g, h) | 1 \leq g \leq h \leq K\}$$

(see (3.4.3)). For this, we refer to the representation of a taxonomic attribute in terms of the ordered partition chain $\mathcal{P}$. If C^{k-2} and D^{k-2} are two classes of the partition P_{k-2} which merge in the partition with one single class P_{k-1}, then every ordered pair (g, h) such that $g \leq h$, where $g \in C^{k-2}$ and $h \in D^{k-2}$ or $g \in D^{k-2}$ and $h \in C^{k-2}$, is assigned to the first preorder class L_1. Analogously, if C^{k-3} and D^{k-3} are two classes of the partition P_{k-3} which merge in one class of the partition P_{k-2}, then every ordered pair (g, h) such that $g \leq h$, where $g \in C^{k-3}$ and $h \in D^{k-3}$ or $g \in D^{k-3}$ and $h \in C^{k-3}$, is assigned to the second preorder class L_2. More generally, L_h results from merging classes which occur in the passage between the partition P_{k-h-1} and P_{k-h}, $1 \leq h \leq k - 1$. The last preorder class L_r ($r = k$ here) is made of ordered pairs of the form (g, g), $1 \leq g \leq K$.

As previously indicated, (see (3.4.15)) the ultrametric preordonance on the category set associated with a taxonomic attribute is coded by means of the *mean rank function*. Thus, comparing objects described by a taxonomic attribute comes down to the above case of comparing objects described by a preordonance attribute. Nevertheless, the mean rank function used here has a particular meaning. More precisely, let o_i and $o_{i'}$ be two objects to be compared with respect to a taxonomic attribute that we denote here by τ. As above, the raw similarity index can be written $s(o_i, o_{i'}) = r(\tau(o_i), \tau(o_{i'}))$. Denote by e and g the respective categories possessed by o_i and $o_{i'}$, such that $e \leq g$. If (e, g) belongs to L_h, $1 \leq h \leq k - 1$, the (3.4.8) can be taken up again as follows

$$r(e, g) = \sum_{1 \leq j < h} l_j + \frac{l_h + 1}{2}$$

A mathematical formula can be provided for $l_h = card(L_h)$, $1 \leq h \leq k - 1$. This is obtained from a generalization of the notion of a type of a partition to a type of an ordered sequence of partitions [11].

Clearly, the development above (Sect. 7.2.4) can be taken up again exactly as it is, if the set $\mathcal{A}$ is constituted by categorical taxonomic attributes. The extension and generalization of this approach has been carried out in the case where description concerns categories and made of taxonomic preordonance attributes (see Sect. 3.4.3) [13].

The *LLA* hierarchical clustering of objects described by taxonomic attributes enables a simple and efficient solution to the classification *consensus* problem to be proposed. Given a finite set of hierarchical classification trees

$$\mathcal{H}_1, \mathcal{H}_2, \ldots, \mathcal{H}_j, \ldots, \mathcal{H}_p$$

on a finite object set $\mathcal{O}$, the matter consists of summarizing these in a *single* hierarchical classification tree $\mathcal{H}$ which has to be situated with respect to the different initial trees. A substantial literature was devoted to this problem [4, 5, 18].

In the solution we propose, the construction of an *LLA* similarity index between objects is come down to the previous treatment, by interpreting each classification tree $\mathcal{H}_j$ as a specific taxonomic categorical attribute, $1 \leq j \leq p$. The only difference with above is that, the tree leaves are associated with *singleton* categories, including each one unique object.

The real problem can be introduced in the case where a given attribute set $\mathcal{A}$ is observed on a given object set $\mathcal{O}$ at different dates : $t_1, t_2, \ldots, t_j, \ldots, t_p$. For each of the latter, say t_j, a data table, indexed by $\mathcal{O} \times \mathcal{A}(j)$ is built. It leads— via a fixed hierarchical clustering method—to a hierarchical classification tree $\mathcal{H}_j$, $1 \leq j \leq p$. Now, imagine the sequence $(\mathcal{H}_1, \mathcal{H}_2, \ldots, \mathcal{H}_j, \ldots, \mathcal{H}_p)$ corresponding to a *stable* evolution period. Then, it may be interesting to summarize the different classification trees of this sequence into a unique one, which defines a consensus between the different $\mathcal{H}_j$, $1 \leq j \leq p$.

However, for this problem, a second alternative can be considered. It requires the establishment of a global data table resulting from the *horizontal* juxtaposition of the different data tables. More precisely, the new data table is indexed by

$$\mathcal{O} \times \bigcup_{1 \leq j \leq p} \mathcal{A}(j).$$

In these conditions, a similarity index on $\mathcal{O}$ can be built according to the general principle described above and concerned by the following section, of *the sum of normalized contributions* of the different descriptive attributes indexing the columns of the global data table. There are in all $p \times card(\mathcal{A})$ attributes: if a is a given attribute of $\mathcal{A}$, "a at date t_j" is a generic new attribute.

An innovator of the classification consensus problem was S. Régnier [17] (see Sect. 2.2 of Chap. 2). In his work the problem considered consists of summarizing a set of partitions of a given set into one single partition called "central" partition.

In our development above the problem addressed is richer and more general. The data can be assimilated to a set of ordered partition chains and the "consensus" is provided as a partition chain.

7.2.6 Similarity Index Between Objects Described by a Mixed Attribute Types: Heterogenous Description

Up to now, for simplicity and consistent reasons, all of the descriptive attributes in a data table are supposed to be of a unique type. Five main types can be distinguished (see Chap. 3): numerical, Boolean, categorical nominal, categorical ordinal and pre-ordonance. However, object description made of different types of attributes may occur in real data. In this respect, let us consider the following decomposition of the attribute set $\mathcal{A}$:

$$\mathcal{A} = \mathcal{A}_{num} + \mathcal{A}_{bool} + \mathcal{A}_{nom} + \mathcal{A}_{ord} + \mathcal{A}_{pre}$$

where $\mathcal{A}_{num}$, $\mathcal{A}_{bool}$, $\mathcal{A}_{nom}$, $\mathcal{A}_{ord}$ and $\mathcal{A}_{pre}$ are formed of numerical, Boolean, categorical nominal, categorical ordinal and categorical preordonance attributes, respectively. The last type includes as subtypes of the categorical taxonomic and the categorical preordonance taxonomic attributes (see Sect. 3.4 of Chap. 3, Sects. 7.2.4 and 7.2.5 above). $\pi(\mathcal{A}) = \{\mathcal{A}_{num}, \mathcal{A}_{bool}, \mathcal{A}_{nom}, \mathcal{A}_{ord}, \mathcal{A}_{pre}\}$ is a partition of $\mathcal{A}$ into five classes.

Given two objects o_i and $o_{i'}$ of $\mathcal{O}$ and a descriptive attribute a^j of $\mathcal{A}$, the definition of the raw contribution of a^j to the similarity between o_i and $o_{i'}$, $s^j(o_i, o_{i'})$, depends on the partition class of $\pi(\mathcal{A})$ to which a^j belongs. The following procedure is proposed in order to define $s^j(o_i, o_{i'})$ and to integrate all of the raw similarity contributions, in a sum—possibly weighted—over $\mathcal{A}$. This sum is in fact decomposed into five parts associated with the five classes of the partition $\pi(\mathcal{A})$.

- For a^j belonging to $\mathcal{A}_{num}$, $s^j(o_i, o_{i'})$ is defined according to (7.2.9) and relatively to the set $\mathcal{A}_{num}$, considered as the whole attribute set;

- For a^j belonging to $\mathcal{A}_{bool}$, $s^j(o_i, o_{i'})$ is defined according to (7.2.9), adapted for description by Boolean attributes (see the end of Sect. 7.2.2) and relatively to the set $\mathcal{A}_{bool}$, considered as the whole attribute set;

- For a^j belonging to $\mathcal{A}_{nom}$, $s^j(o_i, o_{i'})$ is defined according to (7.2.12);

- For a^j belonging to $\mathcal{A}_{ord}$, $s^j(o_i, o_{i'})$ is defined according to (7.2.17);

- For a^j belonging to $\mathcal{A}_{pre}$, $s^j(o_i, o_{i'})$ is defined according to (7.2.25).

Notice that for every of the first four classes of the decomposition of $\mathcal{A}$, the semantic of the value scale of the attributes included in the class concerned, enables the raw similarity contribution to be specified. But, for the attributes belonging to $\mathcal{A}_{pre}$, their respective preordonances on their value sets are needed for this purpose. Nevertheless, notice that for taxonomic attributes, these preordonances can be directly deduced from their respective taxonomic structures.

In these conditions, denoting as above by $S^j(o_i, o_{i'})$ the normalized contribution of the attribute a^j to the similarity between o_i and $o_{i'}$ $1 \leq j \leq p, 1 \leq i < i' \leq n$ (see for example (7.2.1)), the following sums are established

$$S_{num}(o_i, o_{i'}) = \sum_{a^j \in \mathcal{A}_{num}} S^j(o_i, o_{i'})$$

$$S_{bool}(o_i, o_{i'}) = \sum_{a^j \in \mathcal{A}_{bool}} S^j(o_i, o_{i'})$$

$$S_{nom}(o_i, o_{i'}) = \sum_{a^j \in \mathcal{A}_{nom}} S^j(o_i, o_{i'})$$

$$S_{ord}(o_i, o_{i'}) = \sum_{a^j \in \mathcal{A}_{ord}} S^j(o_i, o_{i'})$$

$$S_{pre}(o_i, o_{i'}) = \sum_{a^j \in \mathcal{A}_{pre}} S^j(o_i, o_{i'}) \tag{7.2.30}$$

They are integrated in the global sum

$$S(o_i, o_{i'}) = S_{num}(o_i, o_{i'}) + S_{bool}(o_i, o_{i'}) +$$
$$S_{nom}(o_i, o_{i'}) + S_{ord}(o_i, o_{i'}) + S_{pre}(o_i, o_{i'}) \tag{7.2.31}$$

And then, the rest of the process defined by (7.2.3)–(7.2.6), follows in order to get the table of probabilistic similarities between objects described by a mixing of different attribute types. Notice that in the proposed solution all of the attributes belonging to $\mathcal{A}$ are considered as equally important. Nevertheless, weighting differently each group of attributes characterized by a given type might be envisaged. For this alternative we define $S(o_i, o_{i'})$ as follows

$$S(o_i, o_{i'}) = \alpha_{num} S_{num}(o_i, o_{i'}) + \alpha_{bool} S_{bool}(o_i, o_{i'}) +$$
$$\alpha_{ord} S_{ord}(o_i, o_{i'}) + \alpha_{pre} S_{pre}(o_i, o_{i'}) \tag{7.2.32}$$

where $\alpha_{num}, \alpha_{bool}, \alpha_{nom}, \alpha_{ord}$ and α_{pre} are positive numbers of unity sum.

7.2.7 The Goodall Similarity Index

The probabilistic Goodall similarity index [6] is a *likelihood of the link* similarity index between objects. When we introduced this type of index [9] we were not aware of the latter contribution. However, as we will realize below, our approach is very different of [7], both in its principle and development. The Goodall similarity index concerns comparison between objects described by nominal categorical attributes. In fact, these attributes are transformed in a very particular fashion—depending on their respective statistical distributions on the object set $\mathcal{O}$—into categorical attributes valuated by a numerical similarity (see Sect. 3.3.4 of Chap. 3). In these conditions, let us take up again our notations of Sect. 7.2.3 where $\mathcal{A} = \{a^1, a^2, \ldots, a^j, \ldots, a^p\}$ is a set of nominal categorical attributes and where

$$\mathcal{C}_j = \left\{ d_1^j, d_2^j, \ldots, d_h^j, \ldots, d_{K_j}^j \right\}$$

designates the category set of a^j, $1 \leq j \leq p$. The partition of $\mathcal{O}$ induced by a^j is indicated as

$$\pi(a^j) = \left\{ O_1^j, O_2^j, \ldots, O_h^j, \ldots, O_{K_j}^j \right\}$$

where $O_h^j = a^{j-1}(d_h^j)$. We denote here by n_h^j the cardinality of O_h^j, $1 \leq j \leq p$, $1 \leq h \leq K_j$, the sum of the n_h^j, for j fixed, being $n = card(\mathcal{O})$.

For introducing the Goodall index, let us consider the case of a single attribute a^j and define the relative frequency (proportion) of distinct object pairs possessing the value d_h^j. This can be written as

$$p_h = \frac{n_h^j \left(n_h^j - 1 \right)}{n(n-1)} = \frac{card \left(O_h^{j^{[2]}} \right)}{card(\mathcal{O}^{[2]})}$$

where $X^{[2]}$ denotes the set of distinct ordered pairs of X, $1 \leq h \leq K_j$.

In these conditions, the similarity valuation is introduced on the category set of a^j as follows:

$$\xi(d_g^j, d_h^j) = \begin{cases} 0 & \text{if } g \neq h \\ \chi_{hh}^j = \frac{1}{p_h} & \text{if } g = h \end{cases}$$

Consequently, with this valuation, two objects possessing the same category are considered as much similar as the category concerned is rarely observed. This view is not necessarily admitted by the practitioners. In their book, Sneath and Sokal [19] wrote

"Godall claims that the working taxonomist prefers to enhance the weights of rarer characteristics, but this does not seem to us to be generally accepted dictum of taxonomic practice".

By defining the raw similarity index between two given objects o_i and $o_{i'}$ of $\mathcal{O}$ as

$$s_G^j(i, i') = \xi[a^j(o_i), a^j(o_{i'})],$$

the unlikelihood of bigness of $s^j_G(i, i')$ can be expressed as follows

$$Prob\{s^j_G(i^\star, i'^\star) \geq s^j_G(i, i')\} = \begin{cases} 1 & \text{if } a^j(o_i) \neq a^j(o_{i'}) \\ \sum_{\left\{g \mid \chi^j_{gg} \geq \chi^j_{hh}\right\}} p_g & \text{if } a^j(o_i) = a^j(o_{i'}) = h \end{cases}$$

Let us denote by $P^j_G(i, i')$ this probability index. The latter can be put under the form

$$P^j_G(i, i') = \sum_{\{g \mid p_g \leq p_h\}} p_g$$

if $a^j(o_i) = a^j(o_{i'}) = h$. Hence, this index value is defined by the proportion of distinct object pairs whose components are included in a same class O^j_g, such that $card(O^j_g) \leq card(O^j_h)$.

In these conditions, the probabilistic similarity index can be written as

$$P^G_j(i, i') = 1 - P^j_G(i, i') = \begin{cases} 0 & \text{if } a^j(o_i) \neq a^j(o_{i'}) \\ 1 - \sum_{\left\{g \mid \chi^j_{gg} \geq \chi^j_{hh}\right\}} p_g & \text{if } a^j(o_i) = a^j(o_{i'}) = h \end{cases}$$

This approach appears to us as very particular. A realistic manner to enrich a nominal categorical attribute consists of transforming it into a categorical preordonance attribute by requiring the expert knowledge. For this purpose, let us take up again the set $\mathcal{K}^j_2$ of category pairs (see (7.2.24)). Starting with this set, the expert is asked to select at each step, the category pairs constituted by the most similar components and to take off them of the remaining set of category pairs. More formally, the algorithm process is the following:

Initialization $H \longleftarrow \mathcal{K}^j_2; l \longleftarrow 1;$
While $H \neq \emptyset$ Do
Select in H the set *MSP* of the most similar category pairs;
$L(l) \longleftarrow MSP; l \longleftarrow l + 1;$
$H \longleftarrow H - MSP;$
EndDo

As a result, we obtain the decreasing sequence of the preorder classes. If m is the total number of classes, we have

$$L(1) > L(2) > \cdots > L(l) > \cdots > L(m)$$

It is obvious to make correspondence with (3.4.3) of Chap. 3: $L_1 = L(m), L_2 = L(m - 1), \ldots, L_r = L(1)$.

Let us now consider the general multidimensional case where description is made of a sequence $(a^1, a^2, \ldots, a^j, \ldots, a^p)$ of categorical attributes. Then, associate with two given objects o_i and $o_{i'}$ the probabilistic index

$$P_G^0(i, i') = \Pi_{1 \leq j \leq p} P_G^j(i, i') \tag{7.2.33}$$

where $P_G^j(i, i')$ has been defined above. It corresponds to the unlikelihood degree of how big are the values of $s_G^j(i, i')$, $1 \leq j \leq p$, under a probabilistic independence hypothesis, for which the random raw indices $s_G^j(i^\star, i'\star)$, $1 \leq j \leq p$, are mutually independent. This random model can be obtained by associating with the observed attributes $a^1, a^2, \ldots, a^j, \ldots, a^p$, independent random ones $a^{1\star}, a^{2\star}, \ldots, a^{j\star}, \ldots, a^{p\star}$. The following probabilistic index

$$P_0^G(i, i') = 1 - P_G^0(i, i') \tag{7.2.34}$$

defined as the complement to unity of the previous index, can be interpreted as a specific version of a likelihood of the link index. In order to compare its behaviour with the *LLA* probabilistic index resulting from coding nominal categorical attribute in terms of preordonance categorical attribute (see the above algorithm), we tested it in [12]. This comparison was performed through hierarchical *LLA* clustering applied on an example of real data. Much more consistent results were obtained with the preordonance coding obtained directly by expert knowledge.

As a matter of fact, the exact Goodall's probability similarity index is much more complicated than (7.2.34). The final index is defined by the cumulative empirical distribution of $P_0^G(i, i')$ on $\mathcal{O} \times \mathcal{O}$. It is calculated from the distribution of the attribute vector $(a^1, a^2, \ldots, a^j, \ldots, a^p)$ on $\mathcal{O}$. The latter vector comprises $\Pi_{1 \leq j \leq p} K_j$ points, each corresponding to one value of $(a^1, a^2, \ldots, a^j, \ldots, a^p)$. Therefore, there are in all $\Pi_{1 \leq j \leq p} K_j$ no null values of $P_0^G(i, i')$, each corresponding to a common value of the attribute vector on both o_i and $o_{i'}$. In these conditions, computing the final probabilistic index requires to sort all these values and to calculate their respective frequencies. Consequently the computational complexity becomes of order p whereas, it is of order 2 in our *LLA* index and all of classical indices. Of course, the computational complexity remains polynomial, but, is too large to be acceptable for large databases.

To conclude, let us compare the main features of the Goodall index and that we propose in the *LLA* approach

The Goodall Index

1. For a given categorical attribute a^j, $1 \leq j \leq p$, a valuated similarity on the category set $\{a_1^j, a_2^j, \ldots, a_h^j, \ldots, a_{K_j}^j\}$ is defined such that

   - The similarity between two distinct categories is null;
   - The similarity between one category and itself is as much high as the category is rare;

2. The probabilistic index between objects is based on the empirical distribution of the vector attribute $(a^1, a^2, \ldots, a^j, \ldots, a^p)$ on $\mathcal{O}$. The exact computing is too complex. A more realistic index can be substituted for it. The latter is calculated by considering the marginal distributions of the different attributes and under the independence hypothesis between these (see (7.2.34)).

Our *LLA* Probabilistic Similarity Index

1. With each nominal categorical attribute of $\mathcal{A}$, is associated a preordonance categorical attribute. Only semantic considerations, provided by expert knowledge take part for establishing these preordonances. Each of them is coded by the mean rank function. This gives rise to a similarity index between the categories of the attribute concerned. Let us point out that in this construction, the similarity between two distinct categories is not constant. It varies from one category pair to another one. Besides, the ordinal similarity between one category and itself is neither constant. Nevertheless, it is greater than the ordinal similarity between two distinct categories;

2. The probabilistic index between two objects o_i and $o_{i'}$ follows an index obtained by centring and standardizing a sum of normalized contributions attribute by attribute of a raw similarity index between o_i and $o_{i'}$; the latter being provided as expressed in the previous item. Thus, this index considers the marginal empirical distributions of the different attributes on $\mathcal{O}$. However, from each empirical distribution associated with a given attribute a^j, $1 \leq j \leq p$, only the mean and variance of the raw similarity index are retained. The built probabilistic index can be justified with respect independence hypothesis between attributes, from the "central limit theorem".

7.2.8 Similarity Index Between Rows of a Juxtaposition of Contingency Tables

7.2.8.1 Case of a Single Contingency Table

We return to the data structure defined in Sect. 5.3.2.1 and the associated notations. More particularly, we reconsider the cloud $\mathcal{N}(\mathbb{I})$ of weighted points situated in the geometrical space $\mathbb{R}^J$ and associated with the row set of the contingency table (see (5.3.3)). Recall that this space is endowed with the Chi-square metric (see (5.3.4) and above). In the latter section the elements indexing the columns are represented in a very specific manner as numerical attributes. A correlation coefficient between column attributes was proposed (see (5.3.5)). As emphasized in this section, the perfect symmetry of a contingency table enables the respective roles of the rows and columns, to be reversed. Thus a correlation coefficient between elements of $\mathbb{I}$ can be proposed by referring to the dual cloud

$$\mathcal{N}(\mathbb{J}) = \left\{ \left(f_{\mathbb{I}}^j, p_j \right) | j \in \mathbb{J} \right\} \tag{7.2.35}$$

which is situated in the Euclidean space $\mathbb{R}^I$, endowed with the Chi-square metric. Specifying this metric and the formulas corresponding to (5.3.4) and (5.3.5) is left for the reader.

Here, we are about to apply the approach developed in this section—building a similarity index between objects—to the construction of a similarity index on $\mathbb{I}$. For this, we refer to the representation of $\mathbb{I}$ by means of the cloud $\mathcal{N}(\mathbb{I})$ (see (5.3.3)). In these conditions, the constructive basis consists of setting up a similarity index between objects described by numerical attributes (see Sect. 7.2.2). As for numerical description, we intend to specify the cosine index. By taking into account the Chi-square metric, this index can be written as follows

$$Cos_0(i, i') = Cos(\widehat{f_\mathbb{J}^i Of_\mathbb{J}^{i'}}) = \frac{< f_\mathbb{J}^i - O, f_\mathbb{J}^{i'} - O >}{\| f_\mathbb{J}^i - O \| \cdot \| f_\mathbb{J}^{i'} - O \|} \qquad (7.2.36)$$

where O designates the origin of the space $\mathbb{R}^J$ and where $< \bullet, \bullet >$ and $\| \bullet \|$ designate the inner product and the norm in this space. In these conditions, we get

$$Cos_0(i, i') = \frac{\sum_j (f_{ij} f_{i'j})/p_{.j}}{\sqrt{\left[\sum_j \left(f_{ij}^2 \right) /p_{.j} \right] \left[\sum_h \left(f_{i'h}^2 \right) /p_{.h} \right]}} \qquad (7.2.37)$$

To begin with, let us notice the following important property

Proposition 43 *The cosine index between two elements of* $\mathbb{I}$ *(resp.,* $\mathbb{J}$*) referring to the representation space of* $\mathcal{N}(\mathbb{I})$ *(resp.,* $\mathcal{N}(\mathbb{J})$*) is identical to the correlation coefficient between these two elements relative to the representation space of the cloud* $\mathcal{N}(\mathbb{J})$ *(resp.,* $\mathcal{N}(\mathbb{I})$*).*

Proof We will distinguish two cases depending on the reference point, with respect to which the cosine and the correlation coefficient are defined. For the first case, the reference point is the origin of the representation space and, for the second case, it is the centre of gravity of the cloud concerned. In this respect, let us recall the expressions of the centres of gravity $g_\mathbb{I}$ and $g_\mathbb{J}$ of the clouds $\mathcal{N}(\mathbb{I})$ and $\mathcal{N}(\mathbb{J})$, respectively. We have

$$g_\mathbb{I} = \sum_{i \in \mathbb{I}} p_i f_\mathbb{J}^i$$

$$g_\mathbb{J} = \sum_{j \in \mathbb{J}} p_j f_\mathbb{I}^j \qquad (7.2.38)$$

It is easy to verify that $p_{.1}, p_{.2}, \ldots, p_{.j}, \ldots, p_{.J}$ (resp., $p_{1.}, p_{2.}, \ldots, p_{i.}, \ldots, p_{I.}$) are the components of the vector $g_\mathbb{I} - O$ (resp., $g_\mathbb{J} - O$). O is the origin of the space concerned.

By referring to the cloud $\mathcal{N}(\mathbb{J})$, situated in the space $\mathbb{R}^I$, the correlation coefficient between i and i' with respect to the origin of the space, can be reduced to

$$Cor_0(i, i') = \frac{\sum_j (f_{ij} f_{i'j})/p_{.j}}{\sqrt{\left[\sum_j \left(f_{ij}^2 \right) /p_{.j} \right] \left[\sum_h \left(f_{i'h}^2 \right) /p_{.h} \right]}} \qquad (7.2.39)$$

Obviously, it is identical to the cosine index $Cos_0(i, i')$ (see (7.2.37)), calculated in the dual space $\mathbb{R}^J$.

Now, two indices have to be compared

$$Cos_g(i, i') = Cos(\widehat{f^i_{\mathbb{J}} g_{\mathbb{J}} f^{i'}_{\mathbb{J}}}) = \frac{<f^i_{\mathbb{J}} - g_{\mathbb{J}}, f^{i'}_{\mathbb{J}} - g_{\mathbb{J}}>}{\| f^i_{\mathbb{J}} - g_{\mathbb{J}} \| \cdot \| f^{i'}_{\mathbb{J}} - g_{\mathbb{J}} \|}$$

and

$$Cor_g(i, i') \qquad (7.2.40)$$

where the reference point is the centre of gravity of the cloud concerned: $\mathcal{N}(\mathbb{I})$ for $Cos_g(i, i')$ and $\mathcal{N}(\mathbb{J})$ for $Cor_g(i, i')$. Notice that in the dual case of comparing elements of $\mathbb{J}$ as column attributes (see (5.3.5)), $Cor_g(j, k)$ has been denoted by $\rho(j, k)$.

After simplification, we obtain

$$Cos_g(i, i') = \frac{\sum_j (f_{ij} f_{i'j}/p_{.j}) - p_{i.} p_{i'.}}{\sqrt{\left[\sum_j \left(f_{ij}^2/p_{.j} \right) - p_{i.}^2 \right] \left[\sum_j \left(f_{i'j}^2/p_{.j} \right) - p_{i'.}^2 \right]}} \qquad (7.2.41)$$

It is straightforward to verify that this index comes down to (5.3.5) by transposing the respective roles of $\mathbb{I}$ and $\mathbb{J}$. This accomplishes the proof. $\qquad \square$

We shall now apply the method considered in this section for building a similarity index between objects described by attributes. As mentioned above, the reference concerns the case where the descriptive attributes are numerical (see Sect. 7.2.2). In these conditions, the raw contribution of the jth component, $1 \leq j \leq J$, to the cosine index $Cos_0(i, i')$, has to be centred and standardized with respect to the set $\mathbb{I} \times \mathbb{I}$ of ordered pairs of $\mathbb{I}$. As usually, this contribution is designated by $s^j(i, i')$. By considering the (7.2.37), $s^j(i, i')$ can be written as follows

$$s^j(i, i') = \frac{f^i_j f^{i'}_j}{p_{.j}} = \frac{1}{p_{.j}} \frac{f_{ij}}{p_{i.}} \frac{f_{i'j}}{p_{i'.}} \qquad (7.2.42)$$

Empirical Mean and Variance of $s^j(i^\star, i'^\star)$ Over $\mathbb{I} \times \mathbb{I}$

As above, $s^j(i^\star, i'^\star)$ designates the random index associated with $s^j(i, i')$. We have

$$mean_e\big(s^j(i^\star, i'^\star)\big) = \sum_{(i,i') \in \mathbb{I} \times \mathbb{I}} p_{i.} p_{i'.} \frac{1}{p_{.j}} \frac{f_{ij}}{p_{i.}} \frac{f_{i'j}}{p_{i'.}} = p_{.j} \qquad (7.2.43)$$

The absolute moment of order 2 of $s^j(i^\star, i'^\star)$ can be expressed by

$$M2_e\left(s^j(i^\star, i'^\star)\right) = \sum_{(i,i')\in\mathbb{I}\times\mathbb{I}} p_i.p_{i'}.\frac{1}{p_{.j}^2}\left(\frac{f_{ij}}{p_{i.}}\right)^2\left(\frac{f_{i'j}}{p_{i'.}}\right)^2 = \sum_{(i,i')\in\mathbb{I}\times\mathbb{I}} \frac{f_{ij}^2}{p_{i.}p_{.j}}\frac{f_{i'j}^2}{p_{i'.}p_{.j}}$$

Therefore

$$var_e\left(s^j(i^\star, i'^\star)\right) = \frac{1}{p_{.j}^2}\left(\sum_{(i,i')\in\mathbb{I}\times\mathbb{I}} \frac{f_{ij}^2}{p_{i.}}\frac{f_{i'j}^2}{p_{i'.}}\right) \tag{7.2.44}$$

From (7.2.43) and (7.2.44) we derive the normalized contribution $S^j(i, i')$ of j to the similarity between i and i' (see (7.2.1)). The rest of the process for establishing a probabilistic similarity index on $\mathbb{I}$ remains the same and follows the Equation sequence (7.2.1)–(7.2.6).

It is important to notice that a coefficient of the form (7.2.10) can be proposed in order to define the raw contribution of the jth component to the similarity index between i and i'. In this case, it can be shown (the calculation is left for the reader) that ξ_i^j must be expressed as follows

$$\xi_i^j = \frac{\frac{1}{\sqrt{p_j}}f_j^i}{\sqrt{\sum_{1\le g\le J}\frac{1}{p_j}\left(f_g^i\right)^2}}$$

$1 \le i \le n, 1 \le j \le J.$

7.2.8.2   Case of an "Horizontal" Juxtaposition of Contingency Tables

Remember (see Sect. 5.3.2.2) that the data table is here indexed by a set of the form

$$\mathbb{I} \times (\mathbb{J}_1 + \mathbb{J}_2 + \cdots + \mathbb{J}_l + \cdots + \mathbb{J}_p) \tag{7.2.45}$$

where the sign $+$ indicates union of disjoint subsets.

For a given l, $\mathbb{I} \times \mathbb{J}_l$ indexes a single contingency table. In fact, $\mathbb{I}$ and $\mathbb{J}_l$ are associated, respectively, with the category sets of two categorical attributes. However and generally, in applications, $I = card(\mathbb{I})$ is relatively "large" and $J_l = card(\mathbb{J}_l)$ is relatively "small". This type of data structure occurs frequently in geographical and economical problems.

Now, let us consider the comparison of two distinct elements i and i' of $\mathbb{I}$ through $\mathbb{J}_1 + \mathbb{J}_2 + \cdots + \mathbb{J}_l + \cdots + \mathbb{J}_p$. The additive nature of the similarity index (7.2.2) enables us to integrate all of the $\mathbb{J}_l$, $1 \le l \le p,$, in a consistent way. The contribution of $\mathbb{J}_l$ to the similarity between i and i' can be expressed by

$$Q_l(i, i') = \sum_{j\in\mathbb{J}_l} S^j(i, i') \tag{7.2.46}$$

where $S^j(i, i')$ is the normalized contribution to the similarity $Q_l(i, i')$ of the column attribute j, $j \in \mathbb{J}_l$ (see (7.2.1)). Recall, that we have given above two forms of the raw contribution $s^j(i, i')$ of j to the similarity index between i and i'.

Now, in order to take into account the respective contributions of all of the $\mathbb{J}_l$, $1 \le l \le p$, the following index can be proposed

$$Q(i, i') = \sum_{1 \le l \le p} \frac{1}{J_l} Q_l(i, i') \tag{7.2.47}$$

where the multiplicative factors $1/J_l$, $1 \le l \le p$, make the different $\mathbb{J}_l$ to be of the same importance for establishing the similarity between i and i'.

7.2.9 Other Similarity Indices on the Row Set $\mathbb{I}$ of a Contingency Table

The similarity coefficients we shall present in this section are very general. They enable two probability distributions on a finite set to be compared. The purpose here consists of applying them to compare the elements of the row set $\mathbb{I}$ of a contingency table indexed by the cartesian product $\mathbb{I} \times \mathbb{J}$ (see Sect. 7.2.8.1).

7.2.9.1 The Matusita–Nicolau Affinity Coefficient

Let $f_{\mathbb{J}}$ be a probability (or relative frequency) distribution on a finite set $\mathbb{J}$, whose cardinality being denoted by J. If f_j is the probability assigned to the element j of $\mathbb{J}$, $f_{\mathbb{J}}$ can be represented geometrically by the point of $\mathbb{R}^J$ whose jth component is f_j. Without risk of ambiguity, we can denote this point by $f_{\mathbb{J}}$. We have

$$\text{For any } j, 1 \le j \le J, f_j \ge 0 \text{ and } \sum_{1 \le j \le J} f_j = 1$$

Clearly, the elements of the cloud $\mathcal{N}(\mathbb{I})$ have been represented in this manner (see Sect. 7.2.8.1).

As mentioned in Sect. 5.3.2.1, the set of probability distributions on $\mathbb{J}$ is represented geometrically by the portion of the hyperplane of $\mathbb{R}^J$ defined as follows

$$\left\{ (x_1, x_2, \ldots, x_j, \ldots, x_J) \in \mathbb{R}^J \,\middle|\, \text{For any } j, 1 \le j \le J, x_j \ge 0 \text{ and } \sum_{1 \le j \le J} x_j = 1 \right\}$$

For the coefficient concerned here, the probability distribution $f_\mathbb{J}$ is represented geometrically by the point whose jth component is $\sqrt{f_j}$ and that we denote by $\tilde{f_J}$, $1 \le j \le J$. The set of all these points is the following portion of the hypersphere of $\mathbb{R}^J$ of radius 1.

$$\left\{ (x_1, x_2, \ldots, x_j, \ldots, x_J) \in \mathbb{R}^J \,|\, \text{For any } j, \, 1 \le j \le J, x_j \ge 0 \text{ and} \sum_{1 \le j \le J} x_j^2 = 1 \right\}$$

In these conditions, the Matusita–Nicolau index between two profiles $f_{i\mathbb{J}}$ and $f{i'}_\mathbb{J}$ [1–3, 16] is simply the inner product

$$< f_{i\mathbb{J}} - O, f{i'}_\mathbb{J} - O > = \sum_{1 \le j \le J} \sqrt{f_j^i \cdot f_j^{i'}} \tag{7.2.48}$$

where i and i' are two distinct elements of $\mathbb{I}$. Due to the common value unity of the norm of both component vectors in the inner product, this index coincides with the cosine index between the two vectors concerned.

This index is in fact too general. It does not take into account the specificity of the geometrical representation of $\mathbb{I}$ through $\mathbb{J}$, defined by the cloud $\mathcal{N}(\mathbb{I})$ (see Sect. 7.2.8.1). More precisely, the Chi-square metric endowing the geometrical space $\mathbb{R}^J$ does not intervene. Notice that [1–3] consider, equivalently the comparison between elements of $\mathbb{J}$ and for this, we have to refer to the geometrical representation of $\mathcal{N}(\mathbb{J})$ (see (7.2.35)). Moreover, an adequate *likelihood linkage* version of (7.2.48) is studied in the mentioned papers.

7.2.9.2 The Goodall Proposal

In order to present this coefficient we shall consider—without loss of generality— mutual comparison between rows of a contingency table. For this, let us take up again the notations above (see (5.3.1)), where $\mathbb{I}$ and $\mathbb{J}$ designate the row set and the column set, respectively. Recall that in the cloud $\mathcal{N}(\mathbb{I})$ (see (5.3.3)) a given element i of $\mathbb{I}$ is defined by the relative frequency distribution $f_\mathbb{J}^i$. For comparing two elements i and i' of $\mathbb{I}$, $1 \le i < i' \le n$, Goodall [7] introduces the following parameter set[1]

$$\mathcal{D} = \{d(j^i, h^{i'}) | (i, i') \in \mathbb{I} \times \mathbb{I}, (j, h) \in \mathbb{J} \times \mathbb{J}\} \tag{7.2.49}$$

where $d(j^i, h^{i'})$ is the dissimilarity between j in i and h in i'. In these conditions, the basic index is of the form

---

[1]The notations used here are somewhat different from those of [7].

$$D(i, i') = \sum_{1 \leq j, h \leq J} f_j^i f_h^{i'} d(j^i, h^{i'})$$ (7.2.50)

$1 \leq i < i' \leq n$. Its distribution

$$Dist = \{D(i^\star, i'^\star) | \{i^\star, i'^\star\} \in P_2(\mathbb{I})\}$$

on the set of unordered element pairs of $\mathbb{I}$ is considered for the probabilistic evaluation of the similarity between i and i'. The final index can be expressed as follows

$$P(i, i') = Prob\{D(i^\star, i'^\star) \leq D(i, i')\}$$ (7.2.51)

It is exactly the proportion (or relative frequency) of elements $\{i^\star, i'^\star\}$ of $P_2(\mathbb{I})$ for which the value of the dissimilarity $D(i^\star, i'^\star)$ is lower or equal to that $D(i, i')$, observed.

This index appears as very problematic and complex to calculate. The following questions arise

1. How can we semantically determine the values of the dissimilarity function in table $\mathcal{D}$?
2. How can we handle a table as large as $\mathcal{D}$? this table comprises $I^2 \times J^2$ values (for $I = 1000$ and $J = 100, I^2 \times J^2 = 10^{10}$!)
3. How to extend the index $P(i, i')$ to the case of a juxtaposition of contingency tables.

References

1. Bacelar Nicolau, H.: Two probabilistic models for classification of variables. In: Bock, H.H. (ed.) Classification and Related Methods of Data Analysis, pp. 1981–1986. North-Holland, Amsterdam (1988)
2. Bacelar Nicolau, H.: Classifying integer scale data by the affinity coefficient. Methods Oper. Res. **60**, 587–595 (1990)
3. Bacelar Nicolau, H., Costa Nicolau, F., Sousa, F., Bacelar Nicolau, L.: Clustering variables with a three-way approach for health sciences. Test. Methodol. Psychom. Appl. Psychol. **4**, 435–447 (2014)
4. Barthélemy, P., Leclerc, B., Monjardet, B.: Quelques aspects du consensus en classification. In: Data Analysis and Informatics. North-Holland, Amsterdam (1984)
5. Diday, E.: Croisements, ordres et ultramétriques : application à la rechercherche de consensus en classification automatique. Rapport de recherche 144, INRIA (1982)
6. Goodall, W.D.: A new similarity index based on probability. Biom. **22**(4), 882–890 (1966)
7. Goodall, W.D.: Probabilistic indices for classification—some extensions. Abstr. Bot. **17**(1–2), 125–132 (1993)
8. Gower, J., Legendre, P.: Metric and euclidean properties of dissimilarity coefficients. J. Classif. **3**, 5–48 (1986)
9. Lerman, I.C.: Sur l'analyse des données préalable à une classification automatique; proposition d'une nouvelle mesure de similarité. Mathématiques et Sciences Humaines **32**, 5–15 (1970)

10. Lerman, I.C.: Classification et analyse ordinale des données. Dunod and http://www.brclasssoc. org.uk/books/index.html (1981)
11. Lerman, I.C., PETER, P.: Classification en présence de variables préordonnance taxonomiques à choix multiple. application à la structuration de phlébotomes de la guyanne francaise. Publication Interne 426, IRISA-INRIA (1988)
12. Lerman, I.C., Peter, P.: Indice probabiliste de vraisemblance du lien entre objets quelconques; analyse comparative entre deux approches. Revue de Statistique Appliquée, (LI(1)), pp. 5–35 (2003)
13. Lerman, I.C., Peter, P.: Representation of concept description by multivalued taxonomic preordonance variables. In: Cucumel, G., Brito, P., Bertrand, P., Carvalho, F. (eds.) Selected Contributions in Data Analysis and Classification, pp. 271–284. Springer, Heidelberg (2007)
14. Lesot, M.-J., Rifqi, M., Benhada, H.: Similarity measures for binary and numerical data. Int. J. Knowl. Eng.Soft Data Paradig. 1, 63–84 (2009)
15. Massé, J.R.: Classes de tableaux équivalents en analyse descriptive multidimensionnelle des données. Application à l' étude de mesures statiques sur circuits intégrés logiques. Ph.D. thesis, Université de Rennes 1, Octobre (1978)
16. Matusita, K.: Cluster analysis and affinity of distributions. In: Barra, J.R. et al. (ed.) Recent developments in Statistics, pp. 537–544. North-Holland, Amsterdam (1977)
17. Régnier, S.: Sur quelques aspects mathématiques des problèmes de classification automatique. I.C.C. Bull. 4, 175–191 (1965)
18. Rohlf, F.J.: Consensus indices for comparing classifications. Math. Biosci. 59, 131–144 (1982)
19. Sneath, P.H.A., Sokal, R.R.: Numerical Taxonomy. Freeman, San Francisco (1973)

Chapter 8
The Notion of "Natural" Class, Tools for Its Interpretation. The Classifiability Concept

8.1 Introduction; Monothetic Class and Polythetic Class

Numerical, Statistical and Algorithmic approaches in traditional Biology Classification appear around the 50 years. This introduction of new investigating methods was due to an English and American school of Biologists or Naturalists taxonomists open to metrical statistical and computer science approaches. The principal objective consists of discovering a system of "natural" classes organizing the vast set of living beings. The books "Principles of numerical taxonomy", by R. Sokal and P.H.A. Sneath (1963) and "Numerical taxonomy" (1973) [29, 30] have played a decisive role to assess the importance of the new types of techniques. Given the very broad classification of living, it is not surprising that these advances were due to life scientists.

The main goal of an expert establishing descriptive attributes of the population $\mathcal{P}$ of objects he studies consists of discovering "natural" classes and "natural" classifications (Gilmour 1937) [8]. In order to introduce the notion of a "natural" *class* (one can also say "natural" *cluster*), let us suppose Boolean the description of $\mathcal{P}$ (see Chap. 3, Sects. 3.1 and 3.2). $\mathcal{A} = \{a^1, a^2, \ldots, a^j, \ldots, a^p\}$ will indicate the attribute set. Before going further, note that the terms population $\mathcal{P}$ and universe $\mathcal{U}$ (used in Chap. 3) are interchangeable.

An object class M of $\mathcal{P}$ is said *monothetic* if it can be characterized by an attribute conjunction of the form $A = a^{j1} \wedge a^{j2} \wedge \cdots \wedge a^{jk}$, $1 \leq j1 < j2 < \cdots < jk \leq p$, where an object x of $\mathcal{P}$ belongs to M, if and only if A is $TRUE$ on x. Quickly the notion of monothetic class has appeared too rigid in order to reflect natural classes in Biology and the sciences of living. Already, Vicq d'Azyr [32] emphasized on the fact that a class might be very natural without possible common attribute to all of its elements. Danser [6] realizes the difficulty to assign a scientific status to the notion of a natural class. In fact, there does not exist a mathematical definition of this concept. Its intuitive expression is necessarily relative and has a statistical nature [27]. Even so, this concept founds all the data analysis based on *Clustering*.

© Springer-Verlag London 2016
I.C. Lerman, *Foundations and Methods in Combinatorial and Statistical Data Analysis and Clustering*, Advanced Information and Knowledge Processing, DOI 10.1007/978-1-4471-6793-8_8

By contrast with the notion of a monothetic class, a natural class is also called a *polythetic* class. Although a certain arbitrariness in the formulation, the Beckner (1959) definition [3] of a polythetic class represents a significant advance. According to above, assuming a Boolean description, a polythetic class P of $\mathcal{P}$ refers to a set A of descriptive features ($A \subset \mathcal{A}$), such that every element of P possesses a "large", but not specified part of the elements of A and, conversely, each attribute of A is present in a "large", but not specified part of the elements of P. For this "definition" it is not required to have attributes of A present in all of the elements of P. This definition will be taken up again in Sect. 8.1.1.

In fact, discovering a hierarchical system of natural classes and subclasses underlies the origin of the Adanson (1757) approach [1]. He wrote: "Je me contenterai de rapprocher les objets suivant le plus grand nombre de degrés de leurs rapports et de leur ressemblances ... . Les objets ainsi réunis formeront plusieurs petites familles que je réunirai encore ensemble, afin d'en faire un tout dont les parties soient unies et liées intimement.". A possible translation of this citation might be: "I will content myself with gathering together objects, according to the more large number of their relationships and resemblances. The objects joined will form small families that I put together again, in order to make a whole, whose parts being linked intimately."

The Adanson strategy leads to the *Ascendant Agglomerative Hierarchical Clustering* methods. And, this book is dominated by this approach. Its algorithmic facets, including ordinal, metrical and statistical components, are studied in Chap. 10. The tools developed in Chap. 9 enable clusters or clustering to be validated. Pairwise comparison of the elements of the set to be clustered (Attributes, Objects or Categories) is studied in depth in Chaps. 4–7.

The Beckner intuitive definition as well as the Adanson constructive process are oriented towards clustering an object set. This is also the case of most works in Clustering [33]. One implicit reason for this, might be, in general, the respective size orders of the object set and the attribute set. The size of the object set may reach several millions, whereas that of the attribute set exceeds rarely some thousands. However, clustering the attribute set constitutes a fundamental task. Actually, building an *Ascendant Agglomerative Hierarchical Clustering* on the descriptive attribute set $\mathcal{A}$, according to resemblance notions between attributes or attributes clusters leads to the discovery of the principal components and subcomponents of the population under study. Moreover, the latter clustering enables an object set clustering to be better controlled. In a sense we can say that while clustering attributes contributes to understanding the behaviour of the population studied $\mathcal{P}$, clustering objects enables us (by using attribute clustering) to manage and to make decision relative to $\mathcal{P}$. Otherwise, this approach of attribute clustering gives a nonlinear alternative to *Factorial Analysis* [2, 10]. While factorial analysis methodology is restricted to data where attributes have to be represented geometrically by *linear forms* [16], the method we propose adapt to any type of attributes (see Chap. 3 for the formalization of the different types of attributes).

Besides, such a clustering of the attribute set $\mathcal{A}$ gives rise to *fuzzy* clustering of the object set $\mathcal{O}$. This can be obtained by assigning to each element x of $\mathcal{O}$, numerical coefficients (one by attribute class) indicating to what extent x justifies the different tendencies corresponding to the different attribute classes. In these conditions, it is of importance to situate mutually two fuzzy clusterings on the same set of objects. This is performed in Sect. 8.6, in the case where the attribute set $\mathcal{A}$ is composed of Boolean attributes. This development has required a specific adaptation of the contingency chi-square theory.

Notice that the Beckner formulation considered in the case of Boolean description is perfectly symmetrical with respect to the ordered pair $(\mathcal{P}, \mathcal{A})$. This is due to the formal symmetry of a Boolean incidence data. Nevertheless, we saw in Chap. 7 that comparing Boolean attributes is not statistically equivalent to comparing objects described by Boolean attributes (Compare Sects. 4.2.1, 5.2.1 and 7.2).

A contingency table defines a perfect symmetrical data table (see Sect. 5.3.2.1). In Sect. 8.4, we will analyze the notions and tools developed in this chapter about the latter data structure.

The Beckner view of a natural class leads naturally to a direct approach in constructing clusters. This will be illustrated in Sect. 8.1.1 and more substantially developed in Sect. 9.2 of Chap. 9.

In any case, the objective here is by no means to build clusterings. Chapters 2 and 10 are devoted to study such constructions. We suppose in the first part of this chapter, "good" classifications established on an object set $\mathcal{O}$ (representing the population studied) and on an attribute set $\mathcal{A}$ associated with description of $\mathcal{O}$ by $\mathcal{A}$. And the matter is to "explain", to "understand" the cluster formation of the object classes (resp., the attribute classes) with respect to attributes or clusters of attributes in $\mathcal{A}$ (resp., with respect to objects or object clusters in $\mathcal{O}$).

In fact, this chapter provides gripping tools once the classification built (see Sects. 8.2–8.4 and 8.6) or previously to this construction (see Sects. 8.5 and 8.7).

Consider now a set E endowed with a proximity function p (ordinal or numerical). The *classifiability* of E with respect to p is intuitively defined by the ability of E to be organized in a system of classes and subclasses, respecting to a large extent the proximities between the elements of E. Precise status will be given for this notion in Sect. 8.7, where classifiability measures corresponding to Lerman and Murtagh works [12, 22] are examined. These measures evaluate the *ultrametricity* degree of the proximity function p (see Chap. 1), and then statistical testing hypothesis of non-ultrametricity is proposed, the latter being not dependent on any clustering algorithm.

Interesting remarks and properties are given in [28]. In particular, a new notion quantifying the ultrametricity of a dissimilarity is proposed. Transformations that increase or diminish ultrametricity can be envisaged with respect to various clustering algorithms. The representation of the ultrametric space in terms of equivalence classes of spheres is also considered in this book (see Sects. 1.4.1 and 1.4.2 of Chap. 1). The latter development is provided from Sect. 4 of Chap. 1 of [16].

The classifiability of (E, p) is closely related to the proximity distributions to every element of E. More precisely, denoting by

$$\mathcal{F} = \{P(x) | x \in E\} \tag{8.1.1}$$

the family of these distributions we have

$$P(x) = \{p(x, y) | y \in E - \{x\}\} \tag{8.1.2}$$

For a given x in E the variance $V(x)$ of the distribution $P(x)$ indicates what we call the "projective" importance of x. A small (resp., large) value of $V(x)$ corresponds to a small (resp., large) discrimination of x in its comparison with the other elements of E. Prior to a clustering of E it might be interesting to pinpoint the most "discriminating" (resp., "neutral") elements of E corresponding to the biggest (resp., smallest) values of the variance $V(x)$. Moreover, it is also interesting for a given cluster C—obtained by a clustering algorithm applied on E—to order the elements of C according to their respective projective importances.

The notion of projective importance is conceptually different from that of "classificatory" importance. If x is an element of E, the latter notion concerning x is defined with respect to a partition (classification) of the dual set of E. More clearly, if E is the attribute set (resp., the object set), the dual set of E is the object set $\mathcal{O}$ (resp., the attribute set $\mathcal{A}$). These two dual notions will be addressed in Sects. 8.2 and 8.3 and compared in Sect. 8.5.

8.1.1   The Intuitive Approaches of Beckner and Adanson; from Beckner to Adanson

Let us take up again the intuitive Beckner formulation of a natural (polythetic) class P of objects ($P \subset \mathcal{O} \subset \mathcal{P}$), in the case of Boolean description. Such a class refers to an attribute subset A of the attribute set $\mathcal{A}$, satisfying the conditions

1. Each element of P possesses a "large", but not fixed, proportion of attributes in A;
2. Each attribute of A is present in a "large", but not fixed, part of objects of P;
3. There does not exist necessarily an attribute of A which is present in all of the elements of P.

The impact of this intuitive definition on pairwise comparison between elements of P (resp., A) may be expressed as follows:

1. Two objects of P possess in common a large proportion of attributes of A;
2. Two attributes of A are present simultaneously in a large proportion of objects of P.

Relative to the concept of natural class, going from the first characterization to the second one corresponds to the passage between the Beckner view and the implicit Adanson one. Notice that in the latter the notion of a reference attribute class does not exist; moreover, for this view, comparing clusters is required. This comparison is examined in detail in Chap. 10.

Consider now a partition $\pi(\mathcal{O}) = \{O_1, O_2, \ldots, O_k, \ldots, O_K\}$ into natural classes of an object set $\mathcal{O}$ provided by a representative sample of the population $\mathcal{P}$ studied.

To fix ideas, $\mathcal{O}$ may be supposed large enough and randomly chosen. $\pi(\mathcal{O})$ is generally obtained by means of a clustering algorithm, the latter might be a non-hierarchical one (see Chap. 2) or a hierarchical one (see Chap. 10). In the latter case, the partition $\pi(\mathcal{O})$ may correspond to one of the most significant levels of a classification tree on $\mathcal{O}$ (see Sect. 9.3.6 of Chap. 9). According to the Beckner definition, a sequence $\chi(\mathcal{A}) = (A^1, A^2, \ldots, A^k, \ldots, A^K)$ of attribute $\mathcal{A}$ subsets can be associated with $\pi(\mathcal{O})$, such that O_k refers to A^k, $1 \leq k \leq K$. Contrary to what one might think, the case where each of the descriptive attributes is principally present in only one object class is very particular. In the latter case $\chi(\mathcal{A})$ forms a partition of $\mathcal{A}$. This situation can be illustrated by Fig. 8.1 which represents an incidence data table whose rows and columns are permuted according to the ordered pair of classifications $(\{O_1, O_2, \ldots, O_k, \ldots, O_K\}, \{A^1, A^2, \ldots, A^k, \ldots, A^K\})$. The number of '+' in agiven rectangle of the form $O_h \times A^k$, $1 \leq h, k \leq 5$ reflects the density of presence of the value 1.

The associated classification trees on $\mathcal{O}$ and $\mathcal{A}$ are given by Fig. 8.2.

More general situation of crossing a "good" partition

$$\pi(\mathcal{O}) = \{O_1, O_2, \ldots, O_j, \ldots, O_J\}$$

Fig. 8.1 Cross-polythetic classifications I

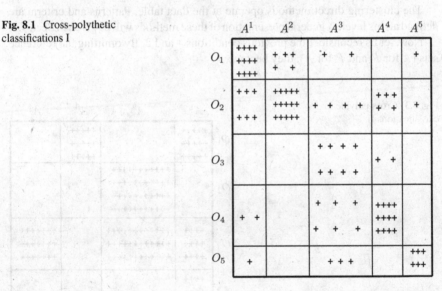

	A^1	A^2	A^3	A^4	A^5
O_1	++++ ++++ ++++	+ + + + +	+ +		+
O_2	+ + + + + +	+++++ +++++ +++++	+ + + +	+ + + + +	+
O_3			+ + + + + + + +	+ +	
O_4	+ +		+ + + + + +	++++ ++++ ++++	
O_5	+		+ + +		+++ +++

Fig. 8.2 Classification trees associated with Fig. 8.1

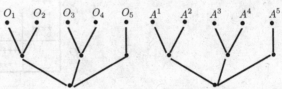

of $\mathcal{O}$ with a "good" partition

$$\chi(\mathcal{A}) = (A^1, A^2, \ldots, A^k, \ldots, A^K)$$

of $\mathcal{A}$ satisfying the Beckner formulation is:

1. To every cluster O_j of $\pi(\mathcal{O})$, $1 \le j \le J$, corresponds a union B^j of $\chi(\mathcal{A})$, where (O_j, B^j) can be substituted for (P, A) in the Beckner definition above;
2. To every cluster A^k of $\chi(\mathcal{A})$, $1 \le k \le K$, corresponds a union G_k of $\pi(\mathcal{O})$, where (G_k, A^k) can be substituted for (P, A) in the Beckner definition above.

Notice that

$$\sum_{1 \le j \le J} O_j \times B^j = \sum_{1 \le k \le K} G_k \times A^k \qquad (8.1.3)$$

where the sums are defined by dijoint unions where "$\times$" is the Cartesian product.

Both associated partitions $\pi(\mathcal{O})$ and $\chi(\mathcal{A})$ make that the proportion of 1 values is significantly high in the subrectangles of the incidence data table concerned by each of the decompositions given in (8.1.3). On the contrary, this proportion is low in the complementary part of (8.1.3) in the incidence data table (Fig. 8.3).

The clustering direct methods operate at the data table. Patterns and criteria are defined at this level. A general presentation of these methods will be given in Sect. 9.2.

Now, let us reconsider the previous conditions 1 and 2. By omitting the reference sets (A for P and P for A), they become

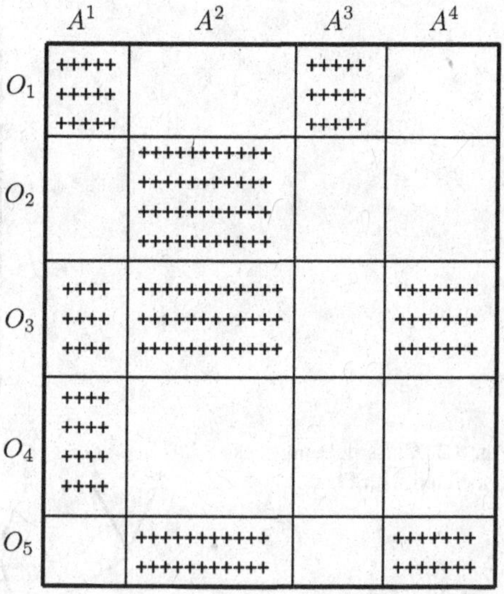

Fig. 8.3 Cross polythetic classifications II

	A^1	A^2	A^3	A^4
O_1	+++++ +++++ +++++		+++++ +++++ +++++	
O_2		++++++++++++ ++++++++++++ ++++++++++++ ++++++++++++		
O_3	++++ ++++ ++++	++++++++++++ ++++++++++++ ++++++++++++		+++++++ +++++++ +++++++
O_4	++++ ++++ ++++ ++++			
O_5		++++++++++++ ++++++++++++		+++++++ +++++++

1. The objects x and y of P possess in common a large number of attributes of the whole attribute set $\mathcal{A}$;
2. The attributes a and b of A are simultaneously present in a large number of objects of the whole object set $\mathcal{O}$.

In Chap. 4, we have designated by $s(x, y)$ (resp., $s(a, b)$) the number of attributes simultaneously present in x and y (resp., the number of objects possessing the attributes a and b) (see Sects. 4.2, 5.2 and 7.2). $s(x, y)$ and $s(a, b)$ were called *raw* similarity indices. We have seen how to formalize the scales evaluation of $s(x, y)$ and $s(a, b)$. Moreover, we have extended this formalization in order to compare descriptive attributes or described objects (resp., categories) for a large range of data types. These have been outlined in Sect. 3.5.4.

8.2 Discriminating a Cluster of Objects by a Descriptive Attribute

8.2.1 Introduction

From the description of an object set $\mathcal{O}$ by an attribute set $\mathcal{A}$, we assume established a partition $\pi(\mathcal{O}) = \{O_1, O_2, \ldots, O_h, \ldots, O_H\}$ into homogeneous classes respecting the resemblances between the elements of $\mathcal{O}$, according to the description. As expressed above in Sect. 8.1, this homogeneity notion is statistical and has to be understood intuitively. To illustrate this, imagine that you take at random two objects in $\mathcal{O}$, their similarity has tendency to be high (resp., low) if both objects belong to the same class (resp., if the two objects belong, respectively, to two different classes). As said above, this partition (classification) is generally built by means of a clustering algorithm (see Chaps. 2 and 10). As usually, if $n = card(\mathcal{O})$ and $p = card(\mathcal{A})$, we may denote $\mathcal{O} = \{o_i | 1 \leq i \leq n\}$ and $\mathcal{A} = \{a^j | 1 \leq j \leq p\}$.

As emphasized in the general introduction above (Sect. 8.1) a fundamental problem of clustering consists of "understanding" the partition $\pi(\mathcal{O})$ and interpreting each of its classes in terms of the initial description (by $\mathcal{A}$). We shall consider in Sect. 8.2.2 the case where $\mathcal{A}$ is composed of attributes of type I, that is, numerical or Boolean (see Sect. 3.2 of Chap. 3). To each element a of $\mathcal{A}$ we will associate the value of a discriminating function of the partition $\pi(\mathcal{O})$ by a. This function will be denoted by $D(\pi(\mathcal{O})|a)$. By associating a specific transformed attribute a_π with a, such that a_π is constant on every class O_k, $1 \leq k \leq K$, $D(\pi(\mathcal{O})|a)$ takes the form of a ratio between two variances: $var(a_\pi)$ and $var(a)$. $var(a_\pi)$ corresponds to the so called "explained" or "retained" variance. $D(\pi(\mathcal{O})|a)$ can also be expressed as a "correlation ratio", that is to say, the square of a correlation coefficient between a_π and a.

Given that generalization of a correlation coefficient was built in Chaps. 5 and 6, under the form defined by (6.2.5) (see also (6.2.92)) for all types of categorical attributes, the coefficient $D(\pi(\mathcal{O})|a)$ can be generalized in the latter extent. More precisely, we will consider the following cases for the attribute structure of a: *binary valuated, nominal categorical, ordinal categorical and categorical valuated by a numerical similarity*. Relative to the development of Chap. 6, one of both attributes concerned in the discriminating coefficient $D(\pi(\mathcal{O})|a)$ is associated with the partition $\pi(\mathcal{O})$, and then it defines a nominal categorical attribute on $\mathcal{O}$.

To end this introduction, let us indicate that this section (Sect. 8.2) reports the article [18] to which the reader may refer for more details.

8.2.2   Case of Attributes of Type I: Numerical and Boolean

Numerical data are the most classical ones. For this, let us recall the data table T_{num} expressed in (7.1.1) of Chap. 7:

$$T_{num} = \{x_i^j = v^j(o_i)|1 \le i \le n, 1 \le j \le p\} \tag{8.2.1}$$

associated with the ordered pair $(\mathcal{O}, \mathcal{A})$, where $\mathcal{O}$ and $\mathcal{A}$ are the described object set and the descriptive attribute set, respectively. The latter descriptive attribute set is here composed of numerical attributes. As usually, $\mathcal{O}$ and $\mathcal{A}$ can be written as

$$\mathcal{O} = \{o_i|1 \le i \le n\} \text{ and } \mathcal{A} = \{v^j|1 \le j \le p\} \tag{8.2.2}$$

According to above

$$\pi(\mathcal{O}) = \{O_1, O_2, \ldots, O_h, \ldots, O_H\} \tag{8.2.3}$$

indicates a partition of $\mathcal{O}$ into homogeneous clusters. $\mathcal{H} = \{1, 2, \ldots, h, \ldots, H\}$ is the class label set. Now, let us desigante by $\mathbb{I} = \{1, 2, \ldots, i, \ldots, n\}$ the index set of the $\mathcal{O}$ elements. $\mathbb{I}_h$ is the $\mathbb{I}$ subset concerned by the elements of O_h. In the following, $\mathbb{I}$ and $\mathcal{O}$ (resp., $\mathbb{I}_h$ and O_h) are interchangeable. We set $n_h = card(O_h) = card(\mathbb{I}_h)$, $1 \le h \le H$. Obviously,

$$n = \sum_{1 \le h \le H} n_h$$

Let v be a generic element of $\mathcal{A}$ and let

$$\{x_i|1 \le i \le n\} \tag{8.2.4}$$

denote the sequence value of v on $\mathbb{I}$. The general empirical mean of v on $\mathbb{I}$ is

$$x_{\cdot} = \frac{1}{n} \sum_{i \in \mathbb{I}} x_i \tag{8.2.5}$$

The class partial means can be expressed by

$$x^h_. = \frac{1}{n_h} \sum_{i \in \mathbb{I}_h} x_i \tag{8.2.6}$$

$1 \leq h \leq H$.

Now, we will associate the attribute v_π with v, such that v_π is constant on each of the classes of $\pi(\mathcal{O})$, the general empirical mean of v_π being equal to that of v. More precisely,

$$(\forall i \in \mathbb{I}),\, v_\pi(i) = x^h_. \text{ if and only iff } i \in \mathbb{I}_h \tag{8.2.7}$$

The variance $var(v)$ of the attribute v can be decomposed—according to the well-known formula—into the intra-class (*Within*) variance and the inter-class (*Between*) variance, namely

$$\frac{1}{n} \sum_{1 \leq i \leq n} (x_i - x_.)^2 = \sum_{1 \leq h \leq H} \frac{n_h}{n} \left(\frac{1}{n_h} \sum_{i \in \mathbb{I}_h} (x_i - x^h_.)^2 \right) + \sum_{1 \leq h \leq H} \frac{n_h}{n} (x^h_. - x_.)^2 \tag{8.2.8}$$

The left member represents $var(v)$ and the second term of the right member, $var(v_\pi)$. The discriminating measure of the partition $\pi(\mathcal{O})$ by the attribute v is given by the ratio

$$D(\pi|v) = \frac{var(v_\pi)}{var(v)} \tag{8.2.9}$$

which is comprised between 0 and 1, 0 if the respective means of v on the differents classes $\mathbb{I}_h$, $1 \leq h \leq H$, are equal, 1, if v is constant on each of the classes $\mathbb{I}_h$, $1 \leq h \leq H$.

It is easy to see that the correlation coefficient between v_π and v is given by

$$Corr(v_\pi, v) = \left(\frac{var(v_\pi)}{var(v)} \right)^{1/2} \tag{8.2.10}$$

Hence, the "Correlation ratio" between v_π and v satisfies the equality

$$\big(Corr(v_\pi, v)\big)^2 = D(\pi|v) \tag{8.2.11}$$

This interpretation of the discrimination coefficient $D(\pi|v)$ of a partition by an attribute will constitute the guiding principle in the case of discriminating a partition by categorical attribute of type II (see Sect. 3.3 of Chap. 3).

Case where the Discriminating Attribute is Boolean

Let us denote by a a Boolean attribute defined on $\mathcal{O}$. As usual, its logical values $TRUE$ and $FALSE$ are coded by 1 an 0. The mean notion considered in the case of numerical attribute becomes a proportion notion. In these conditions, denote by p_a

the proportion in the object set $\mathcal{O}$, of elements for which a is $TRUE$ and by p_a^h this same proportion in the class O_h, $1 \leq h \leq H$. Equations (8.2.9) and (8.2.11) become

$$D(\pi|a) = \frac{var(a_\pi)}{var(a)} = \frac{\sum_{1 \leq h \leq H} \frac{n_h}{n}(p_a^h)^2 - p_a^2}{p_a - p_a^2} \qquad (8.2.12)$$

and

$$\left(Corr(a_\pi, a)\right)^2 = D(\pi|a) \qquad (8.2.13)$$

Now, if we consider the contingency table crossing the partition π with that into two classes $TRUE$ and $FALSE$ associated with the Boolean attribute a. We have

$$D(\pi|a) = \phi^2 \qquad (8.2.14)$$

where $\phi^2 = \chi^2/n$ (see Sect. 6.2.3.5 of Chap. 6).

8.2.3   Discrimination a Partition by a Categorical Attribute

8.2.3.1   Introduction

We have now to propose a generalization of the discrimination coefficient $D(c|\alpha)$ in the case where α is a categorical attribute. Different types of the latter will be considered. In any case and systematically, this coefficient can be obtained by interpreting α as a valuated binary relation on the object set $\mathcal{O}$ [18]. As mentioned at the end of the previous section, three types will be considered for α: *nominal categorical*, *ordinal categorical* and *categorical valuated by a numerical similarity* (see Sect. 3.3 of Chap. 3). Notice that the latter case includes the *categorical preordonance attribute* when it is coded by means of the "mean rank" function (see (3.4.8) of Sect. 3.4.1). Clearly, to each categorical structure of α (i.e. structure on the value set of α), will correspond a specific form of the binary relation on $\mathcal{O}$. Let us denote here by $\mathbb{A} = \{a_1, a_2, \ldots, a_j, \ldots, a_J\}$ the value set of α. On the other hand, $\mathcal{C} = \{c_1, c_2, \ldots, c_h, \ldots, c_H\}$ will designate the value set of the nominal categorical attribute to discriminate by α. As above (see (8.2.3)), the partition $\pi(\mathcal{O})$, induced by c is denoted by

$$\pi(\mathcal{O}) = \{O_1, O_2, \ldots, O_h, \ldots, O_H\}$$

where $O_h = c^{-1}(c_h)$, $1 \leq h \leq H$. $t(\pi) = \{n_1, n_2, \ldots, n_h, \ldots, n_H\}$ indicates the *type* of $\pi(\mathcal{O})$ ($n_h = card(O_h)$, $1 \leq h \leq H$). Moreover, we will designate by

$$\chi(\mathcal{O}) = \{A_1, A_2, \ldots, A_j, \ldots, A_J\} \qquad (8.2.15)$$

the partition of $\mathcal{O}$ induced by α. $A_j = \alpha^{-1}(a_j)$, $1 \leq j \leq J$. $t(\chi) = \{m_1, m_2, \ldots, m_j, \ldots, m_J\}$ indicates the *type* of $\chi(\mathcal{O})$ ($m_j = card(A_j)$, $1 \leq j \leq J$).

In the development below, we shall begin by considering the case where α is a nominal categorical attribute. The case where α is an ordinal categorical attribute will be interpreted as a particular case of that for which α is a categorical attribute valuated by a numerical similarity (see Sect. 3.3.4 of Chap. 3). In these conditions, the latter case will be addressed first in Sect. 8.2.3.3. And consequently, the case where α is an ordinal categorical attribute will follow in Sect. 8.2.3.4.

8.2.3.2 The Discriminant Attribute α Is Nominal Categorical

To begin, we may consider a discriminant attribute referring to the formalism of the chi-square theory for association between nominal categorical attributes [9]. In [26] it is proved that the square of the Tshuprow coefficient between c and α

$$T_{c\alpha}^2 = \frac{\phi^2(c, \alpha)}{\sqrt{(H-1)(J-1)}} \qquad (8.2.16)$$

(see (6.2.44) of Sect. 6.2.3.5) can be geometrically expressed in terms of the square of a correlation coefficient between specific numerical attributes. Therefore, $T_{c\alpha}^2$ can be used as a discrimination coefficient of c by α.

The general method we propose refers to the relational representation of nominal categorical attribute as defined in Sect. 3.3.1 of Chap. 3. The coefficient having a correlation sense that we consider here is given by (6.2.35) of Sect. 6.2.3.5. Let us designate it by $R(c, \alpha)$ and recall its writing

$$R(c, \alpha) = \frac{Q_1(c, \alpha)}{\sqrt{Q_1(c, c) \cdot Q_1(\alpha, \alpha)}}$$

For the construction of the coefficient Q_1 between two nominal categorical attributes, the reader will take up again the development of Sect. 6.2.3. In these conditions, the discrimination coefficient can be written as

$$D(c|\alpha) = R(c, \alpha)^2 \qquad (8.2.17)$$

8.2.3.3 Case Where the Discriminant Attribute Is Categorical Valuated by a Numerical Similarity

In Sect. 6.2.5.5 we have specified how to compare two categorical attributes valuated, respectively, by a numerical similarity function, on the object set $\mathcal{O}$, according to the *LLA* normalization method. This study corresponds to a particular case of that, general, concerning the comparison between two attributes defining valuated binary relations on $\mathcal{O}$ (see Sects. 6.2.5.1–6.2.5.4 of Chap. 6). In order to apply the calculation results of Sect. 6.2.5.5, we shall begin by defining a binary valuation γ on the category set $\mathcal{C}$ of the attribute c as follows:

$$\gamma(c_g, c_h) = \begin{cases} 1 & \text{if } g = h \\ 0 & \text{if } g \neq h \end{cases}$$

$1 \leq g, h \leq H$.

The valuation γ being symmetrical, we have to assume equally a symmetrical structure of the discrimination attribute α, that is, $\alpha_{jk} = \alpha_{kj}$ for $1 \leq j, h \leq J$.

The expressions A_1, A_2 and A_3 of (6.2.125) become for γ

$$A_1 = \left(\sum_{1 \leq h \leq H} n_h(n_h - 1) \right)^2$$

$$A_2 = \sum_{1 \leq h \leq H} n_h(n_h - 1)$$

$$A_3 = \sum_{1 \leq h \leq H} n_h(n_h - 1) \tag{8.2.18}$$

Consequently,

$$A_2 = A_3 \text{ and } A_1 = A_2^2 = A_3^2$$

The parameters A_1, A_2 and A_3 associated with α are calculated according to (6.2.125). The symmetry of α simplifies somewhat the caculations concerned.

Let us now designate by $Q_1(\gamma, \alpha)$ the standardized coefficient between γ and α having the same form as that (6.2.86). The coefficient of correlative nature becomes

$$R(\gamma, \alpha) = \frac{Q_1(\gamma, \alpha)}{\sqrt{Q_1(\gamma, \gamma) \cdot Q_1(\alpha, \alpha)}}$$

Therefore, the discrimination coefficient of γ by α can be written as

$$D(\gamma|\alpha) = R(\gamma, \alpha)^2 \tag{8.2.19}$$

8.2.3.4   Case Where the Discriminant Attribute α Is Ordinal Categorical

The formal representation of this type of attribute was given in Sect. 3.3.2 of Chap. 3. The comparison between two ordinal categorial attributes was addressed in Sect. 6.2.4 of Chap. 6, with the *LLA* view. However, we have here to compare a *nominal* categorical attribute c with an *ordinal* categorical attribute α. The symmetrical character of c makes that we have to retain a symmetrical structure from α. In these conditions, let us define the following proximity index p on the value set of α.

$$(\forall (a_j, a_k) \in \mathbb{A} \times \mathbb{A}), \; p(a_j, a_k) = J - |j - k| \tag{8.2.20}$$

where $| \bullet |$ indicates the absolute value of $\bullet$. By this way the symmetry relation $p(a_j, a_k) = p(a_k, a_j)$, for all $1 \leq j \leq k \leq J$ is acquired.

In these conditions, we define a total a *total preordonance* on $\mathbb{A}$ as a total preorder on $\mathbb{A} \times \mathbb{A}$, according to this index. More precisely, the class $\mathbb{B}_i$ of this total preorder is defined as follows:

$$\{(a_j, a_k) | (a_j, a_k) \in \mathbb{A} \times \mathbb{A} \text{ and } J - |j - k| = i\} \tag{8.2.21}$$

Thus, for the quotient order we have

$$\mathbb{B}_1 < \mathbb{B}_2 < \cdots < \mathbb{B}_i < \cdots < \mathbb{B}_J \tag{8.2.22}$$

Obviously, the class $\mathbb{B}_i$ comprises $2i$ elements for $i = 1, 2, \ldots, (J - 1)$ and J elements for $i = J$.

By denoting $\alpha_{jk} = p(a_j, a_k)$, $1 \leq j \leq k \leq J$, the parameters A_1, A_2 and A_3, associated with the attribute α, become (see (6.2.125)):

$$A_1 = \left(2 \sum_{1 \leq j < k \leq J} m_j m_k \alpha_{jk} - \sum_{1 \leq j \leq J} m_j \alpha_{jj} \right)^2$$

$$A_2 = \sum_{1 \leq j \leq J} \left(n_j (n_j - 1) \alpha_{jj} \right) + \sum_{\{k | k \neq j\}} n_j n_k \alpha_{jk}$$

$$A_3 = 2 \sum_{1 \leq j < k \leq J} m_j m_k \alpha_{jk}^2 \tag{8.2.23}$$

Besides, the definition of the γ function representing the attribute c, has to be considered with respect to the Cartesian product $\mathcal{C} \times \mathcal{C}$ (see above).

In these conditions, when α is an ordinal categorial attribute, the coefficients $Q_1(\gamma, \alpha)$, $Q_1(\gamma, \gamma)$ and $Q_1(\alpha, \alpha)$ can be determined, and therefore the correlative coefficient $R(\gamma, \alpha)$ and its square corresponding to the discrimination coefficient $D(\gamma | \alpha)$, can be established (see (8.2.19) and above).

8.3 "Responsibility" Degree of an Object in an Attribute Cluster Formation

Preamble

In this section, we address the dual problem of that considered in Sect. 8.2. Let $\mathcal{A} = \{a^j | 1 \leq j \leq p\}$ be a set of descriptive attributes of the same type. Consider a partition $\chi(\mathcal{A}) = \{A^1, A^2, \ldots, A^k, \ldots, A^K\}$ of $\mathcal{A}$ into homogeneous clusters based on the description by $\mathcal{A}$ of an object set $\mathcal{O}$. In these conditions, given an element x of $\mathcal{O}$, the matter consists of measuring to what extent x justifies the "tendencies" appeared in the different clusters A^k, $1 \leq k \leq K$. We call the coefficient established for this purpose "Responsability degree" of x with respect to the cluster formation in $\chi(\mathcal{A})$. This coefficient will be denoted by $D(\chi(\mathcal{A})|x)$. The latter will be easier to understand in the case where $\mathcal{A}$ is composed of attributes of type I, that is to say, Boolean or numerical attributes (see Sect. 3.2 of Chap. 3). However, cases where $\mathcal{A}$ comprises attributes of type II (see Sect. 3.3 of Chap. 3), such that ranking attributes or ordinal categorical attributes, will also be examined.

8.3.1 $\mathcal{A}$ is Composed of Attributes of Type I

8.3.1.1   The Attributes are Boolean

The development above will inspire us for specifying $D(\chi(\mathcal{A})|x)$. Roughly, the latter coefficient will correspond to a ratio between two variances defined on the object x: the variance of the respective presences of the different attribute classes A^k ($1 \leq k \leq K$) on x and the variance of the presence of the whole attribute set $\mathcal{A}$ on x. Nevertheless, the development here cannot correspond to a direct transposition of that considered in Sect. 8.2. In fact, the structure induced by a given attribute on the object set concerned, has, essentially, a different nature of that defined by a given object on the attribute set describing it.

Let φ^k designate the proportion of attributes of A^k which are present on a given object x:

$$\varphi^k(x) = \frac{\sum_{a \in A^k} a(x)}{card(A^k)} \tag{8.3.1}$$

where $a(x) = 1$ (resp., $a(x) = 0$) if a is present (resp., absent) on x, $1 \leq k \leq K$. $(\varphi^1(x), \varphi^2(x), \ldots, \varphi^k(x), \ldots, \varphi^K(x))$ will define the distribution of the relative presence frequency of the different classes A^k on x, $1 \leq k \leq K$.

A first coefficient can be proposed for which the cardinalities of the classses A^k are not considered. It can be written as

$$D_1(\chi(\mathcal{A})|x) = \frac{\frac{1}{K} \sum_{1 \leq k \leq K} (\varphi^k(x) - \overline{\varphi}_1)^2}{\overline{\varphi}_1(1 - \overline{\varphi}_1)} \tag{8.3.2}$$

where

$$\overline{\varphi}_1(x) = \frac{1}{K} \sum_{1 \leq k \leq K} \varphi^k(x) \tag{8.3.3}$$

is the equally weighted mean of the sequence $(\varphi^1(x), \varphi^2(x), \ldots, \varphi^k(x), \ldots, \varphi^K(x))$.

The significancy of a given value of $\varphi^k(x)$ (see (8.3.1)) may depend on the size of A^k, $1 \leq k \leq K$. In order to take into account the respective importances of the profiles A^k, $1 \leq k \leq K$, measured by their respective cardinalities, we may introduce a second coefficient:

$$D_2(\chi(\mathcal{A})|x) = \frac{\sum_{1 \leq k \leq K} \pi_k (\varphi^k(x) - \overline{\varphi}_2)^2}{\overline{\varphi}_2(1 - \overline{\varphi}_2)} \tag{8.3.4}$$

where $\pi_k = p_k/p$, $p_k = card(A^k)$ and $p = card(\mathcal{A})$, $1 \leq k \leq K$.

On the other hand,

$$\overline{\varphi}_2(x) = \sum_{1 \leq k \leq K} \pi_k \varphi^k(x) \tag{8.3.5}$$

One might want to consider an additional significancy element related to the marginal distributions of the attributes of $\mathcal{A}$ on the whole object set $\mathcal{O}$. In order to understand intuitively this element, consider two objects x and y dominated by a given profile A^k. Now, imagine identical descriptions of x and y with the exception of two components associated with two attributes a and b of A^k: x (resp., y) possesses a but not b (resp., possesses b but not a). If a is more rarely possessed than b, it is "natural" to consider x *more representative* than y of the type defined by the profile A^k. To integrate this intuitive condition in the "Responsability degree" of a given object x with respect to $\chi(\mathcal{A})$, we may adopt as a measure of the presence of a given attribute a on a given object of $\mathcal{O}$, $(1 - n(a)/n)$ instead of 1, where $n(a)$ is the number of objects of $\mathcal{O}$ where a is present. In these conditions, the new coefficient to propose can be written as

$$D_3(\chi(\mathcal{A})|x) = \frac{\sum_{1 \leq k \leq K} \pi_k (\psi^k(x) - \overline{\psi}(x))^2}{\frac{1}{p} \sum_{a \in \mathcal{A}} (\psi_a(x) - \overline{\psi}(x))^2} \tag{8.3.6}$$

where $\psi_a(x) = (1 - n(a)/n)$ (resp., 0) if the Boolean attribute a is present (resp., absent) in the object x. On the other hand

$$\overline{\psi}(x) = \frac{1}{p} \sum_{a \in \mathcal{A}} \psi_a(x)$$

$$\psi^k(x) = \frac{1}{p_k} \sum_{a \in A^k} \psi_a(x) \tag{8.3.7}$$

Among the three indices (8.3.2), (8.3.4) and (8.3.7), the latter is the more developed. Notice that by interpreting a Boolean attribute as an ordinal categorical attribute whose value set being $\{\neg a, a\}$, ψ_a can be compared with the cumulative distribution function, in a strict sense, of a on $\mathcal{O}$.

Mutual and relative comparisons of the respective behaviors of the three coefficients $D_1(\chi(\mathcal{A})|x)$, $D_2(\chi(\mathcal{A})|x)$ and $D_3(\chi(\mathcal{A})|x)$ on real data require a minute experimentation. Nevertheless, $D_2(\chi(\mathcal{A})|x)$ and its additive components were validated on the basis of a large European survey in 1976 [5, 19] and Chap. 13 of [16]. The questionary submitted concerns

- Economic observations;
- Living conditions and personal satisfaction levels;
- Image and perception of the misery.

From about a hundred of questions corresponding to ordinal categorical attributes, 228 Boolean attributes were set up, constituting the attribute set $\mathcal{A}$. Representative people samples of nine countries were chosen at random. They are distributed as follows. 1232 for *France*, 1340 for *United Kingdom*, 1004 for *Germany*, 923 for *Italy*, 905 for *The Netherlands*, 980 for *Denmark*, 963 for *Belgium*, 268 for *Luxemburg* and 1007 for *Ireland*. Thus, the whole people sample comprises 8622 individuals.

The corresponding partition $\chi(\mathcal{A}) = \{A^1, A^2, \ldots, A^k, \ldots, A^K\}$ of $\mathcal{A}$ into $K = 7$ homogeneous clusters was determined at the most significant level of the *AAHC* classification tree on $\mathcal{A}$ by the *LLA* method (see Sect. 9.3.6 of Chaps. 9 and 10). The behavioral highlighted seven profiles were very clearly interpreted by the expert [5, 19] and Chap. 13 of [16]. They were called *Good economic and religious integration*, *Privileged people*, *Hard-working middle class*, *Passive and egoistic people*, *Christian moral and apolitical attitude*, *Democrat chiristian and liberal viewpoint* and *Unhappy people*.

The empirical distribution of $D_2(\chi(\mathcal{A})|x)$ and its components enable the responsability degrees of the different countries with respect to the behavioral profiles associated with $A^1, A^2, \ldots, A^k, \ldots$ and A^K, respectively, to be specified. In fact, the empirical distribution of $D_2(\chi(\mathcal{A})|x)$ on the whole people sample, defining the object set $\mathcal{O}$, permits us to detect a global threshold ρ such that a given object x is considered as "typical" if and only if $D_2(\chi(\mathcal{A})|x) \geq \rho$ [19]. Let us designate by $\mathcal{O}^t$ the following $\mathcal{O}$ subset

$$\mathcal{O}^t = \{x | x \in \mathcal{O} \text{ and } D_2(\chi(\mathcal{A})|x) \geq \rho\} \tag{8.3.8}$$

This set is divided into nine subsets according to the different countries:

$$\mathcal{O}^t = \sum_{1 \leq c \leq 9} \mathcal{O}^t_c \tag{8.3.9}$$

where $\mathcal{O}^t_c$ is the $\mathcal{O}^t$ subset composed of the elements belonging to the country c.

Consider now the kth component of $D_2(\chi(\mathcal{A})|x)$, namely

$$d_2(A^k|x) = \frac{\pi_k(\varphi^k(x) - \overline{\varphi}_2(x))^2}{\overline{\varphi}_2(x)(1 - \overline{\varphi}_2(x))} \tag{8.3.10}$$

For a given c the empirical distribution of $d_2(A^k|x)$ on $\mathcal{O}_c^t$ shows, generally, the same threshold phenomenon. By denoting ρ_{ck}^t the value of the latter, the set

$$\mathcal{O}_{ck}^t = \{x|x \in \mathcal{O}_c^t \text{ and } d_2(A^k|x) \geq \rho_{ck}^t\} \qquad (8.3.11)$$

is considered as composed of all elements of the country c, taking significant part in the cluster formation of $\chi(\mathcal{A})$ and, additionally, "typical" of the behavioral profile associated with A^k, $1 \leq k \leq K$. In these conditions, the index

$$\rho(A^k|\mathcal{O}_c) = \frac{card(\mathcal{O}_{ck}^t)}{card(\mathcal{O}_c)} \qquad (8.3.12)$$

can be interpreted as the responsibility part of $\mathcal{O}_c$ in the behavioral profile corresponding to the attribute class A^k, $1 \leq k \leq K$. c is here a value of the categorical attribute "nationality" and $\mathcal{O}_c$ is the sample people provided from the country c. By this way, the different nationalities can be scaled according to their respective responsability degrees with respect to the profile formation of the A^k, $1 \leq k \leq K$.

The calculations and the data analysis were carried out by T. Chantrel in the framework of his 3rd cycle thesis [5]. The most significant results are reported in [19] and Chap. 13 of [16]. The software concerned enables the fundamental following question to be answered:

- How many people in a given set S are responsible of a given behavioral profile associated with an attribute class B?

Thus, knowing that—due to the polythetic nature of B—it might be that any element of S possesses all of the B attributes (polythetic nature of B).

The development above is based on the variance analysis of the relative frequency of the B attributes present on a given object x of S. In Sect. 8.6, a different approach will be studied on the basis of an association coefficient between a classical Boolean attribute whose representation in $\mathcal{O}$ is the set S and a fuzzy Boolean attribute associated with the cluster of Boolean attributes B. The latter analysis will be consistent with the chi-square theory.

8.3.1.2   The Attributes Are Numerical

Let us designate here by $\mathcal{V} = \{v^j|1 \leq j \leq p\}$ the set of numerical descriptive attributes (see Sect. 3.2.2 of Chap. 3). On the other hand and as above, the data table is indicated by

$$T_{num} = \{x_i^j = v^j(o_i)|(i, j) \in \mathbb{I} \times \mathbb{J}\} \qquad (8.3.13)$$

where $\mathbb{I}$ (resp., $\mathbb{J}$) labels the object set (resp., the attribute set) (see (7.1.1) of Chap. 7 and (8.2.1)).

The previous development can be transposed easily to the data structure considered here. For this, two techniques can be proposed. The first one follows the transformation proposed in Sect. 7.2.2 of Chap. 7 where to each numerical attribute v^j corresponds a numerical attribute a^j whose scale is the interval $[0, 1]$. Thus, Formula (8.3.1) and all the consequent development can be taken up again in the same conditions. In this first technique, the value scale normalization of a given numerical attribute v^k ignores the set $\mathcal{V} - v^k$ of the other attributes. The second technique is inspired by the fact that in the standardized *Principal Component Analysis* processing rows and columns of T_{num} are equivalent if the values x_i^j are replaced by the standardized values

$$\xi_i^j = \frac{x_i^j - x^j}{s_j} \qquad (8.3.14)$$

where $(i, j) \in \mathbb{I} \times \mathbb{J}$ and

$$x^j = \frac{1}{n} \sum_{i \in \mathbb{I}} x_i^j \text{ and } s_j^2 = \frac{1}{n} \sum_{i \in \mathbb{I}} (x_i^j - x^j)^2 \qquad (8.3.15)$$

[10] and Chap. 6 of [16].

Let us consider now a partition

$$\chi(\mathcal{V}) = \{V^1, V^2, \ldots, V^k, \ldots, V^K\} \qquad (8.3.16)$$

into K homogeneous classes respecting the mutual associations between the attributes of $\mathcal{V}$. $\mathbb{J}_k$ designating the $\mathbb{J}$ subset indexing V^k, we denote by $p_k = card(\mathbb{J}_k)$. Then, $\pi_k = p_k/p$ is the proportion of attributes belonging to V^k, $1 \leq k \leq K$.

The dual decomposition of (8.2.8), with the measures ξ_i^j, relatively to a given object o_i, takes the form

$$\frac{1}{p} \sum_{1 \leq j \leq p} (\xi_i^j - \xi_i)^2 = \sum_{1 \leq k \leq K} \frac{p_k}{p} \left(\frac{1}{p_k} (\xi_i^j - \xi_i^k)^2 \right) + \sum_{1 \leq k \leq K} \frac{p_k}{p} (\xi_i^k - \xi_i)^2 \qquad (8.3.17)$$

In these conditions, the "Responsability degree" of a given object o_i with respect to the partition $\chi(\mathcal{V})$ is defined by

$$D_2(\chi(\mathcal{V})|o_i) = \frac{\sum_{1 \leq k \leq K} \pi_k (\xi_i^k - \xi_i)^2}{\frac{1}{p} \sum_{1 \leq j \leq p} (\xi_i^j - \xi_i)^2} \qquad (8.3.18)$$

Moreover, for a given k, the tendency associated with V^k is all the more justified by an object o_i, that

$$d_2(V^k|o_i) = \frac{\pi_k (\xi_i^k - \xi_i)^2}{\frac{1}{p} \sum_{1 \leq j \leq p} (\xi_i^j - \xi_i)^2} \qquad (8.3.19)$$

is larger, $1 \leq k \leq K, i \in \mathbb{I}$.

New coefficients corresponding to (8.3.18) and (8.3.19) can be considered for the Boolean case. These have the same analytical forms as that in (8.3.4) and (8.3.10), respectively. However, the value of a Boolean attribute a on a given object x is no longer 1 or 0, but defined by a numerical function that we denote by ϕ_a, for which

$$\phi_a(x) = \frac{(1 - f_a)}{\sqrt{f_a(1 - f_a)}} \text{ if the initial value of } a \text{ on } x \text{ is } 1$$

and

$$\phi_a(x) = \frac{-f_a}{\sqrt{f_a(1 - f_a)}} \text{ if the initial value of } a \text{ on } x \text{ is } 0. \quad (8.3.20)$$

where f_a is the proportion of objects possessing a. In this situation the initial Boolean attribute, taking the values 0 and 1, is interpreted as a numerical attribute taking its values in the interval [0, 1]. Therefore, the mean and variance of a on $\mathcal{O}$ are

$$x_{.j} = f_a \text{ and } s_j^2 = f_a(1 - f_a) \quad (8.3.21)$$

these values being used for establishing the function ϕ_a.

8.3.2 The Attribute Set $\mathcal{A}$ is Composed of Categorical or Ranking Attributes

To begin, we may assume that $\mathcal{A}$ is composed of attributes of the same type: nominal categorial, ordinal categorial or ranking attributes. As we will see below both latter cases will be processed in the same way, and then for the ordinal case, we may suppose $\mathcal{A}$ composed of ordinal categorical and ranking attributes. The case where $\mathcal{A}$ is composed of nominal categorical attributes is specific and essentially different.

8.3.2.1 $\mathcal{A}$ is Composed of *Nominal* Categorical Attributes

Each class A^k of a partition $\chi(\mathcal{A}) = \{A^1, A^2, \ldots, A^k, \ldots, A^K\}$ of $\mathcal{A}$ into K homogeneous clusters determines a general aspect of the description (see Chap. 6). At this global level, it is difficult to concieve a coefficient evaluating the extent to which $\chi(\mathcal{A})$ and in particular a given class A^k, is justified by the description of a given object x. In fact, the associations between the attribute categories which have led to the emergence of the A^k, $1 \leq k \leq K$, are hidden. In this situation, it might be recommended to come down to the Boolean attributes associated with the different attribute categories. More precisely, to each nominal categorical attribute c of $\mathcal{A}$, we will associate the set of Boolean attributes defined by the value set of c. This transformation is usually called "complete disjunctive coding". By this way, all the development of Sect. 8.3.1.1 is to take up again.

8.3.2.2 Case Where $\mathcal{A}$ is Composed of *Ordinal* Categorical or *Ranking* Attributes

Let us reconsider the interpretation of a Boolean attribute a in terms of an ordinal categorical attribute with two values $\neg a$ and a. For this expression (8.3.6) and (8.3.7) have been proposed. Recall that the value denoted by $\psi_a(x)$ of a on x is equal to 0 if a is absent in x and $1 - p(a)$ ($p(a) = n(a)/n$) if a is present in x. Notice that $1 - p(a)$ is the proportion of objects for which a is absent. This cumulative interpretation can be generalized in the case where the attribute concerned c is ordinal categorical. More precisely, $\{c_1, c_2, \ldots, c_j, \ldots, c_J\}$ designating the value set of c, where

$$c_1 < c_2 < \cdots < c_j < \cdots < c_J,$$

we set

$$f_j = \frac{card(\mathcal{O}_j)}{card(\mathcal{O})} = \frac{n_j}{n} \tag{8.3.22}$$

as the proportion of objects having the value c_j of c, $\mathcal{O}_j = c^{-1}(c_j)$ (see Sect. 3.3 of Chap. 3). In these conditions, the value $\psi_c(x)$ of the cumulative function ψ_c, generalizing ψ_a, is given by

$$\psi_c(x) = f_1 + f_2 + \cdots + f_{j-1} \tag{8.3.23}$$

if $c(x) = c_j$, $1 \le j \le J$, f_0 is set equal to 0.

Therefore, relative to the partition $\chi(\mathcal{A})$ (see above), Eqs. (8.3.6) and (8.3.7) can be retaken:

$$D_3(\chi(\mathcal{A})|x) = \frac{\sum_{1 \le k \le K} \pi_k(\psi_c^k(x) - \overline{\psi}_c(x))^2}{\frac{1}{p}\sum_{a \in \mathcal{A}}(\psi_c(x) - \overline{\psi}_c(x))^2} \tag{8.3.24}$$

where $\overline{\psi}_c = \frac{1}{p}\sum_{c \in \mathcal{A}} \psi_c(x)$. On the other hand,

$$\overline{\psi}_c(x) = \frac{1}{p}\sum_{a \in \mathcal{A}} \psi_c(x)$$

$$\overline{\psi}_c^k(x) = \frac{1}{p_k}\sum_{a \in A^k} \psi_c(x) \tag{8.3.25}$$

The contribution of x to the kth class is then defined by

$$d_{\mathcal{A}^k|x} = \frac{\pi_k(\psi_c^k(x) - \overline{\psi}_c(x))^2}{\frac{1}{p}\sum_{a \in \mathcal{A}}(\psi_c(x) - \overline{\psi}_c(x))^2} \tag{8.3.26}$$

Clearly, there is no need to distinguish the case where $\mathcal{A}$ is composed partly or totally of ranking attributes. If c is a ranking attribute, Eq. (8.3.23) becomes

$$\psi_c(x) = \frac{r-1}{n}$$

if the rank of x for c is r.

8.4 Rows or Columns of Contingency Tables

Preamble

This section concerns cluster analysis of the row set $\mathbb{I}$ through the column set $\mathbb{J}$ of a contingency table $\mathbb{I} \times \mathbb{J}$ or an "horizontal" juxtaposition of contingency tables $\mathbb{I} \times (\mathbb{J}_1 + \mathbb{J}_2 + \cdots + \mathbb{J}_l + \cdots + \mathbb{J}_L)$. This data structure was defined in Sect. 3.5.2 of Chap. 3 and Sect. 5.3.2 of Chap. 5 (see also Sect. 7.2.8 of Chap. 7). Here, we suppose given an homogeneous partition $\pi(\mathbb{I}) = \{\mathbb{I}_r | 1 \leq r \leq p\}$ of the row set $\mathbb{I}$ of the contingency table and it comes to quantify the discriminating ability of a given column j $(j \in \mathbb{J})$ or a given cluster $\mathbb{J}_s$ of columns $(\mathbb{J}_s \subset \mathbb{J})$ with respect to the partition $\pi(\mathbb{I})$. The geometric representation of the cloud $\mathcal{N}(\mathbb{I})$ (see (5.3.3) of Chap. 5) enables this objective to be accomplished. Clearly, the dual case of discriminating a partition of the column set $\mathbb{J}$ by elements or clusters in $\mathbb{I}$, employs the geometric representation of the cloud $\mathcal{N}(\mathbb{J})$ associated with the columns of the contingency table.

8.4.1 Case of a Single Contingency Table

The notations of Sect. 5.3.2.1 are taken up again here in the same terms. Let $\mathbb{I}$ and $\mathbb{J}$ be the respective sets of subscripts indexing the rows and columns of the contingency table:

$$\{n_{ij} | (i, j) \in \mathbb{I} \times \mathbb{J}\} \tag{8.4.1}$$

$I = card(\mathbb{I})$ and $J = card(\mathbb{J})$ will designate the number of rows and columns of this table, respectively. Writing $i \in \mathbb{I}$ (resp., $j \in \mathbb{J}$) is equivalent to write $1 \leq i \leq I$ (resp., $1 \leq j \leq J$). We put

$$n_{i.} = \sum_{j \in \mathbb{J}} n_{ij}, n_{.j} = \sum_{i \in \mathbb{I}} n_{ij}, 1 \leq i \leq I, 1 \leq j \leq J$$

$$n = n_{..} = \sum_{i \in \mathbb{I}} \sum_{j \in \mathbb{J}} n_{ij} \tag{8.4.2}$$

The following proportions or relative frequencies are naturally associated with the parameters introduced in (8.4.2)

$$f_{ij} = \frac{n_{ij}}{n}, \; p_{i.} = \frac{n_{i.}}{n} \text{ and } p_{.j} = \frac{n_{.j}}{n}, \; 1 \leq i \leq I, 1 \leq j \leq J.$$

The mathematical structure of a contingency table is perfectly symmetrical and the respective roles of $\mathbb{I}$ and $\mathbb{J}$ are interchangeable. We consider here the discrimination of a partition of $\mathbb{I}$ by a single column or a cluster of columns. For this purpose, we will interpret $\mathbb{I}$ as a set of objects and $\mathbb{J}$ as a set of attributes (numerical). More precisely, we will associate the cloud

$$\mathcal{N}(\mathbb{I}) = \{(f_{\mathbb{J}}^i, p_{i.}) | i \in \mathbb{I}\} \tag{8.4.3}$$

in the euclidean space $\mathbb{R}^J$, endowed with the following metric q

$$q(e_j, e_k) = \frac{\delta_{jk}}{p_{.j}}$$

where $\{e_j | 1 \leq j \leq J\}$ designates the canonical basis of $\mathbb{R}^J$ and δ_{jk} designates the δ of Kronecker ($\delta_{jk} = 0$ (resp. 1) if $j \neq k$ (resp., $j = k$)).

$f_{\mathbb{J}}^i$ is the point of $\mathbb{R}^J$ whose coordinates are $f_j^i = f_{ij}/p_{i.}, 1 \leq i \leq I$.

With the metric above q, called the chi-square metric, the square distance between two summits $f_{\mathbb{J}}^i$ and $f_{\mathbb{J}}^{i'}$ is the chi-square distance between the distributions concerned with respect to the distribution $\{p_{.j} | 1 \leq j \leq J\}$, namely

$$d_{\chi^2}^2(f_{\mathbb{J}}^i, f_{\mathbb{J}}^{i'}) = \sum_{j \in \mathbb{J}} \frac{1}{p_{.j}} (f_j^i - f_j^{i'})^2 \tag{8.4.4}$$

Otherwise, in Sect. 7.2.8.1 we have built a similarity index on $\mathbb{I}$, relatively to $\mathbb{J}$, according to the *LLA* methodology.

Let us consider now a partition

$$\pi(\mathbb{I}) = \{\mathbb{I}_1, \mathbb{I}_2, \ldots, \mathbb{I}_h, \ldots, \mathbb{I}_H\}$$

of $\mathbb{I}$ into homogeneous classes constituted each of mutually similar elements with respect to $\mathbb{J}$. Therefore, we shall specify coefficients measuring the discrimination degre of $\pi(\mathbb{I})$ taken globally, or one of its classes $\mathbb{I}_h, 1 \leq h \leq H$, by a given element j of $\mathbb{J}$, or by a subset, that we can denote by $\mathbb{J}^k$, of $\mathbb{J}$.

8.4.1.1   Discriminating a Cluster of Rows by a Given Column

As in the numerical case (see Sect. 8.2.2) the discrimination index is based on the Huygens formula of inertia decomposition (see (8.2.8)). This formula becomes here

$$\sum_{i \in \mathbb{I}} p_{i.} \| f_{\mathbb{J}}^i - g_{\mathbb{J}} \|^2 = \sum_{1 \leq h \leq k} \sum_{i \in \mathbb{I}_h} p_{i.} \| f_{\mathbb{J}}^i - g_{\mathbb{J}}^h \|^2 + \sum_{1 \leq h \leq k} p_{\mathbb{I}_h.} \| g_{\mathbb{J}}^h - g_{\mathbb{J}} \|^2 \qquad (8.4.5)$$

where the norm $\| \bullet \|$ is associated with the q metric above and

$$p_{\mathbb{I}_h.} = \sum_{i \in \mathbb{I}_h} p_{i.}$$

and

$$g_{\mathbb{J}}^h = \frac{1}{p_{\mathbb{I}_h.}} \sum_{i \in \mathbb{I}_h} p_{i.} f_{\mathbb{J}}^i \qquad (8.4.6)$$

The general *center of gravity* of the cloud $\mathcal{N}(\mathbb{I})$ is defined by

$$g_{\mathbb{J}} = p_{.\mathbb{J}} = (p_{.1}, p_{.2}, \ldots, p_{.j}, \ldots, p_{.J})^T \ (T \text{ for Transpose})$$

and we have

$$g_{\mathbb{J}} = \sum_{1 \leq h \leq H} p_{\mathbb{I}_h.} g_{\mathbb{J}}^h$$

where $p_{\mathbb{I}_h.} = \sum_{i \in \mathbb{I}_h} p_{i.}$.

The inertia part "explained" by the partition $\pi(\mathbb{I})$ is defined by the ratio of the *Between* class inertia on the global one:

$$D(\pi(\mathbb{I})) = \frac{\sum_{1 \leq h \leq H} p_{\mathbb{I}_h.} \| g_{\mathbb{J}}^h - g_{\mathbb{J}} \|^2}{\sum_{i \in \mathbb{I}} p_{i.} \| f_{\mathbb{J}}^i - g_{\mathbb{J}} \|^2} \qquad (8.4.7)$$

The discrimination of $\pi(\mathbb{I})$ by a given element j of $\mathbb{J}$ is measured by the contribution of j to the latter ratio. It is also the part of the explained inertia by $\pi(\mathbb{I})$ for the q-orthogonal projection of $\mathcal{N}(\mathbb{I})$ on the axis sustained by the jth vector e_j of the canonical basis of $\mathbb{R}^J$. More precisely, the coefficient concerned can be written as

$$D(\pi(\mathbb{I})|j) = \frac{\sum_{1 \leq h \leq H} p_{\mathbb{I}_h.} \left[\frac{f(\mathbb{I}_h, j)}{p(\mathbb{I}_h.)} - p_{.j} \right]^2}{\sum_{i \in \mathbb{I}} p_{i.} \left[\frac{f_{ij}}{p_{i.}} - p_{.j} \right]^2} \qquad (8.4.8)$$

where $f(\mathbb{I}_h, j) = \sum_{i \in \mathbb{I}_h} f_{ij}$.

In these conditions, the discrimination coefficient of $\mathbb{I}_h$ by j ($j \in \mathbb{J}$) is

$$d(\mathbb{I}_h|j) = \frac{p_{\mathbb{I}_h.} \left[\frac{f(\mathbb{I}_h, j)}{p(\mathbb{I}_h.)} - p_{.j} \right]^2}{\sum_{i \in \mathbb{I}} p_{i.} \left[\frac{f_{ij}}{p_{i.}} - p_{.j} \right]^2} = \frac{p_{\mathbb{I}_h.} \left[\frac{f(\mathbb{I}_h, j)}{p(\mathbb{I}_h.) p_{.j}} - 1 \right]^2}{\sum_{i \in \mathbb{I}} p_{i.} \left[\frac{f_{ij}}{p_{i.} p_{.j}} - 1 \right]^2} \qquad (8.4.9)$$

8.4.1.2   Discriminating a Cluster of Rows by a Cluster of Columns

Two approaches will be considered below for establishing a discrimination coefficient of a row cluster $\mathbb{I}_h$ ($\mathbb{I}_h \subset \mathbb{I}$) by a column cluster $\mathbb{J}^k$ ($\mathbb{J}^k \subset \mathbb{J}$). They will be designated by α and β. The associated coefficients will be denoted below $d_\alpha(\mathbb{I}_h | \mathbb{J}^k)$ and $d_\beta(\mathbb{I}_h | \mathbb{J}^k)$, respectively.

The α approach

By bringing together all of the categories of $\mathbb{J}^k$ and by considering $\mathbb{J}^k$ as a single category, formula of (8.4.9) can be immediatly transposed. We obtain

$$d_\alpha(\mathbb{I}_h | \mathbb{J}^k) = \frac{p_{\mathbb{I}_h \cdot} \left[\frac{f(\mathbb{I}_h, \mathbb{J}^k)}{p(\mathbb{I}_h \cdot) p_{\cdot \mathbb{J}^k}} - 1 \right]^2}{\sum_{i \in \mathbb{I}} p_{i \cdot} \left[\frac{f_{i, \mathbb{J}^k}}{p_{i \cdot} p_{\cdot \mathbb{J}^k}} - 1 \right]^2} \tag{8.4.10}$$

where

$$p_{\cdot \mathbb{J}^k} = \sum_{j \in \mathbb{J}^k} p_{\cdot j} \, , \, f_{i, \mathbb{J}^k} = \sum_{j \in \mathbb{J}^k} f_{ij} \text{ and } f(\mathbb{I}_h, \mathbb{J}^k) = \sum_{(i,j) \in \mathbb{I}_h \times \mathbb{J}^k} f_{ij}$$

The β approach

In this, we will be guided by the connection established in Chap. 7 of [16], between factorial component analysis and inertia methods in cluster analysis. It is proved that a given class $\mathbb{J}^k$ can be interpreted as a "constrained" *factor* of the cloud $\mathcal{N}(\mathbb{I})$. The latter factor is defined by the q-orthogonal projection on the vector

$$\vec{b}_k = \sum_{j \in \mathbb{J}^k} \frac{1}{\sqrt{p_{\cdot \mathbb{J}^k}}} \vec{e}_j$$

With this formalism, the discrimination coefficient of the partition $\pi(\mathbb{I}) = \{\mathbb{I}_h | 1 \leq h \leq H\}$ by $\mathbb{J}^k$ is defined by the explained inertia of this $\mathcal{N}(\mathbb{I})$ partition projected on the axis $\vec{b}_k$ according to the q chi-square metric. More precisely, with the notations above, this coefficient is

$$D_\beta(\pi(\mathbb{I}) | \mathbb{J}^k) = \frac{\sum_{1 \leq h \leq H} p(\mathbb{I}_h \cdot) [q(\vec{b}_k, g_{\mathbb{J}}^h - g_{\mathbb{J}})]^2}{\sum_{i \in \mathbb{I}} p_{i \cdot} [q(\vec{b}_k, f_{\mathbb{J}}^i - g_{\mathbb{J}})]^2} \tag{8.4.11}$$

where $g_{\mathbb{J}}^h$ (resp., $g_{\mathbb{J}}$) designates the center of gravity of the cloud $\mathcal{N}(\mathbb{I}_h)$ (resp., $\mathcal{N}(\mathbb{I})$).

A straightforward calculation leads to the following result

$$D_\beta(\pi(\mathbb{I}) | \mathbb{J}^k) = \frac{\sum_{1 \leq h \leq H} p(\mathbb{I}_h \cdot) \left[\sum_{j \in \mathbb{J}^k} \left(\frac{f(\mathbb{I}_h, j)}{p(\mathbb{I}_h \cdot) p_{\cdot j}} - 1 \right) \right]^2}{\sum_{i \in \mathbb{I}} p_{i \cdot} \left[\sum_{j \in \mathbb{J}^k} \left(\frac{f(i, j)}{p_{i \cdot} p_{\cdot j}} - 1 \right) \right]^2} \tag{8.4.12}$$

Therefore, the discrimination coefficient of $\mathbb{I}_h$ by $\mathbb{J}^k$ is given by

$$d_\beta(\mathbb{I}_h | \mathbb{J}^k) = \frac{p(\mathbb{I}_h.) \left[\sum_{j \in \mathbb{J}^k} \left(\frac{f(\mathbb{I}_h, j)}{p(\mathbb{I}_h.)p.j} - 1 \right) \right]^2}{\sum_{i \in \mathbb{I}} p_i. \left[\sum_{j \in \mathbb{J}^k} \left(\frac{f(i,j)}{pi.p.j} - 1 \right) \right]^2} \qquad (8.4.13)$$

which is to be compared with (8.4.10).

8.4.2 Case of an Horizontal Juxtaposition of Contingency Tables

An horizontal juxtaposition of contingency tables is indexed by a set of the form

$$\mathbb{I} \times \left(\mathbb{J}(1) \cup \mathbb{J}(2) \cup \cdots \cup \mathbb{J}(l) \cup \cdots \cup \mathbb{J}(L) \right) \qquad (8.4.14)$$

where $\mathbb{I} \times \mathbb{J}(l)$ indexes a single and ordinary contingency table, $l \in \mathbb{L}$, where $\mathbb{L}$ designates the label set $\{1, 2, \ldots, l, \ldots, L\}$. The relative frequency table associated with a generic contingency table can be denoted by

$$f_{\mathbb{I} \times \mathbb{J}(l)} = \{ f_{ij} | (i, j) \in \mathbb{I} \times \mathbb{J}(l) \} \qquad (8.4.15)$$

This data structure occurs frequently. A classical family of real examples concerns description of geographic entities by categorical attributes. In Chap. 16 of [16] a real case is given where $\mathbb{I}$ is defined by a set of geographic areas. Three categorical attributes are considered in this case:

1. Type of cultivation;
2. Size of farmland;
3. Size of livestock.

Attributes 2 and 3 are initially numerical. They have been transformed into categorical attributes (see Sect. 3.2.3 of Chap. 3).

If $\mathbb{I}_h$ is a cluster of near elements of $\mathbb{I}$, the discrimination of $\mathbb{I}_h$ by a given element j or a subset $\mathbb{J}(l)^k$ of $\mathbb{J}$ follows Eqs. (8.4.9) and (8.4.10) (resp., (8.4.13)), respectively.

A new situation occurs when we have to evaluate the discrimination part of a set $\mathbb{K}$ of categories, provided from more than a single set $\mathbb{J}(l)$. In this case there exists at least two distinct l and l', such that

$$\mathbb{K} \cap \mathbb{J}(l) \neq \emptyset \text{ and } \mathbb{K} \cap \mathbb{J}(l') \neq \emptyset$$

Now, let $\mathbb{M}$ designate the $\mathbb{L}$ subset, for which

$$l \in \mathbb{M} \Leftrightarrow \mathbb{K} \cap \mathbb{J}(l) \neq \emptyset$$

Then, consider the restriction of the horizontal juxtaposition of contingency tables to $\mathbb{I} \times (\cup_{l \in \mathbb{M}} \mathbb{J}(l))$. In these conditions, two techniques that we denote by (A) and (B) can be proposed in order to evaluate to what extent $\mathbb{K}$ discriminates $\mathbb{I}_h$. For the former, we substitute for the sequence of contingency tables

$$\{f_{\mathbb{I} \times \mathbb{J}(l)} | l \in \mathbb{M}\} \tag{8.4.16}$$

a single one

$$f_{\mathbb{I} \times \mathbb{J}} = \{f'_{ij} | (i, j) \in \mathbb{I} \times (\cup_{l \in \mathbb{M}} \mathbb{J}(l))\} \tag{8.4.17}$$

where

$$f'_{ij} = \frac{1}{M} \sum_{j \in \cup_{l \in \mathbb{M}} \mathbb{J}(l)} f_{ij} \tag{8.4.18}$$

Therefore, Eqs. (8.4.10) and (8.4.13) apply.

In the second method (B), to each l belonging to $\mathbb{M}$, we will associate the contingency table

$$f_{\mathbb{I} \times \mathbb{J}(l)} = \{f_{ij} | (i, j) \in \mathbb{I} \times \mathbb{J}(l)\} \tag{8.4.19}$$

For the latter Eqs. (8.4.10) and (8.4.13) apply, relative to the set $\mathbb{K} \cap \mathbb{J}(l)$. In these conditions, the mean of the M values obtained by (8.4.10) (resp., (8.4.13)), for l describing $\mathbb{M}$ can be proposed for the discrimination evaluation of $\mathbb{I}_h$ by $\mathbb{K}$.

The behavior of the coefficients studied in this section (Sect. 8.4) was analyzed in the third thesis of A. Prod'homme [23].

8.5   On Two Ways of Measuring the "Importance" of a Descriptive Attribute

8.5.1   Introduction

The development considered here is situated in the context of mutual comparison and clustering descriptive attributes. However, this process can be transposed to the case of comparing and clustering described objects. In order to highlight the ideas presented here, we consider the data as Boolean. Nevertheless, generalization to all types of data can be considered from the formulas that we propose below. In fact, these depend only on the association coefficient (similarity index) endowing the set to be clustered. Let $\chi(\mathcal{A}) = \{\mathcal{A}^1, \mathcal{A}^2, \ldots, \mathcal{A}^k, \ldots, \mathcal{A}^K\}$ be a partition of a set $\mathcal{A}$ of descriptive attributes into homogeneous classes generally obtained by a given clustering method, from an association coefficient Q on $\mathcal{A}$. Consider now an arbitrary element a of $\mathcal{A}$ and suppose, without loss of generality, that a belongs to $\mathcal{A}^1$. Let β be a second element of $\mathcal{A}$ ($\beta \neq a$). If a is well integrated in its class ($\mathcal{A}^1$) and in addition, constitutes a "leader" element in $\mathcal{A}^1$, then $Q(a, \beta)$ has some tendency to

Fig. 8.4 Strong variability
of the Qa distribution

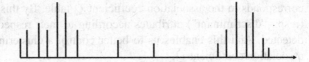

be *large* or *small* whether β belongs or not to $\mathcal{A}^1$. In these conditions, we consider
for every attribute a of $\mathcal{A}$, the distribution of the association coefficient Q on the set,
denoted by $\mathcal{A}_a$, of unordered attribute pairs including the component a, namely

$$\mathcal{A}_a = \{\{a, \beta\} | \beta \in \mathcal{A} - \{a\}\} \tag{8.5.1}$$

We have $card(\mathcal{A}_a) = p - 1$, where $p = card(\mathcal{A})$.

Intuitively, the more a is a leader element in its class, the more the shape of the
distribution

$$Q_a = \{Q(a, \beta) | \beta \in \mathcal{A} - \{a\}\} \tag{8.5.2}$$

is such that the most frequent values occur towards the limits of the variation interval
of $Q(a, \beta)$. The Fig. 8.4, representing the frequency histogram of Q_a suggests this
tendency.

Therefore, a measure of the variability of the distribution Q_a might indicate in
what extent a takes part in the interpretation of the class where it appears. Two related
measures can be proposed. The first one is the variance of the distribution Q_a and
the second one, simpler, is the absolute moment of order 2 of this distribution. These
can be written as

$$\mathcal{V}(a) = \frac{1}{p-1} \sum_{\beta \in \mathcal{A} - \{a\}} (Q(a, \beta) - Q(a))^2 \tag{8.5.3}$$

and

$$\mathcal{M}_2(a) = \frac{1}{p-1} \sum_{\beta \in \mathcal{A} - \{a\}} (Q(a, \beta))^2 \tag{8.5.4}$$

where $Q(a)$ is the empirical mean of the distribution Q_a:

$$Q(a) = \frac{1}{p-1} \sum_{\beta \in \mathcal{A} - \{a\}} Q(a, \beta)$$

Thus, the "*neutrality*" degree of a is measured by the smallness of $\mathcal{V}(a)$ (resp.,
$\mathcal{M}_2(a)$). We will concentrate below on the first measure (8.5.3).

In these conditions, the value table

$$\mathcal{V} = \{\mathcal{V}(a) | a \in \mathcal{A}\}$$

corresponds to the association coefficient Q table. By this way, the most "neutral" (resp., "discriminant") attributes according to their respective associations can be detected. And this enables us to better control a clustering result on $\mathcal{A}$ using the association coefficient Q.

By considering the same approach for the dual problem of object clustering, using a similarity table S on the object set $\mathcal{O}$, we may reach the notion of *strongly* (*weakly*) characterized object.

The formalization of this notion was the starting point for the development of a rich family of data seriation and clustering [11, 14, 20], reported in Chap. 8 of [16]. A basic formula for this development is given by the variance analysis formula of the similarity table on $\mathcal{O}$

$$\{S(i, j) | 1 \le i \ne j \le n\}$$

where $\mathbb{I} = \{1, 2, \ldots, i, \ldots, n\}$ codes $\mathcal{O}$ -. This formula can be written as

$$\frac{1}{n(n-1)} \sum_{(i,j) \in \mathbb{I}^{[2]}} \left(S(i, j) - \overline{S}\right)^2$$

$$= \frac{1}{n} \sum_{1 \le i \le n} \frac{1}{(n-1)} \sum_{\{j | j \ne i\}} \left(S(i, j) - \overline{S}_i\right)^2 + \frac{1}{n} \sum_{1 \le i \le n} \left(\overline{S}_i - \overline{S}\right)^2 \quad (8.5.5)$$

where

$$\overline{S} = \frac{1}{n(n-1)} \sum_{(i,j) \in \mathbb{I}^{[2]}} S(i, j) \text{ and } \overline{S}_i = \frac{1}{(n-1)} \sum_{j | j \ne i} S(i, j), 1 \le i \le n$$

An Illustration in the Case of Boolean Attributes Describing a Geometrical Diagram

Consider a surface composed of two square areas deduced each from the other by an horizontal translation and related with an oblique line having a very small thickness. Uniform density is distributed on the whole surface (see Fig. 8.5). Boolean attributes are associated with horizontal and vertical slices having a fixed thickness. In Fig. 8.5,

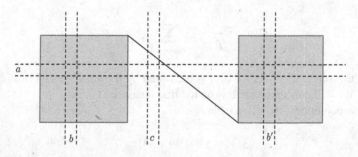

Fig. 8.5 Geometrical illustration of a "neutrality" case

a represents an horizontal slice, whereas b, c and b' represent vertical slices. If h (resp., k) designates the number of horizontal (resp., vertical) slices covering the whole surface, a given point P of the latter is described by a Boolean vector with $h + k$ components. A value 1 (resp., 0) for a given component means that the point P belongs (resp., does not belong) to the slice concerned by this component.

It is easy to show (this is left for the reader) that attributes associated with slices as c are equally the most neutral, and then they can be deleted before a clustering process. This deletion might avoid the well-known chain effect in applying "Single Linkage" or ordinal "Lexicographic" $AAHC$ algorithms (see Chap. 10).

As mentioned above, description by Boolean attributes has been considered in order to introduce the discrimination indices $\mathcal{V}(a)$ and $\mathcal{M}_2(a)$ (see (8.5.3)–(8.5.4)). Let us consider now the LLA approach where a *global* normalization of the association coefficients (resp., similarity indices) is applied (5.2.51)–(5.2.52), following a *local* one (5.2.10). We propose to use the table (5.2.51) obtained in order to calculate the discrimination coefficients $\mathcal{V}(a)$ or $\mathcal{M}_2(a)$ for every element a of $\mathcal{A}$. Due to the fact that the global mean and variance of the table (5.2.51) are equal to 0 and 1, the respective values of the table $\mathcal{V}$ (8.5.5), are more significantly comparable.

In Chap. 6 we have seen how to generalize the association coefficient Q between Boolean attributes, to the case of comparing relational attributes of the same type, whatever the latter is. Hence, building a table as (5.2.51) of Chap. 5 can be considered for mutual comparison of a set of attributes of the same relational type.

On the other hand, a normalized similarity index was built for mutually comparing a set of objects described by attributes of different types (7.2.31)–(7.2.32). Here also the globally reduced table of similarities (7.2.4) is recommended for establishing the discrimination coefficients $\mathcal{V}(a)$ and $\mathcal{M}_2(a)$.

Now, let us return to the basic and guiding case for which the attributes are Boolean where we have to evaluate the "importance" of a given attribute a. The method proposed in Sect. 8.2.2 quantifies this "importance" with respect to a "natural" classification defined by a partition π of the object set $\mathcal{O}$. The latter notion can be called *clustering* importance of a. The "importance" notion presented in the following is independent of a particular clustering. It can be called *projective* importance of a. It is consistent with the fact that the basic version of an association coefficient $Q(v, w)$ between two numerical attributes, v and w, is—up to a multiplicative factor depending on the size of the object set—the classical correlation coefficient between v and w, that is, the cosine of two vectors associated with v and w in a specific way.

Clustering importance and *projective* importance are somewhat connected. In the following, we shall try to compare the respective behaviors of the associated respective coefficients defined on a given attribute a. $\mathcal{C}(a)$ and $\mathcal{V}(a)$ will designate these coefficients.

8.5.2 Comparing Clustering "Importance" and Projective "Importance" of a Descriptive Attribute

8.5.2.1 Introduction

We shall compare the respective behaviors of the coefficients $\mathcal{C}(a)$ (clustering importance) and $\mathcal{V}(a)$ (projective importance) in the framework of a random incidence data table reflecting a pair of clusterings on the object set and on the attribute set, into K classes each, these clusterings being associated bijectively. Let us designate by

$$T^* = \{\alpha_i^j | 1 \le i \le n, 1 \le j \le p\} \qquad (8.5.6)$$

this random incidence table. On the other hand,

$$\pi(\mathcal{O}^*) = \{\mathcal{O}_k^* | 1 \le k \le K\}$$
$$\text{and}$$
$$\chi(\mathcal{A}^*) = \{\mathcal{A}^{*k} | 1 \le k \le K\} \qquad (8.5.7)$$

designate the respective partitions of the object set (represented by the table rows) and the attribute set (represented by the table columns).

For simplicity reasons we suppose

$$\forall k, 1 \le k \le K, card(\mathcal{O}_k^*) = \frac{n}{K} \text{ and } card(\mathcal{A}^{*k}) = \frac{p}{K} \qquad (8.5.8)$$

and this assumes that n and p are multiples of K. By setting $r = n/K$ and $s = p/K$, r (resp., s) is the common size of the classes of $\pi(\mathcal{O}^*)$ (resp., $\chi(\mathcal{A}^*)$).

The random model is defined as follows:

$$Prob\{\alpha_i^j = 1\} = \mathbf{p} \text{ if } (o_i^*, a^{*j}) \in \cup_{1 \le k \le K} \mathcal{O}_k^* \times \mathcal{A}^{*k}$$
$$\text{and}$$
$$Prob\{\alpha_i^j = 0\} = q \text{ if } (o_i^*, a^{*j}) \in (\cup_{1 \le k \le K} \mathcal{O}_k^* \times \mathcal{A}^{*k})^c \qquad (8.5.9)$$

where $0 < q < \mathbf{p} < 1$ and c indicates the complementary operation. Notice that the symbol $\mathbf{p}$—which represents a positive real strictly lower than 1—has to be distinguiched from $p = card(\mathcal{A})$ (Table 8.1).

Now, let us designate by λ the ratio $1/K$. A random component α of the random incidence data table is a Bernouilli variable (attribute) resulting from a mixing of two random Bernouilli variables, whose respective parameters being $\mathbf{p}$ and q. The probabilities of this mixing are λ and $(1 - \lambda)$. In these conditions, the parameter of α is

$$f = \lambda \times \mathbf{p} + (1 - \lambda) \times q \qquad (8.5.10)$$

Table 8.1 A particular random one to one biclustering

$\pi(\mathcal{O}^\star)\backslash\chi(\mathcal{A}^\star)$	$\mathcal{A}^{\star 1}$	$\mathcal{A}^{\star 2}$	•	•	•	$\mathcal{A}^{\star K}$
$\mathcal{O}_1^\star$	p	q	q	q	q	q
$\mathcal{O}_2^\star$	q	p	q	q	q	q
•	q	q	p	q	q	q
•	q	q	q	p	q	q
•	q	q	q	q	p	q
$\mathcal{O}_K^\star$	q	q	q	q	q	p

The random model (8.5.9) corresponds to a very simplified and particular expression of natural classification according to the Beckner definition. Nevertheless, this simplicity enables analytical calculation to be performed.

8.5.2.2 Comparing Variations of $\mathcal{C}(\gamma)$ and $\mathcal{V}(\gamma)$

We shall reason with respect to a random Boolean attribute γ referring to $\mathcal{A}^{\star 1}$. The development below is equal if γ refers to $\mathcal{A}^{\star k}$ for all $k = 1, 2, \ldots, K$. The associated random Boolean vector can be expressed by

$$\gamma^1 = (\gamma_1^1, \gamma_2^1, \ldots, \gamma_r^1, \gamma_{r+1}^1, \gamma_{r+2}^1, \ldots, \gamma_n^1)^T \qquad (8.5.11)$$

where T indicates Transpose. Given two positive reals u and v, such that $0 < u < v < 1$, we set

$$Prob\{\gamma_i^1 = u\} \text{ for } 1 \leq i \leq r$$
$$and$$
$$Prob\{\gamma_i^1 = v\} \text{ for } r + 1 \leq i \leq n \qquad (8.5.12)$$

β designating the random Boolean attribute defined by a random component of γ_1, the probability law of β as that of α, considered above, is defined by mixing two Bernouilli probability laws of parameters u and v, respectively. We have

$$Prob\{\beta = 1\} = \lambda \times u + (1 - \lambda) \times v \qquad (8.5.13)$$

The respective variations of the functions $\mathcal{C}(\gamma)$ and $\mathcal{V}(\gamma)$, introduced above, will be analyzed when the parameters u and v of γ^1 vary. For consistent and simplicity reasons, we put the condition:

$$f = \lambda \times u + (1 - \lambda) \times v \qquad (8.5.14)$$

corresponding to (8.5.10), where f is taken as a constant strictly comprised between 0 and 1. Thus, the only parameter variation is u. Notice that $u > v$ is equivalent to $u > f$.

In these conditions, the global variance of a random component γ_ι^1 of γ^1 is

$$var(\gamma_\iota^1) = f(1 - f) \qquad (8.5.15)$$

Therefore, by considering (8.5.11) and (8.5.14), the *Between* (interclasses) variance is

$$\mathcal{C}(\gamma_\iota^1) = \lambda(u - f)^2 + (1 - \lambda)(v - f)^2 \qquad (8.5.16)$$

We obtain

$$\mathcal{C}(\gamma_\iota^1) = \lambda(1 - \lambda)(u - v)^2 \qquad (8.5.17)$$

whose derivative with respect to u—use (8.5.14)—is

$$\frac{2(u - v)}{(1 - \lambda)}$$

Consequently, γ^1 discriminates all the more the partition $\pi(\mathcal{O}^\star)$, that the difference $(u - v)$ is larger.

Let us consider now the following basic similarity index between γ^1 and a random attribute α^j associated with a random column of the random data table $\mathcal{T}^\star$ (8.5.6):

$$\sigma(\gamma^1, \alpha^j) = \sum_{1 \le i \le n} \gamma_i^1 \cdot \alpha_i^j \qquad (8.5.18)$$

The probability law of this random index is different according to $j = 1$ or $j \ne 1$. Let us denote by L_1 and L_2 the first and the second probability laws, respectively. To fix ideas, take $j = 1$ and $j = 2$. The two corresponding random variables can be written as

$$\sigma(\gamma^1, \alpha^1) = \sum_{i \in \mathbb{I}_1} \gamma_i^1 \cdot \alpha_i^1 + \sum_{i \in \mathbb{I}_1^c} \gamma_i^1 \cdot \alpha_i^1$$

$$\sigma(\gamma^1, \alpha^2) = \sum_{i \in \mathbb{I}_1} \gamma_i^1 \cdot \alpha_i^2 + \sum_{i \in \mathbb{I}_2} \gamma_i^1 \cdot \alpha_i^2 + \sum_{i \in (\mathbb{I}_1 + \mathbb{I}_2)^c} \gamma_i^1 \cdot \alpha_i^2 \qquad (8.5.19)$$

where as usual c indicates the complementary operation and $\mathbb{I}_1$ and $\mathbb{I}_2$ index the classes $\mathcal{O}_1^\star$ and $\mathcal{O}_2^\star$: $\mathbb{I}_1 = \{1, 2, \ldots, r\}$ and $\mathbb{I}_2 = \{r + 1, r + 2, \ldots, 2r\}$.

$\sigma(\gamma^1, \alpha^1)$ takes the form of a sum of two binomial independent random variables whose respective parameters are $(r, \mathbf{p}u)$ and $(n - r, qv)$. $\sigma(\gamma^1, \alpha^2)$ takes the form of a sum of three binomial independent random variables whose respective parameters are $(r, \mathbf{p}v)$, (r, qu) and $(n - 2r, qv)$. Clearly, the probability law of $\sigma(\gamma^1, \alpha^2)$ is identical as that of $\sigma(\gamma^1, \alpha^k)$, for $k \ne 1$.

The expectations of $\sigma(\gamma^1, \alpha^1)$ and $\sigma(\gamma^1, \alpha^2)$ can be written as

$$\mathcal{E}(\sigma(\gamma^1, \alpha^1)) = r[\mathbf{p}u + (K-1)qv]$$
$$\mathcal{E}(\sigma(\gamma^1, \alpha^2)) = r[\mathbf{p}v + qu + (K-2)qv] \qquad (8.5.20)$$

Similarly,

$$var(\sigma(\gamma^1, \alpha^1)) = r[\mathbf{p}u(1 - \mathbf{p}u) + (K-1)qv(1-qv)]$$
$$var(\sigma(\gamma^1, \alpha^2))$$
$$= r[qu(1-qu) + \mathbf{p}v(1-\mathbf{p}v) + (K-2)qv(1-qv)] \qquad (8.5.21)$$

In these expressions, Kr is substituted for n. Now, as said above, for α^j uniformly distributed on $\mathcal{A}^*$, $\sigma(\gamma^1, \alpha^j)$ is a random variable whose law is a mixing of L_1 and L_2 (8.5.19), weighted by $\lambda = \frac{1}{K}$ and $(1 - \lambda) = \frac{K-1}{K}$. Therefore,

$$\mathcal{E}(\sigma(\gamma^1, \alpha^j)) = \frac{r}{K}[\mathbf{p}u + (K-1)qv] + r\frac{K-1}{K}[\mathbf{p}v + qu + (K-2)qv] \qquad (8.5.22)$$

We have

$$\mathcal{E}[(\sigma(\gamma^1, \alpha^1))^2] = r[\mathbf{p}u(1 - \mathbf{p}u) + (K-1)qv(1-qv)] + r^2[\mathbf{p}u + (K-1)qv]^2$$
$$\mathcal{E}[(\sigma(\gamma^1, \alpha^2))^2] = r[qu(1-qu) + \mathbf{p}v(1-\mathbf{p}v) + (K-2)qv(1-qv)] +$$
$$r^2[\mathbf{p}v + qu + (K-2)qv]^2 \qquad (8.5.23)$$

$\mathcal{E}[(\sigma(\gamma^1, \alpha^j))^2]$ is the average, weighted by λ and $(1 - \lambda)$, respectively, of both previous absolute moments of order 2. After assembling the different terms, we obtain

$$\mathcal{E}[(\sigma(\gamma^1, \alpha^j))^2] = \frac{r}{K} \times \mathbf{p}u(1 - \mathbf{p}u)$$
$$+ \frac{r(K-1)}{K} \times [qu(1-qu) + qv(1-qv) + \mathbf{p}v(1-\mathbf{p}v)]$$
$$+ \frac{r(K-1)(K-2)}{K} \times qv(1-qv)$$
$$+ \frac{r^2}{K} \times [\mathbf{p}u + (K-1)qv]^2 + \frac{r^2(K-1)}{K} \times [\mathbf{p}v + qu + (K-2)qv]^2 \qquad (8.5.24)$$

By introducing the new parameters

$$\phi = \mathbf{p}u + (K-1)qv$$
$$\psi = \mathbf{p}v + qu + (K-2)qv$$
$$\theta = \mathbf{p}^2u^2 + (K-1)q^2v^2$$
$$\zeta = \mathbf{p}^2v2 + q^2u^2 + (K-2)q^2v^2 \qquad (8.5.25)$$

We obtain

$$\mathcal{E}[(\sigma(\gamma^1, \alpha^j))] = r[\lambda \times \phi + (1 - \lambda) \times \psi]$$
$$\mathcal{E}[(\sigma(\gamma^1, \alpha^j))^2] = r\{[\lambda \times \phi + (1 - \lambda) \times \psi] - [\lambda \times \theta + (1 - \lambda) \times \zeta]\}$$
$$+ r^2[\lambda \times \phi^2 + (1 - \lambda) \times \psi^2] \tag{8.5.26}$$

It remains now to determine the derivative of the function

$$\mathcal{V}(\gamma^1) = \mathcal{E}[(\sigma(\gamma^1, \alpha^j))^2] - \{\mathcal{E}[(\sigma(\gamma^1, \alpha^j))]\}^2 \tag{8.5.27}$$

with respect to u. For this purpose, we will use the following relations (to be verified by the reader), where the symbol " ' " indicates the derivative with respect to u.

$$\phi' = \mathbf{p} - q$$
$$\psi' = -\frac{1}{K - 1}(\mathbf{p} - q)$$
$$\theta' = 2(\mathbf{p}^2 - q^2 v)$$
$$\zeta' = 2\left\{q^2 u - \frac{1}{K - 1}[\mathbf{p}^2 + (K - 2)q^2]v\right\}$$
$$\lambda \cdot \phi' + (1 - \lambda) \cdot \psi' = 0 \tag{8.5.28}$$

From these equations we deduce

$$\lambda\theta' + (1 - \lambda)\zeta' = 2(u - v)[\lambda\mathbf{p}^2 + (1 - \lambda)q^2]$$
$$\lambda\phi\phi' + (1 - \lambda)\psi\psi' = \lambda(\mathbf{p} - q)^2(u - v)^2 \tag{8.5.29}$$

The second of the previous equations is obtained from

$$\phi\phi' = (\mathbf{p} - q)[u\mathbf{p} + (K - 1)vq]$$
$$\psi\psi' = -\frac{(\mathbf{p} - q)}{K - 1}[uq + v\mathbf{p} + (K - 2)vq] \tag{8.5.30}$$

$$\frac{\partial(\mathcal{V}(\gamma^1))}{\partial u} = -r[\lambda\theta' + (1 - \lambda)\zeta'] + 2r^2[\lambda\phi\phi' + (1 - \lambda)\psi\psi']$$
$$= 2r(u - v)\{r\lambda(\mathbf{p} - q)^2 - [\lambda\mathbf{p}^2 + (1 - \lambda)q^2]\} \tag{8.5.31}$$

For $u > f$, the latter value is positive if

$$(\mathbf{p} - q)^2 > \frac{\lambda\mathbf{p}^2 + (1 - \lambda)q^2}{n\lambda^2} \tag{8.5.32}$$

In conclusion, while $C(\gamma^1)$ is a strictly increasing function with respect to u, the variance of the association coefficients $V(\gamma^1)$ is increasing, but only if the absolute value of the difference between **p** and q is not negligible. It is interesting to notice that the variation of $V(\gamma^1)$ does not depend on the relation **p** $> q$ and this emphasizes the *projective* nature of $V(\gamma^1)$.

8.6   Crossing Fuzzy Categorical Attributes or Fuzzy Classifications (Clusterings)

8.6.1   General Introduction

A very important case applying the results obtained in this section concern Boolean data. For the most part of this introduction and the following development, we suppose Boolean the set $\mathcal{A}$ of descriptive attributes. A description by nominal categorical attributes can be reduced to the latter case by substituting for each categorical attribute the set of its values and by associating with each of them a Boolean attribute (complete disjunctive coding, already mentioned above, Sect. 8.3.2.1).

We shall introduce here the notion of *fuzzy* Boolean attribute. The latter is defined by a class A of ordinary Boolean attributes, restricted to be mutually logically independent. This logical independence means that any distinct Boolean attributes of A are not mutually exclusive. Thus, two Boolean attributes of A defined by two distinct categories of a categorical attribute cannot belong simultaneously to A. A is said constituted by "directed" attributes, it defines a "profile" (one can also say "type"). A is generally obtained as a proximity cluster provided from a clustering of a set $\mathcal{A}$ of Boolean attributes. A fuzzy Boolean attribute takes its values in the real interval [0, 1]. Nevertheless, the formalism introduced by L.A. Zadeh [34] is not required in our development below.

The notion of fuzzy Boolean attribute is clearly needed in the data analysis of behavioral data. In this case, generally, the whole set of descriptive attributes, that we denote here by $\mathcal{B}$ can be splitted into two disjoint subsets $\mathcal{A}$ and $\mathcal{S}$ ($\mathcal{B} = \mathcal{A} + \mathcal{S}$) where $\mathcal{A}$ is the set of behavioral attributes and $\mathcal{S}$ is the set of identification attributes. Representative examples are given by psycho-sociological surveys where the $\mathcal{A}$ attributes comprise psychological facets and $\mathcal{S}$ is composed of the sociological attributes.

As a case relative to the mentioned survey on the European people perception concerning living conditions in 1976, let us indicate two questions coming under $\mathcal{A}$ and two others, under $\mathcal{S}$. The two former questions are:

1. Are you satisfied or not satisfied by the way by which the democracy works in your country?;
2. Are you satisfied or not satisfied by the amount of your earnings?.

The response categories to each of the questions above, are:

1. Very satisfied;
2. Rather satisfied;
3. Rather not satisfied;
4. Not satisfied at all.

The two categorical attributes coming under S might be: "Nationality" and "Religion".

As mentioned above, each behavioral question defines a categorical attribute with four values, and then gives rise to four Boolean attributes. Clustering the set $\mathcal{A}$ of all of them lead to a partition

$$\gamma(\mathcal{A}) = \{A^k | 1 \leq k \leq K\} \tag{8.6.1}$$

into $K = 7$ classes (see Sect. 8.3.1.1 and the references mentioned in this Section). Let us assume, possibly by deleting a few elements, that each of the clusters A^k is composed of mutually non-exclusive Boolean attributes (i.e. for any pair $\{a, b\}$ of elements of A^k, a and b are non-exclusive), $1 \leq k \leq K$. In these conditions, we get a fuzzy categorical attribute whose value set is $\gamma(\mathcal{A})$. Let us designate by α this attribute.

It is therefore interesting to know how different people categories induced by an ordinary identification attribute (e.g. sociological one) are related to the different behavioral profiles associated with the A^k, $1 \leq k \leq K$.

An *ordinary* Boolean or categorical attribute will be said "net" by difference with the fuzzy case, illustrated by α. Designating by $\{c^h | 1 \leq h \leq H\}$ the value set of a net categorical attribute c, it comes to establish a table

$$\chi(c, \boldsymbol{a}) = \{\chi(c^h, A^k) | 1 \leq h \leq H, 1 \leq k \leq K\} \tag{8.6.2}$$

of significant association coefficients on the set $\{c^h | 1 \leq h \leq H\} \times \{A^k | 1 \leq k \leq K\}$. More generally, if

$$\gamma(\mathcal{S}) = \{S^h | 1 \leq h \leq H\} \tag{8.6.3}$$

is a partition of $\mathcal{S}$ determining identification profiles and obtained, for example, from applying clustering algorithm on $\mathcal{S}$, a table such as

$$\chi(\boldsymbol{s}, \boldsymbol{a}) = \{\chi(S^h, A^k) | 1 \leq h \leq H, 1 \leq k \leq K\} \tag{8.6.4}$$

enables us to situate the identification and the behavioral profiles, respectively. s in the left side of (8.6.4) designates the fuzzy attribute whose values are defined by the partition $\gamma(\mathcal{S})$.

A natural thought might be to consider for (8.6.4) a table of similarity indices of the form

$$Sim(\gamma(\mathcal{S}), \gamma(\mathcal{A})) = \{S^\epsilon_{max}(S^h, A^k) | 1 \leq h \leq H, 1 \leq k \leq K\} \tag{8.6.5}$$

where for a given real ϵ, $0 \leq \epsilon \leq 1$, $S^\epsilon_{max}(S^h, A^k)$ is the similarity index between S^h and A^k, concieved in the framework of the $AAHC$ method (see (10.4.8)) of Sect. 10.4. However, this solution provides a weak answer to the problem submitted of situating $\gamma(S)$ with respect to $\gamma(A)$. In fact, the index $S^\epsilon_{max}(S^h, A^k)$ does not take into account intimately enough the internal nature of the structures to be associated, the latter being two polythetic classifications on two disjoint sets of Boolean attributes, respectively. As a matter of fact, a good index devoted to cluster emergence is not necessarily appropriate for associating respective clusters provided from formed classifications. Moreover, the discrimination scale of the table values (8.6.5) cannot be controlled enough.

Otherwise, we may envisage a global clustering treatment of the whole set $\mathcal{B}$ ($\mathcal{B} = \mathcal{A} + \mathcal{S}$) of the Boolean attributes. But this process is not recommended, because the respective links between identification attributes (belonging to $\mathcal{S}$) have different nature are more contrasted than those between behavioral attributes (belonging to $\mathcal{A}$). It follows that the most part of the $\mathcal{S}$ attributes might appear separated in a single cluster, the rest of them mixing up attributes of $\mathcal{A}$, and then disrupting clear and coherent organization of the behavioral attributes taken solely. Consequently, it is essential to separate the respective clustering treatment of $\mathcal{A}$ and $\mathcal{S}$.

Therefore, we will establish a table of the form

$$\chi(\gamma(S), \gamma(A)) = \{\chi(S^h, A^k) | 1 \leq h \leq H, 1 \leq k \leq K\} \qquad (8.6.6)$$

where $\chi(S^h, A^k)$ is obtained by generalizing the association coefficient between Boolean attributes obtained under the Poisson model of the hypothesis of no-relation $\mathcal{H}_3$ (see Sect. 5.2.1.2). This model is motivated by consistency reasons. In fact, the square sum of these coefficients calculated for the cells of a contingency table refers to a chi-square Statistic (see (5.2.14) of Sect. 5.2.1.2). The development of the theory for this new formal situation will be given in Sects. 8.6.2–8.6.5.

To begin, in Sect. 8.6.2, we shall specify the formal expression of a fuzzy Boolean attribute, the latter gives rise to a dichotomous (binary) fuzzy partition of the object set $\mathcal{O}$. The calculations leading to the chi-square statistic χ^2, associated with a classical contingency table have to be taken up again in a specific manner. This enables these calculations to be extended in the case of crossing two fuzzy categorical attributes.

In Sect. 8.6.3, we will make explicit the calculations concerned by the case of crossing net dichotomous attribute and fuzzy dichotomous attribute. The limit distribution of a resulting statistic having the formal structure of the chi-square statistic for 2×2 contingency table is established.

The analysis led in Sect. 8.6.4 is of the same nature as that of Sect. 8.6.3, but, it concerns *two* fuzzy dichotomous attributes.

In Sect. 8.6.5, we will consider the most general case of crossing two fuzzy categorical attributes with K and H fuzzy categories, respectively (see (8.6.1), (8.6.3) and (8.6.6)). Each of them determines a typology on the object set $\mathcal{O}$. We have mentioned above that one of both typologies $\gamma(A)$ is concerned by behavioral psychological attributes and the other one $\gamma(S)$, sociological identification attributes.

In this section the table (8.6.6) will be built and its components justified with respect to the chi-square theory [9].

All the development indicated above is relative to generalization of association coefficients between Boolean or nominal categorical attributes in the fuzzy case and this by referring to the chi-square theory. A natural question may be asked concerning the extension to the fuzzy case of association coefficients already established between relational attributes in the net case (see Sect. 6.2 of Chap. 6). More precisely, the matter concerns comparing fuzzy *nominal* or *ordinal* categorical attributes. The latter case cannot be solved with a chi-square approach. In Sect. 8.6.6, we will indicate how these extensions have been processed. The treatment of the ordinal case has led to an original method of qualitative regression. It was applied very interestingly in an epidemiological survey [17, 21].

8.6.2 Crossing Net Classifications; Introduction to Other Crossings

8.6.2.1   The Net (Classical) Case

We refer here to the notations adopted in Sect. 6.2.3.1. Let

$$P = \{P_h | 1 \leq h \leq H\} \text{ and } Q = \{Q_k | 1 \leq k \leq K\} \tag{8.6.7}$$

two partitions on a set $\mathcal{O}$ of objects, representing a pair $\{c, d\}$ of nominal categorical attributes defined on $\mathcal{O}$. P (resp., Q) is associated with c (resp., d). The cardinalities of the crossing partition

$$P \wedge Q = \{P_h \cap Q_k | 1 \leq h \leq H\}, 1 \leq k \leq K\} \tag{8.6.8}$$

define the contingency table crossing c and d that we disignate by

$$n(P \wedge Q) = \{n_{hk} = card(P_h \cap Q_k) | 1 \leq h \leq H\}, 1 \leq k \leq K\} \tag{8.6.9}$$

Now, denote by $\mathbb{I} = \{i | 1 \leq i \leq n\}$ the integer set coding $\mathcal{O}$ and introduce the Boolean indicator functions ϕ_h and ψ_k of P_h and Q_k, respectively, $1 \leq h \leq H$, $1 \leq k \leq K$. Thus, the entry of the cell (h, k) of $n(P \wedge Q)$ can be written as

$$n_{hk} = \sum_{i \in \mathbb{I}} \phi_h(i) \cdot \psi_k(i) \tag{8.6.10}$$

$1 \leq h \leq H, 1 \leq k \leq K.$

Similarly, the elements of the margin column and the margin row can be expressed as follows:

$$n_{h.} = \sum_{i \in \mathbb{I}} \phi_h(i) \text{ and } n_{.k} = \sum_{i \in \mathbb{I}} \psi_k(i) \tag{8.6.11}$$

The problem consists of defining oriented link measures between the classes partition of P and those of Q. To begin, let us consider the important case for which c and d are binary: $H = K = 2$. In this case, given the margins, the content of a single cell, for example $(1, 1)$, determines the content of the three other cells of the 2×2 contingency table.

By associating a pair $(P^\star, Q^\star)$ of independent random partitions with (P, Q), according to the random model $\mathcal{H}_1$ (see Sect. 6.2.3), the considerations of the *hypergeometric* model $\mathcal{N}_1$ in Sect. 5.2.1.2 apply. In fact, the definition of the random partition pair $(P^\star, Q^\star)$ can be reduced to that of random independent subset pair $(P_1^\star, Q_1^\star)$ of $\mathcal{O}$, associated with the pair (P_1, Q_1), under the hypergeometric model. The random index

$$N_{11} = card(P_1^\star \cap Q_1^\star)$$

associated with the content n_{11} of the cell $(1, 1)$ of the 2×2 contingency table, follows a hypergeometric law, whose mean and variance are

$$\mu_1 = \frac{n_{1.} \cdot n_{.1}}{n} \text{ and } \sigma_1^2 = \frac{n_{1.} \cdot n_{.1} \cdot n_{2.} \cdot n_{.2}}{n^2 \cdot (n-1)} \tag{8.6.12}$$

(see (5.2.9) of Chap. 5).

According to the development of Sect. 5.2.1, the probability law of $card(P_1^\star \cap Q_1^\star)$ is the same as those of the random indices

$$card(P_1^\star \cap Q_1) \text{ and } card(P_1 \cap Q_1^\star)$$

for which only one of both arguments $P_1^\star$ or $Q_1^\star$ is random.

Analytical forms of the three versions of the random index concerned are

$$\sum_{i \in \mathbb{I}} \phi_1(\sigma(i)) \cdot \psi_1(\tau(i)), \ \sum_{i \in \mathbb{I}} \phi_1(\sigma(i)) \cdot \psi_1(i) \text{ and } \sum_{i \in \mathbb{I}} \phi_1(i) \cdot \psi_1(\tau(i)) \tag{8.6.13}$$

where σ (resp., τ) is a random element in the set G_n, provided with equal chances, of all permutations on $\mathbb{I}$ ($card(G_n) = n!$). When σ (resp., τ) describes G_n, each instance of $P_1^\star$ (resp. $Q_1^\star$) is described $n_{1.}! \cdot n_{2.}!$ (resp., $n_{.1}! \cdot n_{.2}!$) times.

Due to (8.6.12), the standardized form of the index n_{11} can be written as

$$Q(1, 1) = \frac{(n_{11} - \frac{n_{1.} \cdot n_{.1}}{n})}{\sqrt{\frac{n_{1.} \cdot n_{2.} \cdot n_{.1} \cdot n_{.2}}{n^2 \cdot (n-1)}}} \tag{8.6.14}$$

Recall that the associated random variable follows under the independence relation $\mathcal{H}_1$ and very general limit conditions, the standardized normal distribution $\mathcal{N}(0, 1)$. The square of $Q(1, 1)$ is nothing else than the chi-square statistic associated with the 2×2 contingency table. It can be decomposed as follows

$$\chi^2 = \sum_{1 \leq h \leq 2} \sum_{1 \leq k \leq 2} \chi_{hk}^2 \tag{8.6.15}$$

where

$$\chi_{hk} = \frac{\left(n_{hk} - \frac{n_h . n_{.k}}{n}\right)}{\sqrt{\frac{n_h . n_{.k}}{n}}} \tag{8.6.16}$$

(see (5.2.14) of Chap. 5).

Clearly, we would have obtained the same previous development if instead of referring to the cell $(1, 1)$, we consider any of the four cells of the 2×2 contingency table. On the other hand and importantly, notice that the coefficient χ_{hk}, $1 \leq h \leq H$, $1 \leq k \leq K$, is obtained by centring and reducing n_{hk} with respect to the hypothesis of no relation $\mathcal{H}_3$ (see (5.2.32) of 5.2.1.2 of Chap. 5).

According to Definition 20 of Chap. 5, χ_{hk} corresponds to the *oriented* contribution of the cell (h, k) to the chi-square statistic associated with the 2×2 contigency table. In the following, we shall begin by extending these coefficients in the classical case of a general contingency table crossing two categorical attributes with H and K values, respectively (8.6.7)–(8.6.11). Next, we will carry out this extension in the case where at least one of both categorical attributes is *fuzzy*.

In the classical general case (see (8.6.7)–(8.6.11)), the probability of a contingency table such (8.6.8), according to the hypothesis of no relation $\mathcal{H}_1$ (see Sect. 6.2.3) is given by

$$\frac{\Pi_{1 \leq h \leq H} n_h. ! \cdot \Pi_{1 \leq k \leq K} n_{.k}!}{n! \cdot \Pi_{1 \leq h \leq H} \Pi_{1 \leq k \leq K} n_{hk}!}$$

A given $H \times K$ contingency table can be reduced to a sequence of $(H-1) \times (K-1)$ 2×2 contingency tables such that the previous probability of the whole contingency table can be expressed as the product of the probabilities associated with these 2×2 contingency tables. To each of the latter corresponds a chi-square statistic as (5.2.14) of Chap. 5, corresponding to the square of a statistic such $Q(1, 1)$ (8.6.14).

In these conditions, the global chi-square statistic can be written as

$$\chi^2(c, d) = \sum_{1 \leq t \leq (H-1) \times (K-1)} \chi_t^2 \tag{8.6.17}$$

where χ_t^2 is attached to the tth contingency table.

Equations (8.6.15) and (8.6.16) hold for the general case of a $H \times K$ contingency table. In this case the summation has to be extended to $\{(h, k) | 1 \leq h \leq H, 1 \leq k \leq K\}$. And, χ_{hk} defines the oriented measure of the link between the hth category of c and the kth category of d.

The associated random version of $\chi^2(c, d)$ follows—under permutational no relation random model $\mathcal{H}_1$ (see Sect. 6.2.3)—a chi-square distribution with $(H - 1) \times (K - 1)$ degrees of freedom.

Let us express that the paradigm of the alternatives of rejecting or not the independence hypothesis between the categorical attributes c and d, according to the chi-square distribution does not correspond to the data analysis philosophy. In fact a non-negligible value of $\chi^2(c, d)$ (8.6.17) may reveal local significant links between certain associated categories of c and d. These are indicated by some of the contingency table cells for which the values of χ_{hk} are pronounced enough, positively or negatively. Consequently, in order to discover these links, we will associate the table

$$\{\chi_{hk} | 1 \leq h \leq H, 1 \leq k \leq K\} \tag{8.6.18}$$

with the contingency table $n(P \wedge Q)$ (see (8.6.9)).

As expressed above, the purpose consists of generalizing this table in the following two cases:

1. Crossing net classification with a fuzzy one (see (8.6.2));
2. Crossing two fuzzy classifications (see (8.6.4)).

The new coefficients will be denoted as in (8.6.18). They will be justified according to the chi-square theory by the property: The sum of their squares—under the permutational random model of no relation $\mathcal{H}_1$—follows a χ^2 distribution with $(H - 1) \times (K - 1)$ degrees of freedom.

8.6.2.2 Introduction to the Fuzzy Cases

To begin, let us give the formal definition of a *fuzzy* Boolean attribute. This will be specified by comparison with the definition of an ordinary, said net, Boolean attribute. Some of the introductive considerations above will be retrieved here.

Let a be a net Boolean attribute. a is represented by the subset $\mathcal{O}(a)$ of the object set $\mathcal{O}$, where a is present (see Chaps. 4 and 5). a induces a partition $\pi(a)$ on $\mathcal{O}$ into two classes $\mathcal{O}(a)$ and $\mathcal{O}(\overline{a})$, where $\mathcal{O}(a)$ (resp., $\mathcal{O}(\overline{a})$) is the set of objects possessing a (resp., not possessing a). $\overline{a}$ designates the negated a attribute. Thus, a determines a net dichotomous partition on $\mathcal{O}$. By associating the indicator functions ϕ_a and $\phi_{\overline{a}}$ of $\mathcal{O}(a)$ and $\mathcal{O}(\overline{a})$, we have

- $(\forall x \in \mathcal{O})$, $\phi_a(x) = 1$ (resp., $\phi_a(x) = 0$) $\Leftrightarrow x \in \mathcal{O}(a)$ (resp., $x \in \mathcal{O}(\overline{a})$);
- and $\phi_{\overline{a}}(x) = 1$ (resp., $\phi_{\overline{a}}(x) = 0$) $\Leftrightarrow x \in \mathcal{O}(\overline{a})$ (resp., $x \in \mathcal{O}(a)$).

  and we have

$$(\forall x \in \mathcal{O}), \phi_a(x) + \phi_{\overline{a}}(x) = 1 \tag{8.6.19}$$

Now, let A be a set of "directed" Boolean attributes, that is to say, mutually non-exclusive (see the begining of Sect. 8.6.1). Thus, for any element a of A, the negated attribute $\overline{a}$ does not belong to A. A is generally obtained as a cluster from a clustering of a set $\mathcal{A}$ of descriptive Boolean attributes, observed on an object set $\mathcal{O}$. Therefore, we may assume A as composed of statistically near attributes. A defines a "profile", also said "type" in $\mathcal{O}$. A given object of $\mathcal{O}$ may be more or less concerned by such a profile. It is more concerned that it possesses a greater number of attributes of A. In these conditions, to each element x of $\mathcal{O}$, we will associate the value of the proportion of A attributes possessed by x, namely

$$\phi_A(x) = \frac{1}{c(A)} \sum_{a \in A} \phi_a(x) \qquad (8.6.20)$$

where $c(A)$ designates the cardinality of A. Analogously, we will associate with x

$$\phi_{\overline{A}}(x) = \frac{1}{c(A)} \sum_{a \in A} \phi_{\overline{a}}(x) \qquad (8.6.21)$$

where $\overline{A}$ is constituted by the negated attributes of A:

$$\overline{A} = \{\overline{a} | a \in A\}$$

Hence, an equivalent relation to (8.6.19) is acquired:

$$(\forall x \in \mathcal{O}), \phi_A(x) + \phi_{\overline{A}}(x) = 1 \qquad (8.6.22)$$

In these conditions, for any x in $\mathcal{O}$, $\phi_A(x)$ defines the belonging degree of x to the *profile* (*type*) associated with A.

As indicated at the end of the general introduction above (Sect. 8.6.1), we will consider first the crossing of a net dichotomous classification with a fuzzy one. Next, we will study the crossing of two fuzzy dichotomous classifications. The analysis of these two cases is sufficient to provide a complete answer for establishing association coefficients between the respective values of

• A net and a fuzzy categorical attributes (see (8.6.2));
• Two fuzzy categorical attributes (see (8.6.4)).

In fact, relative to the first item, let $S = \{s^h | 1 \leq h \leq H\}$ denote the value set of a net categorical attribute s and let $\gamma(\mathcal{A}) = \{A^k | 1 \leq k \leq K\}$ be a partition by proximity of a set $\mathcal{A}$ of Boolean attributes (see (8.6.1)). $\gamma(\mathcal{A})$ is generally obtained by a clustering algorithm. For simplicity but without restricting generality, we suppose that each A^k is composed of directed attributes (i.e. logical exclusivity does not exist between any two elements of A^k), $1 \leq k \leq K$. Then, we have to situate the category set S with respect to the fuzzy category set $\gamma(\mathcal{A})$ (see (8.6.2)). The size of the crossing table is $H \times K$ (H rows and K columns). The content χ_{hk} of the cell

Fig. 8.6 Crossing
dichotomous net an fuzzy
classifications

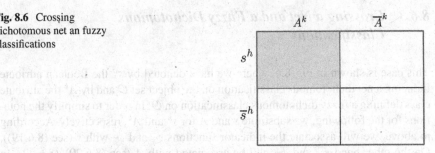

(h, k), $1 \leq h \leq H$, $1 \leq k \leq K$, corresponds to the content of the cell $(1, 1)$ of the 2×2 table depicted in Fig. 8.6, where $\overline{s}^h$ is the s^h negated Boolean attribute and $\overline{A}^k$ is the set of Boolean attributes obtained by associating to each attribute a of A^k its negation $\overline{a}$.

Let us consider now a pair $(\mathcal{S}, \mathcal{A})$ of two disjoint sets of Boolean attributes (see the General Introduction Sect. 8.6.1) and consider the partitions $\gamma(\mathcal{S})$ and $\gamma(\mathcal{A})$ defined by (8.6.1) and (8.6.3), that we recall here as

$$\gamma(\mathcal{S}) = \{S^h | 1 \leq h \leq H\} \text{ and } \gamma(\mathcal{A}) = \{A^k | 1 \leq k \leq K\}$$

As above, we suppose that two different elements of S^h (resp., A^k) cannot be logically exclusive, $1 \leq h \leq H$ (resp., $1 \leq k \leq K$). The crossing of these two fuzzy clusterings leads to a table $H \times K$ of the χ coefficients (see (8.6.6)), such that the content $\chi_{hk} = \chi(S^h, A^k)$ is the same as that of the cell $(1, 1)$ of the table depicted in Fig. 8.7. This table corresponds to the crossing of the dichotomous classifications $\{S^h, \overline{S}^h\}$ and $\{A^k, \overline{A}^k\}$, where $\overline{S}^h$ (resp., $\overline{A}^k$) is defined by the negated attributes of $\{S^h$ (resp., $\{A^k\}$), $1 \leq h \leq H$, $1 \leq k \leq K$.

Nevertheless, for this last case of crossing two fuzzy classifications into H and K clusters, respectively, we will consider in Sect. 8.6.5 a global permutational hypothesis of no relation for studying the statistics corresponding to the integer numbers n_{hk}, $1 \leq h \leq H$, $1 \leq k \leq K$, defined in the net case (see (8.6.9)).

Fig. 8.7 Crossing two fuzzy
dichotomous classifications

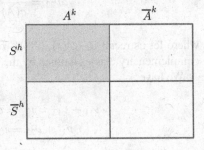

8.6.3 Crossing a Net and a Fuzzy Dichotomous Classifications

This case is shown in Fig. 8.6, where we have denoted by s^h the Boolean attribute inducing a net dichotomous classification on the object set $\mathcal{O}$ and by A^k the attribute class defining a fuzzy dichotomous classification on $\mathcal{O}$. In order to simplify the notations for the following, we substitute s and A for s^h and A^k, respectively. According to above, we will associate the indicator functions ϕ_s and $\phi_{\overline{s}}$ with s (see (8.6.19)). On the other hand ϕ_A and $\phi_{\overline{A}}$ will be associated with A (see (8.6.20)–(8.6.22)). In these conditions, the numbers ν_{11}, ν_{12}, ν_{21} and ν_{22} corresponding to the integers n_{11}, n_{12}, n_{21} and n_{22}, become

$$\nu_{11} = \sum_{i\in\mathbb{I}} \phi_s(i) \cdot \phi_A(i) = \frac{1}{c(A)} \sum_{i\in\mathbb{I}} \sum_{a\in A} \phi_s(i) \cdot \phi_a(i)$$

$$\nu_{12} = \sum_{i\in\mathbb{I}} \phi_s(i) \cdot \phi_{\overline{A}}(i) = \frac{1}{c(A)} \sum_{i\in\mathbb{I}} \sum_{a\in A} \phi_s(i) \cdot \phi_{\overline{a}}(i)$$

$$\nu_{21} = \sum_{i\in\mathbb{I}} \phi_{\overline{s}}(i) \cdot \phi_A(i) = \frac{1}{c(A)} \sum_{i\in\mathbb{I}} \sum_{a\in A} \phi_{\overline{s}}(i) \cdot \phi_a(i)$$

$$\nu_{22} = \sum_{i\in\mathbb{I}} \phi_{\overline{s}}(i) \cdot \phi_{\overline{A}}(i) = \frac{1}{c(A)} \sum_{i\in\mathbb{I}} \sum_{a\in A} \phi_{\overline{s}}(i) \cdot \phi_{\overline{a}}(i) \qquad (8.6.23)$$

where—as set above—$c(A)$ is the cardinality of A. Equally, we can write

$$\nu_{11} = \frac{1}{c(A)} \sum_{a\in A} card\big(\mathcal{O}(s) \cap \mathcal{O}(a)\big)$$

$$\nu_{12} = \frac{1}{c(A)} \sum_{a\in A} card\big(\mathcal{O}(s) \cap \mathcal{O}(\overline{a})\big)$$

$$\nu_{21} = \frac{1}{c(A)} \sum_{a\in A} card\big(\mathcal{O}(\overline{s}) \cap \mathcal{O}(a)\big)$$

$$\nu_{22} = \frac{1}{c(A)} \sum_{a\in A} card\big(\mathcal{O}(\overline{s}) \cap \mathcal{O}(\overline{a})\big) \qquad (8.6.24)$$

where let us recall it, $\{\mathcal{O}(s), \mathcal{O}(\overline{s})\}$ (resp., $\{\mathcal{O}(a), \mathcal{O}(\overline{a})\}$) is the partition into two complementary classes defined by the Boolean attribute s (resp., a).

We have

$$\nu_{11} + \nu_{12} = \nu_{1.} = card\big(\mathcal{O}(s)\big)$$

$$\nu_{21} + \nu_{22} = \nu_{2.} = card\big(\mathcal{O}(\overline{s})\big)$$

$$\nu_{11} + \nu_{21} = \nu_{.1} = \frac{1}{c(A)} \sum_{a \in A} card\left(\mathcal{O}(a)\right)$$

$$\nu_{12} + \nu_{22} = \nu_{.2} = \frac{1}{c(A)} \sum_{a \in A} card\left(\mathcal{O}(\overline{a})\right) \qquad (8.6.25)$$

where the right member of the third (resp., fourth) of Eq. (8.6.25) is the average over A of the absolute frequency of a (resp., $\overline{a}$).

Also, we have

$$\nu_{1.} + \nu_{2.} + \nu_{.1} + \nu_{.2} = n = card(\mathcal{O}) \qquad (8.6.26)$$

Therefore, the content of one of the cells of the 2×2 table

$$\{\nu_{lm} | 1 \le l \le 2, 1 \le m \le 2\} \qquad (8.6.27)$$

determines the content of the three other cells of this table.

In these conditions, the coefficient table

$$\left\{ \chi_{lm} = \frac{\left(\nu_{lm} - (\nu_{l.} \cdot \nu_{.m}/n)\right)}{\sqrt{(\nu_{l.} \cdot \nu_{.m}/n)}} | 1 \le l \le 2, 1 \le m \le 2 \right\} \qquad (8.6.28)$$

corresponds to the table (8.6.18) established in the net case. It defines the directed measures of association between the attributes s ($l = 1$) or $\overline{s}$ and the class of attributes A ($m = 1$) or $\overline{A}$ ($m = 2$).

Statistical justification of these coefficients will be now given according to the permutational hypothesis of no relation, expressed in the net case in (8.6.13). In these conditions, we consider the following dual random variables:

$$X(\sigma) = \sum_{i \in \mathbb{I}} \phi_s\left(\sigma(i)\right) \cdot \phi_A(i) \text{ and } Y(\sigma) = \sum_{i \in \mathbb{I}} \phi_s(i) \cdot \phi_A\left(\sigma(i)\right) \qquad (8.6.29)$$

where σ is a random permutation in the set—denoted above G_n—equally distributed, of all permutations on $\mathbb{I} = \{1, 2, \dots, i, \dots, n\}$. $\phi_A(i)$ is the *fuzzy* indicator function on $\mathbb{I}$ defined in (8.6.20), above.

The probability distributions of $X(\sigma)$ and $Y(\sigma)$ are identical. According to the Wald, Wolfowitz and Noether theorem (see Theorem 8 of Sect. 5.2.2.3), the common distribution tends, under general sufficient conditions, to the normal distribution. These conditions can be easily verified for the case concerned. In fact, the convergence towards the normal distribution is, the most often, very quick.

The common mean and variance are

$$\mathcal{E}\left(X(\sigma)\right) = \mathcal{E}\left(Y(\sigma)\right) = n.\mu\left(\phi_s\right) \cdot \mu\left(\phi_A\right)$$

$$var\left(X(\sigma)\right) = var\left(Y(\sigma)\right) = \frac{n^2}{(n-1)} \cdot var(\phi_s) \cdot var(\phi_A) \qquad (8.6.30)$$

where $\mu(\phi_s)$ and $\mu(\phi_A)$ (resp., $var(\phi_s)$ and $var(\phi_A)$) are the means (resp., the variances) of the functions (ϕ_s) and (ϕ_A) on $\mathbb{I}$. More explicitly,

$$\mu(\phi_s) = \frac{1}{n}\sum_{i\in\mathbb{I}}\phi_s(i) = p(s)$$

$$\mu(\phi_A) = \frac{1}{n}\sum_{i\in\mathbb{I}}\phi_A(i) = \frac{1}{c(A)}\sum_{a\in A}p(a) \qquad (8.6.31)$$

where $p(s)$ (resp., $p(a)$) designates the proportion of elements of the object set $\mathcal{O}$ (coded with $\mathbb{I}$) where the Boolean attribute s (resp., a) is present. Therefore, we have

$$var(\phi_s) = p(s) \cdot (1 - p(s))$$

$$var(\phi_A) = \frac{1}{n}\sum_{i\in\mathbb{I}}[\phi_A(i)]^2 - \left[\frac{1}{c(A)}\sum_{a\in A}p(a)\right]^2 \qquad (8.6.32)$$

where $[\phi_A(i)]^2$ can be written as follows:

$$\frac{1}{c(A)^2}\left[\sum_{a\in A}\phi_a(i) + \sum_{(a,a')\in A^{[2]}}\phi_{a\wedge a'}(i)\right]$$

where $A^{[2]}$ designates the set of ordered pairs of A elements, whose components are distinct ($A^{[2]} = A\times A - \Delta(A\times A)$, Δ indicating the diagonal). $\phi_{a\wedge a'}$ is the indicator function of $\mathcal{O}(a)\cap\mathcal{O}(a')$. Thus, we get

$$var(\phi_A) = \frac{1}{c(A)^2}\left(\sum_{(a,a')\in A\times A}(p(a\wedge a') - p(a)\cdot p(a'))\right) \qquad (8.6.33)$$

where $p(a\wedge a') = card(\mathcal{O}(a)\cap\mathcal{O}(a'))/n$ is the proportion of the $\mathcal{O}$ elements where a and a' are simultanously present. Consequently,

$$\mathcal{E}(X(\sigma)) = \mathcal{E}(Y(\sigma)) = p(s)\cdot\left(\frac{n}{c(A)}\sum_{a\in A}p(a)\right)$$

$$var(X(\sigma)) = var(Y(\sigma)) =$$

$$p(s)(1-p(s))\cdot\frac{1}{n-1}\cdot\left(\frac{n}{c(A)}\right)^2\left(\sum_{(a,a')\in A\times A}(p(a\wedge a') - p(a)\cdot p(a'))\right)$$

$$(8.6.34)$$

The centred and reduced version of ν_{11}, according to the latter model, can therefore be expressed by

$$\frac{\sqrt{(n-1)}\sum_{a\in A}\left(p(a\wedge s)-p(a)p(s)\right)}{p(s)(1-p(s))\left(\sum_{(a,a')\in A\times A}(p(a\wedge a')-p(a)\cdot p(a'))\right)} \tag{8.6.35}$$

whose square can be considered in the permutational hypothesis of no relation described above, as a realization (an instance) of a chi-square random variable with one degree of freedom.

Now, we shall verify that the sum of the squares of the elements of the 2×2 table (8.6.28), namely

$$\sum_{1\le l\le 2,1\le m\le 2}\chi^2_{lm} \tag{8.6.36}$$

is up to a multiplicative factor, the square of (8.6.35). This factor depends only on the margins of the 2×2 crossing table; that is to say, on the marginal distributions of s and A.

We can write

$$\chi_{11}=\chi_{sA}=\frac{\sqrt{\frac{n}{c(A)}}\left(\sum_{a\in A}(p(s\wedge a)-p(s)p(a))\right)}{\sqrt{\sum_{a\in A}p(s)p(a)}}$$

$$\chi_{12}=\chi_{s\overline{A}}=\frac{\sqrt{\frac{n}{c(A)}}\left(\sum_{a\in A}(p(s\wedge\overline{a})-p(s)p(\overline{a}))\right)}{\sqrt{\sum_{a\in A}p(s)p(\overline{a})}}$$

$$\chi_{21}=\chi_{\overline{s}A}=\frac{\sqrt{\frac{n}{c(A)}}\left(\sum_{a\in A}(p(\overline{s}\wedge a)-p(\overline{s})p(a))\right)}{\sqrt{\sum_{a\in A}p(\overline{s})p(a)}}$$

$$\chi_{22}=\chi_{\overline{s}\,\overline{A}}=\frac{\sqrt{\frac{n}{c(A)}}\left(\sum_{a\in A}(p(\overline{s}\wedge\overline{a})-p(\overline{s})p(\overline{a}))\right)}{\sqrt{\sum_{a\in A}p(\overline{s})p(\overline{a})}} \tag{8.6.37}$$

and verify that the square of the numerator of χ_{lm} is the same for all values of l and m, $(1\le l\le 2,1\le m\le 2)$. For this, take into account the following identities

$$p(\overline{a})=1-p(a)\,,\,p(\overline{s})=1-p(s)\,,\,p(\overline{a}\wedge s)=p(s)-p(a\wedge s)\text{ and}$$

$$p(\overline{a}\wedge\overline{s})=1-p(a)-p(s)+p(a\wedge s)$$

Hence, the sum of the squares (8.6.36) can be put in the form

$$\frac{n}{c(A)}\left(\sum_{a\in A}(p(a\wedge s)-p(a)p(s))\right)^2$$

$$\times\left(\left(\sum_{a\in A}p(a)p(s)\right)^{-1}+\left(\sum_{a\in A}p(a)p(\overline{s})\right)^{-1}+\left(\sum_{a\in A}p(\overline{a})p(s)\right)^{-1}+\left(\sum_{a\in A}p(\overline{a})p(\overline{s})\right)^{-1}\right) \tag{8.6.38}$$

which, after simplification, becomes

$$\frac{n\left(\sum_{a \in A}(p(a \wedge s) - p(a)p(s))\right)^2}{\left(\sum_{a \in A} p(a)\right) \cdot \left(\sum_{a \in A} p(\overline{a})\right) \cdot p(s) \cdot p(\overline{s})} \tag{8.6.39}$$

By identifying $(n - 1)$ with n, (8.6.39) is up to the multiplicative factor

$$\frac{\left(\sum_{(a,a') \in A \times A}(p(a \wedge a') - p(a)p(a'))\right)}{\left(\sum_{(a,a') \in A \times A} p(a) \cdot p(\overline{a'})\right)}$$

the square of (8.6.35). In these conditions we get

Theorem 21 *The sum of the squares of the statistics* χ_{lm}, $1 \leq l \leq 2$, $1 \leq m \leq 2$, *(8.6.28) is, in the permutational model of no relation (8.6.29), the realization of a chi-square random variable with one degree of freedom.* χ_{lm}, *represents the directed contribution*, $1 \leq l \leq 2$, $1 \leq m \leq 2$, *of* (l, m), *to the dependence chi-square measure between a net and a fuzzy dichotomous classifications.*

8.6.4 Crossing Two Fuzzy Dichotomous Classifications

The situation we shall consider now is shown in of Fig. 8.7 of the previous section. The development concerned here is analogous to that of Sect. 8.6.3, however it is more general. In fact, we wished to distinguish these two cases because the associated data analysis objectives are different in nature. Moreover, the gradual passage between crossing two nets, to crossing one net and one fuzzy and finally, to crossing two fuzzy dichotomous classifications, highlights the nature of the formal and statistical calculations.

According to Fig. 8.7, let S and A be two disjoint classes of directed attributes. As said above, generally, in real examples, S (resp., A) is constituted by identification attributes (resp., behavioral attributes). The two fuzzy dichotomous classifications to be crossed are associated with $\{S, \overline{S}\}$ and $\{A, \overline{A}\}$, where $\overline{S}$ (resp., $\overline{A}$) indicates the negated attributes of S (resp., A). As above, the fuzzy indicator functions of the profiles S and A (resp., $\overline{S}$ and $\overline{A}$) are designated by ϕ_S and ϕ_A (resp., $\phi_{\overline{S}}$ and $\phi_{\overline{A}}$). The corresponding parameters ν_{11}, ν_{12}, ν_{21} and ν_{22} of (8.6.23) become

$$\nu_{11} = \nu_{SA} = \sum_{i \in \mathbb{I}} \phi_S(i) \cdot \phi_A(i) = \frac{1}{c(S) \cdot c(A)} \sum_{i \in \mathbb{I}} \sum_{(s,a) \in S \times A} \phi_s(i) \cdot \phi_a(i)$$

$$\nu_{12} = \nu_{S\overline{A}} = \sum_{i \in \mathbb{I}} \phi_S(i) \cdot \phi_{\overline{A}}(i) = \frac{1}{c(S) \cdot c(A)} \sum_{i \in \mathbb{I}} \sum_{(s,a) \in S \times A} \phi_s(i) \cdot \phi_{\overline{a}}(i)$$

$$\nu_{21} = \nu_{\overline{S}A} = \sum_{i \in \mathbb{I}} \phi_{\overline{S}}(i) \cdot \phi_A(i) = \frac{1}{c(S) \cdot c(A)} \sum_{i \in \mathbb{I}} \sum_{(s,a) \in S \times A} \phi_{\overline{s}}(i) \cdot \phi_a(i)$$

$$\nu_{22} = \nu_{\overline{S}\overline{A}} = \sum_{i \in \mathbb{I}} \phi_{\overline{s}}(i) \cdot \phi_{\overline{a}}(i) = \frac{1}{c(S) \cdot c(A)} \sum_{i \in \mathbb{I}} \sum_{(s,a) \in S \times A} \phi_{\overline{s}}(i) \cdot \phi_{\overline{a}}(i)$$

(8.6.40)

where—as set above—$c(A)$ is the cardinality of A. Notice that an expression that $\phi_s(i) \cdot \phi_a(i)$ can also be written as $\phi_{s \wedge a}(i)$ where $s \wedge a$ is the Boolean attribute defined by the conjunction of s and a. $\phi_{s \wedge a}$ is the indicator function of the $\mathcal{O}$ subset, $\mathcal{O}(s) \cap \mathcal{O}(a)$. Hence, the previous equations can be expressed as follows:

$$\nu_{11} = \frac{1}{c(S) \cdot c(A)} \sum_{(s,a) \in S \times A} card\big(\mathcal{O}(s) \cap \mathcal{O}(a)\big)$$

$$\nu_{12} = \frac{1}{c(S) \cdot c(A)} \sum_{(s,a) \in S \times A} card\big(\mathcal{O}(s) \cap \mathcal{O}(\overline{a})\big)$$

$$\nu_{21} = \frac{1}{c(S) \cdot c(A)} \sum_{(s,a) \in S \times A} card\big(\mathcal{O}(\overline{s}) \cap \mathcal{O}(a)\big)$$

$$\nu_{22} = \frac{1}{c(S) \cdot c(A)} \sum_{(s,a) \in S \times A} card\big(\mathcal{O}(\overline{s}) \cap \mathcal{O}(\overline{a})\big) \qquad (8.6.41)$$

Moreover, we obtain

$$\nu_{1.} = \nu_{S.} = \nu_{11} + \nu_{12} = \frac{1}{c(S)} \sum_{s \in S} n(s) = \frac{n}{c(S)} \sum_{s \in S} p(s)$$

$$\nu_{2.} = \nu_{\overline{S}.} = \nu_{21} + \nu_{22} = \frac{1}{c(S)} \sum_{s \in S} n(\overline{s}) = \frac{n}{c(S)} \sum_{s \in S} p(\overline{s})$$

$$\nu_{.1} = \nu_{.A} = \nu_{11} + \nu_{21} = \frac{1}{c(A)} \sum_{a \in A} n(a) = \frac{n}{c(A)} \sum_{a \in A} p(a)$$

$$\nu_{.2} = \nu_{.\overline{A}} = \nu_{12} + \nu_{22} = \frac{1}{c(A)} \sum_{a \in A} n(\overline{a}) = \frac{n}{c(A)} \sum_{a \in A} p(\overline{a}) \qquad (8.6.42)$$

And we have

$$\nu_{1.} + \nu_{2.} + \nu_{.1} + \nu_{.2} = n \qquad (8.6.43)$$

Here also, the value of one of the numbers $\nu_{11}, \nu_{12}, \nu_{21}$ or ν_{22} determines the values of all of the other ones.

Therefore, we may use exactly the same formal expression for the table (8.6.28), namely

$$\left\{ \chi_{lm} = \frac{\big(\nu_{lm} - (\nu_{l.} \cdot \nu_{.m}/n)\big)}{\sqrt{(\nu_{l.} \cdot \nu_{.m}/n)}} \,\middle|\, 1 \le l \le 2, 1 \le m \le 2 \right\} \qquad (8.6.44)$$

with an extended interpretation of the coefficients concerned.

As in the previous section (see (8.6.29)), in order to justify statistically the coefficients of (8.6.44), we will associate the dual random variables

$$X(\sigma) = \sum_{i \in \mathbb{I}} \phi_S(\sigma(i)) \cdot \phi_A(i) \text{ and } Y(\sigma) = \sum_{i \in \mathbb{I}} \phi_S(i) \cdot \phi_A(\sigma(i)) \qquad (8.6.45)$$

with ν_{11}. σ is a random permutation in the set of all permutations on $\mathbb{I}$, equally distributed. The same conclusion of the previous section—concerning the normal character of the common limit distribution of $X(\sigma)$ and $Y(\sigma)$—holds. The comparable development of the calculations of $\mathcal{E}(X(\sigma))$, $\mathcal{E}(Y(\sigma))$, $var(X(\sigma))$ and $var(Y(\sigma))$ can be carried out. It leads to the following result

$$\mathcal{E}(X(\sigma)) = \mathcal{E}(Y(\sigma)) = \frac{n}{c(S) \cdot c(A)} \sum_{(s,a) \in S \times A} p(s) \cdot p(a)$$

$$var(X(\sigma)) = var(Y(\sigma)) = \frac{1}{n-1} \cdot \left(\frac{n}{c(S) \cdot c(A)} \right)^2$$

$$\times \sum_{(s,s') \in S \times S} \left(p(s \wedge s') - p(s)p(s') \right) \times \sum_{(a,a') \in A \times A} \left(p(a \wedge a') - p(a)p(a') \right) \quad (8.6.46)$$

Consequently, the centred and reduced version of ν_{11} can be written as

$$\frac{\sqrt{(n-1)} \sum_{(s,a) \in S \times A} \left(p(a \wedge s) - p(a)p(s) \right)}{\sqrt{[\sum_{(s,s') \in S \times S} \left(p(s \wedge s') - p(s)p(s') \right)][\sum_{(a,a') \in A \times A} \left(p(a \wedge a') - p(a)p(a') \right)]}}$$

$$(8.6.47)$$

whose square can be considered, under the permutational hypothesis of no relation, as a realization of a chi-square statistic with one degree of freedom.

As in the previous section, we shall now verify that the extended version of

$$\sum_{1 \leq l \leq 2, 1 \leq m \leq 2} \chi_{lm}^2 \qquad (8.6.48)$$

is, up to a multiplicative factor, the square of (8.6.47), the latter factor depending only, separately, on the distributions of the classes S and A on $\mathcal{O}$.

The new expressions of the χ_{lm}, $1 \leq l \leq 2$, $1 \leq m \leq 2$, become

$$\chi_{11} = \chi_{SA} = \frac{\sqrt{\frac{n}{c(S) \cdot c(A)}} \left(\sum_{(s,a) \in S \times A} (p(s \wedge a) - p(s)p(a)) \right)}{\sqrt{\sum_{(s,a) \in S \times A} p(s)p(a)}}$$

$$\chi_{12} = \chi_{S\overline{A}} = \frac{\sqrt{\frac{n}{c(S) \cdot c(A)}} \left(\sum_{(s,a) \in S \times A} (p(s \wedge \overline{a}) - p(s)p(\overline{a})) \right)}{\sqrt{\sum_{(s,a) \in S \times A} p(s)p(\overline{a})}}$$

$$\chi_{21} = \chi_{\overline{S}A} = \frac{\sqrt{\frac{n}{c(S) \cdot c(A)}} \left(\sum_{(s,a) \in S \times A} (p(\overline{s} \wedge a) - p(\overline{s})p(a)) \right)}{\sqrt{\sum_{(s,a) \in S \times A} p(\overline{s})p(a)}}$$

$$\chi_{22} = \chi_{\overline{S}\overline{A}} = \frac{\sqrt{\frac{n}{c(S) \cdot c(A)}} \left(\sum_{(s,a) \in S \times A} (p(\overline{s} \wedge \overline{a}) - p(\overline{s}) \cdot p(\overline{a})) \right)}{\sqrt{\sum_{(s,a) \in S \times A} p(\overline{s})p(\overline{a})}} \qquad (8.6.49)$$

Once again, by taking into account the relations

$$p(\overline{s}) = 1 - p(s), \ p(\overline{a}) = 1 - p(a), \ p(\overline{s} \wedge a) = p(a) - p(s \wedge a)$$

and

$$p(\overline{s} \wedge \overline{a}) = 1 - p(s) - p(a) + p(s \wedge a)$$

we verify that the square of the numerator of χ_{lm} is the same for all $l = 1, 2$ and for all $m = 1, 2$. Therefore, the square sum (8.6.48) can be expressed by the formula

$$\frac{n}{c(S) \cdot c(A)} \left(\sum_{(s,a) \in S \times A} (p(s \wedge a) - p(s)p(a)) \right)^2$$

$$\times \left(\left(\sum_{(s,a) \in S \times A} p(s)p(a) \right)^{-1} + \left(\sum_{(s,a) \in S \times A} p(s)p(\overline{a}) \right)^{-1} \right.$$

$$\left. + \left(\sum_{(s,a) \in S \times A} p(\overline{s})p(a) \right)^{-1} + \left(\sum_{(s,a) \in S \times A} p(\overline{s})p(\overline{a}) \right)^{-1} \right) \qquad (8.6.50)$$

which can be written after simplification,

$$\frac{n \left(\sum_{(s,a) \in S \times A} (p(s \wedge a) - p(s)p(a)) \right)^2}{\left(\sum_{s \in S} p(s) \right) \cdot \left(\sum_{s \in S} p(\overline{s}) \right) \cdot \left(\sum_{a \in A} p(a) \right) \cdot \left(\sum_{a \in A} p(\overline{a}) \right)} \qquad (8.6.51)$$

By identifying $(n - 1)$ with n, (8.6.51) is up to the multiplicative factor

$$\frac{\left(\sum_{(s,s') \in S \times S} (p(s \wedge s') - p(s)p(s')) \right)}{\left(\sum_{(s,s') \in S \times S} p(s) \cdot p(\overline{s'}) \right)} \times \frac{\left(\sum_{(a,a') \in A \times A} (p(a \wedge a') - p(a)p(a')) \right)}{\left(\sum_{(a,a') \in A \times A} p(a) \cdot p(\overline{a'}) \right)}$$

the square of (8.6.35). In these conditions we get

Theorem 22 *The sum of the squares of the statistics χ_{lm}, $1 \leq l \leq 2$, $1 \leq m \leq 2$, (8.6.48) is, in the permutational model of no relation (8.6.45), the realization of a chi-square random variable with one degree of freedom. χ_{lm}, represents the directed contribution, $1 \leq l \leq 2$, $1 \leq m \leq 2$, of (l, m) to the dependence chi-square measure between two fuzzy dichotomous classifications.*

8.6.5 Crossing Two Typologies

Some elements of the previous general introduction (Sect. 8.6.1) will be taken up again here. Let $\mathcal{S}$ and $\mathcal{A}$ be two disjoint sets of directed Boolean attributes and let $\gamma(\mathcal{S}) = \{S^h | 1 \leq h \leq H\}$ and $\gamma(\mathcal{A}) = \{A^k | 1 \leq k \leq K\}$ be two partitions on $\mathcal{S}$ and $\mathcal{A}$, respectively. $\gamma(\mathcal{S})$ and $\gamma(\mathcal{A})$ define two *typologies* of the object set $\mathcal{O}$ and can be interpreted as the sets values of two *fuzzy* categorical attributes s and a, with H and K values, respectively. In the general introduction mentioned, we have referred—as an example—to the psycho-sociological surveys where $\mathcal{S}$ is composed of identification sociological attributes and $\mathcal{A}$ is defined by behavioral psychological attributes. More generally, $\mathcal{S}$ and $\mathcal{A}$ correspond to two facets of the description of $\mathcal{O}$. The situation can be schematized in Fig. 8.8.

To begin, let us built the table

$$\nu\big(\gamma(\mathcal{S}), \gamma(\mathcal{A})\big) = \left\{ \nu_{hk} = \sum_{i \in \mathbb{I}} \phi_{S^h}(i) \cdot \phi_{A^k}(i) | 1 \leq h \leq H, 1 \leq k \leq K \right\}$$

$$(8.6.52)$$

This table can be considered as a *fuzzy* contingency table crossing the two *fuzzy* categorical attributes that we have just designated by s and a.

Fig. 8.8 Crossing two global fuzzy classifications

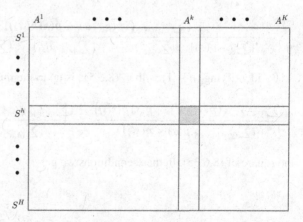

Now, from (8.6.52) we derive the table

$$\chi(s, a) = \{\chi_{hk} = \chi(S^h, A^k) | 1 \le h \le H, 1 \le k \le K\} \tag{8.6.53}$$

where

$$\chi_{hk} = \frac{\nu_{hk} - (\nu_{h.} \cdot \nu_{.k}/n)}{\sqrt{(\nu_{h.} \cdot \nu_{.k}/n)}} \tag{8.6.54}$$

is obtained by centring and reducing ν_{hk} with respect to the Poisson model of the hypothesis of no relation $\mathcal{H}_3$ (see (5.2.32) of 5.2.1.2 of Chap. 5), $1 \le h \le H$, $1 \le k \le K$.

In fact, for a given (h, k), $1 \le h \le H$, $1 \le k \le K$, ν_{hk} corresponds to the cell $(l = 1, m = 1)$ of the fuzzy 2×2 contingency table, crossing $(S^h, \overline{S}^h)$ with $(A^k, \overline{A}^k)$, where $\overline{S}^h$ (resp., $\overline{A}^k$) is the negated S^h (resp., A^k) attributes.

Transposing the statistical study developed above under the permutational model $\mathcal{H}_1$ (see Sect. 5.2.1.2 of Chap. 5) leads to associate the following random table with that obtained from (8.6.52):

$$\nu\big(\gamma(\mathcal{S}^\star), \gamma(\mathcal{A}^\star)\big) = \left\{ \nu_{hk}^\star = \sum_{i \in \mathbb{I}} \phi_{S^h}\big(\sigma_{hk}(i)\big) \cdot \phi_{A^k}\big(\sigma_{hk}(i)\big) | 1 \le h \le H, 1 \le k \le K \right\}$$
$$\tag{8.6.55}$$

where $\{\sigma_{hk} | 1 \le h \le H, 1 \le k \le K\}$ are $H \times K$ independent random permutations on $\mathbb{I}$.

The probability law of the table $\nu\big(\gamma(\mathcal{S}^\star), \gamma(\mathcal{A}^\star)\big)$ is identical to that of $\nu\big(\gamma(\mathcal{S}), \gamma(\mathcal{A}^\star)\big)$ (resp., $\nu\big(\gamma(\mathcal{S}^\star), \gamma(\mathcal{A})\big)$). Therefore, it is sufficient to consider

$$\nu\big(\gamma(\mathcal{S}), \gamma(\mathcal{A}^\star)\big) = \left\{ X_{hk}(\sigma_{hk}) = \sum_{i \in \mathbb{I}} \phi_{S^h}(i) \cdot \phi_{A^k}\big(\sigma_{hk}(i)\big) | 1 \le h \le H, 1 \le k \le K \right\}$$
$$\tag{8.6.56}$$

Since the random model addresses independently each of the cells of the fuzzy contingency table and due to the mutual independence of the random permutations σ_{hk}, $1 \le h \le H$, $1 \le k \le K$, this justificative study is *local*. As announced in the general introduction, we shall consider now a *global* permutational hypothesis of no relation defined at the level of the entire table (8.6.52). For this, a single random permutation σ is involved in all of the contingency table cells. In these conditions, the table (8.6.56) is substituted with the table

$$\nu\big(\gamma(\mathcal{S}), \gamma(\mathcal{A}^\star)\big) = \left\{ X_{hk}(\sigma) = \sum_{i \in \mathbb{I}} \phi_{S^h}(i) \cdot \phi_{A^k}\big(\sigma(i)\big) | 1 \le h \le H, 1 \le k \le K \right\}$$
$$\tag{8.6.57}$$

We shall not detail here the calculations and proofs required by this global statistical analysis. These can be consulted in [15] or Chap. 3 of [16]. We will content ourselves in giving the outline and the result obtained.

Let W be the covariance matrix of the sequence

$$\{X_{hk}(\sigma)|1 \leq h \leq H, 1 \leq k \leq K\} \tag{8.6.58}$$

of permutational random variables described in a certain order (for h, from 1 to H and for a given h, for k, from 1 to K). It can be shown that the element $hkh'k'$ of W can be expressed as follows

$$
\begin{aligned}
w_{hh'kk'} &= cov(X_{hk}(\sigma), X_{h'k'}(\sigma)) \\
&= \frac{\sqrt{n}}{c(S^h) \cdot c(S^{h'})} \sum_{(s,s')\in S^h \times S^{h'}} \left(p(s \wedge s') - p(s)p(s')\right) \\
&\times \frac{\sqrt{n}}{c(A^k) \cdot c(A^{k'})} \sum_{(a,a')\in A^k \times A^{k'}} \left(p(a \wedge a') - p(a)p(a')\right) \tag{8.6.59}
\end{aligned}
$$

Now, let b be a general fuzzy categorial attribute whose value set $\gamma(\mathcal{B}) = \{B^1, B^2, \ldots, B^l, \ldots, B^L\}$, corresponding to a partition of a set $\mathcal{B}$ of directed Boolean attributes. Consider b for the following condition (C):

Condition (C) : The dimension of the set of vectors

$$\left\{\left(\phi_{B^1(i)}, \phi_{B^2(i)}, \ldots, \phi_{B^l(i)}, \ldots, \phi_{B^L(i)}\right)|i \in \mathbb{I}\right\}$$

in the geometrical space $\mathbb{R}^L$, is L.

Notice that this condition is satisfied if there exists a subset $\{i_j|1 \leq j \leq m\}$ $(m < n)$, such that the set of the m vectors

$$\left\{\left(\phi_{B^1(i_j)}, \phi_{B^2(i_j)}, \ldots, \phi_{B^l(i_j)}, \ldots, \phi_{B^L(i_j)}\right)|1 \leq j \leq m\right\}$$

constitutes an independent system of vectors in $\mathbb{R}^L$.

The condition (C) is in fact very little restrictive. Generally, $\gamma(\mathcal{B})$ corresponds to a polythetic classification provided from a clustering algorithm and corresponding, more or less, to the Beckner approach (see Sect. 8.1.1). In these conditions, the respective profiles defined by $B^1, B^2, \ldots, B^l, \ldots$ and B^L are differenciated enough in order to satisfy the condition (C). It suffices, without being necessary, to find a subset of L elements of $\mathcal{O}$, such that each of them is specific of a single profile; that is to say, possesses attributes of exactly one attribute set B^l, $1 \leq l \leq L$.

In these conditions, we have the following

Lemma 6 *If* $\gamma(\mathcal{B}) = \{B^1, B^2, \ldots, B^l, \ldots, B^L\}$ *satisfies the condition* (C), *then the square matrix of the general term*

$$\beta_{j,j'} = \frac{1}{c(B^j) \cdot c(B^{j'})} \sum_{(b,b') \in B^j \times B^{j'}} \left(n(b \wedge b') - \frac{n(b)n(b')}{n} \right)$$

is definite positive.

This implies

Theorem 23 *If both typologies*

$$\gamma(\mathcal{S}) = \{S^1, S^2, \ldots, S^h, \ldots, S^H\} \text{ and } \gamma(\mathcal{A}) = \{A^1, A^2, \ldots, A^k, \ldots, A^K\}$$

satisfy the condition (C), then the covariance matrix W is definite positive.

A lemma (Lemma 2.3) given in [9], Chap. 5, entails the

Corollary 9 *If both typologies*

$$\gamma(\mathcal{S}) = \{S^1, S^2, \ldots, S^h, \ldots, S^H\} \text{ and } \gamma(\mathcal{A}) = \{A^1, A^2, \ldots, A^k, \ldots, A^K\}$$

satisfy the condition (C), then the joint distribution of $\{X_{hk}(\sigma)|1 \leq h \leq H, 1 \leq k \leq K\}$ is asymptotically normal.

For a given Boolean data table indexed by $\mathcal{O} \times \mathcal{A}$ where $\mathcal{O}$ is the object set and $\mathcal{A}$ the attribute set, let us consider a couple of partitions $\pi(\mathcal{O})$ on $\mathcal{O}$ and $\chi(\mathcal{A})$ on $\mathcal{A}$. These are supposed to get separately by Clustering. $\pi(\mathcal{O})$ can be represented by a *net* nominal categorical attribute defined on $\mathcal{O}$ and $\chi(\mathcal{A})$, by a *fuzzy* nominal categorical attribute on $\mathcal{O}$. Crossing the two attributes (see Sects. 8.6.3 and 8.6.5) enable us to interpret in a relative manner one clustering with respect to the other one. In Sect. 9.2 of Chap. 9 *Biclustering* methods are introduced. The objective of the latter methods consists of carrying out Clustering directly at the level of the Cartesian product $\mathcal{O} \times \mathcal{A}$. However, managing these techniques in the case of large data is much more difficult, particularly for computational reasons.

8.6.6 Extension to Crossing Fuzzy Relational Categorical Attributes

8.6.6.1 Introduction

In both versions of the hypothesis of no relation: *local* as in Sects. 8.6.3, 8.6.4 or *global* as in Sect. 8.6.5, we have built and justified—according to the chi-square theory—association coefficients between respective values of *net* or *fuzzy* categorical attributes.

We shall consider in this section the construction of association coefficients between *fuzzy* categorical attributes, taken globally. For this we will refer to the analysis led in Chap. 6 for establishing association coefficients between classical (net) categorical attributes of different types. Two cases will be outlined below for the generalization concerned: *nominal* and *ordinal*.

For comparing globally two fuzzy nominal categorical attributes, the development of Sect. 8.6.5 leads in a natural way to an association coefficient conform to the chi-square statistic. In fact, it is straightforward to adapt each of the coefficients ϕ^2 (6.2.44), T of Tshuprow (6.2.45) and C of Cramer (6.2.46) considered in Sect. 6.2.3.5 of Chap. 6. For this adaptation, the "net" contingency table (8.6.9) is replaced by a "fuzzy" one (8.6.52), in order to carry out the same calculations.

The association coefficients between categorical attributes interpreted as binary relations, studied in Sects. 6.2.3 and 6.2.4 of Chap. 6, can also be generalized in the case of fuzzy categorical attributes. As just mentioned, the two following cases will be considered for this generalization:

1. Association between two fuzzy *nominal* categorical attributes;
2. Association between two fuzzy *ordinal* categorical attributes.

In either of both cases, the method performed in the classical net case is transposed to the fuzzy one. This transposition is obtained from a specific form of the *raw* association index, which uses the belonging function of the object set elements to the different categories. More precisely, if b^l is a generic value of a categorical attribute b with L values, b^l will be expressed using the indicator function ϕ_l of the subset $\mathcal{O}_l$, composed of objects possessing the category b^l of b. In these conditions, by considering the code set $\mathbb{I} = \{1, 2, \ldots, i, \ldots, n\}$ of $\mathcal{O}$, ϕ_l is defined by

$$(\forall i \in \mathbb{I}), \phi_l(i) = 1 \text{ if } i \in \mathbb{I}_l$$
$$\forall i \in \mathbb{I}), \phi_l(i) = 0 \text{ if } i \notin \mathbb{I}_l \qquad (8.6.60)$$

where ϕ_l stands for ϕ_{b^l} and $\mathbb{I}_l$ is the $\mathbb{I}$ subset which codes $\mathcal{O}_l$, $1 \leq l \leq L$.

Now, if we consider a fuzzy categorical attribute b, whose value set is designated by $\{B^l | 1 \leq l \leq L\}$, we substitute the following *fuzzy* belonging functions $\tilde{\phi}_l$, for the *net* belonging functions ϕ_l, $1 \leq l \leq L$.

$$(\forall i \in \mathbb{I}), \tilde{\phi}_l(i) = \frac{1}{card(B^l)} \sum_{b \in B^l} \phi_b(i) \qquad (8.6.61)$$

To begin, we will give the expressions required of the raw association coefficient and the associated random one in the classical case of *net* categorical attributes. Next, the comparison between *fuzzy* categorical attributes can be expressed in the same terms. The intermediary case for which one of the both attributes is net and the other one is fuzzy can immediatly be derived from the previous two cases.

We will end this section by describing—on the basis of a real example in epidemiology—an original method of ordinal regression where the association coefficient between a net ordinal categorical attribute and a fuzzy one enables the regression fit to be evaluated.

8.6.6.2   Comparing Net or Fuzzy Nominal Categorical Attributes

As just mentioned above, we shall begin by giving specific expressions for the net case. These will be extended to the fuzzy case. Let c and d be two nominal categorical attributes, whose value sets are denoted as $\{c^1, c^2, \ldots, c^h, \ldots, c^H\}$ and $\{d^1, d^2, \ldots, d^k, \ldots, d^K\}$. ϕ_h (resp., ψ_k) will designate below the indicator function of the object subset $\mathcal{O}_h = c^{-1}(c^h)$ (resp., $\mathcal{O}_k = d^{-1}(d^k)$).

As for comparing two binary numerical valuations on the object set $\mathcal{O}$ (see Sect. 6.2.5), we refer to the set $\mathbb{I}^{[2]}$ which codes the set of ordered distinct object pairs (6.2.69). Thus, the *raw* coefficient index $s(c, d)$ between c and d (6.2.18) can be—up to the multiplicative factor $1/2$—put under the form

$$s(c, d) = \sum_{1 \leq h \leq H, 1 \leq k \leq K} \sum_{(i,j) \in \mathbb{I}^{[2]}} \phi_h(i) \cdot \phi_h(j) \cdot \psi_k(i) \cdot \psi_k(j) \qquad (8.6.62)$$

The multiplicative factor $1/2$ is not involved in the statistical normalization process, and then it can be dropped.

There are three equivalent versions of the permutational model of the hypothesis of no relation (6.2.77) and (6.2.78). We consider the left one of (6.2.77) for centring and reducing $s(c, d)$. Therefore, the vector $(\phi_h(i) | i \in \mathbb{I})$ is fixed and a random vector $(\psi_k(\tau(i)) | i \in \mathbb{I})$ is associated with $(\psi_k(i) | i \in \mathbb{I})$. τ is a random permutation in the set G_n—provided with equal chances—of all permutations on $\mathbb{I} = \{1, 2, \ldots, i, \ldots, n\}$. In these conditions, the form of the raw *random* index is associated with $s(c, d)$ is

$$s(c, d^\star) = \sum_{1 \leq h \leq H, 1 \leq k \leq K} \sum_{(i,j) \in \mathbb{I}^{[2]}} \phi_h(i) \cdot \phi_h(j) \cdot \psi_k(\tau(i)) \cdot \psi_k(\tau(j)) \qquad (8.6.63)$$

The situation considered corresponds to a particular aspect of the general treatment of Sect. 6.2.5.2. By taking up again the notations of the latter section, we have

$$(\forall (i, j) \in \mathbb{I}^{[2]}) \xi_{ij} = \sum_{1 \leq h \leq H} \phi_h(i) \cdot \phi_h(j) \text{ and } \eta_{ij} = \sum_{1 \leq k \leq K} \psi_k(i) \cdot \psi_k(j) \quad (8.6.64)$$

Clearly, Eq. (6.2.79) applies for the mathematical expectation of $s(c, d^\star)$. Moreover, we are in the case of comparing two *symmetrical* valuated relations

$$\xi_{ij} = \xi_{ji} \text{ and } \eta_{ij} = \eta_{ji} \text{ for any } (i, j) \in \mathbb{I}^{[2]},$$

Then, Eqs. (6.2.83) or (6.2.85) apply for the variance of $s(c, d^\star)$. Nevertheless, the parameters calculated have a very specific interpretation. Thus, for example such an expression

$$\sum_G \xi_{ij} \cdot \xi_{ij'}$$

where $G = \{(i, j, j') | i \in \mathbb{I}, j \in \mathbb{I}, j' \in \mathbb{I}, i \neq j, j \neq j' \text{ and } j \neq j'\}$—which appears in (6.2.83), becomes

$$\sum_G \left(\sum_{1 \leq h \leq H} \phi_h(i) \cdot \phi_h(j) \right) \cdot \left(\sum_{1 \leq h \leq H} \phi_h(i) \cdot \phi_h(j') \right) \qquad (8.6.65)$$

For the fuzzy case, we have to compare two fuzzy *nominal* categorical attributes. Let γ and δ designate these attributes. We denote by

$$\{C^1, C^2, \ldots, C^h, \ldots, C^H\} \text{ and } \{D^1, D^2, \ldots, D^k, \ldots, D^K\}$$

their respective value sets. C^h (resp., D^k) is a set of directed Boolean attributes, $1 \leq h \leq H$ (resp., $1 \leq k \leq K$). These sets are assumed mutually disjoint. The passage from the net case to the fuzzy one is obtained by substituting for the Boolean, fuzzy belonging functions (see (8.6.61)). Let us specify, once more these, in the case of comparing γ and δ.

$$(\forall i \in \mathbb{I}), \tilde{\phi}_h(i) = \frac{1}{card(C^h)} \sum_{c \in C^h} \phi_c(i)$$

$$\text{and } \tilde{\psi}_k(i) = \frac{1}{card(D^k)} \sum_{d \in D^k} \phi_d(i) \qquad (8.6.66)$$

Consequently, the raw association coefficient and the random corresponding one, between γ and δ become

$$s(\gamma, \delta) = \sum_{i \in \mathbb{I}} \tilde{\phi}_h(i) \cdot \tilde{\psi}_k(i)$$

$$s(\gamma, \delta^\star) = \sum_{i \in \mathbb{I}} \tilde{\phi}_h(i) \cdot \tilde{\psi}_k(\tau(i)) \qquad (8.6.67)$$

where τ is a random permutation in the set G_n of all permutations on $\mathbb{I}$ (see above).

Clearly, all the calculations carried out and the results obtained in the net case, hold in the fuzzy one, except that the functions $\{\phi_h | 1 \leq h \leq H\}$ and $\{\psi_k | 1 \leq k \leq K\}$ have to be replaced by $\{\tilde{\phi}_h | 1 \leq h \leq H\}$ and $\{\tilde{\psi}_k | 1 \leq k \leq K\}$, respectively.

8.6.6.3 Comparing Net or Fuzzy Ordinal Categorical Attributes

As for the nominal case, we shall begin by considering the classical net version. Extension to the fuzzy one will be provided next. Let c and d be two *ordinal* categorical attributes, describing a set of objects $\mathcal{O}$. As for the nominal case, their respective value sets are denoted by $\mathcal{C} = \{c^1, c^2, \ldots, c^h, \ldots, c^H\}$ and $\mathcal{D} = \{d^1, d^2, \ldots, d^k, \ldots, d^K\}$, respectively. Moreover, we assume here total orders relations (rankings) on $\mathcal{C}$ and $\mathcal{D}$:

$$c^1 < c^2 < \cdots < c^h < \cdots < c^H \text{ and } d^1 < d^2 < \cdots < d^k < \cdots < d^K$$

As in Sect. 6.2.4.1 of Chap. 6, the total preorders on $\mathcal{O}$ induced by c and d can be designated by ω and ϖ, respectively. Then, $(O_1, O_2, \ldots, O_h, \ldots, O_H)$ and $(P_1, P_2, \ldots, P_k, \ldots, P_K)$ will denote the ordered $\mathcal{O}$ subset sequences associated with ω and ϖ, respectively.

Specific analytical form of the raw association coefficient $s(c, d)$ has to be emphasized in order to make explicit the passage to the fuzzy case. According to (6.2.49) of Chap. 6 and by reffering to (6.2.75), $s(c, d)$ can be reformulated as follows:

$$s(c, d) = \sum_{(i,j) \in \mathbb{I}^{[2]}} \xi_{ij} \cdot \eta_{ij} \tag{8.6.68}$$

where

$$\xi_{ij} = \sum_{1 \leq h < h' \leq H} \phi_h(i)\phi_{h'}(j)$$

and

$$\eta_{ij} = \sum_{1 \leq k < k' \leq K} \psi_k(i)\psi_{k'}(j) \tag{8.6.69}$$

where with the notations adopted previously (see (8.6.60)), ϕ_h (resp., ψ_k) is the indicator function of the class O_h (resp., P_k) of the total preorder ω (resp., ϖ), $1 \leq h \leq H$ (resp., $1 \leq k \leq K$). In these conditions, the random raw association coefficient $s(c, d^\star)$ associated with $s(c, d)$, can be written as

$$s(c, d^\star) = \sum_{(i,j) \in \mathbb{I}^{[2]}} \left(\sum_{1 \leq h < h' \leq H} \phi_h(i)\phi_{h'}(j) \right) \cdot \left(\sum_{1 \leq k < k' \leq K} \psi_k(\tau(i))\psi_{k'}(\tau(j)) \right) \tag{8.6.70}$$

where τ is a random permutation in the set G_n of all permutations on $\mathbb{I}$ (see above).

For a given (i, j) in $\mathbb{I}^{[2]}$, if i and j correspond to two objects belonging to two distinct classes of ω (resp., ϖ), the vectors $(\phi_h(i)|1 \leq h \leq H)$ and $(\phi_h(j)|1 \leq h \leq H)$ (resp., $(\psi_k(i)|1 \leq k \leq K)$ and $(\psi_k(j)|1 \leq k \leq K)$) are different, and then we have neither symmetry nor antisymmetry satisfied for the ξ (resp., η) function on $\mathbb{I}^{[2]}$. Therefore, we have to refer to the general expression (6.2.81), for the expression of the variance of $s(c, d^\star)$. As an example, an expression such

$$\sum_{G_1} \xi_{ij} \cdot \xi_{ij'}$$

becomes

$$\sum_{G_1} \left(\sum_{1 \le h < h' \le H} \phi_h(i)\phi_{h'}(j) \right) \cdot \left(\sum_{1 \le h < h' \le H} \phi_h(i)\phi_{h'}(j') \right)$$

For the fuzzy case, we have to compare two fuzzy *ordinal* categorical attributes. The passage between the net case and the fuzzy one is carried out as for the nominal case. However, we have to take into account the ordinal aspect. So, our expression above will be taken up again here. Let γ and δ designate these categorical attributes (ordinal here). We denote by $\{C^1, C^2, \ldots, C^h, \ldots, C^H\}$ and $\{D^1, D^2, \ldots, D^k, \ldots, D^K\}$ their respective value sets. C^h (resp., D^k) is a set of directed Boolean attributes, $1 \le h \le H$ (resp., $1 \le k \le K$). These sets are assumed mutually disjoint. Moreover, we have:

$$C^1 < C^2 < \cdots < C^h < \cdots < C^H$$
$$\text{and } D^1 < D^2 < \cdots < D^k < \cdots < D^K \qquad (8.6.71)$$

The passage from the net case to the fuzzy one is obtained by substituting for the Boolean attributes, fuzzy belonging functions (see (8.6.61) and (8.6.66)). Let us specify, once more these, in the case of comparing γ and δ.

The raw association coefficient and the random corresponding one, between γ and δ become

$$s(\gamma, \delta) = \sum_{(i,j)\in\mathbb{I}^{[2]}} \left(\sum_{1 \le h < h' \le H} \tilde{\phi}_h(i)\tilde{\phi}_{h'}(j) \right) \cdot \left(\sum_{1 \le k < k' \le K} \tilde{\psi}_k(\tau(i))\tilde{\psi}_{k'}(j) \right)$$

$$s(\gamma, \delta^\star) = \sum_{(i,j)\in\mathbb{I}^{[2]}} \left(\sum_{1 \le h < h' \le H} \tilde{\phi}_h(i)\tilde{\phi}_{h'}(j) \right) \cdot \left(\sum_{1 \le k < k' \le K} \tilde{\psi}_k(\tau(i))\tilde{\psi}_{k'}(\tau(j)) \right) \quad (8.6.72)$$

where τ is a random permutation in the set G_n of all permutations on $\mathbb{I}$ (see above).

Clearly, as for the nominal case, all the calculations carried out and the results obtained in the net case, hold in the fuzzy one, except that the functions $\{\phi_h | 1 \le h \le H\}$ and $\{\psi_k | 1 \le k \le K\}$ have to be replaced by $\{\tilde{\phi}_h | 1 \le h \le H\}$ and $\{\tilde{\psi}_k | 1 \le k \le K\}$, respectively.

8.6.6.4 A Technique of Ordinal Qualitative Regression

This method was applied in a large epidemiological survey, and then we will refer to the latter in order to ilustrate it [4, 17, 21, 31].

Formally, assume a set $\mathcal{A}$ of ordinal categorical attributes describing the behavior of a set $\mathcal{O}$ of objects. In the application mentioned, $\mathcal{O}$ corresponds to a specified sample of living persons. We suppose that associations between the respective categories of the different attributes of $\mathcal{A}$ enable behavioral profiles to be setup. For this purpose, clustering will be required, here, in a particular manner depending on the values of a target ordinal categorical attribute c ($c \notin \mathcal{A}$). These values have to be associated with the different profiles emerged.

In the real example mentioned, $\mathcal{A}$ is obtained from the different parameters:

1. Alcohol consumption (drinking);
2. Ferritin level;
3. Mean corpuscular volume;
4. Triglycerides;
5. Serum iron;
6. Hemathomegaly.

When the initial measure scales are numerical—as the case is for the attribites 1 to 5, above—preliminary studies by the expert, associated with discretization algorithms (see Sect. 3.2.3 of Chap. 3) enable us to substitute for the numerical attributes, ordinal categorical ones. Furthermore, in many data analysis studies, the level value of a given numerical attribute is more significant than the value itself. The value set of a given ordinal categorical attribute a of $\mathcal{A}$, can be designated by $\{a^s | 1 \le s \le S\}$, where a^s is the category ranked s, $1 \le s \le S$.

As just said, let c be a target attribute $c \notin \mathcal{A}$, whose values have to be associated with behavioral profiles emerging from the description by $\mathcal{A}$. In the study mentioned, c is defined by the systolic blood pressure (SBP). Due to the fact that the initial measure scale of c is numerical, c is discretized and reduced to an ordinal categorical attribute. We can designate by $\{c^t | 1 \le t \le T\}$ the category set of c, c^t is the category of rank t, $1 \le t \le T$.

Consider now the completed set $\overline{\mathcal{A}} = \mathcal{A} + \{c\}$ and establish a clustering by proximity of the set , that we denote by $\mathcal{B}$, of all the Boolean attributes corresponding to the categories of $\overline{\mathcal{A}}$. As in [4, 17, 21, 31], this clustering might be obtained from the most "significant" level of the classification tree built by the LLA method (see Sect. 10.4.2 of Chap. 10 and Sect. 9.3.6 of Chap. 9). Let us designate by $P = \{B^h | 1 \le h \le H\}$ the partition obtained.

Two different categories a^s and $a^{s'}$ of a same ordinal categorical attribute a cannot be directly linked in a same B^h, $1 \le h \le H$. In fact, they are exclusive and their raw association index is null. However, a^s and $a^{s'}$ can be found in a same B^h, because of their respective associations with other Boolean categories belonging to B^h, $1 \le h \le H$.

In these conditions, the formal principle of the ordinal qualitative regression considered in [31], can be stated as follows. First, cluster B^h is converted in an attribute Boolean set, that we designate by $\tilde{B}^h$. For this, all categories in B^h, referring to the same ordinal categorical attribute, are replaced by a *single* Boolean attribute which is defined by a specific concatenation of certain of the latter categories. More precisely, let $a_h^s = \{a_1^s, a_2^s, \ldots, a_u^s, \ldots, a_v^s\}$ be the value set of a given ordinal categorical attribute appearing in $\tilde{B}^h$. If $v = 1$, clearly, a_1^s is retained in $\tilde{B}^h$. If $v > 1$, we determine in a_h^s, the set of all connex ordinal intervals and we retain one of them that includes the *less neutral* element a_w^s (see Sect. 8.5.2). The retained ordinal interval of the form $[a_u^s, a_{u+1}^s, \ldots, a_w^s, \ldots, a_{u+l}^s]$ will define a new Boolean attribute belonging to $\tilde{B}^h$. Clearly, the same process is applied to the target attribute c (the SBP in our example). We can designate by $\tilde{P} = \{\tilde{B}^h | 1 \le h \le H\}$ the new partition obtained. Notice that in a given $\tilde{B}^h$ of the latter, $1 \le h \le H$, each of the categorical attribute of $\overline{A}$ appears at most once and is represented by an ordinal interval of its value set. Some of the initial Boolean categories which do not belong to the connex ordinal intervals retained in the definition of $\tilde{P}$, are dropped. Thus, if $[a_u^s, a_{u+1}^s, \ldots, a_{u+l}^s]$ is the retained interval for the attribute a^s, in a given B^h, the attribute set $a_h^s - \{a_u^s, a_{u+1}^s, \ldots, a_{u+l}^s\}$ is deleted in order to establish $\tilde{B}^h$, $1 \le h \le H$.

In these conditions, let us consider the restriction of the partition $\tilde{P}$ to its classes where the target attribute c appears by an ordinal interval of its values. We designate this partition by $\tilde{P}_c = \{\tilde{B}_c^g | 1 \le g \le G\}$ $(G \le H)$ and we assume its classes ordered according to the respective values of c. That is to say, if $\tilde{c}^g$ is the ordinal interval of c, belonging to $\tilde{B}_c^g$, we suppose that

$$\tilde{c}^1 < \tilde{c}^2 < \cdots < \tilde{c}^g < \cdots < \tilde{c}^G$$

Designate by c the new target attribute—that whose value set is $\{\tilde{c}^g | 1 \le g \le G\}$—and consider the partition

$$Q = \{C^g | 1 \le g \le G\}$$

where

$$(\forall g, 1 \le g \le G), C^g = \tilde{B}_c^g - \{\tilde{c}^g\} \tag{8.6.73}$$

The ordinal qualitative regression concerned, implies (predicts) $\tilde{c}^g$ from C^g, $1 \le g \le G$.

The association between the ordered sequence of behavioral profiles

$$(C^1, C^2, \ldots, C^g, \ldots, C^G) \text{ and } (\tilde{c}^1, \tilde{c}^2, \ldots, \tilde{c}^g, \ldots, \tilde{c}^G)$$

is between a fuzzy ordinal categorical attribute and a net one. Consequently, the development of Sect. 8.6.6.3 applies.

8.7 Classifiability

8.7.1 Introduction

Let E be a finite set endowed with a similarity index S (see Sect. 4.1 of Chap. 4). The *Classifiability* of the ordered pair (E, S) is defined by the aptitude of E to be organized according to a hierarchical clustering (classification) scheme (more briefly, we can say, to be h-clustered), respecting the mutual resemblances between the elements of E measured by S. Obviously, S can be derived from a distance or dissimilarity function defined on E.

In order to illustrate this intuitive notion, let us consider an example of non-classifiability. Imagine E to be a sample of points obtained randomly in an euclidean rectangle, according to a uniform distribution. In this case, seeking for a hierarchical classification of E endowed with the euclidean distance is an imaginary problem. The single admissible partition chain (see Chaps. 1 and 10) is the roughest one with two levels, the first one corresponding to the finest partition with $card(E)$ classes (each class is a singleton) and the second one, to the coarsest partition, with a single class, including all of the elements of E.

The general statistical nature of the classifiability problem, consisting of testing the hypothesis of existence of a specific structure in the data, is not new. A very classical case concerns linear regression which is relative to a sequence of observed values $((x_1, y_1), (x_2, y_2), \ldots, (x_i, y_i), \ldots, (x_n, y_n))$ of an ordered pair (v, w) of numerical variables (attributes), a *colinearity* measure is proposed for the points P_i, whose coordinates are (x_i, y_i), $1 \le i \le n$. The sample random variable associated with the latter measure is employed in order to test independence hypothesis between v and w, against the alternative defined by linear regression of w with respect to v. This test is to be carried out prealably to adjusting a line on the set $\{P_i | 1 \le i \le n\}$.

The prior question the taxonomist might ask consists of evaluating the extent to which it is natural to adjust a partition chain structure on (E, S), respecting mutual similarities on E. Mostly, in real data, hierarchical decomposition of the data units set, turns out to be relevant. Nonetheless, this decomposition into clusters and subclusters can be more or less stressed. For a long time, this problem has not preoccupied the taxonomist practicer, possibly because, in his mind, the principal problem consists of discovering the "best" classification!

Intuitively speaking, if (E, S) is sufficiently well classifiable, we can expect from a reasonable hierarchical clustering algorithm (see Chap. 10), to provide nearly the best partition chain with respect several criteria. This intuitive considerationis confirmed by the fact that often "near" hierarchical clusterings are get from different $AAHC$ methods.

Several approaches might be envisaged in order to establish a measure of classifiability degree. Clearly, the latter is related to the frequency of more or less neutral elements in E, with respect to S. Notice that the first term of the right member of Equation (8.5.5) of Sect. 8.5.1, is a statistic defined by the mean of the respective similarity variances of each of the elements of the set concerned (the object set in the formula expression) to the others ones.

Otherwise, in the case where E is representable by a cloud of points in an euclidean space (see Sect. 10.3.3 of Chap. 10), for a a given cloud projection on a factorial subspace, with a few dimensions, comparison can be provided between the explained inertia of the projected cloud and that of an appropriate clustering, optimizing (locally) the inertia criterion (see Chap. 7 of [16]). This comparison might yield a classifiability measure.

The latter possible coefficient would be dependent on applying prealably data analysis or clustering methods. This is also the case of the Rammal's measure based on the subdominant ultrametric [24, 25]. In the latter single linkage method is concerned (see Sect. 10.3.2). In fact, in this coefficient, the initial distances on E—possibly derived from similarities on E—are compared with the ultrametric ones associated with the application of *Single Linkage* method (see Sect. 10.3.2 of Chap. 10).

By contrast, in the approach, we present below, initiated in [13], worked on again in [12] Chap. 4 and included in Chap. 3 of [16], the classifiability measure proposed is independent of applying a specific data analysis or clustering methods. Moreover, the coefficient proposed is get out from metrical considerations. It is concieved at the general level for which the similarity data on E is *ordinal*. That is to say the data similarity is a *preordonance on E* (see Sects. 4.1 and 4.2.2 of Chap. 4). One reason arguing for the relevance of this general structural level is due to the property that a partition chain is equivalently representable by an *ultrametric preordonance* (see Sect. 1.4.3 of Chap. 1).

If E is endowed with a numerical similarity S, we assume the associated preordonance ω_S established according to the first of Eq. (4.1.1) of Chap. 4.

For the Murtagh ultrametricity coefficient, termed α [22], the set E to be clustered is Euclidean spatially represented. Then it is endowed with a distance d. Nonetheless, in order to avoid as far as possible, the influence of d, α is based on the angles. more precisely, α is defined by the proportion in E of triangles which are nearly isoceles— the basis being the lowest side—or equilateral. The Murtagh coefficient will be briefly presented in Sect. 8.7.4.

According to the usage, we will present our index in the case where E is a set $\mathcal{O}$ of objects. However, clearly, E may be a set of categories or a set of attributes.

8.7.2 Discrepancy Between the Preordonance Structure and that Ultrametric, on a Data Set

8.7.2.1 Introduction

Consider a finite set $\mathcal{O}$ endowed with a dissimilarity function d. d defines a *ultrametric* distance on $\mathcal{O}$, if and only if for every triangle $\{x, y, z\}$ in $\mathcal{O}$ $(x, y, z \in \mathcal{O})$

is isosceles; its basis being the lowest side according to d (see Sect. 1.4 of Chap. 1). In the case where $\mathcal{O}$ is provided with an ordinal similarity, a *preordonance* ω, this property becomes:

$$(\forall\{x, y, z\} \in P_3(\mathcal{O})), \{x, y\} \leq \{y, z\} \leq \{x, z\} \Rightarrow \{x, z\} \leq \{y, z\}$$

The total preorder on $F = P_2(\mathcal{O})$ is generally associated with a dissimilarity function d. We assume the latter preorder established according to the first of Eq. 4.1.1 of Chap. 4. In the notations above $P_k(\mathcal{O})$ designates the set of all k-subsets of $\mathcal{O}$. Thus, $F = P_2(\mathcal{O})$ and $J = P_3(\mathcal{O})$ will indicate the set of all 2-subsets and that of 3-subsets of $\mathcal{O}$, respectively. We have

$$card(F) = \frac{n(n-1)}{2} \text{ and } card(J) = \frac{n(n-1)(n-2)}{6}$$

where $n = card(\mathcal{O})$.

The preordonance ω on $\mathcal{O}$, being formally represented by a total preorder on F, we consider the mapping τ of J into the set, designated by $Int(F)$ of open intervals of F, with respect to ω:

$$\tau : J \longrightarrow Int(F) \tag{8.7.1}$$

associating with a given element $\{x, y, z\}$ of J, the open interval

$$]M(x, y, z), S(x, y, z)[$$

where $M(x, y, z)$ and $S(x, y, z)$ are the median and the highest ranked pair among $\{\{x, y\}, \{y, z\}, \{z, x\}\}$, respectively.

In practice, the mapping τ can be graphically represented by a table (see Table 8.2) whose columns are indexed by F (ranked according to ω) and whose rows, by J. The columns associated with a given preorder class of ω are grouped together and

Table 8.2 Classifiability table

$J\backslash F$	ad	ac	ae	ce	bd	cd	bc	de	ab	be
abc		•					•	X	•	
abd	•				•		X	X	•	
abe			•						•	•
acd	•	•	X	X		•				
ace		•	•	•						
ade	•		•	X	X	X	X	•		
bcd					•	•	•			
bce				•			•	X	X	•
bde					•			•	X	•
cde				•		•	X	•		

separated from the following preorder class. In Table 8.2, for a triplet $\{x, y, z\}$ of J, indexing a given row, the elements $\{x, y\}$, $\{y, z\}$ and $\{z, x\}$ are marked with the symbol $\bullet$ and those of the interval $]M(x, y, z), S(x, y, z)[$, with the symbol X.

Example Let $\mathcal{O} = \{a, b, c, d, e\}$ be a set of five elements provided with the following preordonance, $\omega(\mathcal{O})$:

$$ad \sim ac < ae < ce < bd \sim cd < bc < de < ab < be$$

where xy stands for the pair $\{x, y\}$ ($x, y \in \mathcal{O}$). Similarly, in the Table 8.2, a word such xyz represents the triplet $\{x, y, z\}$ ($x, y, z \in \mathcal{O}$).

Now, in the general case, for a given finite set $\mathcal{O}$, endowed with a preordonance $\omega(\mathcal{O})$, we shall define an index of discrepancy between the $\omega(\mathcal{O})$ structure and that of a ultrametric preordonance, based on the mapping τ, introduced above. This index can be interpreted as resulting from a measure on J or a measure on F.

8.7.2.2 Statistics Evaluating the Discrepancy Between the $\omega(\mathcal{O})$ Structure and Ultrametricity

As expressed at the beginning of the previous section, if $\omega(\mathcal{O})$ is ultrametric, for any triplet $\{x, y, z\}$, belonging to J, such that $\{x, y\} \leq \{y, z\} \leq \{x, z\}$, the interval $]\{y, z\}, \{x, z\}[$ is empty. Relative to a fixed triplet $\{x_1, y_1, z_1\}$ such that $\{x_1, y_1\} \leq \{y_1, z_1\} \leq \{x_1, z_1\}$, the preorder $\omega(\mathcal{O})$ is "less ultrametric" to the extent that the cardinal of $]M(x_1, y_1, z_1) = \{y_1, z_1\}, S(x_1, y_1, z_1) = \{x_1, z_1\}[$, defined on $\omega(\mathcal{O})$ is large. In order to take into account the entire set J, we propose as a *raw* measure of discrepancy between the structure of $\omega(\mathcal{O})$ and a ultrametric preordonance

$$HC(\omega) = \sum_{\{x,y,z\}\in J} card(]M(x, y, z), S(x, y, z)[) = \sum_{\{x,y,z\}\in J} \Delta(x, y, z) \quad (8.7.2)$$

where $\Delta(x, y, z)$ is a brief notation for $card(]M(x, y, z), S(x, y, z)[)$.

Now, let us refer to the table representation of τ, which can be called the τ-table (see Table 8.2). $HC(\omega)$ is the number of X in the τ-table. Its expression in (8.7.2) is conformed to describing the τ-table row by row. Its value for Table 8.2 is:

$$HC(\omega) = 1 + 2 + 0 + 2 + 0 + 4 + 0 + 2 + 1 + 1 = 13$$

For a column scanning of the τ-table (column by column), $HC(\omega)$ appears as a measure on F, assigning to each element q of F, the cardinal $card(J_q)$, where J_q is defined as the J subset for which, each of its elements $\{x, y, z\}$ is such that

$$M(x, y, z) < q < S(x, y, z) \text{ for } \omega(\mathcal{O})$$

In other words, $card(J_q)$ is the number of triplets $\{x, y, z\}$ for which the median pair and the highest one emprison strictly q. Differently said, it is the number of times for which the pair q intervenes to separate strictly $M(x, y, z)$ and $S(x, y, z)$. Notice that if q and q' belong to the same preordonance class, $J_q = J_{q'}$.

In order to derive from $HC(\omega)$ a coefficient independent of the cardinality of the set concerned, we reduce $HC(\omega)$ by substituting for an absolute value—defined by a cardinal—a relative value defined by a proportion. Thus, we get

$$H(\omega) = \frac{1}{card(J)} \sum_{\{x,y,z\} \in J} \frac{card(]M(x, y, z), S(x, y, z)[)}{card(F)} = \frac{HC(\omega)}{card(F) \times card(J)}$$
(8.7.3)

Scanning the τ-table, column by column, we can write

$$H(\omega) = \frac{1}{card(F)} \sum_{q \in F} \frac{card(J_q)}{card(J)}$$
(8.7.4)

Therefore, $H(\omega)$ can be interpreted as the empirical mean of the distribution on F of the variable

$$m_q = \frac{card(J_q)}{card(J)}$$

In these conditions, the classifiability of $\mathcal{O}$ can be characterized by the latter distribution; more precisely, by the decreasing sequence of the numbers $\{m_p | p \in F\}$. Let us designate by $D(\omega) = (m_{(1)}, m_{(2)}, \ldots, m_{(p)}, \ldots, m_{(f)})$ $(f = card(F))$ this sequence. It becomes in the example of Table 8.2

$$D(\omega) = (0.3, 0.3, 0.2, 0.2, 0.1, 0.1, 0.1, 0, 0)$$

If ω is ultrametric, $H(\omega)$ vanishes and the vector $D(\omega)$ has all of its components nulls. $H(\omega)$ enables the set $\Omega(\mathcal{O})$ of all of the preordonances which can be defined on $\mathcal{O}$ to be ranked (ranking with ties), according to their respective classifiabilities:

$$\omega(\mathcal{O}) \leq \omega'(\mathcal{O}) \text{ if and only if } H(\omega(\mathcal{O})) \leq H(\omega'(\mathcal{O}))$$

The distribution $D(\omega)$ defines also a preorder (ranking with ties) on $\Omega(\mathcal{O})$. However, this preorder is *partial*:

$$(\forall(\omega, \omega') \in \Omega(\mathcal{O}) \times \Omega(\mathcal{O})), D(\omega) \leq D(\omega') \Leftrightarrow m_{(q)} \leq m'_{(q)}$$
(8.7.5)

where, respectively,

$$D(\omega) = (m_{(1)}, m_{(2)}, \ldots, m_{(q)}, \ldots, m_{(f)})$$

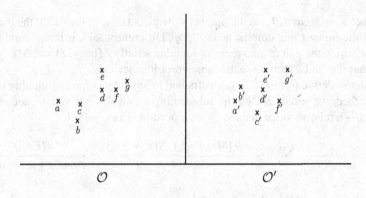

Fig. 8.9 Comparing classifiabilities

and

$$D(\omega') = (m'_{(1)}, m'_{(2)}, \ldots, m'_{(q)}, \ldots, m'_{(f)}).$$

The preorder (8.7.5) is more demanding than that associated with $H(\omega)$:

$$\left(\forall(\omega, \omega') \in \Omega(\mathcal{O}) \times \Omega(\mathcal{O})\right), D(\omega) \leq D(\omega') \Rightarrow H(\omega) \leq H(\omega')$$

Geometrical illustration

In Fig. 8.9,

$$\mathcal{O} = \{a, b, c, d, e, f, g\} \text{ and } \mathcal{O}' = \{a', b', c', d', e', f', g'\}$$

are two sets of seven points in the geometrical plan. In both sets, the number of distinct pairs is $card(F) = 21$ and that of distinct triplets is $card(J) = 35$. The set $\mathcal{O}$ (resp., $\mathcal{O}'$) is endowed with the preordonance $\omega(\mathcal{O})$ (resp., $\omega(\mathcal{O}')$) associated with the geometrical distance in the plan. We have for the preordonances $\omega(\mathcal{O})$ and $\omega(\mathcal{O}')$:

$$\omega(\mathcal{O}) : df \sim fg < bc < de \sim dg < eg < ef < ac < cd < ab <$$

$$bd < ce \sim cf < bf \sim cg < ad < be < ae < bg < af < ag$$

$$\omega(\mathcal{O}') : a'b' < e'g' < d'e' < c'f' \sim d'f' < b'd' < a'd' < b'e' \sim c'd' < d'g' <$$

$$a'c' < f'g' < e'f' \sim a'e' < b'c' < a'f' < b'f' \sim b'g' < c'e' < a'g' \sim c'g'$$

where the word xy stands for the pair $\{x, y\}$, $x \neq y$, x and y, belonging to $\mathcal{O}$ (resp., $\mathcal{O}'$).

Visually, $(\mathcal{O}, \omega(\mathcal{O}))$ appears "more classifiable" than $(\mathcal{O}', \omega(\mathcal{O}'))$. Effectively, we can distinguish for $(\mathcal{O}, \omega(\mathcal{O}))$ the "natural" classification $\{\{a, b, c\}, \{d, e, f, g\}\}$. For $(\mathcal{O}, \omega(\mathcal{O}))$ and $(\mathcal{O}', \omega(\mathcal{O})')$, we obtain for $H(\omega)$ (see (8.7.3)), $77/(21 \times 35) = 0.105$ and $128/(21 \times 35 = 0.174)$, respectively. Otherwise,

$$D(\omega(\mathcal{O})) = \left(\frac{10}{35}, \frac{10}{35}, \frac{9}{35}, \frac{7}{35}, \frac{7}{35}, \frac{7}{35}, \frac{6}{35}, \frac{6}{35}, \frac{4}{35}, \frac{4}{35}, \frac{4}{35}, \frac{1}{35}, \frac{1}{35}, \frac{1}{35}, 0, 0, \ldots, 0\right)$$

$$D(\omega(\mathcal{O}')) = \left(\frac{14}{35}, \frac{13}{35}, \frac{13}{35}, \frac{12}{35}, \frac{11}{35}, \frac{11}{35}, \frac{9}{35}, \frac{9}{35}, \frac{9}{35}, \frac{7}{35}, \frac{5}{35}, \frac{5}{35}, \frac{5}{35}, \frac{3}{35}, \frac{1}{35}, \frac{1}{35}, 0, 0, 0\right)$$

Statistical Simulation of $H(\omega)$

If $n = card(\mathcal{O})$ is large enough, $card(J)$ might be too large. A uniform sampling in J—without replacement—can be envisaged in order to estimate $H(\omega)$. If G is the sample obtained, with $g = card(G)$, the estimation of $H(\omega)$ is given by

$$\hat{H}(\omega) = \frac{1}{g} \sum_{\{x,y,z\} \in G} \frac{\Delta(x, y, z)}{f} \qquad (8.7.6)$$

where $f = card(F)$

Let $\mathcal{H}_g(\omega)$ be the random sampling variable associated with $\hat{H}(\omega)$. For g large enough, due to the central limit theorem [7],

$$\sqrt{k} \times \left(\mathcal{H}_g(\omega) - H(\omega)\right)$$

follows a centered normal distribution of variance

$$V = variance\left(\frac{\Delta(x, y, z)}{f}\right)$$

which can be estimated by the empirical variance of the distribution

$$\left\{\frac{\Delta(x, y, z)}{f} | \{x, y, z\} \in G\right\}$$

This property enables us to establish a confidence interval for $H(\omega)$. For a given risk level, the latter can be yield as narrow as we wish, provided that g is large enough.

To end this section, let us indicate that in [16], Chap. 11, we compare the distribution $D(\omega)$ corresponding to a real case of well-classifiable data, with that of associated random data under a hypothesis of no relation. More precisely, if T is the given data table, two simulations T_1 and T_2 of a random data table T^*, associated with T, under an appropriate hypothesis of no relation, were yield. The corresponding distributions $D(\omega)_1$ and $D(\omega)_2$ turn out to be clearly on the right of $D(\omega)$:

$$D(\omega) < D(\omega)_1 \text{ and } D(\omega) < D(\omega)_2$$

that is: in the graphical representation of $D(\omega)$, $D(\omega)_1$ and $D(\omega)_2$, the respective graphs of $D(\omega)_1$ and $D(\omega)_2$ are strictly higher than that of $D(\omega)$.

In the framework of the latter simulation, the *cohesion* concept of clusters obtained by a given clustering algorithm (see Chap. 9) and the *classifiability* concept, are discussed. The two latter notions are conceptually *independent*.

8.7.3 Classifiability Distribution Under a Random Hypothesis of Non-ultrametricity

To begin, let us specify in set theoretic terms the correspondence between an observed Boolean data table $\mathcal{T}$ and a random one $\mathcal{T}^*$. This association will be concerned here in a particular but important case.

Let $\mathcal{A} = \{a^1, a^2, \ldots, a^k, \ldots, a^K\}$ a set of Boolean attributes observed on a set $\mathcal{O}$ of objects, this description giving rise to a data table $\mathcal{T}$. As usually, the rows (resp., columns) of $\mathcal{T}$ are indexed by $\mathcal{O}$ (resp., $\mathcal{A}$). $n = card(\mathcal{O})$ and $K = card(\mathcal{A})$.

Consider now the decomposition of $\mathcal{O}$ according to the number of attributes possessed by an object. Assuming there is no empty object, we have:

$$\mathcal{O} = \mathcal{O}_1 + \mathcal{O}_2 + \cdots + \mathcal{O}_k + \cdots + \mathcal{O}_K \tag{8.7.7}$$

where $\mathcal{O}_k$ is constituted of all the objects of $\mathcal{O}$ where, exactly, k attributes of $\mathcal{A}$ are present. Clearly, some of the sets of the right member of (8.7.7) may be empty. We have

$$n = n_1 + n_2 + \cdots + n_k + \cdots + n_K \tag{8.7.8}$$

where $n_k = card(\mathcal{O}_k)$, $1 \leq k \leq K$.

The set representation of $\mathcal{O}$ is the set of all subsets of $\mathcal{A}$, that we designate by $P(\mathcal{A})$ (see Sect. 4.2.1.1 of Chap. 4). $P(\mathcal{A})$ can be decomposed into K levels, the kth one is composed of all $\mathcal{A}$ subsets having k elements:

$$P(\mathcal{A}) = P(\mathcal{A})_0 + P(\mathcal{A})_1 + \cdots + P(\mathcal{A})_k + \cdots + P(\mathcal{A})_K \tag{8.7.9}$$

We have

$$card(P(\mathcal{A})) = 2^K \text{ and } card(P(\mathcal{A})_k) = \binom{K}{k}, 0 \leq k \leq K$$

The random model of the hypothesis of no relation considered has the same nature as that of $\mathcal{N}_1$ (see Sect. 5.2.2.1 of Chap. 5), but defined by interverting the respective roles of $\mathcal{O}$ and $\mathcal{A}$. It consists of associating a sequence of random subsets $\{\mathcal{O}_k^* | 1 \leq k \leq K\}$ with the observed one (see (8.7.7)). $\mathcal{O}_k^*$ is a set of n_k independent distinct random objects, corresponding to an element taken at random in $P(\mathcal{A})_k$, with equal chances. To each of the latter objects corresponds a random row of a random table $\mathcal{T}^*$ associated with a row of $\mathcal{T}$, comprising k components equal to 1.

Now, we shall consider the important case for which all of the objects of $\mathcal{O}$ are situated at the same level $\mathcal{L}_h$ of $P(\mathcal{A})$, that means, for all x in $\mathcal{O}$,

$$\sum_{1 \leq k \leq K} x^k = h \qquad (8.7.10)$$

where x^k is the kth component of the description of x with respect to $\mathcal{A}$: $x^k = 1$ (resp., $x^k = 0$) if the attribute a^k is present (resp., absent) in x, $1 \leq k \leq K$.

This case is very important because of the invariance property of the preordonance on $\mathcal{O}$, when it is associated with a similarity index (see Proposition 30 of Chap. 4). This data structure occurs when complete disjunctive coding is considered for a description by categorical attributes. In the case of a data questionnaire, this coding is very often needed for a specific analysis.

In the random model we will consider, $\mathcal{L}_h$ is endowed with a uniform probability. Hence each of its elements has the probability $1/\binom{K}{h}$ to appear.

In order to study the distribution of the classifiability of a random set $\mathcal{O}^*$ corresponding to the object set $\mathcal{O}$, we consider the preordonance defined on $\mathcal{O}$, as associated with a similarity index. Due to the invariance property—just mentioned above—of this preordonance with respect to the analytical form of the similarity index, we consider for the latter, the simplest one, namely

$$s(x, y) = \sum_{1 \leq k \leq K} x^k \cdot y^k \qquad (8.7.11)$$

where $(x^k | 1 \leq k \leq K)$ and $(y^k | 1 \leq k \leq K)$ are the Boolean vectors describing the objects x and y, respectively, with respect to $\mathcal{A}$. Notice that $s = s(x, y)$ ranges between $max(2h - K, 0)$ and h. To simplify and without really restricting generality, we suppose $2h < K$ and then $0 \leq s \leq h$.

The different preordonance classes are defined from the different values of the similarity index $s = s(x, y)$. Therefore, there are at most $h + 1$ preordonance classes that we can denote by

$$\{\omega_s | 0 \leq s \leq h\}$$

where ω_s is the subset of F constituted by the pairs $\{x, y\}$ for which $s(x, y) = s$, $0 \leq s \leq h$.

The evaluation of the classifiability index $H(\omega)$ in a real case is based on two distributions which can be denoted by

$$p = \{(s, p(s)) | 0 \leq s \leq h\}$$
and
$$m = \{(s, m(s)) | 0 \leq s \leq h\} \qquad (8.7.12)$$

where $p(s)$ is the *proportion* of the elements of the class preorder ω_s and $m(s)$ is the *proportion* in J of the triplets whose median and highest pairs emprison strictly the class ω_s.

Therefore, the observed classifiability index can be written as

$$H(\omega) = \sum_{0 \leq s \leq h} m(s) \cdot p(s) \tag{8.7.13}$$

and the associated distribution $D(\omega)$ is defined by the decreasing sequence of the $m(s)$ values, weighted by the corresponding $p(s)$ values.

In these conditions, the theoretical value of $H(\omega)$, denoted in the following by $\mathcal{H}(\omega^*)$, in the framework of the random model of no relation, is

$$\mathcal{H}(\omega^*) = \sum_{0 \leq s \leq h} M(s) \cdot P(s) \tag{8.7.14}$$

where $P(s)$ is the probability for a given pair $\{x^*, y^*\}$ in $F^* = P_2(O^*)$ to have the value s for the similarity index $s(x^*, y^*)$ and $M(s)$ is the probability for a triplet $\{x^*, y^*, z^*\}$ in $J^* = P_3(O^*)$ to have its median and highest pairs emprisonning strictly the class ω_s^* of the preordonance on O^*.

In the following, let $(X^k | 1 \leq k \leq K)$, $(Y^k | 1 \leq k \leq K)$ and $(Z^k | 1 \leq k \leq K)$ designate the respective random Boolean vectors describing three distinct random objects in O^*. On the other hand, X, Y and Z will indicate the respective random O^* subsets.

We shall begin by specifying the distribution

$$\{(s, P(s)) | 0 \leq s \leq h\}$$

$P(s)$ can be interpreted as the proportion of pairs $\{X, Y\}$ of elements taken randomly in $\mathcal{L}_h$ for which $\sum_{1 \leq k \leq K} X^k \cdot Y^k = s$. We have

$$P(s) = Prob\left\{ \sum_{1 \leq k \leq K} X^k \cdot Y^k = s \right\} = \frac{\binom{h}{s} \cdot \binom{K-h}{h-s}}{\binom{K}{h}} \tag{8.7.15}$$

Now, let us consider the distribution

$$\{(s, M(s)) | 0 \leq s \leq h\}$$

The probability $M(s)$ can be put as follows

$$M(s) = \sum_{T} 3 \times Prob\left\{ \sum_{1 \leq k \leq K} X^k \cdot Y^k = p, \sum_{1 \leq k \leq K} Y^k \cdot Z^k = p, \sum_{1 \leq k \leq K} X^k \cdot Z^k = r \right\} \tag{8.7.16}$$

where T is the triplet set

$$T = \{(p, q, r) | p < s, q < s, r > s\}$$

for which we have

$$p + q \leq r + h, q + r \leq p + h \text{ and } r + p \leq q + h$$

An inequality of the type $p + q \leq r + h$ is implied by that

$$\sum_{1 \leq k \leq K} (X^k - Z^k)^2 \leq \sum_{1 \leq k \leq K} (X^k - Y^k)^2 + \sum_{1 \leq k \leq K} (Y^k - Z^k)^2$$

with

$$\sum_{1 \leq k \leq K} X^k = \sum_{1 \leq k \leq K} Y^k = \sum_{1 \leq k \leq K} Z^k = h$$

The multiplicative factor 3 in the $M(s)$ expression is caused by the choice of the pair concerned by the value r. Clearly, $M(s) = 0$ for $s = 0$ and for $s = h$. The probability in (8.7.16) can be put as

$$Prob\left\{ \sum_{1 \leq k \leq K} X^k Y^k = p, \sum_{1 \leq k \leq K} Y^k Z^k = p \Big| \sum_{1 \leq k \leq K} X^k Z^k = r \right\} \cdot Prob\left\{ \sum_{1 \leq k \leq K} X^k Z^k = r \right\} \tag{8.7.17}$$

This probability of the first factor can be decomposed as follows

$$\frac{\sum_{0 \leq l \leq r} \binom{h-r}{p-l} \cdot \binom{h-r}{q-l} \cdot \binom{r}{l} \cdot \binom{K-(2h-r)}{h-(p+q-l)}}{\binom{K}{h}} \tag{8.7.18}$$

This expression is obtained by decomposing the event concerned, relative to the position of Y with respect to the pair (X, Z). For a generic component of (8.7.18) l elements occur in $X \cap Z$, $(p-l)$, in $X - Z$, $(q-l)$ in $Z - X$ and the rest, $h - (p+q-l)$ in $\mathcal{A} - X \cup Z$ (draw a diagram to better see that).

Consequently, we have

$$M(s) = \sum_T 3 \times \frac{\sum_{0 \leq l \leq r} \binom{h-r}{p-l} \cdot \binom{h-r}{q-l} \cdot \binom{r}{l} \cdot \binom{K-(2h-r)}{h-(p+q-l)}}{\binom{K}{h}} \times \frac{\binom{h}{r} \cdot \binom{K-h}{h-r}}{\binom{K}{h}} \tag{8.7.19}$$

The calculation order of this sum is 4. Its computational complexity depends mainly on the binomial coefficient $\binom{K}{h}$. In these conditions, it is interesting to propose a formula approximating (8.7.19) which avoids the sum over $0 \leq l \leq r$, the latter being obtained analytically. The sum concerned can be written as

$$\frac{\binom{h}{p} \cdot \binom{K-h}{h-p}}{\binom{K}{h}} \times \sum_{0 \leq l \leq r} \frac{\binom{h-r}{p-l} \cdot \binom{r}{l}}{\binom{h}{p}} \times \frac{\binom{h-r}{q-l} \cdot \binom{K-(2h-r)}{h-(p+q-l)}}{\binom{K-h}{h-p}} \tag{8.7.20}$$

A given term of the sum above is the product of two hypergeometric probabilities. Due to the approximation of the hypergeometric probability law by that of the Poisson law [7], by putting

$$\mu = \frac{p \cdot r}{h} \text{ and } \nu = \frac{(h-p) \cdot (h-r)}{K-h}$$

we have

$$\frac{\binom{h-r}{p-l} \cdot \binom{r}{l}}{\binom{h}{p}} \simeq \frac{\mu^l}{l!} e^{-\mu} \text{ and } \frac{\binom{h-r}{q-l} \cdot \binom{K-(2h-r)}{h-(p+q-l)}}{\binom{K-h}{h-p}} \simeq \frac{\nu^{q-l}}{(q-l)!} e^{-\nu}$$

The sum in (8.7.20) becomes

$$\sum_{1 \leq l \leq r} \frac{\mu^l}{l!} e^{-\mu} \times \frac{\nu^{q-l}}{(q-l)!} e^{-\nu} = \frac{e^{-(\mu+\nu)}}{q!} \sum_{1 \leq l \leq r} \binom{q}{l} \mu^l \nu^{q-l}$$

Because the latter sum has to be considered only for $r > q$, the right member of the previous equation is equal to

$$\frac{(\mu+\nu)^q}{q!} \cdot e^{-(\mu+\nu)}$$

Therefore, an approximate value for $M(s)$ (see (8.7.19)) is given by

$$M'(s) = \sum_T 3 \frac{\binom{h}{p} \cdot \binom{K-h}{h-p}}{\binom{K}{h}} \times \frac{\binom{h}{s} \cdot \binom{K-h}{h-s}}{\binom{K}{h}} \times \frac{(\mu+\nu)^q}{q!} \cdot e^{-(\mu+\nu)} \tag{8.7.21}$$

Many experiments proved this approximation to be very accurate. Table 8.3 gives the comparison between the distributions $\{M(s)|s \geq 0\}$ and $\{M'(s)|s \geq 0\}$, for $K = 200$ and $h = 40$. We obtain the associated classifiability indices as $H = 0.149$ and $H' = 0.155$. Otherwise,

$$\sum_{s \geq 0} |M(s) - M'(s)| = 0.091$$

Experiment shows that the distribution of $M(s)$ (resp., $M'(s)$) is unimodal. It begins by increasing and next decreases rapidly, the values of $M(s)$ an $M'(s)$ becoming negligible. This observation might be established theoritically.

Table 8.3 Comparing M and M' for $K = 200$ and $h = 40$

s	$M(s)$	$M'(s)$	$P(s)$
1	0.000	0.000	0.001
2	0.000	0.000	0.004
3	0.000	0.000	0.014
4	0.001	0.002	0.038
5	0.008	0.015	0.078
6	0.042	0.060	0.126
7	0.125	0.148	0.164
8	0.229	0.242	0.174
9	0.277	0.273	0.154
10	0.234	0.223	0.114
11	0.146	0.138	0.071
12	0.071	0.067	0.038
13	0.028	0.027	0.017
14	0.009	0.009	0.007
Total	1.170	1.203	0.997

8.7.4 The Murtagh Contribution

In [22], comparing the index $H(\omega)$ (8.7.3) with that proposed by Rammal et al. [24, 25], is considered. The latter index is defined for an object set $\mathcal{O}$, provided with a distance or a dissimilarity function d. It is written as

$$R(d) = \frac{\sum_{(x,y) \in \mathcal{O} \times \mathcal{O}} \left(d(x, y) - d_c(x, y) \right)}{\sum_{(x,y) \in \mathcal{O} \times \mathcal{O}} d(x, y)} \qquad (8.7.22)$$

where d_c is the so-called subdominant ultrametric associated with d. In fact, d_c is obtained from applying *Single Linkage* algorithm on $(\mathcal{O}, d)$ (see Sect. 10.3.2 of Chap. 10). More precisely, in the increasing level sequence of building the classification tree by this algorithm, for a new merging cluster agregating two clusters C and D ($C \subset \mathcal{O}$ and $D \subset \mathcal{O}$), we have

$$\left(\forall (x, y) \in C \times D \right), d(x, y) = min_{(x',y') \in C \times D} d(x', y') \qquad (8.7.23)$$

The $R(d)$ index which is bounded by 0 (complete ultrametricity) and 1, depends on

1. The definition of a distance function d;
2. Applying a particular clustering algorithm, known for its chaning effect.

The new index proposed by Murtagh in [22] concerns, in fact, a clouds of points in an Euclidean space (see (10.3.3) of Sect. 10.3.3.1). Let us denote it by M(d). In order to avoid as far as possible, lack of invariance due to use distances, $M(d)$ is based on angles. Nonetheless, a question might be asked relative to the influence of the distance adopted on the calculation angles from distances.

In the $M(d)$ index, the ultrametricity notion of a given triangle in the space concerned is relaxed. More precisely, for $H(\omega)$ and $R(d)$, a triangle can be said ultrametric if it is isoceles, the basis being (in a large sense) the lowest lenght side. In the Murtagh approach, an *almost* ultrametric triangle is defined. For this purpose, for a given triangle, we consider the vertex for which the underlied angle is the smallest one and then is lower or equal to 60 degrees. Let us designate by X the vertex concerned in a triangle that we denote by (X, Y, Z). Such a triangle is considered as almost ultrametric if the absolute value of the difference between the angles underlied by Y and Z is strictly lower than a fixed threshold. This is taken equal to 2 degrees.

Mainly, the index $M(d)$ is a *proportion*. It is defined as the ratio of the number of triangles almost ultrametric with respect to the total number of triangles. In the latter counting, without clear justification, aligned triangles (X, Y and Z are on a same line) are excluded a priori.

Now the number of triplets being too large (its order is $O(n^3)$), the calculation of the proportion $M(d)$ is carried out on the basis of a random sample taken in the set of all triplets.

In [22] the following two interesting phenomena, simultaneously observed for $H(\omega)$ and $M(d)$ are hightlighted: For a random sample of vertices, associated with a uniform distribution in a hypercube of the geometrical space $\mathbb{R}^p$

1. The respective values of $H(\omega)$ and $M(d)$ do not depend on the sample size;
2. The classifiability (for both indices) increases when the dimensionality p of the geometrical space increases.

References

1. Adanson, M.: Histoire naturelle du Sénégal, Coquillages. Bauche, Paris (1757)
2. Anderson, T.W.: An Introduction to Multivariate Statistical Analysis. Wiley, New York (1958)
3. Beckner, M.: The Biological Way of Thought. Columbia University Press, New York (1959)
4. Caillet, M., Massé, L., Courcoux, H., Coste, E., Abou, E., Tallur, B.: Importance du niveau de la tension artérielle systolique dans la sélection de population-cibles en médecine préventive. In 5-ème Colloque National des Centres d'Examens de Santé. Bordeaux (1981)
5. Chantrel, T.: Nouvelle approche dans la classification et représentation d'un vaste ensemble d'échelles et profils d'attitude. Application à des données en économie rurale et en psychosociologie. Thèse de 3-èse cycle. Ph.D. thesis, Université de Rennes (1979). Accessed 1 May 1979
6. Danser, B.H.: A theory of systematics. Bibl. Biotheor. **4**, 113–180 (1950)
7. Feller, W.: An Introduction to Probability Theory and Its Applications. Wiley, New York (1968)
8. Gilmour, J.S.L.: A taxonomic problem. Nature **139**, 1040–1042 (1937)
9. Lancaster, H.O.: The Chi-squared Distribution. Wiley, New York (1969)
10. Lebart, L., Fenelon, J.-P.: Statistique et Informatique appliquées. Dunod (1973)

11. Leredde, H.: La méthode des pôles d'attraction; la méthode des pôles d'aggrégation : deux nouvelles familles d'algorithmes en classification automatique et sériation. Ph.D. thesis, Université de Paris (1979). Accessed 6 Oct 1979

12. Lerman, I.-C.: Les bases de la classification automatique. Gauthier-Villars (1970)

13. Lerman, I.C.: H-classificabilité. Rev. Math. Sci. Hum. **27**, 21–28 (1969)

14. Lerman, I.C.: Analyse du phénomène de la sériation. Rev. Math. Sci. Hum. **38**, 39–57 (1972)

15. Lerman, I.C.: Croisement de classifications floues. Publications de l'Institut de Statistique des Universités de Paris (XXIV, fasc. 1–2), pp. 13–46 (1979)

16. Lerman, I.C.: Classification et analyse ordinale des données. Dunod (1981). http://www.brclasssoc.org.uk/books/index.html

17. Lerman, I.C.: Analyse classificatoire d'une correspondance multiple; typologie et régression. In: Diday, E. et al. (eds.) Data Analysis and Informatics III, pp. 193–212. North-Holland, Amsterdam (1984)

18. Lerman, I.C.: Coefficient général de discrimination de classes d'objets par des variables de types quelconques. Application à des données génotypiques. Revue de Statistique Appliquée, **54**(2), 33–63 (2006)

19. Lerman, I.C., Hardouin, M., Chantrel, T.: Analyse de la situation relative entre deux classifications floues. In: Diday, E. et al. (eds.) Data Analysis and Informatics, pp. 523–552. North-Holland, Amsterdam (1980)

20. Lerman, I.C., Leredde, H.:. La méthode des pôles d'attraction. In: IRIA, editor, Analyse des Données et Informatique. IRIA (1977)

21. Moreau, A.: Élaboration et calcul d'indices d'association entre variables qualitatives "nettes" ou "floues", Application à une forme d'interprétation d'une classification de paramètres épidémiologiques. Thèse de 3-èse cycle. Ph.D. thesis, Université de Rennes (1985). Accessed 1, June 1985

22. Murtagh, F.: On ultrametricity, data coding and computation. J. Classif. **21**(2), 167–184 (2004)

23. Prod'homme, A.: Indices d'explication des classes obtenues par une méthode de classification hiérarchique respectant la contrainte de contiguïté spatiale. Application à la viticulture Girondine et à la construction de logements dans les Bouches du Rhône. Thèse de 3-èse cycle. Ph.D. thesis, Université de Rennes (1980). Accessed 1 Dec 1980

24. Rammal, R., Angles d'Auriac, J.C., Doucot, B.: On the degree of ultrametricity. J. Phys. Lett. **46**, 945–952 (1985)

25. Rammal, R., Toulouse, G., Virasoro, M.A.: Ultrametricity for physicists. Rev. Mod. Phys. **58**, 765–788 (1986)

26. Saporta, G.: Quelques applications des opérateurs d'Escoufier au traitement des variables qualitatives. Statistique et Analyse des Données **1**, 38–46 (1976)

27. Sattler, R.: Methodological problems in taxonomy. Syst. Zool. **13**(1), 19–27 (1964)

28. Simovici, D.A.: Several remarks on dissimilarities and ultrametrics. Sci. Ann. Comput. Sci. **25**(1), 155–170 (2015)

29. Sneath, P.H.A., Sokal, R.: Numerical Taxonomy. Freeman, San Francisco (1972)

30. Sokal, R., Sneath, P.H.A.: Principles of Numerical Taxonomy. Freeman, San Francisco (1963)

31. Tallur, B.: Méthode d'interprétation d'une classification hiérarchique d'attributs-modalités pour l'explication d'une variable; application à la recherche d'un seuil critique de la tension systolique et des indicateurs de risques cardiovasculaires. Revue de Statistique Appliquée **31**(1), 25–43 (1983)

32. Vicq d'Azyr, F.: Quadrupèpdes, discours préliminaire, encyclopédie méthodique vol. 2. Panckoucke, Paris (1792)

33. Xu, R., Wunsch, D.: Survey of clustering algorithms. IEEE Trans. Neural Netw. **16**(3), 645–678 (2005)

34. Zadeh, L.A.: Fuzzy sets. Inf. Control **8**, 338–353 (1965)

Chapter 9
Quality Measures in Clustering

9.1 Introduction

The construction of a clustering criterion depends on the nature of the data and the mathematical structure retained for its representation. We saw in Chap. 2 two criteria associated with two different methods: the "Central partition" and the "Dynamic adaptative" methods, respectively. A formal definition of a criterion in *Data Analysis* may be expressed as follows:

"The structure σ—on the set E concerned—to which the data representation belongs (e.g. similarity coefficient on E) is more general than that τ of the structure sought (partition or ordered chain of partitions on E). The general strategy consists of injecting the family Θ of the τ structures on E into the family Σ of the σ structures on E. The objective is then to determine that or those of the elements of Θ which are the most comparable with $\sigma(E)$ (σ calculated on E). For this purpose, a criterion is built. Quite generally, a criterion is a ranking (preorder) relation (possibly partial) on the set Θ. For the latter, $\tau(E)$ is preferred to $\tau'(E)$ if and only if, in a given sense, $\tau(E)$ is nearer $\sigma(E)$ than $\tau'(E)$".

Mostly, the criterion is defined from a metric or a dissimilarity coefficient on Σ. Nevertheless, there are cases where it is not so.

To begin, we may consider the case where σ is defined by a rectangular data table crossing two sets. Denote by $\mathbb{I} = \{1, 2, \ldots, i, \ldots, n\}$ and $\mathbb{J} = \{1, 2, \ldots, j, \ldots, p\}$ the respective sets of indices coding the rows and columns of such a table. x_{ij} is the value—taken in a given scale—situated in the cell at the intersection of the ith row and the jth column ($1 \leq i \leq n$, $1 \leq j \leq p$). x_{ij} is a measure relating i to j, $1 \leq i \leq n$, $1 \leq j \leq p$. To fix ideas, assume for the moment x_{ij} as a Boolean measure. More general scales will be considered below (see Chap. 3). Consequently, σ is defined here by a valuated unary incident relation on the cross product $\mathbb{I} \times \mathbb{J}$. The *direct* clustering methods (see [12, 19] for the first uses of this expression) intend to work

© Springer-Verlag London 2016

I.C. Lerman, *Foundations and Methods in Combinatorial and Statistical Data Analysis and Clustering*, Advanced Information and Knowledge Processing, DOI 10.1007/978-1-4471-6793-8_9

at this data level as symmetrically as possible, with respect to rows and columns of the data table. Their objective consists of setting up a system of subrectangles

$$\{\mathbb{I}_k \times \mathbb{J}_k | 1 \leq k \leq K\}$$

where $\mathbb{I}_k \subset \mathbb{I}$ and $\mathbb{J}_k \subset \mathbb{J}$, such that a specific model (logical or statistical) for the distribution of the x_{ij} values inside $\mathbb{I}_k \times \mathbb{J}_k$ is revealed, $1 \leq k \leq K$.

As expressed at the beginning of this introduction, the clustering criterion is established more often than not from a distance or a numerical dissimilarity index on the set E to be clustered. And then, the σ structure—on which the criterion is defined—is a positive numerical binary relation on E. In the literature the case where E is a set of objects $\mathcal{O}$ described by numerical attributes is especially developed. In this case, $\mathcal{O}$ can be represented by a cloud of points in an Euclidean space (see Sects. 7.2.1 and 10.3.3.1 of Chaps. 7 and 10, respectively). For this situation, let us refer to the criteria examined in the Milligan and Cooper paper [32]. In relation to the evaluation of a given partition $\pi = \{\mathcal{O}_1, \mathcal{O}_2, \ldots, \mathcal{O}_k, \ldots, \mathcal{O}_K\}$ of $\mathcal{O}$, in most of the metrical criteria, the *residual* inertia

$$W = \sum_{1 \leq k \leq K} \sum_{i \in \mathbb{I}_k} d^2(O_i, G_k)$$

and the *explained* one

$$B = \sum_{1 \leq k \leq K} n_k d^2(G_k, G)$$

take important roles. In these equations $\mathbb{I}_k$ designates the subscript set indexing the class $\mathcal{O}_k$, $1 \leq k \leq K$. Unity weight is assigned to every element of $\mathcal{O}$. G is the global centre of gravity of the cloud associated with $\mathcal{O}$ and G_k is the centre of gravity of $\mathcal{O}_k$, whose cardinality being n_k, $1 \leq k \leq K$. W and B correspond to the *Within* and *Between* variances associated with the partition π.

Notice that W and B can be expressed as functions of the following valuated binary relation on $\mathcal{O}$

$$\{d^2(O_i, O_{i'}) | \{O_i, O_{i'}\} \in P_2(\mathcal{O})\}$$

and the partition π (see (10.3.5) of Chap. 10). Specifying these expressions is left for the reader.

In the list of criteria examined in [32] the Goodman and Kruskal γ criterion is considered. As seen in Sect. 4.3.3 this coefficient applies for comparing two total preorders (ranking with ties) on a given set E. In the case concerned here these pre-orders are defined on the set $P = P_2(\mathcal{O})$. The first one is the preordonance $\omega_d(\mathcal{O})$ on $\mathcal{O}$ associated with the distance function d (see Sect. 4.2.2 of Chap. 4) and the second one corresponds to the total preorder into two classes $S(\pi)$ and $R(\pi)$ (see Sects. 3.3.1, 4.3.2) associated with the partition π to be evaluated. If the preordonance mentioned is established in such a way, the higher the rank of a given object pair $\{O_i, O_{i'}\}$, the greater is the distance $d(O_i, O_{i'})$, then $R(\pi)$ precedes $S(\pi)$. This comparison structure is considered in Sect. 9.2. To begin, we consider the case where the ordinal similarity and the partition are represented by total preorders on P. In this framework, the adaptation of classical coefficients and therefore the γ coefficient will be specified (see Sect. 9.3.1). Section 9.3.3 justifies the development given in Sect. 9.3.4 where the LLA strategy is applied. The latter development constitutes the most important part of this chapter. The normalized coefficient obtained leads to our methods—worked in collaboration with [35]—of detecting the most "significant" levels and the most "significant" nodes of a classification tree (see Sect. 9.3.6) and Chap. 10.

All these coefficients are established in the case of a set theoretic representation defined in the Cartesian product $P \times P$. We might think that these developments might take place in Chap. 6; this is because they are near comparing two preordonance categorical attributes. However, the specificity of the problem concerned (evaluation of the fit of a partition to an ordinal similarity), the historic of this research and the results obtained (1970–1983) make necessary this subject to be treated separately.

The specific case where the total preorder on $\mathcal{O}$ comprises two classes is associated with a similarity defined by symmetric binary relation on $\mathcal{O}$. This case, for which the representation set can be reduced to P, was studied in the famous Zahn paper [41]. It will be reported and discussed in Sect. 9.3.2.

In [15], comparing two binary numerical valuations on the object set is considered first. Next, it is shown how to apply this general way of comparison—for example by means of the Γ Daniels coefficient [5]—to different scenarios. However, in the case of comparing specific combinatorial structures, this strategy hides the true nature of the comparison problem.

The most important concerns the evaluation of the fitting quality of a partition to a similarity (ordinal or numerical) on a given set. Nevertheless, the problem of evaluating the adequation of a total partition chain to an ordinal or numerical similarity on a given set is equally interesting. The latter partition chain might be associated with a classification tree built from a similarity, dissimilarity or distance index on the set concerned. We shall see in Sect. 9.3 how to generalize the classical criteria established for evaluating the quality of a partition.

For an ordinal similarity (total preordonance) on $\mathcal{O}$ and for the partition case, the two basic criteria are the respective cardinals of the intersection and the symmetric difference between the graph representations of the preordonance and the partition in $P \times P$ (see Sects. 9.3.3 and 9.3.4).

In the previous ordinal criteria there is a metrical aspect for comparing two total preorders on P, which associated with the total preordonance on $\mathcal{O}$ and that corresponding to a partition (resp., partition chain) on $\mathcal{O}$. Pure ordinal criteria, that is, not depending on any metrical considerations, will be defined in Sect. 9.4.4, for the general case of a partition chain search. They are called "lateral order" and "lexicographic order", respectively (see Sects. 9.4.4.1 and 9.4.4.2). In Sect. 9.4.5, the latter criterion is related to the metrical one defined by the "inversion number", the latter being similar to the symmetrical difference criterion.

In our expression above, criterion is defined with respect to clustering an object set $\mathcal{O}$. It goes without saying that a category set $\mathcal{C}$ or an attribute set $\mathcal{A}$ can be substituted for $\mathcal{O}$ (see Chaps. 5 and 6).

9.2   The Direct Clustering Approach: An Example of a Criterion

9.2.1   General Presentation

Let us take up again the terms of the introduction for this subject (see the beginning of Sect. 9.1). $\mathbb{I} \times \mathbb{J}$ indexes the entire data table

$$\{x_{ij} | (i, j) \in \mathbb{I} \times \mathbb{J}\}$$

where the value x_{ij}, generally defined as the value of the jth attribute a^j on the ith object o_i, is considered as a measure relating i to j, $1 \leq i \leq n$, $1 \leq j \leq p$. The case where x_{ij} is a Boolean measure illustrates perfectly this situation. Nevertheless, more general scales will be considered below. Consequently, the σ structure considered above is defined here by a valuated unary incident relation on the cross product $\mathbb{I} \times \mathbb{J}$. Working at this level, the objective of *direct* clustering methods consists of setting up a system of subrectangles

$$\{\mathbb{I}_k \times \mathbb{J}_k | 1 \leq k \leq K\} \tag{9.2.1}$$

where $\mathbb{I}_k \subset \mathbb{I}$ and $\mathbb{J}_k \subset \mathbb{J}$, such that a specific model (logical or statistical) for the distribution of the x_{ij} values inside $\mathbb{I}_k \times \mathbb{J}_k$ is revealed, $1 \leq k \leq K$.

Mostly, the model requested requires the weakest variation within the rectangles mentioned. When a given rectangle—say $\mathbb{I}_{k0} \times \mathbb{J}_{k0}$—is maximal, it may correspond either to a formal concept in the logical case [8] or a polythetic class in the statistical Beckner sense (see Sect. 8.1.1 of Chap. 8), [17]. Also, this direct approach leads to setup joint clusterings [10, 11], or joint hierarchical clusterings [12, 13] on the row and column sets.

More specific models having an ordinal character, the *seriation* models, were studied in order to reorganize, by permutation rows and columns of $\mathbb{I} \times \mathbb{J}$, thus highlighting a highly dense diagonal form with respect to the x_{ij} values, $(i, j) \in \mathbb{I} \times \mathbb{J}$ [4, 16, 18, 21, 22, 26, 31]. Often, this global diagonal form is decomposed into subdiagonal forms, more or less disconnected.

Now, let us specify the data types concerned by direct biclustering methods. They correspond to the following cases:

1. $\mathbb{I}$ codes an object set $\mathcal{O}$ and $\mathbb{J}$ codes a Boolean attribute set $\mathcal{A}$;
2. $\mathbb{I}$ codes an object set $\mathcal{O}$ and $\mathbb{J}$ codes a numerical quantitative attribute set $\mathcal{A}$;
3. $\mathbb{I}$ (resp., $\mathbb{J}$) codes an exhaustive category set Γ (resp., Λ).

Notice that all of these cases come under descriptive attributes of type I (see Sect. 3.2 of Chap. 3 and Sect. 5.3.2 of Chap. 5). For the item 2 above, the scale to which the values x_{ij} ($1 \leq i \leq n$, $1 \leq j \leq p$) refer corresponds generally to positive reals. It is very important for this data structure to have homogeneity of the distribution

$$\{x_{ij}|(i,j) \in \mathbb{I} \times \mathbb{J}\} \qquad (9.2.2)$$

that is to say, an absolute value of a difference such that $|x_{ij} - x_{i'j'}|$ where $(i, j) \neq (i'j')$, must have the same meaning whatever are the respective positions of the cells (i, j) and (i', j') in $\mathbb{I} \times \mathbb{J}$.

Three types of approaches can be distinguished for direct biclustering methods:

1. Graphical;
2. Formal;
3. Statistical.

The graphical methods introduced and developed by Jacques Bertin and his collaborators [3, 4] illustrate very significantly the matter of item 1 above. These techniques are based on the characteristics and powerful of visual perception. For a given data rectangle $\mathbb{I} \times \mathbb{J}$ coming under one of the three data types considered previously (see above), row and column permutations are processed, according to clear recommendations, in order to setup highly dense forms in the data table. Mostly, these forms correspond to biclusters or seriation fragments [3, 4]. Most often the starting point is a permutational organization according to a couple of hierarchical clusterings on the row and column sets (personal communication). In spite of this treatment, the size of the data table which can be processed is limited. The case of 210 rows and 85 columns is among the biggest sizes mentioned.

Formal Concept Analysis (FCA) methods (see item 2 above) principally concern Boolean data. In these conditions the entries of the rectangle $\mathbb{I} \times \mathbb{J}$ are Boolean values, coded by 0 and 1. A *maximal* subrectangle $\mathbb{I}_k \times \mathbb{J}_k$ ($\mathbb{I}_k \subset \mathbb{I}, \mathbb{J}_k \subset \mathbb{J}$) comprises 1 value exclusively. On the other hand, it is such that the maximal property is lost if $\mathbb{I}_k$ or $\mathbb{J}_k$ is augmented. By denoting $\mathcal{O}_k$ and $\mathcal{A}_k$ the respective object subset and attribute subset indicated by $\mathbb{I}_k$ and $\mathbb{J}_k$, respectively, $\mathcal{O}_k \times \mathcal{A}_k$ determines a bicluster called *concept* in the framework of *FCA*. This notion may correspond to a *monothetic* class in an *absolute* sense (see Chap. 8). That is, a given object belongs to the class $\mathcal{O}_k$ if and only if all the Boolean attributes of $\mathcal{A}_k$ are present in every element of $\mathcal{O}_k$. Therefore, essentially, this notion is *opposite* to that given by the notion of *polythetic* class, given by Beckner (see Chap. 8). In fact, in Clustering we seek for *global* statistical *patterns* from local logical descriptions, whereas, in *FCA*, we seek for *global* logical *patterns* from local logical descriptions.

Among the first works in statistical biclustering we may cite the Govaert and Hartigan contributions [10–13]. In the Govaert method non-hierarchical clusterings on rows and columns of the data table are carried out alternatively this, in order to optimize locally a global criterion defined symmetrically on the whole data table. Dynamic adaptative algorithm (see Chap. 2) is employed for this purpose. In the Hartigan approach [12] bi-hierarchical clustering technique is proposed. It is based on a specific binary splitting algorithm. The data on which it applies is a partition of the entire rectangle $\mathbb{I} \times \mathbb{J}$ into subrectangles. If

$$\{\mathbb{I}_k \times \mathbb{J}_k | 1 \leq k \leq K\} \qquad (9.2.3)$$

is such a partition, a subrectangle $\mathbb{I}_{k0} \times \mathbb{J}_{k0}$ is selected in order to be split. This selection can be provided from the criterion to optimize the variance criterion in this case. The row set $\mathbb{I}_{k0}$ or the column set $\mathbb{J}_{k0}$ is split into two subsets leading to the greatest variance reduction. An heuristic may be proposed for this splitting. All these splittings give rise to a pair of binary classification trees on both the row set and the column set. Comparison was made of the latter with those obtained by the classical *Average Linkage* method (see Chap. 10) on the basis of small examples [12]. No significant difference was observed.

As expressed in Chaps. 3 (Sect. 3.5.4) and 8, whatever is the nature of the data table, it is very instructive to begin by crossing two separate hierarchical clusterings on the object set (resp., category set), represented by the row set for one hand and the attribute set represented by the column set, on the other hand. This can be performed whatever is the common type of the descriptive attributes (see Chaps. 3–7 and 10). A reduced representation of the classification trees concerned, given by their respective "significant levels" (see Sect. 9.3.6), provides clear interpretation of both clusterings, each with respect to the other. Our first experiments on this matter are reported in [23, 36]. The statistical logic of the latter correspondence is essentially different from joint clustering (biclustering) defined in the framework of the rectangle $\mathbb{I} \times \mathbb{J}$ for symmetrical data (see items 1–3 above). For these, in the Bioinformatics field new dense forms are defined in the rectangle mentioned. In this respect, earlier or new biclustering methods have arisen [30, 38].

On the other hand and finally, let us mention a new family of Biclustering algorithms concerning numerical data and based on optimizing a least square criterion [40].

9.2.2 An Example

Let us begin by some formal aspects. We consider a set $\mathcal{O}$ of objects described by a set $\mathcal{A}$ of descriptive Boolean attributes. For this we take again the clustering model of $\mathcal{O}$ shown in Chap. 8 (see Sect. 8.1.1) and represented by

$$\pi(\mathcal{O} \backslash \mathcal{A}) = \{(\mathcal{O}_k, \mathcal{A}^k) | 1 \leq k \leq K\} \tag{9.2.4}$$

where $\{\mathcal{O}_k | 1 \leq k \leq K\}$ is a partition of $\mathcal{O}$; but where $\{\mathcal{A}^k | 1 \leq k \leq K\}$ is not a partition necessarily. However, we may assume that the union of the $\mathcal{A}^k$ recovers the entire set $\mathcal{A}$:

$$\bigcup_{1 \leq k \leq K} \mathcal{A}^k = \mathcal{A}$$

$\mathcal{A}^k$ is the attribute set to which $\mathcal{O}_k$ refers, $1 \leq k \leq K$.

As above, $\mathbb{I}_k$ (resp., $\mathbb{J}_k$) designates the subset of $\mathbb{I}$ which indexes $\mathcal{O}_k$. Thus, the rectangle $\mathbb{I}_k \times \mathbb{J}_k$ indexes $\mathcal{O}_k \times \mathcal{A}^k$. Due to the mutual disjunction of the $\mathcal{O}_k$ subsets, the rectangles mentioned are mutually disjoint:

$$\bigcap_{1 \leq k \leq K} (\mathbb{I}_k \times \mathbb{J}_k) = \emptyset$$

Now, designate by L the union of the rectangles $\mathbb{I}_k \times \mathbb{J}_k$ and by M the complementary subset of L in $\mathbb{I} \times \mathbb{J}$:

$$L = \bigcup_{1 \leq k \leq K} (\mathbb{I}_k \times \mathbb{J}_k) \text{ and } M = L^c$$

where the superscript c indicates the complementary function. In these conditions, we represent the decomposition $\pi(\mathcal{O} \setminus \mathcal{A})$ by means of the following Boolean function ϖ on $\mathbb{I} \times \mathbb{J}$:

$$\varpi_{ij} = 1 \text{ if } (i,j) \in L$$
$$\varpi_{ij} = 0 \text{ if } (i,j) \in M \tag{9.2.5}$$

The τ structure is then here defined by such a binary relation ϖ on $\mathbb{I} \bigcup \mathbb{J}$, namely

$$\forall (i,j) \in (\mathbb{I} \bigcup \mathbb{J})^2, i\varpi j \Leftrightarrow (i,j) \in \mathbf{L} \tag{9.2.6}$$

This relation represents the partition $\{\mathcal{O}_k | 1 \leq k \leq K\}$, where the class $\mathcal{O}_k$ is characterized by the attribute subset $\mathcal{A}^k$ in the following monothetic sense: "a given object x belongs to $\mathcal{O}_k$ if and only if, exactly, all the attributes of $\mathcal{A}^k$ are present in x", $1 \leq k \leq K$. To evaluate the fitting of this Boolean structure to the data, it is natural to consider the following criterion:

$$C(\epsilon, \varpi) = \sum_{(i,j) \in \mathbb{I} \times \mathbb{J}} (\epsilon_{ij} - \varpi_{ij})^2 \tag{9.2.7}$$

where $\{\epsilon_{ij} = a^j(o_i) | (i,j) \in \mathbb{I} \times \mathbb{J}\}$ is the incidence data table. This criterion and the development here are taken again from [19, 22]. As mentioned above, an equivalent version of this criterion with a specific development is considered in [12, 13]. If we formalize the Boolean incidence data table as a binary relation ϵ on $\mathbb{I} \bigcup \mathbb{J}$ defined by

$$\forall (i,j) \in (\mathbb{I} \bigcup \mathbb{J})^2, i\epsilon j \Leftrightarrow (i,j) \in \mathbb{I} \times \mathbb{J} \text{ and } \epsilon_{ij} = 1 \tag{9.2.8}$$

the criterion $C(\epsilon, \varpi)$ (see Eq. (9.2.7)) appears as the symmetrical difference between ϵ and ϖ. The following relation can be easily established:

$$C(\epsilon, \varpi) = \sum_{1 \leq i \leq n, 1 \leq j \leq p} \epsilon_{ij} + \sum_{1 \leq i \leq n, 1 \leq j \leq p} \varpi_{ij}$$
$$-2 \sum_{1 \leq i \leq n, 1 \leq j \leq p} \epsilon_{ij} \cdot \varpi_{ij}$$

Table 9.1 Incidence data table $\mathcal{O} \times \mathcal{A}$

$\mathcal{O} \backslash \mathcal{A}$	a^1	a^2	a^3	a^4	a^5	a^6
o_1	0	0	1	1	0	0
o_2	1	1	0	0	0	0
o_3	0	0	0	1	0	1
o_4	1	1	1	0	0	0
o_5	1	0	1	0	0	0
o_6	0	1	1	1	0	0
o_7	0	0	0	0	1	1
o_8	1	0	1	0	0	0
o_9	0	0	1	0	0	0
o_{10}	1	0	0	1	1	1
o_{11}	1	1	1	0	0	1
o_{12}	0	0	0	0	0	1
o_{13}	1	1	0	0	0	1
o_{14}	0	1	1	1	1	0
o_{15}	0	0	0	0	1	1
o_{16}	1	0	1	0	0	0
o_{17}	1	1	1	0	1	0
o_{18}	0	0	0	0	1	1

$$= \sum_{\mathbf{M}} \epsilon_{ij} - \sum_{\mathbf{L}} \epsilon_{ij} + card(\mathbf{L}) \qquad (9.2.9)$$

The incidence data Table 9.1 is obtained by transposing Table 4.1 of Chap. 4 and by reversing the respective roles of described objects and descriptive attributes. Now, we have a set $\{o_1, o_2, \ldots, o_i, \ldots, o_{18}\}$ of eighteen objects described by a set $\{a^1, a^2, a^3, a^4, a^5, a^6\}$ of six Boolean attributes. Relatively to the development above, we shall consider here the very simplified case for which

$$\mathbf{L} = (\mathcal{O}_1 \times \mathcal{A}^1) \bigcup (\mathcal{O}_2 \times \mathcal{A}^2) \qquad (9.2.10)$$

where $\{\mathcal{O}_1, \mathcal{O}_2\}$ (resp., $\{\mathcal{A}^1, \mathcal{A}^2\}$) is a partition of $\mathcal{O}$ (resp., $\mathcal{A}$) into two classes. Hence, the ϖ function is defined as follows:

$$\varpi_{ij} = 1 \text{ if } (i,j) \in (\mathbf{I}_1 \times \mathbf{J}_1) \bigcup (\mathbf{I}_2 \times \mathbf{J}_2)$$

$$\varpi_{ij} = 0 \text{ if } (i,j) \notin (\mathbf{I}_1 \times \mathbf{J}_1) \bigcup (\mathbf{I}_2 \times \mathbf{J}_2) \qquad (9.2.11)$$

where $\mathbf{I}_1$ and $\mathbf{I}_2$ (resp., $\mathbf{J}_1$ and $\mathbf{J}_2$) index the sets $\mathcal{O}_1$ and $\mathcal{O}_2$ (resp., $\mathcal{A}^1$ and $\mathcal{A}^2$), respectively.

In the technique proposed below we assume given the partition $\{\mathcal{A}^1, \mathcal{A}^2\}$ and we derive from it the partition $\{\mathcal{O}_1, \mathcal{O}_2\}$. For this, a given object x belongs to $\mathcal{O}_1$ if and only if the proportion of the $\mathcal{A}^1$ attributes possessed by x is strictly greater than that of the $\mathcal{A}^2$ attributes.

In these conditions, let

$$\pi_1(\mathcal{A}) = \{\{a^1, a^2, a^3\}, \{a^4, a^5, a^6\}\} \text{ and }$$

$$\pi_2(\mathcal{A}) = \{\{a^1, a^2, a^3, a^4\}, \{a^5, a^6\}\} \tag{9.2.12}$$

be two partitions of $\mathcal{A}$. Begin by noticing that these partitions correspond to those obtained at the before last level of the classification trees shown in Figs. 10.3 and 10.2 of Chap. 10. Now, designate by $\pi_1(\mathcal{O}) = \{\mathcal{O}_{11}, \mathcal{O}_{12}\}$ and $\pi_2(\mathcal{O}) = \{\mathcal{O}_{21}, \mathcal{O}_{22}\}$ the respective partitions associated with $\pi_1(\mathcal{A})$ and $\pi_2(\mathcal{A})$. we have

$$\mathcal{O}_{11} = \{o_2, o_4, o_5, o_6, o_8, o_9, o_{11}, o_{13}, o_{16}, o_{17}\}$$

$$\mathcal{O}_{12} = \{o_1, o_3, o_7, o_{10}, o_{12}, o_{14}, o_{15}, o_{18}\} \tag{9.2.13}$$

$$\mathcal{O}_{21} = \{o_1, o_2, o_4, o_5, o_6, o_8, o_9, o_{11}, o_{14}, o_{16}, o_{17}\}$$

$$\mathcal{O}_{22} = \{o_3, o_7, o_{10}, o_{12}, o_{13}, o_{15}, o_{18}\} \tag{9.2.14}$$

The respective values of the criterion $C(\epsilon, \varpi)$ (see (9.2.7)) for $\pi_1(\mathcal{O})$ and $\pi_2(\mathcal{O})$ are

$$3 + 1 + 1 + 0 + 1 + 2 + 1 + 1 + 2 + 1 + 1 + 2 + 2 + 2 + 1 + 1 + 1 + 1 = 24$$

and

$$2 + 2 + 2 + 1 + 2 + 1 + 0 + 2 + 3 + 2 + 2 + 1 + 3 + 2 + 0 + 2 + 2 + 0 = 29$$

Therefore, $\pi_1(\mathcal{O})$ fits better than $\pi_2(\mathcal{O})$. However, the criterion used is too raw. An *LLA* version might be obtained by studying its distribution under a hypothesis of no relation for which we associate a set of mutually independent Boolean random attributes with the observed ones. The random attribute $a^{j\star}$ may be defined from a random permutation of the entries of the jth column of the incidence data table, $1 \leq j \leq p$. The management of this type of test seems difficult.

9.3 Quality of a Partition Based on the Pairwise Similarities

Preamble

Let us designate here by E the set to be clustered, E may be a set of objects, a set of categories or a set of attributes. F will designate the set of unordered pairs composed of distinct elements of E. In other words, F is the set of subsets with two elements of $E\colon F = P_2(E)$. As indicated in the Introduction (see Sect. 9.1), we shall begin with the case where the similarity data is ordinal (preordonance on E, that is, total preorder on F). The construction analysis of a criterion evaluating how much a given partition on E fits an ordinal similarity on E will enable us to control this construction in the case of a numerical similarity on E.

As expressed in the introductory section, the most important problem consists of evaluating the fitting of a *partition* to a similarity on E, the latter being ordinal or numerical. Nevertheless, the problem of evaluating the fitting of a *partition chain* to a similarity on E is interesting by itself and may be important in real applications.

9.3.1   Criteria Based on a Data Preordonance

Let us take up once again our notations. E is the set on which a given partition $\pi(E)$ has to be evaluated and

$$F = P_2(E) = \{\{x, y\} | x \in E, y \in E, x \neq y\} \tag{9.3.1}$$

is the set of unordered distinct element pairs of E. The similarity data is a *preordonance* on E, that is, a total preorder on F. This preorder is generally associated with the choice of a numerical similarity or dissimilarity functions on E (see Chap. 4, Sect. 4.2.2 for the particular case of Boolean data). Without loss of generality, we suppose in this section the preordonance on E associated with a *similarity* (and not a *dissimilarity*) function on E. There are two ways, mentioned in the introduction of Chap. 4, to establish the preordonance on E (see Eq. (4.1.1)). For the first one (resp., the second one) the greater the rank of a given pair $\{x, y\} \in F$ is, the less (resp., more) similar are the components x and y.[1] We designate by $\omega(E)$ (resp., $\varpi(E)$) the first (resp., the second) total preorder. We have (4.1.1) of Chap. 4,

$$\forall (p, q) \in F \times F, p \leq_{\omega(E)} q \Leftrightarrow \mathcal{S}(p) \geq \mathcal{S}(q)$$
$$\forall (p, q) \in F \times F, p \leq_{\varpi_E} q \Leftrightarrow \mathcal{S}(p) \leq \mathcal{S}(q) \tag{9.3.2}$$

Without risk of ambiguity, we shall omit E in the notations below. We represent ω and ϖ in the Cartesian product $F \times F$ by their respective graphs $gr(\omega)$ and $gr(\varpi)$ in a strict sense, namely,

$$gr(\omega) = \{(p, q) | (p, q) \in F \times F, p \leq_{\omega} q \text{ and } \neg (q \leq_{\omega} p)\}$$
$$gr(\varpi) = \{(p, q) | (p, q) \in F \times F, p \leq_{\varpi} q \text{ and } \neg (q \leq_{\varpi} p)\} \tag{9.3.3}$$

where the symbol $\neg(\bullet)$ expresses the negation of $\bullet$.

---

[1] The rank notion used corresponds to that given in Sect. 3.3.3 of Chap. 3.

Now, let π be a partition on E. As seen in Sect. 3.3.1 of Chap. 3, π induces a partition of F into two classes, denoted by $R(\pi)$ and $S(\pi)$, where $R(\pi)$ (resp., $S(\pi)$) is the set of distinct element pairs of E joined (resp., separated) by the partition π. Let us consider the case for which the data is $gr(\omega)$. For this, the partition π is represented by a total preorder on F with two classes $R(\pi)$ and $S(\pi)$, where $R(\pi)$ precedes strictly $S(\pi)$. The strict graph of this total preorder is then the Cartesian product

$$R(\pi) \times S(\pi) \tag{9.3.4}$$

If the ordinal similarity data is $gr(\varpi)$, the representation of π is

$$S(\pi) \times R(\pi) \tag{9.3.5}$$

Clearly, both criteria

$$card[gr(\omega) \cap (R(\pi) \times S(\pi))]$$
$$\text{and } card[gr(\varpi) \cap (S(\pi) \times R(\pi))] \tag{9.3.6}$$

are equivalent. In fact,

$$(\forall (p, q) \in F \times F)\, (p, q) \in gr(\omega) \cap (R(\pi) \times S(\pi))$$
$$\Leftrightarrow (q, p) \in gr(\varpi) \cap (S(\pi) \times R(\pi))$$

J.-P. Benzécri introduced this criterion in 1965 under the form: "in the system of distance inequalities, the *number* of those which are specified by the partition" (see in [1]).

By introducing the indicator functions I_ω, I_ϖ, $I_{R \times S}$ and $I_{S \times R}$ of the respective subsets $gr(\omega)$, $gr(\varpi)$, $R(\pi) \times S(\pi)$ and $S(\pi) \times R(\pi)$, we have

$$B(\omega, \pi) = card[gr(\omega) \cap (R(\pi) \times S(\pi))] = \sum_{(p,q) \in F \times F} I_\omega(p, q) I_{R \times S}(p, q)$$

$$B(\varpi, \pi) = card[gr(\varpi) \cap (S(\pi) \times R(\pi))] = \sum_{(p,q) \in F \times F} I_\varpi(p, q) I_{S \times R}(p, q)$$

$$\tag{9.3.7}$$

For historical reasons related to the development of this research and due to the manner by which the "lexicographic" algorithm of building a classification tree is defined (see Chap. 10, Sect. 10.2), the version $B(\omega, \pi)$ will be taken here and in Sect. 9.3.3. Instead of maximizing this criterion, W.F. de la Véga has proposed to maximize

$$V(\omega, \pi) = card(F \times F - gr(\omega) \triangle (R(\pi) \times S(\pi))) \tag{9.3.8}$$

where $\triangle$ designates the symmetrical difference operator [6]. The following proposition relates the criteria $B(\omega, \pi)$ and $V(\omega, \pi)$.

Proposition 44 *The criteria*

$$card[F \times F - gr(\omega)\triangle (R(\pi) \times S(\pi))]$$
$$and \; card[gr(\omega) \cap (R(\pi) \times S(\pi))] - \frac{1}{2}card(R(\pi)) \times card(S(\pi))$$

$$(9.3.9)$$

are equivalent.

Proof We have

$$card[gr(\omega)\triangle (R(\pi) \times S(\pi))]$$
$$= card(gr(\omega)) + card(R(\pi) \times card(S(\pi)) - 2card[gr(\omega) \cap (R(\pi) \times S(\pi))]$$

$$(9.3.10)$$

Besides

$$card[F \times F - gr(\omega)\triangle (R(\pi) \times S(\pi))] = card(F \times F)$$
$$- card(gr(\omega)) - card[(R(\pi) \times S(\pi))] + 2card[gr(\omega) \cap (R(\pi) \times S(\pi))]$$
$$= card(F \times F) - card(gr(\omega))$$
$$+ 2 \left(card[gr(\omega) \cap (R(\pi) \times S(\pi))] - \frac{1}{2}card (R(\pi)) \times card (S(\pi)) \right)$$

$card(F \times F)$ and $card(gr(\omega))$ being fix, the equivalence property is acquired. $\square$

Example

With respect to the incidence data Table 9.1, let us consider the following preordonance denoted by $\omega(\mathcal{A})$ and associated with the similarity coefficient $(s + t)$ (see Sect. 4.2.1 of Chap. 4):

$$12 \sim 56 < 13 \sim 23 \sim 45 < 24 < 25 \sim 34 \sim 46 < 15 \sim 16 \sim 26 < 14 \sim 35 < 36$$

where we have denoted by ij the attribute pair $\{a^i, a^j\}$. The respective values of $(s+t)$ are

$$12, 11, 10, 9, 8, 6 \text{ and } 2$$

Now, we shall compare the partitions $\pi_1(\mathcal{A})$ and $\pi_2(\mathcal{A})$ considered in Eq. (9.2.12). According to above, these partitions can be denoted as follows:

$$\pi_1(\mathcal{A}) = \{\{1, 2, 3\}, \{4, 5, 6\}\} \text{ and } \pi_2(\mathcal{A}) = \{\{1, 2, 3, 4\}, \{5, 6\}\}$$

where i stands for a^i. We have

$$R(\pi_1) \times S(\pi_1) = \{12, 13, 23, 45, 46, 56\} \times \{14, 15, 16, 24, 25, 26, 34, 35, 36\}$$

and

$$R(\pi_2) \times S(\pi_2) = \{12, 13, 14, 23, 24, 34, 56\} \times \{15, 16, 25, 26, 35, 36, 45, 46\}$$

In order to visualize the calculation of the criterion $card[gr(\omega) \cap (R(\pi) \times S(\pi))]$ for π_1 and π_2, refer to the diagram shown in Table 9.2 continued with Table 9.2.1, where the symbols I, * and x represent elements of the graph $gr(\omega)$, the sets $R(\pi_1) \times S(\pi_1)$ and $R(\pi_2) \times S(\pi_2)$, respectively. Notice that the first (resp., second) component of an ordered pair (p, q) $((p, q) \in F \times F)$ has to be consulted on the row (resp., column) margin of the Table 9.2.

By counting the number of "*I" in Table 9.2 continued with Table 9.2.1, we obtain

$$card[gr(\omega) \cap (R(\pi_1) \times S(\pi_1))] = 51$$

Similarly, by counting the number of "Ix", we obtain

$$card[gr(\omega) \cap (R(\pi_2) \times S(\pi_2))] = 43$$

Table 9.2 Graph relations

P\P	12	56	45	13	23	24	25	46
12								
56								
45	I x	I x		x	x	x		
13	I	I						
23	I	I						
24	* I	* I	* I	* I	* I			*
25	* I x	* I x	* I	* I x	* I x	I x		*
46	I x	I x	I	I x	I x	I x		
34	* I	* I	* I	* I	* I	I		*
15	* I x	* I x	* I	* I x	* I x	I x	I	* I
16	* I x	* I x	* I	* I x	* I x	I x	I	* I
26	* I x	* I x	* I	* I x	* I x	I x	I	* I
14	* I	* I	* I	* I	* I	I	I	* I
35	* I x	* I x	* I	* I x	* I x	I x	I	* I
36	* I x	* I x	* I	* I x	* I x	I x	I	* I

Table 9.2.1

P\P	34	15	16	26	14	35	36
12							
56							
45	x				x		
13							
23							
24							
25	x				x		
46	x				x		
34							
15	I x				x		
16	I x				x		
26	I x				x		
14	I	I	I	I			
35	I x	I	I	I	x		
36	I x	I	I	I	I x	I	

The centred criteria are equal to

$$51 - \frac{1}{2}6 \times 9 = 24 \text{ and } 43 - \frac{1}{2}7 \times 8 = 15$$

for π_1 and π_2, respectively. Therefore, partition π_1 is preferable to partition π_2 for each of both criteria $B(\omega, \pi)$ and $V(\omega, \pi)$ (see Eqs. (9.3.7) and (9.3.8)). However, after the comparative study led in Sect. 9.3.3, we will see remarkable examples for which $V(\varpi, \pi)$ proves to be a discriminant criterion whereas $B(\varpi, \pi)$ is not.

Now, let us consider the case where ω is a complete and strict order on F. ϖ is then also a complete and strict order on F. Designate by h and k the respective rank functions on F associated with ω and ϖ (see Eq. (3.3.28) of Chap. 3 for the definition of a rank function corresponding to a total order on a finite set). We have

$$(\forall p \in F), h(p) + k(p) = f + 1 \tag{9.3.11}$$

where f denotes $card(F)$. Relative to the criterion $B(\omega, \pi)$ (see (9.3.7)), we can write

$$gr(\omega) \cap (R(\pi) \times S(\pi)) = \sum_{q \in S(\pi)} gr(\omega) \cap (R(\pi) \times \{q\})$$

If q is the ith separated pair encountered by describing F from left to right according to the ranking ω, we have

$$card(gr(\omega) \cap (R(\pi) \times \{q\})) = (h(q) - 1) - (i - 1) = h(q) - i$$

Let $h(q_i)$ denote the rank of the latter ith separated pair, and we obtain

$$card(gr(\omega) \cap (R(\pi) \times S(\pi))) = \sum_{1 \leq i \leq s} h(q_i) - \frac{1}{2}s(s+1)$$

where s denotes the cardinality of $S(\pi)$. By introducing the indicator function $(\sigma(q)|q \in F)$ of $S(\pi)$ in F, the latter equation becomes

$$card(gr(\omega) \cap (R(\pi) \times S(\pi))) = \sum_{q \in F} \sigma(q)h(q) - \frac{1}{2}s(s+1) \qquad (9.3.12)$$

Analogously, by considering the complete and strict order ϖ (ϖ is opposite to ω, see Eq. (9.3.3)), we obtain

$$card(gr(\varpi) \cap (S(\pi) \times R(\pi))) = \sum_{p \in F} \rho(p)k(p) - \frac{1}{2}r(r+1) \qquad (9.3.13)$$

where ρ is the indicator function $(\rho(p)|p \in F)$ of $R(\pi)$ in F and $r = card(R(\pi))$.

In Sect. 9.3.3 we will establish, in the case where ω is a total and strict order on F, that the De La Véga criterion is non-biased. For this purpose, we will study the distribution of this criterion (see the second of equations stated in Proposition (44)), when we substitute for the partition π, a random partition π^* in the set $\mathcal{P}(n; t)$, provided with a uniform probability, of all partitions having the same type t as that of π. The mathematical expectation of

$$card[gr(\omega) \cap (R(\pi^*) \times S(\pi^*))] - \frac{1}{2}card(R(\pi)) \times card(S(\pi))$$

is proved to be null. This result is obtained by referring to the first of Eq. (9.3.7) and then by calculating in $F \times F$. Using the random variable associated with the *linear* version (9.3.12) of the criterion permits this result to be get much more easily (see Sect. 9.3.4). However, the approach examined in Sect. 9.3.3 is very instructive. On the other hand, it enables the mathematical expectation of $card[gr(\omega) \cap (R(\pi^*) \times S(\pi^*))]$ to be calculated in the general case where ω is a total preorder (ranking with ties) and not only a total order (ranking without ties).

The form (9.3.13) is that adopted for the distributional study considered in Sect. 9.3.4 of the *raw* criterion on $\mathcal{P}(n; t)$. Since $r(r+1)/2$ is constant, the random criterion can be expressed as

$$S_\pi(\varpi, \pi^*) = s(k, \rho^*) = \sum_{p \in F} \rho^*(p)k(p) \qquad (9.3.14)$$

where ρ^* is the indicator function of $R(\pi^*)$ in F associated with the random partition π^* and k the ranking function on F coding ϖ.

The analysis of the distribution of the latter random variable is compared with the distribution of

$$S_\beta(\varpi, \beta^*) = s(k, \beta^*) = \sum_{p \in F} \beta^*(p)k(p) \qquad (9.3.15)$$

where β^* is the indicator function of $R(\beta^*)$ in F, associated with a random binary relation on E, in the set $\mathcal{B}(r)$, equally distributed, of all symmetrical binary relations β on E, for which $card[R(\beta)] = r$, $R(\beta)$ denoting the F subset which represents the binary relation β.

Next, these results are extended to the more general case where the ordinal similarity corresponds to a total *preorder* on F (preordonance on E). For this, coding the total preorder in terms of the "mean rank" function (see (3.4.8)) is needed. Finally, we consider the adaptation of this analysis to the case of a *numerical* similarity function. In this case, the raw criterion becomes "sum of the similarities of distinct element pairs joined by the partition π". In this respect, similarity indices or coefficients, proposed or established in Chaps. 4–6, can be employed. Normalized versions of the raw criterion are proposed in Sect. 9.3.5.

Now, return to the case where ω is a total *preorder* (ranking with ties) on F. Relatively to the pair (ω, π), let us introduce the subset of $F \times F$ that we denote by $\Omega(\omega, \pi)$ composed of all ordered distinct pairs (p, q) such that the two components p and q are strictly comparable with respect to ω and to π. That is to say

$$[(p, q) \text{ or } (q, p) \in gr(\omega)] \text{ and } [(p, q) \text{ or } (q, p) \in R(\pi) \times S(\pi)]$$

More globally,

$$\begin{aligned}
\Omega(\omega, \pi) &= [gr(\omega) + gr(\varpi)] \cap [R(\pi) \times S(\pi) + S(\pi) \times R(\pi)] \\
&= gr(\omega) \cap (R(\pi) \times S(\pi)) \\
&\quad + gr(\omega) \cap (S(\pi) \times R(\pi)) \\
&\quad + gr(\varpi) \cap (R(\pi) \times S(\pi)) \\
&\quad + gr(\varpi) \cap (S(\pi) \times R(\pi)) \qquad (9.3.16)
\end{aligned}$$

where the symbol "$+$" expresses a set sum (see (9.3.2) and (9.3.3) for the definitions of $gr(\omega)$ and $gr(\varpi)$). It would have been sufficient to consider the reference set as $F^{[2]} = F \times F - \delta(F \times F)$, where $\delta(F \times F)$ is the diagonal of $F \times F$.

The cardinal parameters associated with the decomposition given in (9.3.16) are

$$\begin{aligned}
s(\omega, \pi) &= card[gr(\omega) \cap (R(\pi) \times S(\pi))] \\
u(\omega, \pi) &= card[gr(\omega) \cap (S(\pi) \times R(\pi))] \\
v(\omega, \pi) &= card[gr(\varpi) \cap (R(\pi) \times S(\pi))] \\
t(\omega, \pi) &= card[gr(\varpi) \cap (S(\pi) \times R(\pi))] \qquad (9.3.17)
\end{aligned}$$

Clearly, we have

$$s(\omega, \pi) = t(\omega, \pi) \text{ and } u(\omega, \pi) = v(\omega, \pi)$$

In the example above we have illustrated the calculation of $s(\omega, \pi)$; the reader may compute $u(\omega, \pi)$.

With these formal expressions the famous Goodman and Kruskal coefficient can be written as

$$\gamma(\omega, \pi) = \frac{s(\omega, \pi) - u(\omega, \pi)}{s(\omega, \pi) + u(\omega, \pi)} = \frac{(s(\omega, \pi) + t(\omega, \pi)) - (u(\omega, \pi) + v(\omega, \pi))}{(s(\omega, \pi) + t(\omega, \pi)) + (u(\omega, \pi) + v(\omega, \pi))}$$

(9.3.18)

This expression is formally identical to that obtained in the framework of comparing ordinal categorical attributes (see (4.3.17) of Chap. 4). As said in Chap. 4, it corresponds to the Hamann similarity index proposed for Boolean data. The latter (see Table 4.3) can be written in the context of comparing objects described by Boolean attributes

$$1 - 2\frac{u + v}{p} = \frac{(s + t) - (u + v)}{(s + t) + (u + v)}$$

Clearly, every similarity index of Table 4.3 can be transposed for comparing ω and π, by substituting $s(\omega, \pi)$, $u(\omega, \pi)$, $v(\omega, \pi)$ and $t(\omega, \pi)$ for s, u, v and t, respectively. More generally, every similarity index respecting Definition 14 of Chap. 4 can be transposed for comparing ω and π.

9.3.2 Approximating a Symmetrical Binary Relation by an Equivalence Relation: The Zahn Problem

The similarity data may be assumed to be defined by a symmetrical binary relation β on E. The latter is a specific case of a valuated binary relation on a finite set (see Sect. 3.3.5 of Chap. 3). Here, the valuation is Boolean. The binary relation being symmetrical, β can be represented by the following subset $R(\beta)$ of $F = P_2(E)$:

$$R(\beta) = \{\{x, y\} | \{x, y\} \in F \text{ and } x\beta y\}$$

(9.3.19)

By considering the complementary subset $S(\beta)$ of $R(\beta)$ in F, a total preorder on F with two classes $S(\beta)$ and $R(\beta)$ can be associated with $R(\beta)$. We have for the latter preorder $S(\beta) < R(\beta)$. By difference with the section above, we will work here in the sets F and $R(\beta)$.

In an article published in 1964, C.T. Zahn has studied the problem of approximating a symmetrical binary relation β on a finite set E by an equivalence relation π [41]. The criterion consists of minimizing the cardinal of the symmetrical difference between the respective graphs of β and π in $E \times E$. Solutions for this problem are

provided in [41] for a specific family of binary relations which cannot interest us
here. However, it is interesting to study the behaviour for any symmetrical binary
relation on E, in the clustering framework.

The symmetric difference cardinal between a partition $\pi(E)$ and a preordonance
$\omega(E)$ on E was considered above (see Sect. 9.3.1). It will also be considered below
(see Sect. 9.3.3). In this case, the number of classes of the total preorder on $F = P_2(E)$
associated with $\omega(E)$ is not restricted. On the other hand, this criterion is conceived
and calculated in the Cartesian product $F \times F$.

According to the notations of the previous section, $R(\pi)$ designates the set of
distinct element pairs of E joined by the partition π. In these conditions, the Zahn
criterion can take the following form (suggested by M. Barbut, Centre de Mathématique Sociale, EHESS, Paris):

$$card\ (R(\beta)\triangle R(\pi)) = 2card[R(\beta) - R(\beta) \cap R(\pi)]$$
$$-card\ (R(\beta)) + card\ (R(\pi))$$
$$= 2card[R(\beta) \cap S(\pi)] + card\ (R(\pi)) - card\ (R(\beta)) \qquad (9.3.20)$$

where $\triangle$ designates the symmetrical difference set and where $S(\pi)$ is the set of
separated pairs by π $(R(\pi) + S(\pi) = F)$.

As usual, let us denote by $(E_1, E_2, \ldots, E_i, \ldots, E_k)$ the sequence of labelled classes
of the partition π. $(n_1, n_2, \ldots, n_i, \ldots, n_k)$ designates the type of π $(n_i = card(E_i)$,
$1 \le i \le k$ and $n = card(E))$ (see in an other context (3.3.10) of Chap. 3). In the
equation mentioned, we have expressed $S(\pi)$ as follows:

$$S(\pi) = \sum_{1 \le i < j \le k} E_i \star E_j$$

where $E_i \star E_j$ is the set of unordered pairs $\{x, y\}$ such that $x \in E_i$ and $y \in E_j$, the
cardinality of the latter set being $n_i \times n_j$.

Denote now by r_{ij} the number of elements $\{x, y\}$ of $R(\beta)$ such that $x \in E_i$ and
$y \in E_j$:

$$r_{ij} = R(\beta) \cap E_i \star E_j,$$

$1 \le i < j \le k$. Using the cardinal parameters, the right member of (9.3.20)
becomes

$$2 \sum_{1 \le i < j \le k} r_{ij} + \sum_{1 \le i \le k} \binom{n_i}{2} - card\ (R(\beta)) \qquad (9.3.21)$$

By adding and substracting $card\ (S(\pi))$ we obtain

$$card\ (R(\beta)\triangle R(\pi)) = 2 \sum_{1 \le i < j \le k} \left(r_{ij} - \frac{n_i \times n_j}{2}\right) + card\ (S(\pi))$$
$$= 2 \sum_{1 \le i < j \le k} d_{ij} + card\ (S(\beta)) \qquad (9.3.22)$$

where $d_{ij} = r_{ij} - \frac{n_i \times n_j}{2}$, $1 \leq i < j \leq k$. Hence, the criterion concerned is reduced to

$$d\,(R(\beta), R(\pi)) = \sum_{1 \leq i < j \leq k} d_{ij} \tag{9.3.23}$$

An Example

Let us illustrate the computing of the criterion $d\,(R(\beta), R(\pi))$ on a small example where the respective fittings of two partitions are compared. Set

- $E = \{e_1, e_2, e_3, e_4, e_5, e_6\}$
- $R(\beta) = \{e_1 e_2, e_1 e_3, e_2 e_3, e_3 e_4, e_4 e_6\}$
- $\pi_1 = \{\{e_1, e_2, e_3\}, \{e_4, e_5, e_6\}\}$
- $\pi_2 = \{\{e_1, e_2, e_3, e_4\}, \{e_5, e_6\}\}$

where we have denoted by $e_i e_j$ $(i < j)$ the unordered pair $\{e_i, e_j\}$, $1 \leq i < j \leq 6$. We have

$$S(\pi_1) = \{e_1 e_4, e_1 e_5, e_1 e_6, e_2 e_4, e_2 e_5, e_2 e_6, e_3 e_4, e_3 e_5, e_3 e_6\}$$

$$S(\pi_2) = \{e_1 e_5, e_1 e_6, e_2 e_5, e_2 e_6, e_3 e_5, e_3 e_6, e_4 e_5, e_4 e_6\}$$

In Table 9.3

- The elements of $R(\beta)$ are represented by the symbol "I";
- The elements of $S(\pi_1)$ are represented by the symbol "*";
- The elements of $S(\pi_2)$ are represented by the symbol "x".

The equation above (9.3.23) gives

$$d\,(R(\beta), R(\pi_1)) = 1 - \frac{9}{2} = -\frac{7}{2} \text{ and } d\,(R(\beta), R(\pi_2)) = 0 - \frac{8}{2}$$

In these conditions

$$card\,(R(\beta) \triangle R(\pi_1)) = 10 - 7 = 3 \text{ and } card\,(R(\beta) \triangle R(\pi_2)) = 10 - 8 = 2$$

Therefore, the partition π_2 fits better than π_1 the binary relation β.

Table 9.3 Graphs representation

$E \setminus E$	e_1	e_2	e_3	e_4	e_5	e_6
e_1						
e_2	I					
e_3	I	I				
e_4	*	*	* I			
e_5	* x	* x	* x	x		
e_6	* x	* x	* x	x	I	

9.3.2.1 An Asymptotic Property

In [34], referring to the Zahn problem, a limit result is established. Its generality will make it interesting for clustering. We shall take up again the latter study and show its impact on clustering logic. The binary symmetrical relation β on a finite set E being fixed designate by $\min[R(\beta)]$ the minimum value of the symmetrical difference cardinal between $R(\beta)$ and the representation $R(\pi)$ of an equivalence relation associated with a partition π on E:

$$m[R(\beta)] = \min\{card\ (R(\beta)\Delta R(\pi))\,|\pi \in \mathcal{P}\} \qquad (9.3.24)$$

where $\mathcal{P}$ indicates the set of all partitions on E. According to above $R(\beta)$ and $R(\pi)$ can be interpreted as subsets of $F = P_2(E)$.

Lemma 7

$$m[R(\beta)] \leq \frac{1}{2}\binom{n}{2} \qquad (9.3.25)$$

Denote by $r(\beta)$ the cardinality of $R(\beta)$: $r(\beta) = card[R(\beta)]$. If $r(\beta) \leq \frac{1}{2}\binom{n}{2}$, consider the finest partition π_0 for which each class is a singleton class containing exactly a single element of E. We have $card[R(\beta)\Delta R(\pi)] = card[R(\beta)]$ and then, the (9.3.25) is verified. Now, if $r(\beta) > \frac{1}{2}\binom{n}{2}$, by taking for π the coarsest partition π_1, comprising a single class including all the E elements, $R(\beta)\Delta R(\pi_1)$ is reduced to $S(\beta)$, whose cardinality is $\leq\frac{1}{2}\binom{n}{2}$. $\qquad\qquad \square$

For simplicity notation, we will denote below $m(\beta)$ for $m[R(\beta)]$.

Theorem 24 *For any positive real ϵ, such that $0 < \epsilon < 1$,*

$$\frac{1-\epsilon}{2}\binom{n}{2} \leq m(\beta) \leq \frac{1}{2}\binom{n}{2} \qquad (9.3.26)$$

for almost all symmetrical binary relations β (i.e. for all symmetrical binary relations, except a fraction of them, tending to zero when n tends to infinity).

Proof Let f denote the cardinality of F ($f = \binom{n}{2}$) and let g_n be the cardinal of all equivalence relations π which can be defined on E. To begin, let us recall the classical inequality $g_n \leq n^n$. Now, for a given integer d, we shall specify an upper bound of the cardinal of the set of symmetrical binary relations for which $m(\beta) = d$. This set can be constructed from sets of the form

$$D = R(\pi)\Delta R(\beta) \qquad (9.3.27)$$

such that $card(D) = d$. More precisely, let D be an arbitrary F subset whose cardinal is d, $card(D) = d$. The following algorithm—expressed informally—allows D to get the form (9.3.27).

1. Start with $R(\beta) = \emptyset$ and $S(\beta) = \emptyset$;

   - For every element p of D

2. If $p \in R(\pi)$ increase $S(\beta)$ with p: $S(\beta) \leftarrow S(\beta) + \{p\}$;
3. If $p \in S(\pi)$ increase $R(\beta)$ with p: $R(\beta) \leftarrow R(\beta) + \{p\}$;

   - For every element q of D^c

4. If $q \in R(\pi)$ increase $R(\beta)$ with q: $R(\beta) \leftarrow R(\beta) + \{q\}$;
5. If $q \in S(\pi)$ increase $S(\beta)$ with q: $S(\beta) \leftarrow S(\beta) + \{q\}$;

where D^c is the complementary subset of D in F.

To summarize, the data is given by the ordered pair $(R(\pi), D)$ and the set $R(\beta)$ (resp., $S(\beta)$) is built consequently, in a unique way. The number of ways of choosing such a pair is $g_n\binom{f}{d}$. Therefore,

$$card(\{\beta | m(\beta) = d\}) \leq g_n \binom{f}{d} \tag{9.3.28}$$

In fact, behind each β for which $m(\beta) = d$, there is at least one ordered pair of the form $(R(\pi), D)$. On the other hand, mutually different binary relations of the form $(R(\pi), D)$ correspond necessarily to mutually different binary relations β. In these conditions, an upper bound for the number of relations β such that $m(\beta) < (1-\epsilon)f/2$ is given by

$$\sum_{d < (1-\epsilon)f/2} n^n \binom{f}{d} = n^n \sum_{d < (1-\epsilon)f/2} \binom{f}{d} < n^n f \binom{f}{[(1-\epsilon)f/2]} \tag{9.3.29}$$

where $[\bullet]$ indicates the integer part of $\bullet$.

The cardinal of the entire set of binary relations β, being 2^f, the proportion of those for which

$$m(\beta) < (1-\epsilon)f/2$$

is strictly lower than

$$\frac{n^n f \binom{f}{[(1-\epsilon)f/2]}}{2^f}$$

which is itself strictly lower than

$$\frac{n^n f \binom{f}{[(1-\epsilon)f/2]}}{\binom{f}{[f/2]y}} \tag{9.3.30}$$

The ratio above between the two binomial coefficients can be written as

$$\frac{[f/2]y([f/2]-1)\dots([(1-\epsilon)f/2]+1)}{(f-[(1-\epsilon)f/2])(f-[(1-\epsilon)f/2]-1)\dots(f-[f/2]y+1)}$$

The common number of factors of both numerator and denominator is $[f/2]y - [(1-\epsilon)f/2]$. Now, decompose the product (9.3.30)—from left to right—into $[f/2]y - [(1-\epsilon)f/2]$ factors, each corresponding to a ratio between a numerator element and the associated denominator óne. Then, by identifying the integer parts $[f/2]y$ and $[(1-\epsilon)f/2]$ with $f/2$ and $(1-\epsilon)f/2$, respectively, it can be easily shown that each ratio is strictly lower than $1/(1+\epsilon)$. Notice that the limit conditions authorize the identifications mentioned. In these conditions, the entire ratio (9.3.30) is strictly lower than

$$n^n f (1 + \epsilon)^{(-\epsilon f)/2}$$

which tends to zero when n tends to infinity.

Conclusion

For $n = card(E)$ large enough, the distribution of $m(\beta)$ is very concentrated immediately on the left of the value $\frac{1}{2}\binom{n}{2}$. Therefore, with respect to a given symmetrical binary relation β, an equivalence relation π—possibly obtained by a clustering algorithm—can be evaluated as a very good approximation of β, if its distance to β is appreciably lower than $\frac{1}{2}\binom{n}{2}$. Consequently, for n large enough, if $card[R(\beta)] < \frac{1}{2}\binom{n}{2}$ (resp., $card[R(\beta)] > \frac{1}{2}\binom{n}{2}$), the equivalence relation π associated with the finest (resp., the coarsest) partition is a very good approximation of β relatively to the symmetrical difference cardinal criterion $card\left(R(\beta)\triangle R(\pi)\right)$. The Moon result [34] shows that too little information was retained in order to obtain significant results in non-hierarchical clustering and a fortiori in hierarchical clustering [7]. Hence, the most general similarity information we can retain without compromising the significativity of the clustering results is a *preordonance* associated with a similarity or dissimilarity index on E (see Chap. 4).

9.3.3   Comparing Two Basic Criteria

The data is given here by an ordered pair (ω, π) where ω is assumed to be a total order on $F = P_2(E)$ (see Eqs. (9.3.2) and (9.3.3)) and $\pi = \{E_1, E_2, \ldots, E_i, \ldots, E_k\}$, a partition of E with labelled classes. As above $t = (n_1, n_2, \ldots, n_i, \ldots, n_k)$ designates the *type* of π. As reported in Sect. 9.3.1 (after the example) we shall show that the mathematical expectation of

$$card\left[gr(\omega) \cap \left(R(\pi^\star) \times S(\pi^\star)\right)\right]$$

is

$$\frac{1}{2} card\left(R(\pi)\right) \times card\left(S(\pi)\right)$$

where $\pi^\star$ is a random partition in the set $\mathcal{P}(n; t)$, uniformly distributed, of all partitions with labelled classes, having the same type t as π. This random model has already been considered in Chap. 6, Sect. 6.2.3, it was, in the framework of comparing two nominal categorical attributes. As mentioned above, one specificity of this section consists of working in the set $F \times F$. Calculating in the context of partitions with labelled classes does not restrict the generality. The results obtained will be extended easily in the case of non-labelled classes. Notice that an element π of $\mathcal{P}(n; t)$ can be formalized by means of a surjective mapping φ of E into the label set $\{1, 2, \ldots, i, \ldots, k\}$, such that $card\left(\varphi^{-1}(i)\right) = n_i$, $1 \leq i \leq k$. Thus, $\mathcal{P}(n; t)$ can be identified with the set Φ of these mappings. It will be more flexible to express the proofs of the following lemmas (see Sect. 9.3.3.1) in terms of Φ. These proofs are of the same nature as those of Sect. 6.2.3.4 of Chap. 6.

Let us recall once more that for any element φ of Φ (π of $\mathcal{P}(n; t)$), we have

$$card\ (R(\pi)) = \frac{1}{2} \sum_{1 \leq i \leq k} n_i(n_i - 1) \text{ and } card\ (S(\pi)) = \sum_{1 \leq i < j \leq k} n_i n_j$$

9.3.3.1 Preliminary Lemmas

Lemma 8 *The proportion of elements φ of Φ for which the two components x and y of a given pair $\{x, y\}$ of $P_2(E)$ are included in a same class of the partition π is* $\sum_{1 \leq i \leq k} n_i(n_i - 1)/n(n - 1)$.

Proof The set $\{\varphi | \varphi \in \Phi \text{ and } \varphi(x) = \varphi(y)\}$ can be decomposed as follows:

$$\sum_{1 \leq i \leq k} \{\varphi | \varphi \in \Phi \text{ and } \varphi(x) = \varphi(y) = i\}$$

(set sum) and then,

$$card\ (\{\varphi | \varphi \in \Phi \text{ and } \varphi(x) = \varphi(y)\}) = \sum_{1 \leq i \leq k} card\ (\{\varphi | \varphi \in \Phi \text{ and } \varphi(x) = \varphi(y) = i\})$$

The subset $\{\varphi | \varphi \in \Phi \text{ and } \varphi(x) = \varphi(y) = i\}$ corresponds bijectively to the labelled class partition set of $E - \{x, y\}$ such that the cardinal classes—provided that $n_i \geq 2$—are

$$n_1, n_2, \ldots, n_{i-1}, n_i - 2, n_{i+1}, \ldots, n_k$$

Therefore,

$$card\ (\{\varphi | \varphi \in \Phi \text{ and } \varphi(x) = \varphi(y) = i\}) = \frac{(n - 2)!}{n_1! n_2! \ldots, n_{i-1}!(n_i - 2)! n_{i+1}! \ldots n_k!}$$

if $n_i \geq 2$ and 0, if not. Hence,

$$card\ (\{\varphi | \varphi \in \Phi \text{ and } \varphi(x) = \varphi(y)\}) = \sum_{\{i | n_i \geq 2\}} \frac{(n - 2)!}{n_1! n_2! \ldots, n_{i-1}!(n_i - 2)! n_{i+1}! \ldots n_k!}$$

The result reported is get by constituting the ratio of this cardinal over $card(\Phi)$.   □

The following proofs, which are similar to that given here, will be abbreviated.

Lemma 9 *The proportion of elements φ of Φ for which the two components x and y of a given pair $\{x, y\}$ of $P_2(E)$ belong, respectively, to two distinct classes of the partition π is $2\sum_{1\leq i<j\leq k} n_i n_j/n(n-1)$.*

$$card\left(\{\varphi|\varphi \in \Phi \text{ and } \varphi(x) \neq \varphi(y)\}\right)$$

$$= \sum_{1\leq i\neq j\leq k} card\left(\{\varphi|\varphi \in \Phi \text{ and } \varphi(x) = i, \varphi(y) = j\}\right)$$

This sum comprises $k(k-1)$ terms.

The set $\{\varphi(x) = i, \varphi(y) = j \text{ and } i \neq j\}$ can be mapped bijectively with the set of class labelled partitions of $E - \{x, y\}$, such that the class cardinals of each of these partitions are

$$n_1, n_2, \ldots, n_{i-1}, n_i - 1, n_{i+1}, \ldots, n_{j-1}, n_j - 1, n_{j+1}, \ldots, n_k$$

It is now easy to continue the calculation and to find out the indicated result.

Lemma 10 *Let x, y and z be the three given distinct elements of E. The proportion of elements of Φ, for which the pair $\{x, y\}$ is joined and the pair $\{x, z\}$ is separated, is*

$$\sum_{1\leq i\neq j\leq k} n_i(n_i - 1)n_j/n(n-1)(n-2)$$

$$card\left(\{\varphi|\varphi \in \Phi \text{ and } \varphi(x) = \varphi(y) \neq \varphi(z)\}\right)$$

$$= \sum_{J} card\left(\{\varphi|\varphi \in \Phi \text{ and } \varphi(x) = \varphi(y) = i \text{ and } \varphi(z) = j\}\right)$$

where $J = \{(i,j)|1 \leq i \neq j \leq k \text{ and } n_i \geq 2\}$.

For a fixed $(i,j) \in J$, the set

$$\{\varphi|\varphi \in \Phi \text{ and } \varphi(x) = \varphi(y) = i \text{ and } \varphi(z) = j\}$$

can be mapped bijectively with the set of labelled class partitions of $(E - \{x, y, z\})$, such that the sizes of the different classes—provided that $n_i \geq 2$—are

$$n_1, n_2, \ldots, n_{i-1}, n_i - 2, n_{i+1}, \ldots, n_{j-1}, n_j - 1, n_{j+1}, \ldots, n_k,$$

respectively. This remark enables the generic term of the previous sum to be expressed and then to establish the expected result.

Lemma 11 *Let x, y, z and t be the four given distinct elements of E. The proportion of elements of Φ for which the pair $\{x, y\}$ is joined and the pair $\{z, t\}$ is separated is*

$$\frac{\sum_K n_i(n_i - 1)n_j n_l + 2\sum_L n_i(n_i - 1)(n_i - 2)n_j}{n(n - 1)(n - 2)(n - 3)}$$

where $K = \{(i, j, l) | i \neq j, i \neq l \text{ and } j \neq l, n_i \geq 2\}$ and $L = \{(i, j) | i \neq j, n_i \geq 3\}$. The first sum and the second sum include, at the most, $k(k - 1)(k - 2)$ and $k(k - 1)$ terms, respectively.

Let φ be an element of Φ and let $\{x, y, z, t\}$ be a subset of four elements of E. $\varphi(x) = \varphi(y)$ and $\varphi(z) \neq \varphi(t)$, if and only if, one of the following alternatives holds:

1. $\varphi(x) = \varphi(y) = i$, $\varphi(z) = j$ and $\varphi(t) = l$ for a triplet (i, j, l) in K;
2. a: $(\varphi(x) = \varphi(y) = \varphi(z) = i$ and $\varphi(t) = j)$ or
   b: $(\varphi(x) = \varphi(y) = \varphi(t) = i$ and $\varphi(z) = j)$ for an ordered pair $(i, j) \in L$.

The cardinal of the subset of Φ whose elements satisfy item 1 can be identified with the set of class labelled partitions of $E - \{x, y, z, t\}$, such that the respective cardinalities of the different classes, provided that $n_i \geq 2$, are

$$n_1, n_2, \ldots, n_{i-1}, n_i - 2, n_{i+1}, \ldots, n_{j-1}, n_j - 1, n_{j+1}, \ldots, n_{l-1}, n_l - 1, n_{l+1}, \ldots, n_k$$

The cardinal of the subset of Φ whose elements satisfy item 2 a (resp., item 2 b) is that of the set of class labelled partitions of $E - \{x, y, z, t\}$, for which the respective cardinalities of the different classes, provided that $n_i \geq 3$, are

$$n_1, n_2, \ldots, n_{i-1}, n_i - 3, n_{i+1}, \ldots, n_{j-1}, n_j - 1, n_{j+1}, \ldots, n_k$$

These cardinal properties enable us to complete the calculation and to find out the announced result.

Let us now designate by Π the set of non-labelled class partitions of type $t = (n_1, n_2, \ldots, n_i, \ldots, n_k)$.

Lemma 12 *The different proportions calculated above with respect to the set of labelled class partitions $\mathcal{P}(n; t)$ (identified with Φ) are identical if they were calculated with respect to the set of non-labelled class partitions Π.*

Suppose that in the type t there are in all r distinct values, where the first one ν_1 is repeated k_1 times, the second ν_2, k_2 times, ...and the rth ν_r, k_r times; $k_1 + k_2 + \cdots + k_r = k$. $k_1! k_2! \ldots k_r!$ different elements of $\mathcal{P}(n; t)$ correspond to a given element $\bar{\pi}$ of Π. These are obtained by labelling into $k_i!$ ways the k_i classes having the same cardinality ν_i, $1 \leq i \leq r$. Suppose calculated one of the preceding proportions involved in one of the four lemmas above, with respect to Π. We obtain the same proportion in $\mathcal{P}(n; t)$ by multiplying the two terms (numerator and denominator) of the proportion concerned, by $k_1! k_2! \ldots k_r!$. In fact, for an element $\bar{\pi}$ intervenes in the calculation of a given term of a given proportion, it is necessary and sufficient that the corresponding $k_1!k_2!\ldots k_r!$ elements of $\mathcal{P}(n; t)$ intervene equally in the calculation of this same term for the same proportion in $\mathcal{P}(n; t)$. From now on, we shall refer to Π.

Important Remarks

(1) For symmetry reasons, the different proportions defined above are invariant whatever the E elements constituting the unordered pair $\{x, y\}$ or the ordered pair of distinct unordered pairs $(\{x, y\}, \{x', y'\})$.
(2) It is necessary to recall here that
(2.a) The cardinal of the set G of the ordered pairs (p, q) of the form $(\{x, y\}, \{x, z\})$ where x, y and z are mutually distinct is $n(n - 1(n - 2))$.
(2.b) The cardinal of the set H of the ordered pairs (p, q) of the form $(\{x, y\}, \{z, t\})$ where x, y, z and t are mutually distinct is $\frac{n(n-1)(n-2)(n-3)}{4}$.

9.3.3.2 The Main Result

Theorem 25 Π *being provided with a uniform probability,* ω *designating a total order on F and π an arbitrary element of Π, the* mean *of*

$$card\left(gr(\omega) \cap \left(R(\pi^\star) \cap S(\pi^\star)\right)\right) - \frac{1}{2}card((R(\pi)) \times card((S(\pi))$$

is null.

According to the development above, $card\,(R(\pi)) \times card\,(S(\pi))$ stands for

$$\frac{1}{4}\left[\sum_{1 \le i \le k} n_i(n_i - 1)\right] \times \left[\sum_{1 \le i < j \le k} n_i n_j\right] \tag{9.3.31}$$

As said previously, we will work here at the level of the set $F \times F$ where $F = P_2(E)$. The expression we refer for $card(gr(\omega) \cap (R(\pi) \cap S(\pi)))$ is given by the first of Eq. (9.3.7), where I_ω and $I_{R \times S}$ are the indicator functions of $gr(\omega)$ and $R(\pi) \times S(\pi)$. In these conditions, the *mean* over Π of $card(gr(\omega) \cap (R(\pi^\star) \cap S(\pi^\star)))$ can be written as

$$\frac{1}{card(\Pi)}\left[\sum_{\pi \in \Pi} \sum_{(p,q) \in F \times F} I_\omega(p, q) I_{R \times S}(p, q)\right] \tag{9.3.32}$$

To begin decompose the sum on $F \times F$ into two parts: the first sum on G and the second one on H (see the remark above). Equation (9.3.32) becomes

$$\frac{1}{card(\Pi)}\left[\sum_{\pi\in\Pi}\sum_{(p,q)\in G}I_\omega(p,q)I_{R\times S}(p,q)\right]$$

$$+\frac{1}{card(\Pi)}\left[\sum_{\pi\in\Pi}\sum_{(p,q)\in H}I_\omega(p,q)I_{R\times S}(p,q)\right]\qquad(9.3.33)$$

Now, reverse the sum signs in order to obtain

$$\sum_{(p,q)\in G}I_\omega(p,q)\frac{1}{card(\Pi)}\left[\sum_{\pi\in\Pi}I_{R\times S}(p,q)\right]+\sum_{(p,q)\in H}I_\omega(p,q)\frac{1}{card(\Pi)}\left[\sum_{\pi\in\Pi}I_{R\times S}(p,q)\right]$$
$$(9.3.34)$$

Since ω is a total order $((p,q)\in gr(\omega))\Leftrightarrow((q,p)\notin gr(\omega))$
On the other hand,

- $(p,q)\in G\Leftrightarrow(q,p)\in G$ and
- $(p,q)\in H\Leftrightarrow(q,p)\in G$

By considering these properties and the Lemmas above (8 and 9), the expression (9.3.34) can be written as

$$\frac{1}{2}n(n-1)(n-2)\frac{\sum_J n_i(n_i-1)n_j}{n(n-1)(n-2)}$$
$$+\frac{1}{8}n(n-1)(n-2)(n-3)\frac{(\sum_K n_i(n_i-1)n_jn_l+2\sum_L n_i(n_i-1)(n_i-2)n_j)}{n(n-1)(n-2)(n-3)}$$
$$(9.3.35)$$

where J, K and L have been specified above. The latter expression can be reduced to

$$\frac{1}{2}\left[\sum_J n_i(n_i-1)n_j+\frac{1}{4}\left(\sum_K n_i(n_i-1)n_jn_l+2\sum_L n_i(n_i-1)(n_i-2)n_j\right)\right]$$
$$(9.3.36)$$

The sum over L can be developed as follows:

$$\sum_J n_i^2(n_i-1)n_j-2\sum_J n_i(n_i-1)n_j$$

Therefore, by simplifying, the whole expression (9.3.36) becomes

$$\frac{1}{2}\left[\frac{1}{4}\left(\sum_K n_i(n_i-1)n_jn_l+2\sum_J n_i^2(n_i-1)n_j\right)\right]$$

that is

$$\frac{1}{8}\left[\sum_{\{1\le i\le k\}} n_i(n_i-1)\left(\sum_{\{(j,l)|j\ne l,j\ne i,l\ne i\}} n_j n_l + 2n_i \sum_{\{j|j\ne i\}} n_j\right)\right]$$

The content of () is equal to $2card(S)$. Finally, the whole is equal to $\frac{1}{2}card(R) \times card(S)$.   □

According to the (9.3.34) above, a generalization of this result in the case where ω is a total *pre*order (ranking with ties) requires to calculate

$$\sum_{(p,q)\in G} I_\omega(p,q) \text{ and } \sum_{(p,q)\in H} I_\omega(p,q)$$

9.3.3.3 Some Comparative Results

We consider here the general case of a partition with non-labelled classes. As seen in Chap. 1 the type of such a partition can be defined by the decreasing sequence of its class cardinals. In the chapter mentioned we have introduced the set $\Psi(n, m)$ of all partition types $(n_1, n_2, \ldots, n_i, \ldots, n_k)$ for which the sum of the class cardinals is equal to m ($\sum_{1\le i\le k} n_i^2 = m$). More particularly, we have studied how m provides information on the partition type.

Let us recall a property we have to keep in mind: Among all partitions of E, those which realize the *maximum* value of $card(R(\pi) \times S(\pi))$ are such that $card(R(\pi))$ is the nearest $n(n-1)/4$. This cardinal corresponds to $m = n(n+1)/2$. Notice that if n is a square of an integer, the corresponding $\Psi(n, m)$ is not empty. Effectively, the type $((n+\sqrt{n})/2, ((n-\sqrt{n})/2))$ performs the value $n(n-1)/4$ for $card(R(\pi))$. Note prealably that $(n+\sqrt{n})/2$ and $((n-\sqrt{n})/2)$ are integers, since n and $\sqrt{n}$ have the same parity.

Now, let us consider the partition set on E, denoted by $\mathcal{R}_r$, for which $card(R(\pi)) = r$, where r is a given integer. This set is empty if and only if the case is for $\Psi(n, 2r + n)$. In these conditions, we have the following properties which are corollaries of the main theorem above.

Proposition 45 *If ω is a total order on F, the mean in $\mathcal{R}_r$ of $card(gr(\omega) \cap R(\pi^\star) \times S(\pi^\star)) - \frac{1}{2}card(R(\pi) \times S(\pi))$ is null.*

In this statement the random partition $\pi^\star$ has the same meaning as in the previous theorem. For t belonging to $\Psi(n, 2r + n)$, denote by Π_t the following partition set on E

$$\Pi_t = \{\pi|\pi \in \Pi \text{ and } card(R(\pi)) = r\} \tag{9.3.37}$$

The sets Π_t, for t describing $\Psi(n, 2r + n)$, form a partition of $\mathcal{R}_r$:

$$\mathcal{R}_r = \sum_{t\in\Psi(n,2r+n)} \Pi_t \tag{9.3.38}$$

(set sum). According to the preceding theorem, the mean of $card[gr(\omega) \cap (R(\pi^*)$ $\times S(\pi^*)) - \frac{1}{2}card\,(R(\pi)) \times card\,(S(\pi))]$ is zero on each of the sets Π_t and then the case is on $\mathcal{R}_r$. □

Proposition 46 *Both criteria $B(\omega, \pi)$ and $V(\omega, \pi)$ (see Eqs. (9.3.7) and (9.3.8)) are equivalent in $\mathcal{R}_r$.*

In fact, the difference between these two criteria depends only on r, which is constant in the set $\mathcal{R}_r$. □

Proposition 47 *ω being a total order on $F = P_2(E)$, if $n = card(E)$ is such that $\Psi(n, n(n+1)/2)$ is non-empty, a partition π for which*

$$card\,(R(\pi)) < \frac{n(n-1)}{4}\left(1 - \frac{\sqrt{2}}{2}\right) \text{ or } card\,(R(\pi)) > \frac{n(n-1)}{4}\left(1 + \frac{\sqrt{2}}{2}\right)$$

cannot be optimal for the criterion $B(\omega, \pi)$.

Since $\Psi(n, n(n+1)/2)$ is not empty, there exists at least one partition π for which $card\,(R(\pi)) \times card\,(S(\pi))$ is equal to $n^2(n-1)^2/2^4$. Due to the previous theorem, for any element t of $\Psi(n, n(n+1)/2)$, there exists in Π_t, at least one partition π_t, for which

$$card[gr(\omega) \cap (R(\pi_t) \times S(\pi_t))] - \frac{1}{2}card\,(R(\pi_t) \times S(\pi_t)) \geq 0$$

that is to say,

$$card[gr(\omega) \cap (R(\pi_t) \times S(\pi_t))] \geq \frac{1}{2^5}n^2(n-1)^2$$

where $R(\pi_t)$ and $S(\pi_t)$ are associated with the partition π_t.

Now, let us denote by π_{opt} an optimal partition for the criterion $B(\omega, \pi)$ (card $(gr(\omega) \cap (R(\pi) \times S(\pi)))$ maximum), we have

$$card\left[gr(\omega) \cap \left(R(\pi_{opt}) \times S(\pi_{opt})\right)\right] \geq card[gr(\omega) \cap (R(\pi_t) \times S(\pi_t))] \geq \frac{1}{2^5}n^2(n-1)^2 \tag{9.3.39}$$

It is evident that

$$card\,(R(\pi_{opt}) \times S(\pi_{opt})) \geq card[gr(\omega) \cap \left(R(\pi_{opt}) \times S(\pi_{opt})\right)]$$

Hence,

$$card\,(R(\pi_{opt}) \times S(\pi_{opt})) \geq \frac{1}{2^5}n^2(n-1)^2$$

In other words, a partition π for which

$$card\,(R(\pi) \times S(\pi)) < \frac{1}{2^5}n^2(n-1)^2 \tag{9.3.40}$$

cannot be optimal for the criterion $B(\omega, \pi)$. By noting that card $(R(\pi) \times S(\pi))$ can be written as $r(\pi)(f - r(\pi))$, where $r(\pi) = card\,(R(\pi))$ and $f = n(n - 1/2)$, the inequality (9.3.40) becomes

$$r(\pi)(f - r(\pi)) < \frac{f^2}{2^3}$$

and then, the latter occurs if and only if

$$r(\pi) < \frac{f}{2}\left(1 - \frac{\sqrt{2}}{2}\right) \;\; or\;\; r(\pi) > \frac{f}{2}\left(1 + \frac{\sqrt{2}}{2}\right) \tag{9.3.41}$$

$\square$

In fact, the condition $\Psi(n, n(n + 1)/2) \neq \emptyset$ is not so restrictive. For any $n = card(E)$ a similar result can be obtained by substituting for the extreme right member of (9.3.39) the maximum value of $r(\pi)(f - r(\pi))/2$.

Proposition 48 *ω being a total order on $F = P_2(E)$, if $n = card(E)$ satisfies the two following conditions:*

1. $\Psi(n, n(n + 1)/2) \neq \emptyset$;
2. n is divisible by k: $n = h \times k$

a partition π of E into k classes of the same cardinal h, cannot be optimal for the criterion $B(\omega, \pi)$.

For a partition π of E into k classes having the same cardinal h,

$$r(\pi) = card\,(R(\pi)) = k \times \frac{h(h-1)}{2} \tag{9.3.42}$$

Following the proposition concerned above, a partition for which

$$card\,(R(\pi)) < \frac{n(n-1)}{4}(1 - \frac{\sqrt{2}}{2}) $$

cannot be optimal for the criterion $B(\omega, \pi)$. Therefore, a partition π satisfying the conditions of the previous proposition cannot be optimal if

$$\frac{kh(h-1)}{2} < \frac{kh(kh-1)}{4}\left(1 - \frac{\sqrt{2}}{2}\right) \tag{9.3.43}$$

that is to say, if

$$k > 4 + 2\sqrt{2} - \frac{3 + 2\sqrt{2}}{h}$$

and this inequality is necessarily held if $k \geq 7$.

On the other hand, always according to the proposition concerned above, a partition π for which

$$card\,(R(\pi)) > \frac{n(n-1)}{4}\left(1 + \frac{\sqrt{2}}{2}\right)$$

that is,

$$k < 2(2 - \sqrt{2}) - \frac{3 - \sqrt{2}}{h}$$

cannot be optimal for $B(\omega, \pi)$ and this inequality is necessarily satisfied for $k = 1$ and $n \geq 10$. $\qquad\square$

Let us illustrate the last proposition with an example. For $n = 49$, the set $\Psi(n, n(n+1)/2) = \Psi(49, 1225)$ is not empty, because n is the square of an integer and the type $\big((n + \sqrt{n})/2, (n - \sqrt{n})/2\big) = (28, 21)$ belongs to $\Psi(49, 1225)$. According to the previous property, a partition of a set E, sized 49, into 7 classes of the same cardinal 7 cannot be optimal for the criterion $B(\omega, \pi)$.

One more example given by W.F. de la Véga (personal communication) consists of four spheres in the geometrical space $\mathbb{R}^3$, e_1, e_2, e_3 and e_4 having the same radius r and centred, respectively, at the summits of a regular tetrahedron, the edge of which a being strictly greater than $4r$ ($a > 4r$). The set E is then the solid formed by these four spheres provided with a uniform density (see Fig. 9.1).

Here, it is more adequate to calculate in the framework of the set E^4, rather than in the set F^2, substituting thus E^2 for F. $\mathbb{R}^3$, being provided with the Lebesgue measure that we denote hereafter by μ, the *cardinal* notion is generalized by the *volume* notion. In these conditions, v will designate the common volume of each of the

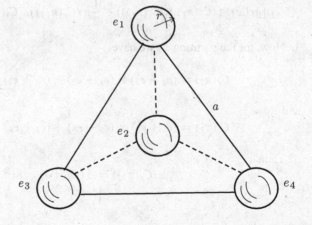

Fig. 9.1 Example given by a solid composed of four spheres

four spheres. The total ordonance ω on E is associated naturally with the Euclidean distance defined on $\mathbb{R}^3$.

Now, let us compare the two following partitions:

$$\pi_1 = \{e_1, e_2, e_3, e_4\} \text{ and } \{e_1 + e_2, e_3 + e_4\}$$

where the symbol "+" indicates a set sum. Denote by $C_r(\pi_1)$ (resp., $C_r(\pi_2)$) the set of ordered point pairs joined by the partition π_1 (resp., π_2). Similarly, $C_s(\pi_1)$ (resp., $C_s(\pi_2)$) will denote the set of ordered point pairs separated by the partition π_1 (resp., π_2).

For π_1, we have

$$C_r(\pi_1) = e_1 \times e_1 + e_2 \times e_2 + e_3 \times e_3 + e_4 \times e_4$$

where the symbol "$\times$" is the Cartesian product.

$$C_s(\pi_1) = e_1 \times e_2 + e_2 \times e_1 + e_1 \times e_3 + e_3 \times e_1 + e_1 \times e_4 + e_4 \times e_1$$

$$+ e_2 \times e_3 + e_3 \times e_2 + e_2 \times e_4 + e_4 \times e_2 + e_3 \times e_4 + e_4 \times e_3$$

Therefore, the measure of $C_r(\pi_1) \times C_s(\pi_1)$ is $4v^2 \times 12v^2 = 48v^4$. On the other hand, due to $a > 4r$, two joined points are strictly closer than two separated points, then $C_r(\pi_1) \times C_s(\pi_1) \subset gr(\omega)$, consequently,

$$\mu[gr(\omega) \cap (C_r(\pi_1) \times C_s(\pi_1))] = \mu(C_r(\pi_1) \times C_s(\pi_1)) = 48v^4$$

Moreover,

$$\frac{1}{2}\mu(C_r(\pi_1))\mu(C_r(\pi_2)) = 24v^4$$

Finally,

$$\mu[gr(\omega) \cap (C_r(\pi_1) \times C_s(\pi_1))] - \frac{1}{2}\mu(C_r(\pi_1))\mu(C_s(\pi_1)) = 24v^4 \qquad (9.3.44)$$

Now, for the partition π_2 we have

$$C_r(\pi_2) = (e_1 + e_2) \times (e_1 + e_2) + (e_3 + e_4) \times (e_3 + e_4)$$

and

$$C_s(\pi_2) = (e_1 + e_2) \times (e_3 + e_4) + (e_3 + e_4) \times (e_1 + e_2)$$

Then

$$\mu(C_r(\pi_2)) = \mu(C_s(\pi_2)) = 8v^2$$

In these conditions,

$$\mu\left(C_r(\pi_2)\right)\mu\left(C_s(\pi_2)\right) = 8v^2 \times 8v^2 = 64v^4$$

Now, decompose the set $C_r(\pi_2) \times C_s(\pi_2)$ into two subsets C and D, where

$$C = \{((x,y),(z,t))|(x,y) \in e_1^2 + e_2^2 + e_3^2 + e_4^2 \text{ and } (z,t) \in C_s(\pi_2)\}$$

and

$$D = \{((x,y),(z,t))|(x,y) \in e_1 \times e_2 + e_2 \times e_1 + e_3 \times e_4 + e_4 \times e_3 \text{ and } (z,t) \in C_s(\pi_2)\}$$

In these conditions

$$\mu[gr(\omega) \cap (C_r(\pi_2) \times C_s(\pi_2))] = \mu\left(gr(\omega) \cap C\right) + \mu\left(gr(\omega) \cap D\right)$$

C is included in $gr(\omega)$. On the other hand, for symmetry reasons, the half part of D is included in $gr(\omega)$. Therefore,

$$\mu[gr(\omega) \cap (C_r(\pi_2) \times C_s(\pi_2))] = 4v^2 \times 8v^2 + \frac{1}{2}4v^2 \times 8v^2 = 48v^4 \qquad (9.3.45)$$

Thus, the intersection criterion $B(gr(\omega), \pi)$ evaluates equally both partitions π_1 and π_2. The centred criterion, which is equivalent to the de la Véga criterion (see Proposition 44) gives for the partition π_2: $48v^4 - 32v^4 = 16v^4$. Consequently, the centred criterion discriminates between π_1 and π_2. For the latter π_1 is preferable to π_2.

Conclusion

To summarize, we have shown that the mean of the random intersection criterion $gr(\omega) \cap (R(\pi^*) \times S(\pi^*))$ on the set Π_t of partitions of a fixed type t is null. Then, the centred intersection criterion is independent of the partition type considered. This property establishes the intrinsic character of the second criteria $V(gr(\omega), \pi)$. Nevertheless, both criteria are equivalent in the set $\mathcal{R}_r$ of partitions π for which $card\,(R(\pi)) = r$. However, we have pointed out (see Proposition 52) and illustrated (see Proposition 48) an a priori impossibility for partitions having specific cardinal structures to be optimal for the first criterion $(B(\omega, \pi))$. The latter is then *biased* and preference will be given to the second criterion $V(\omega, \pi)$. Nevertheless, the latter criterion cannot be the definitive criterion we adopt. In fact, as seen, in another context, in Chap. 6, a relevant reference scale has to be established in order to compare "significantly" different partitions on a same set, provided with a similarity measure. For this purpose, the distribution of the random intersection criterion will be studied with respect to an adequate probabilistic hypothesis of no relation. This will be carried out in the next section.

9.3.4   Distribution of the Intersection Criterion on the Partition Set with a Fixed Type

9.3.4.1   Introduction

As reported in Sect. 9.3.1 we shall work here using the preordonance version $\varpi(E)$ for which, for a given pair $p = \{x, y\}$, the greater is the similarity $S(x, y)$, the higher is the rank $k(p)$ with respect to $\varpi(E)$ (see Eq. (9.3.2)). As indicated in the final part of the section mentioned, we begin by the case where $\varpi(E)$ is a total order on the set $F = P_2(E)$. The case where $\varpi(E)$ is a total *preorder* on F is considered next. Finally, the adaptation of the criterion obtained to the case of a numerical similarity on E is studied.

The general *LLA* strategy defined in Chaps. 5 and 6 is applied here (see Fig. 6.1 of Sect. 6.2.1). The initial form of the *raw* coefficient is

$$s(\varpi, \pi) = card[gr(\varpi) \cap (S(\pi) \times R(\pi))]$$

(see (9.3.6)). The random raw coefficient can be written as

$$s(\varpi, \pi^\star) = card\left[gr(\varpi) \cap \left(S(\pi) \times R(\pi^\star)\right)\right]$$

where, as usually, $\pi^\star$ is a random element in the set Π_t, equally distributed, of all partitions having the same type t as that of π.

Our objective consists of analyzing the distribution of the random coefficient $s(\varpi, \pi^\star)$ and to propose—according to the *LLA* strategy—a statistically normalized coefficient of the following form:

$$S(\varpi, \pi) = \frac{s(\varpi, \pi) - \mathcal{E}[s(\varpi, \pi^\star)]}{\sqrt{var\,(s(\varpi, \pi^\star))}} \tag{9.3.46}$$

(see Fig. 6.1 of Sect. 6.2.1 of Chap. 6).

Let us take up again the notations introduced in the final part of Sect. 9.3.1. ρ designates the indicator function of the subset $R(\pi)$ of F and k the ranking function coding the total order ϖ on F. The difference between the $s(\varpi, \pi)$ index and

$$s(k, \rho) = \sum_{p \in F} \rho(p)k(p) \tag{9.3.47}$$

is a constant depending on $r = card(R(\pi))$ in Π_t (see (9.3.13)). By considering the random indicator function $\rho^\star$ of $R(\pi^\star)$, the distribution of the random raw coefficient

$$s(k, \rho^\star) = \sum_{p \in F} \rho^\star(p)k(p) \tag{9.3.48}$$

is obtained by translating to the right of $\frac{1}{2}r(r+1)$ the distribution of $s(\varpi, \pi^\star)$. Hence, the normalized coefficient $S(\varpi, \pi)$ is the same if we substitute for $s(\varpi, \pi)$ and $s(\varpi, \pi^\star)$, $s(k, \rho)$ and $s(k, \rho^\star)$, respectively.

The distribution of $s(k, \rho^\star)$ will be compared with that of

$$s(k, \beta^\star) = \sum_{p \in F} \beta^\star(p)k(p) \tag{9.3.49}$$

where $\beta^\star$ is the random indicator function associated with the random F subset $R(\beta^\star)$, whose cardinality is r. More precisely, $R(\beta^\star)$ is a random element in the set $\mathcal{B}_r$ of all F subsets, whose cardinal is r, provided with equal probabilities. Each element $R(\beta)$ of $\mathcal{B}_r$ represents a symmetrical binary relation on E. By denoting $\mathcal{R}(n; t)$ the equivalence relations corresponding bijectively to Π_t, $\mathcal{R}(n; t) \subset \mathcal{B}_r$ if and only if the common number of joined pairs of an arbitrary partition π of Π_t is equal to r $(r(\pi) = card(R(\pi)) = r)$.

It will be easy to see that the distribution of the random index $s(k, \beta^\star)$ is identical to that

$$s(k^\star, \beta) = \sum_{p \in F} \beta(p)k^\star(p) \tag{9.3.50}$$

where $k^\star$ is a random ranking on F corresponding to a uniform permutational model.

To begin, the limit form of both dual distributions, those of $s(k, \beta^\star)$ and $s(k^\star, \beta)$, will be analysed. The limit distribution is shown to be the *normal* distribution. Next, we will compare the distribution of $s(k, \rho^\star)$ with that of $s(k, \beta^\star)$. The former distribution is on $\mathcal{B}_r$ and the latter one on $\mathcal{R}(n; t)$, where $\mathcal{R}(n; t) \subset \mathcal{B}_r$. Both distributions have the *same mean*. Otherwise, the asymptotic form of the distribution of $s(k, \rho^\star)$ is under rather general conditions, the same as that of $s(k, \beta^\star)$.

9.3.4.2 The Two Dual Distributions in the Case of a Non-constrained Binary Relation β

In this case, considered already just above, both random coefficients $s(k, \beta^\star)$ and $s(k^\star, \beta)$ (see Eqs. (9.3.49) and (9.3.50)) have exactly the same distribution: that of the sum of r distinct integers taken uniformly at random in the set $\{1, 2, \ldots, g, \ldots, f\}$. In these conditions, we may only consider the random coefficient $s(k, \beta^\star)$ (see (9.3.49)). In fact, $s(k, \beta^\star)$ can also be expressed in the form

$$s(k, \beta^\star) = \sum_{1 \leq g \leq f} g \cdot \beta(\sigma(g)) \tag{9.3.51}$$

where σ is a random permutation in the set G_f of all permutations on $\{1, 2, \ldots, g, \ldots, f\}$, equally distributed. In these conditions, we refer to the Wald and Wolfowitz theorem (see Theorem 8 of Sect. 5.2.2.3 of Chap. 5). For this, we associate the following characteristics:

1. The mean $\mu(k) = \frac{1}{f} \times \frac{f(f+1)}{2} = \frac{f+1}{2}$;

2. The variance $\mu_2(k) = \frac{1}{f} \sum_{1 \leq g \leq f} g^2 - (\frac{f+1}{2})^2 = \frac{(f-1)(f+1)}{2}$;

3. The coefficient W of Wald and Wolfowitz

$$W_q = \frac{\mu_q(k)}{(\mu_2(k))^{q/2}} = \frac{\frac{1}{f}\sum_{1 \leq g \leq f}(g - \frac{f+1}{2})^q}{[\frac{1}{f}\sum_{1 \leq g \leq f}(g - \frac{f+1}{2})^2]^{q/2}}.$$

A simple computing based on the inequality

$$-1 < \frac{2i - f - 1}{f} < 1 \text{ shows that } |W_q| \leq 3^{q/2}$$

this insures the condition W expressed in Theorem 8 of Sect. 5.2.2.3 of Chap. 5. On the other hand, let us consider the Noether condition V of this theorem, expressed with respect to the sequence $(\beta(p) | p \in F)$, which can be written as

$$\frac{\sum_{p \in F} \left(\beta(p) - \frac{r}{f} \right)^q}{f[\frac{r}{f}(1 - \frac{r}{f})]^{q/2}}$$

where $q \geq 3$ and $r = card\,(R(\pi))$. This expression can be written as

$$(-1)^q \left[\left(\frac{r}{f} \right)^{q/2} \times s^{-(h-2)/2} \right] + \left(\frac{s}{f} \right)^{q/2} \times r^{-(h-2)/2}$$

where $s = card\,(S(\pi))$. The latter quantity tends to zero when $n = card(E)$ tends to infinity, provided that the ratio r/f does not tend neither to zero nor to one. In these conditions, the theorem of Wald and Wolfowitz applies and gives the following.

Theorem 26 *The common limiting distribution of $s(k, \beta^\star)$ and $s(k^\star, \beta)$ is normal. The mean and variance of this distribution are $r(f + 1)/2$ and $rs(f + 1)/12$.*

As a direct consequence, the limiting distribution of

$$S(\varpi, \beta^\star) = \frac{card\,(gr(\varpi) \cap (S(\beta^\star) \times R(\beta^\star))) - r \cdot s/2}{\sqrt{r \cdot s(f + 1)/12}} \tag{9.3.52}$$

is the standardized *normal* distribution (mean = 0 and variance = 1), $\beta^\star$ being a random element in $\mathcal{B}(r)$ provided with equal probabilities. By substituting for $k(p)$

$$k'(p) = \frac{k(p) - \mu(k)}{\sqrt{\mu_2(k)}} = \frac{k(p) - \frac{f+1}{2})}{\sqrt{r \cdot s(f + 1)/12}}$$

the form of (9.3.52) becomes

$$S(\varpi, \beta^\star) = \frac{1}{\sqrt{r \cdot s(f - 1)}} \sum_{p \in F} \beta^\star(p)k'(p) \tag{9.3.53}$$

9.3.4.3 Comparing *card* $(gr(\varpi) \cap (S(\beta^\star) \times R(\beta^\star)))$ and *card* $(gr(\varpi) \cap (S(\pi^\star) \times R(\pi^\star)))$ **Distributions**

Due to Eqs. (9.3.48) and (9.3.49), we have to compare the distribution of

$$s(k, \rho^\star) = \sum_{p \in F} \rho^\star(p) k(p) \tag{9.3.54}$$

to that of

$$s(k, \beta^\star) = \sum_{p \in F} \beta^\star(p) k(p) \tag{9.3.55}$$

$\rho^\star$ and $\beta^\star$ were specified above (see what follows Eqs. (9.3.48) and (9.3.49)). According to the notations adopted, we have the following:

Theorem 27 *If the set* $\mathcal{R}(n; t)$ *(associated bijectively to the partition set* Π_t*) is included in* $\mathcal{B}(r)$*, the mean of* $s(k, \rho^\star)$ *in* $\mathcal{R}(n; t)$ *is equal to the mean of* $s(k, \beta^\star)$ *in* $\mathcal{B}(r)$*, and the common value of the mean is* $r(f+1)/2$.

The mean of $s(k, \beta^\star)$ in $\mathcal{B}(r)$ has already been calculated (see Theorem 26). The mean of $s(k, \rho^\star)$ can be expressed by

$$\frac{1}{card\,(\Pi_t)} \sum_{\pi \in \Pi_t} \rho(p) k(p)$$

where ρ is associated univocally to π. By interverting the two sum signs we obtain

$$\sum_{p \in F} \left[\frac{1}{card\,(\Pi_t)} \sum_{\pi \in \Pi_t} \rho(p) \right] k(p)$$

The inside of right brackets is the proportion of partitions in Π_t for which the two components of the pair p are joined. This proportion was calculated above (see Lemma 8), and it is equal to

$$\frac{\sum_{1 \le i \le k} n_i(n_i - 1)}{n(n-1)}$$

Hence, the result stated in the theorem is acquired.   □
This result is obtained here much more easily than in Theorem 25. However, we have pointed out above the own interest of the proof of the latter theorem.
Now, we shall compare the respective moments of order l of $s(k, \beta^\star)$ and $s(k, \rho^\star)$, the condition $\mathcal{R}(n; t) \subset \mathcal{B}(r)$, being satisfied.

Comparing the *l*th Moments of $s(k, \beta^\star)$ and $s(k, \rho^\star)$

$\mathcal{E}$ denoting the mathematical expectation operator, we have for $s(k, \beta^\star)$

$$\mathcal{E}\left[\sum_{p\in F}\beta^{\star}(p)k(p)\right]^{l} = \frac{1}{card\,(\mathcal{B}(r))}\sum_{\beta\in\mathcal{B}(r)}\left(\sum_{p\in F}\beta(p)k(p)\right)^{l} \qquad (9.3.56)$$

By reversing the two sum symbols and by taking into account that β is a Boolean function ($\beta(p) = 0$ or 1), we obtain

$$\mathcal{E}\left(s(k,\beta^{\star})\right)^{l} = \sum \mu(p_{i_1}, p_{i_2}, \ldots, p_{i_m}).c(l; l_1, l_2, \ldots, l_m)k(p_{i_1})^{l_1}k(p_{i_2})^{l_2}\ldots k(p_{i_m})^{l_m}$$
$$(9.3.57)$$

where

- The sum is extended over all permutations $(p_{i_1}, p_{i_2}, \ldots, p_{i_m})$ which can be obtained from a subset of m elements of F (there are in all $\binom{f}{m} \times m!$ such permutations);
- $l = l_1 + l_2 + \cdots + l_m$ and $\mu(p_{i_1}, p_{i_2}, \ldots, p_{i_m})$ is the proportion in $\mathcal{B}(r)$ of symmetrical binary relations for which $\beta(p_{i_1})\beta(p_{i_2})\ldots\beta(p_{i_m}) = 1$;
- $c(l; l_1, l_2, \ldots, l_m) = l!/e_1!e_2!\ldots e_h!l_1!l_2!\ldots l_h!$, where h is the number of distinct l_j, each of them being repeated $e_1, e_2, \ldots, e_h$ times, respectively.

Now, define $\nu(p_{i_1}, p_{i_2}, \ldots, p_{i_m})$ as the proportion of partitions in Π_t which join the pairs p_{i_1}, p_{i_2}, ...and p_{i_m}. The expression of $\mathcal{E}(s(k,\rho^{\star}))^{l}$ is obtained from that $\mathcal{E}(s(k,\beta^{\star}))^{l}$ by substituting $\nu(p_{i_1}, p_{i_2}, \ldots, p_{i_m})$ for $\mu(p_{i_1}, p_{i_2}, \ldots, p_{i_m})$.

Given a pair $p = \{x, y\}$, the proportion in $\mathcal{B}(r)$ of symmetrical binary relations for which $\beta(x, y) = 1$ is

$$\frac{\binom{f-1}{r-1}}{\binom{f}{r}} = \frac{r}{f}$$

In these conditions, for any $(p_{i_1}, p_{i_2}, \ldots, p_{i_m}) \in F^m$, $\mu(p_{i_1}, p_{i_2}, \ldots, p_{i_m})$ is, for n large enough, very approximately equal to

$$\left(\frac{r}{f}\right)^{m} \simeq \left(\sum_{1\leq i\leq k}\pi_i\right)^{2} \text{ where } \pi_i = \frac{n_i}{n} \text{ with } t = (n_1, n_2, \ldots, n_k) \qquad (9.3.58)$$

For a given $(p_{i_1}, p_{i_2}, \ldots, p_{i_m}) \in F^m$ consider the transitive closure of the set of pairs comprised as components of the latter m-uple of pairs. A partition that we indicate by γ, on a E subset, is thus obtained. If $(u_1, u_2, \ldots, u_g)$ is the type of the partition γ, we have

$$\nu(p_{i_1}, p_{i_2}, \ldots, p_{i_m}) \simeq (\pi_1^{u_1}, \pi_2^{u_1}, \ldots, \pi_k^{u_1}) \times \cdots \times (\pi_1^{u_g}, \pi_2^{u_g}, \ldots, \pi_k^{u_g}) \qquad (9.3.59)$$

In fact, every term of the product development of (9.3.59) is associated with a mapping ϕ of the label set $\{1, 2, \ldots, j, \ldots, g\}$ into $\{1, 2, \ldots, i, \ldots, k\}$ that corresponding to the mapping ϕ, for which $\phi(j) = i$, is defined by the proportion of partitions from Π_t, for which the u_j elements of the jth class of γ are joined in the ith class, whose cardinal being n_i, $1 \leq j \leq g$.

Let $F^{[m]}$ ($m \leq f$) be the set of vectors of F having m components mutually distinct:

$$F^{[m]} = \{(p_1, p_2, \ldots, p_i, \ldots, p_m) | p_i \in F \text{ and } p_i \neq p_j \text{ for } i \neq j\}$$

we have

$$card(F^{[m]}) = f(f-1)(f-2)\ldots(f-m+1) \tag{9.3.60}$$

Let us consider now the partition of $F^{[m]}$ into the different subsets G_m^c already introduced in Chap. 3, Sect. 6.2.3.2. $\nu(p_{i_1}, p_{i_2}, \ldots, p_{i_m})$ is invariant when $(p_{i_1}, p_{i_2}, \ldots, p_{i_m})$ describes a given subset G_m^c (see (9.3.59)). Designate by H_m the set of pair m-uples such that any two of them are without a common component. We have

$$card(H_m) = \binom{n}{2} \times \binom{n-2}{2} \times \cdots \times \binom{n-2m-2}{2} \tag{9.3.61}$$

The ratio of this cardinal on the previous one $card(H_m)/card(F^{[m]})$ tends rapidly to 1 when n tends to infinity. Moreover, asymptotically, we have

$$\frac{G_m^c}{card(H_m)} = O\left(\frac{1}{n^q}\right) \tag{9.3.62}$$

where q is an integer greater than 1; q depending on the configuration c which is directly related to the type $(u_1, u_2, \ldots, u_g)$ of the partition deduced from transitive closure of an element of $G_m^{(c)}$.

Now, divide each of both sums such as (9.3.57) defining $\mathcal{E}(s(k, \beta^\star))^l$ and $\mathcal{E}(s(k, \pi^\star))^l$, respectively, into two parts $\Sigma^{(1)}$ and $\Sigma^{(2)}$ where $\Sigma^{(1)}$ is extended over H_m and where $\Sigma^{(2)}$ is extended over the set of vectors $(p_1, p_2, \ldots, p_m)$ for which there exists at least two distinct pairs having a common component; that is to say, extended to the different sets $G_m^{(c)}$, the latter being mutually disjoint. The number of distinct configurations (c) depends only on m.

Equations (9.3.58) and (9.3.59) show that the parts $\Sigma^{(1)}$ of $\mathcal{E}(s(k, \beta^\star))^l$ and $\mathcal{E}(s(k, \pi^\star))^l$ are slightly different, that is, converge quickly to the same limit. On the other hand, (9.3.62) shows that the cardinalities of the basis of $\Sigma^{(2)}$ tend to be negligible with respect to that of $\Sigma^{(1)}$. In order that the value of $\Sigma^{(2)}$ tends to be negligible with respect to that of $\Sigma^{(1)}$ in Equations such (9.3.57), it suffices that a positive measure of the form $k(p_{i_1})^{l_1} k(p_{i_2})^{l_2} \ldots k(p_{i_m})^{l_m}$ not to be particularly high on the sets $G_m^{(c)}$. More precisely, let σ be the positive measure on $F^{[m]}$ defined by the mth power of the positive measure on F, $\{k(p) | p \in F\}$:

$$\sigma(p_1, p_2, \ldots, p_m) = k(p_1) \cdot k(p_2) \cdot \ldots \cdot k(p_m)$$

In these conditions, let us consider the part of a sum such (9.3.57) obtained for a fixed $(l_1, l_2, \ldots, l_m)$. If for the different configurations (c), $\sigma(G_m^{(c)})/\sigma(H_m)$ tends towards 0 when n tends to infinity, the part of the sum concerned, extended over the different sets $G_m^{(c)}$, tends to be negligible with respect to that extended over H_m. Hence, we obtain the following.

Theorem 28 *If $\sigma(G_m^{(c)})/\sigma(H_m)$ tends to 0 when n tends to infinity, for every fixed m and for every configuration (c), then the moments of the distribution of $s(k, \pi^\star)$ in $\mathcal{R}(n; t)$ tend, respectively, to those of the distribution of $s(k, \beta^\star)$ in $\mathcal{B}(r)$, when n tends to infinity.*

In these conditions, the limit distribution of the random coefficient

$$\frac{s(k, \pi^\star) - \frac{r(f+1)}{2}}{\sqrt{\frac{rs(f+1)}{12}}}$$

in $\mathcal{R}(n; t)$, provided with equal probabilities, is the standard normal distribution.

Hence, this random coefficient can put in an analogous form as that of $S(\varpi, \beta^\star)$ (see (9.3.53)), namely

$$S(\varpi, \rho^\star) = \frac{1}{\sqrt{r \cdot s(f - 1)}} \sum_{p \in F} \rho^\star(p)k'(p) \qquad (9.3.63)$$

where $k'(p)$ was defined just above (9.3.53).

9.3.5   Extensions of the Previous Criterion

Preamble

These extensions will turn up by consistent analogy with the case where the similarity data is a total ordonance on E, that is, a total order (i.e. strict ranking) ϖ on $F = P_2(E)$ (see above). According to (9.3.63), the criterion concerned can be written as

$$S(\varpi, \pi) = \frac{1}{\sqrt{r \cdot s(f - 1)}} \sum_{p \in F} \rho(p)k'(p) \qquad (9.3.64)$$

where, as usual, $\{\rho(p) | p \in F\}$ indicates the subset $R(\pi)$ of F and $\{k'(p) | p \in F\}$ the standardized version of the ranking function k on F, associated with ϖ.

Two extensions will be considered. The first one concerns the case for which ϖ is a total *preorder* (ranking with ties) *and not* necessarily a strict total order (ranking without ties) on F; in other words, a *preordonance* on E and not necessarily an *ordonance* on E. The second extension concerns the case for which the criterion uses directly a numerical similarity on E (see Chap. 4).

9.3.5.1 Case Where the Similarity Is Defined by a Total *Preordonance*

The general idea consists of substituting the "mean rank" function for the "rank" function. The mean rank function was presented in Sect. 3.4.1 of Chap. 3 in the context of coding the ordinal similarity of the attribute set of a preordonance categorical attribute. Let us present it here in order to code a total preorder ϖ on F associated with an ordinal similarity on E.

Let $\{F_l | 1 \leq l \leq m\}$ be the class sequence of the total preorder ϖ and denote by $\{f_l | 1 \leq l \leq m\}$ the sequence of the cardinals of the classes F_l, $1 \leq l \leq m$. We have for the quotient order:

$$F_1 < F_2 < \cdots < F_l < \cdots < F_m$$

In these conditions, the mean rank function, designated, without risk of ambiguity, by k is defined as follows:

$$k(p) = \frac{f_1 + 1}{2} \text{ for every } p \in F_1$$

$$k(p) = f_1 + \frac{f_1 + 1}{2} \text{ for every } p \in F_2$$

...

$$k(p) = \sum_{1 \leq j \leq l-1} f_j + \frac{f_l + 1}{2} \text{ for every } p \in F_l$$

...

$$k(p) = \sum_{1 \leq j \leq m-1} f_j + \frac{f_l + 1}{2} \text{ for every } p \in F_m \qquad (9.3.65)$$

In this way, we have the known invariance property:

$$\sum_{p \in F} k(p) = \frac{f(f+1)}{2}$$

In these conditions, the numerator of $k'(p) = (k(p) - \mu(k))/\sqrt{\mu_2(k)}$ is preserved as in the case where ϖ is a strict ranking. However, the denominator is smaller.

9.3.5.2 Case Where the Similarity Is Numerical

A numerical version of a similarity function S on the set E of unit data is dealt here. S can be expressed as the similarity table

$$\mathcal{Q} = \{Q(p) | p = \{x, y\} \in F = P_2(E)\} \qquad (9.3.66)$$

where $Q(p)$ is the value of the similarity function Q on $p = \{x, y\}$.

The starting point in the adaptation process of the coefficient (9.3.64) is the *raw* coefficient considered in Eq. (9.3.47), namely

$$s(k, \rho) = \sum_{p \in F} \rho(p) k(p) \qquad (9.3.67)$$

For this coefficient, the value of the ordinal similarity on p ($p \in F$) is quantified by the integer $k(p)$ which is the rank of p with respect to the total order ϖ on F. In these conditions, in order to obtain the corresponding *raw* coefficient for the numerical case, the similarity numerical function $Q(p)$ is substituted for the ranking function $k(p)$. We obtain

$$s(Q, \rho) = \sum_{p \in F} \rho(p) Q(p) \qquad (9.3.68)$$

Clearly, the distribution study of $s(k, \rho^\star)$ in Sects. 9.3.4.2 and 9.3.4.3 can be transposed for the distribution of $s(S, \rho^\star)$. Notice that in the passage from $s(S, \rho)$ ((9.3.68)) to $s(k, \rho)$ ((9.3.67)), the similarity function S is replaced by its cumulative distribution function, when $\varpi(E)$ is associated with S.

Now, let $mean_e(Q)$ and $var_e(Q)$ be the empirical mean and variance of the distribution of Q (see (9.3.66)):

$$mean_e(Q) = \frac{1}{f} \sum_{p \in F} Q(p) \text{ and } var_e(Q) = \frac{1}{f} \sum_{p \in F} [Q(p) - mean_e(Q)]^2$$

By setting

$$\text{For any } p \in F, c(p) = \frac{Q(p) - mean_e(Q)}{\sqrt{var_e(Q)}} \qquad (9.3.69)$$

the criterion sought, resulting from the adaptation of $S(\varpi, \pi)$ (see (9.3.64)), can be written as

$$S(Q, \pi) = \frac{1}{\sqrt{rs/(f-1)}} \sum_{p \in F} \rho(p) \cdot c(p) \qquad (9.3.70)$$

The similarity coefficient $c(p)$ ($p \in F$) has been introduced in a natural way from a given similarity coefficient $Q(p)$ ($p \in F$). Notice that $c(p)$ corresponds exactly to the coefficient $Q^g(p)$ ($p \in F$) obtained after what we have called the *"Similarity Global Reduction"* (see Sect. 5.2.1.5 of Chap. 5).

Otherwise, it may be easily established that the S. Régnier criterion of the "Central partition method" (see Sect. 2.1 of Chap. 2) can be expressed by

$$C(S, \pi) = \sum_{q \in F} \rho(q) S(q) \qquad (9.3.71)$$

where the indicator function ρ has the same meaning as above and where the similarity index S is up to $-\frac{1}{2}$ the classical Russel and Rao similarity index (see Table 4.3

of Chap. 4). Applying the latter assumes to substitute for categorical attributes the Boolean attributes corresponding to their respective values, this transformation being commonly called *complete disjunctive coding*. In these conditions, according to the principle of our development above, it is easy to see that the coefficient $C(S, \pi)$ is *biased* and does not refer to a statistically relevant measure scale.

If the numerical similarity coefficient Q on E is adequately established (see Chap. 6), a coefficient such as (9.3.70) is generally preferable. In fact, it takes more intimately into account the data similarity structure. On the other hand, its computational complexity is appreciably lower than that (9.3.64), since the latter requires a sorting on F (see Sect. 9.3.5.1).

However, important reasons might justify the use of the criterion form (9.3.64) where $k'(p)$ is defined by a similarity ranking function, strictly ordinal or preordinal (see Sect. 9.3.5.1). The first one is given by the results of Shepard's and Benzécri's works [2, 37] where the following statement is established:

Let $C = \{i_1, i_2, \ldots, i_n\}$ and $C' = \{i'_1, i'_2, \ldots, i'_n\}$ be two configurations in the geometrical space $\mathbb{R}^p$ and let ω and ω' designate the respective ordonances associated with the Euclidean distance, respectively, on C and C'. Suppose that ω' can be deduced from ω by a bijection—which we denote by τ—of C onto C'. In these conditions C' can be obtained from C, by applying on it, translation, rotation, dilatation and a slight distorsion. Now, if π and π' are two respective partitions on C and C', such that $\pi' = \tau(\pi)$, one wishes from a clustering criterion to evaluate equally π and π'. This requirement is clearly satisfied by the ordinal form of the criterion (see (9.3.64)).

Another reason in favour of the ordinal form of the criterion might be given by the stability properties of the preordonance associated with a numerical similarity index (see Sect. 4.2.2 of Chap. 4). Nevertheless, it remains certainly, to compare in depth, experimentally and theoretically both versions of the criterion: ordinal and numerical.

9.3.5.3   Variance Analysis

In Sect. 9.3.4.3 the respective moments of the same order of the random raw coefficients $s(k, \rho^\star)$ and $s(k, \beta^\star)$ (see (9.3.54) and (9.3.55)) were compared. Both concern the ordinal form of the criterion. The first one is that where the random relation $\rho^\star$ is an element of $\mathcal{R}(n; t)$ and the second one is that where the random relation $\beta^\star$ is an element of $\mathcal{B}(r)$ ($\mathcal{R}(n; t)$ and $\mathcal{B}(r)$ have been clearly defined above). This analysis can be naturally extended to the case of comparing the two random criteria:

$$C(c, \rho^\star) = \sum_{p \in F} \rho^\star(p) c(p)$$

and

$$C(c, \beta^\star) = \sum_{p \in F} \beta^\star(p) c(p) \tag{9.3.72}$$

where the coefficient c is associated with a similarity measure Q on E (9.3.69).

It was established that under rather general conditions that the respective moments of the distribution of $C(c, \rho^\star)$ tend towards those of $C(c, \beta^\star)$. We shall consider here the most important case of comparing the respective variances V_ρ and V_β of $C(c, \rho^\star)$ and $C(c, \beta^\star)$. Since these coefficients are centred, we have

$$V_\rho = \sum_{1 \le i \le k} \pi_i^2 \sum_{p \in F} c(p)^2 + \sum_{1 \le i \le k} \pi_i^3 \sum_{(p,q) \in G} c(p)c(q) + \left(\sum_{1 \le i \le k} \pi_i^2 \right)^2 \sum_{(p,q) \in H} c(p)c(q)$$

$$V_\beta = \sum_{1 \le i \le k} \pi_i^2 \sum_{p \in F} c(p)^2 + \left(\sum_{1 \le i \le k} \pi_i^2 \right)^2 \sum_{(p,q) \in G} c(p)c(q) + \left(\sum_{1 \le i \le k} \pi_i^2 \right)^2 \sum_{(p,q) \in H} c(p)c(q)$$

$$(9.3.73)$$

In the equations above, the components of the type $t = (n_1, n_2, \ldots, n_i, \ldots, n_k)$ are assumed large enough in order to approximate very accurately $n_i(n_i - 1) \ldots (n_i - h + 1)/n(n - 1) \ldots (n - h + 1)$ by π^h, where $\pi = n_i/n$ and $h \le 2$, $1 \le i \le k$. It is important to note that the value of V_β is independent of the particular expression of the similarity coefficient Q, whose statistical normalization led to the coefficient c. In effect, by identifying

$$\sum_{1 \le i \le k} \pi^2 \text{ with } \frac{r}{f} \text{ we obtain } V_\beta = \frac{r \cdot s}{f}.$$

This result is obtained by taking into account the identity

$$\sum_{(p,q) \in G} c(p)c(q) + \sum_{(p,q) \in H} c(p)c(q) = \sum_{(p,q) \in (F \times F - \Delta)} c(p)c(q) = -f \qquad (9.3.74)$$

where Δ is the diagonal of $F \times F$.

The difference $V_\rho - V_\beta$ can be written as

$$V_\rho - V_\beta = \left[\sum_{1 \le i \le k} \pi_i^3 - \left(\sum_{1 \le i \le k} \pi_i^2 \right)^2 \right] \sum_{(p,q) \in G} c(p)c(q) \qquad (9.3.75)$$

Proposition 49 *If the different components of the type $t = (n_1, n_2, \ldots, n_i, \ldots, n_k)$ are equal, then the variances V_ρ and V_β are also equal.*

In fact, in this case, for every i, $1 \le i \le k$, $\pi_i = 1/k$. Hence,

$$\sum_{1 \le i \le k} \pi_i^3 = \left(\sum_{1 \le i \le k} \pi_i^2 \right)^2 = \frac{1}{k^2}$$

and then, $V_\rho = V_\beta$.   □

The *Proximity Variance Analysis* formula was introduced in [21] and retaken in [26] and Chap. 8 of [22]. This formula is applied here to the standardized similarity table

$$\{c(x, y) | (x, y) \in E^{[2]}\}$$

where $E^{[2]}$ is the set of ordered distinct element pairs of E. $c(x, y) = c(y, x)$ for $x \neq y \in E$. We have

$$\frac{1}{n(n-1)} \sum_{(x,y) \in E^{[2]}} c(x, y)^2 = \frac{1}{n} \sum_{x \in E} \frac{1}{n-1} \sum_{y | y \in E, y \neq x} [c(x, y) - \bar{c}_x]^2 + \frac{1}{n} \sum_{x \in E} \bar{c}_x^2$$

(9.3.76)

where $\bar{c}_x$ is the mean of the proximities to x, $x \in E$, namely

$$\bar{c}_x = \frac{1}{n-1} \sum_{y | y \neq x} c(x, y)$$

Due to the statistical normalization, the left member is equal to 1. The first (resp., second) term of the right member corresponds to the *within* (resp., *between*) variance.

According to (9.3.75), if $\sum_{1 \leq i \leq k} \pi_i^3$ is different from $(\sum_{1 \leq i \leq k} \pi_i^2)^2$, $(V_\rho - V_\beta)$ is a function of

$$\sum_{(p,q) \in G} c(p)c(q)$$

The following important statement makes that this function can be expressed with respect to the *between* variance:

$$\frac{1}{(n-1)} \left[1 + \frac{1}{n(n-1)} \sum_{(p,q) \in G} c(p)c(q) \right] = \frac{1}{n} \sum_{x \in E} \bar{c}_x^2$$

(9.3.77)

This result is obtained from

$$\sum_{x \in E} \left[\sum_{y | y \in E, y \neq x} c(x, y) \right]^2 = n(n-1) + \sum_{(p,q) \in G} c(p)c(q)$$

In fact, in the development of the square, for every $p \in F$, $c(p)^2$ appears exactly twice and for every $(p, q) \in G$, $c(p)c(q)$ appears exactly once. □

Relation (9.3.74) allows us to obtain

$$1 - \frac{1}{2(n-1)} + \frac{1}{n(n-1)^2} \sum_{(p,q) \in H} [c(p)c(q)]$$

$$= \frac{1}{n} \sum_{x \in E} \frac{1}{n-1} \sum_{y | y \in E, y \neq x} [c(x, y) - \bar{c}_x]^2$$

(9.3.78)

It is interesting to notice that, for a set E provided with a similarity index S, the *neutrality* degree related to the "projective importance" of a given element x with respect to a clustering organization purpose was measured in Sect. 8.5 of Chap. 8, by the smallness of the within variance

$$\mathcal{V}(x) = \frac{1}{(n-1)} \sum_{y|y \in E, y \neq x} [S(x, y) - \overline{S}_x]^2$$

where, as above, $\overline{S}_x$ is the empirical mean of the similarities to x, $x \in E$.

Equation (9.3.78) shows that the within variance of the c similarities depends uniquely on $\sum_{(p,q) \in H} c(p)c(q)$. Formula (9.3.77) can also be written under the form

$$\frac{1}{n(n-1)} \sum_{(p,q) \in G} c(p)c(q) = \frac{1}{n} \sum_{x \in E} \left[\frac{1}{\sqrt{n-1}} \sum_{y|y \in E, y \neq x} c(x, y) \right]^2 - 1 \qquad (9.3.79)$$

Let us now designate by $\sum_{(h)} c(p)$ the sum extended over h elements of $\{c(p)|p \in F\}$. There are in all $\binom{f}{h}$ instances of the previous sum. With this notation the mean and variance of the distribution of the Statistic

$$\frac{1}{\sqrt{n-1}} \sum_{(n-1)} c(p) \qquad (9.3.80)$$

on the set of all $(n-1)$ subsets of F are 0 and 1, respectively. This is because the distribution $\{c(p)|p \in F\}$ is standardized (*mean* $= 0$ and *variance* $= 1$). In these conditions, the following sample of the Statistic (9.3.80)

$$\left\{ \frac{1}{\sqrt{n-1}} \sum_{y|y \neq x)} c(x, y)|x \in E \right\} \qquad (9.3.81)$$

can be considered, in the hypothesis of no relation $\mathcal{H}$, as a sample of n independent realizations of the Statistic (9.3.80). Therefore, under the random model $\mathcal{H}$, the variance of this sample converges almost surely towards unity, for n tending to infinity. That is,

$$Prob\{lim_{n \to \infty} \frac{1}{n} \sum_{x \in E} \left[\frac{1}{\sqrt{n-1}} \sum_{\{y|y \in E, y \neq x\}} c(x, y) \right]^2 = 1\} = 1 \qquad (9.3.82)$$

The expression of the difference $(V_\rho - V_\beta)$ given by (9.3.75) gives the following.

Theorem 29 *Under the hypothesis of no relation model* $\mathcal{H}$, *asymptotically, almost surely,* $(V_\rho - V_\beta) = o(f)$ *when n tends to infinity, o, being the Landau notation and* $f = card(F)$.

9.3.5.4 Summary of the Different Versions of the Standardized Criterion

Let us summarize. A simplified and nevertheless very efficient version of the adequacy criterion of a partition $\pi(E)$ to a similarity $Q(E)$ on a set E is given by

$$C(Q(E), \pi(E)) = \frac{1}{\sqrt{r(\pi) \times s(\pi)/f}} \sum_{p \in F} \rho(p)c(p) \qquad (9.3.83)$$

where

- ρ is the indicator function in F of the subset $R(\pi)$ of the element pairs of E put together by the partition $\pi(E)$;
- $r(\pi) = card[R(\pi)]$ and $s(\pi) = card[S(\pi)]$, where $S(\pi)$ is the set of element pairs separated in different clusters of π $(r(\pi) + s(\pi) = f = card(F))$;
- $c(p)$ is the normalized version of the similarity $Q(E)$.

Consider now the general form of the similarity $Q(E)$ given by the table

$$\{Q(p)|p \in F\} \qquad (9.3.84)$$

The table

$$\{c(p)|p \in F\} \qquad (9.3.85)$$

is obtained from the transformation

$$c(p) = \frac{Q(p) - mean_e(Q)}{\sqrt{var_e(Q)}}$$

where

$$mean_e(Q) = \frac{1}{f} \sum_{p \in F} Q(p)$$

$$var_e(Q) = \frac{1}{f} \sum_{p \in F} [Q(p) - mean_e(Q)]^2$$

thereby, $mean_e(Q)$ and $var_e(Q)$ are the empirical mean and variance of Q on F.

There are two principal alternatives for the definition of Q: *ordinal* and *numerical*. For the ordinal case, Q is defined by the *rank* function k (see the general Sect. 9.3.4 and Sect. 9.3.5.1). There are two subcases for the ordinal version. The first one is associated with the case for which the preordonance $\varpi(E)$ is a total order on $F = P_2(E)$ (ranking without ties) and the second one, to the case, for which $\varpi(E)$ is associated with a total preorder on $F = P_2(E)$ (ranking with ties) (see Sect. 9.3.5.1). In both cases $mean_e(k)$ is the same, namely, $(f + 1)/2$. However, the respective variances are different depending on whether the ranking k is without or with ties. In the former case, this variance is equal to $(f - 1)(f + 1)/12$. In the latter case the variance has to be calculated directly from (9.3.65).

As mentioned above, in the case of numerical similarity index Q on E, whatever the latter is, the associated coefficient c is obtained by what we have called the "Similarity Global Reduction" (see Sect. 5.2.1.5 of Chap. 5).

The reduction of $C(\rho, \pi)$ was obtained using the variance V_β (see Eq. (9.3.73)). A more accurate reduction can be based on V_ρ (see the same equations). We have for the latter

$$V_\rho \simeq \frac{r(\pi)s(\pi)}{f} + \left[\sum_{1 \le i \le k} \pi_i^3 - \left(\sum_{1 \le i \le k} \pi_i^2 \right)^2 \right] \sum_{(p,q) \in G} c(p)c(q) \qquad (9.3.86)$$

where the sum over G can be expressed by a double sum on E elements (see (9.3.77)).

This result can also be retrieved by applying one of the formulas (6.2.83) or (6.2.85) of Sect. 6.2.5.3 of Chap. 6, obtained in the general case of comparing two valuated symmetrical binary relations ξ and η on E. Instead of $F = P_2(E)$, the reference set becomes that $E^{[2]}$ of ordered pairs of distinct elements:

$$E^{[2]} = E \times E - \delta(E \times E)$$

where $\delta(E \times E)$ is the diagonal of $E \times E$.

Relative to above, ξ and η of Sect. 6.2.5.3 will stand for c and ρ, respectively (see Eq. (9.3.72)). Thus, the random coefficient corresponding to $C(c, \rho^\star)$ can be written as $s(\xi, \eta^\star)$. The reader may be interested in making explicit the equivalence between expressions such as (6.2.83) or (6.2.85) and that (9.3.86).

Let us return to the expression (9.3.83) of the criterion. The computational complexity of the latter is $O(n^2)$. In fact, the basic component in its calculation can be reduced to be the sum of the dissimilarities of the element pairs separated by the partition π. In the case of very large data sets it would be interesting to have a linear complexity version $O(n)$ of the criterion. This can be carried out in the case of clustering objects described by numerical and/or categorical attributes and for a specific distance function on the object set [27]. This distance corresponds, for categorical attributes, to the classical distance function between objects described by Boolean attributes, associated with the categorical values. A crucial notion in this linear complexity adaptation of the criterion turns out to be the *centroid* (centre of gravity) of a set of objects described by categorical attributes [14]. Moreover, a transformation such that the passage from (10.3.5) to (10.3.4) of Chap. 10 is employed.

9.3.6   "Significant Levels" and "Significant Nodes" of a Classification Tree

9.3.6.1   Meaning and Detection

The criterion $C(Q, \pi)$ (see (9.3.83)) enables the adequacy between a partition π and a similarity Q on a set E to be evaluated in terms of statistical association. The consistency of the different classes, their number and their respective sizes influence the scale value of the criterion. Let us imagine two clusterings π_1 and π_2 on E; the former with about ten classes and the latter, three classes. Both clusterings are assumed to be coherent and clearly interpretable. Nevertheless, the partition π_2 gives a more global decomposition than π_1. In these conditions, a bigger value of the criterion might be expected for π_2 than for π_1. However, when the class numbers of π_1 and π_2 are near, without extreme size differences, only the respective class consistencies have an effect on the criterion value.

In fact, in real data, the problem of a unique real number of clusters is the most often, a *FALSE* problem. There might be several real numbers of clusters of different partitions mutually related. A typical case is that given by partitions carried out in hierarchical clustering. Clearly, the situation is different in nature for simulated data [32]. Otherwise, for a given clustering, the different clusters have not necessarily the same consistency.

Now, let us anticipate somewhat with respect to the content of Chap. 10. The latter concerns the *Ascendant Agglomerative* construction of a classification tree on a set E endowed with a numerical or ordinal similarity. E may be a set of objects, categories or attributes. Mostly, this construction is based on a dissimilarity function δ between disjoint parts of E. δ is established from the similarity Q on E (see the introductory Sect. 10.1 of Chap. 10). Generally, a required property for δ is to be without reversals, that is to say, the minimal value $\delta_{\min}(l)$ between class pairs of the lth level of the classification tree is greater than that of the preceding level $l - 1$. Therefore, this property is satisfied if and only if from one level to the next, the value of $\delta_{\min}$ increases.

Interpreting directly a classification tree is not an easy task, especially when the size of the set to be clustered increases. Denote by $\tau = (P_1, P_2, \ldots, P_l, \ldots, P_m)$ the partition chain associated with the building of a classification tree on E. This construction, where at each level the nearest class pairs are fusioned, has a systematic character. However, some partitions represent equilibrium states in the automatic synthesis and others represent passages between two achievement states. Otherwise, from one level to the next one, new nodes arise (a single one the most often). These may correspond to cluster completions at a given degree of synthesis. A level giving rise to a partition corresponding to an achievement state will be called "significant level". On the other hand, a node which corresponds to a cluster completion at a given level will be called "significant node". More precise statements of these intuitive notions and specially, the method by which they are discerned, will be given below.

The increasing evolution of $\delta_{\min}(l)$ values, $1 \leq l \leq m$, along the level sequence of the classification tree, cannot really allow "significant" levels or "significant" nodes to be repaired. Many taxonomist researchers propose to examine the distribution of the rate $\delta_{min}(l) - \delta_{min}(l-1)$, $1 \le l \le m$. This practice might be interesting. Nevertheless, how to specify a justified strategy permitting us to compare two consecutive differences of δ_{min} and to give a clear meaning to this comparison. The principal question is: Why the *local* merging criterion concerned in the building of a classification tree can also play the role of evaluator of the quality of the associations and partitions obtained?

Differently, the criterion $C(Q, \pi)$ (see (9.3.83)) is very general and logically independent of the particular *local* merging criterion δ devoted to cluster formation. It only depends on the similarity (resp., dissimilarity) function defined on E. Two main versions were developed above for this criterion: *ordinal* and *numerical*.

As emphasized above, the criterion $C(Q, \pi)$ has a statistical signification. It refers to a hypothesis of no relation (also called "independence hypothesis"). In this, a random partition is associated with an observed one (for example obtained at a given level of the classification tree). For this hypothesis, the similarity Q (ordinal or numerical) is invariant. Thus, the data description does not take part in the definition of the hypothesis of no relation. This is not the case of *Cubic Clustering Criterion* (*CCC*) [39]. First, the latter criterion concerns only numerical data description (i.e. description is made by numerical attributes). Second, the hypothesis of no relation considered for this criterion is much farer from data observation than the case is for the criterion $C(Q, \pi)$. In fact, this hypothesis considers a uniform distribution of the numerical attributes on a hyperbox (a p-dimensional right parallelepiped (p is the number of attributes)). In [33], $C(Q, \pi)$ has proved its better adaptation and ability, with respect to *CCC*, in order to detect, in a hierarchical process, the most relevant partitions. Boolean data associated with categorical attributes description were concerned. In this case, applying *CCC* criterion required correspondence factorial analysis for transforming the qualitative description into a numerical one.

Relative to the partition chain τ considered above, built by means of a merging criterion δ, the empirical distribution of the criterion $C(Q, \pi)$ is examined on the level sequence of the associated classification tree on E. This sequence can be written as

$$\mathcal{C} = \{C(Q, P_l) | 1 \le l \le m\} \tag{9.3.87}$$

The behaviour of the distribution of $C(Q, P_l)$ on the increasing sequence of the levels of the classification tree, observed through manifold real examples, shows a general tendency to go on increasing until *distinctive local maximums* slowly reached. These might be followed by an abrupt decrease (see Fig. 9.2). They determine "significant" levels. A very important one among them corresponds to the level realizing the *global* maximum. Over the beginning sequence of levels, where the general tendency of $C(Q, P_l)$ is to grow, it is possible to observe fluctuations that indicate local maxima and minima as the graph shown in Fig. 9.2 suggests.

A given level l, where $C(Q, P_l)$ shows a diminution, indicates that the corresponding partition P_l fits the similarity Q, less well than the preceding one P_{l-1}. Assuming P_l deduced from P_{l-1} by fusioning two clusters, it is difficult to interpret in this case the link between the two classes of P_{l-1} joined together to form P_l at the synthesis

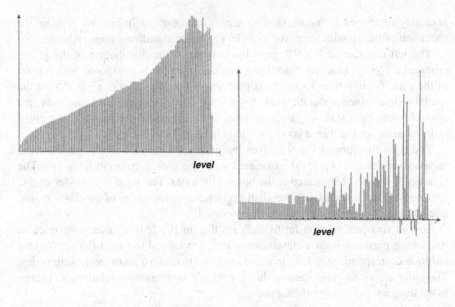

Fig. 9.2 Distribution of the *global* and *local* level criteria

degree given by the classification P_l. Therefore, the data expert who wants a partition with a number of classes around a given value k will retain the tree level associated with a local maximum of $C(Q, P_l)$ corresponding to about k classes.

However, it is not the behaviour of the "global" criterion $C(Q, P_l)$ which allows the reduction of the classification tree to its most pertinent levels, but that of a "local" statistic related to $C(Q, P_l)$ and defined as the variation rate of $C(Q, P_l)$. Designating it by τ_l we have

$$\tau_l = \tau(P_{l-1}, P_l) = C(Q, P_l) - C(Q, P_{l-1}) \tag{9.3.88}$$

$1 \leq l \leq m$.

In these conditions, the "significant" nodes are defined by the local maxima of the distribution

$$\mathcal{T}_{au} = \{\tau_l | 1 \leq l \leq m\} \tag{9.3.89}$$

More precisely, if, for a given k ($1 \leq k \leq m$), τ_k is such a local maximum, the new nodes—very often one—formed at the kth level are considered equally significant. In fact, experiment has shown that the value of $\tau(P_{l-1}, P_l)$ increases when a cluster, having some consistency, which is formed in P_{l-1} grows in the passage from P_{l-1} to P_l and decreases perceptibly, when such a cluster, having some consistency, drops off in favour of a rising embryo of another cluster. Therefore, significant nodes indicate completion stages of the different classes appearing in the tree. Experience shows that a class which comprises several significant nodes is clearly structured for understanding aspect. In these conditions, we propose the reduction of the classification tree to its levels where significant nodes appear.

The left diagram of Fig. 9.2 gives the graphic of the distribution of the global criterion $C(Q, P_l)$ in a real case. Each vertical segment is associated with a level of the classification tree. Its height is proportional to the value of $C(Q, P_l)$ for the partition determined at the lth level, $1 \le l \le m$. The significant levels are indicated at the bottom by a star. As just mentioned above, these levels enable, for a given order of a cluster number, a good partition choice to be done.

The right diagram of Fig. 9.2 gives the graphic of the distribution of the local variation criterion $\tau(P_{l-1}, P_l)$ associated with the global criterion $C(Q, P_l)$. The significant nodes are indicated at the bottom by a star. The significant nodes enable dynamic interpretation of the ascendant agglomerative formation of the classification tree (from the leaves to the root) to be processed.

To end this presentation let us indicate that in [9, 25], in order to extract an interesting partition from a classification tree, we tended to forget the notion of a partition corresponding exactly to a tree level and to retain a set of interesting nodes. The latter are either significant or directly related by branching relation, according to the diagram tree, to significant nodes.

9.3.6.2 Example

Let us refer to the classification tree built in Sect. 10.4.1.3 (see Fig. 10.5). This figure is reconsidered here in Fig. 9.3. This tree is built by a specific merging criterion on a set $\mathcal{A} = \{a^1, a^2, \dots, a^8\}$ of eight Boolean attributes. We shall illustrate the calculation of the criteria $C(Q, \pi)$ and $\tau(P_{l-1}, P_l)$ in the case where the data is an ordinal similarity. The latter, denoted by $\varpi(A)$, is the total preordonance associated with Table 10.7, or equivalently Table 10.8. It can be expressed as follows:

$$\varpi(A) : 58 < 18 < 57 < 13 < 35 < 17 < 34 < 23 < 45 \simeq 47 < 67 \simeq 68 < 26 < 14$$
$$< 46 < 36 < 37 < 16 < 25 \simeq 27 \simeq 28 < 48 < 56 < 24 < 38 < 12 < 78 < 15$$
$$(9.3.90)$$

where jk $(j < k)$ indicates the unordered pair $\{a^j, a^k\}$, $1 \le j < k \le 8$.

The level partition sequences of the classification tree shown in Fig. 9.3 are

$$P_0 = \{\{1\}, \{2\}, \{3\}, \{4\}, \{5\}, \{6\}, \{7\}, \{8\}\}$$
$$P_1 = \{\{1, 5\}, \{2\}, \{3\}, \{4\}, \{6\}, \{7\}, \{8\}\}$$
$$P_2 = \{\{1, 5\}, \{2\}, \{3\}, \{4\}, \{6\}, \{7, 8\}\}$$
$$P_3 = \{\{1, 5\}, \{2, 4\}, \{3\}, \{6\}, \{7, 8\}\}$$
$$P_4 = \{\{1, 5\}, \{2, 4\}, \{3, 7, 8\}, \{6\}\}$$
$$P_5 = \{\{1, 5, 6\}, \{2, 4\}, \{3, 7, 8\}\}$$

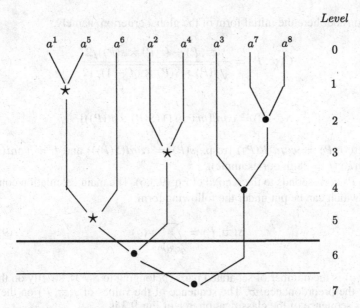

Fig. 9.3 Significant levels and nodes in a classification tree

$$P_6 = \{\{1, 2, 4, 5, 6\}, \{3, 7, 8\}\}$$
$$P_7 = \{\{1, 2, 3, 4, 5, 6, 7, 8\}\} \tag{9.3.91}$$

The associated sequence of the pair subsets joined together by the different partitions is

$$R(P_0) = \emptyset$$
$$R(P_1) = \{15\}$$
$$R(P_2) = \{15, 78\}$$
$$R(P_3) = \{15, 24, 78\}$$
$$R(P_4) = \{15, 24, 37, 38, 78\}$$
$$R(P_5) = \{15, 16, 56, 24, 37, 38, 78\}$$
$$R(P_6) = \{12, 14, 15, 16, 24, 25, 26, 45, 46, 56, 37, 38, 78\}$$
$$R(P_7) = \{12, 13, 14, 15, 16, 17, 18, 23, 24, 25, 26, 27, 28\}$$
$$\cup \{34, 35, 36, 37, 38, 45, 46, 47, 48, 56, 57, 58, 67, 68, 78\} \tag{9.3.92}$$

On the other hand, by considering set theoretic difference, we obtain

$$S(P_l) = F - R(P_l), 0 \le l \le 7 \tag{9.3.93}$$

We consider here the initial form of the global criterion, namely,

$$C(\varpi, P) = \frac{s(\varpi, P) - (r(P) \times s(P))/2}{\sqrt{r(P) \times s(P) \times (f+1)/12}} \qquad (9.3.94)$$

where

$$s(\varpi, P) = card[gr(\varpi) \cap (S(P) \times R(P))]$$

and where $r(P) = card(R(P))$ (resp., $s(P) = card(S(P))$) and $f = card(F) = r(P) + s(P)$ ($f = 28$ in our example).

$C(\varpi, P)$ corresponds to the second of Eq. (9.3.6). The main calculation concerns $s(\varpi, P)$ which can be put under the following form:

$$s(\varpi, P) = \sum_{q \in R(P)} \nu(q) \qquad (9.3.95)$$

where $\nu(q)$ is the number of separated pairs (belonging to $S(P)$), strictly on the left of q, for the preordonance ϖ. The sequence of the values of $s(\varpi, P)$ on the level partitions sequence of the classification tree of Fig. 9.3 is

$s(\varpi, P_0) = \underline{0}$
    $s(\varpi, P_1) = \underline{27}$
    $s(\varpi, P_2) = 26 + 26 = \underline{52}$
    $s(\varpi, P_3) = 25 + 23 + 25 = \underline{73}$
    $s(\varpi, P_4) = 23 + 22 + 16 + 22 + 23 = \underline{106}$
    $s(\varpi, P_5) = 21 + 16 + 20 + 20 + 16 + 20 + 21 = \underline{134}$
    $s(\varpi, P_6) = 15 + 11 + 15 + 12 + 15 + 12 + 11 + 8 + 11 + 15 + 12 + 15 + 15 = \underline{167}$
    $s(\varpi, P_7) = \underline{0}$ (9.3.96)

Notice that for $s(\varpi, P) = 0$, we have

$$C(\varpi, P) = -\sqrt{\frac{3r(P)s(P)}{[r(P) + s(P) + 1]}}$$

On the other hand, $C(\varpi, P) = 0$ if the product $r(P) \times s(P)$ is equal to 0 (Table 9.4).

The most significant level is the fifth one. However, some consideration might be taken into account for the partition of the sixth level, because the drop of the global criterion $C(\varpi, P)$ in the passage from the fifth level to the sixth one is very weak. There are three significant nodes indicated by a star in Fig. 9.3. As expressed above, they correspond to completion stages of clusters at a given synthesis degree.

In the case where the total preorder ϖ on $F = P_2(E)$ is not fine enough, the mean rank function must be used to code ϖ. In the example above the 28 successive values of this function are

Table 9.4 Global and local level criteria distributions

l	$C(\varpi, P)$	$\tau(l-1, l)$
0	0.000	
1	1.671	1.671
2	2.139	0.468
3	2.637	0.498
4	2.909	0.272
5	3.210	0.301
6	3.202	−0.008
7	0.000	−3.202

1, 2, 3, 4, 5, 6, 7, 8, 9.5, 9.5, 11.5, 11.5, 13, 14, 15, 16, 17, 18, 20, 20, 20, 22, 23, 24, 25, 26, 27, 28.

As an exercise the reader is asked to calculate the value of the global criterion, associated with the partition P_5, with this coding of ϖ (see Sects. 9.3.5.1, 9.3.5.4 and (9.3.83)). Let me indicate that in the *CHAVLH* software (see Chap. 11), a numerical version of the similarity on E is directly exploited (see Sect. 9.3.5.4).

9.4 Measuring the Fitting Quality of a Partition Chain (Classification Tree)

9.4.1 Introduction

As previously discussed, we consider a finite set E endowed with an ordinal or numerical similarity. The criteria developed in Sect. 9.3 will be generalized in the case where the matter consists of evaluating the adequacy of a partition chain with a similarity on E, the latter being ordinal or numerical. First, in Sect. 9.4.2.1, we will extend the *intersection* and the *symmetrical difference* criteria of Sect. 9.3.3. In this case, the similarity data considered is a total preordonance $\omega(E)$ on E; that is to say, a total preorder on the set $F = P_2(E)$. In Sect. 9.4.2.2, we will show how to associate a quality measure of a partition chain with an association coefficient between Boolean attributes (see Sect. 4.2.1). The nature of this formal correspondence is the same as that considered in Sect. 4.3.4, and more particularly, in Sect. 4.3.4.1.

The distributional analysis of Sects. 9.3.4 and 9.3.5 and the consequent criteria obtained will be extended in Sect. 9.4.3 for evaluating the fit of a partition chain to a similarity (ordinal or numerical) on E.

Formal ordinal criteria called "lateral order" and "lexicographic order" criteria are defined in Sect. 9.4.4. These could have been expressed above in the context of a partition search. Nevertheless, their presentation for a chain partition search is more general and then, the partition case can be derived from the general treatment. In this, we will establish a mathematical correspondence between the lexicographic order criterion and the classical inversion number criterion which is strongly related to the symmetrical difference criterion.

Let us recall—for the development which follows—that the representation we adopt for a partition chain $\tau = (P_0, P_1, \ldots, P_l, \ldots, P_m)$ is a *ultrametric preordonance* denoted by ω_u (see Chap. 1, Sect. 1.4.3). If $(R_1, R_2, \ldots, R_l, \ldots, R_m)$ is the sequence of the beginning intervals of this preordonance, we have

$$R_1 \subset R_2 \subset \cdots \subset R_l \subset \cdots \subset R_m \tag{9.4.1}$$

where R_l is the set of unordered element pairs of E joined together by the partition P_l, $1 \le l \le m$.

9.4.2 Generalization of the Set Theoretic and Metrical Criteria

These generalizations are the direct consequence of the set theoretic representation of the preordonance $\omega(E)$ and the ultrametric preordonance ω_u associated univocally with a partition chain on E, resulting from building a classification tree on E (see Chap. 10). The latter can be denoted by $\tau(E)$, or more simply, by τ if there is no ambiguity. We assume here the preordonance $\omega(E)$ established in such a way that the rank of a given pair $\{x, y\}$, element of F, is an increasing function of the dissimilarity between x and y (see the second of Eq. (9.3.2)). These representations are of the same nature as those in Sect. 3.4.1 of Chap. 3 (see (3.4.3)–(3.4.6)) and Sect. 4.3.4 of Chap. 4 (see (4.3.34) and (4.3.35)). However, here, both preordonances are defined on the set E, whereas, in the references cited, the preordonances concerned are defined, respectively, on two value sets of two categorical attributes describing the same object set $\mathcal{O}$. Otherwise, we adopt here a strict representation of a total preordonance ω on E by the set of ordered pairs (p, q) of $F \times F$ such that p precedes strictly q for ω, exactly

$$gr(\omega) = \{(p, q) | (p, q) \in F \times F, p < q \text{ and } \neg ((p \sim q) \text{ or } (q > p))\} \tag{9.4.2}$$

where $p \sim q$ indicates that p and q are in the same preorder class.

9.4.2.1 Generalization of the *Intersection* and *Symmetrical Difference* Criteria

Let (ω, τ) be a pair composed of a total preordonance ω and a classification tree τ on a finite set E. The graph representation $gr(\omega)$ of ω has just been given above. Mathematically, τ defines a partition chain that we have denoted by

$\tau = (P_0, P_1, \ldots, P_l, \ldots, P_m)$, where P_l is the partition of the lth level of the classification tree, $1 \leq l \leq m$. According to above, R_l denotes the set of unordered element pairs of E joined by the partition P_l, $1 \leq l \leq m$. The strict graph representation of the ultrametric preordonance associated with τ is

$$gr(\omega_u) = \sum_{1 \leq k < l \leq m} T_k \times T_l \qquad (9.4.3)$$

where $T_k = R_k - R_{k-1}$ is the set of element pairs of E joined "for the first time" at the kth level of the classification tree τ. In these conditions, the *raw* index generalizing $B(\omega, \pi)$ (see the first of Eq. (9.3.7)) can be written as

$$s(gr(\omega), gr(\omega_u)) = card\left(gr(\omega) \cap \sum_{1 \leq k < l \leq m} T_k \times T_l\right)$$

$$= \sum_{1 \leq k < l \leq m} card\left(gr(\omega) \cap (T_k \times T_l)\right) \qquad (9.4.4)$$

By the same way, the symmetrical difference criterion can be written as

$$d(gr(\omega), gr(\omega_u)) = card\left(gr(\omega) \Delta \sum_{1 \leq k < l \leq m} T_k \times T_l\right)$$

$$= \sum_{1 \leq k < l \leq m} card\left(gr(\omega) \Delta (T_k \times T_l)\right) \qquad (9.4.5)$$

where Δ denotes the set symmetrical difference. We have

$$d(gr(\omega), gr(\omega_u)) = card(gr(\omega)) + card\left(\sum_{1 \leq k < l \leq m} T_k \times T_l\right) - 2s(gr(\omega), gr(\omega_u))$$

Introduce now the set $F^{[2]}$ composed of all ordered pairs (p, q) of F whose components are different $(p \neq q)$: $F^{[2]} = F \times F - \delta(F \times F)$, where $\delta(F \times F)$ is the diagonal of $F \times F$. From the previous relation, it follows

$$card(F^{[2]}) - d(gr(\omega), gr(\omega_u))$$

$$= card(F^{[2]}) - card(gr(\omega)) - card\left(\sum_{1 \leq k < l \leq m} T_k \times T_l\right) + 2s(gr(\omega), gr(\omega_u))$$

Therefore, we have the following:

Proposition 50 *The criterion defined by the cardinal the complement in $F^{[2]}$ of the symmetrical difference set between $gr(\omega)$ and $gr(\omega_u)$ is equivalent to the criterion*

$$C(gr(\omega), gr(\omega_u)) = s(gr(\omega), gr(\omega_u)) - \frac{1}{2}card\left(\sum_{1 \le k < l \le m} T_k \times T_l\right)$$

Clearly, this proposition generalizes that Proposition 44. We have preferred here to consider the reference space as $F^{[2]}$ instead of $F \times F$, as considered in the previous proposition. In the latter we have tried to follow De La Véga proposal. By considering the indicator functions in $F^{[2]}$, I_ω of $gr(\omega)$ and I_τ of $gr(\omega_u)$, we may write

$$s(gr(\omega), gr(\omega_u)) = \sum_{(p,q) \in F^{[2]}} I_\omega(p, q) I_\tau(p, q) \qquad (9.4.6)$$

9.4.2.2 Criteria Defined by the Generalization of Classical Association Coefficients

These criteria were defined at the end of Sect. 9.3.1, in the case where the sought structure is a partition π on E. We turn now to a more general case where the structure concerned is a partition chain τ, practically obtained from building a classification tree τ on E (see Chap. 10). All the considerations relative to the evaluation of a partition π will be transposed and adapted for evaluating a partition chain τ. Thus, $\Omega(\omega, \tau)$ will designate the set of ordered pairs (p, q) of $F^{[2]}$, such that p and q are strictly comparable for ω and τ. By referring to the notations above:

$$gr(\omega) = \{(p, q) | (p, q) \in F^{[2]} \text{ and } p < q \text{ for } \omega\}$$
$$gr(\varpi) = \{(p, q) | (p, q) \in F^{[2]} \text{ and } p > q \text{ for } \omega\}$$
$$gr(\omega_u) = \sum_{1 \le k < l \le m} T_k \times T_l$$
$$gr(\overline{\omega_u}) = \sum_{1 \le k < l \le m} T_l \times T_k \qquad (9.4.7)$$

where $\overline{\omega_u}$ is the preordonance ω_u, reversed. We have

$$\Omega(\omega, \tau) = [gr(\omega) + gr(\varpi)] \cap [gr(\omega_u) + gr(\overline{\omega_u})]$$
$$= gr(\omega) \cap gr(\omega_u)$$
$$+ gr(\omega) \cap gr(\overline{\omega_u})$$
$$+ gr(\varpi) \cap gr(\omega_u)$$
$$+ gr(\varpi) \cap gr(\overline{\omega_u}) \qquad (9.4.8)$$

where—as above—the symbol "+" expresses a set sum.

The cardinal parameters associated with this equation decomposition (9.4.8) are

$$s(\omega, \tau) = card[gr(\omega) \cap gr(\omega_u)]$$
$$u(\omega, \tau) = card[gr(\omega) \cap gr(\overline{\omega_u})]$$
$$v(\omega, \tau) = card[gr(\varpi) \cap gr(\omega_u)]$$
$$t(\omega, \tau) = gr(gr(\varpi)) \cap gr(\overline{\omega_u}) \qquad (9.4.9)$$

As for the partition case (see the equation following (9.3.17)), we have

$$s(\omega, \tau) = t(\omega, \tau) \text{ and } u(\omega, \tau) = v(\omega, \tau)$$

Obviously, what is said in the final part of Sect. 9.3.1 applies immediatly for evaluating the adequacy of a partition chain τ to ω. In particular, the Goodman and Kruskal coefficient—which has to be denoted here by $\gamma(\omega, \tau)$—corresponds exactly to (9.3.18), except that τ has to be substituted for π.

Clearly, the parameters of Eq. (9.4.9) can be generalized in order to compare two arbitrary total preorders on a finite set, whatever is the latter. To fix ideas, we may always continue to consider F this set. Designating by α and β these two total preorders on F, Eq. (9.4.9) is generalized as follows:

$$s(\alpha, \beta) = card[gr(\alpha) \cap gr(\beta)]$$
$$u(\alpha, \beta) = card[gr(\alpha) \cap gr(\overline{\beta})]$$
$$v(\alpha, \beta) = card[gr(\overline{\alpha}) \cap gr(\beta)]$$
$$t(\alpha, \beta) = card(gr(\overline{\alpha})) \cap gr(\overline{\beta}) \qquad (9.4.10)$$

where the notations are analogous to those of Eq. (9.4.9). $u(\alpha, \beta)$ (resp., $v(\alpha, \beta)$) $(u(\alpha, \beta) = v(\alpha, \beta))$ is commonly called the *inversion number* of α with respect to β.

9.4.3 Distribution of the Cardinality of the Graph Intersection Criterion

9.4.3.1 Introduction

To begin, we consider here as in Sect. 9.3.3 that the preordonance ω on E is a total ordonance defining a total order on $F = P_2(E)$. Let us recall that the graph representation of ω in $F^{[2]}$ is

$$gr(\omega) = \{(p, q) | (p, q) \in F^{[2]} \text{ and } p < q \text{ for } \omega\} \qquad (9.4.11)$$

(see the first of Eq. (9.3.3)).

For matching a partition π on E with ω, a consistent and realistic hypothesis of no relation was considered in Sect. 9.3.3. In the latter ω is fixed and a random partition $\pi^\star$ on E is associated with π in the set Π of all partitions having the same type t as that of π, the randomness being with equal probabilities. Here, more precisely in Sect. 9.4.3.2, we will begin by a dual form of this hypothesis of no relation. In this, the partition chain τ on E is fixed and a random total order $\omega^\star$ on F is associated with ω in the set of all possible total orders on $F^{[2]}$, with uniformly distributed probability. The expression of this hypothesis, which can be implicitly derived from [20], is more general than that considered in [29] where to the numerical dissimilarity table on E corresponds a random one obtained by permuting the dissimilarity values on $F^{[2]}$. Clearly, the results we will obtain in this framework apply when a single partition is faced with ω.

By coding a total preorder ω on F with the "mean rank function" (see Sect. 3.4.1 of Chap. 3 and Sect. 4.3.4.2 of Chap. 4), the statistical study of Sect. 9.3.4 can be generalized in the case of comparing ω with a partition chain τ (see Sect. 9.4.3.3). An additional extension concerns the case where a numerical similarity is substituted for ω.

9.4.3.2   Distribution of $s(\omega^\star, \tau)$ in the Case Where ω Is a Total Order

As just expressed above, we start with the case where the ordinal similarity data is represented by a total ordonance ω on E; that is to say, a complete order (ranking without ties) on the set $F = P_2(E)$ of unordered element pairs of E. For this ranking, the higher is the rank of a given pair $\{x, y\}$ ($\{x, y\} \in F$), the smaller (resp., the greater) is the similarity value (resp., the dissimilarity value) between the two components x and y. The graph representation of ω in $F^{[2]}$ is given by (9.4.10). Equation (9.4.3) provides the graph representation of the ultrametric preordonance ω_u associated with a partition chain τ, generally obtained from building a classification tree on E (see Chap. 10). Let us rewrite here the *raw* association coefficient $s(\omega, \omega_u)$ (see (9.4.4)) under the form given by (9.4.6):

$$s(\omega, \tau) = \sum_{(p,q)\in F^{[2]}} I_\omega(p, q) I_\tau(p, q) \qquad (9.4.12)$$

where I_ω and I_τ are the indicator functions in $F^{[2]}$ of ω and ω_u, respectively. The latter equation gives for the random raw coefficient

$$s(\omega^\star, \tau) = \sum_{(p,q)\in F^{[2]}} I_{\omega^\star}(p, q) I_\tau(p, q) \qquad (9.4.13)$$

Due to

$$\mathcal{E}\left(I_{\omega^\star}(p, q)\right) = \frac{1}{2}$$

where $\mathcal{E}$ designates mathematical expectation, we have

$$\mathcal{E}\left(s(\omega^\star, \tau)\right) = \frac{1}{2} \times \sum_{1 \le k < l \le m} card(T_k) \cdot card(T_l) \qquad (9.4.14)$$

As for comparing two ordinal categorical attributes (see Sect. 6.2.4.3 of Chap. 6), the calculation of the second absolute moment of $s(\omega^\star, \tau)$ requires the following decomposition of $F^{[2]} \times F^{[2]}$:

$$F^{[2]} \times F^{[2]} = I + I' + G_1 + G'_1 + G_2 + G'_2 + H \text{ set sum}$$

where each set of this decomposition comprises elements of the same form (structure). The respective forms concerned by I, I', G_1, G'_1, G_2, G'_2 and H, are $((p, q), (p, q))$, $((p, q), (p, r))$, $((p, q), (r, p))$, $((p, q), (r, q))$, $((p, q), (q, r))$ and $((p, q), (r, s))$, where distinct letter symbols indicate different elements of F. The respective cardinalities of the sets of this decomposition are $f(f-1)$, $f(f-1)$, $f(f-1)(f-2)$, $f(f-1)(f-2)$, $f(f-1)(f-2)$, $f(f-1)(f-2)$ and $f(f-1)(f-2)(f-3)$. One can verify that the sum of these is equal to $[f(f-1)]^2$.

In these conditions, we have to calculate

1. $\mathcal{E}[I_{\omega^\star}(p, q) \cdot I_{\omega^\star}(p, q)]$
2. $\mathcal{E}[I_{\omega^\star}(p, q) \cdot I_{\omega^\star}(q, p)]$
3. $\mathcal{E}[I_{\omega^\star}(p, q) \cdot I_{\omega^\star}(p, r)]$
4. $\mathcal{E}[I_{\omega^\star}(p, q) \cdot I_{\omega^\star}(r, p)]$
5. $\mathcal{E}[I_{\omega^\star}(p, q) \cdot I_{\omega^\star}(r, q)]$
6. $\mathcal{E}[I_{\omega^\star}(p, q) \cdot I_{\omega^\star}(q, r)]$
7. $\mathcal{E}[I_{\omega^\star}(p, q) \cdot I_{\omega^\star}(r, s)]$

$((p, q), (p', q'))$ being a generic element of $F^{[2]} \times F^{[2]}$, the respective values of $\mathcal{E}[I_{\omega^\star}(p, q) \cdot I_{\omega^\star}(p', q')]$, according to the structure of $((p, q), (p', q'))$ are:

1. $\mathcal{E}[I_{\omega^\star}(p, q) \cdot I_{\omega^\star}(p', q')] = \frac{1}{2}$ if $((p, q), (p', q')) \in I$
2. $\mathcal{E}[I_{\omega^\star}(p, q) \cdot I_{\omega^\star}(p', q')] = 0$ if $((p, q), (p', q')) \in I'$
3. $\mathcal{E}[I_{\omega^\star}(p, q) \cdot I_{\omega^\star}(p', q')] = \frac{1}{3}$ if $((p, q), (p', q')) \in G_1$
4. $\mathcal{E}[I_{\omega^\star}(p, q) \cdot I_{\omega^\star}(p', q')] = \frac{1}{6}$ if $((p, q), (p', q')) \in G'_1$
5. $\mathcal{E}[I_{\omega^\star}(p, q) \cdot I_{\omega^\star}(p', q')] = \frac{1}{3}$ if $((p, q), (p', q')) \in G_2$
6. $\mathcal{E}[I_{\omega^\star}(p, q) \cdot I_{\omega^\star}(p', q')] = \frac{1}{6}$ if $((p, q), (p', q')) \in G'_2$
7. $\mathcal{E}[I_{\omega^\star}(p, q) \cdot I_{\omega^\star}(p', q')] = \frac{1}{4}$ if $((p, q), (p', q')) \in H$

Clearly, as the case is for the *mean* of the random raw coefficient $s(\omega^\star, \tau)$ (see (9.4.13)), the expression of the variance of $s(\omega^\star, \tau)$ will depend on the *type* of the partition chain τ. The latter notion is stated in Sect. 3 of [24]. Now, by restricting the calculations above to the case where the classification tree τ is reduced to a partition π on E (or more exactly, to three levels: the leaf level π and the root level), the following result is obtained.

Proposition 51 *The mean and variance of $s(\omega^{\star}, \pi)$, where π is a partition on E, are equal to*

$$\frac{r(\pi) \cdot s(\pi)}{2}$$

$$\frac{r(\pi) \cdot s(\pi) \cdot (f + 1)}{12} \tag{9.4.15}$$

By referring to Sects. 9.3.4.1 and 9.3.4.2, we obtain the following.

Theorem 30 *The standardized version of $s(\omega, \pi)$ with respect to the distribution of $s(\omega^{\star}, \pi)$ is identical to that obtained with respect to the distribution of $s(\omega, \beta^{\star})$, where $\beta^{\star}$ indicates a random symmetrical binary relation on E, uniformly distributed and conditioned by $card[R(\beta^{\star})] = r(\pi)$ ($R(\beta^{\star})$ is a random subset of F representing $\beta^{\star}$).*

Thereby, the statements of Theorems 26 and 27 are retrieved.

9.4.3.3 Criteria Deduced from the Distribution of $s(\omega, \tau^{\star})$ in the General Case

The most natural version of a random raw coefficient associated with $s(\omega, \tau)$ as defined by formula (9.4.11) is

$$s(\omega, \tau^{\star}) = \sum_{(p,q)\in F^{[2]}} I_{\omega}(p, q)I_{\tau^{\star}}(p, q) \tag{9.4.16}$$

where $\tau^{\star}$ is associated with τ in an adequate *hypothesis of no relation*. The latter is defined by a random permutation on the leaf tree set where equal chance is given for each of the $n!$ permutations. The calculation of the mean and variance of $s(\omega, \tau^{\star})$, assuming such a representation in $F^{[2]}$, requires a direct algorithmic computing where a recursive procedure is needed. More precisely, for a given structure of an ordered pair (p, q) in $F^{[2]}$ (resp., of an ordered pair of two ordered pairs $((p, q), (p', q'))$ in $F^{[2]} \times F^{[2]}$) the tree structure τ has to be explored in order to determine subtree occurrences giving rise to clustering specified components of different pairs [28].

This calculation is feasible but somewhat elaborate. A simpler and nevertheless efficient technique is obtained by considering the representation of a total preorder on F by means of the associated "mean rank" function on F (see Sect. 3.4.1 of Chap. 3). This approach corresponds to that considered in Sect. 9.3.4 in Eqs. (9.3.48) and (9.3.49). Notice that here, the total preorder ω is established in such a way that the rank of a pair $\{x, y\}$ of F is an increasing function of the dissimilarity between its two components x and y. On the other hand, the total preorder on F associated with the partition chain $\tau = (P_0, P_1, \ldots, P_k, \ldots, P_m)$ and that we can also denote by τ, without risk of ambiguity, is defined by

$$\tau : T_1 = R_1 < T_2 = R_2 - R_1 < \cdots < T_k = R_k - R_{k-1} < \cdots < T_m = R_m - R_{m-1}$$

where R_k is the set of element pairs of E joined in the partition P_k and where T_k is the set of element pairs of E joined *for the first "time"* in P_k, $1 \leq k \leq m$. In these conditions, the *raw* association coefficient between ω and τ is written as follows:

$$s1(\omega, \tau) = \sum_{p \in F} r_m^\omega(p) \cdot r_m^\tau(p) \tag{9.4.17}$$

where $r_m^\omega(p)$ (resp., $r_m^\tau(p)$) is the mean rank of p for the total preorder ω (rep., τ). Now, let us make more explicit the value of $r_m^\tau(p)$ when p belongs to T_k. We have

$$r_m^\tau(p) = r_{k-1} + \frac{1}{2}(r_k - r_{k-1} + 1) = \frac{1}{2}(r_k + r_{k-1} + 1) \tag{9.4.18}$$

where $r_k = card(R_k) = r(P_k)$ designates the cardinal of element pairs of E joined by the partition P_k, $1 \leq k \leq m$.

Let now E be coded with the set $\mathbb{I} = \{1, 2, \ldots, i, \ldots, n\}$ of the first n integers, the expression of the raw index becomes

$$s1(\omega, \tau) = \sum_{1 \leq i < j \leq m} r_m^\omega(i, j) \cdot r_m^\tau(i, j) \tag{9.4.19}$$

In these conditions, the associated random raw index can be written as

$$s1(\omega, \tau^\star) = \sum_{1 \leq i < j \leq m} r_m^\omega(i, j) \cdot r_m^\tau(\sigma(i), \sigma(j)) \tag{9.4.20}$$

where σ is a random permutation in the set G_n of all permutations on $\mathbb{I} = \{1, 2, \ldots, i, \ldots, n\}$, provided with a uniform distribution.

By introducing

$$\mathbb{I}^{[2]} = \{(i, j) | 1 \leq i \neq j \leq n\}$$

and by setting

$$r_m^\omega(j, i) = r_m^\omega(i, j) \text{ and } r_m^\tau(j, i) = r_m^\tau(i, j)$$

for $1 \leq i < j \leq m$, we obviously obtain

$$s1(\omega, \tau) = \frac{1}{2} s2(\omega, \tau)$$

where

$$s2(\omega, \tau) = \sum_{(i,j) \in \mathbb{I}^{[2]}} r_m^\omega(i,j) \cdot r_m^\tau(i,j) \qquad (9.4.21)$$

Clearly, the normalized coefficient obtained from $s1(\omega, \tau)$ is identical to that obtained from $s2(\omega, \tau)$. The random raw coefficient associated with the latter one can be written as

$$s2(\omega, \tau^\star) = \sum_{(i,j) \in \mathbb{I}^{[2]}} r_m^\omega(i,j) \cdot r_m^\tau(\sigma(i), \sigma(j)) \qquad (9.4.22)$$

where the random permutation σ has been defined above.

Thus, we bring back to the case considered in Sect. 6.2.5 of Chap. 6 where two valuated symmetrical relations ξ and η have to be compared. By substituting in this section r_m^ω for ξ and r_m^τ for η, we derive the expressions of the mean and variance of $s2(\omega, \tau^\star)$. More precisely, Eqs. (6.2.78) and (6.2.79) will give the *mean* and Eqs. (6.2.84) and (6.2.85), the *variance*. Otherwise, (6.2.86) gives rise to the standardized coefficient

$$Q_1(\omega, \tau) = \frac{s2(\omega, \tau) - n^{[2]} \mu(r_m^\omega) \cdot \mu(r_m^\tau)}{\sqrt{var(s2(\omega, \tau^\star))}} \qquad (9.4.23)$$

where $\mu(r_m^\omega)$ and $\mu(r_m^\tau)$ are the respective means of r_m^ω and r_m^τ on F (or equivalently on $\mathbb{I}^{[2]}$).

Additionally, we may propose, for defining a criterion measuring the adequacy of τ to ω, coefficients of the forms $R_1(\omega, \tau)$ and $R_1^{''}(\omega, \tau)$, according to Eqs. (6.2.91) and (6.2.92).

Clearly, by replacing the *mean rank function*, by a *dissimilarity* function on E, we may derive the same above development and obtain a fit criterion between a dissimilarity and a partition chain on E. Such a dissimilarity might be given by the table $\mathcal{D}$ (see (7.2.7)).

Example

We shall illustrate here the calculations of the graph *intersection* and *symmetrical difference* criteria for comparing the respective adequacies of two trees denoted by τ_1 and τ_2 to a given preordonance ω (see hereafter). To begin, let us recall (see Eq. (9.4.9)) that if ω is a given preordonance on a set E and ω_u a ultrametric preordonance associated with a classification tree on E, then the criteria of the graph intersection and symmetrical difference can be written, respectively, as follows:

$$s(\omega, \omega_u) = card[gr(\omega) \cap gr(\omega_u)]$$
$$d(\omega, \omega_u) = card[gr(\omega) \cap gr(\overline{\omega_u})] + card[gr(\varpi) \cap gr(\omega_u)]$$

$$(9.4.24)$$

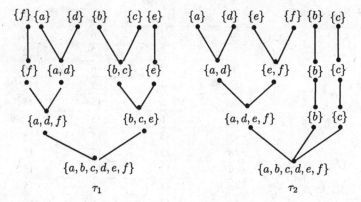

Fig. 9.4 Two trees on the same set

where the different graphs are considered in a strict sense (see Eqs. (9.4.2) and (9.4.3)) and where

$$\left(\forall (p, q) \in F^{[2]}\right), (p, q) \in gr(\omega)[resp., gr(\omega_u)] \Leftrightarrow (q, p) \in gr(\varpi)[resp., gr(\overline{\omega_u})]$$

For ω let us take up again here the preordonance ω_s of Example 1 of Chap. 4. ω_s, established on a set $E = \{a, b, c, d, e, f\}$ of six elements, which is expressed as follows:

$\omega_s : af < ad \sim df < bc < ac \sim ef < ab \sim bd \sim cd \sim be \sim ce \sim de \sim bf < cf \sim ae$

Now, consider the comparison of the respective adequacies of the trees of Fig. 9.4 to ω_s. Both criteria $s(\omega, \omega_u)$ and $d(\omega, \omega_u)$ of Eq. (9.4.23) will be used for this purpose. The ultrametric preordonances, denoted by ω_u^1 and ω_u^2, representing τ^1 and τ^2, respectively, are

$\omega_u^1 : ad \sim bc < af \sim df \sim be \sim ce < ab \sim ac \sim ae \sim bd \sim bf \sim cd \sim cf \sim de \sim ef$

$\omega_u^2 : ad \sim ef < ae \sim af \sim de \sim df < ab \sim ac \sim bc \sim bd \sim be \sim bf \sim cd \sim ce \sim cf$

We obtain Table 9.5 by representing each of the total preorders ω_s, ω_u^1 and ω_u^2, on $F = P_2(E)$, by means of the rank function, as defined in (3.3.28) of Chap. 3. In the following, ω_s will be denoted more simply by ω.

Evaluation of the Intersection Criterion $s(\omega, \omega_u)$

To obtain the value of $s(\omega, \omega_u^1)$ (see Eq. (9.4.23)), we consider the sequence of the elements of F according to the order compatible with ω, as stated in the first column of Table 9.5. For a given element p of F, we describe in the order the elements q of F for which $p < q$ for ω, strictly. When $p < q$ for ω_u^1, the contribution of p to $s(\omega, \omega_u^1)$

Table 9.5 Rank coding of three preordonances

ω	$r(\omega)$	$r(\omega_u^1)$	ω_u^2
af	1	6	6
ad	3	2	2
df	3	6	6
bc	4	2	15
ac	6	15	15
ef	6	15	2
ab	13	15	15
bd	13	15	15
cd	13	15	15
be	13	6	15
ce	13	6	15
de	13	15	6
bf	13	15	15
cf	15	15	15
ae	15	15	6

is increased by one unit. Thus the contribution of *af* is 9, that of *ad* is 11, etc. ..... In these conditions, we have

$$s(\omega, \omega_u^1) = 9 + 11 + 9 + 11 + 3 + 2 + 2 = 47$$

Similarly, we obtain

$$s(\omega, \omega_u^2) = 9 + 11 + 9 + 9 + 1 = 39$$

Evaluation of the Symmetrical Difference Criterion $d(\omega, \omega_u)$

Let ω_u indicate one of the preordonances ω_u^1 and ω_u^2. To calculate the symmetrical difference criterion between ω and ω_u, which we denote by $d(\omega, \omega_u)$, we proceed in the same manner as above. To begin, notice that due to

$$card[gr(\omega) \cap gr(\overline{\omega_u})] = card[gr(\varpi) \cap gr(\omega_u)]$$

we have

$$d(\omega, \omega_u) = 2card[gr(\omega) \cap gr(\overline{\omega_u})]$$

and then, the following criterion values we consider will be those of

$$card[gr(\omega) \cap gr(\overline{\omega_u^1})] \text{ and } card[gr(\omega) \cap gr(\overline{\omega_u^2})]$$

Table 9.6 Mean rank coding of three preordonances

ω	$r(\omega)$	$r(\omega_u^1)$	ω_u^2
af	1	4.5	4.5
ad	2.5	1.5	1.5
df	2.5	4.5	4.5
bc	4	1.5	11
ac	5.5	11	11
ef	5.5	11	1.5
ab	10	11	11
bd	10	11	11
cd	10	11	11
be	10	4.5	11
ce	10	4.5	11
de	10	11	4.5
bf	10	11	11
cf	14.5	11	11
ae	14.5	11	4.5

For a given ω_u ($\omega_u = \omega_u^1$ or ω_u^2) and for each element p in the first column of Table 9.5, the successive elements below p are examined. For each q of them, if $r(\omega)(p) < r(\omega)(q)$ and $r(\omega_u)(p) > r(\omega_u)(q)$ (strictly), the ordered pair (p, q) contributes for one unity to the criterion concerned.

In these conditions, the respective contributions of af, ad, ..., bf and cf to $d'(\omega, \omega_u^1) = \frac{1}{2}d(\omega, \omega_u^1)$ are 2, 0, 1, 0, 2, 2, 0, 0, ..., 0 and 0. Hence,

$$d'(\omega, \omega_u^1) = 2 + 1 + 2 + 2 = 7$$

Similarly, we obtain

$$d'(\omega, \omega_u^2) = 2 + 1 + 3 + 2 + 1 + 1 + 1 + 1 + 1 + 1 = 14$$

Therefore, the tree τ_1 is preferred to the tree τ_1, for both of the criteria.

Table 9.6 is deduced from Table 9.5 by substituting for the rank function the *mean rank* function (see (3.4.8) of Chap. 3).

With this table of values the coefficient $Q_1(\omega, \tau)$ (see (9.4.22)) applies and then $Q_1(\omega, \tau_1)$ and $Q_1(\omega, \tau_2)$ can be calculated—according to the development above—and compared. This is left for the reader interested.

9.4.4 Pure Ordinal Criteria: The Lateral Order and the Lexicographic Order Criteria

We assume given a total preordonance $\omega(E)$ on a finite set E; that is to say, a total preorder on the set $P_2(E)$ of unordered element pairs of E. To fix idea, in the case where $\omega(E)$ follows the definition of a numerical similarity S on E, we suppose $\omega(E)$ established according to the first of Eq. (9.3.2). The *lateral order* criterion is a *partial order* introduced in [6] and reported in [1]. The purpose consists of comparing the respective fits of two partitions π_1 and π_2 to $\omega(E)$. For this comparison π_1 and π_2 are represented by the F subsets $R(\pi_1)$ and $R(\pi_2)$, composed of the unordered distinct element pairs joined together by π_1 and π_2, respectively. Additionally, in this comparison, $card(R(\pi_1))$ and $card(R(\pi_2))$ are supposed to be equal.

As it will appear below, this criterion is very demanding. Its interest is principally formal and theoretical. However, it may be practically interesting in local searches. We shall present it here in a very general way, independently of the matching evaluation problem between a partition and partition chain, to $\omega(E)$. Next, we will illustrate its application for comparing partitions of level 2 of the trees τ_1 and τ_2, with respect to the preordonance ω_s (see the previous example). Finally, we will also consider (using this criterion) the respective global associations of τ_1 and τ_2, with ω_s. These illustrations will depend strongly on the comparison of the shapes of τ_1 and τ_2.

9.4.4.1 The Lateral Order Criterion

Here, F is an abstract finite set whose cardinality is designated by f. Let Ω denote the set of all total preorders on F and let Ω_r be the following set of ordered pairs (ω, R_ω) where ω belongs to Ω and where R_ω is an F subset whose cardinality is r ($r \leq f$):

$$\Omega_r = \{(\omega, R_\omega) | \omega \in \Omega, R_\omega \subset F \text{ and } card(R_\omega) = r\} \qquad (9.4.25)$$

For two elements (ω, R_ω) and $(\omega', R_{\omega'})$ of Ω_r, an increasing bijection ϕ of R_ω onto $R_{\omega'}$ is such that

$$\left(\forall(p, q) \in R_\omega^{[2]}\right), p \leq q \text{ for } \omega \Rightarrow \phi(p) \leq \phi(q) \text{ for } \omega'$$

where $X^{[2]}$ indicates the set of ordered distinct element pairs of X.

Now, let ω_0 be a fixed element of Ω and let $r(\omega_0)$ designate the ranking function on F associated with ω_0. The lateral order on Ω_r, with respect to ω_0, is defined as follows:

Definition 21

$$\left(\forall[(\omega, R_\omega), (\omega', R_{\omega'})] \in \Omega_r^{[2]}\right), (\omega, R_\omega) \prec (\omega', R_{\omega'})$$

if and only if there exists an increasing bijection ϕ of R_ω onto $R_{\omega'}$ such that

$$(\forall p \in R_\omega), r(\omega_0)(p) \leq r(\omega_0)(\phi(p))$$

where $\prec$ denotes the lateral order relation.

Example

Let us illustrate this notion by comparing the partitions obtained at level 2 of the trees τ_1 and τ_2 (see Fig. 9.4). The partitions formed in τ_1 and τ_2 are

$$\pi_2^1 = \{\{a, d, f\}, \{b, c, e\}\} \text{ and } \pi_2^2 = \{\{a, d, e, f\}, \{b\}, \{c\}\},$$

respectively. The respective sets of joined pairs by π_2^1 and π_2^2 are

$$R_2^1 = \{ad, af, bc, be, ce, df\} \text{ and } R_2^2 = \{ad, ae, af, de, def, ef\}$$

Therefore,

$$card(R_2^1) = card(R_2^2) = 6$$

Consider now the ultrametric preordonances ω_u^1 and ω_u^2 representing τ_1 and τ_2 (see the preceding example) and let ϕ be the mapping of R_2^1 onto R_2^2, depicted in Table 9.7 and increasing with respect to the ordered pair (ω_u^1, ω_u^2), that is to say,

$$\left(\forall (p, q) \in F^{[2]}\right) p \leq q \text{ for } \omega_u^1 \Rightarrow \phi(p) \leq \phi(q) \text{ for } \omega_u^2$$

On the other hand, it is immediate to verify that

$$(\forall p \in F), p \leq \phi(p) \text{ for } \omega_s$$

Consequently, π_2^1 is preferred to π_2^2 according to the lateral order criterion $\prec$. This conclusion holds also for comparing the F subset ordered pairs ($R_1^1 = \{ad, bc\}$, $R_1^2 = \{ad, ef\}$) and $\left(T_2^1 = \{af, df, be, ce\}, T_2^2 = \{af, df, de, ae\}\right)$. Hence, the tree τ_1 is preferred to τ_2 with respect to $\prec$.

For a given ordered pair (P, Q) of subsets of F, with equal cardinals, we must be able to prove that if P is preferred to Q with respect to the lateral order criterion, then P is preferred to Q with respect to any of the criteria presented in this chapter. However, the cardinality constraint is difficult to manage. The question arises of how to overcome this restriction.

Table 9.7 The mapping ϕ

p	ad	bc	af	be	ce	df
$\phi(p)$	ad	ef	af	ae	de	df

9.4.4.2   The Lexicographic Order Criterion

Let Ω be the set of all total preorders on an abstract finite set F and let ω_0 be a fixed element of Ω. ω_0 determines a *partial* order on Ω that we call the *lexicographic order* with respect to ω_0. This order will be denoted by $lexico(\omega_0)$. We shall describe it hereafter. To begin, let us define the notion of a *starting interval* $C(\omega_0)$ for ω_0 in F. r_0 designating the ranking function on F for ω_0, $C(\omega_0)$ is defined with respect to a threshold value s of r_0, as follows:

$$C(\omega_0, s) = \{p | p \in F \text{ and } r_0(p) \le s\} \tag{9.4.26}$$

where, clearly, $s \le f = card(F)$.

Definition 22 For any $(\omega^1, \omega^2) \in \Omega^{[2]}$, $\omega^1 \; lexico(\omega_0) \; \omega^2$ if and only if there exists a starting interval (one can also say section) $C(\omega_0, s)$, such that

$$(\forall p \in C(\omega_0, s)), r^1(p) \le r^2(p)$$

and

$$(\exists q \in C(\omega_0, s)), r^1(q) < r^2(q)$$

where r^1 and r^2 are the ranking functions associated with the total preorders ω^1 and ω^2.

In order to illustrate the application of the lexicographic order criterion, let ω, ω_u^1 and ω_u^2 stand for ω, ω^1 and ω^2. We have (see Table 9.5)

$$r^1(af) = r^2(af) = 6$$
$$r^1(ad) = r^2(ad) = 2$$
$$r^1(df) = r^2(df) = 6$$
$$r^1(bc) = 2 < r^2(bc) = 15 \tag{9.4.27}$$

Thus, as expected from Sect. 9.4.4.1, the tree τ_1 is favoured to τ_2 in its association with ω.

9.4.5   Lexicographic Ranking and Inversion Number Criteria

9.4.5.1   Introduction

Refer here to the content of Sect. 9.4.4.2. The simple and natural character of the *lexicographic algorithm* (see Sect. 10.2 of Chap. 10) makes it optimizing the *lexicographic order* global criterion (see also Sect. 10.2.3 of Chap. 10). This criterion is an ordinal one and then, it has not any metrical character. Therefore, it might be interesting to compare it with a classical metrical criterion. Let Ω be the set of all total preorders on F, considered above. The most basic metrical criterion on Ω is defined by the *inversion number* (see Eq. (9.4.10)). In these conditions, by denoting ω_0 a fixed element of Ω, the following question arises: "If ω is an element of Ω near enough ω_0, according to the *lexicographic order* criterion, is it also the case for the *inversion number* criterion?". Consequently, the matter consists of comparing the inversion number and the lexicographic order proximities to ω_0.

This comparison will be carried out below in the case where Ω is the set of all total orders on F. In this way, if F is the set of unordered element pairs of a set E to be clustered, all the necessary generality is preserved when we have to compare the respective associations of two ultrametric preordonances ω_u^1 and ω_u^2 with ω_0 (see Sect. 9.4.4.2). In fact, if ω is a total preorder on F, possibly a ultrametric one, a total order $o(\omega)$ corresponds to it, in such a way that $o(\omega)$ is compatible with ω (without reversals with respect to ω) and the nearest ω_0. More precisely, $o(\omega)$ is obtained by restricting ω_0 to every class preorder of ω. Note that if ω_u^1 and ω_u^2 are two ultrametric preordonances of the same type (i.e. the successive cardinal classes are identical),

$$\omega_u^1 \prec \omega_u^2 \; (lexico \; \omega_0) \Leftrightarrow o(\omega_u^1) \prec o(\omega_u^2) \; (lexico \; \omega_0)$$

Moreover, the inversion number of ω with respect to ω_0 is exactly the same as that of $o(\omega)$ with respect to ω_0. The generality of the subject permits us, below, to abstract from what F does represent.

9.4.5.2 Lexicographic Ranking and Inversion Number

Let F be a finite set, whose cardinality being denoted by f. We consider the set Ω of all total orders on F, $card(\Omega) = f!$. Moreover, we suppose F provided with a fixed total order ω_0. Thus, the elements of F can be designated by their respective ranks for ω_0. In these conditions, a total order o on F is equivalently represented by a permutation on the sequence of the first f integer numbers $(1, 2, \ldots, e, \ldots, f)$. By denoting $(o(1), o(2), \ldots, o(e), \ldots, o(f))$ the latter permutation, $o(i)$, indicates the element of F, whose rank is $o(i)$ for ω_0. The *lexicographic relation* becomes

$$\left(\forall (o, o') \in \Omega^{[2]}\right) o \prec o' (lexico \; \omega_0) \text{ if and only if}$$
$$(\exists j, 1 \leq j \leq f) \text{ such that } (\forall j < i), o(j) = o'(j) \text{ and } o(i) < o'(i) \text{ for } \omega_0$$
$$(9.4.28)$$

This lexicographic order relation $\prec$ is complete. We adopt here the following definition of the ranking function associated with $\prec$.

Definition 23 The lexicographic rank of a given total order o ($o \in \Omega$) is the integer $card([\omega_0, o]) - 1$, where the closed interval $[\omega_0, o]$ is defined in Ω endowed with the lexicographic order relation.

In this way, the first rank is that of ω_0 which is equal to 0.

The notion of *inversion number* between two total preorders on a finite set was given above (see Eq. (9.4.10)). Let us take up again this notion in the context of the set Ω.

Definition 24 The *inversion number* between two total orders o and o' on F is the cardinal of the symmetrical difference of their graphs in $F^{[2]}$, this being commonly denoted by $card\,(gr(o)\Delta gr(o'))$.

Now, referring to ω_0, for a given o in Ω, we denote by $\rho(o)$ the lexicographic order rank of o, with respect to ω_0. On the other hand, $\nu(o)$ will denote the inversion number between o and ω_0. The mapping $o \mapsto \rho(o)$ of Ω onto $\{1, 2, \ldots, f!\}$ being bijective, our purpose in this section is to specify the mapping $\rho \mapsto \nu$ of $\{1, 2, \ldots, f!\}$ onto $\{1, 2, \ldots, f(f-1)/2\}$.

Lemma 13 *The mapping* $\rho \mapsto \nu(\rho)$ *is well defined.*

Let $(o(1), o(2), \ldots, o(f))$ and $\big(o'(1), o'(2), \ldots, o'(g)\big)$ be two permutations, having the same lexicographic rank relative to $(1, 2, \ldots, f)$ and $(1, 2, \ldots, g)$, respectively. Whatever the values of the integers f and g, the inversion number of $(o(1), o(2), \ldots, o(f))$ with respect to $(1, 2, \ldots, f)$ *is the same* as that of

$$\big(o'(1), o'(2), \ldots, o'(g)\big)$$

with respect to $(1, 2, \ldots, g)$. Without any loss of generality, we may suppose $g > f$. Then set $g = f + k$, where $k > 0$. In these conditions, we have, necessarily,

$$\big(o'(1), \ldots, o'(k), o'(k+1), \ldots, o'(g)\big) = (1, 2, \ldots, k, o(1) + k, o(2) + k, \ldots, o(f) + k)$$
$$(9.4.29)$$

that is to say,

$$o'(l) = l \text{ for } l \leq k$$
$$o'(k+m) = k + m \text{ for } 1 \leq m \leq f \tag{9.4.30}$$

Clearly, the permutation $(o(1), o(2), \ldots, o(f))$ has the same inversion number with respect to $(1, 2, \ldots, f)$ as that of $(1, 2, \ldots, k, o(1) + k, o(2) + k, \ldots, o(f) + k)$ with respect to $(1, 2, \ldots, k, 1 + k, 2 + k, \ldots, f + k)$. □

Theorem 31 *Let* $\rho = \rho(o)$ *be the lexicographic rank of the permutation*

$$(o(1), o(2), \ldots, o(f))\,.$$

If s *denotes the integer for which* $s! \leq \rho(o) < (s+1)!$, *consider the following successive divisions:*

$$\rho = q \times s! + \rho_1 , \rho_1 < s!$$
$$\rho_1 = q_1 \times (s - 1)! + \rho_2 , \rho_2 < (s - 1)!$$
$$\rho_2 = q_2 \times (s - 2)! + \rho_3 , \rho_3 < (s - 2)!$$

$$\ldots\ldots\ldots\ldots$$

$$\rho_{s-h} = q_{s-h} \times h! + 0 \tag{9.4.31}$$

We have

$$\nu(\rho) = q + q_1 + q_2 + \cdots + q_{s-h} \tag{9.4.32}$$

Proof In fact the sequence $(\rho_1, \rho_2, \ldots, \rho_i, \ldots)$ is decreasing and each non-null value of ρ_i is followed by a strictly lower value of ρ_j, for a j, strictly greater than i. Thus, the last of Eq. (9.4.31) occurs necessarily.

$\rho < k!$, that is, $0 \le \rho \le k! - 1$, if and only if the starting interval of o, namely $(o(1), o(2), \ldots, o(f - k))$, is

$$(1, 2, \ldots, (f - k)) .$$

In fact, there are $k!$ permutations whose beginning interval is $(1, 2, \ldots, (f - k))$. The first of them, ranked 0, is $(1, 2, \ldots, f)$ and the last, ranked $k! - 1$, is $(1, 2, \ldots, (f - k), f, f - 1, f - 2, \ldots, f - k + 1)$.

Now, consider the permutation associated with a total order o under the form

$$o = (1, 2, \ldots, (f - s + 1), o(f - s), o(f - s + 1), \ldots, o(f))$$

In other words, the first $f - s + 1$ components are the first $f - s + 1$ integer numbers. Note that the number of components between $o(f - s + 1)$ and $o(f)$ is s. There are $s!$ permutations for which $o(f - s) = f - s$, the respective ranks of the first and the last ones being 0 and $(s! - 1)$. There are $2 \times s!$ permutations for which $(f - s) \le o(f - s) \le (f - s + 1)$, the last of them being ranked $(2s! - 1)$. More generally, there are $(k + 1) \times s!$ permutations for which

$$(f - s) \le o(f - s) \le (f - s + k)$$

where the rank of the last one is $(k + 1) \times s! - 1$.

Hence, in order that the quotient of the division of ρ by $s!$ to be q, it is necessary and sufficient that o takes the following form:

$$o = (1, 2, \ldots, f - s - 1, f - s + q, o(f - s + 1), o(f - s + 2), \ldots, o(f)) \tag{9.4.33}$$

and the rest of the division ρ_1 is precisely the lexicographic rank of

$$\theta = (o(f - s + 1), o(f - s + 2), \ldots, o(f))$$

with respect to

$$\tau = ((f - s), (f - s + 1), \ldots, (f - s + q - 1), (f - s + q + 1), \ldots, f)$$

The inversion number of o with respect to $(1, 2, \ldots, f)$ is that resulting from the term $(f - s + q)$ plus that of θ with respect to τ. For o, the element $o(f - s) = (f - s + q)$ gives rise to q inversions with the following terms; these, in fact, include

$$(f - s), (f - s + 1), \ldots, (f - s + q - 1)$$

Moreover, the inversion number of $(o(f - s + 1), o(f - s + 2), \ldots, o(f))$ with respect to τ is exactly $\nu(\rho_1)$. In these conditions, step by step, we obtain the announced result.   □

Corollary 10
$$\rho < k! \Longrightarrow \nu(\rho) \le \frac{k(k - 1)}{2}$$

This result is clearly derived from the beginning of the previous theorem proof. In fact, in the set of all permutations for which $\rho < k!$, that associated with the highest value of $\nu(\rho)$ is whose lexicographic rank is $\rho = (k! - 1)$, namely,

$$(1, 2, \ldots, f - k, f, f - 1, f - 2, \ldots, f - k + 1)$$

Now, let us reconsider the following situation:

1. F is the set of unordered distinct element pairs of a set E to be clustered ($F = P_2(E)$);
2. ω_0 is a total order on F, expressing an ordinal similarity on E;
3. o is associated with a ultrametric preordonance, the nearest ω_0.

In these conditions, the (Corollary 10) gives an answer to the question submitted in the introductory Sect. 9.4.5.1. One may say, if the lexicographic rank function of o is lower than a given threshold, then it is also the case for the inversion number between o and ω_0.

Lexicographic Ranking Successor

In this paragraph we shall specify an algorithm which determines the successor—according to the lexicographic ranking—of a given total order o of the set Ω of all total orders on an abstract set F, whose cardinality is f. According to the above o is expressed as the permutation

$$(r(1), r(2), \ldots, r(e), \ldots, r(f)) \tag{9.4.34}$$

Let i designate the greatest index for which $r(i-1) < r(i)$. Thus, the rank sequence $(r(f), r(f-1), \ldots, r(i))$ is strictly increasing: $r(f) < r(f-1) < \cdots < r(i)$. In this sequence, consider, by describing the latter sequence from right to left, the greatest index $(i+h)$ for which $r(i+h) > r(i-1)$. It is the smallest number (in this sequence) which is greater than $r(i-1)$. In these conditions, transpose $r(i-1)$ and $r(i+h)$. The permutation (9.4.34) becomes

$$(r(1), \ldots, r(i-2), r(i+h), r(i), r(i+1), \ldots, r(i+h-1), r(i-1), r(i+h+1), \ldots, r(f))$$

$$(9.4.35)$$

We have

$$r(i+h-1) > r(i+h) > r(i-1) > r(i+h+1)$$

Therefore, the sequence

$$(r(i), r(i+1), \ldots, r(i+h-1), r(i-1), r(i+h+1), \ldots, r(f))$$

remains strictly decreasing. Then, replace this finishing section by its reverse in the permutation (9.4.35). The latter becomes

$$(r(1), \ldots, r(i-2), r(i+h), r(f), r(f-1), \ldots, r(i+h+1), r(i-1), r(i+h-1), \ldots, r(i))$$

$$(9.4.36)$$

and in this way, the finishing section $(r(f), r(f-1), \ldots, r(i+h+1), r(i-1), r(i+h-1), \ldots, r(i))$ is strictly increasing.

Proposition 52 *The permutation (9.4.36) is the lexicographic immediate successor of the initial permutation (9.4.34).*

Let us retake the initial permutation in the following form:

$$(r(1), r(2), \ldots, r(i-1), r(i), \ldots, r(i+h), \ldots, r(f)) \qquad (9.4.37)$$

where $r(i+h)$ was defined above. The finishing section $(r(i), \ldots, r(i+h), \ldots, r(f))$ being decreasing $((r(i) > r(i+2) > \cdots > r(i+h) > \cdots > r(f)))$, this permutation is the greatest one (for the lexicographic order) among all those for which the beginning section $(r(1), r(2), \ldots, r(i-1))$ is invariant. Therefore, every successor of (9.4.37) differs from the latter permutation on at least one element of this beginning section.

Moreover, every permutation whose starting section is $(r(1), r(2), \ldots, r(i-2))$ is nearer (9.4.37) than all of those for which it is not the case. Consequently, the immediate successor of (9.4.37) is such that the first component where it differs from (9.4.37) is precisely $(i-1)$. The $(i-1)$th component of the sought successor has to be taken in the finishing section $(r(i), r(i+1), \ldots, r(i+h), \ldots, r(f))$. Necessarily, it

Table 9.8 The mapping $\rho \mapsto \nu(\rho)$

					ρ	$\nu(\rho)$
1	2	3	4	5	0	0
1	2	3	5	4	1	1
1	2	4	3	5	2	1
1	2	4	5	3	3	2
1	2	5	3	4	4	2
1	2	5	4	3	5	3
1	3	2	4	5	6	1
1	3	2	5	4	7	2
1	3	4	2	5	8	2
1	3	4	5	2	9	3
1	3	5	2	4	10	3
1	3	5	4	2	11	4
1	4	2	3	5	12	2
1	4	2	5	3	13	3
1	4	3	2	5	14	3
1	4	3	5	2	15	4
1	4	5	2	3	16	4
1	4	5	3	2	17	5
1	5	2	3	4	18	3
1	5	2	4	3	19	4
1	5	3	2	4	20	4
1	5	3	4	2	21	5
1	5	4	2	3	22	5
1	5	4	3	2	23	6

is among the numbers greater than $r(i-1)$, the smallest one, that is $r(i+h)$. Therefore, the immediate successor concerned has the beginning section

$$(r(1), r(2), \ldots, r(i-2), r(i+h))$$

The $(f - i + 1)$ remaining elements are ordered increasingly in the finishing section of (9.4.36). The latter precedes every permutation whose beginning section is the previous one.   □

Example

Applying the previous algorithm with $f = 5$, we obtain $5! = 120$ permutations, lexicographically ranked. The first one, ranked 0, is $(1, 2, 3, 4, 5)$ and the last one, ranked 119, is $(5, 4, 3, 2, 1)$. The sequence of all these permutations can be decomposed into five subsequences of 24 permutations each. Let $(S_1, S_2, S_3, S_4, S_5)$ be the sequence of the 5 subsequences. The first common number of the elements of S_i is i, $1 \le i \le 5$. Table 9.8 gives the first 24 permutations, where, for each of them,

Fig. 9.5 Graph of the
mapping $\rho \mapsto \nu(\rho)$

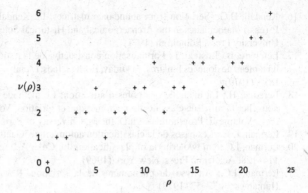

the inversion number $\nu(\rho)$ with respect to the initial permutation $(1, 2, 3, 4, 5)$ is associated with its lexicographic rank ρ. The graph of this correspondence is given in Fig. 9.5. Now, imagine the graph representation of all the permutations (there are in all 120). It is easy to prove that this graph restricted to S_{i+1} can be derived from that concerning S_i by applying to the latter a vector translation $(5, 1)$.

References

1. Benzécri, J.-P.: L'Analyse des Données, 1 La Taxinomie. Dunod (1973)
2. Benzécri, J.P.: Analyse factorielle des proximités. Publications de l'Institut de Statistique de l'Université de Paris, (13–14):13:235–282, 14:219–246 (1964–1965)
3. Bertin, J.: La Graphique et le traitement graphique de l'Information. Flammarion (1977)
4. Bertin, J.: Graphic and graphic information process, translated by William J. Berg and Paul Scott. de Gruyter (1981)
5. Daniels, H.E.: The relation between measures of correlation in the universe of sample permutations. Biometrika **33**, 129–135 (1944)
6. de la Vega, W.F.: Techniques de classification automatique utilisant un indice de ressemblance. Revue Francaise de Sociologie, (8–4):506–520 (1967)
7. de la Vega, W.F.: Quelques propriétés des hiérarchies de classification. In Gardin, J.-C. (ed.) Archéologie et Calculateurs, pp. 329–343. Centre National de la Recherche Scientifique (1970)
8. Ganter, B., Wille, R.: Formal Concept Analysis, Mathematical Foundations. Springer, New York (1999)
9. Ghazzali, N.: Comparaison et réduction d'arbres de classification, en relation avec des problèmes de quantification en imagerie numérique. Ph.D. thesis, Université de Rennes 1, mai 1992
10. Govaert, G.: Classification croisée, Doctorat d'Etat. Ph.D. thesis, University of Paris 6 (1983)
11. Govaert, G.: La classification croisée. La Revue de Modulad 4:9–36 (1989)
12. Hartigan, J.A.: Direct clustering of a data matrix. J. Am. Stat. Assoc. **67**(337), 123–129 (1972)
13. Hartigan, J.A.: Clustering Algorithms. Wiley, New York (1975)
14. Huang, Z.: Extensions to the k-means algorithm for clustering large data sets with categorical values. Data Min. Knowl. Discov. **2**, 283–304 (1998)
15. Hubert, L.J.: Inference procedures for the evaluation and comparison of proximity matrices. In: Felsenstein, J. (ed.) Numerical Taxonomy, pp. 209–228. Springer, New York (1983)

16. Kendall, D.G.: Seriation from abundance matrices. In: Kendall, D.G., Hodson, F.R., Tautu, P. (eds.) Mathematics in the Archaeological and Historical Sciences, pp. 215–252. Edinburgh University Press, Edinburgh (1971)

17. Lefebvre, B., Losfeld, J.: Formalisation constructive de la notion de classe polythétique pour un tableau de données binaires. In Diday, E., et al. (eds.) Analyse des Données et Informatique. IRIA (1979)

18. Leredde, H.: La méthode des pôles d'attraction; La méthode des pôles d'agrégation: deux nouvelles familles de classification automatique et sériation, Volume I: méthodes et exemples réels, Volume II: Programmes. Ph.D. thesis, University of Paris 6 (1979)

19. Lerman, I.-C.: Les bases de la classification automatique. Gauthier-Villars (1970)

20. Lerman, I.C.: On two criteria of classification. In: Cole, A.J. (ed.) Numerical Taxonomy, pp. 114–128. Academic Press, New York (1969)

21. Lerman, I.C.: Analyse du phénomène de la sériation. Revue Mathématique et Sciences Humaines 38:39–57 (1972)

22. Lerman, I.C.: Classification et analyse ordinale des données. Dunod (1981). http://www.brclasssoc.org.uk/books/index.html

23. Lerman, I.C.: Group methodology in production management. Appl. Stoch. Models Data Anal. **2**, 153–165 (1986)

24. Lerman, I.C.: Comparing classification tree structures: a special case of comparing q-ary relations. RAIRO Oper. Res. **33**, 339–365 (1999)

25. Lerman, I.C., Ghazzali, N.: What do we retain from a classification tree? In: Diday, E., Lechevallier, Y. (eds.) Symbolic-Numeric Data Analysis and Learning, pp. 27–42. Nova Science, New York (1991)

26. Lerman, I.C., Leredde, H.: La méthode des pôles d'attraction. In Diday, E., et al. (eds.) Analyse des Données et Informatique, pp. 37–50. IRIA (1977)

27. Lerman, I.C., Pinto da Costa, J., Silva, H.: Validation of very large data sets clustering by means of a nonparametric linear criterion. In: Bock, H.-H., Jajuga, K., Sokolowski, A. (eds.) Classification, Clustering and Data Analysis, pp. 147–157. Springer, New York (2002)

28. Lerman, I.C., Rouxel, F.: Comparing classification tree structures: a special case of comparing q-ary relations ii. RAIRO Oper. Res. **34**, 251–281 (2000)

29. Ling, R.F.: A probability theory of cluster analysis. J. Am. Stat. Assoc. **341**, 159–164 (1973)

30. Madeira, S.C., Oliveira, A.L.: Biclustering algorithms for biological data analysis. IEEE Trans. Comput. Biol. Bioinf. **1**, 24–45 (2004)

31. Mannila, H.: Finding total and partial orders from data for seriation. Lect. Notes Comput. Sci. **5255**, 16–25 (2008)

32. Milligan, G.W., Cooper, M.C.: An examination of procedures for determining the number of clusters in a data set. Psychometrika **2**:159–179 (1985)

33. Mollière, J.-L.: What's the real number of clusters. In: Gaul, W., Schader, M. (eds.) Classification as a Tool of Research, pp. 311–320. North-Holland, Amsterdam (1986)

34. Moon, J.K.: A note on approximating symmetric relations by equivalence relations. S.I.A.M. J. Appl. Math. **14**, 226–227 (1966)

35. Nicolau, M.H.: Analyse d'un algorithme de classification. Ph.D. thesis, University of Paris 6 (1972)

36. Pieraut-le, G., van Meter, K.: Étude génétique de la construction d'une propriété relationnelle: La propriété de passage. CNRS (1976)

37. Shepard, R.N.: The analysis of proximities: multidimensional scaling with unknown distance function. Psychometrika **27**, 219–246 (1962)

38. Tanay, A., Sharan, R., Shamir, R.: Biclustering algorithms: a survey. In: Aluru, S. (ed.) Handbook of Computational Molecular Biology, pp. 1–20. Chapman, Boca Raton (2004)

39. Warren, S.S.: Cubic clustering criterion. Report A - 108, SAS Institute Inc. (1983)

40. Wilderjans, T.F., Depril, D., Mechelen, I.V.: Additive biclustering: a comparison of one new and two existing ALS algorithms. J. Classif. **30**, 56–74 (2013)

41. Zahn, C.T.: Approximating symmetric relations by equivalence relations. S.I.A.M. J. Appl. Math. **12**, 840–847 (1964)

Chapter 10
Building a Classification Tree

10.1 Introduction

The basic data consists of a finite set E provided with a dissimilarity or a similarity function. The elements of E are of the same nature. As seen in Chap. 3, E can be a set of attributes, a set of objects or a set of categories. Tables 3.4 and 3.5 of Chap. 3 give a precise idea of the possible different versions of E. On the other hand, the set E may be weighted by a positive numerical measure μ_E, assigning to each of its elements x ($x \in E$) a weight μ_x. μ_x defines the "importance" with which x has to be considered. The dissimilarity or similarity function on E is mostly numerical. However, it might be ordinal. In Chaps. 4–7, facets of building a similarity index or association coefficient on E have been minutely examined in relation to the descriptive nature of E.

As expressed previously (see Chap. 1), a classification tree on E is defined by an ordered partition chain $\Pi = (P_0, P_1, \ldots, P_{l-1}, P_l, \ldots, P_m)$ on E, where the partition P_l is obtained from P_{l-1} by merging classes of P_{l-1}. P_0 designates the finest partition for which each of its classes is a singleton class, comprising exactly one element of E. P_m is the "rough" partition including exactly a single class corresponding to the whole set E. Let us denote by $\tau(\Pi)$ the classification tree associated with Π. The graphical representation of $\tau(\Pi)$ was considered in Chap. 1. Let us take up again this representation. The leaves of this tree represent singleton subsets of E, comprising each exactly one element of E. These are placed horizontally on an initial level labelled by 0. The partition represented at this level is P_0. The nodes formed at a given level labelled l of this tree are obtained from the nodes of the preceding levels labelled from 0 to $l - 1$ (a leaf is interpreted as a terminal node). This formation is done by aggregating some of the latter and by preserving all of the others. Thus, the last nodes constituted just before level $l + 1$ represent disjoint subsets of E and all of them determine a partition of E which is P_l. The root of the classification tree is located at its last level. It represents a node grouping all of the E elements. It corresponds to the partition denoted above by P_m into one single class.

© Springer-Verlag London 2016
I.C. Lerman, *Foundations and Methods in Combinatorial and Statistical
Data Analysis and Clustering*, Advanced Information and Knowledge Processing,
DOI 10.1007/978-1-4471-6793-8_10

Fig. 10.1 Example of a classification tree

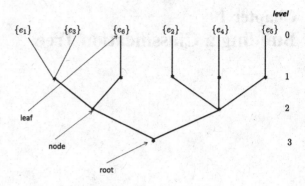

In the example given in Fig. 10.1, the partition sequence Π is

$$P_0 = \{\{e_1\}, \{e_2\}, \{e_3\}, \{e_4\}, \{e_5\}, \{e_6\}\}$$
$$P_1 = \{\{e_1, e_3\}, \{e_2\}, \{e_4\}, \{e_5\}, \{e_6\}\}$$
$$P_2 = \{\{e_1, e_3, e_6\}, \{e_2, e_4, e_5\}\}$$
$$P_3 = \{\{e_1, e_2, e_3, e_4, e_5, e_6\}\} \tag{10.1.1}$$

We assume that set E is to be endowed with a similarity function S or a dissimilarity function $\mathcal{D}$ (see Chaps. 4–7). The objective in this chapter consists of building a classification tree on E which respects as much as possible the similarities between the elements of E. That is, such that, globally and statistically, the joining level of two current elements x and y of E is as much low as the value of $S(x, y)$ (respectively, $\mathcal{D}(x, y)$) is high (respectively, low). In Fig. 10.1, we adopt the same drawing as in Chap. 1 of the classification tree associated with an ordered partition chain, that is, the successive levels increase from top to bottom: from the lowest level 0 to the highest one m.

The similarity function might be ordinal (see Sect. 4.2.2 of Chap. 4). Precisely, we shall begin by this case. The data is then defined by a preordonance w_S on E associated with a similarity index S on E. We assume that w_S is established according to the first of (4.1.1), that is to say

$$\big(\forall (p, q) \in P_2(E) \times P_2(E)\big) \; p \le q \Leftrightarrow S(p) \ge S(q)$$

where—as usually—$P_2(E)$ designates the set of unordered element pairs of E and where $S(p)$ (respectively, $S(q)$)—for $p = \{x, y\}$ (respectively, $q = \{z, t\}$)—is the similarity value between x and y (resp., z and t).

In Sect. 10.2.2, we define an algorithm which builds a classification tree on E directly from this ordinal data (w_S), without any reference to the similarity index S from which w_S is deduced. The natural character of this ordinal algorithm is expressed by a property of optimality relative to the lexographic order criterion seen in the preceding chapter (see Chap. 9). For this reason, the latter algorithm is called "lexicographic algorithm" [33]. It works at the level of set $P_2(E)$. This is equally the case for "Maxima Lower Algorithm" defined by M. Roux [53]. However, the latter has a metrical nature. It is conceived very differently from the "lexicographic" algorithm and can be associated with the N. Jardine and R. Sibson work [25]. In fact and relative to the practical results obtained, both algorithms are equivalent to the famous "Single Linkage" algorithm, most often attributed to Sneath [55, 56] (see below). The latter can be expressed at the level of E subset set. More precisely, it is defined from an extension of the similarity index S between elements of E to a similarity index σ between disjoint subsets of E. Clearly, this extension can be expressed in terms of dissimilarity indices: $\mathcal{D}$ between elements of E and δ, between disjoint subsets of E. For reasons of frequency of use in the taxonomist community, we employ the latter expression. The informal algorithmic definition of the *Ascendant Agglomerative Construction of a Hierarchical Clustering (Classification)* on the set E, can be stated as follows

"At each step merge the nearest class pairs according to the dissimilarity δ"

If C and D are two distinct classes of a partition of E, the dissimilarity notion between C and D underlying the "Single Linkage" algorithm is defined by the minimum function applied to the table of positive numbers.

$$\mathcal{D}(C, D) = \{\mathcal{D}(x, y) | (x, y) \in C \times D\} \tag{10.1.2}$$

where $C \times D$ is the Cartesian product of C by D. This index will be designated by $\delta_{\min}(C, D)$. Clearly, functions other than minimum function can apply to $\mathcal{D}(C, D)$. The most classical ones are the maximum and the average functions. The associated algorithms are called "Complete Linkage" and "Average Linkage" algorithms, respectively (see Sect. 10.3.2). The *inertia* Ward index, which we designate by $\delta_{iner}(C, D)$ is conceived in the framework of clustering a cloud of points situated in an Euclidean space. It can be expressed as a function of $\mathcal{D}(C, D)$ and the respective weights of the elements of $C \cup D$. In this, $\mathcal{D}(x, y)$ is defined by the Euclidean distance between the summits x and y of the cloud of points concerned. In any case, for all of the aggregative hierarchical clustering methods, $\delta(C, D)$ can be formally written as follows:

$$(\forall(C, D) \in \mathcal{P}(E) \times \mathcal{P}(E), C \cap D = \emptyset)$$
$$\delta(C, D) = f(\{\mathcal{D}(x, y) | (x, y) \in (C \cup D) \times (C \cup D)\}, \mu_{C \cup D}) \tag{10.1.3}$$

where $\mathcal{P}(E)$ is the subset of E and where the function f is to be defined on the set of mutual dissimilarities between weighted elements of $C \cup D$. On the other hand, $\mu_{C \cup D}$ is the restriction to $C \cup D$ of a weighting μ_E defined on E

$$\mu_{C \cup D} = \{\mu_x | x \in C \cup D\} \tag{10.1.4}$$

In the absence of a specific weighting, the different elements of E are considered as equally weighted. Thus, in the dissimilarity indices $\delta_{\min}(C, D)$, $\delta_{\max}(C, D)$, $\delta_{ave}(C, D)$ and $\delta_{iner}(C, D)$, the implicit weighting considered assigns the value 1 to each element of E. $\delta(C, D)$ will be called the "merging criterion" defined by f.

Notice that the formal expression (10.1.3) entails that the function f is defined locally with respect to $C \cup D$. More precisely, the complementary of $C \cup D$ in E does not take place in (10.1.3). However, some merging criteria might be defined differently as variations of a partition quality measure between two consecutive partitions P_{l-1} and P_l of Π (see above). So is the case of the Ward criterion based on the inertia variation. However, this variation can be formally reduced to a particular case of formula (10.1.3) (see Sect. 10.3.3). Indeed, intuitively, the merging criterion must be defined locally for comparing pairs of clusters. This enables the association nature between two clusters C and D to be better understood. Nevertheless, there might be global criteria measuring partition quality, whose variations between P_{l-1} and P_l cannot be reduced to local criteria defined at the level of the two clusters C and D. This is the case of the global criterion defined in Sect. 9.3.5 of Chap. 9, in terms of an LLA association coefficient between ordinal or numerical similarity on E and a partition of E. However, the extreme generality of the latter criterion makes it more relevant to compare respective associations between different partitions of E and the similarity (numerical or ordinal) by which E is provided. Thus, this criterion proves to be very important in order to evaluate the evolution of the partition sequence Π of a classification tree on E, obtained step by step by means of a local criterion. "Significant" levels and nodes of the classification tree can be pointed out and this guides and simplifies considerably the interpretation of the hierarchical clustering (see Chap. 9).

Clustering rows of a contingency table, or more generally, horizontal juxtaposition of contingency tables, is a specific problem (see Sect. 7.2.8). The development of the previous merging criteria, as well as new ones, will be considered in Sect. 10.5.

Now, whatever may be the nature of the set E to be clustered: set of objects, categories or attributes, we saw in Chaps. 5–7, how the LLA approach leads to a matrix (table) of discriminant probabilistic indices that we denote as follows:

$$\mathcal{P}^g = \{P^g(x, y) | \{x, y\} \in P_2(E)\} \tag{10.1.5}$$

The following dissimilarity matrix, called "Informational Dissimilarity Matrix"

$$\mathcal{D} = \{D_{inf}(x, y) = -\text{Log}_2(P^g(x, y)) | \{x, y\} \in P_2(E)\} \tag{10.1.6}$$

was associated with $\mathcal{P}^g$ (see for example (7.2.6) and (7.2.7) where E is defined by the object set). In these conditions, each of the previous criteria $\delta_{\min}$, $\delta_{\max}$ and δ_{ave} can apply with the distance matrix $\mathcal{D}$. However, this direct adaptation does not respect the *Likelihood of the Link* principle for comparing disjoint subsets of E. In order to express the application of this principle, let us begin with a merging criterion equivalent to

$$\delta_{min}(C, D) = \min\{D_{inf}(x, y)|(x, y) \in C \times D\} \qquad (10.1.7)$$

whose expression being defined as a function of the probabilistic similarity indices $P^g(x, y)$, $\{x, y\} \in P_2(E)$ (see 10.1.5), namely

$$\sigma_{max}(C, D) = \max\{P^g(x, y)|(x, y) \in C \times D\} \qquad (10.1.8)$$

This index is interpreted as a *raw* probabilistic similarity in the *LLA* approach. The final index called "Maximum Link Likelihood", refers to the *unlikelihood* of "how large is $\sigma_{max}(C, D)$", with respect to a mutual independence hypothesis between the elements of C and D. Alternatives of this hypothesis will be discussed and specified in Sect. 10.4.1. It will lead to a rich family of probabilistic similarity indices for comparing C and D on the basis of $\sigma_{max}(C, D)$. This family is implemented in the *CHAVL* (*C*lassification *H*iérarchique par *A*nalyse de la *V*raisemblance des *L*iens) software (see Chap. 11). Very great number of applied researchs have been carried out by employing *CHAVL*.

Several *raw* probabilistic similarity indices can be proposed from the table

$$\mathcal{P}^g(C, D) = \{P^g(x, y)|(x, y) \in C \times D\} \qquad (10.1.9)$$

the most obvious are obtained by substituting for the maximum operation in (10.1.8), the minimum and the average operations. These indices can be denoted as $\sigma_{min}(C, D)$ and $\sigma_{ave}(C, D)$, respectively. They will be discussed according to the *LLA* approach in Sect. 10.4.2 [11, 26, 49].

This chapter is exclusively devoted to the ascendant agglomerative hierarchical construction of a classification tree on a finite set E endowed with a similarity or a dissimilarity coefficient. Mostly, hierarchical clustering refers to this algorithmic approach. Nevertheless, other algorithmic techniques can be envisaged for building a classification tree on E. In this respect, divisive methods may be mentioned. Generally, these start with the "rough" partition including a single class and proceed by successive binary divisions. At a given step of the building process, the most heterogenous cluster is divided into two disjoint subsets as homogeneous as possible. As mentioned in the interesting review [21] "Algorithm that find the globally optimal division [15, 54] are computationally very demanding". However, mostly, a local optimization is accepted. A classical solution for this consists of using K-means algorithm with $K = 2$ (see Sect. 2.2) Chap. 2 for dividing the most heterogenous cluster into two subclusters. Clearly, a measure of heterogeneity of a given cluster is required for such an algorithm. This measure will be necessarily established from the dissimilarity distribution over the cluster to divide. In any case, conceptually, classification trees obtained by divisive methods are conceptually very different from those obtained by agglomerative methods.

In Chap. 8 of [35], an original descendant, but non-hierarchic, clustering method is presented [29, 34, 41]. This proceeds by successive separations. The initial partition is the *rough* partition into a single class. The partition obtained at the lth step divides E

into l classes. These are obtained by assigning, according to nearest neighbor strategy the E elements around l specific elements, statistically determined as "attraction poles".

Now, let us return to ascendant agglomerative hierarchical clustering. For the algorithm implementation, we may compare the hierarchical ascendant construction of a classification tree to the evolution of a system. If l is a tree level, we characterize the state of the system by the couple (T_l, μ^l) where T_l is the table of the δ indices between the classes formed at the level l and where μ^l is a weighting on the set of these classes. The initial state (T_0, μ^0) is defined by the matrix T_0 of the δ indices between singleton classes, comprising each exactly one element, and by μ^0 the weighting on the set E to be clustered. By denoting L the last level of the classification tree, corresponding to the root ($L = 3$ in Fig. 10.1), the different states of the system for $0 \le l \le L - 1$ have to be considered. Therefore, it is of first importance from computational point of view to have a formula, called "reactualization formula" of the general following form:

$$(T_l, \mu^l) = \varphi(T_{l-1}, \mu^{l-1}) \qquad (10.1.10)$$

where the left and right members define the states of the system at the levels l and $l - 1$, respectively, $1 \le l \le L - 1$. φ is a function to be determined depending on the dissimilarity function δ considered for the merging criterion.

The reactualization formulas for the different merging criteria above considered ("Single Linkage", "Complete Linkage", "Average Linkage", "Ward Inertia Variation", "Maximum Link Likelihood") will be detailed in Sect. 10.6.2. To begin, we express these formulas in the commonly and simplest case where a binary merging of two classes is considered. In this respect, the very known general reactualization formula of Lance and Williams [27] (see also [24]) will be recalled. This formula enables a family of dissimilarity indices δ between classes to be mutually organized. However, not all of the δ indices can appear as particular cases of the latter formula. For example, the reactualization formula associated with the "Maximum Link Likelihood" criterion does not follow the Lance and Williams formula. On the other hand, the latter cannot compare the respective interests of different δ indices and even less provides help for devising new δ indices.

In Sect. 10.6.3, we provide the extension of the reactualization formula in the general case of multiple aggregations at the same level of the classification tree. In this case more than two classes merge in passing from a given level to the next one [36]. This occurs when several class pairs produce simultaneously at a given level the minimal value of the dissimilarity δ (resp., the maximal value of the similarity σ (see above)) between the E classes. This circumstance is frequent in the case of clustering a large object set $\mathcal{O}$ described by categorical attributes such that the total number of categories of all of the attributes is not large enough in order to mutually discriminate the similarities between the different objects. Reactualization formula provides noteworthy reduction of the computational complexity.

Precisely, in order to reduce the computational complexity of a classification tree on large data sets very important algorithmic ideas arise in the eighty years. We discuss them in Sects. 10.6.4–10.6.6. In Sect. 10.6.6, the proposed complexity reduction is carried out thanks to a specific form of *parallel* computing which provides a statistical approximation of the classification tree.

There are clustering problems of object sets endowed with contiguity relations. In this case, it may be needed for the set E concerned, to form clusters which respect the contiguity relation. In other words, relative to agglomerative formation of a classification tree, a necessary condition for merging two clusters C and D, is that there exists two elements x and y belonging to C and D, respectively, such that x and y are related by the contiguity relation. This algorithmic requirement, supported by a real example will be examined in Sect. 10.6.5.

In practice and principally, the algorithmic development of Sect. 10.6 concerns clustering a "large" *object* set. Obviously, this applies also for clustering an *attribute* set. When the problem addressed consists of hierarchically organizing a set of Boolean attributes endowed with an asymmetrical association coefficient (see Sect. 4.2.1.3 and especially Sect. 5.2.2.2), a specific structure was proposed in [23]. This specificity is defined by assigning an orientation to binary merges between leaves and nodes of the classification tree. It was studied in depth in [39, 40].

10.2 "Lexicographic" Ordinal Algorithm

10.2.1 Definition of an Ultrametric Preordonance Associated with a Preordonance Data

As previously stated, E will designate the set on which a classification tree has to be built. We assume E provided with a total preordonance $\omega(E)$ associated with a similarity index S on E. We suppose that $\omega(E)$ is established according to the equation given at the begining of the preceding section. That is, the higher is the similarity value of S between two components of a given pair p, the lower is the rank of p for $\omega(E)$. As usually, $F = P_2(E)$ will designate the set of unordered distinct element pairs of E. The only structure retained here is $\omega(E)$. The latter is defined by a total preorder (ranking with ties) on F. By denoting m the cardinality of E, $card(F) = m(m-1)/2$.

We saw in Chap. 1, Sect. 1.4.3 that an ordered partition chain on E can be represented by a ultrametric preordonance $\omega_u(E)$ on E. By considering an associated ranking function H on F, H is characterized by the following property:

$$\left(\forall x, y \text{ and } z \in E\right) H(x, y) \le r \text{ and } H(y, z) \le r \Rightarrow H(x, z) \le r \quad (10.2.1)$$

where r is a positive integer number, $0 \le r \le m(m-1)/2$.

In the following, we shall define an ordinal algorithm for getting a ultrametric total preordonance $\omega_u(E)$ from $\omega(E)$. $\omega_u(E)$ which approximates in a certain sense $\omega(E)$, is obtained by successive transitive closures of the begining sections (ordinal intervals) of $\omega(E)$. The optimal character of $\omega_u(E)$ will be shown. A specific and formal case is that where the data is a total ordonance on E, that is, a total order on F.

Let us denote here by ρ the ranking function on F associated with the given preordonance $\omega_u(E)$, recall that

$$(\forall p \in F), \; \rho(p) = card\{q|q \leq p\}$$

On the other hand, as just expressed above, H designates the ranking function on F compatible with the ultrametric total preordonance $\omega_u(E)$ to be calculated in order to approximate $\omega(E)$:

$$(\forall p, q \in F), \; H(p) \leq H(q) \Leftrightarrow p \leq q \text{ for } \omega_u(E)$$

Formal Definition of H

$c(x, y) = (z_1, z_2, \ldots, z_j, \ldots, z_k)$ will designate a finite ordered chain of E elements of origin and extremity, x and y, respectively: $z_1 = x$, $z_k = y$ and $z_i \neq z_j$ for all $i \neq j$. On the other hand, we denote by $\mathcal{C}(x, y)$ the set of all ordered chains of origin x and extremity y. The *maximum* and *minimum* operations in the integers will be represented by $\vee$ and $\wedge$, we define

$$\phi(c(x, y)) = \rho(z_1, z_2) \vee \rho(z_2, z_3) \vee \cdots \vee \rho(z_{k-1}, z_k)$$

where $c(x, y) = (z_1, z_2, \ldots, z_j, \ldots, z_k)$. We set

$$H(x, y) = \bigwedge_{c(x,y) \in \mathcal{C}(x,y)} \phi(c(x, y)) \tag{10.2.2}$$

Notice that, if the minimum of $\phi(c(x, y))$ is reached for the chain reduced to (x, y), then $H(x, y) = \rho(x, y)$. Now, we shall prove that the ranking function H determines effectively a ultrametric preordonance, that is, as defined by (10.2.1).

Proposition 53

$$(\forall x, y \text{ and } z \in E) \; H(x, y) \leq r \text{ and } H(y, z) \leq r \Rightarrow H(x, z) \leq r$$

$H(x, y) \leq r$ *means that there exists a chain* $c(x, y)$ *for which* $\phi(c(x, y)) \leq r$. *Let us denote it by* $c(x, y) = (u_1, u_2, \ldots, u_h)$, *where* $u_1 = x$, $u_h = y$ *and* $u_i \neq u_j$, *for* $i \neq j$.

$H(y, z) \leq r$ *means that there exists a chain* $c(y, z)$ *for which* $\phi(c(y, z)) \leq r$. *Let us denote it by* $c(y, z) = (v_1, v_2, \ldots, v_k)$, *where* $v_1 = y$, $v_k = z$ *and* $v_i \neq v_j$, *for* $i \neq j$.

Designate by u_l the first element of the sequence $c(x, y)$ which is found in the sequence $c(y, z)$. This element exits necessarily, since $c(x, y)$ and $c(y, z)$ have y in common. Therefore, assume $u_l = v_m$, where $1 \leq l \leq h$ and $1 \leq m \leq k$. In these conditions, the sequence

$$(u_1, u_2, \ldots, u_l, v_{m+1}, v_{m+2}, \ldots, v_k)$$

defines an element $c(x, z)$ of $\mathcal{C}(x, z)$. Due to the definition of ϕ, we have for this $c(x, z)$, $\phi(c(x, z)) \leq r$. Then a fortiori $H(x, z) \leq r$.   □

10.2.2 Algorithm for Determining ω_u Defined by the H Function

Let $(B_1, B_2, \ldots, B_p)$ be the class sequence of the total preorder ω. We have $B_1 < B_2 < \cdots < B_p$ for the quotient total order on the respective classes. Now, let $(C_1, C_2, \ldots, C_p)$ be the sequence deduced from $(B_1, B_2, \ldots, B_p)$ as follows:

$$C_1 = B_1, \; C_2 = B_1 + B_2, \ldots, \; C_p = B_1 + B_2 + \cdots + B_p$$

where the symbol $+$ expresses disjoint union. For all i, $1 \leq i \leq p$, C_i is a begining section of F for ω.

Now, denote by $\overline{C_i}$ the transitive closure of C_i (recall that if A is a F subset, the transitive closure $\bar{A}$ is the set of unordered element pairs $\{x, y\}$ of E, such that there exists a sequence $(z_1, z_2, \ldots, z_j, \ldots, z_k)$ of E elements for which $x = z_1, y = z_k$ and $\{z_j, z_{j+1}\} \in A$, $1 \leq j \leq k - 1$). Then, let m be the lowest subscript for which $\overline{C_m} = F$. In these conditions, the sequence of transitive closures $(\overline{C_1}, \overline{C_2}, \ldots, \overline{C_m})$ is an increasing sequence with respect to the set inclusion relation. Indeed, $C_i \subset C_j$ implies $\overline{C_i} \subseteq \overline{C_j}$, for $1 \leq i < j \leq m$. This sequence is precisely that of the different begining sections of the ultrametric preordonance ω_u, defined by the function H.

In order to illustrate how this algorithm works, let us apply it to each of the preor-donances ω_s and ω_{s+t} associated with the incidence data Table 4.1 and Sect. 4.2.1.2 of Chap. 4. See Example 1 of Chap. 4 and let us start with

$\omega_s : af < ad \sim df < bc < ac \sim ef$
$< ab \sim bd \sim cd \sim be \sim ce \sim de \sim bf < cf \sim ae$

We get Table 10.1.

The ith row of this table is a Boolean vector with 15 components. It represents the characteristic fonction of $\overline{C_i}$, $1 \leq i \leq 4$. The last row, under the horizontal line, is a ranking (ordinal) function on F, which defines the ultrametric preordonance associated by the *Lexicographic* algorithm with ω_s. A given component of this vector is obtained by substracting to 5 (maximum number of levels -1) the number of 1 components of the corresponding column. In these conditions, the associated partition chain is given in Fig. 10.2.

Table 10.1 Lexicographic algorithm applied on omega(s)

1	0	0	0	0	0	0	0	0	0	0	0	0	0	0
1	1	1	0	0	0	0	0	0	0	0	0	0	0	0
1	1	1	1	0	0	0	0	0	0	0	0	0	0	0
1	1	1	1	1	1	1	1	1	1	1	1	1	1	1
1	2	2	3	4	4	4	4	4	4	4	4	4	4	4

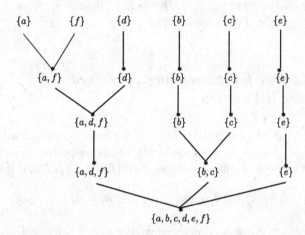

Fig. 10.2 Classification tree associated with the Table 10.1

Similarly, by applying the *Lexicographic* algorithm to

$$\omega_{s+t} : ad \sim bc < be \sim af \sim df < de < bd \sim ce \sim ef$$
$$< ab \sim ac \sim cd < ae \sim bf < cf$$

We obtain Table 10.2.
The associated partition chain is given in Fig. 10.3.

By way of an exercise, the reader is invited to apply the *Lexicographic* algorithm to the *most discriminant* and to the *least discriminant* preordonances deduced from Table 4.2 and associated with Table 4.1 and Sect. 4.2.1.2 (see Sect. 4.2.2.3).

Table 10.2 Lexicographic algorithm applied on omega(s + t)

1	1	0	0	0	0	0	0	0	0	0	0	0	0	0
1	1	1	1	1	0	0	1	0	0	0	0	0	0	0
1	1	1	1	1	1	1	1	1	1	1	1	1	1	1
2	2	3	3	3	4	4	3	4	4	4	4	4	4	4

Fig. 10.3 Classification tree associated with Table 10.2

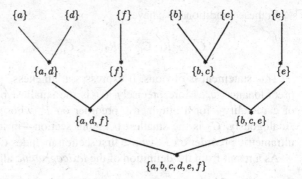

10.2.3   Property of Optimality

The notations are those of the preceding section.

Theorem 32 *Among the ultrametric preordonances of the same type as that of ω_u, ω_u is the nearest to ω, according to the lexicographic order criterion (see Sect. 9.3.2 of Chap. 9).*

Proof Let $t = (t_1, t_2, \ldots, t_i, \ldots, t_m)$ denote the type of ω_u, we have

$$\sum_{1 \leq i \leq j} t_i = card(\overline{C_j}) \geq card(C_j)$$

for $1 \leq j \leq m$.

Let us designate by $(T_1, T_2, \ldots, T_j, \ldots, T_m)$ the sequence of begining sections of a given ultrametric preorder of type t on F. Every ultrametric preordonance on E (defining a ultrametric preorder on F) of type t, for which $C_1 \subseteq T_1$ is nearer ω than that for which the latter inclusion is not satisfied. On the other hand, T_1, being closed transitively, $C_1 \subseteq T_1$ implies $\overline{C_1} \subseteq T_1$. Therefore, due to $card(\overline{C_1}) = card(T_1)$, $\overline{C_1} = T_1$.

Similarly, among the ultrametric preordonances of type t such that $\overline{C_1} = T_1$, each one for which $C_2 \subseteq T_2$ is nearer ω than that for which the latter inclusion is not satisfied. On the other hand, T_2, being closed transitively, $C_2 \subseteq T_2$ implies $\overline{C_2} \subseteq T_2$. Therefore, due to $card(\overline{C_2}) = card(T_2)$, $\overline{C_2} = T_2$. Recursively, we obtain the announced result.   □

Now, let $(R_1, R_2, \ldots, R_k, \ldots, R_l)$ be the sequence of begining sections of a given ultrametric preordonance on E. Assume that there exists an increasing sequence of subscripts $(i_1, i_2, \ldots, i_k, \ldots, i_l)$, where $1 \leq i_1 < i_2 < \cdots < i_k < \cdots < i_l = p$, such that

$$C_{i_1} \subseteq R_1, C_{i_2} \subseteq R_2, \ldots, C_{i_k} \subseteq R_k, \ldots, C_{i_l} \subseteq R_l$$

In these conditions, we have

$$\overline{C_{i_1}} \subseteqq R_1, \overline{C_{i_2}} \subseteqq R_2, \ldots, \overline{C_{i_k}} \subseteqq R_k, \ldots, \overline{C_{i_l}} \subseteqq R_l$$

This statement is obvious. It expresses a "fineness" property of the ultrametric preordonance ω_u. More precisely, $\overline{C_1}$ is the smallest begining section—in terms of cardinality—for a ultrametric preorder on F, whose first section includes C_1. Analogously, $\overline{C_2}$ is the smallest begining section—in terms of cardinality—for a ultrametric preorder on F, whose first section includes C_2 and so on.

As a result from the definition of the *lexicographic* algorithm, we have

Proposition 54 *A necessary and sufficient condition for the ultrametric preordonance ω_u, obtained by the* lexicographic *algorithm from ω, to be comparable with ω,*[1] *is that the transitive closure of each begining section C_i of ω, $1 \leq i \leq p$, is included in a begining section of ω_u.*

This property, which is easy to see, can be formally expressed as follows:
For any i, $1 \leq i \leq p - 1$, there exists an integer h such that $i + h + 1 \leq p$ for which

$$\overline{C_i} = C_{i+h} \cup A_{i+h+1}$$

where A_{i+h+1} is a subset of B_{i+h+1}.

Now, we shall setup some properties of the *lexicographic* algorithm in the case where the data ω is a total ordonance on E (total (linear) order on F).

10.2.4 Case Where ω Is a Total Ordonance

Proposition 55 *In the case where the data ω is a total ordonance on E, the partition chain built by the algorithm is elementary (or maximal) (see Sect. 1.1.2 of Chap. 1)*

The case concerned is characterized by the property: "Each of the classes B_i (see above for its definition), $1 \leq i \leq p$, includes exactly a single element of F". In these conditions, there are $f = n(n-1)/2$ classes B_i, where n designates here $card(E)$. Now, let us compare $\overline{C_{i+1}}$ and $\overline{C_i}$. We have

$$\overline{C_{i+1}} = \overline{\overline{C_i} \cup B_{i+1}}$$

To be more explicit on this formula, let us denote by $\{x, y\}$ the unique element of B_{i+1}. Two alternatives are possible concerning the derivation of $\overline{C_{i+1}}$ from $\overline{C_i}$ and $\{x, y\}$.

---

[1] Two total preorders ω and ω' on a given set are comparable if the graph of one of both total preorders is included in the other one.

1. $\{x, y\} \in \overline{C_i}$ and then $\overline{C_{i+1}} = \overline{C_i}$;
2. $\{x, y\} \notin \overline{C_i}$, then the partition defined by $\overline{C_{i+1}}$ is deduced from that defined by $\overline{C_i}$, by aggregating exactly two classes, those of x and y, respectively.

According to Proposition 54, ω is an ordonance for which the associated ultrametric preordonance ω_u—by the lexicographic algorithm—is *comparable* with ω if and only if the transitive closure of every ω begining section, is an ω begining section. In these conditions, the following question arises "Is this property scarce?" In order to answer this question, let us denote by Ω the set of all total ordonances on E (i.e. all linear orders on F) and by Ω_u the set of total ordonances on E for which the property concerned is satisfied. In these conditions, the problem consists of studying the behavior of the proportion $card(\Omega_u)/card(\Omega)$, when $n = card(E)$ increases.

The cardinality of Ω is $f! = \left(n(n-1)/2\right)!$. The cardinality of Ω_u depends only on $n = card(E)$. Then, we set $\gamma(n) = card(\Omega_u)$. Therefore, we are interested in the ratio $\tau(n) = \gamma(n)/f!$.

10.2.4.1 Enumerating Ω_u

As just expressed above (see Propositions 54 and 55) an ordonance belongs to Ω_u if and only if ω is comparable with a ultrametric preordonance ω_u associated with an elementary partition chain. Thus, ω_u viewed as a total preorder on $F = P_2(E)$, comprises $(n-1)$ classes. By denoting $t_1, t_2, \ldots, t_{n-1}$ the cardinalities of these, taken from left to right, $(t_1, t_2, \ldots, t_{n-1})$ defines the type of ω_u. There are in all $t_1! t_2! \ldots t_{n-1}!$ ordonances in Ω_u comparable with ω. A direct computing of $\gamma(n)$ requires the determination of all types of ultrametric preordonances corresponding to elementary (or maximal) partition chains. Rather than proceeding directly, we shall establish recursive a relation for $\gamma(n)$. We have

Theorem 33

$$\gamma(n) = \sum_{1 \le k \le \frac{n}{2}} \binom{n}{k}\binom{n-2}{k-1}(k(n-k))!\gamma(k)\gamma(n-k) \qquad (10.2.3)$$

Proof The level $n - 2$ of the elementary partition chain defines a partition of E into two classes. Then, the type of this partition is of the form $(n - k, k)$ where $1 \le k \le [n/2]$ (recall that $[n/2]$ is the integer part of $n/2$).

In these conditions, let us consider the decomposition of Ω_u according the type of the partition obtained at the level $n - 2$. There are $\binom{n}{k}$ partitions of E into two classes having the type $(n - k, k)$. Designate by $\Omega_u(2, k)$ the subset of Ω_u for which the partition obtained at the level $n - 2$ is of type $(n - k, k)$. By denoting $\gamma(n, k)$ the cardinality of this subset, we have

$$\gamma(n) = \sum_{1 \le k \le \frac{n}{2}} \gamma(n, k) \qquad (10.2.4)$$

Let us consider a fixed partition of type $(n - k, k)$ that we denote by (E_{n-k}, E_k) and let $\delta(n, k)$ be the number of ordonances for which the partition obtained by the *lexicographic* algorithm at the $n - 2$ level is precisely (E_{n-k}, E_k). Therefore,

$$\gamma(n, k) = \binom{n}{k} \delta(n, k)$$

$$\text{and } \gamma(n) = \sum_{1 \le k \le \frac{n}{2}} \binom{n}{k} \delta(n, k) \tag{10.2.5}$$

Now, let us designate by L and M, the sets of unordered distinct object pairs of E_{n-k} and E_k, respectively. If ω is a total order on F (ordonance on E), we consider the respective restrictions ω_L and ω_M of ω to L and M. ω_{Lu} and ω_{Mu} will designate the ultrametric preordonances associated by the *lexicographic* algorithm applied on E_{n-k} and E_k, respectively.

The following two conditions

1. $\omega \in \Omega_u$;
2. The $n - 2$ level of the partition chain on E—obtained by the *lexicographic* algorithm—is (E_{n-k}, E_k).

are satisfied, if and only if

1. $\omega_L \in \Omega_u(E_{n-k})$ and $\omega_M \in \Omega_u(E_k)$,
2. $L \cup M$ is a begining section for ω,
3. and each preorder class of ω_{Lu} (resp., ω_{Mu}) is an ordinal interval of F for ω.

More precisely, relative to the third item, the different intervals of the sequence of intervals defined by ω_{Lu} are coming between the different intervals of the sequence of intervals defined by ω_{Mu}. In this respect, let us recall that ω_{Lu} and ω_{Mu} include $n - k - 1$ and $k - 1$ classes, respectively.

$\delta(n, k)$ is the number of ways by which we can constitute an ordonance on E satisfying the above three conditions. To build ω, it is necessary to

1. Determine an ordonance ω_L on (E_{n-k}) belonging to $\Omega_u(E_{n-k})$, this can be done in $\gamma(n - k)$ ways;
2. Determine an ordonance ω_M on (E_k) belonging to $\Omega_u(E_k)$, this can be done in $\gamma(k)$ ways;
3. Insert the different ordinal intervals of the sequence of intervals defined by ω_{Lu} with respect to the sequence of intervals defined by ω_{Mu}, this can be done in $\binom{n-2}{k-1}$ different ways (indeed, ω_{Lu} and ω_{Mu} include $n - k - 1$ and $k - 1$ classes, respectively);
4. Finally, establish a total (linear) order on the $(n - k) \times k$ pairs $\{x, y\}$, for which x and y belong to (E_{n-k}) and (E_k), respectively, this can be done in $(k(n - k))!$ different ways (Table 10.3 continued with Table 10.4).

Table 10.3 Behavior of $\gamma(n)$

n	3	4	5	6	7	8
$\gamma(n) \geq$	6	10^2	10^5	10^{10}	10^{14}	10^{21}
$\frac{\gamma(n)}{f!} \leq$	1	$6/10$	$5/10^2$	$3/10^3$	$1/10^5$	$3/10^8$

Table 10.4 Number of ordonances comparable with a ultrametric preordonance

n	9	10	11	12	13	14	15
$\gamma(n) \geq$	10^{30}	10^{40}	10^{51}	10^{65}	10^{81}	10^{99}	10^{118}
$\frac{\gamma(n)}{f!} \leq$	$3/10^{12}$	$2/10^{16}$	$8/10^{22}$	$2/10^{27}$	$2/10^{34}$	$9/10^{42}$	$3/10^{50}$

Consequently, we have the relation

$$\delta(n, k) = \gamma(n - k)\gamma(k)\binom{n - 2}{k - 1}(k(n - k))!$$

which accomplishes the proof. □

As shown in Table 10.3 continued with Table 10.4, $\gamma(n)$ increases quickly when n increases. On the other hand, the proportion $\frac{\gamma(n)}{f!}$ tends quickly towards 0 when n increases.

10.3 Ascendant Agglomerative Hierarchical Clustering Algorithm; Classical Aggregation Criteria

10.3.1 Preamble

Let E be a finite set, possibly weighted and provided with a dissimilarity or distance function $\mathcal{D}$. The data is then defined by a triplet $(E, \mathcal{D}, \mu_E)$, where μ_E is a weighting on E. We refer here to Ascendant Agglomerative Hierarchical Clustering (Classification), following the algorithmic principle expressed in the Introduction (see Sect. 10.1):

"At each step merge the nearest class pairs according to a dissimilarity function δ between disjoint E subsets"

As expressed in Sect. 10.1, the initial state of the algorithm is the finest partition of E, for which each class is a *singleton* class including exactly one element and the final state, is the "rough" partition into one single class, namely E. The passage from a given level to the next one requires the formal definition of a dissimilarity function δ between two disjoint subsets C and D of E. There are many such functions. We referred in the Introduction (see Sect. 10.1) to the most classical of them that we have denoted by $\delta_{\min}$, $\delta_{\max}$, δ_{ave} and δ_{iner}. The associated methods with the three first criteria are called "Single Linkage method", "Complete Linkage method" and "Average Linkage method", respectively. These criteria are defined locally with respect dissimilarities between elements as that of C and those of D (see (10.1.2))

$$\mathcal{D}(C, D) = \{\mathcal{D}(x, y) | (x, y) \in C \times D\}$$

They will be developed in Sect. 10.3.2. The Ward criterion δ_{iner} has a specific global interpretation. It concerns clustering an object set which can be represented as a cloud of points in an euclidean space. It will be detailed in Sect. 10.3.3. Section 10.3.4 is devoted to an interesting result showing equivalence property between "lexicographic" and "single linkage" algorithms.

10.3.2 "Single Linkage", "Complete Linkage" and "Average Linkage" Criteria

10.3.2.1 Single Linkage and Complete Linkage Criteria

In Topology, a metric space is a set E provided with a distance function $\mathcal{D}$ on E. In this context, the distance between two nonempty E subsets C and D is defined as $\delta(C, D) = inf\{\mathcal{D}(x, y) | (x, y) \in C \times D\}$, where inf operation becomes min operation in the case where E is finite. "Single Linkage" criterion is defined exactly in the same way for a more general case where $\mathcal{D}$ is a dissimilarity function. For the latter *triangle inequality*, is not necessarily required. In these conditions, by considering a binary construction of the classification tree on E, at each step of "Single Linkage" algorithm, relative to the classes formed (including singleton classes), a class pair $\{C, D\}$ minimizing

$$\delta_{min}(C, D) = min\{\mathcal{D}(x, y) | (x, y) \in C \times D\}$$

is merged. According to Jardine and Sibson [25], this method was first introduced in the clustering framework by Florek and coworkers [17, 18]. It was introduced independently by McQuitty [45] and by Sneath [55].

Let us consider—for a binary construction of the classification tree—the case, mentioned above, where several class pairs achieve simultaneously the same minimal value of $\delta_{min}(C, D)$. Assume that following each merging class at a given level, calculation of class dissimilarities between the new class and those not affected by the aggregation, is processed. Denote by $\{C_1, D_1\}$ the first merged pair at a given level and by $\{C_2, D_2\}$ a second class pair with the same value of $\delta_{min}(C, D)$. If $\{C_2, D_2\} \cap \{C_1, D_1\} = \emptyset$, the class $\{C_2 \cup D_2\}$ is necessarily constituted at the next level. If $\{C_2, D_2\} \cap \{C_1, D_1\} \neq \emptyset$, by supposing—without loss of generality— $C_1 = C_2$, the class D_2 is aggregated to the new cluster $C_1 \cup D_1$. Indeed, $\delta_{min}(C_1 \cup D_1, D_2) = \delta_{min}(C_2, D_2)$. This invariance property is not preserved for the merging criteria considered below. For any one of them denoted $\delta_\star$ it can be established (this is left for the reader) that

$$\delta_\star(C_1 \cup D_1, D_2) > \delta_\star(C_2, D_2)$$

For this case, the adequate technique consists of merging in the following successive levels of the *binary* tree the nearest class pairs, and calling at each level for a reactualization formula defined in the binary case (see Sects. 10.6.3.1 and 10.6.3.2). More precisely, if $\{C, D\}$ is a class pair among the nearest class pairs appearing at a given level l, C (resp., D) has to join D (resp., C) or the cluster including D (resp., C) which may appear at one of the next levels from l. In these conditions, the minimal value of the dissimilarity function between the new clusters formed is only required after exploiting all the equally nearest class pairs of the level concerned l.

A new solution is expressed in Sect. 10.6.2 [36]. In this, all of the nearest class pairs at a given level l are merged in a single step at the next level $l + 1$. However, and clearly, the classification tree obtained is not necessarily binary.

According to the introduced notations above, the merging criterion sustaining the "Complete Linkage method" is defined by

$$\delta_{max}(C, D) = \max\{\mathcal{D}(x, y) | (x, y) \in C \times D\}.$$

It corresponds to the *diameter* of the cluster $C \cup D$. Thus, the "Complete Linkage Algorithm" consists of agglomerating at each step the class pairs $\{C, D\}$ for which the diameter value of $C \cup D$ is minimal. When this minimal value occurs more than one time at a given level l and for a binary construction of the classification tree, a similar solution as that considered above can be adopted. That is to say, the class pairs concerned by the minimal value of $\delta_{max}(C, D)$ are merged successively without calling for a reactualization formula of class dissimilarities. Each merging is derived from a junction between two clusters including, respectively, two classes C' and D' where $\{C', D'\}$ is one of the class pairs achieving the minimal value of δ_{max} at the tree level l.

According to Jardine and Sibson [25], the original form of Complete Linkage method is due to Sorensen [58].

10.3.2.2 Average Linkage Criterion

This type of merging criterion was suggested by Sokal and Michener (1958) [57]. Relative to the rectangular table of dissimilarities $\mathcal{D}(C, D)$ crossing two clusters C and D (see above and (10.1.2)) This criterion is expressed by

$$\delta_{ave}(C, D) = \frac{1}{card(C) \times card(D)} \sum_{(x,y) \in C \times D} \mathcal{D}(x, y) \qquad (10.3.1)$$

In the Single and Complete Linkage methods, only ordinal operations take part, namely, min and max, respectively. On the other hand, in the Average Linkage criterion, operations concerning the reals are exploited.

Now, let us assume the whole set E provided with a positive weighting $\{\mu_x | x \in E\}$, the Average Linkage criterion becomes

$$\delta_{ave}^w(C, D) = \frac{1}{\mu(C) \times \mu(D)} \sum_{(x,y) \in C \times D} \mu_x \mu_y \mathcal{D}(x, y) \qquad (10.3.2)$$

where, for a subset X of E, $\mu(X)$ is the sum of the weights of the different elements of X.

A particular case of this equation occurs when E corresponds to a set of classes or categories and where μ_x ($x \in E$) is defined by the cardinality of x, $\mathcal{D}$ being a dissimilarity notion between elements of E. $\delta_{ave}(C, D)$ (resp., $\delta_{ave}^w(C, D)$) is commonly called the *Unweighted* (resp., *Weighted*) Average Linkage criterion. We shall work below with the Unweighted version of this criterion, the Weighted version comes done to the Unweighted one.

10.3.3 "Inertia Variation (or Ward) Criterion"

10.3.3.1 Introduction

This criterion addresses clustering of a finite set E provided with a positive weighting $\{\mu_x | x \in E\}$ and represented by a cloud of points in an euclidean space. The classical case was considered in Sect. 7.2.1, it concerns clustering of a set $\mathcal{O} = \{o_1, o_2, \ldots, o_i, \ldots, o_n\}$ of elementary objects described by a set $\mathcal{A} = \{a^1, a^2, \ldots, a^j, \ldots, a^p\}$ of numerical attributes. Thus, $\mathcal{O}$ is represented by a cloud of points in the geometrical space $\mathbb{R}^p$, see (7.1.2) which is taken up again here:

$$\mathcal{N}(\mathbb{I}) = \{(O_i, \mu_i) | i \in \mathbb{I}\} \qquad (10.3.3)$$

where $\mathbb{I}$ indexes the object set $\mathcal{O}$. By denoting O the $\mathbb{R}^p$ space origin, we have

$$\overrightarrow{OO_i} = (O_i - O) = \sum_{1 \leq j \leq p} x_i^j \vec{e}_j$$

where $x_i^j = a^j(o_i)$ and where $\{\vec{e}_j | 1 \leq j \leq p\}$ is the canonical basis of $\mathbb{R}^p$.

The geometrical space $\mathbb{R}^p$ is supposed endowed with a definite positive quadratic form, defining a *metric* q on $\mathbb{R}^p$ (see Chap. 11). q is generally represented by the definite positive symmetrical matrix

$$\{q(e_j, e_k) | 1 \leq j, k \leq p\}$$

The notations of Sect. 7.2.1 are retaken here. Thus

$$G = \frac{1}{\mu} \sum_{i \in \mathbb{I}} \mu_i O_i$$

where $\mu = \sum_{i \in \mathbb{I}} \mu_i$ and G is the *center of gravity* (centroid) of $\mathcal{N}(\mathbb{I})$.

Recall what we saw in Sect. 5.3.2 how to adapt this representation for mutual comparison between rows of a contingency table.

The *inertia moment* of the cloud $\mathcal{N}(\mathbb{I})$, that we denote by $\mathcal{M}(\mathcal{N}(\mathbb{I}))$ is defined by

$$\mathcal{M}(\mathcal{N}(\mathbb{I})) = \sum_{i \in \mathbb{I}} \mu_i \|O_i - G\|^2 \qquad (10.3.4)$$

where $\|O_i - G\|^2$ designates the square distance $d^2(G, O_i)$, according to the q metric between the center of gravity G and the summit O_i, namely, $q(\overrightarrow{GO_i}, \overrightarrow{GO_i})$, $1 \leq i \leq n$.

Using the centroid properties, it is easy to establish that

$$\mathcal{M}(\mathcal{N}(\mathbb{I})) = \sum_{(i,i') \in \mathbb{I} \times \mathbb{I}} \mu_i \mu_{i'} d^2(O_i, O_{i'}) \qquad (10.3.5)$$

In this manner, $\mathcal{M}(\mathcal{N}(\mathbb{I}))$ appears as a measure of how much are mutually distinguishable the different elements of $\mathcal{O}$, or in other terms, a measure of the heterogeneity of the cloud $\mathcal{N}(\mathbb{I})$.

Now, let us consider a partition $P_k = \{\mathcal{O}_j | 1 \leq j \leq k\}$ of $\mathcal{O}$ and denote by $\{\mathbb{I}_j | 1 \leq j \leq k\}$ the corresponding partition of the associated subscript set $\mathbb{I} = \{1, 2, \ldots, i, \ldots, n\}$. More precisely, $\mathbb{I}_j$ is the subscript set of $\mathcal{O}_j$, $1 \leq j \leq k$. In this framework the reduced representation of a given class $\mathcal{O}_j$ is the ordered pair (G_j, M_j), where (G_j and M_j) are the center of gravity and the weight of the subcloud

$$\mathcal{N}(\mathbb{I}_j) = \{(O_i, \mu_i) | i \in \mathbb{I}_j\} \qquad (10.3.6)$$

We have

$$M_j = \sum_{i \in \mathbb{I}_j} \mu_i \quad \text{and} \quad G_j = \frac{1}{M_j} \sum_{i \in \mathbb{I}_j} \mu_i O_i$$

Consequently, the class set $\{\mathcal{O}_j | 1 \leq j \leq k\}$ is represented by the cloud

$$\mathcal{N}(\mathcal{G}) = \{(G_j, M_j) | 1 \leq j \leq k\} \qquad (10.3.7)$$

The inertia moment of this cloud is

$$\mathcal{M}(\mathcal{N}(\mathcal{G})) = \sum_{1 \leq j \leq k} M_j \|G_j - G\|^2 \qquad (10.3.8)$$

where

$$G = \frac{1}{\mu} \sum_{1 \leq j \leq k} M_j G_j$$

It defines the "explained" (or "retained") inertia by the partition P_k. By denoting it $\mathcal{I}_e(P_k)$, we can write

$$\mathcal{I}_e(P_k) = \mathcal{M}(\mathcal{N}(\mathcal{G})) \qquad (10.3.9)$$

since the general center of gravity G is the center of gravity of $\mathcal{N}(\mathcal{G})$.

The passage from the moment of inertia of $\mathcal{N}(\mathbb{I})$ to that of $\mathcal{N}(\mathcal{G})$ is the matter of the famous Huyghens (1629–1695) theorem. The latter is expressed by the following relation:

$$\mathcal{M}(\mathcal{N}(\mathbb{I})) = \mathcal{M}(\mathcal{N}(\mathcal{G})) + \sum_{1 \leq j \leq k} \mathcal{M}(\mathcal{N}(\mathbb{I}_j)) \qquad (10.3.10)$$

The second term of the right member is called the "lost" (or "residual") inertia. For the inertia criterion, the fitting quality of the partition P_k is defined by the value of $\mathcal{M}(\mathcal{N}(\mathcal{G}))$. For a fixed number of classes k, a non-hierarchical clustering algorithm— such as the *K-means* algorithm (see Chap. 2)—seeks to maximize $\mathcal{M}(\mathcal{N}(\mathcal{G}))$.

10.3.3.2 Criterion of Explained Inertia in Hierarchical Clustering

The inertia variation criterion, commonly called *Ward criterion* [59] is defined as the variation of the global explained inertia between two consecutive partition levels of the classification tree associated with a hierarchical clustering. Let us consider a binary version of the classification tree on an object set $\mathcal{O}$. If $\mathcal{O}$ comprises n elements—by counting the leaves and the root level—there are in all n levels. These are labelled $0, 1, 2, \ldots, k, \ldots, n - 1$, respectively. Denote by $P_k = \{\mathcal{O}_1, \mathcal{O}_2, \ldots, \mathcal{O}_j, \ldots, \mathcal{O}_{n-k}\}$, the partition of level k, $0 \leq k \leq n - 1$. The classes of P_k correspond to last nodes formed up to level k, with terminal nodes (leaves) which have not participated to the aggregation process.

Let us now consider the passage between the levels k and $k + 1$, $0 \leq k \leq n - 2$. The partitions concerned can be denoted by $P_k = \{\mathcal{O}_1, \mathcal{O}_2, \ldots, \mathcal{O}_j, \ldots, \mathcal{O}_{n-k}\}$ and $P_{k+1} = \{\mathcal{O}'_1, \mathcal{O}'_2, \ldots, \mathcal{O}'_j, \ldots, \mathcal{O}'_{n-k-1}\}$. Suppose that the classes $\mathcal{O}_h$ and $\mathcal{O}_{h'}$ merge in P_{k+1}, $1 \leq h < h' \leq n - k$, the variation (decrease) of the explained inertia becomes

$$\Delta(k, k+1) = \mathcal{I}_e(P_k) - \mathcal{I}_e(P_{k+1})$$
$$= M_h\|G_h - G\|^2 + M_{h'}\|G_{h'} - G\|^2 - (M_h + M_{h'})\|G_{h\cup h'} - G\|^2$$

$$(10.3.11)$$

where $G_{h\cup h'}$ is the center of gravity of the class $\mathcal{O}_h \cup \mathcal{O}_{h'}$. The property

$$G_{h\cup h'} = \frac{M_h G_h + M_{h'} G_{h'}}{M_h + M_{h'}}$$

enables the relation

$$\Delta(k, k+1) = \frac{M_h M_{h'}}{M_h + M_{h'}} \times \|G_h - G_{h'}\|^2 \qquad (10.3.12)$$

to be established.

It is of importance to notice that this index is exactly the moment of inertia of the following cloud, comprising exactly two weighted summits

$$\{(G_h, M_h), (G_{h'}, M_{h'})\} \qquad (10.3.13)$$

Therefore, the merging criterion of explained inertia variation $\Delta(k, k+1)$—which was presented as the portion concerned of a global criterion in merging two classes—is come down to a *local* criterion depending only on the two classes to aggregate. This criterion has an additive property with respect to the internal nodes (including the root) formed in the classification tree. In order to see that, begin by considering the passage between levels 0 and 1. Without loss of generality, assume that the elements O_1 and O_2 of $\mathcal{O}$ merge at the level 1. The decrease of the moment of inertia between the partitions P_0 and P_1, according to their representations (see (10.3.7)), can be written as

$$\mu_1\|O_1 - G\|^2 + \mu_2\|O_2 - G\|^2 - (\mu_1 + \mu_2)\|G_{1\vee 2} - G\|^2 = \frac{\mu_1\mu_2}{\mu_1 + \mu_2}\|O_1 - O_2\|^2$$

In fact, this decrease is the value of the merging criterion for $\{(O_1, \mu_1), (O_2, \mu_2)\}$.

Now, let us designate by $\mathcal{N}(\mathcal{G}_k)$ the centers of gravity of the P_k partition classes. The total inertia moment $\mathcal{M}(\mathcal{N}(\mathbb{I}))$ corresponds to the inertia of $\mathcal{N}(\mathcal{G}_0)$. In the passage between P_0 and P_1 the lost inertia can be written as

$$\Delta(0, 1) = \mathcal{M}(\mathcal{N}(\mathcal{G}_0)) - \mathcal{M}(\mathcal{N}(\mathcal{G}_1))$$

More generally, in the passage between P_k and P_{k+1}, the lost inertia is

$$\Delta(k, k+1) = \mathcal{M}(\mathcal{N}(\mathcal{G}_k)) - \mathcal{M}(\mathcal{N}(\mathcal{G}_{k+1}))$$

$0 \leq k \leq n - 2$. By noticing that $\mathcal{N}(\mathcal{G}_{n-1})$ is reduced to the single weighted point (G, μ), we obtain

$$\mathcal{M}(\mathcal{N}(\mathbb{I})) = \mathcal{M}(\mathcal{N}(\mathcal{G}_0)) = \sum_{0 \leq k \leq n-2} \Delta(k, k + 1)$$

It follows the

Proposition 56 *The total moment of inertia* $\mathcal{M}(\mathcal{N}(\mathbb{I}))$ *is equal to the sum of merging criteria, assigned respectively to the internal nodes (including the root) of the binary classification tree.*

This result can be extended in the case of non-binary tree (see Sect. 10.6.2). Specification of this extension is left for the reader.

As expressed in the introductory Sects. 10.1 and 10.3.1 the passage from P_k to P_{k+1} in an Ascendant Agglomerative Hierarchical Clustering, is carried out by merging those class pairs for which $\Delta(k, k + 1)$ is minimal.

10.3.4 From "Lexicographic" Ordinal Algorithm to "Single Linkage" or "Maximal Link" Algorithm

We return now to our notations of Sect. 10.2.2: $(B_1, B_2, \ldots, B_p)$ is the class sequence of the total preorder ω on the set $F = P_2(E)$ of unordered element pairs of E. The preordonance on E, represented by ω is supposed associated with a dissimilarity index (function) $\mathcal{D}$ on E. The sequence of begining sections of ω is defined by

$$C_1 = B_1, \ C_2 = B_1 + B_2, \ldots, \ C_p = B_1 + B_2 + \cdots + B_p$$

where the symbol $+$ expresses disjoint union. The transitive closure $\overline{C_i}$ of C_i determines a partition on E. In fact, $\overline{C_i}$ can be interpreted as the graph of a binary equivalence relation on E, $1 \leq i \leq p$. We have $\overline{C_i} \subseteq \overline{C_{i+1}}$, $1 \leq i \leq p - 1$. Thus, two consecutive F subsets $\overline{C_i}$ and $\overline{C_{i+1}}$ are not necessarily different. In these conditions, let m be the lowest subscript for which $\overline{C_m} = F$, $m \leq p$. Therefore, the sequence $(\overline{C_1}, \overline{C_2}, \ldots, \overline{C_p})$ determines an ordered partition chain on E that we designate by $(P_0, P_1, \ldots, P_j, \ldots, P_m)$, where P_0 (resp., P_m) is the finest partition into $card(E)$ classes (resp., the rawest partition into one single class). This partition chain is precisely obtained by the *ordinal lexicographic algorithm*.

Let us denote by $\{E_1^j, E_2^j, \ldots, E_g^j, \ldots, E_{k_j}^j\}$ the set class of P_j. The unordered element pairs of E joined by P_j can be expressed by

$$R(P_j) = \sum_{1 \leq g \leq k_j} P_2(E_g^j)$$

(set sum) where $P_2(E_g^j)$ designates the set of element pairs of E_g^j, $1 \leq g \leq k_j$.

Let $h(j)$ be the greatest subscript for which $\overline{C_{h_j}} = R(P_j)$ where $j < m$. Then, $B_{h_j+1} - \overline{C_{h_j}} \neq \emptyset$ (set difference). A given pair $\{x, y\}$ in the latter set, belongs necessarily to the set $S(P_j)$ of unordered pairs separated by the partition P_j. Among these pairs, $\{x, y\}$ is one for which the dissimilarity value of $\mathcal{D}$ is minimal. Assume $x \in E_f^j$ and $y \in E_g^j$ ($f \neq g$), then E_f^j and E_g^j are nearest neighbours classes according the merging criterion $\delta_{\min}$. Necessarily, they will be aggregated at the next step of the *ordinal lexicographic algorithm* and this, by taking the transitive closure of C_{h_j+1}. It follows the

Theorem 34 *For a preordonance ω on E associated with a dissimilarity index $\mathcal{D}$ on E, the classification tree calculated by the* Ordinal Lexicographic Algorithm *is equivalent to that obtained by* Single Linkage Method. *Nevertheless, the built tree is not necessary binary: in the passage from one tree level to the next one, all the class pairs achieving the same value of the merging criterion $\delta_{\min}$ are joined in one step.*

10.4 *AAHC* Algorithms; Likelihood Linkage Criteria

10.4.1 *Family of Criteria of the Maximal Likelihood Linkage*

10.4.1.1 Introduction

In order to understand in an intuitive manner this type of proximity criterion between classes, consider Fig. 10.4, which is placed in the simple and classical context of geometrical two-dimensional cloud. We suggest two class pairs: $\{C_1, \mathcal{D}_1\}$ and $\{C_2, \mathcal{D}_2\}$. In this drawing $\{C_1$ and $\mathcal{D}_1\}$ are assumed strongly dense, but $\{C_2$ and $\mathcal{D}_2\}$ are more weakly

Fig. 10.4 Principle of the Maximal Likelihood Linkage criterion

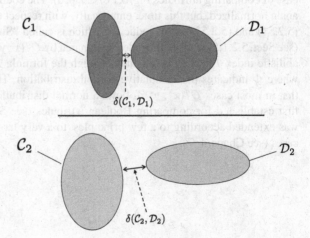

dense. In this figure δ stands for δ_{min} of Single Linkage method (see Sect. 10.3.2.1). As suggested in Fig. 10.4, we have

$$\delta(\mathcal{C}_1, \mathcal{D}_1) < \delta(\mathcal{C}_2, \mathcal{D}_2)$$

Notice that in the forming of an agglomerative ascendant tree, all the construction depends on the aggregation order. In these conditions, the following question arises "Do we have to join $\mathcal{C}_1$ and $\mathcal{D}_1$ before $\mathcal{C}_2$ and $\mathcal{D}_2$?"

In fact, we interpret δ as a *raw* index. The likelihood of the maximal link criterion based directly on δ performs the opposite of the previous proposition: it begins by aggregating $\mathcal{C}_2$ and $\mathcal{D}_2$ before aggregating $\mathcal{C}_1$ and $\mathcal{D}_1$. Indeed, the smallness of $\delta(\mathcal{C}_2, \mathcal{D}_2)$ is considered as more *exceptional* than that of $\delta(\mathcal{C}_1, \mathcal{D}_1)$, by taking into account the point densities of the two classes to be compared.

This idea also appears fundamental in Information theory, where the higher the quantization of an event, the more unlikely it is. The events with which we are concerned here are the observed relations between descriptive attributes (respectively, described objects or categories), or between attribute classes (respectively, object classes or category classes).

As expressed above (see the introductory Sect. 10.1), whatever the nature of the set to be clustered (set of descriptive attributes, set of objects or set of categories), the elaboration of a table of probabilistic similarity indices of the form

$$\mathcal{P}^g = \{P^g(x, y)|\{x, y\} \in P_2(E)\}$$

was achieved in Chaps. 5–7. For a given $\{x, y\}$ of the set of unordered element pairs $P_2(E)$, $P^g(x, y) = \Phi(Q^g(x, y))$, where $Q^g(x, y)$ is obtained from twofold normalizations of a raw similarity index $s(x, y)$. The first one is done with respect to a random choice of an ordered pair $(x^\star, y^\star)$ associated with (x, y) under a suitable random model, where ($x^\star$ and $y^\star$) are independent (see for example for the case of comparing attributes Fig. 6.1 of Chap. 6). The coefficient obtained $Q(x, y)$ is again normalized, but this time, empirically, with respect to $P_2(E)$ (see for example (5.2.51) and (5.2.52)). This reduction which is called "Similarity global reduction" (see Sect. 5.2.1.5) leads to the index designated by $Q^g(x, y)$ (see (5.2.52)). The probabilistic index $P^g(x, y)$ is obtained through the formula $P^g(x, y) = \Phi(Q^g(x, y))$, where Φ indicates the cumulative normal distribution. This is justified by the fact that in most cases $Q^g(x^\star, y^\star)$ follows a normal distribution. This construction was first established for comparing Boolean attributes (see Sect. 5.2.1). Afterwards, it was extended according to a few principles, to a very large and organized scope of cases (see Chaps. 4–7).

10.4.1.2   Analysis of the Independence Hypothesis Between Two Clusters

As expressed above and notably in the introductory Sect. 10.1 the *raw* similarity index between two disjoint subsets C and D of E is defined from the table

$$\mathcal{P}^g(C, D) = \{P^g(x, y)|(x, y) \in C \times D\}$$

(see (10.1.9)). That corresponding to "Maximal Link criterion" is defined by

$$\sigma_{\max}(C, D) = \max\{P^g(x, y)|(x, y) \in C \times D\}$$

(see (10.1.8)). We denote below by l and m the cardinalities of C and D, respectively.

Whatever is the nature of the set E to be clustered, the *Hypothesis of no relation* associates a set $\mathcal{A}^* = \{a^{1*}, a^{2*}, \dots, a^{j*}, \dots, a^{p*}\}$ of independent random attributes with the observed set of attributes $\mathcal{A} = \{a^1, a^2, \dots, a^j, \dots, a^p\}$. The probability law of a^{j*} is associated with the empirical distribution of a^j, $1 \leq j \leq p$. If x^* and y^* are two independent random elements associated with x and y respectively, $P^g(x^*, y^*)$ follows, from a probability integral transformation, a uniform probability distribution on the $[0, 1]$ interval. In these conditions, the random raw index associated with $\sigma_{\max}(C, D)$ can be written as

$$\sigma_{\max}(C^*, D^*) = \max\{P^g_\cdot(x^*, y^*)|(x^*, y^*) \in C^* \times D^*\} \tag{10.4.1}$$

This random index appears as the *maximum* of $l \times m$ random variables uniformly distributed on $[0, 1]$. Strictly, all of these random variables are not mutually independent. Two random elements (x^*, y^*) and (z^*, t^*) of $C^* \times D^*$, where $x^* \neq z^*$ and $y^* \neq t^*$ are independent. However, two random elements of the form (x^*, y^*) and (x^*, t^*) (resp., (x^*, y^*) and (z^*, y^*)) where $x^* \in C^*$ and $\{y^*, t^*\} \subset D^*$ (resp., $\{x^*, z^*\} \subset C^*$ and $y^* \in D^*$) are not in general independent. Nevertheless, the vast majority of pairs $\{(x^*, y^*), (x'^*, y'^*)\}$ of $(C^* \times D^*)^{\{2\}}$ (set of 2 subsets of $C^* \times D^*$) are of the form $\{(x^*, y^*), (z^*, t^*)\}$ where different letters indicate different elements; that is, $x^* \neq z^*$ and $x^* \neq t^*$ (resp., $y^* \neq z^*$ and $y^* \neq t^*$). To realize this statement, let us enumerate these pairs. Their number is equal to $h = l \times (l-1) \times m \times (m-1)$. It remains for the number of pairs $\{(x^*, y^*), (x'^*, y'^*)\}$ such that x^* or y^* is repeated in (x'^*, y'^*), the value $g = l \times m \times (l + m - 2)$.

Notice that if E is a set of attributes of any type, we can establish for the permutational no relation random model, the independence between (x^*, y^*) and (x^*, t^*) [resp., (z^*, y^*)]. Thus, for numerical attributes (see Sect. 5.2.2.1), we get the probabilistic independence between $(x(\sigma), y(\tau))$ and $(x(\sigma), t(\theta))$ [resp., $(z(\theta), y(\tau))$], where σ, τ and θ are three independent permutations taken in the set G_n—provided with a uniform probability measure—of all permutations on the subscript set $\mathbb{L} = \{1, 2, \dots, i, \dots, n\}$, coding the object set $\mathcal{O}$.

Let us mention here the claim of [26] for which "the likelihood linkage analysis approach can be regarded as strongly related to the combination of independence tests" [43].

Now, relaxation of the *Hypothesis of no relation* provides perfect mutual independence between random probabilistic indices associated with $C \times D$. For this purpose and to begin, let us express C and D as follows:

$$C = \{x_1, x_2, \ldots, x_i, \ldots, x_l\}, C = \{y_1, y_2, \ldots, y_j, \ldots, y_m\}$$

Consider a sequence $\left(x_i^{*1}, x_i^{*2}, \ldots, x_i^{*j}, \ldots, x_i^{*m}\right)$ of m independent copies of x_i^*, where x_i^* is the random element associated with x_i, according to the previous version of the hypothesis of no relation.

Analogously, consider a sequence $\left(y_j^{*1}, y_j^{*2}, \ldots, y_j^{*i}, \ldots, y_j^{*l}\right)$ of l independent copies of y_j^*, where y_j^* is the random element associated with y_j, according to the previous version of the hypothesis of no relation.

In these conditions,

$$\{P^g(x_i^{*j}, y_j^{*i})|1 \le i \le l, 1 \le j \le m\} \tag{10.4.2}$$

define a set of $l \times m$ independent random variables, uniformly distributed on the interval $[0, 1]$. Therefore, by considering the random index

$$\sigma_{max}^r(C^*, D^*) = \max\{P^g(x_i^{*j}, y_j^{*i})|1 \le i \le l, 1 \le j \le m\} \tag{10.4.3}$$

where the upper index r in σ_{max}^r refers to the relaxed independence random model. Thus, we obtain the expression of the pure form of the *Maximum Likelihood Linkage* merging criterion

$$P(C, D) = Prob\{\sigma_{max}^r(C^*, D^*) \le \sigma_{max}(C, D)\} = p(C, D)^{l \times m} \tag{10.4.4}$$

where $p(C, D)$ stands for $\sigma_{max}(C, D)$.

Let us express this merging criterion with respect to the table $\mathcal{D}$ of *Informational Dissimilarity matrix* (see (10.1.6)), we obtain

$$\delta_{lla}(C, D) = l \times m \times \min\{-\text{Log}_2(P^g(x, y))|(x, y) \in C \times D\}$$
$$= l \times m \times (-\text{Log}_2(p(C, D))) \tag{10.4.5}$$

By merging—from one level to the next one—the class pairs achieving the value of $\delta_{lla}(C, D)$, we get exactly the same classification tree as that for which the maximal value of $P(C, D)$ is required.

In order to enlarge the formal type of criterion given by $P(C, D)$ (see (10.4.4)), F. and H. Bacelar Nicolau [11, 49] have introduced the following family of criteria

$$P_{f(l,m)}(C, D) = p(C, D)^{f(l,m)} \tag{10.4.6}$$

where $f(l, m)$ is an increasing function of l and m. One specific function proposed in [11, 49] is $\sqrt{l \times m}$. The associated criterion can be written as

$$\mathcal{D}_{(0.5,0.5)}(C, D) = \sqrt{l \times m} \times \min\{-\mathrm{Log}_2(P^g(x, y)) | (x, y) \in C \times D\}$$

Some formal analogy might be noticed between the latter and the Ward criterion (see (10.3.12)). In fact, the passage from the Ward criterion to $\mathcal{D}_{(0.5,0.5)}(C, D)$ is obtained by substituting for the *harmonic* mean between l and m the geometric one and for the square distance between the centers of gravity of C and D, $\min\{-\mathrm{Log}_2(P^g(x, y)) | (x, y) \in C \times D\}$.

Following the latter Nicolau proposal $(\mathcal{D}_{(0.5,0.5)}(C, D))$, we suggested the family of criteria:

$$LL_{\max}^{\epsilon}(C, D) = \big(p(C, D)\big)^{(l \times m)^{\epsilon}} \tag{10.4.7}$$

where ϵ is a positive real parameter between 0 and 1 [37, 38]. This family goes from the maximal link ($\epsilon = 0$) to the pure form of maximal link likelihood criterion ($\epsilon = 1$). Notice that for this value, the merging criterion was designated above by $P(C, D)$ (see (10.4.4)). For reasons of accuracy in numerical computing, we consider the strictly increasing function

$$S_{\max}^{\epsilon}(C, D) = -\log_2\big(-\log_2(LL_{\max}^{\epsilon}(C, D))\big) \tag{10.4.8}$$

which leads to exactly the same classification tree. Notice that using the logarithm function $\log_e$ instead of $\log_2$ does not affect the tree formation.

Influenced by the above formal comparison with the Ward criterion in the case where $\epsilon = 0.5$, this value can be taken as a reference value. Around it and beyond, there is a great stability in the hierarchical clustering results. Nevertheless, an adjusted value of ϵ might lead to a progress and enables the finest results in accordance with the expert knowledge, to be obtained. This fitting depends on the nature of the data and the application domain concerned. The "optimal" value of ϵ is directly related to a clear interpretation in agglomerative cluster formation. In the most recent experiments concerning clustering proteic structures, a weak value of ϵ such as $\epsilon = 0.2$ has led to the best results [1]. The same observation occured in experiments in image quantization [19, 20]

I. Kojadinovic [26] considers that an accurate random model of no relation (independence) between two classes C and D must preserve the mutual relations between the elements of both classes taken separately. In the case studied in [26]—where C and D are two classes of numerical attributes, the permutational form of such an independence hypothesis can be expressed as follows: Fix C and associate a random set $D^\star$ with D depending on a single random permutation $\pi^\star$ on the subscript set $\mathbb{I} = \{1, 2, \ldots, i, \ldots, n\}$ of the object set $\mathcal{O}$. $\pi^\star$ has a simultaneous effect on all attributes of D. More precisely, if $d^\star$ is the random attribute associated with the observed attribute d of D, $d^\star$ is defined—with understandable notations—by

$$(\forall i \in \mathbb{I})d^\star(o_i) = d(o_{\pi^\star(i)})$$

where $\pi^\star$ is a random permutation of the set G_n of all permutations on $\mathbb{I}$, equally distributed.

By this way, the inter-relations between the random attributes of $D^\star$ are the same as those between the attributes of D. Clearly, in defining the independence hypothesis, it is equivalent to fix D and to associate in the same manner a random set $C^\star$ with C. If, according to the notations above, σ designates a similarity index between classes, the distribution of $\sigma(C, D^\star)$ is the same as that of $\sigma(C^\star, D)$. Due to the difficulty of computing analytically this distribution, an extensive study by simulations is carried out in [26]. The analyzed criteria concentrate on $\sigma_{\min}(C, D)$, $\sigma_{ave}(C, D)$ and (10.3.1) and their Likelihood Linkage versions. In terms of dissimilarity functions between C and D they correspond directly to $\delta_{\min}(C, D)$ (Single Linkage criterion) and $\delta_{ave}(C, D)$ (Average Linkage criterion) (see (10.1.7) and (10.3.1)). The Likelihood Linkage version of $\sigma_{ave}(C, D)$ was introduced in [49]. It has been taken up again in a more general framework in [26].

The conclusion of the Kojadinovic paper favouring the behavior of the *Average Likelihood Linkage* criterion with respect to the *Maximal Likelihood Linkage* one, for clustering numerical attributes, might be reconsidered for two reasons

1. The probabilistic similarity table $\mathcal{P}^g$, assuming *Similarity Global Reduction* (see (5.2.53) in Chaps. 5 and 7) is not involved in the simulations;
2. A single value of ϵ ($\epsilon = 1$) is retained and experimented from the rich family of criteria of the *Maximal Likelihood Linkage*.

Our experience of the behaviour of the *Average Likelihood Linkage* criterion leads us to express two general remarks

1. In the *AAHC* Construction of the classification tree, quick convergence towards large and main clusters is observed. However, relevant and fine sub-clusters are not necessarily distinguished.
2. The analytical form of the Average criterion makes that a partition quality measure based on the sum of joined pair similarities (see Chap. 9) has some tendency to produce more large values compared with a merging criterion such as deduced from Single Linkage method.

In Sect. 10.6.3, the reactualization formulas for the most classical merging criteria—including Likelihood Linkage criterion—will be expressed in the general case of multiple aggregations occuring at a given level of the classification tree (see (10.1.10)). Let us consider here, the classical form of this formula in the case of binary aggregation, for $S_{\max}^\epsilon(C, D)$. C_1, C_2 and D being three classes, we have

$$S_{max}^{\epsilon}(C_1 \cup C_2, D) = -\log_2(l_1 + l_2)$$
$$+ \max\{S_{max}^{\epsilon}(C_1, D) + \log_2(l_1), S_{max}^{\epsilon}(C_2, D) + \log_2(l_2)\} \qquad (10.4.9)$$

where $l_1 = card(C_1)$ and $l_2 = card(C_2)$. This formula reactualizes the similarities between classes after aggregation of *only two classes*. It will be employed in the next example with $\epsilon = 1$.

10.4.1.3 An Illustrative Example

This example is taken up from [38]. It concerns the following data Table 10.5 giving the description of 10 objects by 8 Boolean attributes. The addressed facet here concerns hierarchical clustering of the set of Boolean attributes.

In the following square table 8×8 (see Table 10.6), the attribute raw indices of the form $s(a, b)$ (see (5.2.1)) are situated below, but including, the main diagonal. The *Binomial* random model of no relation is adopted here. For this, the random

Table 10.5 Data table

$\mathcal{O} \backslash \mathcal{A}$	a^1	a^2	a^3	a^4	a^5	a^6	a^7	a^8
o_1	0	0	1	1	0	1	0	1
o_2	0	0	0	1	0	0	1	1
o_3	1	1	0	1	1	0	0	0
o_4	0	0	1	0	0	0	1	1
o_5	1	0	1	0	1	0	0	0
o_6	1	0	0	1	1	0	0	0
o_7	1	0	0	0	1	1	1	0
o_8	0	0	1	1	0	0	1	1
o_9	1	0	0	1	1	1	0	0
o_{10}	1	1	0	1	0	0	1	1

Table 10.6 Raw indices\ p parameter

$\mathcal{A} \backslash \mathcal{A}$	a^1	a^2	a^3	a^4	a^5	a^6	a^7	a^8
a^1	6	0.12	0.24	0.42	0.30	0.18	0.30	0.30
a^2	2	2	0.08	0.14	0.10	0.06	0.10	0.10
a^3	1	0	4	0.28	0.20	0.12	0.20	0.20
a^4	4	2	2	7	0.35	0.21	0.35	0.35
a^5	5	1	1	3	5	0.15	0.25	0.25
a^6	2	0	1	2	2	3	0.15	0.15
a^7	2	1	2	3	1	1	5	0.25
a^8	1	1	3	4	0	1	4	5

Table 10.7 Probability indices $\backslash -\log_e(-\log_e(P(a^j, a^k)))$

$\mathcal{A} \backslash \mathcal{A}$	a^1	a^2	a^3	a^4	a^5	a^6	a^7	a^8
a^1	–	0.886	0.270	0.582	**0.953**	0.735	0.389	0.149
a^2	2.112	–	0.449	0.842	0.736	0.549	0.736	0.736
a^3	−0.270	0.222	–	0.440	0.376	0.659	0.678	0.879
a^4	0.614	1.760	0.197	–	0.514	0.647	0.514	0.752
a^5	3.034	1.182	0.022	0.407	–	0.820	0.244	0.056
a^6	1.178	0.511	0.875	0.831	1.617	–	0.544	0.544
a^7	0.041	1.182	0.945	0.407	−0.344	0.496	–	0.922
a^8	−0.644	1.182	2.048	1.255	−1.059	0.496	2.511	–

Table 10.8 Standardized coefficients

$\mathcal{A} \backslash \mathcal{A}$	a^1	a^2	a^3	a^4	a^5	a^6	a^7	a^8
a^1								
a^2	0.778							
a^3	−1.037	−0.933						
a^4	−0.128	0.547	−0.563					
a^5	**1.380**	0.000	−0.791	−0.331				
a^6	0.165	−0.799	−0.195	−0.078	0.443			
a^7	−0.690	0.000	0.000	−0.331	−1.095	−0.443		
a^8	−1.1380	0.000	0.791	0.331	−1.826	−0.443	1.095	

raw index $s(a^\star, b^\star)$ follows a binomial probability law, whose parameter being $p = p(a) \times p(b)$, (see (5.2.1)) for the notations. The respective values of this parameter for the different ordered pairs (a^j, a^k) of attributes, $1 \leq j < k \leq 8$ are given above the main diagonal of Table 10.6. The respective values of the *local* form of the probabilistic index $P_l(a, b) = Prob\{s(a^\star, b^\star) \leq s(a, b)\}$ are given in Table 10.7, above the main diagonal, $1 \leq j < k \leq 8$. These values refer directly to the Binomial cumulative distribution function $B(n = 10, p)$. Thus, for example, $P_l((a^1, a^5)) = Prob\{K \leq 5\}$, where K follows $B(n = 10, p = 0.30)$. By considering the small size of the object set $(n = 10)$, calculating the standardized index $Q_2(a, b)$ is not needed. Nevertheless, Table 10.8 gives the values of $Q_2(a^j, a^k)$ for $1 \leq j < k \leq 8$.

Tables 10.7, 10.9, 10.10, 10.11, 10.12, 10.13 and 10.14 give the sequence of the states of the couple designated by (T_l, μ^l) in the first section above (see (10.1.10)), $0 \leq l \leq 6$. Table 10.7 is associated with (T_0, μ^0) and Table 10.14 with (T_6, μ^6). Therefore by including the zero level (see Fig. 10.5), the obtained tree comprises eight levels. The weighting μ^l is defined here by the class cardinalities. These integers take place in the last column at the right side (see Tables 10.9, 10.10, 10.11, 10.12, 10.13 and 10.14).

Table 10.9 Second aggregation

$1 \leftarrow 1 \vee 5$	$2 \leftarrow 2$	$3 \leftarrow 3$	$4 \leftarrow 4$	$5 \leftarrow 6$	$6 \leftarrow 7$	$7 \leftarrow 8$	
							2
1.419							1
−0.617	0.222						1
−0.079	1.760	0.197					1
0.924	0.511	0.875	0.831				1
−0.652	1.182	0.945	0.407	0.496			1
−1.337	1.182	2.048	1.255	0.496	**2.511**		1

Table 10.10 Third aggregation

$1 \leftarrow 1$	$2 \leftarrow 2$	$3 \leftarrow 3$	$4 \leftarrow 4$	$5 \leftarrow 5$	$6 \leftarrow 6 \vee 7$	
						2
1.419						1
−0.617	0.222					1
−0.079	**1.760**	0.197				1
0.924	0.511	0.875	0.831			1
−0.1345	0.489	1.355	0.562	−0.197		2

Table 10.11 Fourth aggregation

$1 \leftarrow 1$	$2 \leftarrow 2 \vee 4$	$3 \leftarrow 3$	$4 \leftarrow 5$	$5 \leftarrow 6$	
					2
0.726					2
−0.617	−0.471				1
0.924	0.138	0.875			1
−1.345	0.131	**1.355**	−0.197		2

Table 10.12 Fifth aggregation

$1 \leftarrow 1$	$2 \leftarrow 2 \vee 4$	$3 \leftarrow 3 \vee 5$	$4 \leftarrow 4$	
				2
0.726				2
−1.751	−0.537			3
0.924	0.138	−0.224		1

For each similarity matrix T_l, $0 \leq l \leq 6$, we have marked the maximum value with a bold type, indicating by this way the class pair to merge. In our example, for each step, there is only one couple of classes which realizes the latter maximum value of the similarity between classes. Thus, the tree is binary (see Fig. 10.5). As expressed above, this particular case is directly related to the example, but does not correspond to a general fact.

Table 10.13 Sixth
aggregation

$1 \leftarrow 1 \vee 4$	$2 \leftarrow 2$	$3 \leftarrow 3$	
			3
0.32			2
-1.323	-0.537		3

Table 10.14 Seventh
aggregation

$1 \leftarrow 1 \vee 2$	$2 \leftarrow 3$	
		5
-1.453		3

Fig. 10.5 Classification tree
on the attribute set

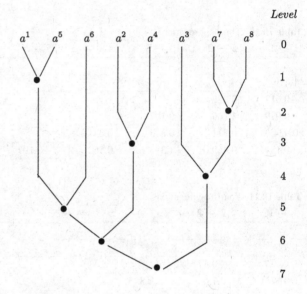

Level

When l increases, $0 \leq l \leq 6$, the number of classes formed decreases. We have in the Tables 10.9, 10.10, 10.11, 10.12, 10.13 and 10.14, clearly indicated the evolution of a given labelled class, according to the column (respectively, row) concerned in the matrix T_l, $0 \leq l \leq 6$. By this way, it is possible to retreive the different elements included in a given class which takes part in a given table T_l, $0 \leq l \leq 6$.

Now, let us mention that [38] provides also hierarchical clustering of the object set according to the development given in Sect. 7.2.2. On the other hand, important aspects are illustrated, namely:

- The most "significant" levels and nodes of a classification tree (see Chap. 9, Sect. 9.3.6);
- "Neutrality" degree of the elements of the set to be clustered;
- Crossing between a significant partition on the object set and a significant partition on the attribute set.

10.4.2   *Minimal Likelihood Linkage and Average Likelihood Linkage in the* LLA *Analysis*

The Likelihood Linkage merging criteria defined from *Complete Linkage* and *Average Linkage* methods were initiated in [49] and reported in [11]. They have been worked on again—with more references with respect developments in Mathematical Statistics—in [26]. The idea was to extend the *Maximal Likelihood* criterion defined from the *Single Linkage* method to other classical merging criteria. Our development here brings some particular elements in this extension. By referring to table $\mathcal{P}^g$ of probabilistic similarities on the set E to be clustered (see Sect. 10.4.1.1), the respective criteria $\sigma_{\min}(C, D)$ and $\sigma_{ave}(C, D)$ associated with the *Complete Linkage* and *Average Linkage* methods, for comparing two disjoint subsets C and D of E can be written as

$$\sigma_{\min}(C, D) = \min\{\mathcal{P}(x, y) | (x, y) \in C \times D\} \qquad (10.4.10)$$

$$\sigma_{ave}(C, D) = \frac{1}{card(C) \times card(D)} \sum_{(x,y) \in C \times D} \mathcal{P}(x, y) \qquad (10.4.11)$$

10.4.2.1   Minimal Likelihood Linkage in the *LLA* Analysis

According to the basic definition of the *Maximal Likelihood Linkage* (see (10.4.4)), the *Minimal Likelihood Linkage* criterion can be expressed by

$$LL_{\min}(C, D) = Prob\{\sigma^\star_{\min}(C, D) \le \sigma_{\min}(C, D)\} \qquad (10.4.12)$$

where $\sigma^\star_{\min}(C, D)$ is defined by the *minimum* of $l \times m$ ($l = card(C)$ and $m = card(D)$) independent random variables uniformly distributed on the [0, 1] interval. We have

$$LL_{\min}(C, D) = 1 - Prob\{\sigma^\star_{\min}(C, D) > \sigma_{\min}(C, D)\} = 1 - \left(1 - \sigma_{\min}(C, D)\right)^{l \times m}$$
$$(10.4.13)$$

As for the Likelihood of Maximal Linkage criterion, a numerical accuracy in the calculation of $LL_{\min}(C, D)$ has to be solved. A similar solution to the previous one consists of substituting for $LL_{\min}(C, D)$ the strictly increasing function

$$S_{\min}(C, D) = -\text{Log}\Big(-\text{Log}\big(LL_{\min}(C, D)\big)\Big)$$

where for simplicity, Log refers to neperian logarithm. In fact, mathematically, $S_{min}(C, D)$ leads to exactly the same classification tree as that obtained by LL_{min} (C, D). By using $Log(1 - u) = -u + o(u)$, where $0 < u < 1$ and o the Landau notation, we obtain the approximation

$$S_{min}(C, D) = (l \times m)\sigma_{min}(C, D) \tag{10.4.14}$$

Now, introducing a family of criteria of the *Minimal Likelihood Linkage* can be proposed in the same way as for the family of criteria of the *Maximal Likelihood Linkage*. For this, we consider an exponential weighting of $l \times m$ by a parameter ϵ, comprised between 0 and 1 ($0 \le \epsilon \le 1$). We obtain

$$S_{min}^{\epsilon}(C, D) = (l \times m)^{\epsilon}\sigma_{min}(C, D) \tag{10.4.15}$$

We have no experience relative to the behaviour of this family of criteria. The basic form from which this family has been derived was studied in [26, 49]. It was not considered as realizing a high performance comparative to with other criteria. However, the new version given here (see (10.4.15)) is worth to carry out experiments.

10.4.2.2 Average Likelihood Linkage in the *LLA* Analysis

The *Likelihood of the Average Linkage* criterion can be stated as follows

$$LL_{ave}(C, D) = Prob\{\sigma_{ave}^{\star}(C, D) \le \sigma_{ave}(C, D)\} \tag{10.4.16}$$

where $\sigma_{ave}(C, D)$ is recalled in (10.4.11). $\sigma_{ave}^{\star}(C, D)$ corresponds to the mean of $l \times m$ random variables uniformly distributed on the interval $[0, 1]$ ($l = card(C)$ and $m = card(D)$). The exact cumulative distribution function is established in [49] and reported in [11]. By denoting, for simplicity, $\sigma = \sigma_{ave}(C, D)$, it can be written as

$$LL_{ave}(C, D) = \frac{1}{(l \times m)!} \sum_{0 \le i \le l \cdot m\sigma} (-1)^i \binom{l \cdot m}{i}(l \cdot m\sigma - i)^{l \cdot m}$$

$$\text{if } 0 \le \sigma < 0.5$$

$$= 1 - \frac{1}{(l \times m)!} \sum_{0 \le i \le l \cdot m(1-\sigma)} (-1)^i \binom{l \cdot m}{i}(l \cdot m(1 - \sigma) - i)^{l \cdot m}$$

$$\text{if } 0.5 \le \sigma \le 1 \tag{10.4.17}$$

This result can be related to a classical one giving the distribution of the sum of independent random variables uniformly distributed on [0, 1] [16].

If the set E to be clustered is large enough, (10.4.17) cannot be exploited for building an ascendant agglomerative hierarchical classification tree on E. In practice, *Central limit theorem* enables Normal approximation of the distribution of $\sigma^{\star}_{ave}(C, D)$ to be used. The mean and variance of the latter being 0.5 and $1/12l \cdot m$, we have

$$LL_{ave}(C, D) = \Phi\left(\sqrt{3l \cdot m}(2\sigma - 1)\right) \qquad (10.4.18)$$

where, as above, σ stands for $\sigma_{ave}(C, D)$.

However, due to the multiplicative factor $\sqrt{3l \cdot m}$ the probability scale associated with the latter function becomes no discriminant if l and m are large enough. This may be the reason for which the merging criterion (10.4.18) has some tendency to produce large clusters in the hierarchical agglomerative forming of the classification tree.

10.4.2.3   *Inversion or Reversal* Phenomenon for the Likelihood Linkage Criteria

An important property of a merging criterion is to not produce *reversal* in the *AAHC* construction of a classification tree. In order to define this notion, let us consider a binary construction of a classification tree using a merging criterion σ corresponding to a similarity notion between classes (disjoint subsets) of the set E to be clustered. Let k and $k + 1$ be two consecutive levels of the binary tree on a given set E ($1 \leq k \leq card(E) - 2$) obtained from σ. Designate by C_1 and C_2 the two nearest classes of the partition P_k of the k level. Thus, the partition P_{k+1} of the $k + 1$ level is obtained by merging C_1 and C_2. *reversal* occurs in the passage between P_k and P_{k+1} if there exists a class D appearing in both partitions P_k and P_{k+1} such that $(C_1 \cup C_2) \cap D = \emptyset$ and $\sigma(C_1 \cup C_2, D) > \sigma(C_1, C_2)$.

In order to control the reversal phenomenon for a given merging criterion σ, an efficient technique consists of applying a reactualization formula for σ. Relative to the merging of C_1 and C_2, this expresses $\sigma((C_1 \cup C_2), D)$ with respect to $\sigma(C_1, D)$ $\sigma(C_2, D)$. This formula was given above for $\sigma(C, D) = S^\epsilon_{max}(C, D)$ (10.4.9). It is easy to derive these formulas for the merging criteria $\sigma_{max}(C, D)$, $\sigma_{min}(C, D)$ and $\sigma_{ave}(C, D)$. We have

$$\sigma_{max}((C_1 \cup C_2), D) = \max\{\sigma_{max}(C_1, D), \sigma_{max}(C_2, D)\}$$

$$\sigma_{min}((C_1 \cup C_2), D) = \min\{\sigma_{min}(C_1, D), \sigma_{min}(C_2, D)\}$$

$$\sigma_{ave}((C_1 \cup C_2), D) = \frac{l_1}{l_1 + l_2}\sigma_{ave}(C_1, D) + \frac{l_2}{l_1 + l_2}\sigma_{ave}(C_2, D)$$

$$(10.4.19)$$

where $l_1 = card(C_1)$ and $l_2 = card(C_2)$. These equations enable us to see that no reversal can occur with these criteria.

It is also obvious to see from (10.4.9) that every element of the family of the *Maximal Likelihood Linkage* criteria $\{S_{\max}^\epsilon(C, D)|0 \le \epsilon \le 1\}$ cannot produce *reversals* in the classification tree building.

However, the criteria of the *Minimal and Average Likelihood* criteria might cause reversals. In the case of a given element $S_{\min}^\epsilon(C, D)$ of the family of the Minimal Likelihood criteria parametrized by ϵ (see (10.4.15)), the reactualization formula can be written as

$$S_{\min}^\epsilon((C_1 \cup C_2), D) = \min \left\{ \left(\frac{l_1 + l_2}{l_1}\right)^\epsilon S_{\min}^\epsilon(C_1, D), \left(\frac{l_1 + l_2}{l_2}\right)^\epsilon S_{\min}^\epsilon(C_2, D) \right\}$$

(10.4.20)

where $l_1 = card(C_1)$ and $l_2 = card(C_2)$. By considering a class D of the partition P_k different from C_1 and C_2 $((C_1 \cup C_1) \cap D = \emptyset)$, due to the multiplicative factors $\left(\frac{l_1+l_2}{l_1}\right)^\epsilon$ and $\left(\frac{l_1+l_2}{l_2}\right)^\epsilon$ there is no garantee to have

$$\left(\frac{l_1 + l_2}{l_1}\right)^\epsilon S_{\min}^\epsilon(C_1, D) < S_{\min}^\epsilon(C_1, D)$$

and

$$\left(\frac{l_1 + l_2}{l_2}\right)^\epsilon S_{\min}^\epsilon(C_2, D) < S_{\min}^\epsilon(C_2, D)$$

Now, the reactualization formula for the *Average Likelihood Linkage* (see (10.4.18)) is somewhat complicated to detail. In these conditions, we shall directly show why this criterion may cause reversals. As above, let D be a class of the partition P_k different from C_1 and C_2 $((C_1 \cup C_1) \cap D = \emptyset)$. For simplicity, but without restricting the generality of the argument, let us assume $l_1 = l_2 = l$. The question which needs to be addressed is whether or not the following equations

1. $LL_{ave}(C_1, D) < LL_{ave}(C_1, C_2)$
2. $LL_{ave}(C_2, D) < LL_{ave}(C_1, C_2)$

entail

$$LL_{ave}(C_1 \cup C_2, D) < LL_{ave}(C_1, C_2)$$

By referring to (10.4.18), the previous equations become

1. $\sqrt{3l \cdot m}(2\sigma_{ave}(C_1, D) - 1) < \sqrt{3l}(2\sigma_{ave}(C_1, C_2) - 1)$
2. $\sqrt{3l \cdot m}(2\sigma_{ave}(C_2, D) - 1) < \sqrt{3l}(2\sigma_{ave}(C_1, C_2) - 1)$

Since $\sqrt{6l \cdot m} > \sqrt{3l \cdot m}$, this implication cannot hold.

We did not give above the reactualization formula for the *Ward criterion* (see Sect. 10.3.3) in the case of a binary construction of a classification tree. This will be more adequately inserted in Sect. 10.6.2. We can already indicate that no inversion (reversal) phenomenon can occur for the latter criterion. Let us recall that the philosophy of the Ward criterion is essentially different from a criterion belonging to the Likelihood Linkage family. For the Ward criterion the basic dissimilarity index is the euclidean distance between weighted points in a geometrical space; whereas, the basic similarity index in a Likelihood Linkage method refers to a very general notion of *correlation* between combinatorial structures defined on the object set.

10.5 *AAHC* for Clustering Rows or Columns of a Contingency Table

10.5.1 *Introduction*

Definition of a *contingency table* was given in Sect. 5.3.2.1 of Chap. 5 (see (5.3.1) and (5.3.2)). The reader is asked to take it up again here with the notations concerned. The perfect formal symmetry of this table enables two dual geometrical representations to be given for it. The first one, $\mathcal{N}(\mathbb{I})$ (see (5.3.3)) is a cloud of weighted points in the geometrical space $\mathbb{R}^J$, endowed with the Chi-square metric (see (5.3.4)). The second one, $\mathcal{N}(\mathbb{J})$ (see (7.2.35) in Sect. 7.2.8.1 of Chap. 7) is situated in the space $\mathbb{R}^I$.

For $\mathcal{N}(\mathbb{I})$, $\mathbb{I}$ represents the object set and $\mathbb{J}$ can be assimilated to a set of numerical descriptive attributes. For $\mathcal{N}(\mathbb{J})$, the respective roles of the object set and the attribute set are inverted. Clearly, all that will be expressed for clustering in the framework of $\mathcal{N}(\mathbb{I})$ can be transposed in the framework of $\mathcal{N}(\mathbb{J})$. Our development below will be in the context of $\mathcal{N}(\mathbb{I})$. By this way, clustering the row set of an horizontal juxtaposition of contingency tables may be considered.

Solution was given for clustering columns or rows of a contingency table in the framework of the *Likelihood Linkage clustering* family. In fact these methods require the construction of an association coefficient, having a correlative form, between the elements of the set to be clustered (see Sects. 5.3.2.1 of Chap. 5 and 7.2.8.1 of Chap. 7). In Sect. 5.3.2.1 the set concerned is $\mathbb{J}$ interpreted as a set of attributes with respect to the geometrical representation of $\mathcal{N}(\mathbb{I})$ (see (5.3.5) and (5.3.6)). In Sect. 7.2.8.1 the set concerned is $\mathbb{I}$ interpreted as a set of objects, with respect the same representation (see (7.2.36) and all that follows in Sects. 7.2.8.1 and 7.2.8.2).

In the following two sections, two new types of criteria will be developed: an adaptation of the *Ward* criterion and the *Mutual Information* criterion. It will be shown that there is connection between them.

10.5.2 Chi Square Criterion: A Transposition of the Ward Criterion

J.P. Benzécri and Collaborators [6] have extensively developed the metrical and geometrical analysis of statistical dependency between two nominal categorical attributes, defining each a partition with a large number of classes on a given object set. In these conditions, the matter consists of applying the general development of Sect. 10.3.3—concerning clustering a cloud of points situated in a geometrical space—to the case of the cloud $\mathcal{N}(\mathbb{I})$ associated with the geometrical representation of the *row* set of a contingency table considered in Sect. 5.3.2.1 of Chap. 5. As mentioned above, elements and notations of the latter will be taken up again here for the purpose of this adaptation. Thus, the cloud $\mathcal{N}(\mathbb{I})$ given by (5.3.3) was expressed as follows

$$\mathcal{N}(\mathbb{I}) = \{(f_{\mathbb{J}}^i, p_{i.}) | i \in \mathbb{I}\} \tag{10.5.1}$$

where $f_{\mathbb{J}}^i$ is the point of the geometrical space $\mathbb{R}^J$ defined by

$$\overrightarrow{Of_{\mathbb{J}}^i} = (f_{\mathbb{J}}^i - O) = \sum_{1 \leq j \leq J} f_j^i \vec{e}_j \tag{10.5.2}$$

where f_j^i is the relative frequency n_{ij}/n_i. (See the first paragraph of Sect. 5.3.2.1) and where $\{\vec{e}_j | 1 \leq j \leq J\}$ is the canonical base of the space $\mathbb{R}^J$, $1 \leq i \leq I$.

The *center of gravity* (centroid) of $\mathcal{N}(\mathbb{I})$ is the point

$$g_{\mathbb{I}} = p_{.\mathbb{J}}$$

whose components are the marginal proportions $p_{.j}$, $1 \leq j \leq J$.

$\mathbb{R}^J$ is provided with the Chi-square metric q:

$$q(e_j, e_k) = \frac{\delta_{jk}}{p_{.j}}$$

where δ_{jk} designates the δ of Kronecker ($\delta_{jk} = 0$ (resp. 1) if $j \neq k$ (resp., $j = k$)). For this, the square distance between two summits $f_{\mathbb{J}}^i$ and $f_{\mathbb{J}}^{i'}$ is

$$d_{\chi^2}^2(f_{\mathbb{J}}^i, f_{\mathbb{J}}^{i'}) = \sum_{j \in \mathbb{J}} \frac{1}{p_{.j}} (f_j^i - f_j^{i'})^2 \tag{10.5.3}$$

In these conditions, the *total inertia moment* can be written as

$$\mathcal{M}(\mathcal{N}(\mathbb{I})) = \sum_{i \in \mathbb{I}} \sum_{i \in \mathbb{I}} p_i \cdot d_{\chi^2}^2(f_{\mathbb{J}}^i, g_{\mathbb{J}}) \tag{10.5.4}$$

where

$$d_{\chi^2}^2(f_{\mathbb{J}}^i, g_{\mathbb{J}}) = \sum_{j \in \mathbb{J}} \frac{1}{p_{\cdot j}}(f_j^i - p_{\cdot j})^2$$

Remember that the *total inertia moment* is nothing else than the coefficient χ^2/n calculated for the contingency table concerned. Thus, the heterogeneity measure defined in the general case by the total moment of inertia, becomes here a measure of statistical deviation from independence between the two categorical attributes whose crossing corresponds to the contingency table studied.

Now, let us consider a partition $\{\mathbb{I}_h | 1 \le h \le k\}$ of $\mathbb{I}$ to which corresponds a partition $\{\mathcal{N}(\mathbb{I}_h) | 1 \le h \le k\}$ of the cloud $\mathcal{N}(\mathbb{I})$. This partition into k classes can be designated by P_k. The weight and the center of gravity of $\mathcal{N}(\mathbb{I}_h)$ are

$$p_{\mathbb{I}_h \cdot} = \sum_{i \in \mathbb{I}_h} p_{i \cdot}$$

$$g_{\mathbb{J}}^h = \frac{1}{p_{\mathbb{I}_h \cdot}} \sum_{i \in \mathbb{I}_h} p_{i \cdot} \cdot f_{\mathbb{J}}^i \qquad (10.5.5)$$

More precisely, the jth component of $g_{\mathbb{J}}^h$ can be written as

$$f_j^{\mathbb{I}_h} = \frac{f(\mathbb{I}_h, j)}{p_{\mathbb{I}_h \cdot}} = \frac{\sum_{i \in \mathbb{I}_h} f_{ij}}{p_{\mathbb{I}_h \cdot}}$$

Thus, the ordered pair $(g_{\mathbb{J}}^h, p_{\mathbb{I}_h \cdot})$ defines the reduced representation of the subcloud $\mathcal{N}(\mathbb{I}_h), 1 \le h \le k$. Consequently, the class set $\{\mathcal{N}(\mathbb{I}_h) | 1 \le h \le k\}$ formed by disjoint subclouds of $\mathcal{N}(\mathbb{I})$ is represented by the cloud

$$\mathcal{N}(\mathcal{G}) = \{(g_{\mathbb{J}}^h, p_{\mathbb{I}_h \cdot}) | 1 \le h \le k\} \qquad (10.5.6)$$

The inertia moment of this cloud is

$$\mathcal{M}(\mathcal{N}(\mathcal{G})) = \sum_{1 \le h \le k} p_{\mathbb{I}_h} \cdot \|g_{\mathbb{J}}^h - g_{\mathbb{J}}\|^2 \qquad (10.5.7)$$

where the metric concerned is the Chi-square metric and where we have

$$g_{\mathbb{J}} = \sum_{1 \le h \le k} p_{\mathbb{I}_h} \cdot g_{\mathbb{J}}^h$$

since the general center of gravity $g_{\mathbb{J}}$ is the center of gravity of $\mathcal{N}(\mathcal{G})$.

$\mathcal{M}(\mathcal{N}(\mathcal{G}))$ defines the "explained" (or "retained") inertia by the partition P_k.

The general formula of Huyghens inertia decomposition (10.3.10) becomes here

$$\sum_{i\in\mathbb{I}} p_i \cdot \|f_{\mathbb{J}}^i - g_{\mathbb{J}}\|^2 = \sum_{1\le h\le k} \sum_{i\in\mathbb{I}_h} p_i \cdot \|f_{\mathbb{J}}^i - g_{\mathbb{J}}^h\|^2 + \sum_{1\le h\le k} p_{\mathbb{I}_h} \cdot \|g_{\mathbb{J}}^h - g_{\mathbb{J}}\|^2 \quad (10.5.8)$$

The first term of the right member is called the "lost" (or "residual") inertia. Notice that the order of both terms in (10.5.8) is inverted with respect to that in (10.3.10).

Finally, the clustering methods, mentioned in Sect. 10.3.3, using the inertia variation criterion can be used for clustering the row set $\mathbb{I}$ of a contingency table on the basis of its geometrical representation by the cloud $\mathcal{N}(\mathbb{I})$ (see (10.5.1)). Let us end this section by giving in this specific case, the inertia variation between two consecutive levels k and $k+1$ of an *AAHC* construction of a classification tree (see (10.3.12) concerning the general case). By denoting $\mathbb{I}_h$ and $\mathbb{I}_{h'}$ two candidate clusters for merging, we have

$$\Delta(k, k+1) = \frac{p_{\mathbb{I}_h} \cdot p_{\mathbb{I}_{h'}}}{p_{\mathbb{I}_h} + p_{\mathbb{I}_{h'}}} \|g_{\mathbb{J}}^h - g_{\mathbb{J}}^{h'}\|^2 \quad (10.5.9)$$

10.5.3 Mutual Information Criterion

To my knowledge, this approach which introduces *Information theory models* in clustering was proposed by L. Orloci [51], rediscovered and worked on again by J.P. Benzécri [6], Volume I. The latter development is taken up again here.

Let $\mathbb{K}$ be a category set of a nominal categorical attribute c having k values. Without ambiguity we may denote $\mathbb{K}$ by $\{1, 2, \ldots, h, \ldots, k\}$. The distribution of c on a given object set $\mathcal{O}$ is defined by

$$f_{\mathbb{K}} = (f_1, f_2, \ldots, f_h, \ldots, f_k)$$

where f_h is the relative frequency $\frac{n_h}{n}$, $n = card(\mathcal{O})$, $n_h = card(\mathcal{O}_h)$ and $\mathcal{O}_h = c^{-1}(h)$ (Sect. 3.3.1).

Now, consider the experiment which consists of randomly choosing an element x of $\mathcal{O}$ with equal chances. The *Shannon* entropy associated with this random experiment is defined as follows:

$$H(f_{\mathbb{K}}) = - \sum_{1\le h\le k} f_h \cdot \log_2(f_h) \quad (10.5.10)$$

This is the mathematical expectation of the quantity of information $-\log_2(f_h)$ brought by the outcome: "x possesses the hth category of the attribute concerned".

Several axiomatic characterizations enable the Shannon entropy to be presented as satisfying "natural" conditions [2]. We will not discuss this subject any further.

Now, let us return to the framework of cluster analysis of the rows through the columns (resp., the columns through the rows) of a contingency table. Consider the following table of proportions

$$f_{\mathbb{I} \times \mathbb{J}} = \{ f_{ij} | (i, j) \in \mathbb{I} \times \mathbb{J} \} \tag{10.5.11}$$

corresponding to a contingency table as that given in (5.3.1) of Chap. 5. As expressed above, this can be considered defined from crossing two partitions π and χ on the object set $\mathcal{O}$. The entropy of the latter distribution can be written as

$$H(f_{\mathbb{I} \times \mathbb{J}}) = \sum_{(i,j) \in \mathbb{I} \times \mathbb{J}} f_{ij} \cdot \log_2(f_{ij}) \tag{10.5.12}$$

We have the very important relation

$$H(f_{\mathbb{I} \times \mathbb{J}}) \le H(p_{\mathbb{I}.}) + H(p_{.\mathbb{J}}) \tag{10.5.13}$$

(see notations in Sect. 5.3.2.1 an above). Equality in (10.5.13) characterizes statistical independence between the partitions π and χ, which is expressed by

$$\forall((i, j) \in \mathbb{I} \times \mathbb{J}), \, f_{ij} = p_{i.} \times p_{.j} \tag{10.5.14}$$

Let us indicate that the inequality (10.5.13) can be proved by using the concavity property of the function $u\log_2(u)$ for positive real u.

The positive difference

$$H(p_{\mathbb{I}.}) + H(p_{.\mathbb{J}}) - H(f_{\mathbb{I} \times \mathbb{J}}) \tag{10.5.15}$$

can be put equivalently under one of the following both forms

$$H(p_{\mathbb{I}.}) - H(\mathbb{I}/\mathbb{J})$$
$$H(p_{.\mathbb{J}}) - H(\mathbb{J}/\mathbb{I}) \tag{10.5.16}$$

where the conditional entropies $H(\mathbb{I}/\mathbb{J})$ and $H(\mathbb{J}/\mathbb{I})$ are defined as follows

$$H(\mathbb{I}/\mathbb{J}) = \sum_{j \in \mathbb{J}} H(f_{\mathbb{I}}^j)$$
$$H(\mathbb{J}/\mathbb{I}) = \sum_{i \in \mathbb{I}} H(f_{\mathbb{J}}^i) \tag{10.5.17}$$

where, clearly,

$$H(f_{\mathbb{I}}^j) = -\sum_{i \in \mathbb{I}} f_i^j \log_2(f_i^j) \quad \text{and} \quad H(f_{\mathbb{J}}^i) = -\sum_{j \in \mathbb{J}} f_j^i \log_2(f_j^i)$$

The measure defined in (10.5.15) is called *"Mutual Information"* between the two categorical attributes giving rise to the contingency table concerned. It can be designated by $\mathcal{H}(f_{\mathbb{I}\times\mathbb{J}}; p_{\mathbb{I}.} \times p_{.\mathbb{J}})$, where $p_{\mathbb{I}.} \times p_{.\mathbb{J}} = \{p_{i.} \cdot p_{.j} | (i, j) \in \mathbb{I} \times \mathbb{J}\}$ is the distribution in the case of independence. This notation is taken from the chapter "Théorie de l'Information et Classification d'après un tableau de contingence" in [6], Volume I. We have

$$\mathcal{H}(f_{\mathbb{I}\times\mathbb{J}}; p_{\mathbb{I}.} \times p_{.\mathbb{J}})$$

$$= \sum_{(i,j)\in\mathbb{I}\times\mathbb{J}} f_{ij} \log_2\left(\frac{f_{ij}}{p_{i.} \cdot p_{.j}}\right)$$

$$= \sum_{(i,j)\in\mathbb{I}\times\mathbb{J}} p_{i.} \cdot p_{.j} \phi\left(\frac{f_{ij}}{p_{i.} \cdot p_{.j}}\right) \qquad (10.5.18)$$

where, for positive real u, $\phi(u) = u log_2 u$.

Therefore, the *Mutual Information* criterion appears as the mathematical expectation with respect to the distribution $p_{\mathbb{I}.} \times p_{.\mathbb{J}}$ of the function ϕ applied to the *density* function $f_{\mathbb{I}\times\mathbb{J}/p_{\mathbb{I}.}\times p_{.\mathbb{J}}}$. As mentioned, the strict concavity of the function ϕ on the positive reals enables the (10.5.13) to be established. Effectively, the value of such a function on a weighted mean of a sequence of numbers, is lower or equal to this weighted mean on these numbers.

Notice that the Chi-square measure χ^2/n associated with a contingency table can be expressed by the same general analytic form as that in the last equation of (10.5.18), namely,

$$\frac{\chi^2}{n}(\mathbb{I} \times \mathbb{J}) = \sum_{(i,j)\in\mathbb{I}\times\mathbb{J}} p_{i.} \cdot p_{.j} \psi\left(\frac{f_{ij}}{p_{i.} \cdot p_{.j}}\right) \qquad (10.5.19)$$

where $\psi(u) = u^2 - 1$ for u positive real.

Let us now consider the merging of two categories $i1$ and $i2$ and their replacement by a single category $i0$, representing the union of $i1$ and $i2$. Remember that this corresponds to aggregate both disjoint subsets of the object set $\mathcal{O}$, possessing the categories $i1$ and $i2$, respectively. The decrease of the *Mutual Information* expressed as in (10.5.18) is

$$\sum_{j\in\mathbb{J}} \left[-p_{i0.} \cdot p_{.j} \phi\left(\frac{f_{i0j}}{p_{i0.} p_{.j}}\right) + p_{i1.} p_{.j} \phi\left(\frac{f_{i1j}}{p_{i1.} p_{.j}}\right) + p_{i2.} p_{.j} \phi\left(\frac{f_{i2j}}{p_{i2.} p_{.j}}\right) \right]$$

$$(10.5.20)$$

where $f_{i0j} = f_{i1j} + f_{i2j}$ and $p_{i0.} = p_{i1.} + p_{i2.}$. Due to the cocavity of ϕ, the latter quantity is positive.

In these conditions, the first step of a binary construction of a classification tree on $\mathbb{I}$ by an ascendant agglomerative procedure consists of merging the category pair $\{i1, i2\}$ for which the quantity (10.5.20) is *minimum*. Thus, in substituting $\mathbb{I}' = \mathbb{I} - \{i1, i2\} + \{i0\}$ for $\mathbb{I}$,

$$\mathcal{H}(f_{\mathbb{I}' \times \mathbb{J}}; p_{\mathbb{I}'.} \times p_{.\mathbb{J}}) \text{ is the nearest } \mathcal{H}(f_{\mathbb{I} \times \mathbb{J}}; p_{\mathbb{I}.} \times p_{.\mathbb{J}})$$

The principle of the passage from level k to level $k + 1$ in the construction of the binary classification tree ($0 \leq k \leq I - 2$) is identical to the previous one where the passage concerned is between levels 0 and 1. Let $P_k = \{\mathbb{I}_h | 1 \leq h \leq k\}$ the k level partition of $\mathbb{I}$. In a natural way, we can associate a categorical attribute a^k with this partition, having k values, where the hth one corresponds to $\mathbb{I}_h$, $1 \leq h \leq k$. Let us designate by $\mathbb{K} = \{1, 2, \ldots, h, \ldots, k\}$ this value set. The step of the construction algorithm enabling the passage from P_k to P_{k+1}, consists of merging the pair of classes $\{\mathbb{I}_h, \mathbb{I}_{h'}\}$ which minimizes the decrease of the mutual information between $\mathbb{K}$ and $\mathbb{J}$; that is

$$\mathcal{H}(f_{\mathbb{K} \times \mathbb{J}}; p_{\mathbb{K}.} \times p_{.\mathbb{J}}) - \mathcal{H}(f_{\mathbb{K}' \times \mathbb{J}}; p_{\mathbb{K}'.} \times p_{.\mathbb{J}}) \qquad (10.5.21)$$

which is calculated by a formula analogous to (10.5.20) and where $\mathbb{K}'$ is deduced from $\mathbb{K}$ by concatenating h and h'.

It is interesting to notice that: as it was for the Ward and then, for the Chi-square criterion, in the *local* part of the *global* criterion (the *Mutual Information* here) concerned by merging two given classes, only these classes intervene. Notice also that, in substituting the function ψ for ϕ in (10.5.20), we obtain another version of the expression of the Chi-square criterion (compare (10.5.9) and (10.5.19)).

Clearly, the problem of clustering $\mathbb{J}$ through $\mathbb{I}$ by using mutual information criterion, is equivalent to that of clustering $\mathbb{I}$ through $\mathbb{J}$. For this the respective roles of $\mathbb{I}$ and $\mathbb{J}$ have to be reversed. Nevertheless, the problem of finding a "good" simultaneous clustering of rows and columns of a contingency table has a different nature [22].

10.6 Efficient Algorithms in Ascendant Agglomerative Hierarchical Classification (Clustering)

10.6.1 Introduction

This section is devoted to the presentation and analysis of efficient *AAHC* algorithms for processing large data sets. Mostly, the size problem occurs for clustering a set of objects $\mathcal{O}$, or possibly, a set $\mathcal{C}$ of described categories; but rarely for a set $\mathcal{A}$ of descriptive attributes. The majority of the complexity problems arised for clustering a large object set $\mathcal{O}$, are not concerned by the different types of descriptive attributes participating to the establishment of $\mathcal{A}$. More precisely, if the starting point of a hierarchical clustering algorithm is a similarity function S or a dissimilarity function $\mathcal{D}$, on the object set $\mathcal{O}$, the nature of the descriptive attributes giving rise to S (resp., $\mathcal{D}$) is generally immaterial. However, as mentioned in Sect. 10.1, when the set $\mathcal{A}$ of descriptive attributes is composed of categorical attributes having in all a not large enough number of categories, there might be a large number of tied values for S (resp., $\mathcal{D}$) on $\mathcal{O}$. On the other hand, the nature of $\mathcal{A}$ might intervene in the case where the clustering algorithm works at the level of the data table $\mathcal{O} \times \mathcal{A}$, crossing the object set $\mathcal{O}$ with the attribute set $\mathcal{A}$ (see Sect. 10.6.2). In fact, this algorithm may require a mathematical calculation of a cluster representation by a single element in the representation space of the object set defined by $\mathcal{A}$ and this cannot be performed in the same way whatever is the structure of the latter space. More clearly, for a representation in the geometrical space $\mathbb{R}^p$ corresponding to a description by numerical attributes, the most usual practice consists of representing a given cluster by the *center of gravity* (centroid) of the subcloud associated with the cluster representation. This practice cannot hold in the case of description by categorical attributes. In [42, 52] an original representation is considered in the case where $\mathcal{A}$ is constituted by categorical preordonance attributes (see Sect. 3.4.1 of Chap. 3).

To fix ideas and to simplify, but without restricting generality for the algorithmic study, we assume $\mathcal{A}$ composed of numerical attributes. In these conditions, $\mathcal{O}$ is represented by a cloud of points $\mathcal{N}(\mathbb{I})$ in the geometrical space $\mathbb{R}^p$ ($p = card(\mathcal{A})$) (see Sect. 10.3.3.1). On the other hand, without loss of generality, we suppose the different elements of $\mathcal{O}$, equally weighted.

For homogeneity reasons, all the algorithmic development below will be expressed in terms of a dissimilarity or distance function $\mathcal{D}$ established on the object set $\mathcal{O}$. In the case where $\mathcal{O}$ is provided with a similarity function S (built or given), we transform S prealably into a dissimilarity function $\mathcal{D}$ as consistently as possible. Thus and for example, in the case where an *LLA* probabilistic similarity index is established for comparing objects (see Chap. 7), table $\mathcal{D}$ of informational dissimilarities (see (7.2.7) of Chap. 7) is substituted for the table $\mathcal{P}^g$ of probabilistic indices (see (7.2.6) of Chap. 7).

In the classical literature concerning *AAHC*, building *binary* classification trees is nearly exclusively considered. The complexity facets of the algorithms presented below can be discussed without loss of generality in this binary framework. As expressed in Sects. 10.1 and 10.3, the following items are required in the classical version of the construction of a binary classification tree on $\mathcal{O}$

1. Definition of a dissimilarity or distance index (function) $\mathcal{D}$ between elements of $\mathcal{O}$;
2. Deriving from $\mathcal{D}$ a dissimilarity function δ between disjoint subsets of $\mathcal{O}$, giving rise to the merging criterion.

In Sect. 10.6.2, some computing properties of the basic and traditional *AAHC* algorithm will be expressed. By considering the application of a reactualization formula as those introduced by Lance and Williams [27] it will be seen clearly that the spatial complexity is $O(n^2)$. Besides, the time complexity of the part of the clustering process consisting of choosing at a given level, the pair of classes to be fusioned, is about $n^3/6$ (time complexity $O(n^3)$).

In these conditions, since the end of the 1970s, taxonomist researchers observed that these orders of complexity cannot be compatible with processing cluster "large" data sets. Very original ideas arised and became operational in order to decrease substantially

1. The number of distances (or dissimilarities) calculated between the cluster formed;
2. The number of comparisons between class distances;
3. The size of the memory space needed.

Two types of objectives were considered for the new hierarchical clustering algorithms, answering the previous requirements. For the first one, the classification tree has to be exactly the same as that obtained by applying the traditional algorithm. In the framework of the second objective, an approximation of the exact tree, having a certain form is proposed. A third and specific type of algorithms concerning clustering large data sets, occurs when a contiguity constraint is given or defined on the object set and when this constraint has to be satisfied at each level of the cluster formation.

The major headings associated with the first objective are

1. Reactualization formulas in the case of simultaneous multiple aggregations [36] (see Sect. 10.6.3);
2. Reducibility, monotonicity and the reducible neighborhoods clustering algorithm [9, 10] (see Sect. 10.6.4);
3. Reciprocal Nearest Neighbors (RNN) clusters and chain search of RNN [5, 9, 13, 46] (see Sect. 10.6.4).

All of these techniques can apply in the case mentioned above where, additionally, a contiguity constraint has to be satisfied in the cluster formation [4, 30, 31, 47] (see Sect. 10.6.5).

The approximation needed of the classification tree in *Information Retreivial* follows the subsequent stages

1. Random splitting of the entire set $\mathcal{O}$ of the elements concerned (documents) into a partition $\{\mathcal{O}^1, \mathcal{O}^2, \ldots, \mathcal{O}^k, \ldots, \mathcal{O}^K\}$ where the disjoint and complementary subsets $\mathcal{O}^k$, $1 \leq k \leq K$, are, as much as possible, of equal size;
2. Respective partial ascendant hierarchical classifications of the different subsets $\mathcal{O}^k$, $1 \leq k \leq K$, each of the hierarchical classification construction being stopped at the first relevant level, giving rise to a "significant partition" for which the number of classes reaches a given order of size;
3. Joining by the same hierarchical clustering method the different classes obtained through the K partial hierarchical clusterings, after representing each class by a single element.

This method was established in [32] and developed in [42, 52]. The idea of this fractionization technique was retrieved by Cutting et al. (1992) [12].

10.6.2 Complexity Considerations of the Basic AAHC Algorithm

Let us consider the construction of a classification tree on an object set $\mathcal{O}$ associated with a classical version of an *AAHC* algorithm. As expressed above, this construction requires the definition and computing a dissimilarity function δ for pairwise comparisons between disjoint $\mathcal{O}$ subsets. Generally, δ is calculated from a dissimilarity or distance function $\mathcal{D}$ defined on $\mathcal{O}$. $\mathcal{D}$ can be given by a table, denoted also by $\mathcal{D}$, of the form

$$\mathcal{D} = \{\mathcal{D}(x, y) | \{x, y\} \in P_2(\mathcal{O})\} \qquad (10.6.1)$$

where, as usually, $P_2(\mathcal{O})$ designates the set of unordered object pairs of $\mathcal{O}$. On the other hand, δ is formally expressed by

$$\delta = \{\delta(X, Y) | X \in \mathcal{P}(\mathcal{O}), Y \in \mathcal{P}(\mathcal{O}) \quad \text{and} \quad X \cap Y = \emptyset\} \qquad (10.6.2)$$

$\mathcal{D}$ gives rise to the subtable of δ concerning dissimilarities between singleton classes, namely,

$$\delta^0 = \{\delta(\{x\}, \{y\}) | \{x, y\} \in P_2(\mathcal{O})\}$$

If $\mathcal{O}$ is endowed with a weighting, the passage from $\mathcal{D}$ to δ^0 may require the latter weighting (e.g. Ward criterion). However, as specified in the previous section complexities are not affected, and then, uniform weighting is assumed. In these conditions, The respective values of δ^0 are those of $\mathcal{D}$. On the other hand, as said in the previous section, we do not restrict generality if we address the complexity problems in the case where $\mathcal{O}$ is described by p numerical attributes and then, represented by a cloud of points in the geometrical space $\mathbb{R}^p$. The latter cloud was denoted above by $\mathcal{N}(\mathbb{I})$. Therefore, $\mathcal{D}$ (resp., δ^0) can be obtained from a distance defined in $\mathbb{R}^p$ (see for example Sect. 4.2.1.3).

There are in all $(n-1)^2$ values of the table δ which have to be calculated in order to build the binary classification tree concerned. $n(n-1/2)$ values are comprised in δ^0 for the passage from the 0 level to the 1 level. Besides, the passage from the $k-1$ level to the k level requires the caculation of the dissimilarity between the new class formed in the partition P_k with the $n - k - 1$ classes remained from the previous partition P_{k-1}. We have

$$\frac{n(n-1)}{2} + \sum_{1 \le k \le n-2} (n - k - 1) = (n-1)^2$$

Let us denote by $C_1 \cup C_2$ the new class formed in the passage from P_{k-1} to P_k, resulting of merging the classes C_1 and C_2 of P_{k-1}. If D is a common class of P_{k-1} and P_k, the calculation of $\delta(C_1 \cup C_2, D)$, using a general analytical equation such (10.1.3) requires to turn to table $\mathcal{D}$ and to select $(card(C_1) + card(C_2)) \times card(D)$ values. While, using a reactualization formula such (10.1.10), requires to pick up only three values in the dissimilarity table between classes of P_{k-1}, namely, $\delta(C_1, D), \delta(C_2, D)$ and $\delta(C_1, C_2)$ (see 10.6.3).

Due to the uniform weighting adopted on $\mathcal{O}$ for the complexity considerations, the general reactualization formula of (10.1.10) becomes

$$(T_k, t^k) = \varphi(T_{k-1}, t^{k-1}), \ 1 \leq k \leq n-2$$

where T_k is the dissimilarity δ table (subtable of the global δ table above) between the classes of the partition P_k and where t^k is the type of P_k (see Sect. 1.3 of Chap. 1), $0 \leq k \leq n-2$. T_k comprises $(n-k)(n-k-1)/2$ values. Hence, the time complexity for sorting T_k in order to determine its minimum value and the pair of classes concerned is $(n-k)(n-k-1)/2$. Recall that the time complexity is evaluated in the case of the construction of a binary tree, for which the minimal value of T_k is reached once for every k, $0 \leq k \leq n-2$. Therefore, the total time complexity for this computational part of the classification tree is

$$\frac{1}{2} \sum_{0 \leq k \leq n-2} (n-k)(n-k-1) = \frac{1}{6}(n-1)n(n+1)$$

The space complexity of an *AAHC* algorithm using a reactualization formula is $n^2/2$, it corresponds to the size of δ^0 which is exactly $n(n-1)/2$. In these conditions, processing very large data sets with this algorithm might be difficult. An alternative consists of carrying out all the calculations needed for building the classification tree in the framework of the data table indexed by $\mathcal{O} \times \mathcal{A}$. In this case, which is all the better when p is small enough, the space complexity becomes linear with respect to the number of objects n.

For this data structure, the different partitions of the chain partition are stored at the level of the data table Objects by Attributes. If C and D are two classes of the lth partition P_l of this chain, $0 \leq l \leq n-2$, the following table $\mathcal{D}(C, D)$ of dissimilarities between respective elements of C and D, has to be calculated. Recall that dissimilarities or distances between classes or elements are lost after determining a given partition in the partition chain.

$$\mathcal{D}(C, D) = \{\mathcal{D}(x, y) | (x, y) \in C \times D\}$$

This table, already considered in (10.1.2), comprises $card(C) \times card(D)$ unit distances or dissimilarities in $\mathbb{R}^p$. Now, let us regard the passage from the partition P_{k-1} to P_k, $1 \leq k \leq n-3$. At a given step of the comparison process of class pair dissimilarities of P_{k-1}, consider the last class pair $\{C, D\}$ for which δ has the lowest value. The next step consists of selecting a class pair $\{C', D'\}$ of P_{k-1}, not yet examined in order to calculate $\delta(C', D')$ and to compare $\delta(C', D')$ with $\delta(C, D)$, retaining the class pair ($\{C, D\}$ or $\{C', D'\}$) for which δ is minimum and the associated value of δ. The complexity of this stage will depend closely on the nature of the merging criterion δ. For a criterion such that $\delta_{\min}$ (Single Linkage), $\delta_{\max}$ (Complete

Linkage), or δ_{lla} (Maximum Likelihood Linkage), a sort of the values of the table $\mathcal{D}(C, D)$ is needed. The time complexity of the latter is $card(C) \times card(D)$. A criterion such that δ_{ave} requires an arithmetic summation on $card(C) \times card(D)$ numerical values. Thus, we have also in this case the same linear complexity.

Interesting mathematical property occurs in the case of *Inertia variation* (or Ward) criterion (see Sect. 10.3.3). The expression of the latter depends only on the representation of a given class C by the ordered pair (G_C, μ_C) where the first argument is the center of gravity of the cloud in $\mathbb{R}^p$ representing C and its second argument is the weight of C (see (10.3.12) and (10.3.13)). Recall that for studying complexity we have considered—without loss of generality—μ_C given by $card(C)$.

For studying complexity in the framework of the implementation considered, in terms of calculation of unit distances in $\mathbb{R}^p$, we shall consider that the binary classification tree has a comb structure. For the latter, in the passage from P_{k-1} to P_k, $2 \le k \le n - 1$, only a single element joins the class of P_{k-1} which includes more than one object. Considering this specific structure simplifies but not restrict the generality (the proof of this latter point is left for the reader).

In the case of Ward criterion or either one for which a class formed can be represented by a single point—generally weighted—obtained linearly with respect to the class size, the number of unit distance calculations in $\mathbb{R}^p$ can be written as

$$\frac{n \times (n - 1)}{2} + \sum_{2 \le k \le n-1} (n - k) = (n - 1)^2$$

In the other mentioned cases (δ_{ave}, $\delta_{\min}$, $\delta_{\max}$ and δ_{lla}), the number of unit distance calculations becomes

$$\frac{n \times (n - 1)}{2} + \sum_{2 \le k \le n-1} (n - k) \times k = \frac{1}{6}(n - 1)(n^2 + 4n - 6)$$

Thus, the time complexity in $O(n^2)$ for the first type of criterion and $O(n^3)$ for the second family of criteria.

10.6.3   Reactualization Formulas in the Cases of Binary and Multiple Aggregations

First, in Sect. 10.6.3.1, we address the classical case for which a single pair of classes merge from a given level to the next one. That is, in the construction of the classification tree, for a given partition determined at a given level, there exists exactly a single class pair which realizes the minimum value of the dissimilarity δ between its classes. In the next Sect. 10.6.3.2, we study the case where the minimal dissimilarity δ value may occur several times at a given level. In this case, the built classification tree is not binary necessarily.

10.6.3.1 Reactualization Formulas in the Case of Binary Aggregations

The merging criteria we shall consider are presented and largely discussed in Sects. 10.3 and 10.4. For frequency of use, they will be expressed in terms of distances or dissimilarities (and not of similarities) between two disjoint subsets C and D of the object set $\mathcal{O}$, provided with a dissimilar or distance function $\mathcal{D}$, on the basis of the table $\mathcal{D}(C, D)$ of $\mathcal{D}(x, y)$ between x and y $((x, y) \in C \times D)$ (see Sect. 10.3.1 ·and (10.1.2) and (10.6.2)). These criteria are

1. $\delta_{\min}$ (Single Linkage criterion)
2. $\delta_{\max}$ (Complete Linkage criterion)
3. δ_{ave} (Average Linkage criterion)
4. δ_{iner} (Inertia variation (Ward) criterion)
5. Δ_{lla}^{ϵ} (Maximal Likelihood Linkage family criteria)

Let us recapitulate their respective formulations

$$\delta_{\min}(C, D) = \min\{\mathcal{D}(x, y)|(x, y) \in C \times D\}$$
$$\delta_{\max}(C, D) = \max\{\mathcal{D}(x, y)|(x, y) \in C \times D\}$$
$$\delta_{ave}(C, D) = \frac{1}{card(C) \times card(D)} \sum_{(x,y) \in C \times D} \mathcal{D}(x, y)$$
$$\delta_{iner}(C, D) = \frac{card(C) \times card(D)}{card(C) + card(D)} \|G_C - G_D\|^2$$
$$\Delta_{lla}^{\epsilon}(C, D) = \epsilon[\text{Log}_2(card(C)) + \text{Log}_2(card(D))]$$
$$\times \min\{\text{Log}_2(D_{inf}(x, y))|(x, y) \in C \times D\} \qquad (10.6.3)$$

For the items 1, 2 and 3, $\mathcal{D}(x, y)$ is a dissimilarity function calculated on (x, y) in the attribute representation space. In 4, the norm $\|\bullet\|$ corresponds to an euclidean distance in the geometrical space $\mathbb{R}^p$, on the other hand G_C and G_D are the centers of gravity of the classes C and D. In 5, $D_{inf}(x, y)$ refers to the table of *informational dissimilarities* (see (7.2.7) of Chap. 7) and (10.1.6). We have

$$\Delta_{lla}^{\epsilon}(C, D) = -S_{\max}^{\epsilon}(C, D)$$

of (10.4.8).

These merging criteria are the most employed. On the other hand, the reversal phenomenon, defined in Sect. 10.4.2.3, is impossible for any of them. This property was shown in the latter section for the criteria $\sigma_{\max}$, $\sigma_{\min}$, σ_{ave} and $S_{\max}^{\epsilon}$, these, expressed in terms of similarities are equivalent to the $\delta_{\min}$, $\delta_{\max}$, δ_{ave} and Δ_{lla}^{ϵ} criteria, respectively. The non-inversion, said also *monotonic* property, for these criteria were seen from the reactualization formula in the binary case. The same technique will be considered in order to establish this property for δ_{iner} (Ward criterion). All of the merging criteria do not have this property. For example, no reversal cannot be garantee for the *centroid* criterion defined by the square distance between the centers of gravity in $\mathbb{R}^p$ of the subclouds associated with the clusters to be compared. We have also seen that inversion can occur for criteria such that *Minimal and Average Likelihood* criteria (see Sect. 10.4.2.3).

Up untill now, in the binary construction of a classification tree we referred to a reactualization formula of the form

$$\delta(C_1 \cup C_2, D) = \varphi\big(\delta(C_1, D), \delta(C_2, D), \delta(C_1, C_2), l_1, l_2, m\big)$$

where $l_1 = card(C_1), l_2 = card(C_2)$ and $m = card(D)$ (see (10.4.9), Sects. 10.4.1.3 and 10.6.2). This general recurrence formula is very important. Consider the passage between the partition P_{k-1} of the $(k-1)$th level to the partition P_k of the kth level, in the binary classification tree. For a candidate class pair $\{C_1, C_2\}$ of P_{k-1} to be merged, there are only twice $(n - k - 1)$ values of $\delta(C_i, D)$ $(i = 1$ and $i = 2)$ to consult in the dissimilarity table T_{k-1}. Thus, this formula avoids the direct calculation of $\delta(C_1 \cup C_2, D)$ for a class D of P_{k-1}, distinct from C_1 and C_2.

The importance of the latter equation led Lance and Williams [27] to give a parametrized form of this recurrence formula including several dissimilarity indices δ. This can be written as follows:

$$\delta(C_1 \cup C_2, D) = \alpha_1\delta(C_1, D) + \alpha_2\delta(C_2, D) + \beta\delta(C_1, C_2) + \gamma|\delta(C_1, D) - \delta(C_2, D)| \tag{10.6.4}$$

Some classical merging criteria can be expressed in the framework of this formula. Thus, $\delta_{min}, \delta_{max}, \delta_{ave}$ and δ_{iner} can be represented by specific parametrizations. However and in spite of the generalization provided in [24], not all of the possible criteria can be covered by such a recurrence formula. Thus, δ_{lla} cannot be considered as a particular case of the Lance and Williams formula (see below).

To summarize and as expressed in the introductory Sect. 10.1, the general parametrization of binary reactualization formulas provided by the Lance and Williams recurrence formula is theoretically interesting but useless in practice. The reasons of our claim are

1. Every merging criterion δ cannot be interpreted as a particular case of the Lance and Williams recurrence formula;
2. This formula enables differences between different criteria δ, to be observed but not to be understood;
3. It is more profitable to devise freely a new criterion than to be constrained by an a priori specific form;
4. This formula does not include the case of multiple aggregations occuring at the same time at a given level of the classification tree.

To be completely explicit, let us make clear the reactualization formulas in the binary case for each of the criteria $\delta_{min}, \delta_{max}, \delta_{ave}, \delta_{iner}$ and Δ_{lla}^ϵ.

$$\delta_{\min}((C_1 \cup C_2), D) = \min\{\delta_{\min}(C_1, D), \delta_{\min}(C_2, D)\}$$

$$\delta_{\max}((C_1 \cup C_2), D) = \max\{\delta_{\max}(C_1, D), \delta_{\max}(C_2, D)\}$$

$$\delta_{ave}((C_1 \cup C_2), D) = \frac{l_1}{l_1 + l_2}\delta_{ave}(C_1, D) + \frac{l_2}{l_1 + l_2}\delta_{ave}(C_2, D)$$

$$\delta_{iner}((C_1 \cup C_2), D) = \frac{1}{(l_1 + l_2 + m)}$$

$$\times \big((l_1 + m)\delta_{iner}(C_1, D) + (l_2 + m)\delta_{iner}(C_2, D) - m\delta_{iner}(C_1, C_2)\big)$$

$$\Delta_{lla}^{\epsilon}((C_1 \cup C_2), D) = \epsilon\log_2(l_1 + l_2)$$

$$+ \min\{\Delta_{lla}^{\epsilon}(C_1, D) - \epsilon\log_2(l_1), \Delta_{lla}^{\epsilon}(C_2, D) - \epsilon\log_2(l_2)\} \qquad (10.6.5)$$

where, for notation simplicity, we have denoted as previously, $l_1 = card(C_1)$, $l_2 = card(C_2)$ and $m = card(D)$. Let us indicate the following relation which is useful in order to establish the reactualization formula for δ_{iner}:

$$\|G_1 - G_2\|^2 = \|G_1 - G\|^2 + \|G_2 - G\|^2 - 2q(G_1 - G, G_2 - G)$$

where G is the general center of gravity of the total cloud representing the object set $\mathcal{O}$ and where G_1 and G_2 are the respective centers of gravity of the subclouds representing the classes C_1 and C_2, q being the metric (scalar product) of which the geometrical representation space $\mathbb{R}^p$ is provided.

It is straightforward to establish that all of the five criteria listed in (10.6.5) are *monotonic* (no possible reversal), that is,

$$\delta(C_1, C_2) \leq \delta(C_1, D) \quad \text{and} \quad \delta(C_1, C_2) \leq \delta(C_2, D)$$

$$\Rightarrow \delta(C_1, C_2) \leq \delta(C_1 \cup C_2, D)$$

However, and as an example, reversal is possible for the *centroid* criterion δ_{cent}, for which the binary reactualization formula is

$$\delta_{cent}(C_1 \cup C_2, D)$$

$$= \frac{l_1}{l_1 + l_2}\delta_{cent}(C_1, D) + \frac{l_2}{l_1 + l_2}\delta_{cent}(C_2, D)$$

$$- \frac{l_1 l_2}{(l_1 + l_2)^2}\delta_{cent}(C_1, C_2) \qquad (10.6.6)$$

The reactualization formulas (10.6.5) can be specified in the case of clustering a set $\mathcal{O}$ of weighted objects described by p numerical attributes and represented in the geometrical space $\mathbb{R}$ by a cloud of points $\mathcal{N}(\mathbb{I})$ (see Eq. 10.3.3). In this case, the Ward criterion is clearly justified as defined by the *inertia* of $\mathcal{N}(\mathbb{I})$. Consider now an arbitrary subset $\mathcal{Q}$ of $\mathcal{O}$ ($\mathcal{Q} \subseteq \mathcal{O}$).

Let $\mathbb{K}$ denote the subset of $\mathbb{I}$ which indexes $\mathcal{Q}$, the inertia of $\mathcal{Q}$ representation can be written as follows

$$\mathcal{M}(\mathbb{K}) = \frac{1}{2\mu(\mathbb{K})} \sum_{(k,k') \in \mathbb{K} \times \mathbb{K}} \mu_k \cdot \mu_k d^2(k, k')$$

where $\mu(\mathbb{K})$ is the total weight of the subcloud $\mathcal{N}(\mathbb{K})$ and where $d^2(k, k')$ is the square of the euclidean distance between the two vertices of of $\mathcal{N}(\mathbb{K})$, labelled k and k'.

The expression above of the inertia $\mathcal{M}(\mathbb{K})$ and the fourth reactualization formula in (10.6.5) enable the Ward method to be applied in the case where a non-euclidean distance or even a dissimilarity is substituted for d. This proposition is studied with some detail in [28].

10.6.3.2 Reactualization Formulas in the Case of Multiple Aggregations

We consider the case mentioned at the end of the introductory Sect. 10.1 and also expressed in Sect. 10.3.2.1, where several class pairs achieve "at the same time", at a given level of the construction of the classification tree, the minimal value of the class dissimilarity concerned δ. As claimed in Sect. 10.1, this case occurs frequently for some data descriptive structures. A binary solution for building the classification tree in this case was outlined in Sect. 10.3.2.1. For the latter, let us denote by $P_{k-1} = \{E_1, E_2, \ldots, E_j, \ldots, E_{n-k+1}\}$ the partition determined by the $(k-1)$th level of the latter classification tree, on an object set of size n. On P_{k-1} define the binary relation that we designate by $Min\delta$, for which $E_{j0}Min\delta E_{j1}$, if and only if $\delta(E_{j0}, E_{j1})$ is *minimal* in the sequence $\{\delta(E_j, E_{j'})|1 \leq j < j' \leq n-k+1\}$ of class dissimilarities. Now, consider a given connected component $\mathcal{C} = \{C_1, C_2, \ldots, C_h, \ldots, C_r\}$ in P_{k-1}. The elements of $\mathcal{C}$ are mutually distinct classes of P_{k-1}, such that for any two elements C_h and $C_{h'}$ in $\mathcal{C}$, there is a chain for $Min\delta$ connecting C_h to $C_{h'}$. On the other hand, if E_j of P_{k-1} does not belong to $\mathcal{C}$, then E_j is disconnected from any element of $\mathcal{C}$.

As pointed out in Sect. 10.3.2, in order to retreive the cluster $C_1 \cup C_2 \cup \cdots \cup C_h \cup \cdots \cup C_r$ by using a binary version of the construction of a classification tree, it is necessary to proceed to a series of $r - 1$ binary aggregations, followed each by applying the binary reactualization formula concerned by the class dissimilarity δ. In the multiple aggregation method, only a single multiple aggregation is carried out in order to pass from the set $\mathcal{C}$ of r classes to one class defined by their union: $C_1 \cup C_2 \cup \cdots \cup C_h \cup \cdots \cup C_r$.

Now, let us describe in a somewhat more general way the multiple classification tree construction. We are interested in the passage between two consecutive levels $i - 1$ and i of this classification tree ($i \geq 1$). The associated partitions are designated—as usually—by P_{i-1} and P_i. The new clusters formed in P_i correspond to connected components of the binary relation $Min\delta$ defined on P_{i-1} (see above). If $\mathcal{C} = \{C_1, C_2, \ldots, C_h, \ldots, C_r\}$ is a connected component in P_{i-1}, then $C_1 \cup C_2 \cup \cdots \cup C_h \cup \cdots \cup C_r$ is a new cluster in P_i. Clearly, there are as many new clusters in P_i, as connected components in P_{i-1}. If c is the number of connected components in P_{i-1}, including $r_1, r_2, \ldots, r_b, \ldots, r_{c-1}$ and r_c elements, respectively, a binary version of the tree classification construction would require $(r_1 - 1) + (r_2 - 1) + \cdots + (r_b - 1) + \cdots + (r_c - 1)$ binary aggregations, each of two classes belonging to a same connected component; each aggregation being followed by a calling to the reactualization formula in the binary case. In the multiple aggregation procedure, only a single new level is built. In the latter, a cluster such $C_1 \cup C_2 \cup \cdots \cup C_h \cup \cdots \cup C_r$ is directly built. This is possible provided to have available a *multiple* reactualization formula of the form

$$\delta(C_1 \cup C_2 \cup \cdots \cup C_h \cup \cdots \cup C_r, D_1 \cup D_2 \cup \cdots \cup D_j \cup \cdots \cup D_s)$$
$$= \varphi\big(\delta(C_h, D_j), l_h, m_j | 1 \leq h \leq r, 1 \leq j \leq s\big) \tag{10.6.7}$$

where, in the context, $\{C_1, C_2, \ldots, C_h, \ldots, C_r\}$ and $\{D_1, D_2, \ldots, D_j, \ldots, D_s\}$ are two distinct connected components in P_{i-1} and where $l_h = card(C_h)$ (resp., $m_j = card(D_j)$), $1 \leq h \leq r$ (resp., $1 \leq j \leq s$).

The different versions of this formula were established for the five criteria $\delta_{\min}$, $\delta_{\max}$, δ_{ave}, δ_{iner} and Δ_{lla}^{ϵ} in [36]. Let us give here these equations:

$\delta_{\min}$ Single Linkage criterion

$$\delta_{\min}(C_1 \cup C_2 \cup \cdots \cup C_h \cup \cdots \cup C_r, D_1 \cup D_2 \cup \cdots \cup D_j \ldots \cup \cdots \cup D_s)$$
$$= \min\big(\delta_{\min}(C_h, D_j) | 1 \leq h \leq r, 1 \leq j \leq s\big) \tag{10.6.8}$$

$\delta_{\max}$ Complete Linkage criterion

$$\delta_{\max}(C_1 \cup C_2 \cdots \cup C_h \cdots \cup C_r, D_1 \cup D_2 \cdots \cup D_j \cdots \cup D_s)$$
$$= \max\big(\delta_{\min}(C_h, D_j) | 1 \leq h \leq r, 1 \leq j \leq s\big) \tag{10.6.9}$$

δ_{ave} Average Linkage criterion

$$\delta_{ave}(C_1 \cup C_2 \cdots \cup C_h \cdots \cup C_r, D_1 \cup D_2 \cdots \cup D_j \cdots \cup D_s)$$
$$\sum_{1 \leq h \leq r} \sum_{1 \leq j \leq s} \nu_{hl} \delta_{ave}(C_h, D_j) \tag{10.6.10}$$

where

$$\nu_{hl} = \frac{l_h . m_j}{(l_1 + l_2 + \cdots + l_p + \cdots + l_r) . (m_1 + m_2 + \cdots + m_q + \cdots + m_s)}$$
$$1 \leq h \leq r, 1 \leq j \leq s$$

$$\delta_{iner} \text{ Variation Inertia or Ward criterion}$$

$$\delta_{iner}(C_1 \cup C_2 \cdots \cup C_h \cdots \cup C_r, D_1 \cup D_2 \cdots \cup D_j \cdots \cup D_s)$$

$$= \frac{1}{N_{rs}} \left(\sum_{1 \le h \le r} \sum_{1 \le j \le s} (l_h + m_j)\delta_{iner}(C_h, D_j) \right.$$

$$- \sum_{1 \le h < h' \le r} (l_h + l_{h'}) \frac{M_s}{L_r} \delta_{iner}(C_h, C_{h'})$$

$$\left. - \sum_{1 \le j < j' \le s} (m_j + m_{j'}) \frac{L_r}{M_s} \delta_{iner}(D_j, D_{j'}) \right) \qquad (10.6.11)$$

where

$$L_r = \sum_{1 \le h \le r} l_h, M_s = \sum_{1 \le j \le s} m_j \text{ and } N_{rs} = L_r + M_s$$

This equation is proved recursively with respect to (r, s) in [36].

$$\Delta_{lla}^{\epsilon} \text{ Maximal Likelihood Linkage family criteria}$$

With the notations defined just above, we have

$$\Delta_{lla}^{\epsilon}(C_1 \cup C_2 \cdots \cup C_h \cdots \cup C_r, D_1 \cup D_2 \cdots \cup D_j \cdots \cup D_s)$$
$$= \epsilon\log_2(L_r) + \epsilon\log_2(M_s)$$
$$+ \min\{\Delta_{lla}^{\epsilon}(C_h, D_j - \epsilon\log_2(l_h) - \epsilon\log_2(m_j)|$$
$$1 \le h \le r, 1 \le j \le s\} \qquad (10.6.12)$$

10.6.4   Reducibility, Monotonic Criterion, Reducible Neighborhoods and Reciprocal Nearest Neighborhoods

In the *AAHC*, the "reductibility" property of a fusion class criterion δ—defined in the following—enables a large reduction of complexity time to be obtained through specific hierarchical algorithms. We specify below the principle of the latter. They are called "The Reducible Neighborhoods (*RedN*)" and "The Reciprocal Nearest Neighbors (*RNN*)" algorithms, respectively. With these, the number of δ dissimilarities (distances) calculations and their mutual comparisons, are drastically decreased. In fact, this decrease is observed in practice. Nevertheless, theoretical results strengthen this empirical observation. Processing at the initial level of the data table, for any reducible merging criterion, the maximal theoretical complexity is $O(n^2\log(n))$ for the $RedN$ algorithm, while it is $O(n^3)$ for the classical $AAHC$ algorithm [10]. The reducibility property associated with the Reducible Neighborhoods algorithm were introduced and analyzed in [8–10]. Now, in regard to Reciprocal Nearest Neighbors algorithm, by working with the dissimilarity matrix, the maximal theoretical complexity is $O(n^2 \log(n))$, regardless the reducible merging criterion considered [9, 48]. Concerning the latter algorithm, the Sect. 6 of an interesting article of Murtagh [48] relates an illustrated and well-documented historic concerning the RNN algorithm. Relevant implementations taking into account the nature of the merging criterion δ are mentioned. $O(n^2)$ time and spatial complexities are pointed out for each of the criteria δ_{min}, δ_{max} and δ_{ave} (see Sect. 10.3.2).

Before expressing the principles, which are related, of each of both algorithms, $RedN$ and RNN, we start by defining the reducibility property of a criterion δ. Next, we notice the very important equivalence property between reducibility and monotony.

Let (P, δ) be a couple composed of a partition P on the object set $\mathcal{O}$ and a dissimilarity coefficient δ between the elements of P, these being classes of $\mathcal{O}$. For C defined by a union of P classes and a positive numerical value ρ, corresponding to a possible value of δ, the ball $\mathcal{B}(C, \rho)$ in P, centered in C of radius ρ, is defined as follows

$$\mathcal{B}(C, \rho) = \{X | X \in P \quad \text{and} \quad \delta(C, X) \le \rho\} \tag{10.6.13}$$

Reducibility Property of δ

Definition 25 δ is reducible if for any pair $\{C_1, C_2\}$ of distinct elements of P and for any possible value ρ of δ, we have

$$\delta(C_1, C_2) \le \rho \Rightarrow \mathcal{B}(C_1 \cup C_2, \rho) \subset \mathcal{B}(C_1, \rho) \cup \mathcal{B}(C_2, \rho)$$

In other words, if $\delta(C_1, C_2) \le \rho$, the ball centered in $C_1 \cup C_2$ of radius ρ is included in the union of both balls, having the same radius ρ and centered in C_1 and C_2, respectively. An equivalent expression for Definition 25 is given by

Definition 26 δ is reducible if for any pair $\{C_1, C_2\}$ of distinct elements of P and for any possible value ρ of δ, we have

$$\delta(C_1, C_2) \le \rho \Rightarrow \mathcal{B}^c(C_1, \rho) \cap \mathcal{B}^c(C_2, \rho) \subset \mathcal{B}^c(C_1 \cup C_2, \rho)$$

where $\mathcal{B}^c(C_i)$ ($i = 1$ or $i = 2$) designates the complementary subset of $\mathcal{B}(C_i, \rho)$ ($i = 1$ or $i = 2$) in P.

In other terms, for C_1, C_2 and D belonging to P

$$\delta(C_1, C_2) \le \rho, \delta(C_1, D) > \rho, \delta(C_2, D) > \rho \text{ then } \delta(C_1 \cup C_2, D) > \rho$$

Clearly, if δ is reducible, δ is monotonic. This is obtained from the last expression of reducibility by setting $\rho = \delta(C_1, C_2)$.

Conversely, assume δ monotonic and consider its reactualization formula under the form (see Sect. 10.6.3.1),

$$\delta(C_1 \cup C_2, D) = \varphi\big(\delta(C_1, D), \delta(C_2, D), \delta(C_1, C_2), l_1, l_2, m\big)$$

where—as above—$l_1 = card(C_1)$, $l_2 = card(C_2)$ and $m = card(D)$.

The symmetry of φ with respect to its two first arguments enable us to reduce—without loss of generality —the expression of monotonicity property to

$$\delta(C_1, D) \geq \delta(C_2, D) \geq \delta(C_1, C_2) \Rightarrow$$
$$\varphi\big(\delta(C_1, D), \delta(C_2, D), \delta(C_1, C_2), l_1, l_2, m\big) \geq \delta(C_1, C_2)$$

$$(10.6.14)$$

In fact, the analytical form of the function φ for monotonic criteria makes that the latter equation holds if we substitute for $\delta(C_1, C_2)$, $\alpha \times \delta(C_1, C_2)$, where α is a real positive number comprised between 0 and 1. Consequently, δ is reducible. Notice that in the case given by the left member (premise) of the implication in (10.6.14), all of the four criteria δ_{min}, δ_{max}, δ_{ave} and δ_{iner} become linear combinations of $\delta(C_1, D)$, $\delta(C_2, D)$ and $\delta(C_1, C_2)$.

The above proof of the property "monotonicity $\Rightarrow$ reductibility" reconsiders that given in [14].

Reducibile Neighborhoods (*RedN*) Algorithm

Description of the Reducibile Neighborhoods algorithm requires to fix a strictly increasing sequence of stratification thresholds

$$(\rho_1, \rho_2, \ldots, \rho_i, \ldots, \rho_{max}),$$

$\rho_1 < \rho_2 < \cdots < \rho_i < \rho_{i+1} < \cdots < \rho_{max}$, where ρ_{i+1} has to be determined judiciously, once ρ_i has been exploited. Each of these stratification indices gives rise to a sequence of proximity graphs between classes of an $\mathcal{O}$ partition. Let us designate this sequence, when it is associated with ρ_i, by

$$\mathcal{G}_i = (G_{i1}, G_{i2}, \ldots, G_{i(j-1)}, G_{ij}, \ldots, G_{imax}) \qquad (10.6.15)$$

where G_{ij} is deduced from $G_{i(j-1)}$ by fusioning vertices of $G_{i(j-1)}$. G_{ij} comprises less vertices and less edges than $G_{i(j-1)}$. G_{imax} is the graph for which the edge set becomes empty. The vertices of G_{imax} represent clusters in $\mathcal{O}$ such that any two of them are strictly more dissimilar than ρ_i. ρ_{i+1} ($\rho_{i+1} > \rho_i$) is a threshold for class dissimilarity value enabling the *RedN* algorithmic principle to apply to G_{imax}. To be completely explicit, let us describe the construction of the sequence of graphs $\mathcal{G}_1 = (G_{11}, G_{12}, \ldots, G_{1(j-1)}, G_{1j}, \ldots, G_{1max})$ and also the passage from this sequence to the next one $\mathcal{G}_2$. Clearly, the passage from $\mathcal{G}_i$ to $\mathcal{G}_{i+1}$ will be analogous to that from $\mathcal{G}_1$ to $\mathcal{G}_2$.

The vertices of G_{11} is the set of singleton classes which we denote by

$$C_{11} = \{\{x\}|x \in \mathcal{O}\} \tag{10.6.16}$$

ρ_1 enables the graph G_{11} to be specified:

$$G_{11} = (C_{11}, \Gamma_{11}, \rho_1) \tag{10.6.17}$$

where Γ_{11} is the edge set of G_{11}:

$$\Gamma_{11} = \{\{\{x\}, \{y\}\}|x \in \mathcal{O}, y \in \mathcal{O} \text{ and } \delta(\{\{x\}, \{y\}\}) \leq \rho_1\}$$

For each vertice $\{x\}$ of C_{11}, we associate a neighborhood ball centered in $\{x\}$ of radius ρ_1 of the form

$$B_1(\{x\}, \rho_1) = \{\{y\}|\delta(\{x\}, \{y\}) \leq \rho_1\} \tag{10.6.18}$$

In the binary version of this algorithm as given in [10], one seeks for a pair of vertices $\{\{x\}, \{y\}\}$ in

$$\bigcup \{B_1(\{x\}, \rho_1)|x \in \mathcal{O}\} \tag{10.6.19}$$

which realizes the minimal value of $\delta(\{x\}, \{y\})$. Let us designate by $c_1 = \{x_1, y_1\}$ such a pair. c_1 is then a vertice of the graph G_{12} of which the vertice set C_{12} is obtained by deleting x_1 and y_1 and by adding $c_1 = \{x_1, y_1\}$, namely,

$$C_{12} = C_{11} - \{\{x_1\}, \{y_1\}\} + c_1 \text{ (set sum)} \tag{10.6.20}$$

The neighborhood ball $B(c, \rho_1)$ is then constituted by scanning $B_1(\{x_1\}, \rho_1) \cup B_1(\{y_1\}, \rho_1)$ (reducibility property). Moreover, if $\{x_1\}$ (resp., $\{y_1\}$) appears in a ball of the form $B_1(\{z\}, \rho_1)$, it is replaced by c_1, provided that $\delta(\{z\}, c_1)$ is lower or equal to ρ_1. When all these substitutions are carried out, the graph

$$G_{12} = (C_{12}, \Gamma_{12}, \rho_1)$$

is obtained. Thus, in the binary expression of this algorithm, the passage from G_{11} to G_{12} is the resulting of the fusion of a single pair designated above by $\{\{x_1\}, \{y_1\}\}$. Obviously, the passage from G_{1j} to $G_{1(j+1)}$ is completely analogous to that from G_{11} to G_{12}.

Notice that a "small" value of ρ_1 entails a few number of calculations and comparisons. Progressively to the execution of the algorithm, due to the monotonic property of δ, the dissimilarities between formed classes increase. Therefore, necessarily, there exists a subscript 1max, such that the graph $G_{1\text{max}}$ becomes empty, that is to say, $\Gamma_{1\text{max}} = \emptyset$. In this case

1. A new stratification threshold ρ_2 is chosen ($\rho_2 > \rho_1$) for the definition of the graph G_{21}, whose vertices being the same as those of $G_{1\max}$;
2. The new dissimilarity graph G_{21} is built by associating a ball of radius ρ_2 with each vertice of G_{21}, interpreted as the center of the ball concerned.

As above, G_{21} is followed by $G_{22}, G_{23}, \ldots, G_{2\max}, \ldots$and so on. The process ends when a single class grouping together all elements of $\mathcal{O}$, is obtained. Hence, the complete hierarchy is built.

In the algorithm proposed [10], for spatial complexity reasons, all the calculations and comparisons are achieved by using uniquely the initial data table. These can be performed at the level of the dissimilarity matrix between unit elements of $\mathcal{O}$, when the spatial complexity authorizes.

As emphasized above, in many applications the scales of the descriptive attributes might not be rich enough in order to discriminate finely the dissimilarities between unit elements. In these conditions, a bias might occur, due to the binary expression of the algorithm concerned and the class dissimilarities reactualizations into the different neighborhoods [10]. In fact, several class pairs belonging to a same or different neighborhoods may realize at the same time the minimal value of the class dissimilarities. It is then of importance to introduce and to adapt the method of multiple aggregations in the Reducible Neighborhoods algorithm (see Sect. 10.6.3.2).

Reciprocal Nearest Neighbor (*RNN*) Algorithm

The nature of Reciprocal Nearest Neighbor (RNN) algorithm is essentially binary. We shall describe the first step of this algorithm. The following ones are similar in all respects. To begin, let us associate a singleton class $\{x\}$ with each object x of $\mathcal{O}$. Thus, as in the general case, the starting point is the pair $(\mathcal{C}_1, \delta)$, where $\mathcal{C}_1$ is the set of singleton classes and where δ is a dissimilarity measure between disjoint subsets of $\mathcal{O}$, satisfying the reducibility property. For consistency reasons $\mathcal{C}_1$ was denoted above by $\mathcal{C}_{11}$ (see (10.6.16)). As usual, δ will play the role of the merging criterion. For notation reasons, in the development below, the singleton subset $\{x\}$ will be denoted by the capital letter X. By this way, $\mathcal{C}_1$ can be written as

$$\mathcal{C}_1 = \{X | X = \{x\}, x \in \mathcal{O}\} \tag{10.6.21}$$

Let us now define a pair $\{X, Y\}$ of elements of $\mathcal{C}_1$ which are *reciprocal nearest neighbors*. For each element of $\mathcal{C}_1$, we associate the smallest dissimilarity δ to X. Formally, it is

$$\delta_{\min}(X) = \min\{\delta(X, Y') | Y' \in \mathcal{C}_1 - X\} \tag{10.6.22}$$

Additionally, the neighborhood ball composed of the nearest elements of $\mathcal{C}_1$ to X, is associated with X, namely,

$$NN(X) = \{Y' | Y' \in \mathcal{C}_1 - X \quad \text{and} \quad \delta(X, Y') = \delta_{\min}(X)\} \tag{10.6.23}$$

Definition 27 Let $\{X, Y\}$ be a pair of distinct elements of $\mathcal{C}_1$, $\{X, Y\}$ is a pair of reciprocal nearest neighbors of $\mathcal{C}_1$ if and only if $Y \in NN(X)$ and $X \in NN(Y)$.

For such a pair $\delta(X, Y) = \delta_{\min}(X) = \delta_{\min}(Y)$. If Z is an element of $C_1 - X \cup Y$, we have necessarily

$$\delta(Z, X) \geq \delta_{\min}(X) = \delta_{\min}(Y)$$
$$\text{and } \delta(Z, Y) \geq \delta_{\min}(X) = \delta_{\min}(Y) \tag{10.6.24}$$

if not, $\delta_{\min}(X)$ (resp., $\delta_{\min}(Y)$) cannot be the lowest value reached by the dissimilarity δ to X (resp., to Y). Let us suppose the strict inequalities

$$\delta(Z, X) > \delta(X, Y)$$
$$\text{and } \delta(Z, Y) > \delta(X, Y) \tag{10.6.25}$$

In these conditions, the reducibility property of δ implies

$$\delta(X \cup Y, Z) > \delta(X, Y)$$

Therefore, the cluster $X \cup Y$ will appear necessarily in the classification hierarchy built by the classical version of the *AAHC* algorithm.

The case where there exists Z such that $\delta(Z, X) = \delta(X, Y)$ or $\delta(Z, Y) = \delta(X, Y)$ will be considered hereafter.

In this first step of the algorithm—as besides in the following ones—several class pairs of RNN singleton classes may occur, these are at the same time. More precisely, suppose we have found k pairs of RNN after scanning the element pairs of C_1 (see (10.6.21)) and denote these pairs

$$\{X_1, Y_1\}, \{X_2, Y_2\}, \ldots, \{X_j, Y_j\}, \ldots, \{X_k, Y_k\} \tag{10.6.26}$$

where $X_j = \{x_j\}$ (resp., $Y_j = \{y_j\}$), $1 \leq j \leq k$. In these conditions all of the RNNs of (10.6.26) are fusioned and the partition obtained is

$$C_2 = C_1 - \left(\bigcup_{1 \leq j \leq k} X_j \right) \cup \left(\bigcup_{1 \leq j \leq k} Y_j \right) + \{\{x_j, y_j\} | 1 \leq j \leq k\} \tag{10.6.27}$$

The mutual dissimilarities between the new classes formed and the dissimilarities of the latter with the unchanged classes are calculated. There will be in all $k(k - 1)/2 + k(n - k)$ new class dissimilarities to compute. For this purpose, reactualization formulas are requested (see Sect. 10.6.3.2).

In the case where δ is defined by the Ward criterion, it may be interesting for spatial complexity reasons, to process at the level of the data table and this, by substituting for $X_j \cup Y_j$ the center of gravity of $X_j \cup Y_j$, $1 \leq j \leq k$. In these conditions, the new weighted distances required by the Ward criterion, are calculated in order to pass to the next phase. For the moment, let us continue the presentation of the first phase of the algorithm.

The chain search appear to be the most efficient procedure to determine the $RNNs$ class pairs [5]. From an arbitrary given element $X_{(1)}$ of C_1, we build a sequence of C_1 elements, $X_{(1)}, X_{(2)}, \ldots, X_{(i-1)}, X_{(i)}, \ldots,$, where $X_{(i)}$ is the nearest neighbor of $X_{(i-1)}, i \geq 1$. The sequence construction is stopped at $X_{(r)}, r \geq 2$, when $X_{(r-1)}$ is among the nearest neighbors of $X_{(r)}$: $X_{(r-1)} \in NN(X_{(r)})$, this because $(X_{(r-1)}, X_{(r)})$ is a pair of reciprocal nearest neighbors. As mentioned above, an illustrated and well-documented presentation is given in [48]. A parallel implementation of the RNN algorithm can be studied in connection with [12, 50].

Now, let us envisage the case mentioned above where several reciprocal nearest neighbour class pairs are connected, that is, the set of classes concerned form a connected graph with respect to reciprocal nearest neighbor binary relation. The most elementary example might be expressed by

$$\delta(Z, Y) = \delta(X, Y) = \delta_{\min}(X) = \delta_{\min}(Y) = \delta_{\min}(Z) \qquad (10.6.28)$$

for three classes X, Y and Z. In this case we propose to carry out a multiple aggregation on all vertices of the connected graph concerned. The reactualization formula for multiple aggregation is then required for this situation (see Sect. 10.6.3.2).

The second step of the RNN algorithm is analogous to the first one, but, with respect to the class set C_2 (see (10.6.27)), provided with its class dissimilarities, and so on, passing from C_j to C_{j+1}, $j \geq 1$, until all elements of $\mathcal{O}$ are joined together in a single class. What is obtained by this process is an indexed hierarchy on $\mathcal{O}$ (see Sect. 1.4.4). From this structure, the classification tree on $\mathcal{O}$ can be derived.

The most autonomous and intrinsic expression of the RNN algorithm in its binary version is due to de Rham [13]. Nevertheless, the general principle of this algorithm goes back to Mac Quitty [44]. Moreover, theoretical basis and characteristics of the latter are mentioned in [7] and considered in [8].

To end this presentation, let us mention that the RNN technique search and the multiple aggregation procedure are implemented in the $CHAVL$ software (see Chap. 11).

10.6.5 Ascendant Agglomerative Hierarchical Clustering (AAHC) Under a Contiguity Constraint

A valuable general presentation of contiguity constrained clustering including many reported study cases is given in [47]. In the latter, the criteria discussed are those of Single Linkage, Complete Linkage, Centroid and Inertia Variation (Ward) methods. Whereas, in the development below—which is based on [4, 31]—we focus on the Inertia variation criterion and the family of the Likelihood of the Maximal Link criteria. Formally, the data is defined by two complete graphs, that we denote by Gr_d and Gr_c, on the object set $\mathcal{O}$ to be clustered. Gr_d is a numerically valuated graph associated with a dissimilarity or distance function d defined on $\mathcal{O}$. The graph Gr_c is discrete. It is defined from a binary contiguity relation between the elements of $\mathcal{O}$.

As expressed in Sects. 10.3.3 and 10.5.2, the inertia Ward criterion makes sense when the object set $\mathcal{O}$ is faithfully represented by a cloud of points in an euclidean space. Therefore, d is defined by the euclidean distance associated with this representation. Otherwise, let us recall that in the framework of the likelihood Linkage Analysis approach, the dissimilarity d is given by the *Informational Dissimilarity* (see (7.2.7) of Chap. 7 and (10.1.6)), the latter being very general with respect to the data structure.

Mostly, the binary contiguity relation, that we denote by R_c, has a geographical nature. For example, in agricultural data, $\mathcal{O}$ might be a set of parcels of a given terrain, characterized by their crop yields. If $\{x, y\}$ is a pair of elements of $\mathcal{O}$, $x R_c y$ if and only if x and y has a common border (are adjacent). In image processing, the described unit is a pixel characterized by its luminance in terms of grey level intensity value. For a pair $\{x, y\}$ of pixels of a given image, $x R_c y$ if and only if x and y are adjacent according to one of different directions. We have been concerned by this application domain in [3, 4, 30, 31] where two versions of the contiguity relation were experienced and compared. The first one is the most widely adopted. For a given pixel x, all of the pixels surrounding x in one of the eight directions (est, south-est, south, south-ouest, north-ouest, north and north-est) are contiguous to x. In the second contiguity relation introduced, only four directions are retained (est, south, ouest and north). A formal definition of both contiguity notions given in [30, 31] where first and the second contiguity graphs are designated by G_8 and G_4, respectively.

The contiguity notions between unit elements of $\mathcal{O}$ extends to the contiguity notion between disjoint subsets of $\mathcal{O}$. By adopting the same notation R_c for disjoint subsets contiguity relation, we have for $X \subset \mathcal{O}$, $Y \subset \mathcal{O}$ and $X \cap Y = \emptyset$, $X R_c Y$ if and only if there exists $(x, y) \in X \times Y$, such that $x R_c y$.

In the framework of *AAHC*, the contiguity constrained clustering consists of realizing a classification tree on $\mathcal{O}$, according to one of the merging criteria established independently; but, in respecting the contiguity relation. Therefore, the algorithmic basic principle is:

"At each step merge the contiguous class pairs which are the nearest for the merging criterion concerned".

Recall that we have considered above the merging criteria: δ_{min}, δ_{max}, δ_{cent}, δ_{iner} and Δ_{lla}^{ϵ} (see Sects. 10.3 and 10.4).

In these conditions, the partitions corresponding to the different levels of the classification tree determine segmentations of $\mathcal{O}$, that is to say, partitions whose classes are connected by the contiguity relations, these, respecting in a certain sense, as much as possible the merging criterion. The matter will be then to detect relevant levels giving rise to "good" segmentations of $\mathcal{O}$. For this purpose a criterion as that analyzed in Chap. 9, Sect. 9.3 which serves to pinpoint the most "significant" levels and nodes of a classification tree can be used (see Sect. 9.3.6).

The contiguity constraint may cause reversals in the case of a reducible merging criterion δ, for which—as we know—inversion is impossible in non-constrained hierarchical clustering. In fact, the frequency of inversions will depend on the analytical form of the reactualization formula concerning δ. Let C_1, C_2 and D be three classes

Fig. 10.6 Possible inversion

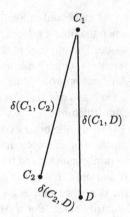

whose cardinalities being denoted as above l_1, l_2 and m, respectively. Suppose as pictured in Fig. 10.6 that (C_1, C_2) and (C_1, D) belong to the contiguity graph, but (C_2, D) does not. Furthermore, suppose

$$\delta(C_2, D) < \delta(C_1, C_2) < \delta(C_1, D) \qquad (10.6.29)$$

In this situation, the cluster $C_1 \cup C_2$ is first formed, followed by the formation of $C_1 \cup C_2 \cup D$. Reversal occurs if

$$\delta(C_1 \cup C_2, D) < \delta(C_1, C_2) \qquad (10.6.30)$$

this might be caused by a small enough value of $\delta(C_2, D)$

Now, we shall compare briefly the inertia Ward criterion δ_{iner} and a generic element of the criteria family Δ_{lla}^{ϵ} of the *Likelihood of Maximal Link* (see above). We also consider a new criterion which refers to the latter family, but which is more accurately adapted to contiguity-constrained clustering [31]. By denoting it Λ_{lla}^{ϵ}, we have

$$\Lambda_{lla}^{\epsilon}(B, C) = a(B, C)^{\epsilon}\left(-\log_2(p_0(B, C))\right) \qquad (10.6.31)$$

where B and C are two connected classes of $\mathcal{O}$, $a(B, C)$ is the *number of edges* in the contiguity graph relating B and C, and

$$p_0(B, C) = \max\{P(x, y)|\{x, y\} \in (B \times C) \cap G_c\} \qquad (10.6.32)$$

where $P(x, y)$ is the probabilistic index between x and y.

As a matter of fact, we have to make explicit how do we obtain the probabilistic similarity indices corresponding to (7.2.6) of Chap. 7. First, the raw similarity index $s(x, y)$—for the only luminance attribute—is defined from the absolute value of the difference between the respective luminances of x and y. On the other hand, the standardization of $s(x, y)$ is done with respect to the set G_c of contiguous pairs instead of all the unordered element pairs $P_2(\mathcal{O})$. This is consistent with the fact that two compared elements determine an edge of G_c. Besides, by this manner, the time complexity for establishing the table

$$\{s(x, y)|(x, y) \in G_c\} \tag{10.6.33}$$

becomes linear.

Relative to an arbitrary class dissimilarity δ, let us consider in an abstract manner two reactualization formulas concerning δ, indicated by R_{lla} and R_{iner} in the following (see (10.6.34) and (10.6.35)):

$$R_{lla} : \delta(C_1 \cup C_2, D) = \min \left\{ \left(1 + \frac{l_2}{l_1}\right)^\epsilon \delta(C_1, D), \left(1 + \frac{l_1}{l_2}\right)^\epsilon \delta(C_2, D) \right\} \tag{10.6.34}$$

$$R_{iner} : \delta(C_1 \cup C_2, D) = \frac{1}{l_1 + l_2 + m} \times$$
$$\left((l_1 + m)\delta(C_1, D) + (l_2 + m)\delta(C_2, D) - m\delta(C_1, C_2) \right) \tag{10.6.35}$$

In the former formula, we have taken the form $(l \cdot m)^\epsilon \left(-\log_2(p(C, D)) \right)$ of the Likelihood Maximal Link criterion; that is to say, before the iterated application of the function $\log_2$. In [30, 31] the following properties are proved

Proposition 57 *If δ satisfies the reactualization formula R_{lla}, the necessary and sufficient condition for inversion is*

$$\delta(C_2, D) < \frac{l_2}{l_1 + l_2} \delta(C_1, C_2) \tag{10.6.36}$$

Proposition 58 *If δ satisfies the reactualization formula R_{iner}, the necessary and sufficient condition for inversion is*

$$\delta(C_2, D) < \frac{(l_1 + m) + (l_2 + m)}{(l_2 + m)} \delta(C_1, C_2) - \frac{(l_1 + m)}{(l_2 + m)} \delta(C_1, D) \tag{10.6.37}$$

Mostly, (10.6.36) implies (10.6.37). To see that, consider the case where $l_1 = l_2$. Equations (10.6.36) and (10.6.37) become

$$\delta(C_2, D) < \frac{1}{2}\delta(C_1, C_2)$$
$$\text{and } \delta(C_2, D) < 2\delta(C_1, C_2) - \delta(C_1, D) \tag{10.6.38}$$

respectively. Consequently, the condition $\delta(C_1, D) < 1.5 \times \delta(C_1, C_2)$ suffices for deriving the implication: (10.6.36) $\Rightarrow$ (10.6.37).

Therefore, a criterion of the family Δ_{lla}^ϵ causes much less inversions than δ_{iner}. This statement was observed experimentally. A very few reversals occured with Δ_{lla}^ϵ, whereas a non negligible number of reversals were produced with δ_{iner}.

In the framework of the Likelihood of the Maximal Link approach, the most interesting experimental results in processing satellite images were obtained with the Λ_{lla}^ϵ version of the merging criterion (see (10.6.31) and (10.6.32)) [4, 30, 31].

Independently, this criterion has the mathematical property to not give rise to inversion. This follows from

$$\Lambda_{lla}^\epsilon(C_1 \cup C_2, D) = \Lambda_{lla}^\epsilon(C_1, D) > \Lambda_{lla}^\epsilon(C_1, C_2) \qquad (10.6.39)$$

(see Fig. 10.6).

Now, let us report a reinforcement of the contiguity constraint in this application. It is obtained from a positive numerical parameter π, comprised between 0 and 0.5, defining a threshold for the probabilistic similarity index $P(x, y)$ between two contiguous elements x and y. More precisely, the value $P(x, y)$ is retained if and only if $P(x, y) \geq \pi$. Otherwise, $P(x, y)$ is set a very low value, in order to not take part in the clustering process.

Experimental results of the inertia (Ward) and Λ_{lla}^ϵ criteria were compared on the basis of the treatment of two satellite pictures. The first one comprises $n = 459 \times 686$ pixels and the second, $n = 758 \times 419$ pixels [4, 31]. As mentioned above, the unique attribute kept for the segmentation is the luminance of the pixel (grey level). The results obtained with both criteria are quite comparable. However, the best reconstructed zones are not the same for either two criteria.

The software concerned $CAHCVR$ (Classification Ascendante Hiérarchique sous contrainte de Contiguité utilisant l'algorithme des Voisins Récoproques) was setup by Kaddour Bachar (ESSCA, Angers, France). It includes the criteria δ_{iner}, Δ_{lla}^ϵ and Λ_{lla}^ϵ. On the other hand, binary and multiple aggregations are two options taken into account in $CAHCVR$.

Theoretical and experimental results [3, 30] prove clearly the linearity of time complexity, with respect to the image size. Theoretical results were obtained for binary aggregation. Relative to the latter, multiple aggregation gives approximately a 50 % gain in time processing. On the other hand, the contiguity constraint defined by the graph G_4 enables 70 % of computing time to be saved with respect to that defined by G_8.

10.6.6 Ascendant Agglomerative Parallel Hierarchical Clustering

The general principle of approximating $AAHC$ from a decomposition of the object set and using a *parallel* approach was given at the end of the introductory section (see Sect. 10.6.1). Here, we shall detail the different phases of this algorithmic approach in order to make them as explicit as possible.

Random Cutting the Object Set $\mathcal{O}$ into Equal Slices

The object set $\mathcal{O}$ is randomly divided according to a partition

$$\{\mathcal{O}^1, \mathcal{O}^2, \ldots, \mathcal{O}^k, \ldots, \mathcal{O}^K\}$$

where the different classes have, to within unity, the same cardinal. As above, n designates the cardinality of $\mathcal{O}$. Let us assume that the maximal size of a set which can be processed by the available *AAHC* algorithm and computer tools, is m, $m < n$. Consider then the lowest integer K such that $[n/K] \le m$ ($[\bullet]$ indicates the integer part of $\bullet$). Thus, for example, if $n = 900{,}000$ and $m = 8{,}000$, $K = 113$ and $[n/K] = 7{,}964$. We have

$$n - [n/K] \times K = r < K$$

For the example considered, this equation is written as

$$900{,}000 - 7{,}964 \times 113 = 68 < 113$$

In these conditions, we constitute a subset $\mathcal{O}'$ of $\mathcal{O}$, sized by $n' = K \times [n/K]$, chosen uniformly at random (with equal chances). In the case of our illustration, $n' = 113 \times 7{,}964 = 899{,}932$. A partition of $\mathcal{O}'$ into K classes having the same cardinaly $[n/K]$ is then randomly sampled, according to a probabilistic uniform model. Let us designate this partition by $\{\mathcal{O}'^1, \mathcal{O}'^2, \ldots, \mathcal{O}'^k, \ldots, \mathcal{O}'^K\}$. Now, determine a random subset $\{k(1), k(2), \ldots, k(q), \ldots, k(r)\}$ of r mutually distinct subscripts in $\{1, 2, \ldots, k, \ldots, K\}$. In the given example, it comes to choose a random subset of 68 subscripts among $\{1, 2, \ldots, k, \ldots, 113\}$. The $r = n - n'$ elements of $\mathcal{O} - \mathcal{O}'$ are then distributed, uniformly at random, each exactly in one of the classes $\mathcal{O}'^{k(q)}$, $1 \le q \le r$. The partition obtained can be denoted by $\{\mathcal{O}^1, \mathcal{O}^2, \ldots, \mathcal{O}^k, \ldots, \mathcal{O}^K\}$ where $\mathcal{O}^k = \mathcal{O}'^k$ if $k \notin \{k(1), k(2), \ldots, k(q), \ldots, k(r)\}$ and $\mathcal{O}^{k(q)} = \mathcal{O}'^{k(q)} + \{x_{k(q)}\}$, where $\{x_{k(q)}\}$ designates the qth element of $\mathcal{O} - \mathcal{O}'$, $1 \le q \le r$. Thus, $K - r$ classes of π include $[n/K]$ elements and r classes of π comprise $[n/K] + 1$ elements.

Partial Hierarchical Clusterings of the Different Slices

If, as usually, $\mathcal{A}$ designates the set of the descriptive attributes of $\mathcal{O}$, for every subset $\mathcal{O}^k$ of the above decomposition of $\mathcal{O}$ (provided by the partition π) we associate a data table indexed by $\mathcal{O}^k \times \mathcal{A}$, $1 \le k \le K$. Relative to these descriptions the same *AAHC* method is applied on each $\mathcal{O}^k$, but in a truncated manner. The tree formation is stopped when the number of classes of the last "interesting" partition obtained satisfies certain specific cardinality conditions (see hereafter).

Depending on whether the criterion used, the *AAHC* method may process either directly at the level of the data table or at that of the dissimilarity or distance table, established prealably on $\mathcal{O}^k$, $1 \le k \le K$. It is of importance for the criterion considered, that the respective sizes of the classes formed, gradually as the algorithm proceeds, are of the same order. For this feature, intuitively expressed, criteria of interest are the inertia variation (Ward) and the family of the Maximal Link Likelihood. For the former criterion, in the case of numerical data (i.e. $\mathcal{A}$ is composed of numerical attributes), it may be useful to process computings at the level of the different data tables $\mathcal{O}^k \times \mathcal{A}$, $1 \le k \le K$. For the Likelihood of the Maximal Link criteria, it is more manageable to proceed at the level of the different dissimilarity tables. These are of the form

$$\mathcal{D}_k = \left\{ D_{inf}(x, y) = -\text{Log}_2\big(P(x, y)|\{x, y\} \in P_2(\mathcal{O}^k)\big)\right\} \qquad (10.6.40)$$

$1 \le k \le K$ (see (7.2.7) of Chap. 7).

Now, let us designate by ClN (class Number) the greatest integer such that $K \times ClN \le m$. In other words $ClN = [m/K]$. Thus, if $m = 8{,}000$ and $K = 113$, $ClN = 70$ and $K \times ClN = 7{,}910$, whereas $K \times (ClN + 1) = 8{,}023$. In these conditions, it is a matter to retain from the starting formation of each of the K classification trees, the kth being on $\mathcal{O}^k$ ($1 \le k \le K$), ClN classes at most. For a given k, $1 \le k \le K$, let $Lev1$ be the first level for which the class number of the partition obtained, is lower or equal to ClN. In general, unless multiple aggregations, we have exactly ClN classes for $Lev1$. The partition of $Lev1$ might be less "significant" (see Sect. 9.3.6 of Chap. 9) than a coarser partition of the immediately following levels. In these conditions, we consider a "small enough" integer l and we introduce $Lev2 = Lev1+l$. Next, we scan the partition sequence of the levels $Lev1$, $Lev1+1$, $Lev1+2$, ..., $Lev2 = Lev1+l$. Among these $l + 1$ partitions, the retained one is that which maximizes the criterion $C(\pi, \mathcal{D}_k)$ (see the previous mentioned reference). For the above considered criteria: $n = 900{,}000$, $m = 8{,}000$, $K = 113$ and $ClN = 70$, l might be taken equal to 5.

Cluster Representation

Relative to the partial hierarchical clustering of $\mathcal{O}^k$, $ClNk$ will designate the number of constituted classes of the last partition retained. These underlie the subtrees formed. We have $ClNk \le ClN$, $1 \le k \le K$. Let us denote this partition by $\{\mathcal{O}_1^k, \mathcal{O}_2^k, \ldots, \mathcal{O}_j^k, \ldots, \mathcal{O}_{ClNk}^k\}$. Each of its classes will be represented by its central element according to the criterion employed in the $AAHC$ method concerned. For the inertia Ward criterion, a given class $\mathcal{O}_j^k$ will be represented by its center of gravity (centroid) g_j^k, possibly endowed with the number of elements of the class represented, namely: $n_j^k = card(\mathcal{O}_j^k)$, $1 \le j \le ClNk$. If one version of the Likelihood of the maximal Link method is used, then $\mathcal{O}_j^k$ will be represented by its most central element c_j^k, the latter being defined as minimizing the sum of its square dissimilarities to the other elements of $\mathcal{O}_j^k$. Here also, as for Ward method, we can attach to c_j^k, $n_j^k = card(\mathcal{O}_j^k)$, $1 \le j \le ClNk$.

Hierarchical Clustering of the Set of Slices

We substitute for the subset $\mathcal{O}^k$ the representatives of the class decomposition $\{\mathcal{O}_j^k | 1 \leq j \leq ClNk\}$, $1 \leq k \leq K$. Thus, a set of n^k(n superscript $k = card(\mathcal{O}^k)$) elements is replaced by a set of $ClNk$ elements, $1 \leq k \leq K$. Let us recall that the magnitude orders of n^k and $ClNk$ are $[n/K]$ and $[m/K]$, respectively. In the illustration considered above, the magnitude orders of n^k and $ClNk$ are 8,000 and 70, respectively. According to the notations above, let us indicate by

$$\{(g_k^j, n_j^k) | 1 \leq j \leq ClNk\} \qquad (10.6.41)$$

the representative set of $\mathcal{O}^k$ when the criterion used is the inertia variation (Ward) and by

$$\{(c_k^j, n_j^k) | 1 \leq j \leq ClNk\} \qquad (10.6.42)$$

this representative set when the criterion used corresponds to a given element (specified by a value of ϵ) of the criteria family of the Maximal Link Likelihood.

In the case of use of the inertia criterion, the representative set to be clustered can be expressed by

$$\bigcup \{(g_k^j, n_j^k) | 1 \leq j \leq ClNk\} \qquad (10.6.43)$$

Reactualization formulas (see (10.6.3) and (10.6.5)) enable this clustering to be carried out.

In the case of use a Likelihood Linkage criterion, according to (10.6.42) the representative set to be clustered is

$$\bigcup \{(c_k^j, n_j^k) | 1 \leq j \leq ClNk\} \qquad (10.6.44)$$

As for the previous criterion, the reactualization formulas (see (10.6.3) and (10.6.5)) enable this clustering method to be performed.

Whatever the aggregative criterion considered the method defined in Sect. 9.3.6 of Chap. 9 apply in order to determine the most significant levels an nodes of the classification tree built.

In practice, it makes no sense to exhibit the entire or even parts of any of the precedings classification trees. In fact, only some "significant" levels (for Examples 3 or 4) have to be retained from the previous global classification tree. The respective ordered partitions of these levels will be then reconstructed with respect to the entire object set $\mathcal{O}$ and characterized by using the tools developed in Chap. 8.

References

1. Abbassi, N.: Identification de familles protéiques. Master 2, Research report, IRISA-INRIA, June 2013
2. Aczel, J., Forte, B., Ng, C.T.: Why the Shannon and Hartley entropies are "natural". Adv. Appl. Probab. **6**, 131–146, Printed in Israel (1974)
3. Bachar, K., Lerman, I.-C.: Statistical conditions for an algorithm of hierarchical classification under constraint of contiguiuty. In: Rizzi, A., Bock, H.-H., Vichi, M. (eds.) Advances in Data Science and Classification, pp. 131–136. Springer, Berlin (1998)
4. Bachar, K., Lerman, I.-C.: Fixing parameters in the constrained hierarchical classification method: application to digital image segmentation. In: Banks, D., et al. (eds.) Classification Clustering and Data Mining Applications, pp. 85–94. Springer, New York (2004)
5. Benzécri, J.-P,: Construction d'une classification ascendante hiérarchique par la recherche en chaîne des voisins réciproques. Les Cahiers de l'Analyse des Données **7**, 209–218 (1982)
6. Benzécri, J.-P.: L'analyse des données, tomes I et II. Dunod (1973)
7. Bock, H.H.: Automatische Klassifikation Vandenhoeck und Rupprecht. Gottingen (1974)
8. Bruynooghe, M.: Classification ascendante hiérarchique des grands ensembles de données; un algorithme rapide fondé sur la construction des voisinages réductibles. Cahiers de l'Analyse des Données **III**(1), 7–33 (1978)
9. Bruynooghe, M.: Nouveaux algorithmes en classification automatique applicables aux très grands ensembles de données, rencontrés en traitement d'images et en reconnaissance de formes. Ph.D. thesis, Thèse d'Etat, Université de Paris 6, Jan 1989
10. Bruynooghe, M.: Recent results in hierarchical clustering. I—the reducible neighborhoods clustering algorithm. Int. J. Pattern Recognit. Artif. Intell. **7**(3), 541–571 (1993)
11. Costa Nicolau, F., Bacelar-Nicolau, H.: Some trends in the classification of variables.In: Hayashi, C., et al. (eds.) Data Science, Classification and Related Methods, pp. 89–98.Springer, New York (1998)
12. Cutting, D., Karger, D., Pederson, J., Tukey, J.: Scatter/gather: a cluster-based approach to browsing large document collections. In: Belkin, N., et al. (eds.) International ACM SIGIR Conference on Research and Development in Information Retrevial, pp. 318–339. ACM Press (1992)
13. de Rham, C.: La classification hiérarchique ascendante selon la méthode des voisins récipro-ques. Les Cahiers de l'Analyse des Données **V**(2), 135–144 (1980)
14. Diday, E.: Inversions en classification hiérarchique: application à la construction adaptative d'indices d'agrégation. Revue de Statistique Appliquée **31**(1), 45–62 (1983)
15. Edwards, W.F., Cavalli-Sforza, L.L.: A method for cluster analysis. Biometrics **21**, 363–375 (1965)
16. Feller, W.: An Introduction to Probability Theory and Its Applications, vol. II, 2nd edn. Wiley, New York (1971)
17. Florek, K.J., Lukaszewick, J., Perkal, J., Steinhaus, H., Zubrzycki, S.: Sur la liaison et la division des points d'un ensemble fini. In: Colloquium Mathematics, vol. 2, pp. 282–285 (1951)
18. Florek, K.J., Lukaszewick, J., Perkal, J., Steinhaus, H., Zubrzycki, S.: Taksonomia wroclawska (in polish with english summary). Przegl. antrop. **17**, 193–217 (1951)
19. Ghazzali, N.: Comparaison et réduction d'arbres de classification, en relation avec des prob-lèmes de quantification en imagerie numérique. Ph.D. thesis, Université de Rennes 1, May 1992
20. Ghazzali, N., Léger, A., Lerman, I.C.: Rôle de la classification statistique dans la compression du signal d'image: Panoram et étude spécifique de cas. La Revue de Modulad **14**, 51–91 (1994)
21. Gordon, A.D.: A review of hierarchical classification. J. R. Stat. Soc. **150**(2), 119–137 (1987)
22. Govaert, G.: La classification croisée. La Revue de Modulad **4**, 9–36 (1989)
23. Gras, R., Larher, A.: L'implication statistique, une nouvelle méthode d'analyse des données. Mathématiques, Informatique et Sciences Humaines **120**, 5–31 (1993)
24. Jambu, M.: Classification Automatique pour l'Analyse des Données. Dunod (1978)

25. Jardine, N., Sibson, R.: Mathematical Taxonomy. Wiley, New York (1971)
26. Kojadinovic, I.: Hierarchical clustering of continuous variables based on the empirical copula process and permutation linkages. Comput. Stat. Data Anal. **54**, 90–108 (2010)
27. Lance, G.N., Williams, W.T.: A general theory of classificatory sorting strategies. I. Hierarchical systems. Comput. J. **9**, 373–380 (1967)
28. Lee, A., Willcox, B.: Minkowski generalizations of Ward's method in hierarchical clustering. J. Classif. **31**, 194–218 (2014)
29. Leredde, H.: La méthode des pôles d'attraction; La méthode des pôles d'agrégation : deux nouvelles familles d'algorithmes en classification automatique et sériation. Ph.D. thesis, Université de Paris 6, Oct 1979
30. Lerman, I.-C., Bachar, K.: Contruction et justification d'une méthode de classification ascendante hiérarchique accélérée fondée fondée sur le critère de la vraisemblance du lien en cas de données de contiguïté. application en imagerie numérique. Publication Interne 1616, IRISA-INRIA, April 2004
31. Lerman, I.-C., Bachar, K.: Comparaison de deux critères en classification ascendante hiérarchique sous contrainte de contiguïté. application en imagerie numérique. Journal de la Société Française de Statistique **149**(2), 45–74 (2008)
32. Lerman, I.-C., Peter, Ph.: Analyse d'un algorithme de classification hiérarchique en parallèle pour le traitement de gros ensembles, aspects méthodologiques et programmation. Publication Interne IRISA et Rapport de Recherche INRIA 232, IRISA-INRIA, Aug 1984
33. Lerman, I.C.: Les bases de la classification automatique. Gauthier-Villars (1970)
34. Lerman, I.C.: Analyse du phénomène de la "sériation". Revue mathématique et Sciences Humaines **38** (1972)
35. Lerman, I.C.: Classification et analyse ordinale des données. Dunod. http://www.brclasssoc.org.uk/books/index.html (1981)
36. Lerman, I.C.: Formules de réactualisation en cas d'agrégations multiples. Recherche Opérationnele, Operations Research **23**(2), 151–163 (1989)
37. Lerman, I.C.: Foundations of the likelihood linkage analysis (LLA) classification method. Appl. Stoch. Models Data Anal. **7**, 63–76 (1991)
38. Lerman, I.C.: Likelihood linkage analysis (LLA) classification method: an example treated by hand. Biochimie **75**, 379–397 (1993)
39. Lerman, I.C.: Analyse logique, combinatoire et statistique de la construction d'une hiérarchie binaire; niveaux et noeuds significatifs. Mathématiques et Sciences Humaines, Mathematics and Social Sciences **184**, 47–103 (2008)
40. Lerman, I.C., Kuntz, P.: Directed binary hierarchies and directed ultrametrics. J. Classif. **28**, 272–296 (2011)
41. Lerman, I.C., Leredde, H.: La méthode des pôles d'attraction. In: Diday, E., et al. (eds.) Journées Analyse des Données et Informatique. IRIA (1977)
42. Lerman, I.C., Peter, Ph.: Organisation et consultation d'une banque de petites annonces à partir d'une méthode de classification hiérarchique en parallèle. In: Data Analysis and Informatics IV, pp. 121–136. North Holland (1986)
43. Loughin, T.M.: A systematic comparison of methods for combining p-values from independent tests. Comput. Stat. Data Anal. **47**, 467–485 (2004)
44. McQuitty, L.L.: Similarity analysis by reciprocal pairs for discrete and continuous data. Educ. Psychol. Meas. **26**, 825–831 (1966)
45. McQuitty, L.L.: Elementary linkage analysis for isolating orthogonal and oblique types and typal relevancies. Educ. Psychol. Meas. **17**, 207–229 (1957)
46. Murtagh, F.: A survey of recent advances in hierarchical clustering. Comput. J. **26**, 354–359 (1983)
47. Murtagh, F.: A survey of algorithms for contiguity constrained clustering and related problems. Comput. J. **28**, 82–88 (1985)
48. Murtagh, F.: Clustering massive data sets. In: Handbook of Massive Data Sets, pp. 501–543. Kluwer Academic Publishers, Norwell (2002)

49. Nicolau, F.: Criterios de análise classificatoria hierarquica baseados na funçao de distribuiçao. Ph.D. thesis, Faculty of Science of Lisboa University, Feb 1981
50. Olson, C.F.: Parallel algorithms for hierarchical clustering. Parallel Comput. **21**, 1313–1325 (1995)
51. Orloci, L.: Information theory models for hierarchic and non-hierarchic classifications. In: Cole, A.J. (ed.) Numerical Taxonomy, pp. 148–164. Academic Press, New York (1968)
52. Peter, Ph.: Méthodes de classification hiérarchique et problèmes de structuration et de recherche d ' informations assistée par ordinateur. Ph.D. thesis, Université de Rennes 1, mars 1987
53. Roux, M.: Deux algorithmes récents en classification automatique. Revue de Statistique Appliquée **18**(4), 35–40 (1970)
54. Scott, A.J., Symons, M.J.: On the Edwards and Cavalli Sforza method of cluster analysis. Biometrics **27**, 217–219 (1971)
55. Sneath, P.H.A.: The application of computers to taxonomy. J. Gen. Microbiol. **17**, 201–226 (1957)
56. Sneath, P.H.A., Sokal, R.R.: Numerical Taxonomy. Freeman, San Francisco (1973)
57. Sokal, R.R., Michener, C.D.: A statistical method for evaluating systematic relationships. Kansas Univ. Sci. Bull. **38**, 1409–1438 (1958)
58. Sorensen, T.: A method of establishing groups of equal amplitude in plant sociology based on similarity of species content. K. danske Vidensk. Selsk. Skr. (biol) **5**, 1–34 (1948)
59. Ward, J.H.: Hierarchical grouping to optimise an objective function. J. Am. Stat. Assoc. **58**, 238–244 (1963)

Chapter 11
Applying the *LLA* Method to Real Data

11.1 Introduction: the *CHAVL* Software (*C*lassification *H*iérarchique par *A*nalyse de la *V*raisemblance des *L*iens)

The theoretical and methodological aspects of the *LLA* approach were highlighted and studied in Chaps. 3, 5–7, 9 and 10. The edification of this methodology enables us an overall view on nonsupervised methods of *Data Analysis* and *clustering methods*.

Many applications work on a large scale, provided by various fields (Bioinformatics, Informatics, Social sciences, Image processing, Natural language processing, …) have been carried out and then, have validated this approach. Any of the logical or mathematical types of data description can be handled in an accurate fashion.

• **The data structures format**

Basically, the data are inserted in a rectangular table called a "data table" (see Sect. 3.5.5 of Chap. 3). The columns of the latter are indexed by the elements of a set $\mathcal{A}$ of descriptive attributes. If the description concerns an object set $\mathcal{O}$, the rows of the data table are labelled by the elements of $\mathcal{O}$. On the other hand, if the description concerns a set $\mathcal{C}$ of categories, then, the rows of the data table represent the elements of $\mathcal{C}$. Clustering can be about the set $\mathcal{A}$ of descriptive attributes or the set $\mathcal{O}$ of objects (resp., $\mathcal{C}$ of categories).

CHAVL performs clustering by the *LLA* method either the set $\mathcal{A}$ or the set $\mathcal{O}$ (resp., $\mathcal{C}$). The respective hierarchical clusterings involved are realized separately. Each of them takes intimately into account the mathematical structure of the data table indexed by $\mathcal{O}$ (resp., $\mathcal{C}$) × $\mathcal{A}$. For this table and relative to the description of a set of objects, the set $\mathcal{A}$ of descriptive attributes may be composed of attributes of a single type or, of a mixing of attributes of different types. The respective types considered are:

– Numerical;
– Boolean;

© Springer-Verlag London 2016
I.C. Lerman, *Foundations and Methods in Combinatorial and Statistical Data Analysis and Clustering*, Advanced Information and Knowledge Processing,
DOI 10.1007/978-1-4471-6793-8_11

– Nominal categorical;
– Ordinal categorical;
– Preordonance or categorical similarity valuated.

Mathematical definitions of these different types are provided in Chap. 3.

All the attributes are supposed to be of the same type in the case of clustering the attribute set $\mathcal{A}$. Clustering a set $\mathcal{O}$ of objects allows description to be made by a *mixing* of different types of attributes (see Sect. 7.2.6 of Chap. 7).

A particular structure dealt in *CHAVL* is defined by an *horizontal juxtaposition* of *contingency tables* (see Sect. 7.2.8 of Chap. 7). In this case, the rows of the data table are indexed by a set $\mathcal{C}$ of categories. The columns correspond to categories associated with nominal categorical attributes.

Thereby, in the *CHAVL* software the respective clusterings of the attribute set on the one hand and the object set (resp., the category set) on the other one, are performed separately.

Correspondence between a classification of the attribute set (resp., of the object set) and the object space (resp., the space of the descriptive attributes) was carried out with specific programs in [20], [14] Part II, Chap. 13 (pp. 563–599), [1, 18]. Let us mention that the set concerned by applying *AAHC* in the latter [14] is constituted by Boolean attributes associated with the categories of ordinal categorical attributes defined from behavioural scales.

In Chap. 8, Sects. 8.2–8.6, methodological (mathematical and statistical) tools are established in order to evaluate

– Associations between objects taken separately, or a classes of objects and a cluster of descriptive attributes, also dually;
– Associations between descriptive attributes taken separately, or attribute classes and a clustering of an object set.

A final, but a very useful structure handled by *CHAVL* is defined as a numerical similarity table between the data units, directly given by the expert. This table might result from a specific calculation. As usually, its general form is

$$S = \{S(x, y) | x, y \in F = P_2(E)\} \tag{11.1.1}$$

where E is the set to be clustered. In this situation—where S is given—E is generally a set of objects. Now, to reach a probability scale for the similarity table, S is substituted for $\mathcal{Q}$ in (7.2.3) of Sect. 7.2.1 of Chap. 7.

To summarize, we can distinguish eight different versions of the data table which can be dealt by the *CHAVL* software:

– *Five* versions where the data table is in the format Objects × Attributes and where the attributes are of a single type;
– *One* version where the data table is in the format Objects × Attributes, but where the descriptive attributes share different types;
– *One* version corresponding to horizontal juxtaposition of contingency tables;
– *One* version corresponding to numerical similarity table on the set to be clustered (generally an object set).

The *CHAVL* software is provided in the site [26]. In this, the directions for use are given. The respective authors of this program are P. Peter (Laboratoire d'Informatique de l'University of Nantes (LINA)), H. Leredde (University of Paris Nord), M. Ouali (IRISA, University of Rennes1, CREC, Saint-Cyr Coëtquidan) and I.C. Lerman (IRISA, University of Rennes1, CREC, Saint-Cyr Coëtquidan). P. Peter has been the architect of the last finalizations.

A prior version of this program—not including clustering preordonance attributes—was registered [1] at the "Agence de Protection des Programmes (*APP*)" in the name of University of Rennes 1, in 2006.

An ergonomic and variant simplified of this program called *LLAhclust*, including the most classical cases of a data table, was implemented in the environment of **R** software [10]. This work has mainly been carried out by I. Kojadinovic and P. Peter of the University of Nantes. Improvements in the latter implementations have been performed by N. Le Meur and B. Tallur (Irisa, University of Rennes 1).

• **The different subroutines and their respective organization**
In *CHAVL* the treatment is realized through different processes. The first one concerns the calculation of the similarity matrix between objects (resp., categories) or the association coefficient matrix between descriptive attributes. The table of similarity indices (resp., association coefficients) is globally reduced according to (5.2.52) of Sect. 5.2.1.5 of Chap. 5 or (7.2.5) of Sect. 7.2.1 of Chap. 7.

The subroutine SIMOB (SIMilarités entre OBjets) intervenes in the case where we have to establish the similarity index table between objects described by a single type of attributes; also, when we have to compare rows of a juxtaposition of contingency tables.

Otherwise, the subroutine SIMHET (SIMilarités en cas d'attributs HÉTérogènes) is considered for calculating the similarity index table between objects described by a mixing of attributes of different types.

The subroutine ASVAR (ASsociations entre VARiables) is devoted to the computing of the association coefficient matrix between descriptive attributes of a same type (*numerical*, *Boolean*, *nominal categorical* and *ordinal categorical*).

AVARE (Associations entre Variables RElationnelles) is placed at the same level as ASVAR. Its function is the same as that of ASVAR. However, the specificity of AVARE consists of coding all types of attributes in terms of *preordonances* (see Sect. 3.4.4) of Chap. 3. In these conditions, it enables association coefficient table between descriptive attributes of a single type to be established. Boolean, numerical, nominal categorical, ordinal categorical and preordonance categorical are involved. Notice that establishing association coefficient table between preordonance attributes is not comprised in ASVAR.

Thus SIMOB, SIMHET, ASVAR and AVARE define the different alternatives of the first step of *CHAVL*. These programs enable the similarity index or association coefficient table on the set concerned, to be established. In a certain extent, we can say that SIMHET plays with respect to SIMOB, the same role as AVARE with respect

---

[1] IDDN.FR.001.240016.000.S.P.2006.000.20700.

to ASVAR. Indeed, SIMOB and ASVAR are relative to data tables with a unique type of attributes, whereas SIMHET and AVARE may address description data table with a mixing of attribute types.

For a given similarity or association coefficient table on a set E to be clustered, the next step called AVLEPS (AVL EPSilon) is devoted to the Ascendant Agglomerative Hierarchical clustering of E, according to the criterion of the Likelihood of the maximal link, parametrized by the numerical parameter ϵ ($0 \leq \epsilon \leq 1$) (see (10.4.7) and (10.4.8) of Sect. 10.4.1.2 of Chap. 10). Reciprocal Nearest Neighbours algorithm is employed in order to build the classification tree (see Sect. 10.6.1 of Chap. 10). The AVLEPS step requires the calculation of the lower triangular matrix (including the diagonal) of the E similarities or association coefficients.

AVLEPS is followed by INTRP (INTRPrétation). The latter step provides numerical indices for guiding interpretation of the results of a hierarchical clustering. In this respect, two data are required:

1. The lower half similarity triangular matrix (including the diagonal terms);
2. The prefixed polish representation of the classification tree.

   The indices calculated are:

 - For each element e of the set E to be clustered $(E = \mathcal{O}$ (resp., $\mathcal{C}$) or $\mathcal{A}$, where as usually, $\mathcal{O}, \mathcal{C}$ or $\mathcal{A}$ designate a set of objects, categories or attributes), the variance $\mathcal{V}(e)$ of its similarities to the other elements of E (see (8.5.2), (8.5.3) and what follows of Sect. 8.5.1 of Chap. 8), the values of $\mathcal{V}(e)$ are listed decreasingly, the stronger of them indicate the leader elements in the cluster formation;
 - The distribution over the increasing level sequence of the adequation criterion between partition and similarity on E, this criterion associated with the partition sequence of a classification tree, is called "*Global* Level Statistic" (see Sect. 9.3.6 and more particularly, Sect. 9.3.6.2 and Table 9.4);
 - The distribution over the increasing level partition sequence of the variation rate of the Global Level Statistic, this index is called "*Local* Level Statistic".

The last step of the *CHAVL* program is DESAB (DESsin de l'ArBre). Its function consists of drawing the classification tree. More precisely, a graphical representation of the classification tree is built from its prefixed polish representation. Prior to this construction, levels are selected in order to propose a condensed form of the tree. The latter levels may correspond to those for which a *significant node* appears (see Sect. 9.3.6).

The general organization of *CHAVL* is given in the diagram depicted in Fig. 11.1, where, $\mathcal{O}$ (resp., $\mathcal{C}$) indicates a set of objects (resp., categories) and where $\mathcal{A}$ designates a set of descriptive attributes.

11.2 Real Data: Outline Presentation of Some Processings

As expressed in the beginning of the previous section, the *LLA* approach has been intensively applied in many domains. It enabled original and powerful results to be obtained.

Fig. 11.1 Organization of
the CHAVL software

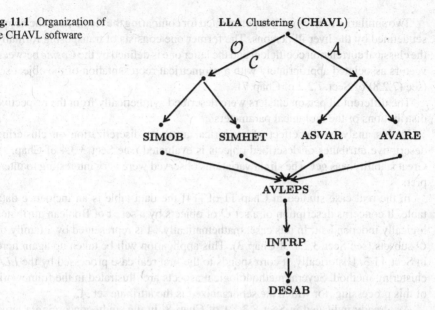

As observed, the data format is always reduced to a *data table* indexed by a Cartesian product $O \times A$ (resp., $C \times A$) where O (resp;, C) is a set of (elementary) objects (resp., C is a set of categories) and where A is a set of descriptive attributes.

As emphasized all along this book, the mathematical representation of a data table is of essential importance. This governs the formal aspects of the similarity function between the O (resp., C) elements and those of the associations coefficients between the A elements. The metrical or statistical facets of these functions (Similarity index or Association coefficient) are derived in a next stage. For this as emphasized, these formal, metrical and statistical features depend intimately on the set to be clustered: O (resp., C) or A (see Chaps. 4–7).

Now, let us refer to Part II of [14]. The application considered in Chap. 15 of the latter reference gives a perfect illustration of the case where A is composed of *numerical* attributes describing a set of objects (individuals). The data concerns the pathology hepato-biliary. As frequently in Medicine, the data table is small. 16 biological parameters are measured on a set of 117 liver ill persons ($card(A) = 16$ and $card(O) = 117$) (the initial data comprising 228 subjects was truncated in order to avoid a missing measure).

Clustering biological parameters enabled the different *syndromes* of the hepato-biliary pathology to be discovered or retrieved.

Clustering the object set composed of liver ill persons led us to distinguish the different illness states (cancer, cirrhosis, …) through clusters obtained at the most significant level of the classification tree (see Sect. 9.3.6 of Chap. 9).

Before referring to the probability scale in the context of the *LLA* clustering method, the association coefficient used for comparing the numerical attributes defined by the biological parameters, is defined by the classical Bravais-Pearson correlation coefficient (see (5.2.69) of Chap. 5).

Two similarity indices were experimented for comparing the elements of the object set defined by the liver ill persons. The former one consists of transposing formally the classical correlation coefficient and the latter one is defined by the *Cosine* between vectors associated appropriately with a geometrical representation of the object set (see (7.2.8) of Sect. 7.2.2 of Chap. 7).

The different ill person clusters were described synthetically from the respective distributions of the biological parameters.

In this analysis, the effect of numerical attribute discretization on clustering descriptive attributes or described objects is evaluated (see Sect. 3.3.4 of Chap. 3). Great stability was get. The slight variations observed were very interesting to interpret.

In the real case studied in Chap. 11 of [14] the data table is an incidence data table. It concerns description of a set $\mathcal{O}$ of objects by a set $\mathcal{A}$ of Boolean attributes logically independent. In this case, mathematically, $\mathcal{A}$ is represented by a family of $\mathcal{O}$ subsets (see Sect. 3.2.1 of Chap. 3). This application will be taken up again here in Sect. 11.3. Historically, it corresponds to the first real case processed by the *LLA* clustering method. Several methodological aspects are illustrated in the framework of this processing, for which the set organized is the attribute set $\mathcal{A}$.

As already indicated in Sect. 8.3.1.1 of Chap. 8, in the study concerning a large European survey (see Chap. 13 of Part II of [14]), clustering is about a set $\mathcal{A}$ of Boolean attributes. However, these are not logically independent since they correspond to the categories of ordinal or nominal categorical attributes (see Sect. 8.6.1 for a recall of the context and examples).

In Chap. 16 of Part II of [14] the data structure is defined by an horizontal juxtaposition of contingency tables. This data structure was introduced in Sect. 5.3.2.2 of Chap. 5 and Sect. 7.2.8 of Chap. 7. Let us recall that the form of this data is

$$\mathbb{I} \times (\mathbb{J}^{(1)} + \mathbb{J}^{(2)} + \cdots + \mathbb{J}^{(l)} + \cdots + \mathbb{J}^{(L)}) \tag{11.2.1}$$

(see (7.2.45) of Sect. 7.2.8.2 of Chap. 7) where the sign + indicates union of disjoint subsets and where $\mathbb{I}$ (resp., $\mathbb{J}^{(l)}$, for any l, $1 \leq l \leq L$) is an exclusive and exhaustive category set associated with a nominal categorical attribute.

The application concerns the analysis of the structure of the French agriculture in 1972. $\mathbb{I}$ codes the set of the 89 French départements. $L = 3$. $\mathbb{J}^{(1)}$, $\mathbb{J}^{(2)}$ and $\mathbb{J}^{(3)}$ are the respective value sets of the categorical attributes: "structure of the farm", "mode of using the land" and "magnitude of the livestock". The reference mentioned above in [14] was preceded by [22] and followed by [16, 28].

Now, let $\mathbb{J}$ stand here for one of the $\mathbb{J}^{(l)}$ ($1 \leq l \leq L$) and consider the clustering of $\mathbb{I}$ through $\mathbb{J}$. Though the contingency table $\mathbb{I} \times \mathbb{J}$ is perfectly symmetrical, the similarity index proposed in Sect. 7.2.8.1 of Chap. 7 is relative to the Euclidean representation of the cloud $\mathcal{N}(\mathbb{I})$ (see (5.3.3) of Sect. 5.3.2.1 of Chap. 5). Thereby, $\mathbb{I}$ is represented by a cloud of points in the geometrical space $\mathbb{R}^{|\mathbb{J}|}$ ($|\mathbb{J}|$ stands for $card(\mathbb{J})$) and the basis of the similarity index between two elements i and i' of $\mathbb{I}$ is the cosine between the vectors joining the origin O to the vertices $f_{\mathbb{J}}^i$ and $f_{\mathbb{J}}^{i'}$, representing i and i' (see (7.2.36) of Sect. 7.2.8.1 of Chap. 7).

The similarity index on $\mathbb{I}$ considered in the application concerned is associated with the geometrical representation of the cloud $\mathcal{N}(\mathbb{J})$ (see (7.2.35) of Sect. 7.2.8.1 of Chap. 7). For two elements i and i' of $\mathbb{I}$, it is defined by the correlation coefficient between i and i' interpreted as column attributes. It corresponds to the dual transposition of the index $\rho(j, k)$ (see (5.3.5) of Sect. 5.3.2.1 of Chap. 5). More precisely, it can be written

$$\rho(i, i') = \frac{\sum_{j \in \mathbb{J}} p_j (f_i^j - p_i.)(f_{i'}^j - p_{i'}.)}{\sqrt{\left(\sum_{j \in \mathbb{J}} p_j (f_i^j - p_i.)^2\right)\left(\sum_{j \in \mathbb{J}} p_j (f_{i'}^j - p_{i'}.)^2\right)}} \tag{11.2.2}$$

Proposition 43 of Chap. 7 shows that this index is equivalent to the Cosine between the vectors $(f_{\mathbb{J}}^i - g_{\mathbb{J}})$ and $(f_{\mathbb{J}}^{i'} - g_{\mathbb{J}})$, defined in the dual representation space $\mathbb{R}^{|\mathbb{J}|}$, where $g_{\mathbb{J}}$) is the centre of gravity of $\mathcal{N}(\mathbb{I})$.

Now, in order to integrate the respective contributions of the different sets $\mathbb{J}^l$, $1 \leq l \leq L$, an average operation is carried out on the sequence of correlation coefficients

$$\{\rho^l(i, i') | 1 \leq l \leq L\} \tag{11.2.3}$$

where $\rho^l(i, i')$ is associated with the representation of the cloud $\mathcal{N}(\mathbb{J}^l)$. Therefore, a global index such

$$R(i, i') = \frac{1}{L} \sum_{1 \leq l \leq L} \rho^l(i, i') \tag{11.2.4}$$

may be proposed.

In these conditions, the similarity index applied (see the references above) is not that developed in Sect. 7.2.8.2 of Chap. 7. However, the starting point of the latter is the Cosine index defined by Eq. (7.2.36). Applying the latter similarity index (that of Sect. 7.2.8.2) would require to introduce the sequence

$$\{\mathcal{N}_l(\mathbb{I}) | 1 \leq l \leq L\} \tag{11.2.5}$$

where $\mathcal{N}_l(\mathbb{I})$ is the cloud representing $\mathbb{I}$ in the Euclidean space $\mathbb{R}^{|\mathbb{J}_l|}$, $1 \leq l \leq L$. Let us precise in passing that it is this alternative which is implemented in *CHAVL* (see Sect. 11.1).

On the other hand, the latter representation (see (11.2.5)) was employed for the Ward method using the Chi square distance (see (5.3.4) of Sect. 5.3.2.1 of Chap. 7 and Sect. 10.5.2 of Chap. 10) in order to compare the practical results obtained with those of the *LLA* method using (11.2.4).

By considering the respective classification trees associated with the inertia and the Likelihood of the maximal link criteria, it turns out that the clusters get before the last levels are almost identical. However, the last cluster merging leading to the largest classes are more consistent in the *LLA* fusions than in the inertial ones.

This clustering is about a set of geographical units: the French départements. We can observe (see Fig. 3 of [22]) that the statistical behaviour through the attributes "structure of the farm", "mode of using the land" and "magnitude of the livestock", considered above, entails that the spatial connexity is satisfied naturally to a large extent in the classes formed.

Differently, in the application concerning pixel classification of a satellite image, in order to perform its segmentation (see [2, 19]), a contiguity constraint is introduced for the cluster formation. By this way, the classes get at each level of the classification tree are necessarily connex. Informally, at each step (level) of the ascendant agglomerative hierarchical building of the classification tree on the pixel set, the geographically adjacent class pairs, which realize the maximal value of the chosen similarity between classes, are merged. To this respect, two types of criteria have been considered: the family of the Likelihood of the maximal link criteria and the Ward criterion (see Sects. 10.4.1 and 10.3.3. of Chap. 10). Several aspects of this research are reported and commented in Sect. 10.6.5 of Chap. 10.

In Sect. 11.4 we consider the classification of a set of aligned proteic sequences. These are defined by the cytochrome C of different living organisms. In this case, the sequence set is considered as a set of objects (and not an attribute set as it would be possible with respect to a dual interpretation (see Sect. 5 of [21])). The descriptive attributes associated with the different sequence sites are defined from a *substitution* matrix between amino acids. The latter matrix is interpreted as a categorical attribute whose category set is valuated by a numerical similarity (see Sect. 3.3.4 of Chap. 3). The results associated with three matrices (Dayhoff, Henikoffs and *LLA*) are defined and compared.

In Sect. 11.5, clusterings sets of categorical attributes for descriptions using specific codings are studied.

11.3 Types of Child Characters Through Children's Literature

11.3.1 Preamble: Technical Data Sheet

- **The data table**
  The data table is a rectangular data table indexed by $\mathcal{O} \times \mathcal{A}$. It concerns the description of objects by Boolean attributes (see Sect. 3.5.4 of Chap. 3). Its dimension is 1500×110 ($card(\mathcal{O}) = 1500$ and $card(\mathcal{A}) = 110$).

- **The set to be clustered**
  Hierarchical clustering concerns the set $\mathcal{A}$ of descriptive attributes. The objective consists of organizing the respective relationships between the elements of $\mathcal{A}$ in order to turn up behavioural profiles and subprofiles.

- **The association coefficient employed**
  This coefficient is specified in (5.2.32) of Sect. 5.2.1.2 of Chap. 5. Let us recall the expression of this index

$$(\forall a, b \in \mathcal{A}), \quad Q_P(a, b) = \frac{n(a \wedge b) - \frac{n(a)n(b)}{n}}{\sqrt{\frac{n(a)n(b)}{n}}} \qquad (11.3.1)$$

where, as usually, $n = card(\mathcal{O})$, $n(a) = card(\mathcal{O}(a))$ (resp., $n(b) = card(\mathcal{O}(b))$) and $n(a \wedge b) = card(\mathcal{O}(a) \cap \mathcal{O}(b))$, $\mathcal{O}(a)$ and $\mathcal{O}(b)$ being the subsets of $\mathcal{O}$ where a and b are present, respectively. The P subscript in $Q_P(a, b)$ indicates that the random model of the hypothesis of no relation considered is the Poisson model (see references mentioned above).

- **Making use**
  M.H. Nicolaü (University of Lisbon) and I.C. Lerman (University of Rennes 1).

- **Psycho-Sociologist experts providing the data and interpreting results**
  M-J. Chombard de Lauwe and her collaborator Cl. Bellan (Centre National de la Recherche Scientifique) [5].

11.3.2 General Objective and Data Description

The real example that we deal with here was reported in several publications. In the first of them Appendix 2 of [5, 11], the "Lexicographic" ordinal algorithm was used (see Sects. 10.2 and 10.3.4). The subject of the thesis work [24] was the analysis of the behaviour of the *LLA* algorithm which was applied the first time on the data considered here. This application is taken up again in [12–14], Chap. 11.

The data was elaborated relatively to a problem in psycho-sociology field, situated in the framework of research led by M-J. Chombard de Lauwe [5]. The purpose is to determine the *models* created by the adults and proposed to the child in the literature concerning the child. The data consists of a set of 1500 child personages provided by a sample of books published on this literature. Notice that the sample unit is a book— selected by criteria put forth by the psycho-sociologist—and all of its personages take part in the constitution of the object set $\mathcal{O}$. These publications are edited between 1880 and 1960; as a matter of fact the author of the research wanted to recover three consecutive periods: before the first war, between the two wars and after the second war. The global analysis was followed by partial analyses associated with the three periods, respectively. This subdivision enables the psychosociologist to study the evolution of the models proposed by the adults to child in the child literature.

Each subject (individual), element of $\mathcal{O}$, is described by a set $\mathcal{A}$ of Boolean attributes (see Sect. 3.2.1 of Chap. 3), which establish the following:

- *The portrait*: Aptitudes (knowledge, memory, particular gifts ...), the fields where the aptitudes are demonstrated (art, sport, command, adventure, first aid, practical jokes, way of life, scholastic results, ...), religion (Christian, believes in God, not specified), temperament (patience, anger, coquetry, impulsiveness, frankness, gaiety, ...), relations with others (submission, equality or command with respect to another child, equality or command with respect to an adult, ...), the role of the subject with respect to action (active, passive), great topics where the subject evolves (childhood, drama, moral teaching, vocation, adventure, …), general appreciation of the subject (positive, firstly negative but becoming positive, essentially negative);

- *The material environment*: Town, castle, habitation (e.g. house on wheels, caravan), vehicle action (e.g. car, aeroplane), wild nature, road track, living abroad;
- *His close surroundings*: Father, mother, substitute father, substitute mother, presence of a rival child, friend of the same or opposite sex, animal presence of a policeman (representative of "law"), presence of a traitor or bandit (representative of "bad"), …;
- *His family atmosphere*: Not described, absence of family, incomplete or replacing family, marked by a type of adult surrounding (e.g. army), community of children;
- *His social environment*: Labourer, aristocracy and personage are impossible to locate in the described social context, …

Thus, 110 Boolean attributes were retained and the basic information is contained in an incidence data table T with 1500 rows indexed by the O elements and 110 columns associated with the elements of the attribute set A.

A personage model as sought by the Psycho-Sociologist corresponds to a *behavioural profile*. This is necessarily a fuzzy notion. It can be covered by the notion of *natural* class. The latter was intuitively expressed relatively to a set O of objects described by a set A of Boolean attributes (see Chap. 8). Thereby, in order to determine different profiles, we may consider classes of a partition of O, where each of them is composed of near elements. Then, for each object class, its more discriminant attributes can be specified.

However, we have observed that, dually, the notion of *natural* class applies with respect to the attribute set A. A cluster of statistically near Boolean attributes gives a direct answer to the question of the Psycho-Sociologist, concerning the specification of personage model. Determining such clusters is the approach which was chosen. By this way, for a given behavioural profile, return to individuals can be carried out by detecting the more responsible individuals of the latter profile (see Sect. 8.3 of Chap. 8).

Moreover, instead of a single partition on the attribute set A, a partition chain, corresponding to a hierarchy of classifications, provides a richer structure on A. In fact, in this fashion, the more emphasized types (profiles) appear from the first levels of the classification tree on A. In these conditions, we may retain partitions of A corresponding to the most significant levels of the classification tree on A (see Sect. 9.3.6).

As examples, we will present below the respective formations of two classes which define two semantically opposite profiles. Also, we will indicate the partition obtained at the most significant level of the classification tree. Finally, we will indicate developments and analyses performed in [5, 12, 13] and Chap. 11 of [14].

11.3.3 Profiles Extracted from the Classification Tree on $\mathcal{A}$

The classification tree is binary, having 109 levels (there are in all 110 elements). Due to the large number of individuals (1500) with respect to the attribute number (110), the association coefficient (11.3.1) is discriminant enough in order to not give rise to multiple aggregations in building the classification tree (see Sect. 10.6.3 of Chap. 11). The levels retained for codensing the tree correspond to the *local maxima* of the local level Statistic defined by the increment rate τ of the global level Statistic C (see (9.3.88) of Sect. 9.3.6.1 and Table 9.4 of Chap. 9). The respective numbers of these levels are 8, 12, 41, 55, 71, 86, 90, 93, 99, 102, 104, 106, 107 (giving the absolute maximum of C) and 108. When a node ν is constituted between two levels retained l and m, where $l < m$, $l \leq level(\nu) < m$, its representation is expressed at the lowest level l. The most often m is near l and to simplify interpretation we shall say that ν is formed at level l.

The "negative" type appears quickly and forcefully. Due to several significant nodes going with its formation, it is very well structured. Each of these nodes is marked by a • in Fig. 11.2. Its formation is achieved at level 106. It comprises three components C_1, C_2 and C_3, where

- $C_1 = \{$ 'negative', 'hypocrisy', 'jealousy', 'anger', 'quarelling', 'pride', 'selfishness', 'non-positive', and 'negative character evolving towards a positive one', 'lack of will'$\}$;
- $C_2 = \{$ 'rebellious', 'highly independent', 'impulsive', 'emotional', 'greediness', 'tendency to lie', 'ugly', 'boy'$\}$;
- and $C_3 = \{$ 'passive in the action', 'lack of courage', 'obeying a child', 'aristocracy', 'great *bourgeoisie*', 'servant', 'castle', 'coquetry', 'mode of life'$\}$.

C_1 and C_2 are the more associated and then, they are fusioned before than C_3 joins the union $C_1 \cup C_2$.

Figure 11.2 gives the formation of C_1. the kernel $\{$ 'negative', 'hypocrisy', jealousy$\}$ represents the *most abject aspect* of *negativeness*. Furthermore, disagreeable external associated manifestations: 'anger' and 'quarelling' join the latter kernel. The cluster obtained: $\{$ 'negative', 'hypocrisy', 'jealousy', 'anger', quarelling$\}$ joins a cluster

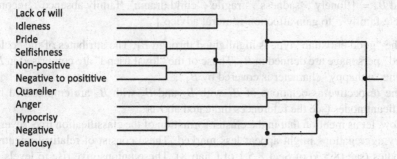

Lack of will
Idleness
Pride
Selfishness
Non-positive
Negative to posititive
Quareller
Anger
Hypocrisy
Negative
Jealousy

Fig. 11.2 Formation of the cluster C_1

already constituted, including 'pride' and 'selfishness', 'non-positive' and 'negative character evolving towards the positive one', 'idleness' and 'lack of will', defining a type of personage whose character is full of defaults and weak, but not fundamentally bad because he evolves towards the good.

The C_2 cluster (class) is composed of two subclusters that we denote by C_{21} and C_{22}, where

- $C_{21} = \big(('\text{rebellious}', '\text{highly independent}'), ('\text{emotive}', '\text{impulsive}')\big)$;

- $C_{22} = \big(('\text{greediness}', '\text{tendency to lie}'), ('\text{boy}', '\text{ugly}')\big)$.

The respective groupings are represented by round brackets.

We will not detail more the cluster C_3 which represents the "*aristocrate bourgeois*". The latter is connoted negatively since this personage comprises features such 'passive', 'lack of courage' and 'submitted to child'.

Now, we shall describe a class (cluster), that we designate by B, which is semantically opposite to class C. The formation of this class ends at level 108. The associations leading to B are defined as follows in the classification tree

$$((B_1, B_2), (B_3, B_4))$$

where

$$B = B_1 + B_2 + B_3 + B_4 \,(\text{union of disjoint subsets})$$

We have

- $B_1 = \{$'christian', 'believes', 'faith', 'modesty', 'patience', 'perfect manners', 'submit to adult', 'sensitive to environment', 'likes the nature', 'animal', 'animal care'$\}$;
- $B_2 = \{$'knowledge', 'memory', 'temperate', 'scholarly success', 'child community', 'plans his future', 'aspiration', 'very gifted', 'art', 'vocation', 'distant past', 'future perspectives'$\}$;
- $B_3 = \{$'honesty', 'frankness', 'generosity', 'positive', 'father substitute', 'gaiety', 'friend opposite sex'$\}$;
- and $B_4 = \{$'lonely', 'sadness', 'fragile', 'child drama', 'family absence', 'incomplete family', 'to gain affection', 'moral advice'$\}$.

The "good christian" type is highlighted through B_1. The attributes of the "celebrated" personage are defined in B_2. Those of the "loyal friend" are comprised in B_3 and the "unhappy" character is covered by B_4.

The respective associations of B_1 with B_2 and B_3 with B_4 are emphasized by significant nodes (see the references indicated above).

Now, let us mention that in the cluster formation of the classification tree, several binary aggregations might appear less marked. These consist of relatively neutral attributes (see (8.5.3) of Sect. 8.5.1 of Chap. 8). These unions give rise to levels at which the global level Statistic C shows a decrease, which means that considered on the whole, the partition of this level is less in accordance with the similarities between units than that of the preceding level. In this case, we prefer not to insist upon the significance of such local and partial groupings, an example of these being the association of greediness with an ugly boy. Generally, the small clusters including those not very marked because of neutral attributes, will join together at the following levels and give birth to some more general types, the meaning of which becoming clearer.

At the most significant level the 107th the classes B and C are completed. They are clearly differentiated. A third class D is also already finished at this level and its achievement is punctuated with a significant node. It can be put in the form of union of four subclasses D_1, D_2, D_3 and D_4, as follows:

$$((D_1, D_2), (D_3, D_4))$$

where D_1 and D_2 (resp., D_3 and D_4) merge before merging $D_1 \cup D_2$ with $D_3 \cup D_4$. We may point out that aggregation of D_1 and D_2 is accompanied with a significant node. The respective compositions of D_1, D_2, D_3 and D_4 are

- $D_1 = \{$'voluntary', 'obstinate', 'initiative', 'controlled', 'life saving', 'first aid', 'justicer', 'courage'$\}$;
- $D_2 = \{$'adventure', 'adventure mission', 'investigation', 'traitor-bandit', 'policeman', 'religion not mentioned', 'tracks', 'wild nature', 'vehicle action', 'living abroad'$\}$;
- $D_3 = \{$'authority', 'commandment', 'command child', 'equal adult', 'command adult', 'rival child', 'sport', 'active', 'particular gifts'$\}$;
- and $D_4 = \{$'comic personage', 'caricatural personage', 'set in time', 'jokes', 'funny', 'family no described', 'out social classes', 'adult surrounding', 'vehicle habitat'$\}$.

The class D_1 contains the attributes of what be called the "scout hero". D_2 represents the milieu and the atmosphere in which the latter personage is living. The class D_3 is around viril characteristics of commandment and competition. The last class D_4 joins together the attributes which set up the "comic personage". Some of these classes such D_1 have a weak internal cohesion, the attributes of which they are constituted are sufficiently neutral.

Finally, at the most significant level 107 a classification with three wide classes appears. The first one regroups the "negative" and the "droll" and thus represents a very general "negative" type. The second one embodies the "comic", the "sportive-authoritarive", the "adventurer" and the "scout hero", defining the usual character of the escapist literature. The third class gathers together the "model child", the "good Christian", the "unhappy" and the "loyal friend" defining a general "normative type". The last levels where these classes are joined by two's lead to the abrupt fall in the global Statistic.

11.3.4 Developments

In [12, 13, 24] and Chap. 11 of [14], the following points are considered:

1. Using a standardized association coefficient between Boolean attributes associated with the random hypergeometric model of the hypothesis of no relation (see Sect. 5.2.1.2 of Chap. 5);
2. Return to individuals;
3. Classifiability, neutral elements and cluster cohesion:

   - Classifiability and neutral elements;
   - Cluster cohesion and neutral elements.

11.3.5 Standardized Association Coefficient with Respect to the Hypergeometric Model

In this case, the coefficient $Q_P(a, b)$ is replaced by

$$(\forall a, b \in \mathcal{A}), \ Q_H(a, b) = \frac{n(a \wedge b) - \frac{n(a)n(b)}{n}}{\sqrt{\frac{n(a)n(\bar{a})n(b)n(\bar{b})}{n^2(n-1)}}} \tag{11.3.2}$$

(see (5.2.9) and (5.2.10) of Sect. 5.2.1.2 of Chap. 5.)

The results obtained are very similar to those obtained with $Q_P(a, b)$. However, some differences can be noticed. They concern associations in which neutral attributes participate in the cluster formation. An experiment where the 75 most discriminant attributes were retained gave rise, very clearly, to the following five types (where marginal attributes are ignored)

1. The "negative": 'non-positive', 'selfishness', 'pride', 'lack of will', 'hypocrisy', 'jealousy', 'quarreller', 'anger', 'highly independent', 'revolting', 'impulsive', 'lie', 'coquetry', 'lack of courage', 'passive';
2. The "comic": 'comic personage', 'caricatural personage', 'set in time', 'funny', 'out social classes', 'family not described', 'jokes';
3. The "adventurer": 'adventure', 'adventure-mission', 'traitor-bandit', 'investigation', 'policeman', 'wild nature', 'tracks', 'vehicle action', 'living abroad';
4. The "unhappy": 'lonely', 'sadness', 'family absence', 'child drama', 'fragile', 'to gain affection', 'incomplete family' 'father substitute', 'mother substitute';
5. The "model": 'distant past', 'vocation', 'art', 'knowledge', 'memory', 'aspiration', 'plans his future', 'very gifted', 'temperate', 'believes', 'christian', 'faith', 'perfect manners'.

11.3.6 Return to Individuals

LLA method applies also to hierarchical clustering of the individual (object) set (see Chap. 7). The general form of the similarity index is given in (7.2.5) of Sect. 7.2.1 of Chap. 7. The specific case where the descriptive attributes are numerical or Boolean is considered in Sect. 7.2.2.

By considering a pair $(\pi(\mathcal{A}), \pi(\mathcal{S}))$ of "significant" partitions on the attribute set $\mathcal{A}$ and the individual set $\mathcal{S}$ (the child personages), respectively, Eq. (8.6.37) of Chap 8 enable us to situate the respective classes of $\pi(\mathcal{S})$ with respect to the behavioural profiles specified through $\pi(\mathcal{A})$.

In fact, what was carried out consists of determining the more responsible personages of the typology into the five classes expressed above. Designating by

$$\pi(\mathcal{A}) = \{A^k | 1 \le k \le K\}$$

such a partition, with each of the 1500 individuals assigned the value of the statistical index

$$D_1 = \frac{\frac{1}{k}\sum_{1 \le k \le K}(\varphi_i^k - \varphi_i)}{\frac{\varphi_i(1-\varphi_i)}{n}} \qquad (11.3.3)$$

where K is the class number and n the number of individuals (1500 in our case). φ_i is the proportion in $\mathcal{A}$ of attributes possessed by (present in) the ith individual (object) o_i and φ_i^k is the proportion of attributes of the class A^k present in the object o_i, $1 \le k \le K$, $1 \le i \le n$.

This coefficient is given in (8.3.2) and (8.3.3) of Sect. 8.3.1.1 of Chap. 8. For the latter coefficient, the respective cardinalities of the classes A^k, $1 \le k \le K$, are equally considered. This is not the case for the coefficient expressed in (8.3.4) and (8.3.5). For the latter, the respective classes A^k, $1 \le k \le K$, are weighted according to their respective cardinalities.

Too large values of (11.3.3) have been observed for personages without real interest, these being reduced to schematic purity of behavioural profiles. This result is due to the fact that the individuals described are obtained by imaginary constructions. On the contrary, in human populations, discovering the *most typical* individuals might be of an extreme interest. Because it obtains to the Sociologist, the possibility to study in depth his research by coming back on the field in order to analyse more accurately each subject determined in the above way.

11.3.6.1 Classifiability, Neutral Elements and Cluster Cohesion

Let us return to Sect. 8.7 and more particularly to Sect. 8.7.2. In the context of the latter the attribute set $\mathcal{A}$ considered above will play the role of the object set $\mathcal{O}$. Thus, $F = P_2(\mathcal{A})$ designates the set of unordered pairs of elements of $\mathcal{A}$ and $J = P_3(\mathcal{A})$

Fig. 11.3 Classifiability
graph

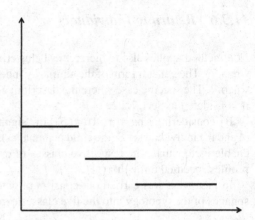

the set of $\mathcal{A}$ subsets comprising three elements. Let $\omega(\mathcal{A})$ denote the preordonance on $\mathcal{A}$ associated with the similarity coefficient chosen on $\mathcal{A}$, for example Q_P defined by (11.3.1).

The classifiability measure of $\mathcal{A}$ via $\omega(\mathcal{A})$ was defined (see Sect. 8.7.2) from the following distribution on F

$$\{m(q)|q \in F\} \tag{11.3.4}$$

where $m(q)$ is the proportion in J of triplets of which the median and upper pairs (according to ω) are strictly separated by q. The decreasing sequence of the values of $m(q)$ was denoted by $D(\omega)$ in Sect. 8.7.2.2. As an illustration $D(\omega)$ can be represented by a graph such that of Fig. 11.3. The latter is associated with the following preordonance

$$\{a, d\} \sim \{a, c\} < \{a, e\} < \{c, e\} < \{b, d\} \sim \{c, d\} < \{b, c\} < \{d, e\} < \{a, b\} < \{b, e\}$$

The classifiability graph consists only of horizontal lines, one of which represents the subset of pairs which intervene to strictly separate the median pair from the upper one of the *same number of triplets*; the length of a given line represents the proportion of these pairs, and its elevation the proportion of triplets of which the median and upper pair are strictly separated by any one of these pairs. The better the classifiability the more flattened the graph consisting of horizontal lines, on the horizontal axis; thus, it can be measured by the smallness of the area between the two perpendicular axes.

In [12] a subset $\mathcal{B}$ of 67 attributes taken in the $\mathcal{A}$ of 110 attributes is retained. $\mathcal{B}$ is composed of very characteristic descriptive attributes of the child personages of the child literature. The classifiability graph of the distribution $D(\omega(\mathcal{B}))$ was established on the basis of the preordonance $\omega(\mathcal{B})$ associated with the coefficient $Q_P(a, b)$ (see (11.3.1)). A random set $\mathcal{B}^*$ is associated with $\mathcal{B}$. $\mathcal{B}^*$ is composed of independent random attributes associated, respectively, with the attributes of $\mathcal{B}$. Writing $\mathcal{B}$ as follows:

$$\mathcal{B} = \{b^j | 1 \leq j \leq 67\} \tag{11.3.5}$$

$\mathcal{B}^\star$ can be written

$$\mathcal{B}^\star = \{b^{j^\star} | 1 \leq j \leq 67\} \tag{11.3.6}$$

where $b^{j^\star}$ is the random attribute associated with b^j. The corresponding random model is defined by H_1 (see Sect. 5.2.1.2 of Chap. 5). More precisely, let us recall that if $\mathcal{O}(b^j)$ is the $\mathcal{O}$ subset where b^j is present and if $n_j = card(\mathcal{O}(b^j))$, $\mathcal{O}(b^{j^\star})$ is a random $\mathcal{O}$ subset in the set, endowed with equal probabilities (chances) of all $\mathcal{O}$ parts, having the same cardinal n_j, $1 \leq j \leq 67$.

Two independent realizations $\mathcal{B}_1$ and $\mathcal{B}_2$ of $\mathcal{B}^\star$ have been carried out. The respective graphs of the classifiability distribution $D(\omega(\mathcal{B}_1))$ and $D(\omega(\mathcal{B}_2))$ show clearly ([12] page 43) that

$$D(\omega(\mathcal{B})) < D(\omega(\mathcal{B}_1))$$
and
$$D(\omega(\mathcal{B})) < D(\omega(\mathcal{B}_2)) \tag{11.3.7}$$

(see (8.7.5) of Sect. 8.7.2.2 of Chap. 8).

- **Classifiability and neutral elements**
  The classifiability of a set $\mathcal{A}$ provided with a similarity Q is closely linked to the relative removal of each of its elements with respect to the sequence of the others, established by increasing dissimilarity. The data analysis prior to a clustering led us to measure the neutral character of a given element a of $\mathcal{A}$, with respect to a classificatory goal by the smallness of the observed variance $\mathcal{V}(a)$ of the distribution of the Q similarities to a. The expression of the latter distribution and the associated variance are given in (8.5.2) and (8.5.3) of Sect. 8.5.1 of Chap. 8. If in $\mathcal{A}$ there are a few elements such that their degrees of neutrality are rather high to disturb the classifiability nature of the whole set, we are able to detect these elements and to remove them before any classification (clustering).
  For a finite set E, endowed with a dissimilarity δ, it might be interesting to study— at least experimentally—the influence of the neutrality degree of a given element e ($e \in E$) with respect to *interiority* level as defined in [4]. On the other hand, *separability* (see the latter reference) of a partition obtained by Clustering is necessarily strongly related to the Classifiability of (E, δ).

- **Classifiability and cluster cohesion**
  The presence of elements in $\mathcal{A}$ with a high neutrality degree (i.e. weakly discriminant with respect to their associations with the others elements) influence the cohesion of the classes formed in a partition. In the *LLA* method, the criterion of a partition quality was defined and studied in Sect. 9.3.4 of Chap. 9. Two versions were established according to either an *ordinal* (see (9.3.46)) or *numerical* (see

Fig. 11.4 Cohesion and
classifiability

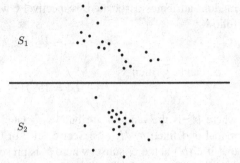

(9.3.70)) similarity on the set to be clustered. We have noticed that in the numerical case, the average of the proximity variance participate to the statistical analysis of the criterion (see (9.3.78) of Chap. 9).

We may be interested in exploring the relation between the degree of classifiability and the cohesion of classes obtained by a good algorithm. All things equal and to a certain extent, class cohesions and classifiability might be associated. However, apart from very extreme cases, we must not believe that a relative amount of classifiability increase will necessarily imply stronger cohesion for the formed classes. We realized this fact from random simulations of an incidence data table, associated with a real case, under the H_1 hypothesis of nonlink. To illustrate this geometrically, let us consider two sets of points S_1 and S_2 in Fig. 11.4. The set S_1 of points seems to be more suitable that than the set S_2 for organizing in classes; nevertheless, the classification of S_2 into seven classes, of which the first groups all the central points and where each of the six others contains exactly one peripheral point, will have a stronger cohesion than any classification of S_1.

11.4   Dayhoff, Henikoffs and *LLA* Matrices for Comparing Proteic Sequences

11.4.1   Preamble: Technical Data Sheet

- **The data table**
  The data table is a rectangular data table indexed by $S \times A$, where S is a set of proteic sequences of a given family (e.g. the cytochrome C) and A is a set of categorical attributes valued by a numerical similarity (see Sect. 7.2.4 of Chap. 7).

- **The set to be clustered**
  Hierarchical clustering concerns the set S interpreted as a set of objects. The objective consists of observing to what extent the species assessed by the evolution theory are recognized.

- **The association coefficient employed**
  Given two sequences s_k and $s_{k'}$ ($1 \leq k \neq k' \leq K$), its construction starts with the respective normalized contributions of the different attributes to the similarity of s_k and $s_{k'}$. Let us recall the expression of this type of similarity index (see (7.2.26)) with the notations adopted below

$$S^l(s_k, s_{k'}) = \frac{s^l(s_k, s_{k'}) - M_1}{\sqrt{M_2 - M_1^2}} \qquad (11.4.1)$$

where $s^l(s_k, s_{k'})$ is defined by the *raw* contribution of the lth attribute to the comparison of s_k and $s_{k'}$ and where M_1 and M_2 are the mean and the absolute moment of order 2 of a random index associated with the raw one under a hypothesis of no relation, $1 \leq l \leq L$.
  The construction of the global similarity index follows the sequence of Eqs. (7.2.1)–(7.2.6).

- **Making use**
  I.C. Lerman (University of Rennes 1), Ph. Peter (University of Nantes) and J.L. Risler (Centre National de la Recherche Scientifique, Centre de Génétique Moléculaire)

- **Biologist expert providing the data and interpreting results**
  J.L. Risler (Centre National de la Recherche Scientifique, Centre de Génétique Moléculaire).

11.4.2   Introduction

Basically, a proteic sequence can be expressed as a *word* using an alphabet $\mathbb{A}$ of 20 letters where each of them represents an amino acid:

$$\mathbb{A} = \{A, R, N, D, C, Q, E, G, H, I, L, K, M, F, P, S, T, W, Y, V\}$$

The set of proteic sequences can be decomposed into families. An example of one of the latters is the "cytochrome C". The respective lengths of the different words associated with the proteic sequences of a given family are, generally, not equals. This is justified in evolution theory by phenomenons of deletion or insertion of amino acids in the proteic sequences during time. A special symbol denoted by "−" represents a deletion. By inserting the symbol "−" at certain positions of the respective words representing the proteic sequences, those of a given family may be mutually aligned (*multiple alignment*). Thereby, the alphabet used becomes $\mathbb{A}' = \mathbb{A} + \{-\}$. It comprises 21 letters. One major objective in molecular biology consists precisely of validating multiple alignments. Recognizing from these consistent subfamilies, according to species evolution, is of great importance, This recognition enables implicitly the multiple alignment to be authenticated.

For a given set of aligned sequences, the approach considered here for recognizing consistent subfamilies is based on clustering. For methodological clarity reasons, we assume the multiple alignment to be carried out by structural methods [23]. Thereby, the data comes down to a rectangular table indexed by the Cartesian product $S \times \mathcal{A}$, where S—which indexes the row set—represents the sequence set and where $\mathcal{A}$—which indexes the column set—the attribute set. The latter will be specified below. Let us put S and $\mathcal{A}$ in the forms

$$S = \{s_k | 1 \leq k \leq K\}$$

and

$$\mathcal{A} = \{a^l | 1 \leq l \leq L\} \qquad (11.4.2)$$

where a^l is a categorical attribute whose value set is $\mathbb{A}'$. Thus a^l is the categorical attribute giving the element of $\mathbb{A}'$ (an amino acid or a deletion) situated at the position l of the sequence. Notice that the only distinction between two attributes a^l and $a^{l'}$, for $l \neq l'$, is due to the fact that a^l and $a^{l'}$ are concerned by two *distinct* positions in the sequence, $1 \leq l \neq l' \leq L$.

If no structure (scale) is retained on the value set $\mathbb{A}'$ of the attributes a^l ($1 \leq l \leq L$); in other terms, if we interpret the a^l attributes as nominal categorical ones (see Sect. 3.3.1 of Chap. 3), the clustering which could be obtained is necessarily very poor with respect to species evolution. Therefore, we shall consider a numerical similarity on $\mathbb{A}'$ depending intimately on evolution. Designating by *Sim* such a similarity function, for Y and Z belonging to $\mathbb{A}$, $Y \neq Z$, $Sim(Y, Z)$ is calculated in order to reflect the propensity for Y (resp., Z) to mutate into Z (resp., Y). The propensity symmetry assuming that the likelihood of amino acid Z replacing Y is the same as that of amino acid Y replacing Z, is generally admitted. Now, similarly, for Y in $\mathbb{A}$, $Sim(Y, Y)$ is related to the persistence of Y. Finally, we will see below how to derive the values of $Sim(Y, -) = Sim(-, Y)$ for $Y \in \mathbb{A}$ and also $Sim(-, -)$.

Thereby, the *Sim* function induces a symmetrical valuation graph on $\mathbb{A}'$ and this makes that for every l, $1 \leq l \leq L$, a^l defines a valuated graph on the sequence set S. Hence, the comparison of two proteic sequences s and s' of S, requires to integrate the vector of numerical similarity values

$$\{Sim(a^l(s), a^l(s'))\} \qquad (11.4.3)$$

A daring assumption on amino acid mutability during time for near and already aligned pairs of proteic sequences leads to the famous Dayhoff matrix [6]. The latter gives a specific estimation of an association matrix between amino acids. Let us designate this matrix by

$$\mathcal{D} = \{D(Y, Z) | Y, Z \in \mathbb{A}\} \qquad (11.4.4)$$

The principles and calculations leading to the Dayhoff matrix will be made explicit in Sect. 11.4.3. To this respect, precise formulas will be established. In fact, this matrix is parametrized by the Biologist. In our experiments, the matrix corresponding to the so called 250 PAM (Point Accepted Mutation) is adopted (see in Sect. 11.4.3). The matrix on which we have worked is provided from [9].

A second matrix family called BLOSUM has been proposed by the Henikoffs [8]. Let us designate it by

$$\mathcal{B} = \{B(Y, Z) | Y, Z \in \mathbb{A}\} \tag{11.4.5}$$

At present, this type of matrix seems to be more employed than that of Dayhoff. However, in our treatments [21] no better performance was observed with the more recent matrices. And in fact, the results obtained with the Dayhoff matrix turn out to be more in accordance with the Biologist knowledge. Clearly, the elaboration of a BLOSUM matrix is different from a Dayhoff one; however, its statistical nature is the same. Indeed, a given element of either a Dayhoff or a BLOSUM matrices represents the logarithm of a probability density. In Sect. 11.4.4 we will detail the elements for calculating a BLOSUM matrix.

In Sect. 11.4.6, we will describe how to derive an *LLA* matrix of probabilistic association indices from a Dayhoff or a BLOSUM matrices. We will designate by *LLA*($\mathcal{D}$) and *LLA*($\mathcal{B}$) these two matrices, respectively. To each of them we may associate the corresponding matrix of "Informational dissimilarity" (see (7.2.7) of Sect. 7.2.1 of Chap. 7).

An association matrix

$$\mathcal{Q} = \{Q(Y, Z) | Y, Z \in \mathbb{A}\} \tag{11.4.6}$$

on $\mathbb{A}$ allows mutual alignments of sequence pairs to be performed. Very popular methods for this are provided by dynamic programming techniques (BLAST, FASTA). Generally, the deletion symbol "−" appears in these alignments. A statistical solution is proposed in order to estimate the respective associations between a deletion and each of the 20 amino acids according to $\mathcal{Q}$.

Now, let $\mathcal{S}$ be a set of sequences of a given family (e.g. the cytochrome C). As just mentioned above, the pairwise alignment of $\mathcal{S}$ leads through several algorithms to a multiple alignment of $\mathcal{S}$ [23]. Thereby, we get a rectangular table indexed by the Cartesian product $\mathcal{S} \times \mathcal{A}$ (see (11.4.2)). In Sect. 11.4.7 the *LLA* probabilistic similarity index on a set $\mathcal{S}$ of aligned sequences is established. In these conditions, the *AAHC LLA* method applies for clustering $\mathcal{S}$, interpreted as a set of objects described by categorical attributes whose value sets are valuated by a numerical similarity. Four valuations might be considered in our context: $\mathcal{D}$, $\mathcal{B}$, *LLA*($\mathcal{D}$) and *LLA*($\mathcal{B}$). As mentioned above, the finest and most consistent results were obtained with *LLA*($\mathcal{D}$). We will give a general idea of these in Sect. 11.4.7.

11.4.3 Construction of the Dayhoff Matrix

In this section, the principles of calculation of the Dayhoff matrix will be made explicit. For this purpose, precise mathematical formulas will be detailed.

11.4.3.1 Preliminaries

The basic data is a set S of proteic sequences provided from a given family (e.g. the cytochrome C) and more or less mutually cotemporary with respect to their respective evolutions. We assume given a pairwise alignment of S. These alignments are essential for the computing. It is not a necessary condition that the common length of two aligned sequences is the same for the different aligned pairs. Otherwise, many sequence families may contribute to the calculation of the statistical indices which will be defined below. In fact, in these, the fundamental operation consists of summation over a set of sequence pairs.

To fix ideas and to simplify in order to make clearer the nature of the following calculations, we assume given a set

$$S = \{s_k | 1 \le k \le K\} \tag{11.4.7}$$

of aligned sequences (multiple alignment) with respect to the alphabet

$$\mathbb{A} = \{Z_i | 1 \le i \le 20\} \tag{11.4.8}$$

of 20 letters corresponding to the 20 amino acids. By difference with above (see what follows (11.4.2)) the deletion symbol "$-$" is not comprised in the alphabet. Nevertheless, the general case where the latter symbol has to be added for alignment, can be deduced without major difficulty. In these conditions, a proteic sequence s_k belonging to S can be formalized by a mapping that we can denote also (without ambiguity) by s_k, of

$$\mathcal{L} = \{1, 2, \ldots, l, \ldots, L\} \tag{11.4.9}$$

into $\mathbb{A}$ (recall that L is the common length of the S sequences):

$$s_k : \mathcal{L} \to \mathbb{A}$$
$$l \mapsto s_k(l) \tag{11.4.10}$$

where $s_k(l)$ is the amino acid occupying the lth site of the kth sequence, $1 \le l \le L$, $1 \le k \le K$.

Now, the computing will refer to the set of S unordered pairs, that is to say to the set $P_2(S)$ of 2 subsets of S:

$$P_2(S) = \left\{\{s_k, s_{k'}\} | 1 \le k < k' \le K\right\} \tag{11.4.11}$$

Let a designate the amino acid attribute whose value set is A. To a given s of S we will associate the absolute frequency distribution of a over s:

$$\mathcal{D}(a) = \left\{n_s(i) = card\left(s^{-1}(Z_i)\right) | 1 \le i \le 20\right\} \tag{11.4.12}$$

$n_s(i)$ is the number of times where the letter Z_i appears in the sequence s. Clearly, we have

$$\sum_{1 \leq i \leq 20} n_s(i) = L \qquad (11.4.13)$$

$$N(i) = \sum_{s \in S} n_s(i) \qquad (11.4.14)$$

is the total number of times where the amino acid Z_i occurs in S. We have

$$\sum_{1 \leq i \leq 20} N(i) = KL \qquad (11.4.15)$$

Now, let us introduce, with respect to the set S of the K sequences, the relative frequencies or proportions

$$f_i = \frac{N(i)}{KL}, \, 1 \leq i \leq 20. \qquad (11.4.16)$$

These proportions are the same if they are calculated on the basis of the set $P_2(S)$ (see (11.4.11)). As a matter of fact, we have

$$f_i = \frac{\sum_{\{s,s'\} \in P_2(S)} \left(n_s(i) + n_{s'}(i) \right)}{\left(K(K-1)L \right)}, \quad 1 \leq i \leq 20. \qquad (11.4.17)$$

where the denominator is the total number of letter symbols occurring in $P_2(S)$. Indeed, each pair of sequences gives rise to $2L$ letter occurrences.

11.4.3.2 *Relative Mutability* of an Amino Acid Through a Set of Pairwise Aligned Sequences

A quantified fundamental notion for elaborating a Dayhoff matrix is the *relative mutability* of an amino acid through a set S of proteic sequences, two by two aligned. To begin, let us introduce with a small example the non-normalized version of this notion. We designate it by $\mu_i(\{s, s'\})$ for a given amino acid Z_i involved in an aligned sequence pair $\{s, s'\}$. This number is the ratio of the number of sites for which Z_i occurs in exactly one of both sequences to the total number of times where Z_i appears. Thus for the aligned sequences

$$A \, C \, D \, E \, F \, L$$
$$A \, G \, D \, E \, A \, L$$

the non-normalized relative mutability of A is $1/3$, that of G or C is 1, that of D is 0, …Trivially, $\mu_i(\{s, s'\})$ is an index comprised between 0 and 1.

More formally, let us designate by $e_i(\{s, s'\})$ (resp.; $d_i(\{s, s'\})$) the number of sites including Z_i in exactly *one* of both sequences (resp., *both* sequences). We have

$$\mu_i(\{s, s'\}) = \frac{e_i(\{s, s'\})}{2d_i(\{s, s'\}) + e_i(\{s, s'\})}, \, 1 \leq i \leq 20. \tag{11.4.18}$$

We can observe that $\mu_i(\{s, s'\})$ is a too raw index in order to render an adequate measure of mutability. Effectively, the significance of the latter index cannot be the same according to a large or a small percentage of mutations between s and s'. In these conditions, let us introduce for a given pair $\{s, s'\}$ of aligned sequences the percentage of mutations $p(\{s, s'\})$, namely,

$$p(\{s, s'\}) = \frac{L_m(\{s, s'\})}{L} \tag{11.4.19}$$

where L is the common length of both sequences and $L_m(\{s, s'\})$ is the number of sites for which mutation occurs; that is to say, for which the respective letters representing amino acids are distinct. The *normalized* mutability or more briefly, the mutability is defined by

$$m_i(\{s, s'\}) = \frac{\mu_i(\{s, s'\})}{p(\{s, s'\})} = \frac{e_i(\{s, s'\})}{(2d_i(\{s, s'\}) + e_i(\{s, s'\}))p(\{s, s'\})}, \, 1 \leq i \leq 20. \tag{11.4.20}$$

$mi(\{s, s'\})$ quantifies the relative mutability with respect to a single pair of aligned sequences. In order to take into account the whole set $\mathcal{S}$, the respective contributions of all of the aligned pairs have to be globalized. The solution proposed for this in [6, 9] consists of defining a ratio whose numerator (resp., denominator) is the sum over the set $P_2(\mathcal{S})$—of unordered sequence pairs—of the numerator (resp., denominator) of $mi(\{s, s'\})$ as expressed in (11.4.20). More precisely, this ratio can be written

$$m_i(\mathcal{S}) = \frac{\sum_{\{s,s'\} \in P_2(\mathcal{S})} e_i(\{s, s'\})}{\sum_{\{s,s'\} \in P_2(\mathcal{S})} (2d_i(\{s, s'\}) + e_i(\{s, s'\}))p(\{s, s'\})}, \, 1 \leq i \leq 20. \tag{11.4.21}$$

Another method for defining globally the relative mutability of an amino acid Z_i through $\mathcal{S}$ consists of averaging $m_i(\{s, s'\})$ over $P_2(\mathcal{S})$. More precisely, by denoting $\mathcal{S}_i$ the sequence set for which Z_i appears in every of its elements, we have

$$\mathcal{S} = \mathcal{S}_i + \mathcal{T}_i \text{ set sum} \tag{11.4.22}$$

where $\mathcal{T}_i$—the complementary subset to $\mathcal{S}_i$ with respect to $\mathcal{S}$—is composed of $\mathcal{S}$ sequences for which the amino acid Z_i does not appear.

The set of sequence pairs $\{s, s'\}$ for which Z_i takes part in at least one of both sequences s and s' can be written as follows:

$$P_2(i) = P_2(\mathcal{S}_i) + \mathcal{S}_i \star \mathcal{T}_i \text{ set sum} \tag{11.4.23}$$

where $P_2(S_i)$ (resp., $S_i \star T_i$) is the set of sequence pairs whose two components belong to S_i (resp., whose one of both components belong to S_i and the other one to T_i). By denoting K_i the size of S_i ($K_i = card(S_i)$), we have

$$card(P_2(i)) = \frac{1}{2}K_i(K_i - 1) + K_i(K - K_i) = \frac{1}{2}K_i(2K - K_i - 1) \qquad (11.4.24)$$

And then,

$$m_i' = \frac{2}{K_i(2K - K_i - 1)} \sum_{\{s,s'\} \in P_2(i)} m_i(\{s, s'\}) \qquad (11.4.25)$$

corresponds to the new interpretation of the global relative mutability.

Notice that in the calculation of $m_i(\{s, s'\})$, for a given pair of aligned sequences s and s', the common length associated with the alignment concerned, does not intervene. Nevertheless, the average considered in (11.4.25) might be weighted according to the respective lengths of the different alignments.

Now, if deletions have to take part in the pairwise alignments, one of the following solutions can be considered:

1. Substitute for $\mathbb{A}$, $\mathbb{A}'$ which includes an additional 21th symbol denoted by "–" and considered as any of the 20 letters of $\mathbb{A}$;
2. For every aligned sequence pair $\{s, s'\}$ of $P_2(S)$, ignore the sites where a deletion intervenes in s or (non exclusively) in s';
3. For every aligned pair of sequences $\{s, s'\}$ ignore the sites where a deletion appears for the calculation of $e_i(\{s, s'\})$ and $d_i(\{s, s'\})$, but retain these sites for the calculation of the percentage p($\{s, s'\}$).

Clearly, many sequence families may contribute to the determination of m_i (see (11.4.21)) or m_i' (see (11.4.25)). For this, consider a set

$$\{S^t | 1 \leq t \leq u\}$$

of sequence sets provided, respectively, from homogeneous families and replace $P_2(S)$ in (11.4.21) by the set sum

$$\sum_{1 \leq t \leq u} P_2(S^t) \qquad (11.4.26)$$

Now, for extending m_i' in (11.4.25), begin by associating $P_2^t(i)$ with S^t in the same way that $P_2(i)$ was associated with S (see (11.4.23)). In these conditions m_i' is defined by a ratio whose numerator has the following form

$$\sum_{1 \leq t \leq u} \sum_{\{s,s'\} \in P_2^t(i)} m_i(\{s, s'\}) \qquad (11.4.27)$$

The denominator of this ratio is the sum of the cardinals of the sets $P_2^t(i)$, that is

$$\sum_{1 \leq t \leq u} \frac{K_i^t (2K^t - K_i^t - 1)}{2} \tag{11.4.28}$$

where K^t is the cardinal of S^t and where K_i^t is the cardinal of the subset S_i^t of S^t for which all of its elements include Z_i.

11.4.3.3   Mutation Probability

The mutation probability of a given amino acid for a unit period time of evolution is a crucial notion in the Dayhoff construction. Let us designate by $M_{\bar{i}i}$ the passage probability from i to negated i, denoted by $\bar{i}$. $M_{\bar{i}i}$ is set proportional to the relative mutability m_i (see (11.4.21)):

$$M_{\bar{i}i} = \lambda m_i(\mathcal{S}) , \ 1 \leq i \leq 20. \tag{11.4.29}$$

This condition would also be considered for m_i' instead of m_i (11.4.25).
According to (11.4.29) the probability of persistence of Z_i can be written

$$M_{ii} = 1 - M_{\bar{i}i} = 1 - \lambda m_i(\mathcal{S}) , \ 1 \leq i \leq 20. \tag{11.4.30}$$

Now, consider

$$p_{\mathcal{S}}(i) = f_i M_{ii}, \quad 1 \leq i \leq 20. \tag{11.4.31}$$

This represents the probability for the amino acid Z_i to occur and to persist in the framework of $\mathcal{S}$. More precisely, it is the probability for a random site of a random aligned sequence pair $\{s, s'\}$ taken in $P_2(\mathcal{S})$, to meet the amino acid Z_i in s and in s' at the site concerned, the randomness being with equal chances.

In these conditions, the global probability of persistence through $\mathcal{S}$ is

$$p_{\mathcal{S}} = \sum_{1 \leq i \leq 20} p_{\mathcal{S}}(i) = \sum_{1 \leq i \leq 20} f_i M_{ii} \tag{11.4.32}$$

$p_{\mathcal{S}}$ represents the probability to meet an amino acid which persists from s to s', whatever is the amino acid. Precisely, this global persistence probability is put equal to 0.99:

$$p_{\mathcal{S}} = \sum_{1 \leq i \leq 20} f_i M_{ii} = 0.99 \tag{11.4.33}$$

and this corresponds—on average—to a single mutation for 100 sites during a unit evolution period. Equation (11.4.33) defines a kind of normalization which can be written

$$\sum_{1 \leq i \leq 20} f_i M_{\bar{i}i} = \lambda \sum_{1 \leq i \leq 20} f_i m_i = 0.01 \tag{11.4.34}$$

Equation (11.4.34) enables the value of the parameter λ to be fixed:

$$\lambda = \frac{0.01}{\sum_{1 \leq i \leq 20} f_i m_i} \quad (11.4.35)$$

Let us summarize before going on. The mutation probability $M_{\bar{i}i}$ of a given amino acid Z_i was defined as proportional to the relative mutability m_i of Z_i. The proportionality coefficient was determined in order to impose the condition that the persistence probability of an amino acid during an evolution unit time is equal to 0.99.

Now, we have to derive from $M_{\bar{i}i}$ the sequence of 19 probabilities

$$\{M_{ji} | 1 \leq j \leq 20, j \neq i\} \quad (11.4.36)$$

For this purpose, we evaluate by counting—on the basis of aligned sequence pairs—the conditional probability that Z_i mutates into Z_j $(j \neq i)$, knowing that Z_i mutates. Thus, this probability can be written as follows:

$$P^i_{ji} = \frac{A_{ji}}{\sum_{k|k \neq i} A_{ki}} \quad (11.4.37)$$

where A_{ki} $(k \neq i)$ is the number of times calculated on a set of aligned sequence pairs, where Z_k is faced to Z_i, $1 \leq k \leq 20$, $k \neq i$.

Therefore, we can deduce the transition probability from i to j:

$$M_{ji} = M_{\bar{i}i} P^i_{ji} = \lambda m_i \frac{A_{ji}}{\sum_{k|k \neq i} A_{ki}} \quad (11.4.38)$$

In these conditions, the verification of the following relations is immediate

$$\sum_{1 \leq j \leq 20} M_{ji} = 1 \quad (11.4.39)$$

In the same way, that the probability to meet a *persistent* amino acid Z_i was defined (see (11.4.31)), the probability $q_S(i)$ to meet a *mutated* amino acid Z_i can be defined. We have

$$q_S(i) = f_i M_{\bar{i}i}, \quad 1 \leq i \leq 20. \quad (11.4.40)$$

Trivially, we have

$$p_S(i) + q_S(i) = f_i$$

and

$$\sum_{1 \leq i \leq 20} (p_S(i) + q_S(i)) = 1 \quad (11.4.41)$$

Let us emphasize once more the principle of the method employed. To begin, the probability (or percentage) of global mutation is fixed equal to 0.01 for a unit evolution time. Then a learning set S composed of contemporary sequences of a given family, aligned by pairs, is considered. In these conditions, the latter probability is distributed on the 20 amino acids proportionally to two factors (see (11.4.29) and (11.4.34)):

1. The respective relative mutabilities m_i, $1 \leq i \leq 20$, of the different amino acids;
2. The respective frequencies of exposure to mutation f_i, $1 \leq i \leq 20$.

Now, the mutation probability of the amino acid Z_i, M_{ii} is distributed proportionally with respect to the set $\{Z_j | j \neq i\}$ of the 19 mutation possibilities, $1 \leq i \leq 20$. This proportionality is calculated according to the empirical observation of S (see (11.4.37)).

The matrix obtained

$$\mathcal{M} = \left(M_{ji} | 1 \leq i,j \leq 20 \right) \tag{11.4.42}$$

is a probability transition matrix relative to one percent of global mutation in an unit period of evolution. For a k units duration period, called k *PAM* (Point Accepted Mutation), we have to raise to the power k the matrix $\mathcal{M}$. As mentioned in the introduction we have worked in [21] with a 250 *PAM* matrix provided from [9]. Without risk of ambiguity, the we will continue to denote the latter matrix as in (11.4.42).

Whatever the evolution period concerned, we shall determine now, for a random site of a given aligned sequence pair $\{s, s'\}$, to meet the amino acid Z_i in one of both pairs and the amino acid Z_j in the other one, $1 \leq i,j \leq 20$.

The site being fixed, we may imagine the random choice in s followed by the random choice in s'. If $i \neq j$, the event in which we are interested can be expressed as follows:

- "meet Z_i in s and Z_i mutates into Z_j in s'" or;
- "meet Z_j in s and Z_j mutates into Z_i in s'".

Therefore, the probability of this event is

$$(f_i M_{ji} + f_j M_{ij}) \tag{11.4.43}$$

If j is identical to i the event in which we are interested becomes

- "meet Z_i in s and Z_i persists in s'".

The probability of this event is

$$f_i M_{ii} \tag{11.4.44}$$

Now, consider the probability of the first of both events under the independence hypothesis. It can be written

$$f_i f_j + f_j f_i = 2 f_i f_j, \quad 1 \le i \ne j \le 20. \tag{11.4.45}$$

Similarly, the probability of the second event is

$$f_i^2, \quad 1 \le i \le 20. \tag{11.4.46}$$

In these conditions, the probability densities of the first and second events can be written, respectively, as follows:

$$R_{ji} = \frac{f_i M_{ji} + f_j M_{ij}}{2 f_i f_j}, R_{ii} = \frac{f_i M_{ii}}{f_i^2}, \quad 1 \le i \ne j \le 20. \tag{11.4.47}$$

Clearly, the matrix

$$\{R_{ij} | 1 \le i, j \le 20\} \tag{11.4.48}$$

is symmetrical. Besides, the symmetry property

$$f_i M_{ji} = f_j M_{ij}, 1 \le i, j \le 20, \tag{11.4.49}$$

is generally admitted. The left (resp., right) member of (11.4.49) is defined by the probability that the amino acid Z_i (resp., Z_j) occurs and mutates into Z_j (resp., Z_i).

In these conditions, the probability density for $1 \le i \ne j \le 20$ is defined by

$$R_{ji} = \frac{f_i M_{ji}}{f_i f_j} = \frac{M_{ji}}{f_j}, \quad 1 \le i, j \le 20. \tag{11.4.50}$$

Nevertheless, the exact formula (11.4.47) would have been used more accurately. Notice that (11.4.49) applies in the case where $i = j$.

Now, instead of an unordered sequence pair $\{s, s'\}$ let us consider at random an ordered sequence pair (s, s'), where $s \ne s'$. For a given site chosen at random, designate by O_{ij} the probability to meet Z_i in s and Z_j in s'. O_{ij} is given by the left member of (11.4.49), though O_{ji} is given by the right member of (11.4.49). Admitting the equality between O_{ij} and O_{ji}, we obtain

$$R_{ij} = \frac{O_{ij}}{f_i f_j}, \quad 1 \le i, j \le 20. \tag{11.4.51}$$

Thereby, the Dayhoff matrix is the matrix of the logarithms to base 10 of the numbers R_{ij}, namely,

$$\mathcal{D} = \{D_{ij} = \text{Log}_{10}(R_{ij}) | 1 \le i, j \le 20\} \tag{11.4.52}$$

which is an upper semi-matrix comprising its diagonal.

$\mathcal{D}$ can be interpreted as an association (similarity) matrix on the set $\mathbb{A}$ of the amino acids, where the similarity between a given amino acid and itself is not constant.

11.4.4 The Henikoffs Matrix: Comparison with the Dayhoff Matrix

In order to introduce this matrix [8], we shall use as much as possible the same notations as previously. To begin, we consider the simplest but general enough case of a set $\mathcal{S}$ of aligned sequences provided by a single family of proteic sequences. For biological reasons, the authors of this matrix consider a more complex case that we will mention below. However, the nature of the statistical computing is exactly the same.

Relative to $\mathcal{S}$ (11.4.2) the Henikoffs matrix is obtained by a *counting* based on the set $P_2(\mathcal{S})$ of unordered sequence pairs. For this purpose, let us consider the following table of integers

$$\{A_{ij} | 1 \le j \le i \le 20\} \tag{11.4.53}$$

where A_{ij} is the number of sites for which the amino acids Z_i and Z_j are face to face in the whole set $P_2(\mathcal{S})$. More precisely, by denoting $A_{ij}(\{s, s'\})$ the number of sites where Z_i and Z_j are face to face, one of them in s and the other one in s', we have

$$A_{ij} = \sum_{\{s,s'\} \in P_2(\mathcal{S})} A_{ij}(\{s, s'\}), \quad 1 \le j \le i \le 20. \tag{11.4.54}$$

Thereby, the empirical probability of a face to face meeting between Z_i and Z_j, is given by

$$Q_{ij} = \frac{A_{ij}}{\sum_{1 \le j' \le i' \le 20} A_{i'j'}}, \quad 1 \le j \le i \le 20. \tag{11.4.55}$$

Notice that the denominator value is $LK(K-1)/2$, where L is the common length of the $\mathcal{S}$ sequences and where K is the number of sequences. In fact, for a given sequence pair $\{s, s'\}$ each site takes part to increment one unity exactly one of the $A_{i'j'}, 1 \le j' \le i' \le 20$.

By difference with (11.4.37), (11.4.55) is a total probability and not a conditional one. Q_{ij} will play the same role as (11.4.43) (resp., (11.4.44)) in the case where $i \ne j$ (resp., $i = j$). Precisely, let us recall again the events concerned by these probabilities with respect to a random site of an aligned sequence pair $\{s, s'\}$.

- If $i \ne j$: "meet face to face Z_i in s and Z_j in s'" or "meet Z_j in s and Z_i in s'";
- If $i = j$: "meet face to face Z_i in s and s'".

The respective probabilities of these two events were specified in (11.4.45) and (11.4.46) under the independence hypothesis. Let us denote the latter probabilities as follows:

$$E_{ij} = 2f_i f_j \text{ for } 1 \leq i \neq j \leq 20$$
$$E_{ii} = f_i^2 \text{ for } 1 \leq i \leq 20 \tag{11.4.56}$$

In order to adopt similar notations to those of [8] let us remark that f_i (see (11.4.16)) goes as

$$2A_{ii} + \sum_{\{j|j \neq i\}} A_{ij}, 1 \leq i \leq 20. \tag{11.4.57}$$

Hence f_i can be written

$$f_i = Q_{ii} + \frac{1}{2} \sum_{\{j|j \neq i\}} Q_{ij}, 1 \leq i \leq 20. \tag{11.4.58}$$

Introduce now the probability density in (i, j)

$$r_{ij} = \frac{Q_{ij}}{E_{ij}}, \quad 1 \leq i, j \leq 20. \tag{11.4.59}$$

And then, consider as in the case of the Dayhoff matrix, the logarithm, now to base 2, of the r_{ij}:

$$s_{ij} = \log_2(r_{ij}), \quad 1 \leq i, j \leq 20. \tag{11.4.60}$$

The matrix

$$\mathcal{H} = \{h_{ij} = 2s_{ij} | 1 \leq i \leq j \leq 20\} \tag{11.4.61}$$

(the Henikoffs matrix) is considered as estimating the respective substitutions degrees between the amino acids.

Our aim now is to make explicit the conceptual difference between the latter matrix and that—finer but more hypothetical—Dayhoff matrix. In the following, we will designate by $\mathcal{D}$ and $\mathcal{H}$ the Dayhoff and Henikoffs cases, respectively.

Whereas we observe for $\mathcal{H}$ a *persistence* empirical probability in the form

$$\sum_{1 \leq i \leq 20} Q_{ii} \tag{11.4.62}$$

this probability is imposed to be equal to 0.99 for $\mathcal{D}$:

$$\sum_{1 \leq i \leq 20} O_{ii} = \sum_{1 \leq i \leq 20} f_i M_{ii} = 0.99 \tag{11.4.63}$$

The original notion of relative mutability of an amino acid (see (11.4.21)) permits us to distribute the complementary of the global persistence probability (see (11.4.34)) proportionally (see (11.4.29)). Then, the mutation probability $M_{\bar{i}i}$ of a given amino acid i is distributed over the different mutation alternatives, according to the observation of the training set $\mathcal{S}$ (see (11.4.37) and (11.4.38))

Consequently, the occurrence probability matrix

$$\{O_{ij} | 1 \leq i \leq j \leq 20\} \tag{11.4.64}$$

is deduced from the transition matrix (11.4.42) and the empirical distribution $\{f_i | 1 \leq i \leq 20\}$ (see (11.4.16), (11.4.17), (11.4.49) and (11.4.50)).

In $\mathcal{H}$, the occurrence probability matrix

$$\{Q_{ij} | 1 \leq i \leq j \leq 20\} \tag{11.4.65}$$

which can correspond to (11.4.64) is directly calculated from the learning set $\mathcal{S}$ (see (11.4.55)).

Now, let us point out that the square matrix,

$$\{O_{ij} | 1 \leq i, j \leq 20\} \tag{11.4.66}$$

which is symmetrical by nature, defines a joint probability distribution on $\mathbb{A} \times \mathbb{A}$. The associated row and column marginal distributions are identical and defined by

$$\{f_i | 1 \leq i \leq 20\} \tag{11.4.67}$$

To obtain from (11.4.65) a matrix of the same nature as (11.4.66), follow the argument between (11.4.50) and (11.4.51) and set

$$(\forall (i, j), 1 \leq i < j \leq 20), O'_{ij} = O'_{ji} = \frac{1}{2} Q_{ij}$$

and

$$(\forall i, 1 \leq i \leq 20), O'_{ii} = Q_{ii} \tag{11.4.68}$$

The matrix

$$\{O'_{ij} | 1 \leq i, j \leq 20\} \tag{11.4.69}$$

is symmetrical by construction. It defines a joint relative frequency distribution on $\mathbb{A} \times \mathbb{A}$ whose common row and column marginal distributions being given by (11.4.67).

Nevertheless, the direct calculation of Q_{ij} in the framework of $\mathcal{H}$ (see (11.4.55)) can be biased in its biological meaning if the set $\mathcal{S}$ can be decomposed into B blocs:

$$\mathcal{S} = \sum_{1 \leq b \leq B} \mathcal{S}_b \qquad (11.4.70)$$

where the respective sizes of the different blocs are too heterogeneous and where each of the different blocs is composed of proteic sequences strongly homologous. Let us notice that a clustering method enables the (11.4.70) decomposition to be obtained. However the Henikoffs [8] provide this decomposition by using a successive chainings algorithm of nearest neighbour type. According to this technique, imagine that at a given step of the algorithm process the bloc $\mathcal{S}_b$ includes r aligned sequences

$$\mathcal{S}_b = \{s_{b1}, s_{b2}, \ldots, s_{bq}, \ldots, s_{br}\} \qquad (11.4.71)$$

Then, consider a new sequence s to be assigned to one of the blocks constituted and suppose that the nearest sequence s is in $\mathcal{S}_b$ for a given b. If for the latter nearest sequence s_{bq}, aligned with s, there are at least a percentage of p sites for which the same residue is in s and s_{bq}, then s will define the $(r + 1)$th element $s_{b(r+1)}$ of $\mathcal{S}_b$. *BLOSUM*62 corresponds to $p = 62$.

Relative to the decomposition (11.4.70), equal weighting will be associated with the different blocks $\mathcal{S}_b$, $1 \leq b \leq B$; each class $\mathcal{S}_b$ will be considered as a single sequence. More precisely, (11.4.54) will be replaced by

$$A_{ij} = \sum_{1 \leq b < b' \leq B} A_{ij}(b, b') \qquad (11.4.72)$$

where

$$A_{ij}(b, b') = \frac{1}{k_b k_{b'}} \sum_{(s,s') \in \mathcal{S}_b \times \mathcal{S}_{b'}} A_{ij}(s, s') \qquad (11.4.73)$$

where $k_b = card(\mathcal{S}_b)$ and $k_{b'} = card(\mathcal{S}_{b'})$, $1 \leq b < b' \leq B$.

Let us return now to the classical case. Two coefficients are associated with the distribution of the entries s_{ij}, $1 \leq i \leq j \leq 20$ (see (11.4.60)). Both consist of averaging the s_{ij} values. The first average is established with respect to the relative frequency distribution (11.4.65) and the second one, with respect to the square of the marginal frequency distribution (11.4.67). The first coefficient called "relative entropy" can be written as follows:

$$H = \sum_{1 \leq i \leq j \leq 20} Q_{ij} \qquad (11.4.74)$$

and the second one,

$$E = \sum_{1 \leq i,j \leq 20} f_i f_j s_{ij}$$ (11.4.75)

where $s_{ij} = s_{ji}$ for all (i, j), $1 \leq i \neq j \leq 20$. In this way, the sum of the probability coefficients is equal to 1.

It is on the basis of the index H that the comparison between BLOSUM and Dayhoff matrices is done [8]. However, the latter index is too global and then somewhat imprecise in order to capture finely the comparison wished.

11.4.5 The LLA Matrices

The elements required to establish an *LLA* matrix on the set of the 20 amino acids are, on the one hand, the joint empirical distribution (11.4.66) or (11.4.69) and, on the other hand, the marginal distribution (11.4.67).

The starting point of the construction of this matrix is the association coefficient between Boolean attributes which refers to the Poisson hypothesis of no relation (see (5.2.32) of Sect. 5.2.1.2 of Chap. 5). Let us recall that if $\{a_i, b_j\}$ is a pair of Boolean attributes observed on an object set $\mathcal{O}$ of size n, the correlational form of such a coefficient can be written

$$\rho_{ij} = \frac{f(i \wedge j) - f_i g_j}{\sqrt{f_i g_j}}$$ (11.4.76)

where $f(i \wedge j)$ is the proportion (relative frequency) of objects where the conjunction $a_i \wedge b_j$ is *TRUE* (a_i and b_j are present simultaneously) and where f_i (resp., g_j) is the proportion of objects where a_i (resp., b_j) is *TRUE* (a_i (resp., b_j) is present), $1 \leq i \leq I$, $1 \leq j \leq J$.

Here, let us recall that—provided that $n f_i g_j$ is not too small—$\sqrt{n} \rho_{ij}$ can be considered in the statistical independence hypothesis between a_i and b_j, as a realization of a standardized normal random variable.

Now, suppose that $\{a_i | 1 \leq i \leq I\}$ (resp., $\{b_j | 1 \leq j \leq J\}$) is the set of categories (values) of a categorical attribute a (resp., b). Crossing a with b leads to a contingency table to which we can associate the following table of relative frequencies

$$\{f(i \wedge j) | 1 \leq i \leq I, 1 \leq j \leq J\}$$ (11.4.77)

whose column (resp., row) margin is defined by $\{f_i | 1 \leq i \leq I\}$ (resp., $\{g_j | 1 \leq j \leq J\}$).

$\sqrt{n} \rho_{ij}$ was called "the *oriented* contribution of the cell (i, j) to the Chi square statistic χ^2 associated with the contingency table" (see Definition 20 in Sect. 5.2.1.2, (6.2.43) in Sect. 6.2.3.5 of Chap. 6 and (8.6.18) in Sect. 8.6.2.1 of Chap. 8). ρ_{ij} is then the contribution of the cell (i, j) to the coefficient

$$\frac{\chi^2}{n} = \sum_{1 \leq i \leq I, 1 \leq j \leq J} \left(\frac{f(i \wedge j) - f_i g_j}{\sqrt{f_i g_j}} \right)^2 \tag{11.4.78}$$

which only depends on the table of proportions (11.4.77).

In the case of interest the categorical attributes a and b are identical. Both represent the "amino acid" attribute. The value set of the latter comprises 20 elements or categories and the value table (11.4.77) becomes (11.4.66) or (11.4.69). Formally, the data table structure is analogous to Burt table [3]. A similar situation is provided by a "confusion matrix". b being identical to a, the matrix (11.4.77) becomes symmetrical.

For a quantified pairwise comparison of the amino acid set $\mathbb{A}$, the index (11.4.76) becomes

$$\rho_{ij} = \frac{f(i \wedge j) - f_i f_j}{\sqrt{f_i f_j}}, \quad 1 \leq i, j \leq 20. \tag{11.4.79}$$

In this case both matrices (11.4.77) and (11.4.79) are symmetrical. The latter one leads to an *LLA* matrix of probabilistic similarity indices (see (5.2.53) which follows (5.2.51) and (5.2.52) of Sect. 5.2.1.5 of Chap. 5). The $-\text{Log}_2$ transformation enables the *LLA* matrix to be substituted by an "Informational" dissimilarity matrix (see (7.2.7) of Sect. 7.2.1 of Chap. 7).

More particularly, we substitute for the matrix

$$\{\rho_{ij} | 1 \leq i, j \leq 20\} \tag{11.4.80}$$

a matrix of *globally* normalized indices

$$\{\rho_{ij}^g | 1 \leq i, j \leq 20\} \tag{11.4.81}$$

where the ρ_{ij}^g are deduced from the ρ_{ij} by a normalization process with respect to an adequate probability distribution on $\mathbb{A} \times \mathbb{A}$. For this reduction purpose we can take the joint distribution

$$\{f(i \wedge j) | 1 \leq i, j \leq 20\} \tag{11.4.82}$$

which corresponds to consider ρ_{ij} as observed with the relative frequency $f(i \wedge j)$, $1 \leq i, j \leq 20$. In these conditions,

$$\left\{ \rho_{ij}^g = \frac{\rho_{ij} - m_1(\rho)}{\sqrt{var_1(\rho)}}, 1 \leq i, j \leq 20 \right\} \tag{11.4.83}$$

where

$$m_1(\rho) = \sum_{1 \le i,j \le 20} f(i \wedge j)\rho_{ij}$$

and

$$var_1(\rho) = \sum_{1 \le i,j \le 20} f(i \wedge j)\big(\rho_{ij} - m_1(\rho)\big)^2$$

$$= \sum_{1 \le i,j \le 20} f(i \wedge j)\rho_{ij}^2 - \big(m_1(\rho)\big)^2 \tag{11.4.84}$$

where—let us recall once again—the role of the matrix (11.4.77) can be played by (11.4.66) if we adopt $\mathcal{D}$ (Dayhoff) or (11.4.69) if we adopt $\mathcal{H}$ (Henikoffs).

An alternative method of global reduction of the similarity table (11.4.80) can be carried out with respect to the marginal law (11.4.67). In this case, instead of (11.4.82), we consider

$$\{f_i f_j | 1 \le i,j \le 20\} \tag{11.4.85}$$

The normalized ρ_{ij} coefficient becomes

$$\left\{\rho_{ij}^h = \frac{\rho_{ij} - m_2(\rho)}{\sqrt{var_2(\rho)}}, \quad 1 \le i,j \le 20\right\} \tag{11.4.86}$$

where

$$m_2(\rho) = \sum_{1 \le i,j \le 20} f_i f_j \rho_{ij}$$

and

$$var_2(\rho) = \sum_{1 \le i,j \le 20} f_i f_j \big(\rho_{ij} - m_2(\rho)\big)^2$$

$$= \sum_{1 \le i,j \le 20} f_i f_j \rho_{ij}^2 - \big(m_2(\rho)\big)^2 \tag{11.4.87}$$

Now, let us designate by

$$\{\sigma_{ij} | 1 \le i,j \le 20\} \tag{11.4.88}$$

the matrix adopted of the globally normalized coefficients such (11.4.83) or (11.4.86). Therefore, the matrix of the probabilistic indices of the "Likelihood of the Link" is expressed as follows:

$$\{P_{ij} = \Phi(\sigma_{ij}) | 1 \le i,j \le 20\} \tag{11.4.89}$$

where Φ is the cumulative distribution function of the $\mathcal{N}(0, 1)$ standardized normal law.

The matrix of the "Informational" dissimilarities is then

$$\{\delta_{ij} = -\text{Log}_2(P_{ij})|1 \leq i, j \leq 20\} \tag{11.4.90}$$

where δ_{ij} defines the information quantity of the event whose probability is P_{ij}, $1 \leq i, j \leq 20$.

In the case where we need to work with a similarity matrix between amino acids, we may consider, according to the nature of the problem concerned, (11.4.89) or that which is deduced monotically from (11.4.90), namely,

$$\{\theta_{ij} = (\delta_{\max} - \delta_{ij})|1 \leq i, j \leq 20\} \tag{11.4.91}$$

11.4.6 LLA Similarity Index on a Set of Proteic Aligned Sequences

11.4.6.1 Introduction

To begin, we shall make explicit how to obtain coefficients of the form (11.4.83) associated with the Dayhoff and Henikoffs matrices, respectively. Let us designate by $Q_g(D)$ and $Q_g(B)$ the matrices of the (11.4.83) indices associated with the Dayhoff (250 PAM) and Henikoffs (BLOSUM62) coefficients:

$$Q_g(D) = \{Q_{gd}(i,j)|1 \leq j \leq i \leq 20\} \tag{11.4.92}$$

and

$$Q_g(B) = \{Q_{gb}(i,j)|1 \leq j \leq i \leq 20\} \tag{11.4.93}$$

The Dayhoff matrix adopted for our experiments is exactly that of Table I of page 279 of [9]. The BLOSUM62 matrix is provided by the matrix of Fig. 2 of page 10917 of [8]. In both cases, for symmetrical reasons, only the diagonal lower part, including the diagonal, is taken into account. Also, let us specify that the marginal relative frequency distribution $\{f_i|1 \leq i \leq 20\}$ (see (11.4.67)) is taken from the third column of Table III of [9]. This distribution will serve to reduce both Dayhoff and Henikoffs matrices. As expressed above, it is the joint frequency $f(i \wedge j)$ which is calculated differently according to the Dayhoff or Henikoffs versions, $1 \leq i, j \leq 20$. Let us designate by $\tau(D)$ and $\tau(B)$ the matrices considered—as just indicated—for the calculation: $\tau(D)$ for Dayhoff 250 PAM and $\tau(B)$ for BLOSUM62.

Firstly, we shall show how the matrices $Q_g(D)$ and $Q_g(B)$ (see (11.4.92) and (11.4.93)) can be derived from $\tau(D)$ and $\tau(B)$. Next, we will associate with $Q_g(D)$ and $Q_g(B)$ the "Informational" *dissimilarity* matrices (11.4.90) $\delta(D)$ and $\delta(B)$, respectively. Hence, we obtain the "Informational" (11.4.91) *similarity* matrices $\theta(D)$ and $\theta(B)$. In the framework of these matrices of similarity indices, comparison values between one deletion and a given amino acid and also, between two deletions will be proposed.

Finally, an *LLA* similarity between proteic sequences will be deduced from $\theta(D)$ and $\theta(B)$.

11.4.6.2 $Q_g(D)$ and $Q_g(B)$ Matrices

The matrix $\tau(D)$ is the matrix (11.4.52) multiplied by the factor 10. Therefore, the matrix

$$\tau(D) = \{S'(i,j)|1 \le j \le i \le 20\} \tag{11.4.94}$$

gives the matrix (11.4.51) of the R_{ij} coefficients by applying

$$R_{ij} = 10^{0.1 \times S'(i,j)}, \quad 1 \le j \le i \le 20. \tag{11.4.95}$$

In these conditions,

$$O_{ij} = f_i f_j R_{ij} = f_i f_j 10^{0.1 \times S'(i,j)}, \quad 1 \le j \le i \le 20. \tag{11.4.96}$$

From the O_{ij}—by taking into account the symmetry with respect to (i,j)—it is easy to obtain the coefficients $\rho_{ij}(D)$ (see (11.4.79)); this, since $f(i \wedge j)$ is identical to O_{ij}, $1 \le i,j \le 20$. The statistical normalization described in (11.4.81)–(11.4.84) leads to the indices (11.4.83) that we have denoted by $Q_{gd}(i,j)$, $1 \le j \le i \le 20$.

The matrix $\tau(B)$ is that already designated by $\mathcal{H}$ (see (11.4.61)). The $f(i \wedge j)$ of $\rho_{ij}(B)$ (see (11.4.79)) corresponds to O'_{ij} (see (11.4.68) and (11.4.69)), $1 \le j \le i \le 20$. By considering the expressions (11.4.59) and (11.4.60), we have for any (i,j), $1 \le j \le i \le 20$,

$$O'_{ij} = f_i f_j 2^{0.5 h_{ij}} \tag{11.4.97}$$

where h_{ij} is the (i,j)th term of the matrix $\tau(B)$ (11.4.61).

From the O'_{ij} we get the ρ_{ij} (see (11.4.79)) associated with B. These give the $\rho_{ij}^g(B)$ (see (11.4.83)) denoted here by $Q_{gb}(i,j)$, where the O'_{ij} are substituted for the $f(i \wedge j)$, $1 \le i,j \le 20$. Thus we obtain the matrix $Q_g(B)$ (see (11.4.93)).

11.4.6.3 The "Probabilistic" or "Informational" Similarities on the Amino Acid Set

According to (11.4.89) the probabilistic similarity indices associated with $Q_g(D)$ and $Q_g(B)$, respectively, are defined in the matrices

$$P(D) = \{P_d(i,j) = \Phi\big(Q_{gd}(i,j)\big| 1 \le i, j \le 20\big)\} \tag{11.4.98}$$

and

$$P(B) = \{P_b(i,j) = \Phi\big(Q_{gb}(i,j)\big| 1 \le i, j \le 20\big)\} \tag{11.4.99}$$

Now, following (11.4.90), the matrices of informational dissimilarities can be written

$$\delta(D) = \{\delta_d(i,j) = -\text{Log}_2\big(P_d(i,j)\big)| 1 \le i, j \le 20\} \tag{11.4.100}$$

and

$$\delta(B) = \{\delta_b(i,j) = -\text{Log}_2\big(P_b(i,j)\big)| 1 \le i, j \le 20\} \tag{11.4.101}$$

By considering, according to (11.4.91) the maximal values $\delta_{\max}(D)$ and $\delta_{\max}(B)$ of each of the previous matrices $\delta(D)$ and $\delta(B)$ (see (11.4.100) and (11.4.101)), we substitute for each of them, respectively, the associated matrices of informational similarities, namely,

$$\theta(D) = \{\theta_d(i,j) = \big(\delta_{\max}(D) - \delta_d(i,j)\big)| 1 \le i, j \le 20\} \tag{11.4.102}$$

and

$$\theta(B) = \{\theta_b(i,j) = \big(\delta_{\max}(B) - \delta_b(i,j)\big)| 1 \le i, j \le 20\} \tag{11.4.103}$$

It is on the basis of these matrices that we will establish our similarity indices between proteic sequences. For this purpose, it is necessary to be able to compare—according to θ_d (resp., θ_b)—a given amino acid with a deletion and also, two deletions between them. In other words it is question to infer from $\theta(D)$ (resp., $\theta(B)$) a similarity matrix on the alphabet A' composed of 21 letters, including the deletion symbol "−".

Let us designate by θ the similarity index adopted on A. $\theta = \theta_d$ (11.4.102) or $\theta = \theta_b$ (11.4.103). On the other hand, let Z_i be one of the 20 amino acids, $1 \le i \le 20$. In order to resolve the inference problem submitted, two comparison cases have to be considered: $(Z_i, -)$ and $(-, -)$.

The estimation of these associations rests on the following interpretation of the substitution event of a given amino acid Z into a deletion $-$:

"Z was about to be transformed into another amino acid, but this did not happen"

In these conditions, for comparing an amino acid Z_i with a deletion $-$, we propose a *weighted* average of the association indices between Z_i and the *others* amino acids. This *weighting*, precisely, considers the passage probabilities between Z_i and the others amino acids. Thereby, we have

$$\theta(Z_i, -) = \frac{1}{(f_i - f(i \wedge j))} \sum_{j|j \neq i} f(i \wedge j)\theta(i, j) \qquad (11.4.104)$$

where $f(i \wedge j)$ has to be replaced either by O_{ij} (11.4.96) if we refer to $\theta(D)$ (11.4.102) or by O'_{ij} (11.4.97) if we refer to $\theta(B)$ (11.4.103). We may indicate here that this point of view is essentially different from that considered in [7].

Now, according to the interpretation above, comparing a deletion with itself has the same nature as that of comparing an amino acid with itself. But, in the case concerned, the latter is not specified. Therefore, we will take the average—weighted by the exposure frequencies—of the similarity indices (θ_d or θ_b) between an amino acid and itself. Thus, we obtain

$$\theta(-, -) = \sum_{1 \leq i \leq 20} f_i\theta(i, i) \qquad (11.4.105)$$

Let us mention that we consider in [21] the case where an amino acid has not been possible to identify. However, this case is extremely rare and then will not be considered here. In these conditions, we end up with an alphabet $\mathbb{A}'$ including 21 symbols:

$$\mathbb{A}' = \{A, C, D, E, F, G, H, I, K, L, M, N, P, Q, R, S, T, V, W, Y, -\}$$
$$(11.4.106)$$

The attribute whose value set is $\mathbb{A}'$ is a categorical attribute a valuated by a numerical similarity (see Sect. 3.3.4 of Chap. 3). Four versions of this similarity are considered in [21]. There are associated with $\tau(D)$ (denoted also by $\mathcal{D}$ (see (11.4.52))), $\tau(B)$ (denoted also by $\mathcal{H}$ (see (11.4.61))), $\theta(D)$ (11.4.102) and $\theta(B)$ (11.4.103), respectively. a may designate the attribute concerned.

To each numerical version of the similarity, we have associated an ordinal version. a becomes a *preordonance* categorical attribute (see Sect. 3.4.1 of Chap. 3). Thereby, a induces a total preorder on the set

$$C = \{(Z_i, Z_j)|1 \leq j \leq i \leq 21\} \qquad (11.4.107)$$

representing the subset of ordered pairs of elements of $\mathbb{A}'$ to be compared. By ranking from left to right these pairs, the position of a given pair (Z_i, Z_j) is all the more to the right since Z_i and Z_j are more similar. This position will be coded by the "mean rank" function (see (3.4.8) of Sect. 3.4.1 of Chap. 3). Notice that the numerical or ordinal valuation concerning the comparison of an amino acid with itself, is not constant. On the other hand, the latter valuation is not necessarily always greater than that associated with the comparison of two different amino acids. This property has no effect on the association method for building similarity indices between proteic sequences.

11.4.6.4 Similarity *LLA* Indices Between Proteic Sequences

Eight *LLA* cluster processings have been considered in [21]. They are grouped two by two. For a given numerical similarity matrix on $\mathbb{A}'$, the first treatment concerns the numerical valuation and the second one, an ordinal similarity in terms of a preordonance on $\mathbb{A}'$, associated with the numerical similarity. As implicitly indicated above, the numerical similarity matrices considered are $\tau(D)$ (denoted also by $\mathcal{D}$ (see (11.4.52))) and $\tau(B)$ (denoted also by $\mathcal{H}$ (see (11.4.61))) on the one hand, $\theta(D)$ (11.4.102) and $\theta(B)$ (11.4.103) on the other hand. Denoting by *NM* for numerical and *PO* for "preordonance", the four pairs of treatments can be indicated as follows:

$$NM\big(\tau(D)\big), PO\big(\tau(D)\big)$$
$$NM\big(\tau(B)\big), PO\big(\tau(B)\big)$$
$$NM\big(\theta(D)\big), PO\big(\theta(D)\big)$$
and
$$NM\big(\theta(B)\big), PO\big(\theta(B)\big) \tag{11.4.108}$$

The only difference between a *PO* type processing and a *NM* one resides in the fact that the numerical valuation assigned to an element of C (see (11.4.107)) is replaced by its rank according to the "mean rank" function. The interest of the latter ordinal coding associated with a numerical similarity consists of observing the stability of the classification results. And, in the case of no perfect stability, the matter is to discuss both codings and possibly to choose between them.

The set of proteic sequences is interpreted here as a set of objects. The multiple alignment makes possible to associate a descriptive attribute with each site (see (11.4.2)). As seen above, this attribute is a categorical one whose value set is $\mathbb{A}'$, where $\mathbb{A}'$ is endowed with a numerical or ordinal similarity. Clearly, as expressed above (see what follows (11.4.2)), it is the variability of the behaviour of this attribute through the different sites which makes the classification.

Although the different sites have not the same biological importance and that there exists complex relationships between them, the similarity index we consider is built such that the same discrimination power is given to each of the attribute sites. This alternative is the most logical one in the case of no knowledge. In fact, it is the hidden statistical relationships between the different sites which enable the classification to emerge.

Let us designate by

$$\{\sigma_{ij} | 1 \le j \le i \le 21\} \tag{11.4.109}$$

one of the similarity matrices on $\mathbb{A}'$. As defined above, the latter is provided from $\tau(D)$, $\tau(B)$, $\theta(D)$, $\theta(B)$ or the associated respective preordonances. In the latter case, as expressed above, the value of the mean rank function is substituted for the numerical similarity value.

Consider now a pair $\{s_k, s_{k'}\}$ of sequences in the set S of aligned sequences (see (11.4.11)). We shall specify the contribution of the lth site to the comparison between s_k and $s_{k'}$, $1 \leq l \leq L$. According to the notations introduced (see (11.4.10)), $\{s_k(l), s_{k'}(l)\}$ is the pair of amino acids encountered. Then following the construction of Sect. 7.2.4 of Chap. 7, begin by defining the *raw* contribution of the site l to the comparison between s_k and $s_{k'}$ as

$$c^l(k, k') = \sigma\big(s_k(l), s_{k'}(l)\big) \qquad (11.4.110)$$

(see (7.2.25) of Sect. 7.2.4 of Chap. 7).

The next step consists of centring and reducing this raw index with respect to the set $S \times S$ of all ordered sequence pairs. Thereby, the *normalized* contribution of the site l to the comparison between s_k and $s_{k'}$, can be written

$$C^l(k, k') = \frac{c^l(k, k') - mean_e(c^l)}{\sqrt{var_e(c^l)}} \qquad (11.4.111)$$

where $mean_e(c^l)$ and $var_e(c^l)$ are the mean and variance of $c^l(k, k')$ (11.4.110) on the ordered sequence pairs; that is, on $\underline{K} \times \underline{K}$ where $\underline{K} = \{1, 2, \ldots, k, \ldots, K\}$. In Sect. 7.2.4 of Chap. 7 a synthetic expression for (11.4.111) is proposed. In the latter the statistical distribution of the lth attribute a^l takes part.

The sum of normalized coefficients for comparing s_k and $s_{k'}$ becomes

$$C(k, k') = \sum_{1 \leq l \leq L} C^l(k, k') \qquad (11.4.112)$$

The latter coefficient is statistically globally normalized in order to obtain

$$Q_s(k, k') = \frac{C(k, k') - mean_e(C)}{\sqrt{var_e(C)}} \qquad (11.4.113)$$

where $mean_e(C)$ and $var_e(C)$ are the mean and variance of C on the set of unordered sequence pairs (see (11.4.11)), that is, on the set $P_2(\underline{K}) = \{(k, k')|1 \leq k < k' \leq K\}$.

In these conditions, the *LLA* probabilistic index can be written as follows:

$$P(k, k') = \Phi(Q_s(k, k')) \qquad (11.4.114)$$

where, as usually, Φ is the cumulative normal function distribution $\mathcal{N}(0, 1)$, $1 \leq k < k' \leq K$.

It is the table of values (11.4.114) which is given as argument to the *LLA* criterion for building a hierarchical clustering on S (see Sect. 10.4.1 of Chap. 10). A parametrization of the family of criteria of the maximal likelihood linkage by a real positive number ϵ, $0 \leq \epsilon \leq 1$, is considered (see (10.4.7) and (10.4.8) of Sect. 10.4.1.2 of Chap. 10). The value 0.5 is often used in our experiments (see in Sect. 10.4.1.2 of Chap. 10 what follows (10.4.8)).

```
2.2785139
1.2574933 3.0485260
1.4461807 1.0511439 2.8774893
1.0902845 0.9095557 2.7170441 2.9560292
0.7150888 1.4552970 0.6127312 0.5225282 3.0453665
1.8570709 1.2616836 1.7922755 1.4450164 0.4210062 3.0091465
1.0631216 1.4570786 1.4536560 1.4528335 1.4547640 1.0715559 2.5683739
1.4460235 1.1587363 0.7603540 0.6872585 1.4506751 0.6052425 1.0143239 2.6716611
1.1000243 1.0221388 1.4482390 1.7846550 0.5489843 1.1076385 1.6645865 0.7086759
2.9363647
1.0048492 0.9048023 0.2785048 0.1478879 2.1351538 0.0000000 1.0255015 2.2156837
0.4351146 3.0401185
1.2370453 1.2620064 1.0085783 0.9630430 1.4545969 0.9122524 1.2468621 2.1428313
1.1082051 2.3037999 2.5975044
1.4476792 1.3108960 2.0106878 1.7405829 0.9218428 1.4479743 1.6361873 1.0048938
1.7342013 0.6063662 1.1630094 2.2519710
1.7953781 1.1648540 0.9655890 0.9153895 0.8685930 1.1341295 1.4537466 0.9603977
0.9300714 1.4449174 1.1346551 1.2159203 2.9969170
1.1637790 1.1003690 1.7128764 2.0415778 0.8171082 1.1699888 1.8362802 0.8474665
2.0294731 0.8631580 1.1704556 1.4517292 1.4507571 2.7392020
1.1271187 1.2969317 1.1900305 1.4480207 0.7333012 1.4466629 1.8750945 0.7679510
2.6944678 0.5182908 1.1346551 1.4504694 1.1927974 1.9950739 2.8674059
1.8447546 1.6662111 1.4470628 1.1114709 0.9546423 1.8379109 1.2542595 1.1395745
1.1206255 0.6603217 1.2495892 1.7548249 1.7789348 1.1805944 1.1461017 2.2118475
2.1980586 1.2839677 1.1682619 1.1395298 0.9961412 1.1076385 1.2705377 1.7620621
1.1479177 1.0660927 1.4528675 1.7342013 1.7567459 1.2028949 1.1712632 1.8009442
2.1229463
1.8371065 1.1212186 0.8901660 0.8316810 1.4492844 0.7662575 0.9575734 2.7432480
0.6087465 2.2825398 1.9342194 0.9468505 1.1534717 0.7675209 0.6766317 1.0995551
1.4464102 2.8104279
0.9503884 1.5566432 0.9884725 0.9407687 1.3409066 1.1623911 1.2413585 1.0443201
1.0989610 1.1276033 1.2363904 1.0348313 1.0528978 1.1629117 1.4557014 1.0671469
1.0193918 1.0764722 3.0506892
0.8029671 1.7753067 1.0762570 0.7446253 2.6495769 0.6692612 2.2285907 1.0722921
0.8934274 1.1762559 1.2061894 1.2691501 0.9371850 1.1223230 1.0802619 1.2150354
0.8934274 0.8570857 1.4571486 3.0472760
1.3585238 0.9379746 1.3880049 1.2226409 1.0659977 0.9782194 1.3493228 1.4569216
1.2125098 1.0078677 1.4522192 1.3398169 1.0926322 1.2838297 1.2843280 1.2411964
1.3399767 1.4014330 0.6181077 1.0512260 2.7385223
```

Fig. 11.5 Matrix $NM(\theta(D))$: $\{(i, j)|1 \le j \le i \le 21\}$

11.4.7 Some Results

11.4.7.1 Preamble

To begin, the *LLA* methodology was applied [21] in order to evaluate and compare the two matrices $\tau(D)$ and $\tau(B)$ associated with the Dayhoffs and the Henikoffs approaches (see Sect. 11.4.7.1). For this purpose the normalized *LLA* association coefficients on $\mathbb{A}$, defined by the matrices $Q_g(D)$ and $Q_g(H)$ (see (11.4.92) and (11.4.93)) are used.

Second, *LLA* clustering was carried out on a set of 89 cytochromes c provided from different living organisms. The general form of the probabilistic similarity coefficient between two sequences s_k and $s_{k'}$ is given by (11.4.114). The latter is obtained from $\{c^l(k, k')|1 \le l \le L\}$ (see (11.4.110)–(11.4.113)) where σ in (11.4.110) is defined from a numerical or ordinal matrix on $\mathbb{A}'$. As expressed above (see (11.4.108)) the numerical similarity matrix may be $\tau(D)$, $\tau(B)$, $\theta(D)$ or $\theta(B)$. For a given one, the associated ordinal similarity is derived by substituting a ranking for the numerical

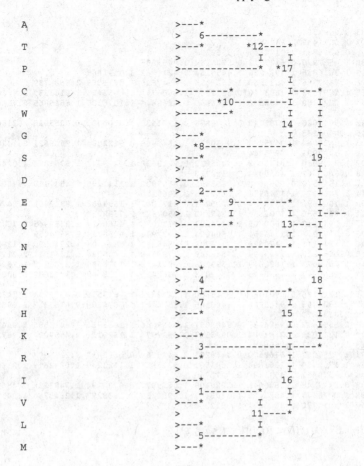

Fig. 11.6 *LLA* classification tree on $\mathbb{A}$ associated with $Q_g(D)$

similarity. Thus, eight versions of the similarity on $\mathbb{A}'$ are considered. To each of them corresponds a classification tree on S. The eight dendrograms were mutually compared [21]. The more consistent result was obtained with $\theta(D)$ (see Fig. 11.5). We will present briefly in Sect. 11.4.7.3 the results obtained.

11.4.7.2 *LLA* Clustering of the Amino Acid Set

The *LLA* hierarchical clusterings on $\mathbb{A}$ get from $Q_g(D)$ and $Q_g(B)$ (see (11.4.92) and (11.4.93)), respectively, are clearly not identical (see Figs. 11.6 and 11.7). They are different with respect to some aspects. We have not at one's disposal available criterion to compare both results and to decide which of them is the best one. The practice and the intuition of the Biologist are the only guides. According to these, we may state that no usual clusters are obtained with $Q_g(B)$. More particularly, the

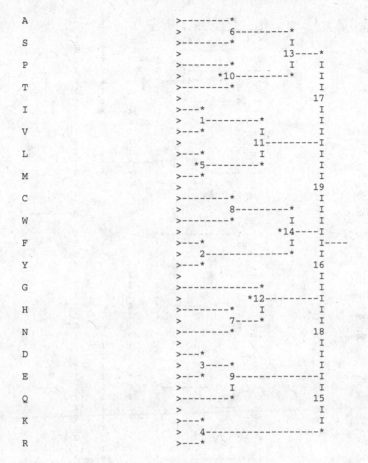

Fig. 11.7 *LLA* classification tree on A associated with $Q_g(B)$

respective associations (P, T) and (H, N) cannot be understood. On the contrary, nothing is shoking in the cluster associations of the classification tree provided from $Q_g(D)$.

In conclusion, there is not an objective criterion enabling us to decide what is the best of the matrices $\tau(D)$ and $\tau(B)$, through their *LLA* treatments. However, the practice and intuition of the biologist makes $Q_g(D)$ preferable.

11.4.7.3 Proteic Sequences Clustering with $NM(\theta(D))$

A selective clustering method is expected to group together sets of affiliated sequences. For example, the "bacteriums" must give rise to a cluster (possibly sub-divided into subclusters). Other clusters have to correspond to the "insects", the "animals", the "plants", etc. Besides, a sensitive method is required to recognize

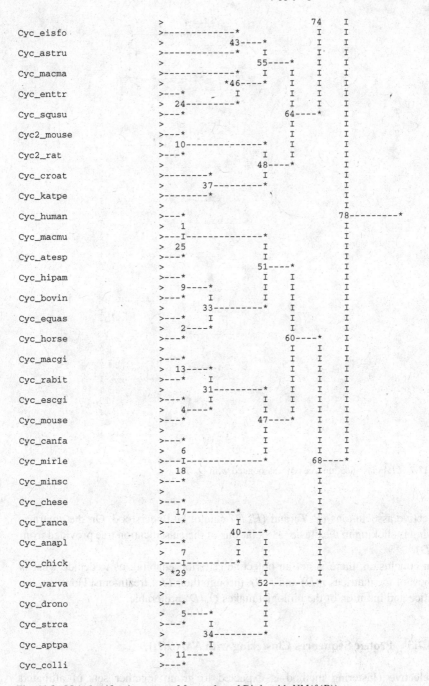

Fig. 11.8 *LLA* classification tree on Mammals and Birds with $NM(\theta(D))$

coherent subclasses. For example, in the cluster of animals, the "birds" must be separated of the "mammals".

Let us notice that the cytochrome sequences are very well preserved. There are only in all 16 differences between the *man* and the *tortoise*. Even more, these differences correspond to substitutions between near amino acids.

Otherwise, we cannot expect a perfect classification tree; because, some species are very weakly represented. For example, there is a single specimen for the "nematodes" (kind of small worm). It follows that nematode sequence might agglomerate its nearest group.

The aim of the *LLA* hierarchical clustering is not to produce a phylogeny of the species. Indeed, there is not direct reason for which the respective lengths of the classification tree branches are proportional to the evolution time. Nevertheless, phylogenetic classifications on the same set of sequences were carried out (see in [21]). The methods employed "Neighbor Joining" and "UPGMA" are provided from the PHYLIP software which is very known in the Biologist community. The results obtained were not satisfactory.

The whole tree on the set of 89 sequences is reported in [21]. By deleting of the latter the microorganisms and the plants, we obtain the tree of Fig. 11.8.

- Two distinct classes are highlighted, the first one includes the "birds" and the second one the "mammals". The "varanus" (lizard) and the "tortoise" are correctly associated with the "birds". The "frog" should not be in this cluster and the "kangaroo" should be more clearly separated from the "mammals". Nevertheless, the separation mentioned into two distinct classes is completely satisfactory.
- A third class including the cytochromes C2 (and not the cytochrome C) of the "rat" and the "mouse", emerges clearly. As a matter of fact it gathers a variety of pluricellular organisms corresponding to sequences which are not associated with neither "birds" nor "mammals". Possibly, we would prefer to observe the "rattlesnake" nearer the "varanus" than the "tortoise".

According to the Biologist view, the clustering of the set of proteic sequences obtained by the *LLA* method employing the similarity matrix $NM(\theta(D))$ on $\mathbb{A}'$ is extremely satisfactory. This approach turns out to be sensitive and selective.

11.5 Specific Results in Clustering Categorical Attributes by *LLA* Methodology

11.5.1 Structuring the Sets of Values of Categorical Attributes

We shall emphasize in this section the importance of the structure retained on the respective value sets of categorical descriptive attributes. For this purpose, we take up again here an illustration provided from [25], which corresponds to a specific aspect of a large survey on public opinion in France, led in 1989. This survey was carried out by an organization called "Agoramétrie" (AESOP/AGORAMETRIE *Les structures de l'opinion 1978 et 1988*) which was created by two essential institutions providing energy in France: the "EDF" (*Electricité De France*) and the "CEA" (*Commissariat à l'Energie Atomique*).

The reason of the complete survey was to analyze the conflicts underlying the public opinion. The data table is defined by the description of 500 individuals representing the French population by means of 250 descriptive attributes. These the more often than not are ordinal categorical attributes. Precisely, we will concerned here by a small part of the questionary, giving rise to categorical attributes which cannot be interpreted as ordinal categorical attributes.

A set of 19 politicians representing the political scene of France in the period concerned, is considered. There are (in alphabetic order):

R. Barre, P. Beregovoy, J. Chirac, J. Delors, V.G. D'Estaing, L. Fabius, L. Jospin, B. Lalonde, J. Lang, F. Léotard, J.M. LePen, G. Marchais, P. Méhaignerie, F. Mitterand, M. Noir, M. Roccard, M. Séguin, S. Veil and A. Waechter.

Relatively to each of these politicians, the following question is asked:
Do you wish this politician play an important role?
The respective choices proposed are

1. **YES**;
2. **NO**;
3. **NO ANSWER**.

This question attached with a given political personality defines a categorical attribute with three values **YES**, **NO** and **NO ANSWER**. Thereby 19 categorical attributes respectively associated with 19 political personalities are obtained.

For these, the classical and mostly used coding consists of no retaining structure on the value set of each of the different attributes. Thus, each of the latter is interpreted as a nominal categorical attribute (see Sect. 3.3.1 of Chap. 3).

A richer interpretation of the semantics sustaining the value set of a given attribute consists of assuming an ordinal similarity on the latter set. In these conditions, the attribute concerned is considered as a *preordonance* categorical attribute (see Sect. 3.3.4 of Chap. 3).

This ordinal similarity is perceived by the expert. Clearly, the categories "**YES**" and "**NO**" are the most dissimilar. Otherwise, the category "**NO ANSWER**", which reflects indifference or even rejection, is estimated nearer "**NO**" than "**YES**". Moreover, we have opt for a maximal ordinal similarity between a category and itself, whatever is the latter. In these conditions, we get the following preordonance

$$\{1, 2\} < \{1, 3\} < \{2, 3\} < \{1, 1\} \sim \{2, 2\} \sim \{3, 3\} \qquad (11.5.1)$$

Notice that the extreme right preorder class composed of tied pairs can be refined. The proposed preordonance becomes

$$\{1, 2\} < \{1, 3\} < \{2, 3\} < \{3, 3\} < \{1, 1\} < \{2, 2\} \qquad (11.5.2)$$

This because an answer "**NO**" is more accentuated than an answer "**YES**". On the other hand, an answer "**YES**" is more specific than an answer "**NO ANSWER**".

In order to cluster the set of 19 categorical attributes, the *LLA* method was employed for each of both representations: that classical for which any structure is assumed on the value set of each of the categorical attributes and that, where we suppose given an ordinal similarity (preordonance) on each of these sets (see (11.5.1)).

For the first coding, the association coefficient used, was denoted by $R''(\pi, \chi)$ (see (6.2.35) of Sect. 6.2.3.5 of Chap. 6). For the latter, the basic coefficient is $Q_1(\pi, \chi)$ (see (6.2.32)). π and χ are the partitions induced by the two categorical attributes on the set of 500 individuals. The partitions to be mutually compared are all into three classes.

For the second coding, the association coefficient between two preordonance categorical attributes corresponds to a particular case of the coefficient $R''(\xi, \eta)$ (see (6.2.92)), established to compare two valued binary relations on an object set. The adaptation to the preordonance case is specified in Sect. 6.2.5.5. Relative to the notations of the latter section (see (6.2.114) and (6.2.115)), we have $K = M = 3$ and the respective preordonances on the value sets of ξ and η are identical. However, clearly, the statistical distributions of ξ and η are generally distinct. The ξ (resp., η) valuation can be instantiated by using the "mean rank" function (see (3.4.8) of Sect. 3.4.1 of Chap. 3).

The first and classical coding leads to the classification tree of Fig. 11.9. In this case, the political right and left have certainly some influence in the groupings observed. However, these two sides of the policy scene are not really separated. The main tendency in the groupings or separations might be explained by a rigor or even a sectarianism attitude around patriotism including nationalism and attachment to land, face to open mindedness. Thereby, we observe associations such [J.M. LePen, G. Marchais], [F. Mitterand, M. Roccard], [P. Méhaignerie, M. Séguin], [P. Beregovoy, J. Delors], [S. Veil, R. Barre] and [B. Lalonde, A. Waechter].

The classification tree provided in Fig. 11.10 corresponds to the case for which the coding adopted is in terms of preordonances (see (11.5.1)). In this case, the two political sides (*right* and *left*) are clearly separated. Besides, in each of these political groups, the traditional generation and the younger rising one are clearly distinguished. Thus, for the right side, the more classical generation is represented by [J. Chirac, V.G. D'Estaing, R. Barre, S. Veil], that younger one with a new focus, by [F. Léotard, M. Noir, M. Séguin, P. Méhaignerie]. Now, for the left political group, the traditional generation is instantiated by [F; Mitterand, M. Roccard, P. Beregovoy, J. Delors] and the younger one by [L. Fabius, L. Jospin, J. Lang]. A last cluster [J.M. LePen, G. Marchais, B. Lalonde, J. Waechter] gathers the extreme facets of the French political scene. In these, *nationalism* and *attachment to land* take an important part.

The comparison between both trees (see Figs. 11.9 and 11.10) shows how much the relational structure retained for mutually comparing the elements of the value set of a given descriptive attribute, is important.

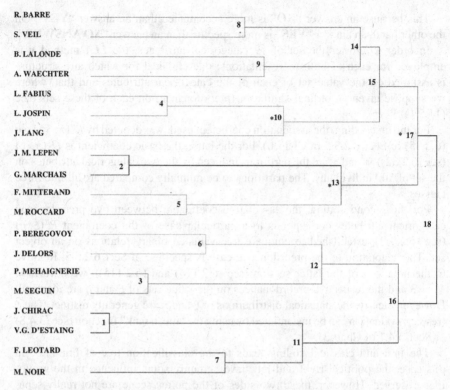

R. BARRE
S. VEIL
B. LALONDE
A. WAECHTER
L. FABIUS
L. JOSPIN
J. LANG
J. M. LEPEN
G. MARCHAIS
F. MITTERAND
M. ROCCARD
P. BEREGOVOY
J. DELORS
P. MEHAIGNERIE
M. SEGUIN
J. CHIRAC
V.G. D'ESTAING
F. LEOTARD
M. NOIR

Fig. 11.9   A first tree organizing the policy opinion of France in 1989

11.5.2   From Total Associations Between Categorical Attributes to Partial Ones

In this section, we shall point out situations for which *partial* (and not *total*) associations coefficients are more relevant for aggregating categorical attributes. In the real case presented below, ordinal categorical attributes are concerned (see Sect. 6.2.6.3). This illustration is provided from the thesis work [27]. It is reported in [17].

Relative to clustering by the *AAHC* method a given set—denoted by $\mathcal{B}$ below—of ordinal categorical attributes, we shall compare two classification trees on $\mathcal{B}$. The first one results from applying the index (6.2.59) of Sect. 6.2.4.3. The second tree is obtained from the partial association coefficients expressed in Sect. 6.2.6 and developed in [15]. It is the *local* version of the hypothesis of no relation which is employed in order to build the association coefficient between categorical attributes (see Sect. 6.2.6). In this framework—as mentioned in Sect. 6.2.6.3—if c and d indicate two ordinal categorical attributes to compare and if a is the ordinal categorical attribute to neutralize through the association between c and d, the ordinal structure underlying the value set of a is not taken into account. More precisely, if ω and ϖ are

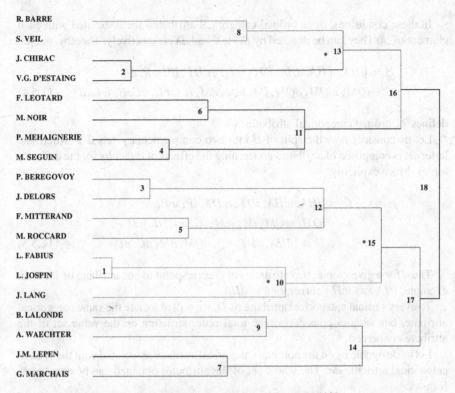

R. BARRE
S. VEIL
J. CHIRAC
V.G. D'ESTAING
F. LEOTARD
M. NOIR
P. MEHAIGNERIE
M. SEGUIN
P. BEREGOVOY
J. DELORS
F. MITTERAND
M. ROCCARD
L. FABIUS
L. JOSPIN
J. LANG
B. LALONDE
A. WAECHTER
J.M. LEPEN
G. MARCHAIS

Fig. 11.10 A second tree organizing the policy opinion of France in 1989

the two total preorders on the set of objects $\mathcal{O}$ induced by c and d (see Sect. 3.3.2) and if α is the total preorder induced by a, we only retain from α the associated partition of $\mathcal{O}$, whose classes are the classes of α.

In the real example considered here, $\mathcal{O}$ corresponds to a sample of students of a given graduate class of Polytechnic School (Palaiseau, France), $n = card(\mathcal{O}) = 198$. The descriptive attributes are ordinal categorical (see Sect. 3.3.2). The number of categories per attribute is not the same for all of the attributes. It is around 4–5. These attributes are associated with questions of ordinal evaluation concerning the following taught topics: "Hilbertian Analysis", "Distribution Theory", "Probability", "Quantum Mecanics", "Mecanics", "Chemistery" and "Informatics". They will be denoted more briefly by HA, DT, P, QM, M, C and I. Their whole set can be designated by $\mathcal{A}$ where

$$\mathcal{A} = \{HA, DT, P, QM, M, C, I\} \qquad (11.5.3)$$

For each of the $\mathcal{A}$ subjects, three principal dimensions are considered:

"*interest*" for, "*work*" and effects of "*difficulty*"

In these conditions, three ordinal categorical attributes are associated with each element of $\mathcal{A}$. They can be denoted by iX, wX and dX, respectively. Thereby, the set

$$\mathcal{B} = \{iHA, wHA, dHA, iDT, wDT, dDT, iP, wP, dP$$
$$iQM, wQM, dQM, iM, wM, dM, iC, wC, dC, iI, wI, dI\} \quad (11.5.4)$$

defines 21 ordinal categorical attributes.

Let us consider now the split of $\mathcal{B}$ into two complementary sets $\mathcal{C}$ $\mathcal{D}$ where the latter set is composed of attributes concerning the effects of *difficulty* for the different topics. More explicitly,

$$\mathcal{C} = \{iHA, wHA, iDT, wDT, iP, wP,$$
$$iQM, wQM, iM, wM, iC, wC, iI, wI\}$$
$$\mathcal{D} = \{dHA, dDT, dP, dQM, dM, dC, dI\} \quad (11.5.5)$$

Thus, for a given topic, two attributes of $\mathcal{C}$ correspond to one attribute of $\mathcal{D}$; as an example, iHA and wHA correspond to dHA.

To every ordinal categorical attribute of $\mathcal{D}$, we will associate the same categorical attribute, but without considering the total order structure on the value set of the attribute concerned.

Let us designate by δa the nominal categorical attribute associated with the ordinal categorical attribute da. The whole set of the attributes obtained can be denoted as follows:

$$\Delta = \{\delta HA, \delta DT, \delta P, \delta QM, \delta M, \delta C, \delta I\} \quad (11.5.6)$$

To each categorical attribute δa belonging to Δ, we will associate the partition π_a of the object set $\mathcal{O}$ which represents it (see (3.3.2)) of Sect. 3.3.1.

Let us consider now partial comparison between two attributes b and c of $\mathcal{C}$, neutralizing the influence of $\mathcal{D}$. Two cases have to be distinguished:

1. Both attributes concern the same topic. These are necessarily of the form ia and wa, where $a \in \mathcal{A}$ (see (11.5.3)). In this case, the nominal categorical attribute δa is neutralized. In these conditions, the respective and separated contributions of the different classes of π_a are calculated. These are globalized by means of a weighted average.
2. The two attributes b and c are relative to two different subjects. The attribute pair $\{b, c\}$ corresponds necessarily to one of the following forms:

$$\{iX, iY\}, \{wX, wY\} \text{ or } \{iX, wY\}$$

where X and Y are two distinct elements of $\mathcal{A}$.

Fig. 11.11 Total association

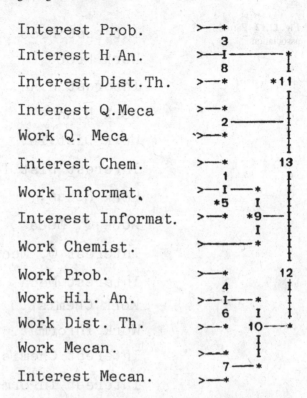

If δX and δY designate the nominal categorical attributes taken in Δ (see (11.5.6)) and respectively associated with X and Y, the neutralization is carried out with respect to the joint partition $\pi_X \wedge \pi_Y$, where π_X (resp., π_Y) is the partition of $\mathcal{O}$ induced by δX (resp., δY).

In both of the classification trees of Figs. 11.11 and 11.12 obtained by the *AAHC* method, the set $\mathcal{D}$ of the attributes concerning the *difficulty* dimension do not appear. The tree of Fig. 11.11 is built on the basis of *total* association coefficients. That of Fig. 11.12, on the basis of *partial* association coefficients, neutralizing the difficulty dimension (see above).

It appears clearly from the tree of Fig. 11.11 that the higher is the place of Calculus in the topic concerned, the lower is the association degree between "interest" and "work" for the same field. This observation is instantiated for "Hilbertian Analysis", "Distribution Theory" and "Probability". On the contrary, "work" follows rather well "interest" for the other subjects: "Quantum Mechanic", "Mechanic", "Chemistry" and "Informatics". Notice that for any of the latter fields, the objective of the discipline involved concerns mainly development of applied science.

In the second tree (see Fig. 11.12), where partial association coefficients are used, "work" follows "interest" for every topic. This correspondence is now very clear, particularly for subjects where calculus takes a large part ("Hilbertian Analysis", "Probability", "Distribution Theory").

Fig. 11.12 Partial
association

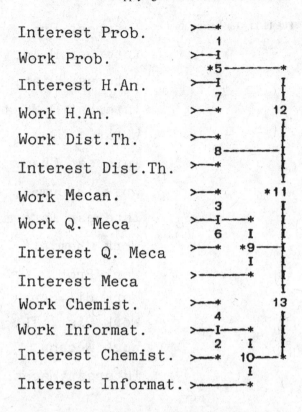

Both trees (see Figs. 11.11 and 11.12) are in fact reduced to their levels where "significant" nodes appear. These nodes are indicated by stars (⋆).

The detection method of "significant" levels and nodes of a classification tree is expressed in Sect. 9.3.6. An illustration of this method is given in Sect. 9.3.6.2.

This reduction technique of the tree levels to those where a "significant" node appears is applied in the tree drawings of Figs. 11.11 and 11.12. The principle of this drawing is the following.

Let $l_1, l_2, \ldots, l_s, l_{s+1}, \ldots, l_S$ be the levels where a "significant" node appears

$$1 \leq l_1 < l_2 < \cdots < l_s < l_{s+1} < \cdots < l_S < L$$

where L is the last tree level.

- The nodes constituted between the levels 1 and $l_2 - 1$ of the tree appear at the first physical level of the tree representation;
- The nodes constituted between the levels l_s and $l_{s+1} - 1$ of the tree appear at the sth physical level of the tree, $1 \leq s \leq S - 1$;
- The nodes constituted between the levels l_S and the last level L, appear at the Sth physical level of the tree.

In the example of Fig. 11.11 (resp., Fig. 11.12), the sequence $(l_1, l_2, \ldots, l_s, l_{s+1}, \ldots, l_S)$ is (5, 9, 11) (resp., (5, 9, 11)).

References

1. Adoue, V.: Élaboration d'un logiciel d'explication de classes pour une classification de données génotypiques. Rapport DESS-CCI, IFSIC-Université de Rennes1, September 2003
2. Bachar, K., Lerman, I.C.: Fixing parameters in the constrained hierarchical classification method: application to digital image segmentation. In: Banks, D., et al. (eds.) Classification Clustering and Data Mining Applications, pp. 85–94. Springer, Berlin (2004)
3. Burt, C.: The factorial analysis of qualitative data. Br. J. Stat. Psychol. **3**(5), 166–185 (1950)
4. Cerdeira, J.O., Martins, J., Silva, P.C.: A combinatorial approach to assess the separability of clusters. J. Classif. **29**(1), 7–22 (2012)
5. de Lauwe, M.-J.C., Bellan, Cl.: Enfants de l'image. PAYOT (1979)
6. Dayhoff, M.O., Eck, R.V., Park, C.M.: A model of evolutionary change in proteins. Atlas Protein Seq. Struct. **5**, 89–99 (1972)
7. Gonnet, G.H.: A tutorial introduction to computational biochemistery using Darwin. Research Report 24, Informatik E.T.H., Zurich, Switzerland, November 1992
8. Henikoff, S., Henikoff, J.G.: Amino acid substitution matrices from protein blocks. Proc. Natl. Acad. Sci. USA Biochem. **89**, 10915–10919 (1992)
9. Jones, D.T., Taylor, W.R., Thornton, J.M.: The rapid generation of mutation data matrices from proteic sequences. Cabios **8**(3), 275–282 (1992)
10. Kojadinovic, I., Lerman, I.C., Peter, P.: http://cran.rproject.org/src/contrib/descriptions/llahclust.html. Software, École Polytechnique de L'Université de Nantes and IRISA (2007)
11. Lerman, I.C.: Les bases de la classification automatique. Gauthier-Villars (1970)
12. Lerman, I.C.: Introduction à une méthode de classification automatique illustrée par la recherche d'une typologie de personnages enfants à travers la littérature enfantine. Revue de Statistique Appliquée **21**(3), 23–49 (1973)
13. Lerman, I.C.: Combinatorial analysis in the statistical treatment of behavioral data. Qual. Quant. **14**, 431–469 (1980)
14. Lerman, I.C.: Classification et analyse ordinale des données. Dunod and http://www.brclasssoc.org.uk/books/index.html (1981)
15. Lerman, I.C.: Interprétation non-linéaire d'un coefficient d'association entre modalités d'une juxtaposition de tables de contingences. Mathématiques et Sciences Humaines **83**, 5–30 (1983)
16. Lerman, I.C.: Analyse classificatoire d'une correspondance multiple, typologie et régression. In: Diday, E., et al. (eds.) Data Analysis and Informatics, III, pp. 193–221. North Holland (1984)
17. Lerman, I.C.: Comparing relational variables according to likelihood of the links classification method. In: Jambu, M., Diday, E., Hayashi, C., Oshumi, N., et al. (eds.) Recent Developments in Clustering and Data Analysis, pp. 187–200. Academic Press (1988)
18. Lerman, I.C.: Coefficient général de discrimination de classes d'ojets par des variables de types quelconques. Application à des données génotypiques. Revue de Statistique Appliquée, (LIV(**2**)):33–63 (2006)
19. Lerman, I.C., Bachar, K.: Comparaison de deux critères en classification ascendante hiérarchique sous contrainte de contigïté; application en imagerie numérique. Journal de la Société Française de Statistique, **149**(2):45–74 (2008)
20. Lerman, I.C., Hardouin, M., Chantrel, T.: Analyse de la situation relative entre deux classifications floues. In: Diday, E., et al. (eds.) Data Analysis and Informatics, pp. 523–533. North Holland (1980)
21. Lerman, I.C., Peter, P., Risler, J.L.: Matrices AVL pour la classification et l'alignement de séquences protéiques. Research Report 2466, IRISA-INRIA, September 1994
22. Lerman, I.C., Tallur, B.: Classification des éléments constitutifs d'une juxtaposition de tableaux de contingence. Revue de Statistique Appliquée **28**, 5–28 (1980)
23. Mount, D.M.: Bioinformatics: Sequence and Genome Analysis, 2nd edn. Cold Spring Harbor Laboratory Press, Cold Spring Harbor (2004)
24. Nicolau, M.H.: Analyse d'un algorithme de classification, Thèse de 3-ème cycle. Ph.D. thesis, Université de Paris 6, November 1972

25. Ouali-allah, M.: Analyse en préordonnance des données qualitatives. Application aux données numériques et symboliques. Ph.D. thesis, Université de Rennes 1, décembre 1991
26. Peter, P., Leredde, H., Ouali, M., Lerman, I.C.: CHAVL software. http://chavl.gforge.inria. fr/CHAVL-11-2015.zip, University of Nantes and University of Rennes 1 IRISA, November 2015
27. Sbii, A.: Validité et Logiciel de coefficients de corrélation partielle entre variables descriptives. Application à la perception de l'enseignement des différentes matières à l'École Polytechnique. Thèse de Troisième Cycle. Ph.D. thesis, Université de Rennes 1, February 1988
28. Tallur, B.: Contribution à l'analyse exploratoire de tableaux de contingence par la classification, Doctorat d' État. Ph.D. thesis, Université de Rennes 1, 1988

Chapter 12
Conclusion and Thoughts for Future Works

12.1 Contribution to Challenges in Cluster Analysis

The work presented in this book is at the heart of *Data Science*. In this general area the *Data Analysis* and more particularly *Data Clustering* plays a fundamental part. Non-supervised Clustering is concerned in our development.

To begin, let us refer to two classical books of general interest on the subject. The first one is the Gordon book "Classification" [11]. The latter runs through a broad range of methods, giving the guiding ideas, describing and briefly illustrating them. This book is an excellent introductory book.

The spirit of the description of methods considered in the book "Data Clustering Theory, Algorithms and Applications" of Gan et al. [8] is very different. Already, the methods concerned are not exacly the same. They are dominated by geometrical and statistical aspects. These are classic methods widely used and for which language codes are provided.

In both books, mainly, clustering a set of objects described by numerical attributes is the only problem of concern. Building association coefficients between descriptive categorical attributes is not really considered. Consequently, the problem of clustering a set of descriptive attributes of different types is no more considered. On the other hand, building similarity indices between objects (respectively, categories of objects) in the case of qualitative descriptions is not studied.

In our book, we propose a constructive theory of Clustering which permits us to analyze in a relative manner all that was proposed in the area concerned. Each of the steps enabling the passage from the initial data to their clustering is minutely studied. Thereby, significant contributions on the following subjects have been obtained:

1. The nature of the data and its logical, mathematical and statistical formalization;
2. Numerical or ordinal quantization of the resemblance notion between the elements of the set E to be structured by Clustering (E may be a set $\mathcal{A}$ of descriptive attributes or a set $\mathcal{O}$ of elementary objects (respectively, a set $\mathcal{C}$ of object categories)) (see Chaps. 4–7);

© Springer-Verlag London 2016

I.C. Lerman, *Foundations and Methods in Combinatorial and Statistical Data Analysis and Clustering*, Advanced Information and Knowledge Processing, DOI 10.1007/978-1-4471-6793-8_12

3. For a given set E, extension of the proximity (resemblance) notion between elements to that between classes of elements (see Chap. 10);

4. Detailed synthetic description of the principal methods of nonhierarchical clustering (see Chap. 2);

5. Efficient algorithms in order to built hierarchical clusterings (see Chap. 10);

6. Statistical validation of partitions, classes or partition chains on the set E clustered, according to the pairwise similarities observed on E (see Chap. 9);

7. Dual interpretation of the correspondence between elements of the descriptive attribute set $\mathcal{A}$ and elements or classes of the set $\mathcal{O}$ of objects (respectively, a set $\mathcal{C}$ of categories of objects) (see Chap. 8);

8. Description and mathematical properties of the sought synthetic structure in Clustering: a partition or an ordered partition chain (see Chap. 1);

9. Specific software performing the application of the Likelihood Linkage Analysis Hierarchical Classification method and real cases processed (see Chap. 11).

In [1], eight challenges are listed for the future scope in the Cluster Analysis research field: *A. Identification of clusters, B. Selection of distance measure, C. Structure of database, D. Types of attributes, E. Selecting the starting cluster, F. Choosing the best method of clustering, G. Cluster validity* and *H. Standard benchmarks*. In fact, these challenges are constant problems in Cluster Analysis. The developments above (1–9), studied minutely in the book have a real impact on the challenges expressed. By denoting $K \longrightarrow L$ such an impact for $K \in \{1, 2, 3, 4, 5, 6, 7, 8, 9\}$ and $L \in \{A, B, C, D, E, F, G, H\}$, we have:

1. $1 \longrightarrow B, 1 \longrightarrow C, 1 \longrightarrow D$ and $1 \longrightarrow F$;
2. $1 \longrightarrow B$;
3. $3 \longrightarrow B$ and $3 \longrightarrow F$;
4. $4 \longrightarrow F$ and $4 \longrightarrow E$;
5. $5 \longrightarrow F$;
6. $6 \longrightarrow G, 6 \longrightarrow F$ and $6 \longrightarrow G$;
7. $7 \longrightarrow G$;
8. $8 \longrightarrow F$;
9. $9 \longrightarrow H$.

Now, relative to the "Recent Advances in Cluster Analysis" (see page 596 of [1]), elegant solutions are proposed in the book for the problem in vogue of *Ensembles* of a clustering algorithms. In fact, the matter consists of clustering a set of objects described by nominal categorical attributes; each one of these being associated with a result of clustering algorithm. A nonhierarchical clustering can be proposed from the "Central partition method" (see Sect. 2.2 of Chap. 2) or associated with a "significant" level of the classification tree obtained by the LLA method, using a similarity index defined in Sect. 7.2.3.1 of Chap. 7. According to the latter, different significant levels can be studied.

On the other hand, parallel implementation of the $AAHC$ algorithm as expressed in Sect. 10.6.6 of Chap. 10 may contribute significantly to the problem of "Distributed clustering" also expressed as a recent advance in Cluster Analysis.

Finally and following the same reference indicated above, for the "Multiway clustering" subject, where heterogeneous data has to be dealt, coding as defined in Sect. 3.4.4 of Chap. 3 provides a flexible and powerful method for converting objects into pooled feature vectors. Also, in the case of semantic relationships within each object, graph representation of an object on the set of elements included in it, might be useful for formalizing the clustering problem.

12.2 Around Two Books Concerning *Relational* Aspects

The first characteristic of the LLA hierarchical clustering method consist of representing descriptive attributes of a set of objects $\mathcal{O}$ in terms of relations on $\mathcal{O}$ (see Chap. 3). Let us now examine two books, mutually very different, concerned by *relational* aspects. Special facets are considered in each of them. These are *Relational Data Clustering* of Bo Long et al. [19] and *Ordinal and Relational Clustering* of M.F. Janowitz [12].

Contrary to what seems to be claimed in [19], the notion of a data table crossing an object set or a set of categories of objects with a set of descriptive attributes is extremely important. However, very general sense must be given to the latter notion (see Tables 3.4 and 3.5 of Chap. 3). Moreover, all the descriptive structures considered in [19] can be reduced to be represented by a single or juxtaposition of data tables such that $\mathcal{T}$ or $\mathcal{S}$ (see Sect. 3.5.5 of Chap. 3).

More precisely, in the description mentioned, a set of sets

$$\mathcal{P} = \{\mathcal{O}^l | 1 \leq l \leq L\} \tag{12.2.1}$$

is considered. Each of them is provided by a valuated graph of the form

$$S^l = \{S^l_{ij} | (i, j) \in \mathbb{I}^l \times \mathbb{I}^l\} \tag{12.2.2}$$

where $\mathbb{I}^l = \{1, 2, \ldots, i, \ldots, n_l\}$ is the set of the indices labelling the elements of $\mathcal{O}^l$.

On the other hand, a valuated biparti graph links each pair $(\mathcal{O}^l, \mathcal{O}^m)$ of sets, $1 \leq l \neq m \leq L$. This graph can be expressed as follows

$$R^{lm} = \{R^{lm}_{ij} | (i, j) \in \mathbb{I}^l \times \mathbb{I}^m\} \tag{12.2.3}$$

According to the terminology used in [19], (12.2.2) corresponds to *homogeneous* data and (12.2.3), to *heterogeneous* data.

A clustering problem as it can be considered in [19] may concern a given set $\mathcal{O}^l$, where the similarity between two objects belonging to $\mathcal{O}^l$ is percieved through all the other sets $\mathcal{O}^m$, $1 \leq m \leq L$, $m \neq l$. For this, all the valuated graphs associating $\mathcal{O}^l$ with the other sets $\mathcal{O}^m$, $1 \leq m \leq L$, $m \neq l$, are taken into account, namely

$$\mathcal{R}^l = \{R^{lm} | 1 \leq m \leq L, m \neq l\} \tag{12.2.4}$$

Each of the $\mathcal{O}^m$ sets ($m \neq l$), gives rise to a data table T_l^m of type T (see Sect. 3.5.5). T_l^m has $n_l = card(\mathcal{O}^l)$ rows and $n_m = card(\mathcal{O}^m)$ columns. The jth one of the latter $(R_{ij}^{lm} | i \in \mathbb{I}^l)^t$ (t as transpose) can be assimilated to the value sequence of a numerical attribute. Thus, the description is defined from the the juxtaposition of numerical data tables $\{T_l^m | 1 \leq m \leq L, m \neq l\}$. Obviously, the homogeneous data defined by the valuated graph S^l (see 12.2.2) might also intervene in establishing similarity indices on $\mathcal{O}^l$. However, S^l has a specific role with respect to the valuated biparti graphs $\mathcal{R}^l$ (see 12.2.4).

For the analysis of the association between *words* and *documents*, often mentioned in [19], the data table structure concerned in our approach is given by a contingency table $T(D, W)$ (see Sects. 5.3.2 and 10.5 of Chaps. 5 and 10, respectively). The row set of such a table is indexed by a set D of documents and the column set, by a set W of words. The number of times n_{wd} for which the word w is present in the document d is given at the intersection of the row d and the column w. For a pair of partitions $\pi(D)$ of D and $\chi(W)$ of W, obtained by clustering, the correspondence between the respective classes of $\pi(D)$ and $\chi(W)$ can be obtained by means of the statistical tools studied in Chap. 8.

The problem often posed for this data structure concerns scarcity. The numbers n_{wd} might be too small and then, densifying the data by grouping turns out to be needed. An important application was carried out in [23] on the subject of determination semantic classes in automatic natural language processing.

In Sect. 4.2.1 of Chap. 4 we have introduced the notion of the preordonance associated with a Similarity index in the case of boolean data. In Sect. 10.2 of Chap. 10, we have proposed and studied an $AAHC$ ordinal algorithm on a set E, expressed with respect to a preordonance on E. The latter preordonance is generally associated with a numerical similarity index $S(E)$, or equivalently, a numerical dissimilarity index $\mathcal{D}(E)$. Thus, the only data on which the algorithm mentioned is based, is a total preorder $\omega(E)$ on the set F of unordered element pairs of E. We suppose here this preorder established such as $\{x, y\} \leq \{z, t\}$ if and only if x and y are more similar (less dissimilar) than z and t.

If $(B_1, B_2, \ldots, B_h, \ldots, B_H)$ is the totally ordered class sequence of $\omega(E)$:

$$B_1 < B_2 < \cdots < B_h < \cdots < B_H,$$

the ordinal dissimilarity (the preordonance) $\omega(E)$ can be represented as a mapping of F into the *totally ordered* set of integers $\{1, 2, \ldots, h, \ldots, H\}$.

In the second book mentioned above (*Ordinal and Relational Clustering* by M.F. Janowitz [12]), the dissimilarity is defined as a mapping of E into a *partially ordered set* (poset). In fact and mainly, only boolean data are considered. For a set of objects $\mathcal{O}$ described by p boolean attributes, the similarity index between objects takes its values in the same space as that representing the object set. Indices such that (1) and (2) of Table 4.3 of Chap. 4 are shown to be reflected in the representation chosen.

Now, let us consider the case where the set $\mathcal{A} = \{a^1, a^2, \ldots, a^j, \ldots, a^p\}$ of the descriptive attributes includes attributes of another type than Boolean ($\mathcal{A}$ might be composed of attributes of different types) (see Chap. 3). According to the representation above, the similarity (respectively, dissimilarity) representation has to be in

$$\mathcal{L} = \mathcal{E}_1 \times \mathcal{E}_2 \times \cdots \times \mathcal{E}_j \times \cdots \times \mathcal{E}_p$$

where $\mathcal{E}_j$ is the value set of a^j, $1 \leq j \leq p$. In these conditions, the question is how to represent a dissimilarity between two given objects of the set $\mathcal{O}$. One more question concerns the tractability of the algorithm proposed in the case of large databases.

Relative to the latter computational complexity problem, notice that in the much more simple case for which the Similarity takes its values in a totally ordered set; that is to say, in practice, when the similarity data handled is a preordonance $\omega\big(\mathcal{S}(E)\big)$ associated with a numerical similarity $\mathcal{S}(E)$ on E, the order of the computational complexity needed to establish $\omega\big(\mathcal{S}(E)\big)$ from $\mathcal{S}(E)$ is—up to a multiplicative coefficient—$n^2 log_2 n$ ($n = card(E)$).

We saw that in the case of boolean data (see Proposition 33) of Chap. 4 the preordonance $\omega\big(\mathcal{S}(E)\big)$ associated with $\mathcal{S}(E)$, has—in certain statistical conditions— stability properties with respect to the choice of the similarity index $\mathcal{S}(E)$.

Defining clustering algorithm on the basis of the ordinal data $\omega\big(\mathcal{S}(E)\big)$ has a theoretical interest. Moreover, this type of data may efficiently intervene in validation clustering problems (see for example Sect. 9.3.1 of Chap. 9). However, the finest result for the cluster emergence are obtained by considering *numerically* the nature of the data; that is, by working directly with $\mathcal{S}(E)$ than with $\omega\big(\mathcal{S}(E)\big)$ [17]. Nevertheless, deep thought is necessary for situating the respective roles of *numerical* and *ordinal* approaches in Data Analysis and Clustering.

M.F. Janowitz [12] percieves a connection between his approach and the FCA (Formal Concept Analysis) method (see [9] and Sect. 9.2.1 of Chap. 9). He states that "… it appears that cluster analysis and FCA are essentially the same thing, only with different notation and terminology…" In these conditions we may propose, in the case of boolean data, to compare the results of a FCA with those obtained by crossing two *LLA* reduced classification trees on the attribute set and on the object set (see Sect. 9.3.6 of Chap. 9).

12.3   Developments in the Framework of the *LLA* Approach

12.3.1   *Principal Component Analysis*

To condense each of the trees just mentioned, two versions are expressed for the level Statistic criterion. The former is based on the preordonance associated with a similarity (see Sect. 9.3.5.1 of Chap. 9) and the latter, directly with the numerical similarity (see Sect. 9.3.5.2 of Chap. 9).

This second Statistic refers in fact—the most general case—to comparing two valuated binary relation attributes (see Sect. 6.2.5 of Chap. 6). This can be used to recognize the "good" number of factorial axes to retain in a Principal Component Analysis (see Chap. 6 of [15] and [14]).

More precisely, let us consider a cloud of points such that $\mathcal{N}(\mathbb{I})$ of Eq. 7.1.2 in Chap. 7 and without restricting generality, suppose the latter to be equiweighted. Now, designate by

$$\mathcal{D} = \{d(i, i') | \{i, i'\} \in P_2(\mathbb{I})\} \qquad (12.3.1)$$

the $\mathbb{I}$ pairwise distances in the whole space $\mathbb{R}^p$ ($P_2(\mathbb{I})$ is the set of unordered pairs of distinct elements of $\mathbb{I}$ and p is the number of numerical attributes describing the set of objects).

Similarly, to $\mathcal{D}$,

$$\mathcal{D}^k = \{d^k(i, i') | \{i, i'\} \in P_2(\mathbb{I})\} \qquad (12.3.2)$$

is the distance table between the projected elements of $\mathbb{I}$ into the subfactorial space sustained by the first k factorial axes.

We consider the association coefficient $C(\mathcal{D}, \mathcal{D}^k)$ between the matrices $\mathcal{D}$ and $\mathcal{D}^k$ in accordance with the coefficient Q_1 instanciated in Eq. 6.2.87 of Sect. 6.2.5.3 of Chap. 6. This coefficient is of the same nature as that defined in Eq. 9.3.70 concerning evaluation of partitions in Chap. 9.

Contrary to that we may think, the distribution of

$$\{C(\mathcal{D}, \mathcal{D}^k) | k \geq 1\} \qquad (12.3.3)$$

is not strictly increasing with respect to k. Therefore, its variation does not follow that of the inertia of the cloud $\mathcal{N}^k(\mathbb{I})$ obtained by projection of $\mathcal{N}(\mathbb{I})$ on the subspace sustained by the first k factorial axes. This statement was observed by Habib Benali through first experiments several years ago. These have to be worked on again.

Thereby, the local maxima of Eq. 12.3.3 may indicate the value k_0 of k to be retained for the factorial representation of $\mathcal{N}(\mathbb{I})$.

12.3.2 Multidimensional Scaling

In the Introduction of Chap. 10 (see Sect. 10.1) we introduced the matrix of "Informational" dissimilarity indices (see 10.1.6) associated with the probabilistic similarity indices (see 10.1.5). This association enables us to insert the LLA approach in the data analysis methods based on using a dissimilarity index. This is the case of Multidimensional Scaling [4, 13]. For a set E endowed with a dissimilarity index δ, the objective of the latter method consists of representing E in a geometrical space of low dimension—mostly the two dimensional space $\mathbb{R}^2$—in such a way that the interdistances d between the representation points reflect as much as possible the dissimilarities δ between the elements represented. More precisely, $R(e)$ denoting the representation of the element e of E, the matter consists of doing in such a way that, globally, the respective distances $d\big(R(x), R(y)\big)$ become as near as possible the dissimilarities $\delta(x, y)$, $\{x, y\} \in P_2(E)$.

This analysis was carried out for a set E of biological sequences provided with an Informational dissimilarity index (see Sect. 11.4 of Chap. 11). The algorithm processed works with respect to a specific computer structure [16, 21, 22].

Equally, we might represent synthetically the Table 10.1.6 of Chap. 10 by means of *Additive Similarity Tree* [2, 24].

12.3.3 In What *LLA* Hierarchical Clustering Method Is a Probabilistic Method?

The *LLA* method is considered as a probabilistic method because similarity indices between objects or association coefficients between attributes refer to a probability scale. In these conditions, the value of a similarity index (respectively, of an association coefficient) between two elements to compare is, under probabilistic hypothesis of no relation, the realization of a random variable uniformly distributed over the [0, 1] interval.

The meaning given to a probabilistic method in [3] or in [5] is very different. In these, clustering concerns a set of objects described by numerical variables where each cluster is expressed in terms of a multidimensional probability distribution on $\mathbb{R}^p$, where p is the variable number.

In the overview article [10], a null hypothesis is defined over the geometrical space $\mathbb{R}^p$. The purpose of the latter hypothesis is not—as in our case—to establish a similarity measure, but to validate the existence of a cluster in a given area of the representation space. An effective definition of such a hypothesis is not really manageable.

12.3.4 Semi-supervised Hierarchical Classification

The *AAHC LLA* method is a non-supervised clustering method. However, it can as in [7] takes part in a supervised classification method. In this way, we obtain more significant results and a considerable reduction of the computational complexity. The whole process leads to a *semi-supervised* classification method.

The meaning we give to the term "semi-supervised" is absolutely not the same as that given in [25] (see also [20]). In the latter work, "semi-supervised hierarchical clustering" corresponds in our expression to "hierarchical clustering *under constraints* provided by knowledge" (see Sect. 10.6.5 of Chap. 10).

12.4   Big Data

A final and very important point for the future. We are living in the *Big Data* era.
An immense challenge concerns exploration of Big Data by Data Analysis methods
[6]. The methods we have set up contribute to data organization. Essential concepts to assess statistical tendencies are *Correlation* between attributes and *Similarity* between objects. The importance of these notions led us to a large development
through the Chaps. 4, 5, 6 and 7.

The research in *Big Data Mining* is in its earliest beginnings. Relative to a problem
of Data Analysis of a fixed type, we have to confront the three Vs difficulties: *Volume*,
Velocity and *Variation*. A computing model and specific technologies have to be built.
The matter will be to be able to apply the general principle of MAP and REDUCE.
Let us notice that philosophically, this principle was applied in [18] (see Sect. 10.6.6
of Chap. 10).

References

1. Ahuja, M.S., Bal, J.S.: Exploring cluster analysis. Int. J. Comput. Inf. Technol. **3**, 594–597
   (2014)
2. Barthélémy, J.P., Guénoche, A.: Trees an proximity representations. Wiley-Interscience (1991)
3. Bock, H.-H.: Probabilistic aspects in classification. In: Yajima, K., Tanaka, Y., Bock, H.-H.,
   Hayashi, C., Ohsumi, N., Baba, Y. (eds.), Data Science, Classification, and Related Methods,
   pp. 3–21. Springer, Berlin (1998)
4. Borg, L., Groenen, P.: Modern Multidimensional Scaling: Theory and Applications, 2nd edn.
   Springer, Berlin (2005)
5. Celeux, G., Govaert, G.: Comparison of the mixture and the classification maximum likelihood
   in cluster analysis. J. Stat. Comput. Simul. **47**, 127–146 (1993)
6. Che, D., Safran, M., Peng, Z.: From big data to big data mining: challenges, issues and opportu-
   nities. In: Hong, B., et al., (ed). Database Systems for Advanced Applications, 2013 DASFAA
   Workshops, LNCS 7827, pp. 1–15. Springer Berlin (2013)
7. Costa, J.F.P., Lerman, I.C.: Arcade: a prediction method for nominal variables. Intell. Data
   Anal. **2**, 265–286 (1998)
8. Gan, G., Ma, C., Wu, J.: Data clustering theory, algorithms and applications. SIAM (2007)
9. Ganter, B., Wille, R.: Formal Concept Analysis, Mathematical Foundations. Springer, Berlin
   (1999)
10. Gordon, A.: Cluster validation. In: Yajima, K., Tanaka, Y., Bock, H.-H., Hayashi, C., Ohsumi,
    N., Baba, Y. (eds). Data Science, Classification, and Related Methods, pp. 22–39. Springer,
    Berlin (1998)
11. Gordon, A.D.: Classification, 2nd edn. Chapman and Hall, London (1999)
12. Janowitz, M.F.: Ordinal and Relational Clustering. World Scientific, Singapore (2010)
13. Kruskal, J.B., Wish, M.: Multidimensional Scaling. Sage Publications, Beverly Hills and Lon-
    don (1978)
14. Lebart, L., Fenelon, J-P.: Statistique et Informatique appliquées. Dunod (1973)
15. Lerman, I.C.: Classification et analyse ordinale des données. Dunod and http://www.brclasssoc.
    org.uk/books/index.html (1981)
16. Lerman, I.C., Ngouenet, R.: Algorithmes génériques séquentiels et parallèles pour une
    représentation affine des proximités. Publication Interne IRISA, Research Report INRIA P.I.
    901 and RR 2570, IRISA-INRIA (1995)

17. Lerman, I.C., Peter, P., Risler, J.L.: Matrices AVL pour la classification et l'alignement de séquences protéiques. Research Report 2466, IRISA-INRIA (1994)
18. Lerman, I.C., PETER, Ph.: Organisation et consultation d ' une banque de petites annonces à partir d ' une méthode de classification hiérarchique en parallèle. In: Data Analysis and Informatics IV, pp. 121–136. North Holland (1986)
19. Long, B., Zhang, Z., Yu, P.S.: Relatinal Data Clustering Models, Algorithms, and Applications. Chapman and Hall/CRC, Boca Raton (2010)
20. Miyamoto, S.: An overview of hierarchical and non-hierarchical algorithms of clustering for semi-supervised classification. In: Torra, V. et al., (ed.), Proceedings of the 9th international conference on modelling decisions for artificial intelligence, pp. 1–10. Springer (2012)
21. Ngouenet, R.: Une nouvelle famille d'indices de dissimilarité pour la mds. Research Report RR-2087, IRISA-INRIA, October 1993
22. Ngouenet, R.: Analyse géométrique des données de dissimilarité par la Multidimentional Scaling. Application aux séquences biologiques. Parallel genetic algorithms for Multidimentional Scaling: Application to statistical analysis of biological sequences. Ph.D. thesis, Université de Rennes 1, December 1995
23. Rossignol, M.: Acquisition automatique de lexiques sémantiques pour la recherche d'information. Ph.D. thesis, Université de Rennes 1, October 2005
24. Sattah, S., Tversky, A.: Additive similarity trees. Psychometrika **42**(3), 319–345 (1977)
25. Zheng, L., Tao, L.: Semi-supervised hierarchical clustering. In: 2011 11th IEEE international conference on data mining, pp. 982–991. IEEE (2011)

Printed in the United States
By Bookmasters